Water Supply in Agriculture

In Focus – Special Book Series

Water Supply in Agriculture

Editor
Nektarios N. Kourgialas

IWA PUBLISHING

Published by **IWA Publishing**
Unit 104–105, Export Building
1 Clove Crescent
London E14 2BA, UK
Telephone: +44 (0)20 7654 5500
Fax: +44 (0)20 7654 5555
Email: publications@iwap.co.uk
Web: www.iwapublishing.com

First published 2021

British Library Cataloguing in Publication Data
A CIP catalogue record for this book is available from the British Library

ISBN: 9781789062946 (paperback)

Contents

doi: 10.2166/ws.2021.244

Editorial: Could advances in geoinformatics, irrigation management and climate adaptive agronomic practices ensure the sustainability of water supply in agriculture?

PROMISING FEATURES

The Food and Agriculture Organization (FAO) predicts that up to 2050, food production will increase by about 70% above today's levels. This increase in food production could be accomplished by developing more agricultural land and transitioning from rainfed to irrigated agriculture. The development of the agricultural sector is interrelated with the effective management of water resources as irrigated agriculture is the world's largest water consumer with trends of increasing irrigation needs and pollution load in the oncoming years (Ricciardi *et al.* 2020). Water supply in the rural sector is expected to be most affected by climate change with reduced crop yields due to declining water availability and abandonment of climate-disadvantaged parcels. Agricultural production stability faces many challenges related to climate-water effects, with the following being of utmost importance: changes in levels and frequency of precipitation, more frequent dry spells, increased extreme weather events including flash droughts/floods, losses of agricultural lands due to water erosion, salinization of coastal cultivated areas and freshwater, as well as surface and groundwater contamination by fertilizers or other agronomically related sources of pollution. Introducing and incorporating efficient use and management practices of irrigation water to new climate-smart production methods is a vital priority for major sectoral policies, notably the Common Agricultural Policy (CAP) (Kourgialas, 2021).

In line with this, of high importance for water policy makers is to establish effective agricultural water management strategies to enhance preparedness and capacity to respond to the impacts of climate change and anthropogenic interventions. These strategies should focus on three key objectives, which, once they scale up, can increase the resilience of agriculture as a whole. These three objectives are related to: advanced geoinformatics tools, proper irrigation management and climatic water – adaptive agronomic practices (Grafton *et al.* 2018).

Thereby, the answer to the question of this Editorial '*Could advances in geoinformatics, irrigation management and climate adaptive agronomic practices ensure the sustainability of water supply in agriculture?*' is certainly YES. The sustainability of water supply in cropping systems could be effectively addressed based on the advantages that raised from the above-mentioned key objectives divided in turn into the following scientific topics:

- Precision agriculture and irrigation technologies – *In situ* and satellite remote sensing
- GIS and advanced decision-making models in water resources under climate change scenarios – Forecasting of environmental risks and pollution loads in agriculture
- Optimization of efficiency of irrigation systems and distribution of water
- Management of water salinity and reuse of treated water for irrigation
- Proper agronomic practices for water saving at farm and catchment scale
- Choosing crops that are genetically more tolerant to environmental stress and extreme weather conditions

THIS SPECIAL ISSUE

In this Special Issue (SI) of *Water Supply*, 33 papers were selected that contribute to a broad discussion and demonstration of state-of-the-art multifunctional role of water resources in agriculture. The aim of this SI is to provide insights into novel modelling (monitoring, analyzing/visualizing and prediction) approaches, irrigation management and agronomic practices to investigate the adaptability of water supply and crop production systems to changing environment.

The inspiration came from the international conference 'Water Efficiency & Climate Resilient Agriculture' organized by the AgroClimaWater project (LIFE14 CCA/GR/000389), '*Promote water efficiency and support the shift towards climate resilient agriculture in Mediterranean countries*' co-funded by the European Commission (http://www.lifeagroclimawater.eu/).

This SI presents characteristic examples of new technologies and decision support systems (e.g., artificial intelligence/optimization modelling approaches, Big Geo data) in water efficiency at different levels, including: water supply hydraulic infrastructure systems, water retention measures, less exposed to evaporation and better adapted to infiltration, solutions to reduce water demand and developing techniques for reusing water. Specifically, this SI brings advances in precision irrigation techniques adapted to the actual needs of plants at farm and extended scale applications (Filintas *et al.* 2021; Markos & Radoglou 2021; Mousabeygi *et al.* 2021). Also, new modelling approaches to predict and/or monitor water dynamics (Kader *et al.* 2021; Mridha & Rahman 2021; Sihag *et al.* 2021; Tziatzios *et al.* 2021; Xu *et al.* 2021) or pollutants transport such as salinity (Daghari *et al.* 2021; Yuan 2021) and nitrates (Bian *et al.* 2021; Nie *et al.* 2021) are provided. An important part of this SI is covered by optimal water allocation and the newest developments in irrigation management (Cortés-Ruiz & Azuz-Adeath 2021; Lyra *et al.* 2021; Mahdavi 2021; Morianou *et al.* 2021a; Wei *et al.* 2021). Successfully tested methods for improving irrigation efficiency for quantity (Hui *et al.* 2021a, 2021b) and quality (Jie *et al.* 2021; Lemos *et al.* 2021; Lima *et al.* 2021; Tao *et al.* 2021; Xie & Hung 2021; Zhang *et al.* 2021) are also addressed as one of the utmost prospectives in agricultural water management. Notable results on hydraulic irrigation networks and equipment/systems (Samarinas & Evangelides 2021; Shi *et al.* 2021; Sun *et al.* 2021) are presented as well. Finally, this SI highlights the benefits in the adaptability of agronomical practices and decision-making systems to release climatic water effects on crops and ensure long-term stability of ecosystems (Biswas *et al.* 2021; Economidou *et al.* 2021; Gkiatas *et al.* 2021; Morianou *et al.* 2021b; Ragkos & Ambas 2021).

Many appreciations to the authors for their particularly interesting papers, which we would like to believe have contributed to the improvement of our scientific knowledge in the field of water supply in agriculture. Furthermore, I would like to express my deepest thanks and gratitude to the reviewers and the *Water Supply* (IWA Publishing) staff who contributed to a successful preparation of this SI.

I hope you enjoy the read and be inspired.

Guest Editor
Nektarios N. Kourgialas
Hellenic Agricultural Organization (ELGO-DIMITRA),
Institute for Olive Tree Subtropical Crops and Viticulture,
Water Resourses-Irrigation & Env. Geoinformatics Lab,
Chania, Greece

REFERENCES

Bian, J., Wang, Q., Nie, S., Wan, H. & Wu, J. 2021 Understanding nitrogen transport in the unsaturated zone with fluctuations in groundwater depth. *Water Supply* **21** (6), 2691–2708.

Biswas, A, Mailapalli, D. R. & Raghuwanshi, N. S. 2021 Consumptive water footprints, water use efficiencies and productivities of rice under alternate wetting and drying for Kharagpur, West Bengal, India. *Water Supply* **21** (6), 2935–2946.

Cortés-Ruiz, A. & Azuz-Adeath, I. 2021 Estimating the future hydric needs of Baja California, Mexico. Assessment of scenarios to stop being a region with water scarcity. *Water Supply* **21** (6), 2760–2771.

Daghari, I., El Zarroug, M. R., Muanda, C., Kompany, J. R., Kanzari, S. & Mimoun, A. B. 2021 Feasibility of water desalination for irrigation: the case of the coastal irrigated area of Dyiar-Al-Hujjej, Tunisia. *Water Supply* **21** (1), 24–45.

Economidou, Y., Doula, M. K. & Zorpas, A. A. 2021 Mitigation of the effects of climate change in the agricultural sector of Cyprus, through optimization of benefit. *Water Supply* **21** (6), 2947–2958.

Filintas, A., Wogiatzi, E. & Gougoulias, N. 2021 Rainfed cultivation with supplemental irrigation modelling on seed yield and oil of *Coriandrum sativum L.* using precision agriculture and GIS moisture mapping. *Water Supply* **21** (6), 2569–2582.

Gkiatas, G., Kasapidis, I., Koutalakis, P., Iakovoglou, V., Savvopoulou, A., Germantzidis, I. & Zaimes, G. N. 2021 Enhancing urban and suburban riparian areas through ecosystem services and ecotourism activities. *Water Supply* **21** (6), 2974–2988.

Grafton, R. Q., Williams, J., Perry, C. J., Molle, F., Ringler, C., Steduto, P., Udall, B., Wheeler, S. A., Wang, Y., Garrick, D. & Allen, R. G. 2018 The paradox of irrigation efficiency. *Science* **361**, 748–750.

Hui, X., Yan, H., Xu, Y. & Tan, H. 2021a Sprinkler droplet impact angle affects shear stress distribution on soil surface – a case study of a ball-driven sprinkler. *Water Supply* **21** (6), 2772–2785.

Hui, X., Yan, H., Zhang, L. & Chen, J. 2021b A simplified method to improve water distribution and application uniformity for sprinkler irrigation on sloping land: adjustment of riser orientation. *Water Supply* **21** (6), 2786–2798.

Jie, R., Xiao-wen, L. & Ying-xin, W. 2021 Characteristics and evaluations of heavy metal contents in irrigation water sources on a suburb farmland in the lower reaches of Xijiang River in the early busy season of farming. *Water Supply* **21** (6), 2811–2824.

Kader, M. A., Nakamura, K., Senge, M. & Mojid, M. A. 2021 Two-dimensional numerical simulations of soil-water and heat flow in a rainfed soybean field under plastic mulching. *Water Supply* **21** (6), 2615–2632.

Kourgialas, N. 2021 Hydroclimatic impact on Mediterranean tree crops area – mapping hydrological extremes (drought/flood) prone parcels. *Journal of Hydrology* **596**, 125684.

Lemos, L. T. O., de Deus, F. P., Thebaldi, M. S., Diotto, A. V., de Andrade Júnior, V. C. & de Almeida, R. C. 2021 Influence of the soil water retention curve type and magnetic water treatment on lettuce irrigation management responses. *Water Supply* **21** (6), 2850–2862.

Lima, M., de Araujo, B. M., Soares, S. R. A., Santos, A. S. P. & Vieira, J. M. P. 2021 Water reuse potential for irrigation in Brazilian hydrographic regions. *Water Supply* **21** (6), 2799–2810.

Lyra, A., Loukas, A. & Sidiropoulos, P. 2021 Impacts of irrigation and nitrate fertilization scenarios on groundwater resources quantity and quality of the Almyros Basin, Greece. *Water Supply* **21** (6), 2748–2759.

Mahdavi, T. 2021 Application of the 'theory of planned behavior' to understand farmers' intentions to accept water policy options using structural equation modeling. *Water Supply* **21** (6), 2720–2734.

Markos, N. & Radoglou, K. 2021 Estimation of ecosystem evapotranspiration in a *Robinia pseudoacacia* L. plantation with the use of the eddy covariance technique and modeling approaches. *Water Supply* **21** (6), 2553–2568.

Morianou, G., Ziogas, V., Kourgialas, N. N. & Karatzas, G. P. 2021a Effect of irrigation practices upon yield and fruit quality of four grapefruit (*Citrus paradisi* Mac.) cultivars. *Water Supply* **21** (6), 2735–2747.

Morianou, G., Kourgialas, N. N., Pisinaras, V., Psarras, G. & Arambatzis, G. 2021b Assessing desertification sensitivity map under climate change and agricultural practices scenarios: the island of Crete case study. *Water Supply* **21** (6), 2916–2934.

Mousabeygi, F., Rezaei, Y. & Akhavan, S. 2021 Assessment of maize water status using a consumer-grade camera and thermal imagery. *Water Supply* **21** (6), 2543–2552.

Mridha, M. A. H. & Rahman, S. H. 2021 Prediction of groundwater trends for irrigation in Northern Bangladesh. *Water Supply* **21** (6), 2596–2614.

Nie, W.-B., Nie, K.-K., Li, Y.-B. & Ma, X.-Y. 2021 HYDRUS-2D Simulations of nitrate nitrogen and potassium transport characteristics under fertilizer solution infiltration of furrow irrigation. *Water Supply* **21** (6), 2665–2680.

Ragkos, A. & Ambas, V. 2021 Examining the potential of an irrigation work to improve sustainability in a rural area. *Water Supply* **21** (6), 2959–2973.

Ricciardi, V., Wane, A., Singh, B. S., Goode, C., Solomon, D., McCullough, E., Diekmann, F., Porciello, J., Jain, M., Randall, N. & Mehrabi, Z. 2020 A scoping review of research funding for small-scale farmers in water scarce regions. *Nature Sustainability* **3**, 836–844.

Samarinas, N. & Evangelides, C. 2021 Discharge estimation for trapezoidal open channels applying fuzzy transformation method to a flow equation. *Water Supply* **21** (6), 2893–2903.

Shi, K., Liu, Z., Xie, Y. & Li, M. 2021 Study on optimal start-up time and water usage volume of blowdown residue discharge of mesh filter. *Water Supply* **21** (6), 2904–2915.

Sihag, P., Kumar, M. & Sammen, S. S. 2021 Predicting the infiltration characteristics for semi-arid regions using regression trees. *Water Supply* **21** (6), 2583–2595.

Sun, B., Yang, L., Zhu, S., Zhang, H., Zhang, C. & Zhang, J. 2021 Investigation of flow measurement mechanism and hydraulic characteristics of the NACA airfoil pillar-shaped flume with different wing lengths in a rectangular channel. *Water Supply* **21** (6), 2875–2892.

Tao, H., Zhou, Y., Aihemaiti, M., Li, Q., Yang, W., Jiang, Y. & Wu, Z. 2021 Water–sediment separation efficiency prediction of gill-piece separation device. *Water Supply* **21** (6), 2837–2849.

Tziatzios, G., Sidiropoulos, P., Vasiliades, L., Lyra, A., Mylopoulos, N. & Loukas, A. 2021 The use of the pilot points method on groundwater modelling for a degraded aquifer with limited field data: the case of Lake Karla aquifer. *Water Supply* **21** (6), 2633–2645.

Wei, C., Cheng, J., Gong, Y. & Gong, Z. 2021 Optimal water allocation method for joint operation of a reservoir and pumping station under insufficient irrigation. *Water Supply* **21** (6), 2709–2719.

Xie, J. & Hung, Y.-C. 2021 Effect of water compounds on photo-disinfection efficacy of TiO_2 NP-embedded cellulose acetate film in natural water. *Water Supply* **21** (6), 2825–2836.

Xu, C., Liu, Z., Zhu, X., Tian, J., Gu, F. & Wang, Y. 2021 Water and soil environmental vulnerability of artificial oases in arid areas and its temporal and spatial differentiation and evolution. *Water Supply* **21** (6), 2646–2664.

Yuan, C. 2021 Simulation soil water–salt dynamics in saline wasteland of Yongji irrigation area in hetao irrigation district of China. *Water Supply* **21** (6), 2681–2690.

Zhang, J., Wei, K., Wang, Q., Sun, Y. & Mu, W. 2021 Effects of magnetized fresh water on seed germination and seeding growth of cotton. *Water Supply* **21** (6), 2863–2874.

Assessment of maize water status using a consumer-grade camera and thermal imagery

Fatemeh Mousabeygi, Yousef Rezaei and Samira Akhavan

ABSTRACT

The analysis of plant response to water deficits can help us to identify appropriate water-saving and irrigation methods. The goal of this study is to assess the potential of eight indices derived from a modified consumer-grade camera and a thermal camera for monitoring the relative water content (RWC) of maize. The study design was a randomized complete block design with three replications and 16 treatments with four levels of irrigation water percentage based on field capacity (100% Fc, 80% Fc, 60% Fc and 40% Fc), and four levels of nitrogen (without nitrogen, 100 kg N/ha, 200 kg N/ha and 300 kg N/ha) were used. RWC values were used to evaluate the performance of the eight crop water stress indicators. The results showed that the best performance results of the studied vegetation indices were DANS, CTSD and CTCV, respectively. As observed, R^2 values were 0.88, 0.76 and 0.67, respectively.

Key words | maize, modified camera, Otsu algorithm, red-green ratio index, relative water content

Fatemeh Mousabeygi
Samira Akhavan
Department of Water Engineering,
Faculty of Agriculture,
Bu-Ali Sina University,
Hamedan,
Iran

Yousef Rezaei (corresponding author)
Department of Civil Engineering, Faculty of Engineering,
Bu-Ali Sina University,
Hamedan, P.O. Box 65178-33131,
Iran
E-mail: *yrezaei@basu.ac.ir*

HIGHLIGHTS

- Low-cost modified camera and thermal camera was used to collect images of maize plants.
- The performance of Tc-based indicators and vegetation indices for monitoring maize water status was assessed.
- High correlations were found between indices based on thermal images with maize water status.
- RGRI-Otsu method was used to improve the accuracy of Tc extraction and background elimination.

INTRODUCTION

As the world population grows, the world population's living standards increase, diets change and the effects of climate change intensify. So, the development of agriculture is critical for food security. Agriculture is both a major cause and casualty of water scarcity, especially in arid and semi-arid areas worldwide (FAO 2019). Consequently, a deeper understanding is necessary of abiotic stress such as drought, salinity and low or high temperatures on productivity demands and plant growth and development, through precision irrigation (Ihuoma & Madramootoo 2017). Drought or crop water stress induces various biochemical and physiological responses in plants. Extensive studies on plant water status based on detecting the crop parameters and soil water status have been reported (Osakabe *et al.* 2014). However, laboratory methods of determining plant water status such as the measurement of the stomatal conductance, leaf water potential and sap flow are costly and time-consuming (Li *et al.* 2010; Pinheiro & Chaves 2011; Gerhards *et al.* 2018). For decades, sensing of crops from platforms such as satellites or unmanned aerial vehicles (UAVs) has presented a more appropriate method to use for extensively planted crops such as maize (Toureiro *et al.* 2017; Zhang *et al.* 2019). All of

doi: 10.2166/ws.2020.270

them are equipped with hyperspectral, multispectral or digital imaging techniques to calculate indices from their spectral bands and also provide crop details such as plant water status (Sugiura *et al.* 2007; Santesteban *et al.* 2017; Zhang *et al.* 2018a, 2018b). However, satellite platforms have limitations such as high cost for high special resolution images, short availability of usable data (because clouds and cloud shadows may hide ground features), low temporal resolution and fixed schedule. On the other hand, most aerial imagery utility and availability are limited by the cost and complexity of the imaging system. So, consumer-grade cameras that have been designed to acquire imagery, reduce costs and increase the spatial resolution of images. Modified consumer-grade cameras, as another type of consumer-grade camera, replace the infrared (IR)-blocking filter in front of the complementary metal oxide semiconductor (CMOS) or charge coupled device (CCD) sensors with a long-pass IR filter to capture IR light spectral information details (Yang *et al.* 2014; Zhang *et al.* 2016). A low-cost modified camera that use external filters does not require special skill for opening the sensors and removing or changing them (Putra & Soni 2017).

Under water-stress, plant leaves display higher leaf temperature. Therefore, an infrared thermographic camera can measure canopy temperature. Hence, vegetation temperature can be used as an indicator of vegetation health. One issue that needs to be eliminated is the soil background from the thermal image. There are two commonly used methods to eliminate background pixels: (i) a threshold-based approach that uses thermal imagery only, and canopy temperature (*Tc*) extraction using algorithms such as Otsu and edge detection (Rud *et al.* 2014; Ludovisi *et al.* 2017; Park *et al.* 2017); (ii) a co-registration approach that uses both thermal and visible (red–green–blue (RGB)) images. These visible images were used to make a mask image to eliminate background pixels (Baluja *et al.* 2012; Han *et al.* 2018; Poblete *et al.* 2018).

The first indicator of crop stress conditions was proposed by Idso *et al.* (1981). They established a relationship between leaf-to-air temperature difference and vapor pressure deficit as the crop water stress index (CWSI). This model has been widely used to monitor plants' water status. However, this approach requires more meteorological data (Taghvaeian *et al.* 2014). Some researchers have attempted to reduce the number of parameters required to calculate crop water stress indicators. Han *et al.* (2016) developed the standard deviation of canopy temperature (CTSD) as a new approach to monitoring the water stress of maize crops, within a thermal image. Also, Zhang *et al.* (2018b) monitored water stress of cotton crops using Tc characteristics obtained from thermal images, including CTSD and coefficient of variation of Tc (CTCV).

In our study, a low-cost modified camera and external filters that included RGB and IR spectrum wavebands and also a thermal camera were used to collect images of maize plants cultivated in a greenhouse under different irrigation and nitrogen treatments. The objectives were to: (i) explore the performance of Tc-based indicators for monitoring maize water status; (ii) explore the use of low-cost modified camera and extracted vegetation indices in response to irrigation treatments and nitrogen variations; and (iii) clarify their correlations with relative water content (RWC). The results would provide guidance for developing non-destructive and rapid monitoring of water status in maize crops.

MATERIALS AND METHODS

Experimental site and crop management

The field experiment was planned with a randomized complete block design and three replicates during 2018. An early maturing variety of maize in a greenhouse located in the western region of Iran (34°47′N, and 48°28′E) was used. Four irrigation treatments (I_1: 100% (control), I_2: 80%, I_3: 60%, and I_4: 40% of field capacity) and also four nitrogen treatments (N_1: without nitrogen, N_2: 100 kg N/ha, N_3: 200 kg N/ha and N_4: 300 kg N/ha) were applied as urea (46% N). Maize seeds were sown, in pot containers of 18.38 L capacity. The pot substrate was made by loam soil and plant nutrient content. The chemical and physical properties were assessed and are presented in Table 1. Soil testing procedures were utilized to determine the level of plant-available nutrients. Also, a soil moisture meter (LUTRON PMS-714) was used for soil moisture analysis and irrigation volume calculated based on field capacity. The experiment was conducted in late June/early July 2018, and the harvesting date was 81 days after sowing (DAS). Plants emerged approximately two weeks prior to initiation of the experiments. We took visible, infrared and thermal images in the

Table 1 | Physical and chemical properties of soil

N (ppm)	P (ppm)	K (ppm)	O.C (%)	TNT (%)	PH	SP (%)	EC (μs/m)	Soil texture	Clay (%)	Silt (%)	Sand (%)
0.18	10.1	205	1.95	8	7.32	35	17.4	Loam	20.01	38.41	41.58

cases of days 23, 27, 36, 43, 48, 55 and 61 after sowing dates. Then one leaf of each plant was cut using scissors; afterwards leaf nitrogen content and RWC were calculated for each plant sample by Kjeldahl and relative turgidity methods, respectively (Kjeldahl 1883; Barrs & Weatherley 1962).

Acquisition and pretreatment of RGB, IR and thermal images

Figure 1 shows some of the equipment and processing steps of extracting plant indices from RGB, IR and thermal images.

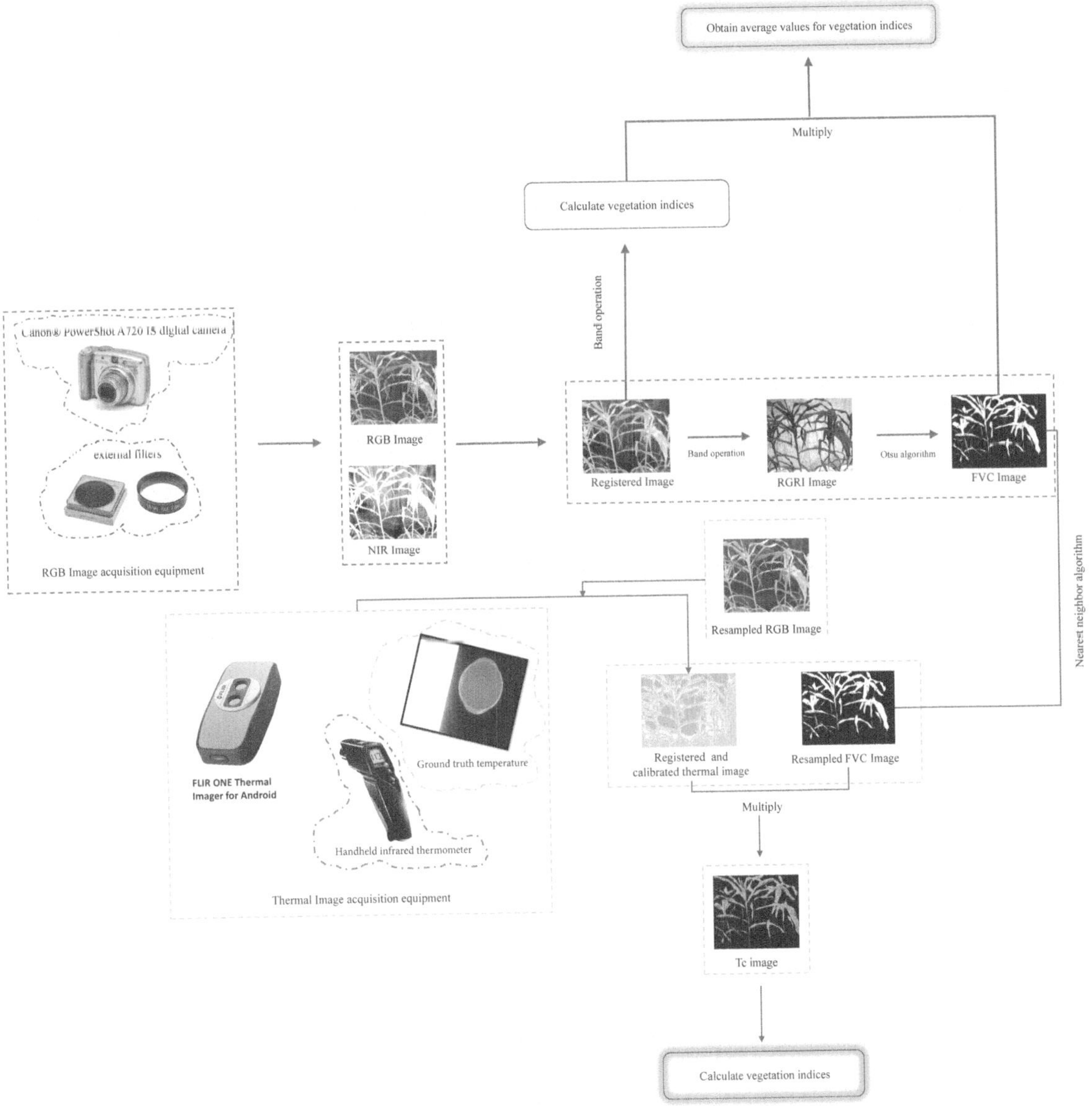

Figure 1 | The acquisition and pre-treatment of RGB, IR and thermal imagery for extracting vegetation indices.

Accordingly, for taking RGB and IR images, a Canon® PowerShot A720 IS digital camera (8 Megapixel CCD sensor) was modified to detect radiation in IR spectral bands and placed in front of each pot. Thereafter, the combination of the IR blocking filter and the IR filter was used. An IR filter (GREEN.L 67 mm IR 720) prevents visible light from passing through while only allowing IR light to strike the camera's sensor (IR720 passes only infrared rays above 720 nm), and a UV/IR cuts filter (Night Sky UV/IR Cut 2' Filter) cuts out IR rays above 700 nm and retains the visible light spectrum. These two circular filters attached to the front of the camera lens. In this way, after taking the image using the IR filter, the IR filter was separated from the lens and UV/IR cut filter placed on the front of the lens, and then the image was taken.

To initiate image processing, we registered visible and infrared images using MATLAB R2017a (Version 9.2). The image registration algorithm in shown in Figure 2. In the next step, for background elimination, the red–green ratio index (RGRI) was calculated using Equation (1), where R and G represent the digital numbers of the red and green bands.

$$RGRI = \frac{R}{G} \tag{1}$$

Then the Otsu algorithm was applied to the RGRI to separate green plants from the background (Verrelst *et al.* 2008). Otsu's algorithm is undoubtedly one of the most suitable thresholding techniques. It uses the image histogram data as input and finds a pixel value (so-called threshold level) that is able to separate pixels into two classes, foreground and background (Bangare *et al.* 2015). Therefore, the maize fractional vegetation cover (FVC) on RGB images was extracted. As shown in Figure 1, the FVC was used as a binary mask to segment the green plants against the background by multiplying FVC in the vegetation indices.

In the next step, to take thermal images, an FLIR ONE thermal imager for Android (resolution 0.1 °C) was used. Afterward, the spatial resolution of the RGB and FVC images was resampled to match the scale of the thermal images by using the nearest-neighbour interpolation algorithm. So, resampled visible and thermal images were registered, using MATLAB R2017a (Version 9.2). At each image, ground-truth temperature measurements were taken. The ground-truth temperature was measured by a gun-style design of the TESTO 830-T1 infrared thermometer from water, maize leaves and white–black boards. To calibrate the thermal imagery, a linear regression model was established between ground-truth temperature and the obtained digital number from the thermal images (Harvey *et al.* 2016; Bian *et al.* 2019; Sagan *et al.* 2019). Then by multiplying the FVC images in registered and calibrated images, crop temperature was extracted. In the following, we chose and calculated eight indices. The formulas for calculating the indices are given in Table 2. We obtained these indices in each plant and compared with RWC by statistical methods.

Statistical analyses

The regression models between vegetation indices and RWC were developed. Then the adjusted determination coefficient

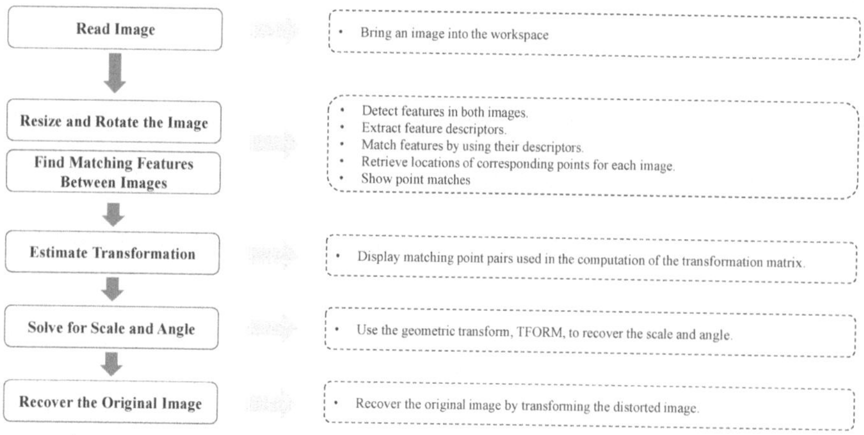

Figure 2 | The image registration algorithm.

Table 2 | Selected vegetation indices for RGB, IR and thermal images

Vegetation indices	Formula	Equation	Reference	Definition
CTCV	$\frac{CTSD}{\overline{Tc}}$	(2)	Zhang *et al.* (2018b)	Coefficient of variation of *Tc*
CTSD	$\sqrt{\left[\frac{(T_{c1}-\overline{Tc})^2+(T_{c2}-\overline{Tc})^2+\ldots+(T_{cn}-\overline{Tc})^2}{n}\right]}$	(3)	Han *et al.* (2016)	Standard deviation of *Tc*
DANS	$\begin{cases} \overline{Tc}-T_{NS} & , \overline{Tc}>28^{\circ}C \\ 0 & , \overline{Tc}<28^{\circ}C \end{cases}$	(4)	Taghvaeian *et al.* (2014)	Degrees above non-stress
DVI	$IR - R$	(5)	Tucker (1979)	Difference vegetation index
EVI	$2.5((IR-R)/(IR+6R-7.5B+1))$	(6)	Huete *et al.* (2002)	Enhanced vegetation Index
NDVI	$(IR-R)/(IR+R)$	(7)	García *et al.* (2017)	Normalized difference vegetation index
SR	IR/R	(8)	Le Maire *et al.* (2004)	Simple ratio
TRVI	$\sqrt{(IR-R)/(IR+R)}+0.5$	(9)	Deering *et al.* (1975)	Transformed vegetation index

Note: *R*, *G*, *B* and *IR* are the mean of the digital number of red, green, blue and infrared bands, respectively. $\overline{Tc}$ is the mean maize *Tc* within a sampling plot derived by the RGRI-Otsu method, T_{NS} is the non-stressed maize *Tc* and 28 °C has been suggested by the ARS Plant Stress and Water Conservation Laboratory (DeJonge *et al.* 2015), T_{ci} ($i = 1, 2, \ldots, n$) is the actual *Tc* of each maize canopy pixel within a sampling plot, and *n* is the number of maize canopy pixels within a sampling plot.

(R^2) (Woebbecke *et al.* 1995), Root Mean Square Error (RMSE) and Mean Bias Error (MBE) (Coelho *et al.* 2018) were calculated for comparisons. RMSE and MBE are given in Equations (10) and (11):

$$RMSE = \sqrt{\frac{\sum_{i=1}^{N}(Yobs_i - Yest_i)^2}{N}} \tag{10}$$

$$MBE = \frac{\sum_{i=1}^{N}(Yobs_i - Yest_i)}{N} \tag{11}$$

where: N is the number of sample data, $Yobs_i$ is the observed values and $Yest_i$ the estimated values.

RESULT AND DISCUSSION

Irrigation water applied, and leaf nitrogen status

Figure 3(a) shows the specific dates and cumulative irrigation depth of irrigation events. The average total amount of water applied was 457 mm for the control treatment, 416 mm for 80% of field capacity (I_2), 360 mm for 60% of field capacity (I_3), and 349 mm for 40% of field capacity (I_4). Figure 3(b) presents the box plot of leaf N concentration during the study. Throughout the growing season, younger leaves have lower leaf N concentrations, but

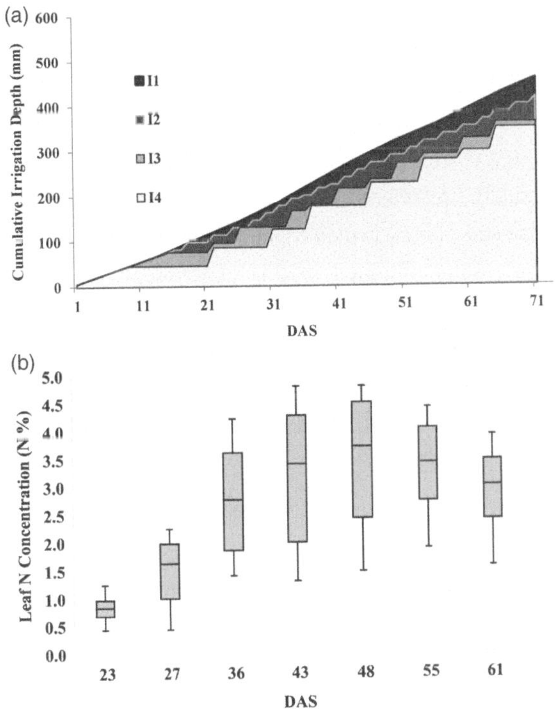

Figure 3 | (a) Cumulative irrigation depth for four irrigation treatments (control (I_1), 80% of field capacity (I_2), 60% of field capacity (I_3), and 40% of field capacity (I_4)). (b) The boxplots of leaf N concentration (N %) for maize samples.

with plant growth, leaf N concentrations increased. Also, because of the rapid growth before silking, for all fertilizer N rates, a gradual decline in leaf N concentration was observed. After silking, N content slightly increased then continued to decline (Schepers *et al.* 1992).

Distribution of ground-truth *Tc* and RWC

In Figure 4, the box plot shows the RWC rate data over the study. Regarding the irrigation scheduling (Figure 3(a)), RWC values were varied in each day of the growth stage and their values were also lower when water stress increased. Figure 5 shows the distributions of ground-truth *Tc* and RWC for I_1, I_2, I_3 and I_4 based on data acquired on DAS 36 and 48, respectively. As shown in Figure 5, there was a difference between the distributions of ground-truth *Tc* and RWC among treatment groups. The results, which are not presented in this paper, showed that there were also not significant interaction effects between nitrogen and irrigation on ground-truth *Tc* and RWC. In I_1 treatment, soil moisture was always maintained at field capacity level, so RWC was higher than 80% and the temperature was lowest. However, with prolonged water deficit among other treatments, the *Tc* increased with decreasing severity of RWC, and generally, as the RWC increases temperature usually decreases or vice versa, and generally, drought-stressed plants display higher *Tc* than well-watered plants. The results obtained in this experiment indicated that drought stress significantly decreased the relative water content of maize and increased leaf temperature due to drought stress that might have occurred due to increased respiration and decreased transpiration resulting from stomatal closure (Siddique *et al.* 2000).

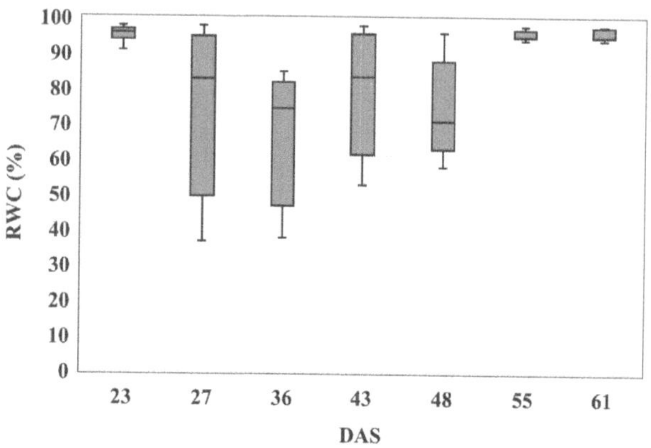

Figure 4 | The boxplots of relative water content (%) for maize samples.

Calibration of temperature derived from thermal images and extraction of maize *Tc*

Figure 6 shows the relationships between temperatures derived from the thermal images and the temperature of water and white and black boards as ground-truth measurements. There are good linear correlations with a slope of about 0.8. However, the intercept of the linear correlation for water and the two diffuse boards is 5.25, 6.29, and 8.51, respectively. Also Figure 7 shows the relationship between ground-truth *Tc* and *Tc* derived from thermal images using linear regression analysis. There is a high correlation with an R^2 value of 0.87 and an RMSE value of 1.41 °C. Meanwhile, the slope and intercept of the linear correlation are 0.91 and 2.16, respectively. In this study, we acquired data at about 2 m height and in the RGRI-Otsu method, the nearest-neighbour algorithm was applied to resample the FVC image resolution to thermal image resolution (from 3,264 × 2,448 to 640 × 480 pixels). Similar to the results of Yang *et al.* (2014), our findings showed that the linear regression model (Figures 6 and 7) had a slope of approximately 1.0. Consequently, the temperatures derived from thermal images were accurate (Zhang *et al.* 2019).

Relationships between crop water stress indicators and RWC of maize

Figure 8 and Table 3 illustrate the relationships between crop water stress indicators and RWC and the detail of the linear regression models. In particular, degrees above non-stress (DANS), CTSD and CTCV provided the best coefficient of determination values with $R^2 = 0.88$, 0.76 and 0.67, respectively. All three methods showed a low RMSE and MBE for RWC estimation compared with the other indices studied. Other used indices like difference vegetation index (DVI), transformed vegetation index (TRVI), normalized difference vegetation index (NDVI), enhanced vegetation index (EVI) and simple ratio (SR) showed coefficients of determination less than 0.50. When comparing RWC estimation through these indices by means of RMSE and MBE, these indices performed worse than the three top indices. Also, there were negative correlations between RWC and the three top indices, that is, the mean DANS, CTSD and CTCV increased with the decrease in relative

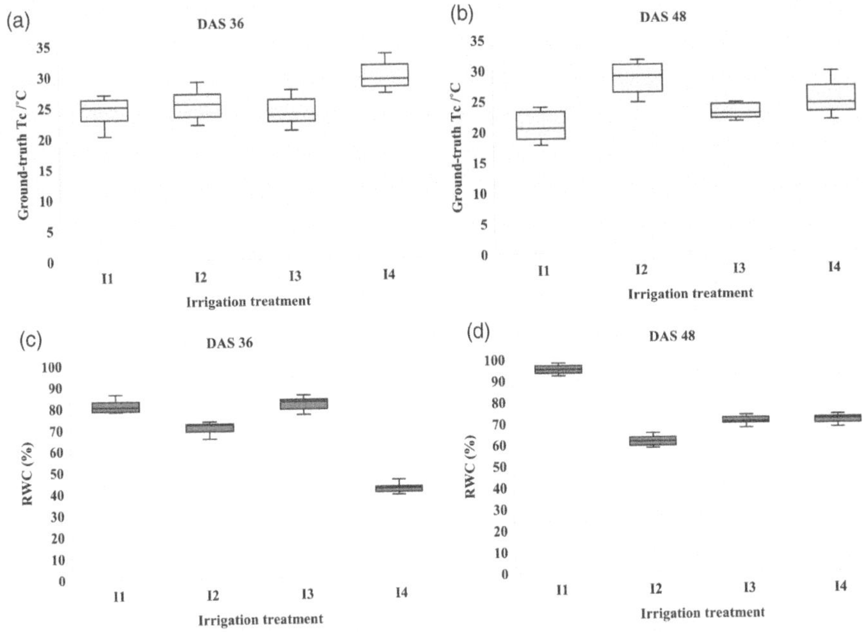

Figure 5 | The distributions of ground-truth *Tc* and RWC for the studied irrigation treatments at four different levels (I_1: 100%, I_2: 80%, I_3: 60%, and I_4: 40% of field capacity). Panels (a) and (c) are ground-truth *Tc* and RWC acquired on DAS 36, respectively, while panels (b) and (d) are ground-truth *Tc* and RWC acquired on DAS 48, respectively.

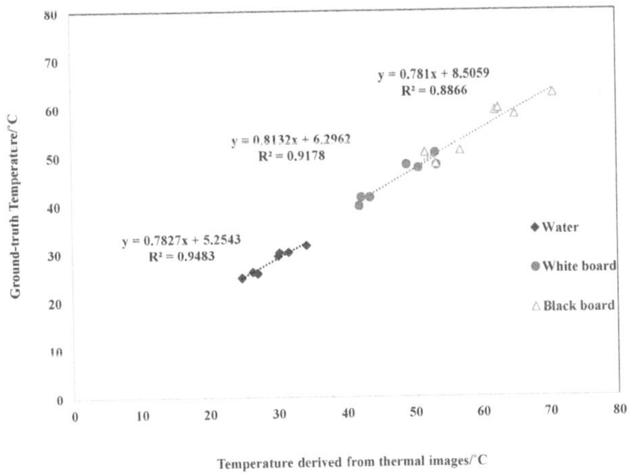

Figure 6 | Regression model between temperatures derived from thermal imagery and temperature of water and black–white boards.

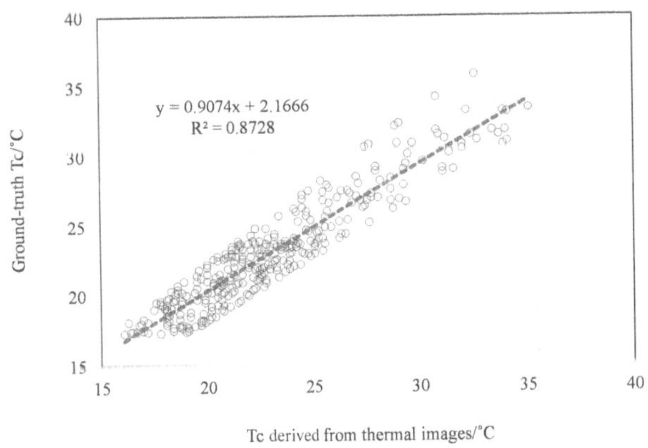

Figure 7 | Linear regression model between maize canopy temperatures (*Tc*) extracted from thermal images and by ground-truth.

water content (Taghvaeian *et al.* 2014), while for other indices, there were positive correlations. In general, compared with the studied indices, DANS was the more effective indicator of RWC in maize and was more sensitive to irrigation treatments, so that we were able to determine when and how much water to apply. Zhang *et al.* (2019) presented the same result in their paper. However, some researchers have investigated that CWSI has great potential to be used in the assessment of crop water stress (Lee & Park 2019; Costa-Filho *et al.* 2020). But the more important point is that DANS greatly simplifies the calculation and does not require any meteorological factors compared with CWSI, and so could be used in the diagnosis of crop water stress (Zhang *et al.* 2018b). After reviewing these studies and studies on CWSI, we strongly recommend additional research to compare the performance of the studied water stress indicators in different locations and under different meteorological conditions.

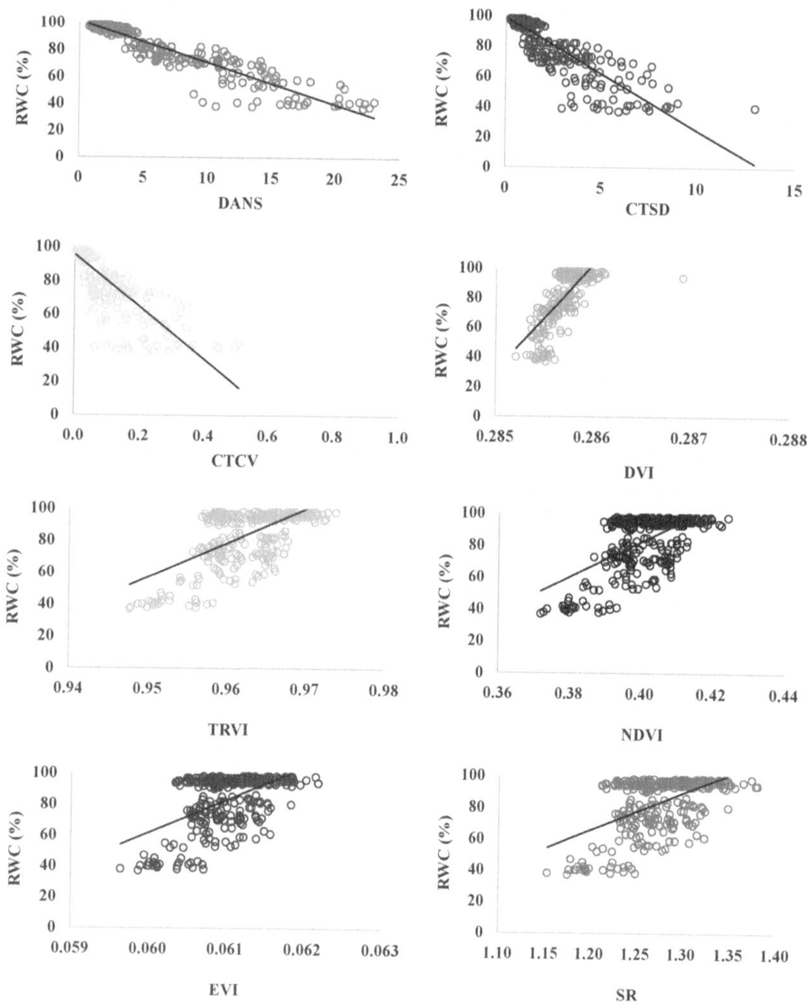

Figure 8 | Relationships of crop water stress indicators with RWC of maize.

Table 3 | The regression equations between RWC and crop water stress indicators

INDEX	RWC	R^2	RMSE	MBE
DANS	$Y = -3.054(X) + 101.02$	0.88	5.794287	−0.00321
CTSD	$Y = -7.3981(X) + 98.795$	0.76	8.395986	−0.00024
CTCV	$Y = -155.44(X) + 95.515$	0.69	9.539177	−0.00024
DVI	$Y = 53{,}056(X) - 15{,}076$	0.41	11.67889	−0.29376
TRVI	$Y = 2{,}145(X) - 1{,}980.6$	0.40	13.28678	−0.04385
NDVI	$Y = 1{,}100.8(X) - 357.99$	0.37	13.52913	0.009812
EVI	$Y = 21{,}039(X) - 1{,}200.6$	0.31	14.1716	0.007598
SR	$Y = 235.91(X) - 217.47$	0.30	14.26647	0.006633

CONCLUSION

The use of both modified consumer-grade cameras and thermal cameras allowed the estimating of RWC, to assess maize water status. In fact, high correlations were found between indices based on thermal images with maize water status. To improve the accuracy of *Tc* extraction and background elimination, we used the RGRI-Otsu method based on the combination of RGB and thermal images. The *Tc* extracted was highly correlated with the ground-truth measurements with R^2 of 0.87 and RMSE of 1.41 °C.

This finding showed that DANS, CTSD and CTCV can estimate maize water stress and were also simpler than the CWSI index. This method is low-cost and easy to use; therefore, there is a huge potential for implementing it on UAVs that allow farmers to observe their fields from the sky. However, further studies are required to test the performance of it for other crops and climatic and farming-practice conditions.

DATA AVAILABILITY STATEMENT

All relevant data are included in the paper or its Supplementary Information.

REFERENCES

Baluja, J., Diago, M. P., Balda, P., Zorer, R., Meggio, F., Morales, F. & Tardaguila, J. 2012 Assessment of vineyard water status variability by thermal and multispectral imagery using an unmanned aerial vehicle (UAV). *Irrigation Science* **30** (6), 511–522. doi:10.1007/s00271-012-0382-9.

Bangare, S. L., Dubal, A., Bangare, P. S. & Patil, S. T. 2015 Reviewing Otsu's method for image thresholding. *International Journal of Applied Engineering Research* **10** (9), 21777–21783.

Barrs, H. D. & Weatherley, P. E. 1962 A re-examination of the relative turgidity technique for estimating water deficits in leaves. *Australian Journal of Biological Sciences* **15** (3), 413–428. doi:10.1071/bi9620413.

Bian, J., Zhang, Z., Chen, J., Chen, H., Cui, C., Li, X. & Fu, Q. 2019 Simplified evaluation of cotton water stress using high resolution unmanned aerial vehicle thermal imagery. *Remote Sensing* **11** (3), 267. doi:10.3390/rs11030267.

Coelho, A. P., Rosalen, D. L. & de Faria, R. T. 2018 Índices de vegetação na predição da produtividade de biomassa e grãos de aveia-branca sob lâminas de irrigação (Vegetation indices in the prediction of biomass and grain yield of white oat under irrigation levels). *Pesquisa Agropecuária Tropical* **48** (2), 109–117. doi:10.5039/agraria.v14i4a7310.

Costa-Filho, E., Chávez, J. L. & Comas, L. 2020 Determining maize water stress through a remote sensing-based surface energy balance approach. *Irrigation Science* **38** (5), 501–518. doi:10.1007/s00271-020-00668-1.

Deering, D. W., Rouse Jr, J. W., Haas, R. H. & Schell, J. A. 1975 Measuring forage production of grazing units from LANDSAT MSS data. In: *Paper Presented at the Proc. of the 10th Intl. Symp. on Remote Sens. of Env Iron. II*.

DeJonge, K. C., Taghvaeian, S., Trout, T. J. & Comas, L. H. 2015 Comparison of canopy temperature-based water stress indices for maize. *Agricultural Water Management* **156**, 51–62. doi:10.1016/j.agwat.2015.03.023.

FAO 2019 *Water scarcity – one of the greatest challenges of our time*. http://www.fao.org/fao-stories/article/en/c/1185405/.

García, C. E., Montero, D. & Chica, H. A. 2017 Evaluation of a NIR camera for monitoring yield and nitrogen effect in sugarcane. *Agronomía Colombiana* **35** (1), 82–91. doi:10.15446/agron.colomb.v35n1.60852.

Gerhards, M., Schlerf, M., Rascher, U., Udelhoven, T., Juszczak, R., Alberti, G., Miglietta, F. & Inoue, Y. 2018 Analysis of airborne optical and thermal imagery for detection of water stress symptoms. *Remote Sensing* **10** (7), 1139. doi:10.3390/rs10071139.

Han, M., Zhang, H., DeJonge, K. C., Comas, L. H. & Trout, T. J. 2016 Estimating maize water stress by standard deviation of canopy temperature in thermal imagery. *Agricultural Water Management* **177**, 400–409. doi:10.1016/j.agwat.2016.08.031.

Han, M., Zhang, H., DeJonge, K. C., Comas, L. H. & Gleason, S. 2018 Comparison of three crop water stress index models with sap flow measurements in maize. *Agricultural Water Management* **203**, 366–375. doi:10.1016/j.agwat.2018.02.030.

Harvey, M. C., Rowland, J. V. & Luketina, K. M. 2016 Drone with thermal infrared camera provides high resolution georeferenced imagery of the Waikite geothermal area, New Zealand. *Journal of Volcanology and Geothermal Research* **325**, 61–69. doi:10.1016/j.jvolgeores.2016.06.014.

Huete, A., Didan, K., Miura, T., Rodriguez, E. P., Gao, X. & Ferreira, L. G. 2002 Overview of the radiometric and biophysical performance of the MODIS vegetation indices. *Remote Sensing of Environment* **83** (1–2), 195–213. doi:10.1016/s0034-4257(02)00096-2.

Idso, S. B., Jackson, R. D., Pinter Jr, P. J., Reginato, R. J. & Hatfield, J. L. 1981 Normalizing the stress-degree-day parameter for environmental variability. *Agricultural Meteorology* **24**, 45–55. doi:10.1016/0002-1571(81)90032-7.

Ihuoma, S. O. & Madramootoo, C. A. 2017 Recent advances in crop water stress detection. *Computers and Electronics in Agriculture* **141**, 267–275. doi:10.1016/j.compag.2017.07.026.

Kjeldahl, J. 1883 Neue Methode zur Bestimmung des Stickstoffs in organischen Körpern. *Zeitschrift für Analytische Chemie* **22** (1), 366–382. doi:10.1007/bf01338151.

Lee, D.-H. & Park, J.-H. 2019 Comparison between NDVI and CWSI for waxy corn growth monitoring in field soil conditions. In: *Remote Sensing for Agriculture, Ecosystems, and Hydrology XXI* (C. M. U. Neale & A. Maltese, eds), Proceedings Volume 11149, SPIE, Bellingham, WA, USA, 111491L. doi:10.1117/12.2533558.

Le Maire, G., François, C. & Dufrêne, E. 2004 Towards universal broad leaf chlorophyll indices using PROSPECT simulated database and hyperspectral reflectance measurements. *Remote Sensing of Environment* **89** (1), 1–28. doi:10.1016/j.rse.2003.09.004.

Li, L., Nielsen, D. C., Yu, Q., Ma, L. & Ahuja, L. R. 2010 Evaluating the crop water stress index and its correlation with latent heat and CO_2 fluxes over winter wheat and maize in the North China plain. *Agricultural Water Management* **97** (8), 1146–1155. doi:10.1016/j.agwat.2008.09.015.

Ludovisi, R., Tauro, F., Salvati, R., Khoury, S., Mugnozza Scarascia, G. & Harfouche, A. 2017 UAV-based thermal imaging for high-throughput field phenotyping of black poplar response to drought. *Frontiers in Plant Science* **8**, 1681. doi:10.3389/fpls.2017.01681.

Osakabe, Y., Osakabe, K., Shinozaki, K. & Tran, L.-S. P. 2014 Response of plants to water stress. *Frontiers in Plant Science* **5**, 86. doi:10.3389/fpls.2014.00086.

Park, S., Ryu, D., Fuentes, S., Chung, H., Hernández-Montes, E. & O'Connell, M. 2017 Adaptive estimation of crop water stress in nectarine and peach orchards using high-resolution imagery from an unmanned aerial vehicle (UAV). *Remote Sensing* **9** (8), 828. doi:10.3390/rs9080828.

Pinheiro, C. & Chaves, M. M. 2011 Photosynthesis and drought: can we make metabolic connections from available data? *Journal of Experimental Botany* **62** (3), 869–882. doi:10.1093/jxb/erq340.

Poblete, T., Ortega-Farías, S. & Ryu, D. 2018 Automatic coregistration algorithm to remove canopy shaded pixels in UAV-borne thermal images to improve the estimation of crop water stress index of a drip-irrigated Cabernet Sauvignon vineyard. *Sensors* **18** (2), 397. doi:10.3390/s18020397.

Putra, B. T. W. & Soni, P. 2017 Evaluating NIR-red and NIR-red edge external filters with digital cameras for assessing vegetation indices under different illumination. *Infrared Physics & Technology* **81**, 148–156. doi:10.1016/j.infrared.2017.01.007.

Rud, R., Cohen, Y., Alchanatis, V., Levi, A., Brikman, R., Shenderey, C., Heuer, B., Markovitch, T., Dar, Z., Rosen, C., Mulla, D. & Nigon, T. 2014 Crop water stress index derived from multi-year ground and aerial thermal images as an indicator of potato water status. *Precision Agriculture* **15** (3), 273–289. doi:10.1007/s11119-014-9351-z.

Sagan, V., Maimaitijiang, M., Sidike, P., Eblimit, K., Peterson, K. T., Hartling, S., Esposito, F., Khanal, K., Newcomb, M., Pauli, D., Ward, R., Fritschi, F., Shakoor, N. & Mockler, T. 2019 UAV-based high resolution thermal imaging for vegetation monitoring, and plant phenotyping using ICI 8640 P, FLIR Vue Pro R 640, and thermoMap cameras. *Remote Sensing* **11** (3), 330. doi:10.3390/rs11030330.

Santesteban, L. G., Di Gennaro, S. F., Herrero-Langreo, A., Miranda, C., Royo, J. B. & Matese, A. 2017 High-resolution UAV-based thermal imaging to estimate the instantaneous and seasonal variability of plant water status within a vineyard. *Agricultural Water Management* **183**, 49–59. doi:10.1016/j.agwat.2016.08.026.

Schepers, J. S., Francis, D. D., Vigil, M. & Below, F. E. 1992 Comparison of corn leaf nitrogen concentration and chlorophyll meter readings. *Communications in Soil Science and Plant Analysis* **23** (17–20), 2173–2187. doi:10.1080/00103629209368733.

Siddique, M. R. B., Hamid, A. & Islam, M. S. 2000 Drought stress effects on water relations of wheat. *Botanical Bulletin of Academia Sinica* **41** (1) 35–39.

Sugiura, R., Noguchi, N. & Ishii, K. 2007 Correction of low-altitude thermal images applied to estimating soil water status. *Biosystems Engineering* **96** (3), 301–313. doi:10.1016/j.biosystemseng.2006.11.006.

Taghvaeian, S., Comas, L., DeJonge, K. C. & Trout, T. J. 2014 Conventional and simplified canopy temperature indices predict water stress in sunflower. *Agricultural Water Management* **144**, 69–80. doi:10.1016/j.agwat.2014.06.003.

Toureiro, C., Serralheiro, R., Shahidian, S. & Sousa, A. 2017 Irrigation management with remote sensing: evaluating irrigation requirement for maize under Mediterranean climate condition. *Agricultural Water Management* **184**, 211–220. doi:10.1016/j.agwat.2016.02.010.

Tucker, C. J. 1979 Red and photographic infrared linear combinations for monitoring vegetation. *Remote Sensing of Environment* **8** (2), 127–150. doi:10.1016/0034-4257(79)90013-0.

Verrelst, J., Schaepman, M. E., Koetz, B. & Kneubühler, M. 2008 Angular sensitivity analysis of vegetation indices derived from CHRIS/PROBA data. *Remote Sensing of Environment* **112** (5), 2341–2353. doi:10.1016/j.rse.2007.11.001.

Woebbecke, D. M., Meyer, G. E., Von Bargen, K. & Mortensen, D. A. 1995 Shape features for identifying young weeds using image analysis. *Transactions of the ASAE* **38** (1), 271–281. doi:10.13031/2013.27839.

Yang, C., Westbrook, J. K., Suh, C. P.-C., Martin, D. E., Hoffmann, W. C., Lan, Y., Fritz, B. K. & Goolsby, J. A. 2014 An airborne multispectral imaging system based on two consumer-grade cameras for agricultural remote sensing. *Remote Sensing* **6** (6), 5257–5278. doi:10.3390/rs6065257.

Zhang, J., Yang, C., Song, H., Hoffmann, W. C., Zhang, D. & Zhang, G. 2016 Evaluation of an airborne remote sensing platform consisting of two consumer-grade cameras for crop identification. *Remote Sensing* **8** (3), 257. doi:10.3390/rs8030257.

Zhang, L., Niu, Y., Han, W. & Liu, Z. 2018a Establishing method of crop water stress index empirical model of field maize. *Nongye Jixie Xuebao/Transactions of the Chinese Society of Agricultural Machinery* **49** (5), 233–239.

Zhang, Z., Bian, J., Han, W., Fu, Q., Chen, S. & Cui, T. 2018b Cotton moisture stress diagnosis based on canopy temperature characteristics calculated from UAV thermal infrared image|. *Transactions of the Chinese Society of Agricultural Engineering* **34** (15), 77–84. doi:10.3390/rs11030267.

Zhang, L., Niu, Y., Zhang, H., Han, W., Li, G., Tang, J. & Peng, X. 2019 Maize canopy temperature extracted from UAV thermal and RGB imagery and its application in water stress monitoring. *Frontiers in Plant Science* **10**, 1270. doi:10.3389/fpls.2019.01270.

First received 11 August 2020; accepted in revised form 5 October 2020. Available online 19 October 2020

Estimation of ecosystem evapotranspiration in a *Robinia pseudoacacia* L. plantation with the use of the eddy covariance technique and modeling approaches

Nikos Markos and Kalliopi Radoglou

Nikos Markos
Kalliopi Radoglou (corresponding author)
Department of Forestry and Management of Environment and Natural Resources,
Democritus University of Thrace,
N. Orestiada,
Greece
E-mail: *kradoglo@fmenr.duth.gr*

ABSTRACT

The eddy covariance technique provides reliable ecosystem-level evapotranspiration (ET) measurements. These measurements, when combined with models and satellite products, could offer high spatiotemporal coverage and valuable mechanistic interpretation of the underlying processes. This study address one-year eddy covariance measurements from a *Robinia pseudoacacia* site in Northern Greece and remote sensing products: we (a) provide a medium-term description of daily ET fluxes for a *R. pseudoacacia* plantation in a degraded land, (b) assess the contribution of environmental drivers (e.g. net radiation, temperature etc.) on ET and (c) evaluate a simple satellite and meteorological driven model for larger-scale applications, based on the Land Surface Water Index (LSWI) and the FAO approach. *R. pseudoacacia* was found to have quite high water consumption, especially during leaf expansion. Net radiation and soil water content had the greatest effect on ecosystem evapotranspiration. LSWI was found to be correlated with both soil water content and evapotranspiration. Its use as an index for water limitation in models leads to high accuracy when compared to ET measurements. Our results (a) provide a significant contribution to the assessment of *R. pseudoacacia* ecophysiology and (b) highlight the potential of accurate ecosystem ET estimation with simple modeling approaches.

Key words | black locust, deciduous forest, eddy covariance, evapotranspiration, Land Surface Water Index, restoration, *Robinia pseudoacacia*, transpiration

HIGHLIGHTS

- The eddy covariance method contributes to the better knowledge of the species ecophysiology, as well as its performance under highly degraded soils.
- An evapotranspiration model, evaluated against the eddy covariance measurements, appears to give a very good performance.
- Land Surface Water Index was found to correlate well with ecosystem evapotranspiration and can be used as a crop coefficient scalar.

doi: 10.2166/ws.2021.142

INTRODUCTION

Evapotranspiration (ET) is one of the greatest components of energy fluxes and water cycling in the atmosphere (Katul *et al.* 2012). It is responsible for the returning of up to 2/3 of precipitation water back to the atmosphere on average globally (Jasechko *et al.* 2013; Wei *et al.* 2017; Yao *et al.* 2017), an amount that can be as high as 95% in dry ecosystems (Raz-Yaseef *et al.* 2010; Stoy *et al.* 2019). The accurate estimation of evapotranspiration is essential for a better understanding of the global water cycle (Cheng *et al.* 2011) and the improvement of water resource management (Mu *et al.* 2013a; Paschalis *et al.* 2018).

For ET estimation, several methods have been developed, including the use of the eddy covariance method (Baldocchi 2014), the Bowen Ratio (Scott *et al.* 2006), sap flow, large aperture scintillometers, lysimeters and others (Allen *et al.* 2011). The eddy covariance method is considered the most reliable and state-of-the-art technique for accurate ET measurement at the field scale (Zhang *et al.* 2014; Saitta *et al.* 2020).

All methods mentioned above can cover small to medium-sized areas. The estimation of ET for larger areas is performed with the use of ET models. Several ET modeling approaches exist with different degrees of complexity (Chen & Liu 2020). The basis of those models is solving the energy balance equation using the well-known Penman-Monteith equation (Allen *et al.* 1998). ET models can be summarized into two different categories: (a) those that are based on the land surface temperature (LST), which is closely linked to ecosystem water status and thus ET (Bastiaanssen *et al.* 1998; Anderson *et al.* 2007; Mu *et al.* 2013b; Chen & Liu 2020) and (b) those that are based on the crop coefficient approach, the so-called FAO approach (Allen *et al.* 1998). The actual evapotranspiration was estimated as the product of reference evapotranspiration (ET0) multiplied by a factor called crop coefficient (Kc).

Both approaches have significantly benefited from the advantages that remote sensing is offering for large scale monitoring. The first category of models uses the thermal bands of Modis or Landsat satellites to estimate LST by applying quite sophisticated approaches (Anderson *et al.* 2012; Mu *et al.* 2013b). However, as satellite thermal bands are provided in lower spatial resolutions when the modeling applications refer to smaller areas, there is a need for downscaling with special algorithms (Agam *et al.* 2007; Gao *et al.* 2012).

The second category has been so far following a more semi-empirical approach. While reference evapotranspiration is estimated as a function of environmental parameters, the crop coefficients are empirical species, specific factors, fluctuating during the year according to the species' growth stage (Allen *et al.* 1998; Pereira *et al.* 2021). In the past few years, satellite vegetation indices have been recognized to offer great advantages in crop coefficient modeling (Pereira *et al.* 2021). So far, several studies have used satellite spectral indices to represent the seasonal fluctuation of crop coefficients, such as NDVI (Hunsaker *et al.* 2005; Drerup *et al.* 2017; French *et al.* 2020), or relative indices, like EVI and Soil Adjusted Vegetation Index (SAVI) (Guerschman *et al.* 2009; Glenn *et al.* 2011; He *et al.* 2019). Most of the studies that follow similar approaches refer to agricultural sites, which means that water availability is not a marginal factor, so NDVI and EVI can represent the phenological status. However, it is difficult to describe the water stress effects. In satellite primary productivity models that use the light use efficiency approach (Monteith 1972), the Land Surface Water Index (LSWI), (Gao 1996)) was widely used as a representative of ecosystem water status (Xiao 2004; Mahadevan *et al.* 2008; Chen *et al.* 2011; Bao *et al.* 2016). The index is related to the total canopy water content (Gao 1996), and thus it could potentially represent both the canopy growth stage and water stress. This approach can provide further benefits; the vegetation indices mentioned can be acquired with a high spatial resolution, such as Sentinel 2, making the applications more suitable for small or medium-sized areas.

Modeling applications have been mainly focused on agricultural sites or natural ecosystems. On the other hand, the applications in forest plantations on disturbed or degraded land, like black locust plantations (Kanzler *et al.* 2020), are generally missing.

Black locust (*Robinia pseudoacacia* L.) is a medium-sized deciduous species. Although it is considered an invasive species in Europe, it has already become a very common component of both economy and culture in many European countries (Vítková *et al.* 2017). Even though the use of the species in forest plantations has met a lot of criticism, the species has a long history of use in restoration projects due to its unique characteristics to adapt to marginal sites. It is a fast-growing, drought-tolerant and low demand species, able to grow in poor-nutrient areas (Nicolescu *et al.* 2020). Because of these unique characteristics, the species has become very common in reclamation projects for disturbed lands, such as former coal mines (Skousen & Zipper 2014; Carl *et al.* 2017; Vítková *et al.* 2017; Nicolescu *et al.* 2020). In Europe, black locust has been studied for a long time, with the studies being focused more on its economic value than ecology (Vítková *et al.* 2017). Over the last few years, many studies of the species have been mainly from the Loess Plateau in China (Wang *et al.* 2015; Jiao *et al.* 2018; Miyazawa *et al.* 2018; Zhao *et al.* 2018). Despite that, there are things about its ecology and ecophysiology that remain elusive.

The aim of this study is (a) to provide a medium-term presentation of measurements of daily ET fluxes for a tree plantation with the use of the eddy covariance method, (b) to assess the contribution of the main environmental driving parameters of the ET fluxes for the specific site and (c) to provide a simple, but accurate methodology for the extrapolation of the eddy tower measurements to greater scales with the use of a hybrid satellite and meteorological driven model, based on high spatial analysis vegetation indices from the Sentinel 2 satellite.

MATERIALS AND METHODS

Study site

The site is located on a flat field of about 5×10^5 m^2 (40.59 °N, 21.65 °E) that is part of a greater planted area of about 24×10^6 m^2 in the restored areas of the Western Macedonia Lignite Center, in Northern Greece (Figure 1). The dominant species of the plantations is *Robinia pseudoacacia*, a deciduous broadleaf species, which accounts for 95% of the total area of the plantations. The plantations consist of mature and younger trees, due to the continuous restoration and plantation process of more than 20 years. The age of the trees at the study site is about 20 years. The mean tree height in the study site is 13.5 m and is relatively homogenous. *Robinia pseudoacacia* is a fast-growing species that can reach its maximum height before the age of 20 years (Nicolescu *et al.* 2020). The seasonal fluctuation of the Enhanced Vegetation Index (EVI) in the plantation, as estimated from the Sentinel 2 imagery, for a five-year period (2016–2020) is steady (see supplementary material). EVI is an index closely related to Leaf Area Index (LAI) (area of transpiration). Understory vegetation consists mainly of perennial grasses, with *Cynodon dactylon* being the dominant one. The growing period of the understory grass starts after leaf fall, as the leaves of black locust are very thin and delicate, resulting in a very quick decomposition, and it is active until the trees' leaf expansion at the beginning of May. The substrate consists of deposition material from the mining process that, in general, is very depleted and poor. It is characterized by alkaline pH and classified from moderately coarse to fine soils (Papadopoulos *et al.* 2015). Mean annual precipitation for the ten-year period in question (2010–2020) is 510 ± 156 mm, and mean annual temperature for the same period is 13.36 ± 0.92 °C, as estimated from the nearby meteorological station of the National Observatory of Athens (http://penteli.meteo.gr/stations/amyntaio/).

Meteorological and eddy covariance measurements

An eddy tower was installed at the end of July 2019, measuring continuously the exchanges of vapor, energy and CO_2 in the study area. The tower was installed in the middle of the study site in order to achieve a unidirectional footprint (no prevailing wind direction). The tower is equipped with an integrated IRGA analyzer and sonic anemometer (Irgason, Campbell Scientific Inc.) with an enhanced barometer (PTB 110, Vaisala Inc.), which perform CO_2/H_2O, three-component wind analysis and barometric pressure respectively and a full meteorological station, including, among others, a net radiometer (NR Lite2, Kipp and Zonen BV), a temperature/humidity probe (HygroClip2 Advanced, Rotronic Inc.) and three soil heat flux plates (HFP01, Hukseflux Thermal Sensors BV). The Irgason analyzer was placed in the top of the tower, at a total height of 17.5 m,

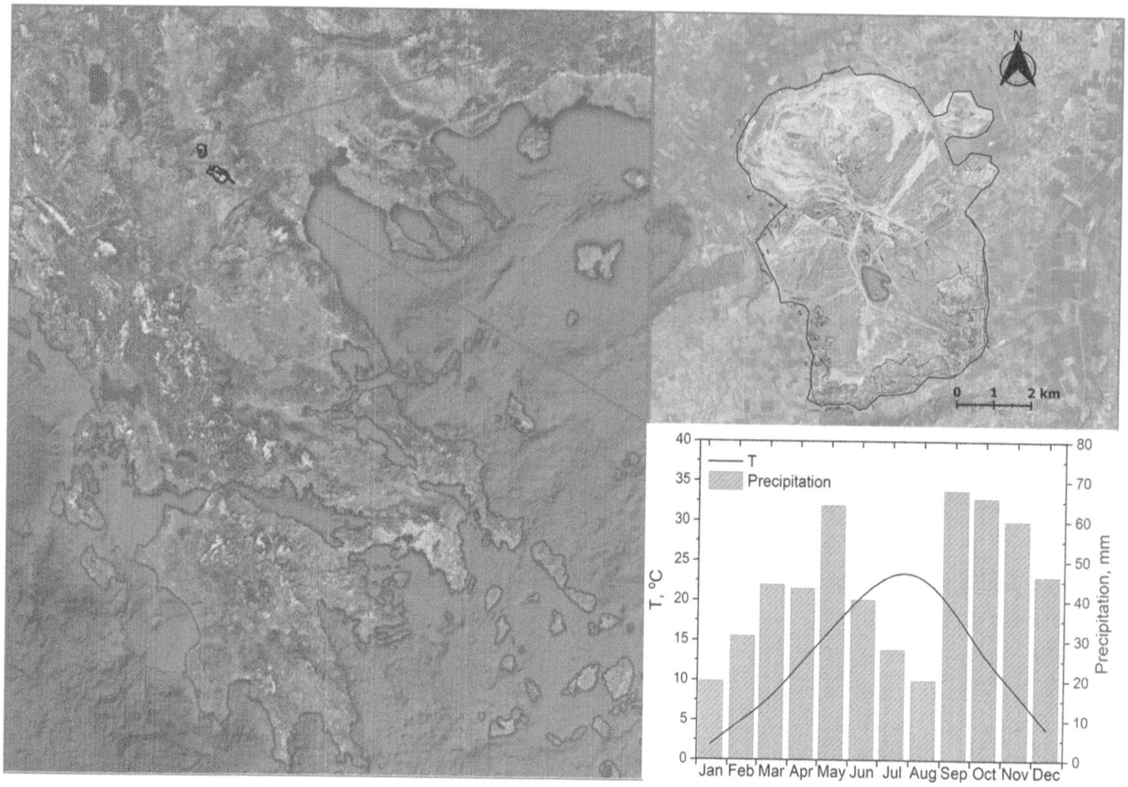

Figure 1 | Map and ombrothermic diagram for the study site. The eddy tower is installed inside the area marked with the red polygon (upper right). The ombrothermic diagram refers to the period 2010–2020. The full color version of this figure is available in the online version of this paper, at http://dx.doi.org/10.2166/ws.2021.142.

providing us an average fetch radius of 200 meters. Soil water content (SWC) is measured at 15 cm depth. Raw data are collected from a datalogger (CR1000X, Campbell Scientific Inc.) at a frequency of 10 Hz and processed on-site via an integrated analysis software (EasyFlux DL, Campbell Inc.) and post-processed with EddyPro 7 (Licor GmBh) with the use of the standardized eddy covariance methodology (Aubinet *et al.* 1999). The default time step of data acquisition was 30 min. Daily values were obtained by integrating or averaging the instantaneous values, depending on the parameter. Data gap-filling and flux partitioning were done using a standardized methodology (Reichstein *et al.* 2005). Data analysis covered the period from August 2019 to September 2020.

Reference evapotranspiration estimation

Reference evapotranspiration for the study site was calculated according to the FAO's methodology (Allen *et al.* 1998), Equation (1)

$$ET_0 = \frac{0.408(R_n - G) + \gamma \frac{900}{T + 273} u(es - ea)}{\Delta + \gamma(1 + 0.34u)} \quad (1)$$

where

ET_0 is the reference evapotranspiration (mm d^{-1})
R_n is the net radiation of the surface (MJ m^{-2} d^{-1})
G is the soil heat flux (MJ m^{-2} d^{-1})
γ is the psychrometric constant (kPa $^oC^{-1}$)
T is the mean daily air temperature (^{o}C)
es is the saturated vapor pressure (kPa)
ea is the actual vapor pressure (kPa)
Δ is the slope vapor pressure curve (kPa $^oC^{-1}$)
u is the wind speed (m s^{-1}).

All the above parameters are estimated by the eddy tower on half-hourly time steps and then scaled to daily time steps. By definition, air temperature and wind speed must be measured at the height of 2 meters or, if measured elsewhere, they have to be converted to that height. Our measurements were performed at the tower height (17.5 m) and no conversions were applied, to have a common reference with the other tower measurements.

Satellite image analysis

For the calculation of vegetation indices, we used the free-distributed products of Copernicus Sentinel 2 (https://sentinel.esa.int/web/sentinel/missions/sentinel-2), which offer the advantage of high resolution (10 × 10 m) and frequent acquisition. For our analyses, we used L2A products. The average acquisition of cloud-free images during the study period for our study site was four days.

Using five bands in total (blue, green, red, nir and swir), we estimated the time series of the vegetation indices Land Surface Water Index (LSWI), Normalized Difference Vegetation Index (NDVI), Enhanced Vegetation Index (EVI) and Normalized Difference Water Index (NDWI) (Table 1) for the whole study period.

All vegetation indices were estimated as the average ± standard deviation for the whole study site for cloud-free days, using the online analysis tool Sentinel-hub (https://www.sentinel-hub.com/). Sentinel 2 provides two SWIR bands, one at 1,565–1,655 nm and the other at 2,100–2,280 nm (https://sentinels.copernicus.eu). In this study, we used the SWIR1 (1,565–1,655 nm) band for the estimation of LSWI.

Table 1 | Definition and formulas for the vegetation indices used in the study

Index	Full name	Formula
NDVI	Normalized Difference Vegetation Index	$\frac{nir - red}{nir + red}$
EVI	Enhanced Vegetation Index	$2.5 * \frac{\text{nir} - \text{red}}{\text{nir} + 6.5 * \text{red} - 7.5 * \text{blue} + 1}$
LSWI	Land Surface Water Index	$\frac{nir - swir}{nir + swir}$
NDWI	Normalized Difference Water Index	$\frac{green - nir}{green + nir}$

Nir, red, green and swir stand for respective bands of Sentinel 2 imagery.

Modeling description and evaluation

For the modeling of ET, three different approaches were followed. The first one was the crop coefficient approach. Daily evapotranspiration was estimated as the product of reference evapotranspiration (as estimated from Equation (1)) multiplied by the crop coefficient Kc (Equation (2)). During this procedure, the daily values of Kc were estimated as the product of the maximum seasonal value (Kc_{max}) multiplied by the normalized fluctuation of the vegetation index (Equation (3)), according to Choudhury (1994).

$$ET = Kc * ET_0 = \text{Kc}_{\text{max}} * f(VI) * ET_0 \tag{2}$$

$$f(VI) = \frac{VI - VI_{min}}{VI_{max} - VI_{min}} \tag{3}$$

where

ET is the actual evapotranspiration (mm d^{-1})
ET_0 is the reference evapotranspiration (mm d^{-1}, Equation (1))
Kc is the crop coefficient (dimensionless)
VI, VI_{min} and VI_{max} are the running, minimum and maximum values of the corresponding spectral index during the study period.

Due to the specific conditions in the study site; that is, the two distinctive states of the vegetation (seasonally dominated either by grass or black locust canopy), we used two separate crop coefficients for different periods of the year. For the period between the leaf fall and new leaf expansion (November to April), during which the site's vegetation was dominated by the understory grass, the value of Kc_{max} equal to 1.2 was used (Allen *et al.* 2011). For the period after leaf expansion until leaf fall (May-October), during which the vegetation was dominated by the canopy, the value of Kc_{max} equal to 0.75 was used, which was the average maximum value of the ecosystem's Kc during the period of full leaf expansion, estimated as the ratio of ET/ET0 during the corresponding period, according to FAO methodology (Allen *et al.* 1998; Pereira *et al.* 2021). The value 0.75 for Kc_{max} used is inside the range of values for Kc of deciduous trees (0.45–1.2) proposed by FAO (Allen *et al.* 1998; Pereira *et al.* 2021). For the intermediate period from the beginning of leaf expansion until full development (April-May), during which both understory vegetation and canopy are active, the average of the two values was used. The vegetation index used as a scalar for Kc_{max} (Equation (3)) was better correlated to the measured ET fluxes (see results). As min and max values in Equation (3) minimum and maximum values of the indices during the study period were used and as the running values, the calculated indices, interpolated for the interval periods.

The second approach was similar to the one described above, with the difference that instead of ET0 we used the H_2O equivalent for net radiation (mm d^{-1}) as input to Equation (2). ET0 is a quantity that depends on many environmental parameters, as extensively described before. All the parameters do not have the same impact on actual ET. The subtraction of soil heat flux from net radiation should ideally be equal to the sum of latent and sensible heat flux (LE + H) which determines the total evaporative demand. We tried to further minimize the input parameters. An alternative version of the model was utilized. Total evaporative demand is not represented by reference evapotranspiration, but by the water equivalent of net radiation (mm d^{-1}). However, as the sum of LE + H are usually less than net radiation – global radiation (Rn-G) (Allen *et al.* 2011) and G is small compared to Rn, the water equivalent of Rn was multiplied by 0.88, a factor that resulted from the solving of the energy balance equation for our study site (see supplementary material).

The third approach was the derivation of an empirical formula from the regression of the measured daily evapotranspiration values with the satellite vegetation indices mentioned in the previous section (Table 1). Although this is the simplest approach, it lacks a mechanistic interpretation. While it can be useful for the specific site, it would be quite uncertain to use it in transferable applications.

For statistical analyses and diagram construction, the packages Origin 2020 and SPSS were used. The models' results were quantified against the measured ET values from the tower. The criteria used were the coefficients of determination from the linear regression analysis (R^2), the root mean square error (RMSE) (Equation (4)) and the Nash–Sutcliffe model coefficient (NSE) (Equation (5)), which determines the relative magnitude of the residual variance compared to the measured data variance (Nash & Sutcliffe 1970).

$$RMSE = \sqrt{\frac{\sum_{i=1}^{N} (x_m^i - x_o^i)^2}{N}} \tag{4}$$

$$NSE = 1 - \frac{\sum_{i=1}^{N} (x_m^i - x_o^i)^2}{\sum_{i=1}^{N} (x_o^i - \overline{x_o})^2} \tag{5}$$

The assessment of the importance of relative contribution of each independent variable (net radiation (Rn), wind speed (WS), temperature (T) and vapor pressure deficit (VPD) on ecosystem evapotranspiration was done by means of *P* values and standardized beta coefficients after the application of multiple regression analysis.

RESULTS AND DISCUSSION

Meteorological parameters

The seasonal fluctuation of meteorological data from the eddy tower is presented in Figure 2 and was used as inputs in Equation (1) to estimate reference evapotranspiration. The peak of net radiation (Rn) was observed during the period between mid-June to mid-July. Average Rn values were slightly higher during August 2020 compared to August 2019. Differences between 8/2019 and 8/2020 were observed at air temperature (T) and vapor pressure deficit (VPD) (Figure 2(b) and 2(c)). Both quantities were lower during 8/2020 compared to 8/2019 by about 2 °C and 0.3 kPa respectively. The precipitation and SWC are presented in Figure 2(d). The total precipitation amount for the whole study period (1/8/2019–27/9/2020) was 622 mm. Summer precipitation for 2020 (June – August) was quite high (170.8 mm), which accounted for 27.5% of total precipitation for the whole study period. SWC was kept high from the end of autumn until the middle of spring, when a rapid decrease was observed from the end of April until the end of May. In general, summer SWC values were quite high, as well as the precipitation. However, interannual differences were also observed, as the average % SWC during 8/2019 was 9.09 ± 1.55, while in

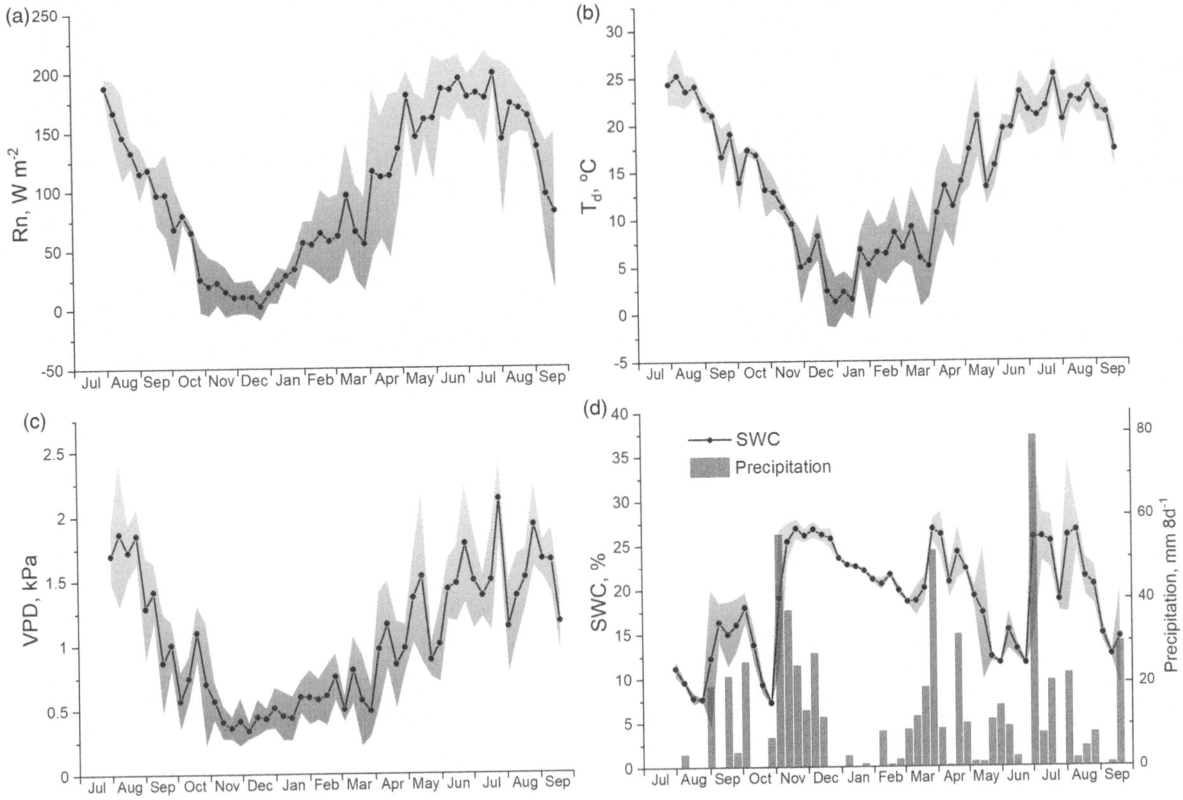

Figure 2 | Seasonal fluctuation of 8-day running averages of daily values of environmental parameters net radiation (a), air temperature (b), VPD (c) and SWC and precipitation (d) for the study period (2019–2020).

8/2020 23.54 ± 4.99 and the corresponding precipitation values were 3.1 and 36.9 mm, respectively. Finally, wind speed values did not appear to follow a clear seasonal pattern or have great differentiations, except for some higher values during the winter (data not shown).

Actual and reference evapotranspiration seasonal fluctuation

The seasonal fluctuation of both reference (ET0) as estimated by Equation (1) and actual evapotranspiration (ET), measured by the tower, is presented in Figure 3. The overall seasonal fluctuation of both quantities is generally following the annual pattern of net radiation and SWC (Figure 2). ET0 value is higher than the ET during the whole measured period. The lower annual rates are met during January, while two peaks can be found, one from May to the beginning of June and the other from July until the middle of August. Between those periods there is an intermediate period of a sharp decline in the activity, which reflects the significant drop in rainfall and the decrease in SWC. The ET rates are considerably high from February to April,

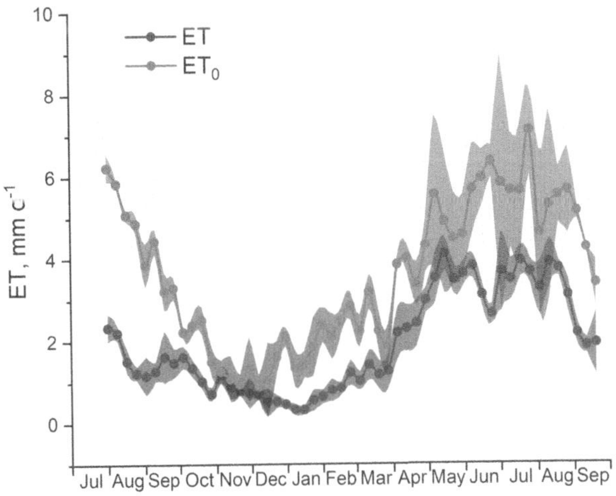

Figure 3 | Seasonal fluctuation of 8-day running averages of daily values of actual evapotranspiration (blue line) and reference evapotranspiration (red line) for the study period.

before the leaf expansion, which can be attributed to the well-developed understory grass.

The period of the first peak (May-June) is similar to the period of ET maximization for the same species in the Loess Plateau in China (Miyazawa *et al.* 2018). The trees of the referred study were grown on an arid to semi-arid region with a different precipitation pattern and that's probably the reason for this very narrow peak. The peak during the beginning of May in our study site could be considered quite early, as during May the canopy was still under development and full leaf expansion is met about the middle of June (data not shown). It seems this species and the whole ecosystem, consequently, follow a strategy of very luxury water consumption during the canopy development. The result of this is rapid soil drying until the beginning of June, despite the considerable amount of precipitation during that period (Figure 2(d)). The excessive water consumption of black locust and the resulting decline in soil moisture has been reported in another study from a relative ecosystem in China (Liang *et al.* 2018). The estimated weighted annual sum of ET with eddy covariance for *Robinia pseudoacacia* is 595 mm, a value that is very close to the respective total sum of precipitation. Although this value is comparable to many deciduous broadleaf sites of the fluxnet network (fluxnet.org) (Fatichi & Pappas 2017), there are not many sites with the same species that are comparable, as was mentioned previously. In a relevant study, the average total ET of a two-year study for *R. pseudoacacia* was estimated to be about 208 mm y^{-1}. However, the average amount of precipitation was quite lower, about 262 mm y^{-1} (Jiao *et al.* 2018). Another study, (Fu *et al.* 2021) estimated the annual evapotranspiration for an *R. pseudoacacia* plantation under non-water stress conditions at 576.88 and 848.55 mm y^{-1} for a dry and a wet year respectively.

The assumptions mentioned before are made with a quite high degree of uncertainty, since our results refer to the total ecosystem fluxes and any partitioning of ecosystem ET to transpiration and soil evaporation was not applied. Although generally the contribution of transpiration to total ecosystem evapotranspiration in closed canopies is estimated to be up to 90% (Jasechko *et al.* 2013; Good *et al.* 2014; Li *et al.* 2019), in a relevant study in a *R. pseudoacacia* plantation, the corresponding proportion was estimated between 9.7 and 28.8% (Jiao *et al.* 2018), which can be considered quite low.

Spectral indices

Measured ET values are strongly correlated with all studied vegetation indices, with coefficients of determination ranging from 0.51 to 0.77. The strongest correlated index was the LSWI ($R^2 = 0.77$, Figure 4(a)). LSWI was used as the input parameter to describe the Kc scalar (Equation (3)) for the models based on reference evapotranspiration (Equation (2)). However, as the initial idea was to use a scalar for Kc that would be representative of both phenological and water status, the performance of LSWI versus ET was investigated under different water status scenarios. Given that daily ET seems to co-fluctuate seasonally with SWC (Figures 2 and 3), the effect of SWC to ET was assessed by comparing the respective values for the period of full leaf expansion (June – August, Figure 5(a)). This period limitation was followed to avoid the misinterpretation or noise that would come from data from the non-growing period. According to our findings, SWC is related to daily ET via an exponential function (Figure 5(a)) with a very high coefficient of determination ($R^2 = 0.76$, $P < 0.001$). Based on this relationship, an apparent threshold of 15% SWC can be identified, below which ET values are strongly increasing with the increase of SWC and reach a plateau at about the value of 15% SWC. Following that threshold, the LSWI – ET values were separated into two sub-datasets, one that corresponded to SWC values greater than 15% and the other to SWC values less than 15% for the whole period of study, including the non-growing period. Regression analysis with the use of an exponential function was performed for the two datasets independently, as well as for the whole dataset together (Figure 5(b)). Based on the results of this analysis, ET is related to LSWI with exponential functions. Although ET values corresponding to SWC > 15% are significantly higher than the relative values that correspond to SWC < 15% for the same LSWI range, the use of a single exponential function applied to the whole dataset can be considered adequate to describe the ET dependence on LSWI for the whole SWC range, as nearly all values are inside the corresponding 95% prediction band (Figure 5(b)). Consequently, we can assume that LSWI is an appropriate index for the description of both the phenological stage and water status of our studied ecosystem. The accurate representation of water availability

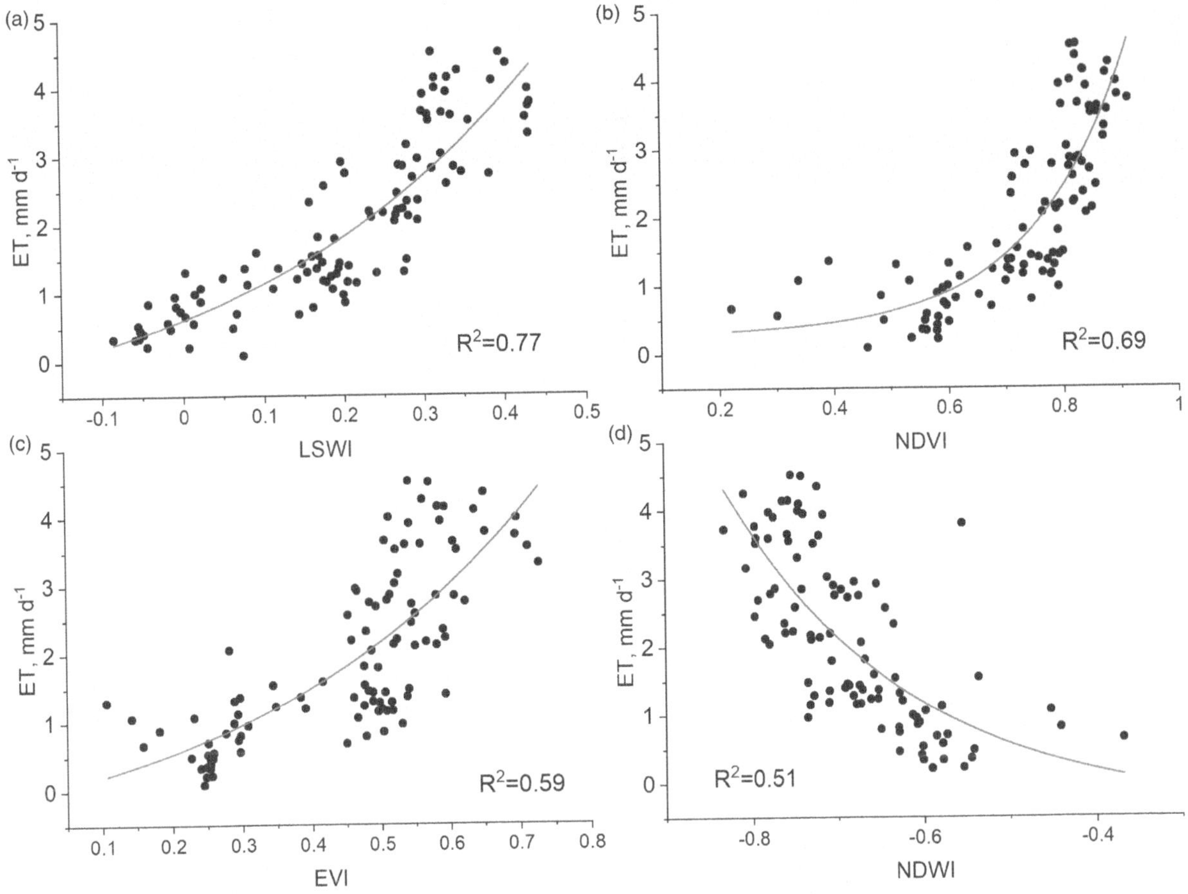

Figure 4 | Diagrams of correlation between daily ET values and and the coresponding satellite spectral indices LSWI (a), NDVI (b), EVI (c) and NDWI (d) for the study period.

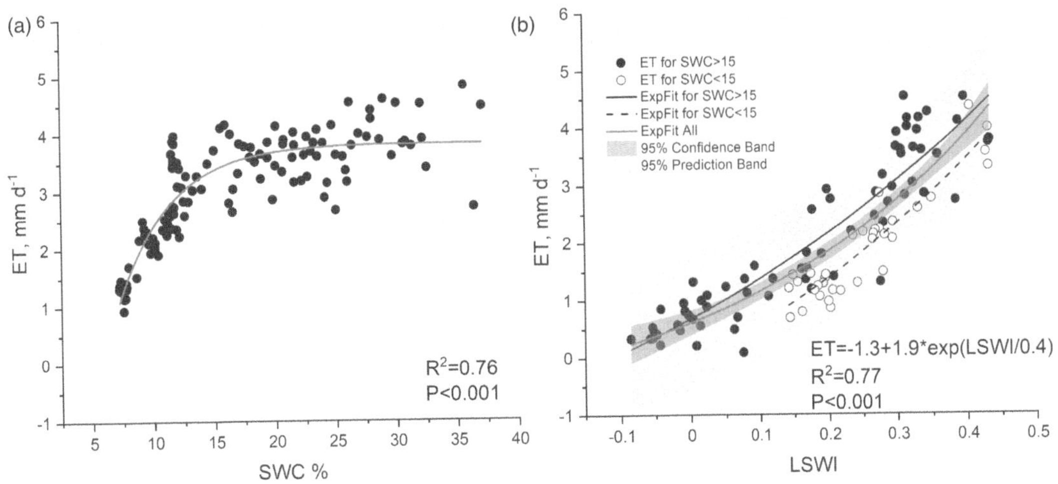

Figure 5 | Correlation of daily ET values with SWC during *Robinia*'s full leaf expansion (June – August, (a)) and correlation of daily ET values with LSWI for the whole study period (b). In (b), the closed circles stand for ET values that correspond to SWC > 15 and the open circles to SWC < 15. The solid black line is the regression of LSWI and ET for the first instance, the dashed black line for the second instance and the red line for the whole dataset. The 95% confidence and prediction bands refer to the regression for the whole dataset.

effects on ecosystems performance is one of the most important, yet demanding and uncertain procedures. Our findings highlight the great potential of LSWI to be used as a proxy for water limitations in our studied ecosystem and could act as a driver for further implications in other ecosystems.

Impact of environmental parameters on ET fluxes

Pearson R and regression coefficients R^2 between ET and ET0 daily values and the corresponding environmental parameters and LSWI are presented in Table 2. ET and ET0 are very strongly correlated with Rn, VPD, T and LSWI, while the correlation with wind speed is quite weaker for both quantities. The latter can be explained from the relative constant wind speed conditions throughout the study period. ET is better correlated with net radiation and LSWI, rather than ET0.

Table 2 | Pearson R coefficients and regression coefficients between the measured environmental parameters and actual (ET) and reference evapotranspiration (ETo)

	ET		ET0	
	Pearson R	R^2	**Pearson R**	R^2
Rn	0.80	0.65	0.95	0.90
VPD	0.60	0.36	0.88	0.78
T	0.64	0.41	0.79	0.62
WS	−0.06	0.00	−0.03	0.00
LSWI	0.87	0.76	0.74	0.55
ET0	0.78	0.61	–	–

Net radiation (Rn), vapor pressure deficit (VPD), air temperature (T), wind speed (WS), Land Surface Water Index (LSWI).

Both ET and ET0 are simultaneously affected by all the independent parameters mentioned above. The one-to-one comparison can lead to false conclusions about the real effect of each parameter on the final estimates. For example, net radiation (Rn) and temperature (T) follow similar seasonal fluctuation patterns, so their true relative impact cannot be distinguished in that way. Additional multiple regression analysis was applied, in order to estimate both the significance and the relative effect of each studied environmental parameter to ET and ET0. This was assessed by the *P* values and the standardized beta coefficients for each parameter. The results of the analysis are presented in Table 3.

According to the multiple regression analysis results, ET0 was significantly affected by all environmental parameters ($P < 0.001$), while the parameters that have greater effect on ET0 are net radiation and VPD, with beta values 0.64 and 0.39, respectively. The effect of temperature and wind speed are quite weaker. All independent parameters were positively correlated to ET0.

For actual ET, when LSWI was not included in the analysis, all environmental parameters were significant ($P < 0.001$) and Rn was the parameter with the strongest effect on ET. The second most important parameter was T. VPD was also found to have a significant effect; however, contrary to ET0 on this occasion, the effect is negative. The results of the analysis were altered when LSWI was included as an independent parameter in the analysis. In this case, both air temperature and VPD had no significant correlation to ET, as their *P* values are 0.27 and 0.71, respectively. The only important parameters are Rn, WS and LSWI. The parameter with the strongest effect was LSWI, followed by net radiation and wind speed.

Table 3 | Beta standardized coefficients and *P* values between actual (ET)/reference evapotranspiration (ET0) and the environmental parameters that are used as inputs to Equation (1)

	ET						ET0		
	LSWI excluded			LSWI included					
	Beta	**Sig.**	**Tolerance**	**Beta**	**Sig.**	**Tolerance**	**Beta**	**Sig.**	**Tolerance**
Rn	0.77	0.00	0.38	0.35	0.00	0.26	0.64	0.00	0.38
T	0.43	0.00	0.18	−0.06	0.27	0.14	0.05	0.00	0.18
VPD	−0.34	0.00	0.18	−0.02	0.71	0.16	0.39	0.00	0.18
WS	0.14	0.00	0.78	0.11	0.00	0.77	0.19	0.00	0.78
LSWI	–	–	–	0.68	0.00	0.29	–	–	–

Regarding Rn, these results are somehow expected, as Rn was the strongest component in the energy balance equation and thus determined the total evaporative demand. However, when T and VPD are analyzed, things are more complicated. Firstly, VPD is dependent on air temperature, so the use of both parameters in the same analysis can create some uncertainty. Secondly, while the increase of VPD increases the total evaporative demand, its effect on actual evapotranspiration could be either positive or negative (Massmann *et al.* 2019) as the increase of VPD leads to decrease in stomatal conductance (Dewar 2002).

According to the above, the model can be further simplified by using the water equivalent of net radiation (Rn) instead of reference evapotranspiration as a proxy for evaporative demand. This led to a model that relies on simplicity, without compromises in accuracy ($R^2 = 0.89$, RMSE = 0.53, NSE = 0.80). This can provide further advantages, as net radiation is a parameter that can be easily measured in the field or estimated from relevant satellite maps. The problem in this approach is the energy closure. The sum of LE + H does not equal to Rn-G, with an uncertainty that can be as high as 30%, with usual values around 10–20% (Wilson *et al.* 2002; Allen *et al.* 2011; Burba 2013). In the case of our studied ecosystem, the relative value is estimated around 12% (see supplementary material).

Minimizing the input parameters, without making great compromises in accuracy, is of great importance in ecosystem modeling. According to our findings, not all environmental parameters have the same impact on ET fluxes for the studied ecosystem. Net radiation and LSWI can be adequate in order to provide an adequate estimation of ET on seasonal basis. Considering that net radiation is a quite easily measured or estimated parameter, that satellite imagery offers many advantages in its estimation (Hou *et al.* 2014; Verma *et al.* 2016; Chen *et al.* 2020). Its use along with LSWI could meet further applications in evapotranspiration modeling (Barr *et al.* 2012).

Model performance

The seasonal fluctuation of measured against modeled *ET* values, according to the three different modeling approaches, as well as the comparison of the measured and modeled values, is presented in Figure 6. The modeled values derived from Equation (2), using ET0 as a proxy for evaporative demand, are presented as Model 1, while the corresponding values that use the water equivalent of net radiation as a proxy for evaporative demand are presented as Model 2. Finally, the modeled ET values that come from the empirical regression of measured ET with LSWI, are presented as Model 3. All the three models show very good seasonal co-fluctuation and agreement with the measured values for their 8-day running averages (Figure 6(a)–6(c)), with comparable R^2 values, ranging from 0.85 to 0.89 (Figure 6(d)–6(f)). The same applies when RMSE and NSE are considered, with values ranging from 0.45–0.53 and 0.80–0.85, respectively. It must be mentioned that typically this comparison can be considered valid only for the cases of Model 1 and Model 2, because the regression analysis in the case of Model 3 comes from the same dataset with which the model was evaluated, so typically that can be considered as autocorrelation. However, our scope in this case is to present how the modeled values of the empirical function co-fluctuate seasonally with the measured values and what are the advantages or disadvantages compared to the more mechanistic models.

In general, ET models have been reported to perform quite satisfactory in several studies with R^2 coefficients ranging from 0.3 to 0.86 for crops (Battude *et al.* 2017; He *et al.* 2019; French *et al.* 2020) and 0.12–0.90 in natural forests (Nagler *et al.* 2005; Verstraeten *et al.* 2005; Cleugh *et al.* 2007; Velpuri *et al.* 2013; Yao *et al.* 2017). According to that, our modeling approach resulted in a model performance among the higher range.

However, the advantages of the developed model are not mainly concerned with its performance, but its simplicity. Although several models have used similar approaches; that is, to use satellite indices to model the crop coefficient, the most commonly used indices were NDVI and EVI (Hunsaker *et al.* 2005; Guerschman *et al.* 2009; Glenn *et al.* 2011; Drerup *et al.* 2017; He *et al.* 2019; French *et al.* 2020), which can adequately describe the phenological stage of the vegetation but are poorer for its water status. On the other hand, LSWI is related to the total vegetation water content, and as a result to both phenological stage and water status (Gao 1996). Additionally, LSWI has been found to have a strong correlation with LST (Mokhtari *et al.* 2019; Alexander 2020) and thus can be used as a substitute of thermal spectrum

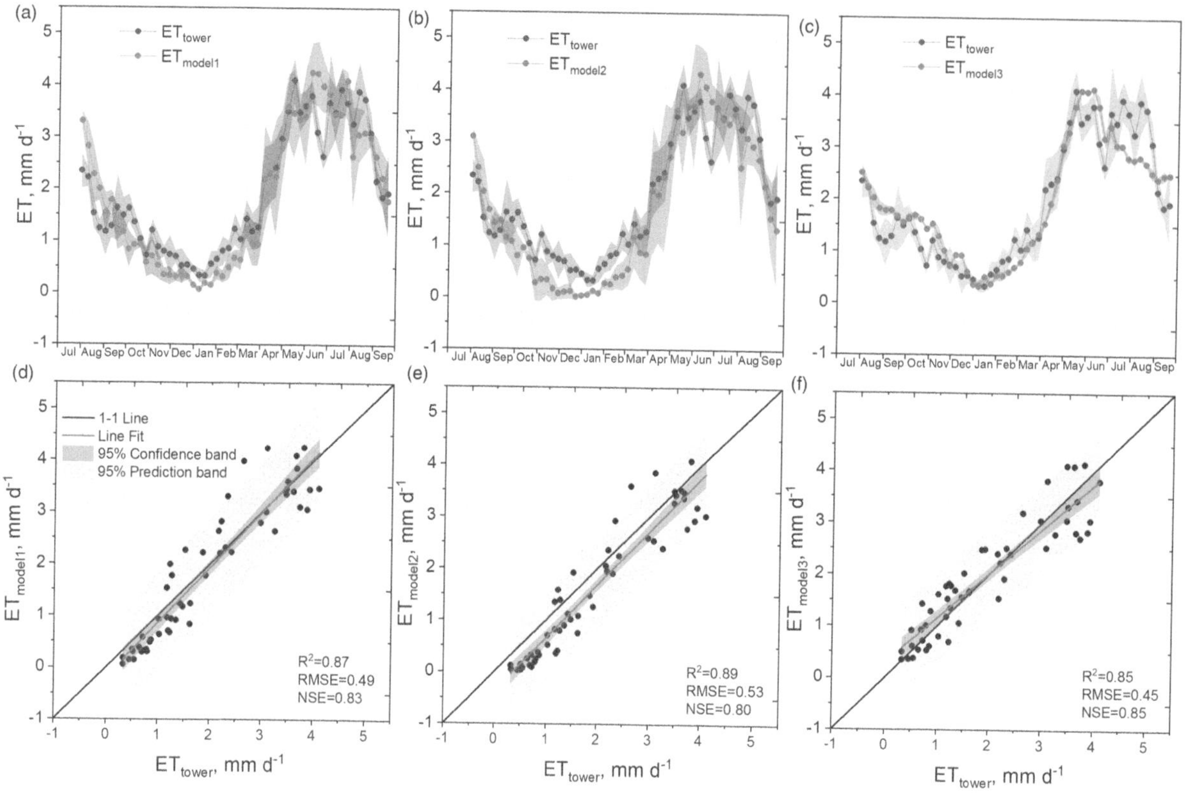

Figure 6 | Seasonal fluctuation of measured and modeled ET values at 8-day running averages; ET tower stands for the measured values from the eddy tower. ETmodel1 (a) and ETmodel2 stand for the modeled values using the reference evapotranspiration (b) and water equivalent of net radiation as proxy for evaporative demand. ETmodel3 stands for modeled ET values with the use of the empirical regression with LSWI. (c). The corresponing correlations of measured against modeled values with the three different modeling approaches (d)–(f), respectively. The coefficients of determination from the linear regression analysis (R2), the root mean square error (RMSE) and the Nash–Sutcliffe model efficiency coefficient (NSE).

(Sadeghi *et al.* 2015). Therefore, the use of LSWI, as a scalar to crop coefficient, could potentially work as a bridge between LST and crop coefficient based models. This is even more important if we considered that the studied ecosystem is not an agricultural site or a natural forest, but a plantation grown on a degraded site.

CONCLUSIONS

In this study we presented the results of more than one-year measurements and modeling of ecosystem evapotranspiration for a *Robinia pseudoacacia* plantation grown after restoration of highly disturbed land.

Reference evapotranspiration estimated from Equation (1) (ET0) has given a higher value from actually measured evapotranspiration (ET) from an eddy tower during the whole study period. A high value of ET was recorded in May to middle August. SWC is related to daily ET via an exponential function. The relationship between satellite spectral indices and the corresponding daily evapotranspiration values showed that LSWI was the most higher correlated index. From the environmental parameters, LSWI and Rn were those which had strong effect on ET. ET is dependent on LSWI and it is an appropriate index for the description of both phenological stage and water status of the studied ecosystem.

The results of this study contribute to the better understanding of the behavior *of Robinia pseudoacacia*, as well as the ecosystem processes of forest plantations grown on disturbed sites, and potentially they can lead to further investigation of the potential of minimization of the input parameters on ET models, without great compromises on model performance and accuracy.

In general, although *R. pseudoacacia* is already a species with an extensive distribution in Europe and worldwide, there are still many things to be studied concerning its ecology and behavior, and its contribution to the global carbon and water cycle. We believe that our findings can have a significant contribution in this direction.

ACKNOWLEDGEMENTS

The authors kindly acknowledge the Hellenic PPC (Public Power Corporation) S.A. for the provision of assisting personnel and necessary infrastructure during field campaigns and data collection. Special thanks are due to Lamprini Patmanidou and Simela Andreadi and their teams.

FUNDING

This research has been co-financed by the European Union and Greek national funds through the Operational Program Competitiveness, Entrepreneurship, and Innovation, under the call RESEARCH–CREATE–INNOVATE (project: COFORMIT. T1EDK-02521).

DATA AVAILABILITY STATEMENT

Data cannot be made publicly available; readers should contact the corresponding author for details.

REFERENCES

Agam, N., Kustas, W. P., Anderson, M. C., Li, F. & Neale, C. M. U. 2007 A vegetation index based technique for spatial sharpening of thermal imagery. *Remote Sensing of Environment* **107**, 545–558.

Alexander, C. 2020 Normalised difference spectral indices and urban land cover as indicators of land surface temperature (LST). *International Journal of Applied Earth Observation and Geoinformation* **86**, 102013.

Allen, R. G., Pereira, L. S., Raes, D. & Smith M, W. A. B. 1998 Crop evapotranspiration – Guidelines for computing crop water requirements – FAO Irrigation and Drainage Paper 56, 300pp.

Allen, R. G., Pereira, L. S., Howell, T. A. & Jensen, M. E. 2011 Evapotranspiration information reporting: I. Factors governing measurement accuracy. *Agricultural Water Management* **98**, 899–920.

Anderson, M. C., Norman, J. M., Mecikalski, J. R., Otkin, J. A. & Kustas, W. P. 2007 A climatological study of evapotranspiration and moisture stress across the continental United States based on thermal remote sensing: 1. Model formulation. *Journal of Geophysical Research Atmospheres* **112**, 1–17.

Anderson, M. C., Allen, R. G., Morse, A. & Kustas, W. P. 2012 Use of Landsat thermal imagery in monitoring evapotranspiration and managing water resources. *Remote Sensing of Environment* **122**, 50–65.

Aubinet, M., Grelle, A., Ibrom, A., Rannik, Ü., Moncrieff, J., Foken, T., Kowalski, A. S., Martin, P. H., Berbigier, P., Bernhofer, C., Clement, R., Elbers, J., Granier, A., Grünwald, T., Morgenstern, K., Pilegaard, K., Rebmann, C., Snijders, W., Valentini, R. & Vesala, T. 1999 Estimates of the annual net carbon and water exchange of forests: the EUROFLUX methodology. *Advances in Ecological Research* **30**, 113–175.

Baldocchi, D. 2014 Measuring fluxes of trace gases and energy between ecosystems and the atmosphere – the state and future of the eddy covariance method. *Global Change Biology* **20**, 3600–3609.

Bao, G., Bao, Y., Qin, Z., Xin, X., Bao Yulong Bayarsaikan, S., Zhou, Y. & Chuntai, B. 2016 Modeling net primary productivity of terrestrial ecosystems in the semi-arid climate of the Mongolian Plateau using LSWI-based CASA ecosystem model. *International Journal of Applied Earth Observation and Geoinformation* **46**, 84–93.

Barr, A. G., van der Kamp, G., Black, T. A., McCaughey, J. H. & Nesic, Z. 2012 Energy balance closure at the BERMS flux towers in relation to the water balance of the White Gull Creek watershed 1999–2009. *Agricultural and Forest Meteorology* **153**, 3–13. https://doi.org/10.1016/j.agrformet.2011.05.017.

Bastiaanssen, W. G. M., Menenti, M., Feddes, R. A. & Holtslag, A. A. M. 1998 A remote sensing surface energy balance algorithm for land (SEBAL): 1. Formulation. *Journal of Hydrology* **212–213**, 198–212.

Battude, M., Al Bitar, A., Brut, A., Tallec, T., Huc, M., Cros, J., Weber, J. J., Lhuissier, L., Simonneaux, V. & Demarez, V. 2017 Modeling water needs and total irrigation depths of maize crop in the south west of France using high spatial and temporal resolution satellite imagery. *Agricultural Water Management* **189**, 123–136.

Burba, G. 2013 *Eddy Covariance Method for Scientific, Industrial, Agricultural and Regulatory Applications: A Field Book on Measuring Ecosystem Gas Exchange and Areal Emission Rates*. LI-COR Biosciences, Lincoln, USA.

Carl, C., Landgraf, D., van der Maaten-Theunissen, M., Biber, P. & Pretzsch, H. 2017 *Robinia pseudoacacia* L. Flower analyzed by using Unmanned Aerial Vehicle (UAV). *Remote Sensing* **9**, 1–19.

Chen, J. M. & Liu, J. 2020 Evolution of evapotranspiration models using thermal and shortwave remote sensing data. *Remote Sensing of Environment* **237**, 111594.

Chen, M., Zhuang, Q., Cook, D. R., Coulter, R., Pekour, M., Scott, R. L., Munger, J. W. & Bible, K. 2011 Quantification of terrestrial ecosystem carbon dynamics in the conterminous United States combining a process-based biogeochemical model and MODIS and AmeriFlux data. *Biogeosciences* **8**, 2665–2688.

Chen, J., He, T., Jiang, B. & Liang, S. 2020 Estimation of all-sky all-wave daily net radiation at high latitudes from MODIS data. *Remote Sensing of Environment* **245**, 111842.

Cheng, L., Xu, Z., Wang, D. & Cai, X. 2011 Assessing interannual variability of evapotranspiration at the catchment scale using satellite-based evapotranspiration data sets. *Water Resources Research* **47** (9), 1–11.

Choudhury, B. J. 1994 Synergism of multispectral satellite observations for estimating regional land surface evaporation. *Remote Sensing of Environment* **49**, 264–274.

Cleugh, H. A., Leuning, R., Mu, Q. & Running, S. W. 2007 Regional evaporation estimates from flux tower and MODIS satellite data. *Remote Sensing of Environment* **106**, 285–304.

Dewar, R. C. 2002 The Ball-Berry-Leuning and Tardieu-Davies stomatal models: synthesis and extension within a spatially aggregated picture of guard cell function. *Plant, Cell and Environment* **25**, 1383–1398.

Drerup, P., Brueck, H. & Scherer, H. W. 2017 Evapotranspiration of winter wheat estimated with the FAO 56 approach and NDVI measurements in a temperate humid climate of NW Europe. *Agricultural Water Management* **192**, 180–188.

Fatichi, S. & Pappas, C. 2017 Constrained variability of modeled T: ET ratio across biomes. *Geophysical Research Letters* **44** (13), 6795–6803. https://doi.org/10.1002/2017GL074041.

French, A. N., Hunsaker, D. J., Sanchez, C. A., Saber, M., Gonzalez, J. R. & Anderson, R. 2020 Satellite-based NDVI crop coefficients and evapotranspiration with eddy covariance validation for multiple durum wheat fields in the US Southwest. *Agricultural Water Management* **239**, 106266.

Fu, C., Song, X., Li, L. & Li, H. 2021 Study on vegetation water requirements of *Robinia pseudoacacia* and its mixed plantation in Loess Gully Region. *IOP Conference Series: Earth and Environmental Science* **643**, 012122.

Gao, B. C. 1996 NDWI – A normalized difference water index for remote sensing of vegetation liquid water from space. *Remote Sensing of Environment* **58**, 257–266.

Gao, F., Kustas, W. P. & Anderson, M. C. 2012 A data mining approach for sharpening thermal satellite imagery over land. *Remote Sensing* **4**, 3287–3319.

Glenn, E. P., Neale, C. M. U., Hunsaker, D. J. & Nagler, P. L. 2011 Vegetation index-based crop coefficients to estimate evapotranspiration by remote sensing in agricultural and natural ecosystems. *Hydrological Processes* **25**, 4050–4062.

Good, S. P., Soderberg, K., Guan, K., King, E. G., Scanlon, T. M. & Caylor, K. K. 2014 δ2 h isotopic flux partitioning of evapotranspiration over a grass field following a water pulse and subsequent dry down. *Water Resources Research* **50**, 1410–1432.

Guerschman, J. P., Van Dijk, A. I. J. M., Mattersdorf, G., Beringer, J., Hutley, L. B., Leuning, R., Pipunic, R. C. & Sherman, B. S. 2009 Scaling of potential evapotranspiration with MODIS data reproduces flux observations and catchment water balance observations across Australia. *Journal of Hydrology* **369**, 107–119.

He, M., Kimball, J. S., Yi, Y., Running, S. W., Guan, K., Moreno, A., Wu, X. & Maneta, M. 2019 Satellite data-driven modeling of field scale evapotranspiration in croplands using the MOD16 algorithm framework. *Remote Sensing of Environment* **230**, 111201.

Hou, J., Jia, G., Zhao, T., Wang, H. & Tang, B. 2014 Satellite-based estimation of daily average net radiation under clear-sky conditions. *Advances in Atmospheric Sciences* **31**, 705–720.

Hunsaker, D. J., Pinter, P. J. & Kimball, B. A. 2005 Wheat basal crop coefficients determined by normalized difference vegetation index. *Irrigation Science* **24**, 1–14.

Jasechko, S., Sharp, Z. D., Gibson, J. J., Birks, S. J., Yi, Y. & Fawcett, P. J. 2013 Terrestrial water fluxes dominated by transpiration. *Nature* **496**, 347–350.

Jiao, L., Lu, N., Fu, B., Wang, J., Li, Z., Fang, W., Liu, J., Wang, C. & Zhang, L. 2018 Evapotranspiration partitioning and its implications for plant water use strategy: evidence from a black locust plantation in the semi-arid Loess Plateau, China. *Forest Ecology and Management* **424**, 428–438.

Kanzler, M., Böhm, C. & Freese, D. 2020 The development of soil organic carbon under young black locust (*Robinia pseudoacacia* L.) trees at a post-mining landscape in eastern Germany. *New Forests* **52**, 47–68.

Katul, G. G., Oren, R., Manzoni, S., Higgins, C. & Parlange, M. B. 2012 Evapotranspiration: a process driving mass transport and energy exchange in the soil-plant-atmosphere-climate system. *Reviews of Geophysics* **50** (3), RG3002.

Li, X., Gentine, P., Lin, C., Zhou, S., Sun, Z., Zheng, Y., Liu, J. & Zheng, C. 2019 A simple and objective method to partition evapotranspiration into transpiration and evaporation at eddy-covariance sites. *Agricultural and Forest Meteorology* **265**, 171–182.

Liang, H., Xue, Y., Li, Z., Wang, S., Wu, X., Gao, G., Liu, G. & Fu, B. 2018 Soil moisture decline following the plantation of *Robinia pseudoacacia* forests: evidence from the loess plateau. *Forest Ecology and Management* **412**, 62–69.

Mahadevan, P., Wofsy, S. C., Matross, D. M., Xiao, X., Dunn, A. L., Lin, J. C., Gerbig, C., Munger, J. W., Chow, V. Y. & Gottlieb, E. W. 2008 A satellite-based biosphere parameterization for net ecosystem CO_2 exchange: Vegetation Photosynthesis and Respiration Model (VPRM). *Global Biogeochemical Cycles* **22** (2), GB2005.

Massmann, A., Gentine, P. & Lin, C. 2019 When does vapor pressure deficit drive or reduce evapotranspiration? *Journal of Advances in Modeling Earth Systems* **11**, 3305–3320.

Miyazawa, Y., Du, S., Taniguchi, T., Yamanaka, N. & Kumagai, T. 2018 Gas exchange by the mesic-origin, arid land plantation species *Robinia pseudoacacia* under annual summer reduction in plant hydraulic conductance. *Tree Physiology* **38**, 1166–1179.

Mokhtari, A., Noory, H., Pourshakouri, F., Haghighatmehr, P., Afrasiabian, Y., Razavi, M., Fereydooni, F. & Sadeghi Naeni, A. 2019 Calculating potential evapotranspiration and single crop coefficient based on energy balance equation using Landsat 8 and Sentinel-2. *ISPRS Journal of Photogrammetry and Remote Sensing* **154**, 231–245.

Monteith, J. 1972 Solar radiation and productivity in tropical ecosystems. *Journal of Applied Ecology* **9**, 747–766.

Mu, Q., Zhao, M., Kimball, J. S., McDowell, N. G. & Running, S. W. 2013a A remotely sensed global terrestrial drought severity index. *Bulletin of the American Meteorological Society* **94**, 83–98.

Mu, Q., Zhao, M. & Running, S. W. 2013b MODIS global Terrestrial Evapotranspiration (ET) Product (NASA MOD16A2/A3) Collection 5. Algorithm Theoritical Basis Document. Available at: https://modis-land.gsfc.nasa.gov/pdf/MOD16ATBD.pdf.

Nagler, P. L., Scott, R. L., Westenburg, C., Cleverly, J. R., Glenn, E. P. & Huete, A. R. 2005 Evapotranspiration on western U.S. rivers estimated using the Enhanced Vegetation Index from MODIS and data from eddy covariance and Bowen ratio flux towers. *Remote Sensing of Environment* **97**, 337–351.

Nash, J. E. & Sutcliffe, J. V. 1970 River flow forecasting through conceptual models part I - a discussion of principles. *Journal of Hydrology* **10** (3), 282–290.

Nicolescu, V. N., Rédei, K., Mason, W. L., Vor, T., Pöetzelsberger, E., Bastien, J. C., Brus, R., Benčať, T., Dodan, M., Cvjetkovic, B., Andrašev, S., La Porta, N., Lavnyy, V., Mandžukovski, D., Petkova, K., Roženbergar, D., Wąsik, R., Mohren, G. M. J., Monteverdi, M. C., Musch, B., Klisz, M., Perić, S., Keça, L., Bartlett, D., Hernea, C. & Pástor, M. 2020 Ecology, growth and management of black locust (*Robinia pseudoacacia* L.), a non-native species integrated into European forests. *Journal of Forestry Research* **31**, 1081–1101.

Papadopoulos, C., Gekaa, C., Pavloudakis, F., Roumpos, C. & Andreadou, S. 2015 Evaluation of the soil quality on the reclaimed lignite mine land in West Macedonia, Greece. *Procedia Earth and Planetary Science* **15**, 928–932.

Paschalis, A., Fatichi, S., Pappas, C. & Or, D. 2018 Covariation of vegetation and climate constrains present and future T/ET variability. *Environmental Research Letters* **13** (10), 0–18. https://doi.org/10.1088/1748-9326/aae267.

Pereira, L. S., Paredes, P., Hunsaker, D. J., López-Urrea, R. & Mohammadi Shad, Z. 2021 Standard single and basal crop coefficients for field crops. Updates and advances to the FAO56 crop water requirements method. *Agricultural Water Management* **243**, 106466.

Raz-Yaseef, N., Rotenberg, E. & Yakir, D. 2010 Effects of spatial variations in soil evaporation caused by tree shading on water flux partitioning in a semi-arid pine forest. *Agricultural and Forest Meteorology* **150**, 454–462.

Reichstein, M., Falge, E., Baldocchi, D. D., Papale, D., Aubinet, M., Berbigier, P., Bernhofer, C., Buchmann, N., Gilmanov, T., Granier, A. A., Grunwald, T., Havrankova, K., Ilvesniemi, H., Janous, D., Knohl, A., Laurila, T., Lohila, A., Loustau, D., Matteucci, G., Meyers, T., Miglietta, F., Ourcival, J.-M. M., Pumpanen, J., Rambal, S., Rotenberg, E., Sanz, M., Tenhunen, J., Seufert, G. G., Vaccari, F., Vesala, T., Yakir, D., Valentini, R., Grünwald, T., Havránková, K., Ilvesniemi, H., Janous, D., Knohl, A., Laurila, T., Lohila, A., Loustau, D., Matteucci, G., Meyers, T., Miglietta, F., Ourcival, J.-M. M., Pumpanen, J., Rambal, S., Rotenberg, E., Sanz, M., Tenhunen, J., Seufert, G. G., Vaccari, F., Vesala, T., Yakir, D. & Valentini, R. 2005 On the separation of net ecosystem exchange into assimilation and ecosystem respiration: review and improved algorithm. *Global Change Biology* **11**, 1424–1439.

Sadeghi, M., Jones, S. B. & Philpot, W. D. 2015 A linear physically-based model for remote sensing of soil moisture using short wave infrared bands. *Remote Sensing of Environment* **164**, 66–76.

Saitta, D., Vanella, D., Ramírez-Cuesta, J. M., Longo-Minnolo, G., Ferlito, F. & Consoli, S. 2020 Comparison of orange orchard evapotranspiration by eddy covariance, sap flow, and FAO-56 methods under different irrigation strategies. *Journal of Irrigation and Drainage Engineering* **146**, 05020002.

Scott, R. L., Huxman, T. E., Cable, W. L. & Emmerich, W. E. 2006 Partitioning of evapotranspiration and its relation to carbon dioxide exchange in a Chihuahuan Desert shrubland. *Hydrological Processes* **20**, 3227–3243.

Skousen, J. & Zipper, C. E. 2014 Post-mining policies and practices in the Eastern USA coal region. *International Journal of Coal Science and Technology* **1**, 135–151.

Stoy, P. C., El-Madany, T. S., Fisher, J. B., Gentine, P., Gerken, T., Good, S. P., Klosterhalfen, A., Liu, S., Miralles, D. G., Perez-Priego, O., Rigden, A. J., Skaggs, T. H., Wohlfahrt, G., Anderson, R. G., Coenders-Gerrits, A. M. J., Jung, M., Maes, W. H., Mammarella, I., Mauder, M., Migliavacca, M., Nelson, J. A., Poyatos, R., Reichstein, M., Scott, R. L. & Wolf, S. 2019 Reviews and syntheses: turning the challenges of partitioning ecosystem evaporation and transpiration into opportunities. *Biogeosciences* **16**, 3747–3775.

Velpuri, N. M., Senay, G. B., Singh, R. K., Bohms, S. & Verdin, J. P. 2013 A comprehensive evaluation of two MODIS evapotranspiration products over the conterminous United States: using point and gridded FLUXNET and water balance ET. *Remote Sensing of Environment* **139**, 35–49.

Verma, M., Fisher, J. B., Mallick, K., Ryu, Y., Kobayashi, H., Guillaume, A., Moore, G., Ramakrishnan, L., Hendrix, V., Wolf, S., Sikka, M., Kiely, G., Wohlfahrt, G., Gielen, B., Roupsard, O., Toscano, P., Arain, A. & Cescatti, A. 2016 Global surface net-radiation at 5 km from MODIS Terra. *Remote Sensing* **8** (9), 739.

Verstraeten, W. W., Veroustraete, F. & Feyen, J. 2005 Estimating evapotranspiration of European forests from NOAA-imagery at satellite overpass time: towards an operational processing chain for integrated optical and thermal sensor data products. *Remote Sensing of Environment* **96**, 256–276.

Vítková, M., Müllerová, J., Sádlo, J., Pergl, J. & Pyšek, P. 2017 Black locust (*Robinia pseudoacacia*) beloved and despised: a story of an invasive tree in Central Europe. *Forest Ecology and Management* **384**, 287–302.

Wang, J. J., Hu, C. X., Bai, J. & Gong, C. M. 2015 Carbon sequestration of mature black locust stands on the Loess Plateau, China. *Plant, Soil and Environment* **61**, 116–121.

Wei, Z., Yoshimura, K., Wang, L., Miralles, D. G., Jasechko, S. & Lee, X. 2017 Revisiting the contribution of transpiration to global terrestrial evapotranspiration. *Geophysical Research Letters* **44**, 2792–2801.

Wilson, K., Goldstein, A., Falge, E., Aubinet, M., Baldocchi, D., Berbigier, P., Bernhofer, C., Ceulemans, R., Dolman, H., Field, C., Grelle, A., Ibrom, A., Law, B. E., Kowalski, A., Meyers, T., Moncrieff, J., Monson, R., Oechel, W., Tenhunen, J., Valentini, R. & Verma, S. 2002 Energy balance closure at FLUXNET sites. *Agricultural and Forest Meteorology* **113**, 223–243.

Xiao, X. 2004 Modeling gross primary production of temperate deciduous broadleaf forest using satellite images and climate data. *Remote Sensing of Environment* **91**, 256–270.

Yao, Y., Liang, S., Li, X., Chen, J., Liu, S., Jia, K., Zhang, X., Xiao, Z., Fisher, J. B., Mu, Q., Pan, M., Liu, M., Cheng, J., Jiang, B., Xie, X., Grünwald, T., Bernhofer, C. & Roupsard, O. 2017 Improving global terrestrial evapotranspiration estimation using support vector machine by integrating three process-based algorithms. *Agricultural and Forest Meteorology* **242**, 55–74.

Zhang, Z., Tian, F., Hu, H. & Yang, P. 2014 A comparison of methods for determining field evapotranspiration: photosynthesis system, sap flow, and eddy covariance. *Hydrology and Earth System Sciences* **18**, 1053–1072.

Zhao, Q., Wang, F., Zhao, J., Zhou, J., Yu, S. & Zhao, Z. 2018 Estimating forest canopy cover in black locust (*Robinia pseudoacacia* L.) plantations on the Loess Plateau using random forest. *Forests* **9**, 1–16.

http://penteli.meteo.gr/stations/amyntaio/

https://sentinel.esa.int/web/sentinel/missions/sentinel-2

fluxnet.org

https://www.sentinel-hub.com/

First received 31 January 2021; accepted in revised form 30 April 2021. Available online 17 May 2021

Rainfed cultivation with supplemental irrigation modelling on seed yield and oil of *Coriandrum sativum* L. using precision agriculture and GIS moisture mapping

Agathos Filintas, Eleni Wogiatzi and Nikolaos Gougoulias

ABSTRACT

The aim of the present study was to determine the effects of rainfed and supplemental irrigation, and sowing period (SP) treatments on coriander (*Coriandrum sativum* L.) yield, essential oil content and umbel heights by applying new agro-technologies (TDR-sensors for soil moisture (SM), geographical information system (GIS), precision agriculture, soil-hydraulic analyses and geostatistical models) for yield and SM root zone geospatial modelling and two-dimensional GIS mapping. Results of laboratory analysis indicated a suitable soil for coriander's growth and revealed that field's soil was characterized sandy clay loam (SCL) with mean values: soil organic matter (SOM) = 1.70%, bulk specific gravity = 1.42 $g{\cdot}cm^{-3}$, plant available water = 0.129 $cm{\cdot}cm^{-1}$, pH = 7.10 and cation-exchange capacity (CEC) = 19.3 $cmol{\cdot}kg^{-1}$. The two-way ANOVA statistical analysis ($P = 0.05$) results revealed that the irrigation treatments (IR1: rainfed, IR2: rainfed plus supplemental irrigation [best]), and the SP treatments (SP1: October's last week, SP2: November's first week [best]) significantly affects coriander's seed yield and essential oil content, but the SP have no significant effect on plant's umbel height ($P = 0.873$). Supplemental irrigation, using a limited amount of water, if applied during the critical crop growth stages, can result in substantial improvement on seed yield (+284.934%), essential oil content (+125.396%) and plant's umbel height (+117.929%).

Key words | 2-D TDR-GIS soil moisture mapping, geostatistical modelling on yield and oil of coriander (*Coriandrum sativum* L.), GIS, precision agriculture, rainfed cultivation with supplemental drip irrigation, soil and hydraulic analyses

Agathos Filintas (corresponding author)
Eleni Wogiatzi
Nikolaos Gougoulias
Department of Agriculture-Agrotechnology,
University of Thessaly,
Gaiopolis Campus, Larisa,
Greece
E-mail: *filintas@uth.gr*

HIGHLIGHTS

- Supplemental irrigation, if applied during the critical crop growth stages (L_{mid} and L_{late}), can result in substantial improvement on coriander's seed and essential oil yields.
- Daily monitoring with TDR sensors, 2-D TDR-GIS geostatistical mapping of soil's moisture root zone profile and calculation of the soil-water-plant-atmosphere balance, can outcome better decisions for supplemental irrigation which is an effective response to alleviate the adverse impact of soil moisture stress during dry spells.
- The significance of the results is strongly indicating that farmers should adopt these practices (supplemental irrigation, November sowing, 2-D GIS moisture monitoring and precision agriculture mapping) in order to achieve better management decisions and yields, environmentally friendly agricultural management and increased income.

doi: 10.2166/ws.2021.108

- In the future, modern farms will work differently in order to be more profitable, primarily because of advancements in technology, such as soil, water and crop sensors, new devices, drones and robots, farm machines and information technology.

GRAPHICAL ABSTRACT

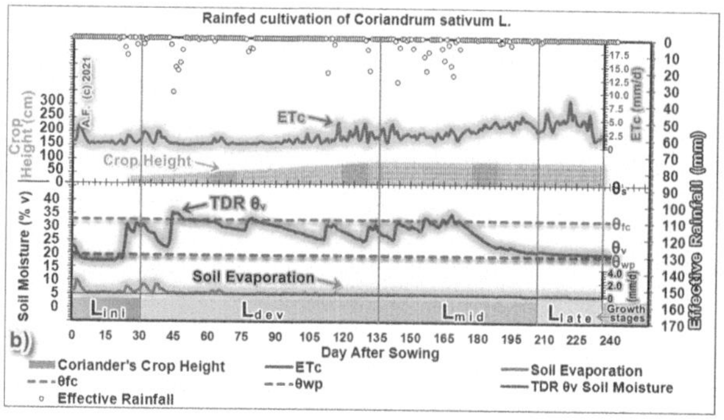

INTRODUCTION

Globally, the herb and spice sector has been enjoying rapid growth for several decades on the back of burgeoning demand from the booming economies, alongside with increased recognition of the health and culinary benefits of herbs and spices (FAO 2018). The worldwide herb and spice trade currently represents a multibillion-dollar industry and one that is booming with international demand and expanding areas of production with a clear need for international standards (FAO 2018). A well known herb and spice plant is coriander (*Coriandrum sativum* L.) which is an annual herbaceous crop of the *Apiaceae* (*Umbellifera*) family and it is known to be a native plant of Mediterranean region, Western Europe and Asia. Coriander is also commonly referred to as cilantro when grown for its herbage, used in many foods.

The erect, glabrous, annual plant coriander is produced throughout the world for culinary, aromatic and medicinal uses (Diederichsen 1996; Evergetis & Haroutounian 2014; Mandal & Mandal 2015; Khodadadi *et al.* 2016; Wei *et al.* 2019). The plant of coriander has been used since ancient times for cooking, medication and flavouring (Nadeem *et al.* 2013).

The essential oil extracted from coriander fruits (commonly termed seeds) has numerous uses (Diederichsen 1996; Lenardis *et al.* 2000; Evergetis & Haroutounian 2014; Mandal & Mandal 2015; Wei *et al.* 2019).

The essential oil from coriander is most commonly isolated from the plant seeds (fruits) usually by either steam or hydrodistillation or by the new microwave-assisted hydrodistillation extraction method (Ghazanfari *et al.* 2020). The essential oil content in coriander seeds has been found to vary between 0.125 and 1.90% (v w^{-1}) essential oil, of which 60–70% is linalool, the compound that gives the pleasant characteristic odor (Diederichsen 1996; Anitescu *et al.* 1997; Jeliazkova *et al.* 1997; Lenardis *et al.* 2000; Smallfield *et al.* 2001; Ayanoglue *et al.* 2002; Msaada *et al.* 2007, 2009; Gil *et al.* 2009; Neffati *et al.* 2011; Mandal & Mandal 2015; Khodadadi *et al.* 2016; Beyzi *et al.* 2017; Ghazanfari *et al.* 2020).

Coriander plant has many therapeutic properties including as an anti-inflammatory, analgesic, anticonvulsant, blood pressure lowering, cholesterol lowering, antioxidant, anti-diabetic, anti-mutagenic, sedative-hypnotic, protective against lead toxicity and heavy metal detoxificier, (Asgharpanah & Kazemivash 2012; Momin *et al.* 2012; Laribi *et al.* 2015; Mandal & Mandal 2015).

Regarding soil moisture is the content of water in the soil, held in the spaces between soil particles. Soil moisture availability is considered essential for nutrient cycling, a prerequisite for primary crop production (Filintas 2011). Moreover, soil moisture affects the land evapotranspiration which is a central process in the climate system and a nexus of the water, energy and carbon cycles (Allen *et al.* 1998; Wan *et al.* 2007; Filintas 2008, 2011; Garten *et al.* 2009; Koutseris *et al.* 2010; Falloon *et al.* 2011, Filintas *et al.* 2019, 2020; Stamatis *et al.* 2011). Soil moisture is the most important factor for enhanced growth and production of coriander under rainfed condition. Among different growth stages of coriander, reproductive stages (L_{mid} and L_{late} growth stages) are much more sensitive than vegetative stage.

As the main source of natural water for agriculture and natural vegetation, soil moisture influences a variety of processes related to plant growth and agricultural production (Allen *et al.* 1998; Rodriguez-Iturbe 2000; Filintas 2011), as well as a range of soil processes (Allen *et al.* 1998; Filintas 2011; Filintas *et al.* 2019). Rainfall and irrigation water supply the moisture in soil and are used for field crop production which will become increasingly limited at Mediterranean latitudes due to climate and land-use changes (Clarke 1993). It is evident that scarce water resources frequently limit crop production in semiarid lands. Decreases in and scarcity of water resources resulting from many environmental effects and sources, and especially from agricultural irrigation consumption, are major environmental issues worldwide (Filintas 2008; Dioudis *et al.* 2009a, 2009b; Koutseris *et al.* 2010).

Regarding precision agriculture (PA), is referred as a management strategy which gathers using various ways, means and processes, and analyzes temporal, spatial and individual geodata and combines it with other digital or analog information in order to support proper crop management decisions according to the estimated field's variability for improved resources use efficiency, productivity, quality, profitability and sustainability of agricultural production.

Properly used, supplemental drip irrigation cultivation vs. rainfed cultivation, in conjunction with soil moisture monitoring and geographical information system (GIS) precision agriculture moisture root zone profile geospatial modelling and two-dimensional (2-D) mapping, may constitute a method for improving seed yield, essential oil and plant's umbel heights and also water use efficiency of coriander's crops for sustainable management of resources, and can contribute to environmental protection and sustainability.

The aim of the present study was to determine the effects of rainfed and supplemental drip irrigation, and of sowing period treatments on coriander (*Coriandrum sativum* L.) (Figure 1) yield, essential oil content and umbel heights by applying new agro-technologies (TDR-sensors measured soil's moisture (SM), GIS, global positioning system (GPS), precision agriculture (PA), soil-hydraulic analyses and geostatistical models) for yield and SM root zone profile geospatial modelling and 2-D GIS mapping), adapted to the specific climatic conditions of Central Greece and also the Mediterranean, in order to increase the harvested seed yield and essential oil without reducing the quality of the final product.

METHODS

Study area

The experiment was conducted at the farmland of Gaiopolis campus–University of Thessaly, in a typical soil climatic

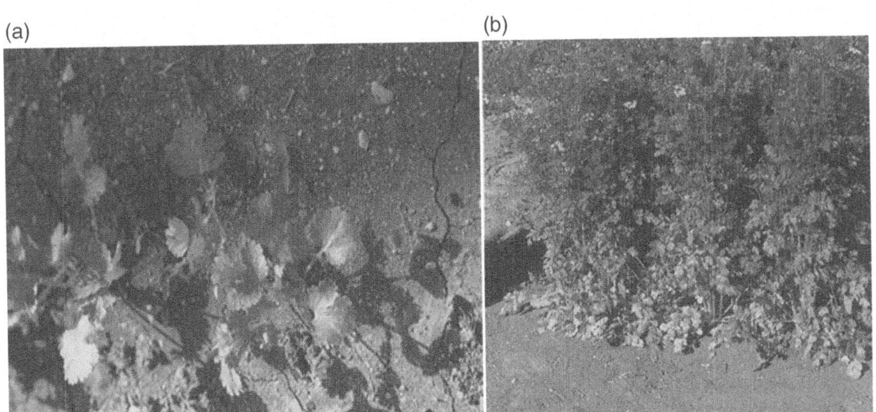

Figure 1 | (a) Young plants of coriander, (b) coriander plants at full umpel growth.

environment of Thessaly plain, at Larisa city in Central Greece (coordinates: latitude 39°62′68″N, longitude 22°38′5 ″E). The study area (Central Greece) is characterized by a typical Mediterranean climate with a cold winter, hot summer and low precipitation in spring and summer. The soil of the experimental site is characterized as sandy clay loam (SCL), soil organic matter (SOM) content was found to be 1.70%, bulk specific gravity of soil was 1.42 g·cm^{-3}, the plant available water (PAW) was 0.129 cm·cm^{-1}, the pH at (1:5) soil/water extract was 7.10 and the cation-exchange capacity (CEC) of soil was 19.3 cmol·kg^{-1}, all at a depth of 0–30 cm.

Experimental plot design and soil sampling

The coriander's (Figure 1) experimental plot design was a factorial split plot design with the main factor of the two irrigation treatments (IR1: rainfed, IR2: rainfed plus supplemental drip irrigation (100% *ETo*) using a surface drip irrigation system) and sub factor the two sowing periods (SP1: last week of October, SP2: first week of November) treatments.

The size of each experimental plot unit was 1.8 m^2 (0.9 m width × 2 m length) containing two rows of plants each and the total area cultivated was 28.80 m^2. The layout and size of the 16 plots were designed that way because it was suitable and easier for the correct application of the drip irrigation system, the water cleaning system (filters) and to achieve a high distribution uniformity of irrigation water.

Control of weeds in all plots was performed manually by hand and no pesticides were used.

The effective root depth of coriander plant is 0–30 cm, where in a typical water extraction pattern the root hairs absorb about 90% of the water (soil moisture) and available nutrients of the soil. So, in order to determine soil's physical, chemical and hydraulic characteristics at the effective root depth zone (0–30 cm) of coriander, a soil sampling survey was carried out at the experimental plot units.

Soil samples at depth 0–30 cm were collected and analysed at the laboratory of Applied Soil Science of Gaiopolis campus–University of Thessaly in order to determine SOM content, soil's texture, structure and classes, soil's hydraulic properties (plant available water, field capacity (FC, θ_{fc}), wilting point (θ_{wp})), soil's pH and CEC. A GPS receiver was used to identify the sampling locations.

Planting season

The ideal season for sowing coriander seeds is the last week of October to the first week of November (Meena *et al.* 2013). So, the seeding took place on two different sowing periods on October (last week) and November (first week) applying 20 kg·ha $^{-1}$ of coriander (*Coriandrum sativum var. microcarpum* L.) seed in a row distance of 40 cm and plant distance of 10 cm. Two doses, basic and surface fertilization (A dose = 66.67 kg P_2O_5 and B dose = 74.44 kg N per hectare) were applied.

Soil moisture determination

Soil moisture of the plots was measured by applying the time domain reflectometry (TDR) method. For the application of the TDR method, a TDR measuring instrument-model MP-917 ESI (Environmental Sensors INC 1997; Dioudis *et al.* 2010; Filintas 2011) and soil moisture TDR probes with five sensors each were used (Figure 3(a)). Each probe have five sensors (segments), placed at 0–15, 15–30, 30–45, 45–60 and 60–75 cm depths and each sensor was used for measuring volumetric soil water content ($\theta v_i, \ldots, \theta v_n$) ($i = 1, 2, \ldots, n$ and $n = 5$) of coriander's root zone.

In TDR method, the propagation velocity (v) of an electromagnetic energy pulse in the soil is determined. In order to determine v in soil, very short electrical pulses are sent through a metallic two rod probe (Figure 2(a)) with one or more sensors (segments). From the travel time (t) and the length (l) of the probe, which has been travelled along twice, the propagation velocity $v = 2\,l/t$ is calculated (Worsching *et al.* 2006). When the pulse velocity changes in the soil, is a result of a change in soil moisture content due to the relatively large dielectric value of water (Muñoz-Carpena 2012). The pulse velocity in the probe sensors is measured by the instrument and correlated to the soil moisture. A lower velocity indicates a wetter soil. TDR has been used for measuring volumetric soil water (θ_V) by several researchers (Yu & Drnevich 2004; Filintas 2008, 2011; Dioudis *et al.* 2010; Filintas *et al.* 2010). TDR is a non-destructive and relatively less labour intensive technique; the TDR instrument used is portable, the sensors are easy to install

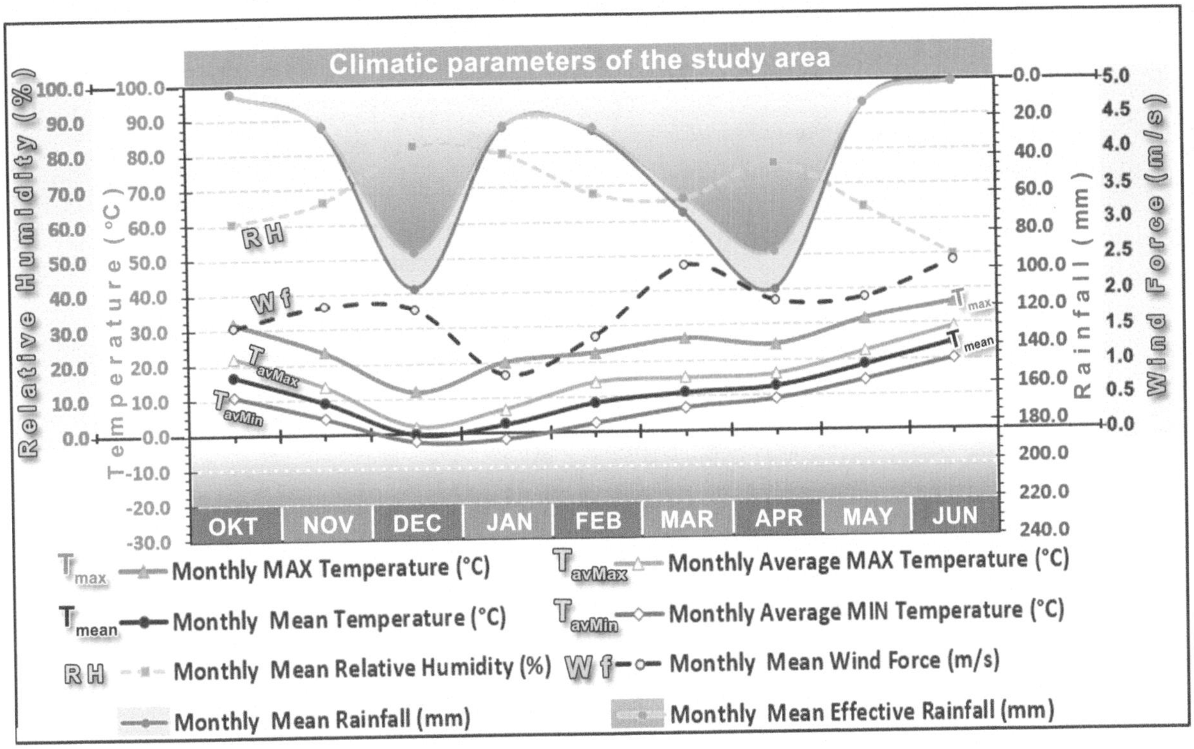

Figure 2 | The various monthly climatic parameters (max air temperature, mean air temperature, average max air temperature, average min air temperature, mean relative humidity, mean precipitation (rainfall), mean effective rainfall, and mean wind force) of the study area.

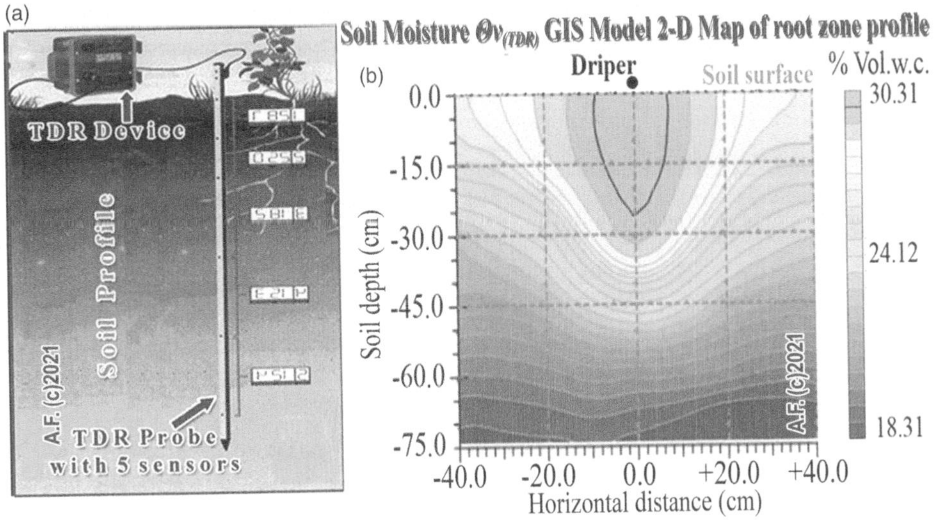

Figure 3 | (a) TDR instrument and probe with five sensors for soil moisture monitoring, (b) an example of the produced digital GIS 2-D maps of coriander's soil moisture root zone profile (supplementary drip irrigation plot).

and safe to operate. This technique allows reliable measurements of volumetric water content (θ_V) within a short time period. No soil specific calibrations are required (Topp & Davis 1985; Ferrara & Flore 2003). TDR technique gives accurate results within an error limit of ±1% and allows continuous measurements to be obtained over the full soil moisture range (profile) (Chandler *et al.* 2004; Filintas 2008, 2011; Dioudis *et al.* 2010; Filintas *et al.* 2010), along with measurements of the electrical conductivity of the soil (Thomsen *et al.* 2007). The TDR measurements data of soil moisture were imported daily in a digital geodatabase and used in a GIS utilizing precision agriculture and geostatistics in order to model and produce digital soil moisture 2-D maps of coriander's root zone soil profile of the experimental plots.

Evapotranspiration measurements and supplemental net irrigation requirement

Evaporation from an open water surface such as the Class 'A' evaporation pan, provides an index of the integrated effect of solar radiation, air temperature, air humidity and wind force on crop evapotranspiration. The Class 'A' evaporation pan method uses a 1.21 m (4 ft) diameter circular pan filled with water. The daily rate of evaporation from the pan is determined from the change in water level (adjusted for rainfall). The crop evapotranspiration method, which is estimated using a Class 'A' evaporation pan that is corrected by the evaporation pan coefficient (*Kp*) and by the crop stage coefficient (*Kc*), has been used successfully by many researchers (Allen *et al.* 1998; Yu & Drnevich 2004; Filintas 2008; Dioudis *et al.* 2010; Filintas *et al.* 2010). In the present study, the following Equations (1) and (2) were used for the calculation of the daily and monthly reference evapotranspiration (*ETo*) and crop evapotranspiration (*ETc*):

$$ETo = Epan \times Kp \tag{1}$$

where:

ETo = reference evapotranspiration (mm·day^{-1}),
$Epan$ = pan evaporation (mm·day^{-1}),
Kp = pan coefficient,

and,

$$ETc = ETo \times Kc \tag{2}$$

where:

ETc = crop evapotranspiration (mm·day^{-1}),
ETo = reference evapotranspiration (mm·day^{-1}),
Kc = crop stage coefficient of each one of the four crop growth stages.

Meteorological data were obtained from a meteorological station near to experimental field. The effective rainfall for the experimental site conditions was calculated according to USDA-SCS (1970).

The net irrigation requirement (*NIR*), that is the irrigation water delivered to the experimental field and available for the coriander crop to be used, was calculated using Equation (3):

$$NIR = ETc - Pe - GW - \Delta\theta_{V(TDR)} \tag{3}$$

where:
NIR = net irrigation requirement (mm),
ETc = evapotranspiration (mm),
Pe = effective rainfall (mm),
GW = groundwater contribution from water table (mm),
$\Delta\theta_{V(TDR)}$ = change in TDR measured soil-water content $\theta_{V(TDR)}$ (mm).

With appropriate daily soil-water content $\theta_{V(TDR)}$ measurement, accurate estimated daily crop *ETc* and monitoring of system inflows (irrigation amount, effective rainfall and groundwater contribution) and estimated surface outflow, a reliable daily soil-water balance can be developed for the crop. It requires appropriate and current monitoring of soil-water content, irrigation water applied, rainfall measurement, runoff, and full climatic data (rainfall, solar radiation, maximum and minimum air temperature, maximum and minimum air humidity and wind force) for daily crop *ETc* determination.

Essential oil content determination

The fruits (seeds) of *Coriandrum sativum var. microcarpum* L. were harvested in June at the stage of full maturity, where only brown fruits were present. Post-harvest processing results found that the fruits' mean diameter was D_f <3 mm, while the mean weight of 1000 fruits was found to be 9.26 g. After air drying fruits in a shady place (humidity 9%) at the laboratory, a commercial grain crusher with two rotating rollers was used to crush the fruits and subsequently were sieved with a 2.0 mm metallic sieve, before distillation.

Essential oil content determination was performed by extraction from coriander seeds with hydro-distillation using a Clevenger-type distillation apparatus. Twelve and a half gram of dried coriander seeds were subjected into 250 mL of water for 105 min of hydrodistillation. The extraction procedure was repeated three times for each sample and the essential oil content was estimated on dry weight basis (DW) of coriander seed material (ml·100 g^{-1}).

Soil sample collection, satellite GPS registration and soil's characteristics laboratory determination

At the experimental cropfield, soil samples (0–30 cm) were collected in an grid pattern of 0.9 m × 2 m to determine exture [(clay content (Cl), silt (Si) content, sand (Sa) content and gravel (Gr) content)], SOM content, pH and CEC, soil's structure and texture classes, bulk specific gravity of soil and PAW (% vol.) that was determined by FC water content (θ_{fc}) (or water holding capacity) (% vol.) and wilting point water content (θ_{wp}) (% vol.) measurements.

A satellite GPS receiver was used to identify and register the sampling locations. A total of 16 soil samples were air-dried and passed through a 2 mm mesh to determine soil texture by the Bouyoucos-hydrometer method (Page *et al.* 1982; Beretta *et al.* 2014).

Soil's pH was determined in a 1:5 soil/water extract, while pH value was measured by using glass electrode and a pH meter. Soil organic matter was analysed by chemical oxidation with 1 mol·L^{-1} $K_2Cr_2O_7$ and titration of the remaining reagent with 0.5 mol·L^{-1} $FeSO_4$. Soil's nitrogen inorganic forms were extracted with 0.5 mol·L^{-1} $CaCl_2$ and estimated by distillation in the presence of MgO and Devarda's alloy, respectively (Page *et al.* 1982).

The PAW for plant growth is the difference between soil's FC (θ_{fc}) and wilting point (WP) (θ_{wp}) water contents. The FC and WP water contents were measured with the porous ceramic plate method placed into a container that is pressurized with 1/3 atmospheres (about 5 psi) for FC and with 15 atmospheres (about 225 psi) for WP (Filintas 2008; Dioudis *et al.* 2009a; Filintas *et al.* 2010, 2019). Soil classification with USDA Soil Classification Triangle was performed according to USDA classification (Soil Survey Staff 1975).

Statistical data analysis

Data analysis was performed using the IBM SPSS v.26 (Norusis 2011) statistical software package. The results are means of the samples and measurements of all measured and derived data groups. Mean separation was made using $LSD_{0.05}$ statistical test as the test criterion when significant differences ($P = 0.05$) between treatments were found (Steel & Torrie 1982).

The Zero Hypothesis (H0) for the main factor (irrigation treatments [IR1: rainfed, IR2: rainfed plus supplemental drip irrigation (100% *ETo*)], for the sub factor [sowing periods treatments (SP1: last week of October, SP2: first week of November)] and for irrigation treatments and sowing period treatments interaction effects on coriander's seed yield, essential oil content and plant's umbel heights were H0ir, H0sp and H0irxsp, respectively, as seen below.

H0ir – irrigation treatments [IR1: (rainfed), IR2: (rainfed plus supplemental drip irrigation)] have no significant effect on coriander's seed yield and/or on essential oil content and/or on plant's umbel heights.

H0sp = sowing period treatments [SP1 (last week of October), SP2 (first week of November)] have no significant effect on coriander's seed yield and/or on essential oil content and/or on plant's umbel heights.

H0irxsp = irrigation treatments and sowing period treatments interaction have no significant effect on coriander's seed yield and/or on essential oil content and/or on plant's umbel heights.

Geostatistical modelling and spatial interpolation methodology

For the experimental farm field, spatial interpolation was used with the geostatistical models of CoKriging, which are used to estimate an unknown value, given the observed values at sampled plots (Lu & Wong 2008; Dioudis *et al.* 2009a; Filintas *et al.* 2010, 2019; Filintas 2011; Stamatis *et al.* 2011). The method is based on the assumptions that the attribution values of measured parameters at the unsampled soil sites are a weighted average of values at sampled soil sites of the experimental farm field. Using the parameters found from measurements and laboratory analyses (which were digitally mapped in a GIS geodatabase environment) as input auxiliary variables, we delineated soil moisture profile digital GIS maps of coriander's root zone with the help of spatial analysis and the use of a GIS software (ArcGIS© version 10.2).

In addition, the evaluation of TDR measured soil moisture, require statistical analysis of residual errors, the difference between predicted and observed values and prediction characterization between over- and underestimates. To that end, we used the statistical parameters described by Loague & Green (1991); Filintas (2011); Filintas *et al.* (2019), such as the equations for the mean prediction error (MPE), mean standardized prediction error (MSPE) and the root mean square standardized error (RMSSE). The MPE and MSPE values should approach zero for an optimal prediction and the RMSSE should approach one.

RESULTS AND DISCUSSION

The study area is characterized by a typical Mediterranean climate with cold humid winters and hot-dry summers. In particular, the average max air temperature ranged during the cultivation period (October to June) from 1.98 °C to 29.25 °C (mean 15.68 °C), during the autumn-winter (October–February) from 1.98 °C to 22.11 °C (mean 11.68 °C), during the spring (March–May) from 15.07 °C to 22.35 °C (mean 17.81 °C) and during the summer months (June and July) had a mean of 29.25 °C. Monthly relative humidity ranged during the cultivation period (October to June) from min 49.76% to max 82.11% (mean 68.07%), during the autumn-winter (October–February) from min 60.26% to max 82.11% (mean 71.29%), during the spring (March–May) from min 63.65% to max 76.48% (mean 68.82%) and during the summer month (June) had a mean of 49.76%.

Precipitation during the autumn-winter (October–February) was 184.00 mm, on spring (March–May) was 192.10 mm and only 1.00 mm on summer (June) with a total precipitation (October to June) of 377.10 mm. The various climatic parameters (max air temperature, mean air temperature, average max air temperature, average min air temperature, mean relative humidity, mean precipitation (rainfall), mean effective rainfall, and mean wind force) of the study area are presented in Figure 2.

The above, mentioned values of climatic data are very important because directly affect crop growth, evapotranspiration and crop water requirements (needs). A certain crop (in our case coriander) grown in a sunny and hot climate needs per day more water than the same crop grown in a cloudy and cooler climate.

However, there are – apart from air temperature and precipitation – other climatic factors which influence the crop water needs. These factors are the relative humidity, the wind force (see Figure 2) and sunshine. When the climate is dry, the crop water needs are higher than when it is humid. Moreover, in windy climates, the crops will use more water than in calm climates.

A first reading of the above climatic values indicates that the air temperature, relative humidity and wind force are acceptable values that permit the coriander proper growth. On the contrary, precipitation amount and its distribution through time on the four crop growth stages seems to be not enough to cover crop water needs as rainfed cultivation on all stages for the coriander's full growth.

According to the USDA (Soil Survey Staff 1975), soil texture was characterized as SCL. Results and statistical analysis of soil laboratory analysis at the laboratory of Applied Soil Science of Gaiopolis campus revealed that the soil of the experimental field (mean values) was characterized as SCL, SOM was 1.70%, bulk specific gravity of soil was 1.42 $g{\cdot}cm^{-3}$, the PAW was 0.129 $cm{\cdot}cm^{-1}$, the pH at (1:5) soil/water extract was 7.10 and the cation-exchange capacity (CEC) of soil was 19.3 $cmol{\cdot}kg^{-1}$.

The TDR sensor (Figure 3(a)) measurements (average of the soil moisture content measurements at five different soil depths [0–15 cm, 15–30 cm, 30–45 cm, 45–60 cm and 60–75 cm]) were used in order to monitor daily soil moisture $\theta_{V(TDR)}$ and the depletion of available soil moisture (ASMD) (Dioudis *et al.* 2009a; Filintas *et al.* 2010; Filintas 2011) that was calculated and evaluated in relation to each rainfed and supplementary drip irrigation treatment.

The TDR measurements datasets (data of five different soil depths) of soil moisture were imported in a digital geodatabase and used in a GIS environment utilizing precision agriculture and geostatistics (Filintas *et al.* 2010, 2019, 2020; Filintas 2011) in order to model and produce digital soil moisture 2-D maps of coriander root zone soil profile of the plots. An example of the produced digital soil moisture GIS Model 2-D maps of coriander root zone soil profile of the supplemental drip irrigation plots is depicted in Figure 3(b).

In Table 1 is presented the rainfall, effective rainfall, supplemental irrigation and crop evapotranspiration *ETc* of the cultivation period of coriander's plant.

In Figure 4 are depicted the results of the daily soil-water balance model (Filintas 2011) with the daily monitoring of soil moisture $\theta_{V(TDR)}$ (measured with TDR instrument and sensors (Dioudis *et al.* 2009a; Filintas *et al.* 2010; Filintas 2011)), effective rainfall, soil evaporation, crop *ETc*, field capacity (θ_{fc}), permanent wilting point (θ_{wp}), saturation θ_S (soil moisture content at saturation), and coriander's crop height, through the four crop growth stages of treatment IR1: rainfed.

As it can be observed in Figure 4, in the middle of one of the most sensitive growth stages [the flowering (L_{mid} stage)] (Allen *et al.* 1998; Filintas 2008, 2011), the crop *ETc* almost gradually increases till the end of the flowering stage, while precipitation is deteriorating to very low amounts.

At the fourth crop growth stage (L_{late} stage) the crop *ETc* continues to increase furthermore, while precipitation is deteriorating to almost zero amounts.

In Figure 4, in the middle of one of the most sensitive growth stages [the flowering (L_{mid} stage)] a shortage of soil moisture $\theta_{V(TDR)}$ occurs in the dry rainfed area of the experimental field's location due to increased crop *ETc* and the prevailing climatic conditions, driving soil moisture at the end of L_{mid} stage to drop down at very low levels of soil water content near the permanent wilting point θ_{wp}, at the end of the flowering stage.

During the grain filling crop growth stage (L_{late} stage) of the coriander's crop, the soil moisture $\theta_{V(TDR)}$ continues to be at low levels of soil water content which is dropping down gradually at the edge of the permanent wilting point θ_{wp} due to increased crop *ETc* and to very low precipitation amount of the study area. This soil water content shortage situation at very low levels caused a serious water stress to the coriander plants and as a result, rainfed crop growth was poor and yield was consequently low (IR1: rainfed = 422.369 (±17.205) Kg·ha^{-1}).

Shortage of soil moisture in the dry rainfed areas often occurs during the most sensitive growth stages [flowering (L_{mid} stage) and grain filling (L_{late} stage)] of the coriander crop and in other crops also. As a result, rainfed crop growth is poor and yield is consequently low.

Productivity results of IR2: rainfed plus supplemental irrigation plots were superior because the soil water content shortage situation did not occur at IR2 and soil moisture distribution (see Figure 3(b)) as it can be observed in digital

Table 1 | Rainfall, effective rainfall, supplemental irrigation and crop evapotranspiration of the coriander's cultivation period

Period of supplemental irrigations **[b]SP1 with DAS 0 = 25 Oct** **[b]SP2 with DAS 0 = 07 Nov**	**Rainfall (mm)**	**Effective rainfall (mm)**	**Supplemental irrigation (mm)**	**Crop ETc (mm/period)**
1st period of supplemental irrigations, 0–30 DAS[a]	27.80	26.56	10.29	36.85
2nd period of supplemental irrigations, 31–60 DAS	104.90	87.29	0.00	14.52
3rd period of supplemental irrigations, 61–90 DAS	21.40	20.67	12.55	33.22
4th period of supplemental irrigations, 91–105 DAS	0.00	0.00	19.80	19.8
5th period of supplemental irrigations, 106–150 DAS	126.40	100.84	0.00	70.73
6th period of supplemental irrigations, 151–180 DAS	78.50	68.64	40.81	109.45
7th period of supplemental irrigations, 181–210 DAS	13.20	12.92	185.34	198.262
Harvest of seeds period, 211–235 DAS	0.40	0.40	140.33	140.734
Total water	372.60	317.32	409.12	623.57

[a]DAS, days after sowing.
[b]Sowing periods (SP1: last week of October, SP2: first week of November).

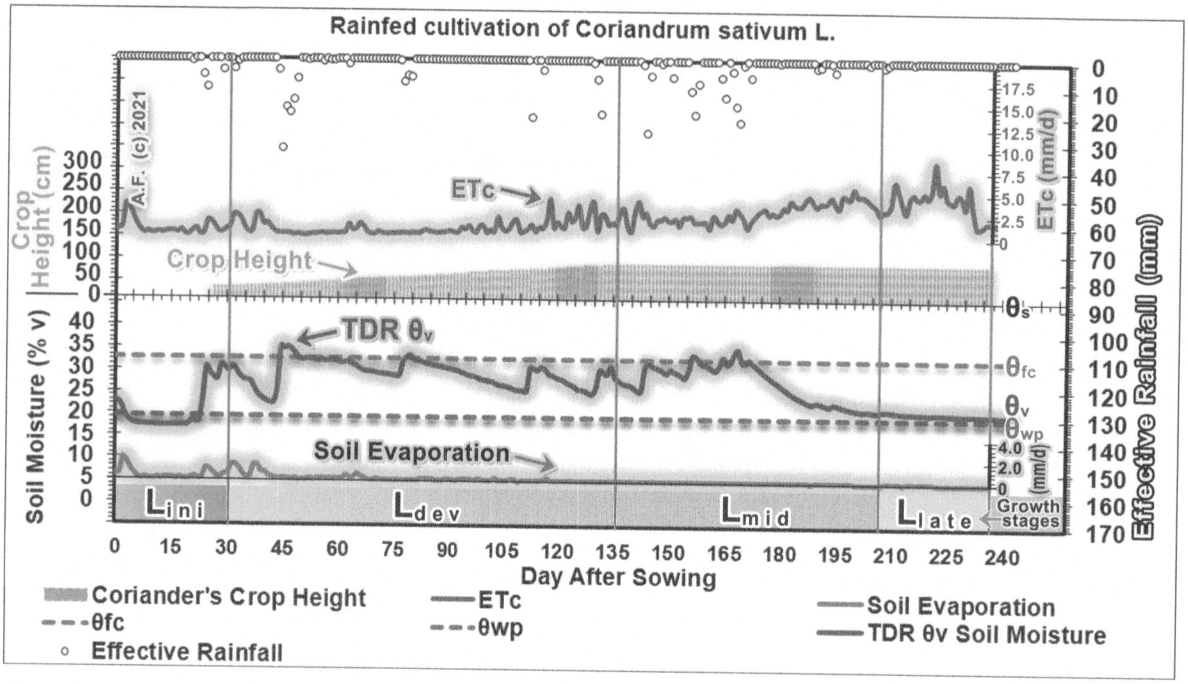

Figure 4 | Daily monitoring of soil moisture $\theta_{V(TDR)}$ (measured with TDR instrument and sensors), effective rainfall, soil evaporation, crop evapotranspiration (*ETc*), field capacity (θ_{fc}), permanent wilting point (θ_{wp}), saturation (θ_S), and coriander's crop height through the four crop growth stages of treatment IR1: rainfed.

GIS 2-D maps of coriander's soil moisture root zone profile was ranged at high volumetric soil water contents, and it did not cause any water stress to the coriander plants. As a result, supplemental irrigation crop growth was increased in comparison to the rainfed and supplemental irrigation treatments yield was consequently high (IR2: rainfed plus supplementary irrigation = 1203.473 (±34.246) Kg·ha^{-1}).

The two-way ANOVA (analysis of variance) (Filintas 2011; Norusis 2011) statistical analysis ($P = 0.05$) results of the irrigation treatments, sowing periods and irrigation treatments*sowing period interactions effects on coriander's dry seed yield, essential oil contents and plant's umbel heights are presented in Table 2.

According to the two-way ANOVA results of Table 2, the zero hypothesis (H0ir) for the main factor irrigation treatments [IR1: rainfed, IR2: rainfed plus supplemental drip irrigation (100% *ETo*)], was rejected for coriander's dry seed yield, essential oil contents and plant's umbel heights. Statistically, this means that the irrigation treatments [IR1: rainfed, IR2: rainfed plus supplemental drip irrigation] have significant effect on coriander's dry seed yield, on essential oil contents and on the plant's umbel heights.

According to the two-way ANOVA results of Table 2, the zero hypothesis (H0sp) for the sub factor sowing period treatments (SP1: last week of October, SP2: first week of November), was rejected for coriander's dry seed

Table 2 | Two-way ANOVA results for the coriander's dry seed yield, essential oil content and plant's umbel height

Dependent variable: source	Dry seed yield		Essential oil content		Plant's umbel height	
	F	Sig.	F	Sig.	F	Sig.
Corrected Model	4550.699	.0000000	103.238	.0000000	30.809	.0000064
Intercept	58951.379	.0000000	22986.407	.0000000	13651.219	.0000000
Irrigation_Treatments	13606.773	.0000000	291.837	.0000000	92.387	.0000005
Sowing_Period	37.300	.0000528	17.877	.0011718	.026	.8734888
[Irrigation_Treatments* Sowing_Period]	8.024	.0150986	.000	.9929850	.014	.9072088

yield and essential oil contents but it was not rejected for plant's umbel heights.

Statistically, this means that the sowing period treatments (SP1: last week of October, SP2: first week of November) have a significant effect on coriander's dry seed yield and on essential oil content but have no significant effect on the plant's umbel heights ($P = 0.873$). There is a 0.005% and a 0.117% chance of getting results by random chance for coriander's dry seed yield and essential oil contents respectively.

According to the two-way ANOVA results of Table 2, the zero hypothesis (H0irxsp) for irrigation's and sowing period's treatments interaction was rejected for coriander's dry seed yield but it was not rejected for essential oil contents and for the plant's umbel heights.

Statistically, this means that the irrigation treatments and sowing period treatments interaction have significant effect on coriander's dry seed yield but have no significant effect on essential oil contents ($P = 0.993$) and on the plant's umbel heights ($P = 0.907$). There is a 1.5% chance of getting results by random chance for coriander's dry seed yield.

The two-way ANOVA (Filintas 2011; Norusis 2011) statistical analysis ($P = 0.05$) results revealed that the irrigation treatments (IR1: rainfed, IR2: rainfed plus supplementary irrigation (100% *ETo*) [best]), significantly affects coriander's dry seed yields (IR1 = 422.369 (±17.205) Kg·ha^{-1} and IR2 = 1203.473 (±34.246) Kg·ha^{-1}) essential oil contents (IR1 = 0.937 (±0.047) % v·w^{-1} and IR2 = 1.175 (±0.034) % v·w^{-1}) and plant's umbel heights (IR1 = 73.746 (±2.849) cm and IR2 = 86.968 (±2.213) cm).

Supplemental irrigation, using a limited amount of water, if applied during the critical crop growth stages (especially on L_{mid} and L_{late} growth stages), can result in substantial improvement on coriander's seed yield (+284.934%), on essential oil contents (+125.396%) and on the plant's umbel heights (+117.929%).

Therefore, daily monitoring with TDR sensors and 2-D GIS geostatistical mapping of soil moisture root zone profile and calculation of the soil-water balance of coriander's root zone profile for the four crop growth stages, can outcome better decisions for supplemental drip irrigation which is an effective response to alleviate the adverse impact of soil moisture stress during dry spells on the yield of rainfed coriander (*Coriandrum sativum L.*) crop and probably in other rainfed crops.

The above advances in agriculture will let businesses be more profitable, efficient, safer, and environmentally friendly.

CONCLUSIONS

The two-way ANOVA statistical analysis ($P = 0.05$) results revealed that the irrigation treatments (IR1: rainfed, IR2: rainfed plus supplementary irrigation (100% *ETo*) [best]), significantly affects coriander's dry seed yields (IR1 = 422.369 (±17.205) Kg·ha^{-1} and IR2 = 1203.473 (±34.246) Kg·ha^{-1}) essential oil contents (IR1 = 0.937 (±0.047) % v·w^{-1} and IR2 = 1.175 (±0.034) % v·w^{-1}) and plant's umbel heights (IR1 = 73.746 (±2.849) cm and IR2 = 86.968 (±2.213) cm).

Statistical results revealed that the sowing period treatments (SP1: last week of October, SP2: first week of November [best]) have significant effect on coriander's dry seed yield and on essential oil content but have no significant effect on plant's umbel heights ($P = 0.873$).

Moreover, the two-way ANOVA statistical analysis also revealed that the irrigation treatments and sowing period treatments interaction have significant effect on coriander's dry seed yield but have no significant effect on essential oil contents (P = 0.993) and on plant's umbel heights (P = 0.907). There is a 1.5% chance of getting results by random chance for coriander's dry seed yield.

Supplemental irrigation, using a limited amount of water, if applied during the critical crop growth stages (especially on L_{mid} and L_{late} growth stages), can result in substantial improvement on coriander's seed yield (+284.934%), on essential oil contents (+125.396%) and on the plant's umbel heights (+117.929%).

Therefore, daily monitoring with TDR sensors and 2-D GIS geostatistical mapping of soil moisture root zone profile and calculation of the soil-water balance of coriander's root zone profile for the four crop growth stages, can provide better decisions for supplemental drip irrigation which is an effective response to alleviate the adverse impact of soil moisture stress during dry spells on the yield of rainfed coriander (*Coriandrum sativum L.*) crop and probably in other rainfed crops.

The significance of the results achieved is strongly indicative that farmers should adopt these management practises (rainfed plus supplementary irrigation, November sowing and 2-D TDR-GIS SM monitoring and precision agriculture GIS mapping) in order to achieve better management decisions, better dry seed yields, better essential oil contents, environmentally friendly agricultural management and increased income.

In the future, modern farms and various agricultural operations will work differently, primarily because of advancements in technology, such as soil, water and crop sensors, devices, farm machines, and information technology. Future agriculture will use sophisticated technologies such as GIS, precision agriculture, robots, temperature and moisture sensors, aerial and satellite images, drones equipped with various sensors, GPS technology, etc. These advances in agriculture will let businesses be more profitable, efficient, safer, and environmentally friendly.

DATA AVAILABILITY STATEMENT

Data cannot be made publicly available; readers should contact the corresponding author for details.

REFERENCES

Allen, R., Pereira, L., Raes, D. & Smith, M. 1998 *Crop evapotranspiration*. FAO Drainage and Irrigation paper No. 56. FAO, Rome, Italy, p. 300.

Anitescu, G., Doneanu, C. & Radulescu, V. 1997 Isolation of coriander oil: comparison between steam distillation and supercritical CO2 extraction. *Flavour Fragance J.* **12** (3), 73–176. https://doi.org/10.1002/(SICI)1099-1026(199705) 12:3 < 173::AID-FFJ630 > 3.0.CO;2-1.

Asgharpanah, J. & Kazemivash, N. 2012 Phytochemistry, pharmacology and medicinal properties of Coriandrum sativum L. *Afr. J. Pharm. Pharmacol.* **6** (31), 2340–2345. https://doi.org/10.5897/AJPP12.901.

Ayanoglue, F., Ahmet, M., Neset, A. & Bilal, B. 2002 Seed yields, yield components and essential oil of selected coriander (*Coriandrum sativum* L.) lines. *J. Herbs Spices Med. Plants* **9** (2–3), 71–76. https://doi.org/10.1300/J044v09n02_10.

Beretta, N. A., Silbermann, V. A., Paladino, L., Torres, D., Bassahun, D., Musselli, R. & García-Lamohte, A. 2014 Soil texture analyses using a hydrometer: modifiation of the Bouyoucos method. *Cien. Inv. Agr.* **41** (2), 263–271. doi: 10. 4067/s0718-16202014000200013.

Beyzi, E., Karaman, K., Gunes, A. & Buyukkilic-Beyzi, S. 2017 Change in some biochemical and bioactive properties and essential oil composition of coriander seed (*Coriandrum sativum L.*) varieties from Turkey. *Ind. Crops Prod.* **109**, 74–78. https://doi.org/10.1016/j.indcrop.2017.08.008.

Chandler, D. G., Seyfried, M., Murdock, M. & McNamara, J. P. 2004 Field calibration of water content reflectometers. *Soil Sci. Soc. Am. J.* **68**, 1501–1507.

Clarke, R. 1993 *Water: The International Crisis*. Earthscan Publications Ltd, London, UK.

Diederichsen, A. 1996 *Coriander, Coriandrum Sativum L.* International Plant Genetic Resources Institute (IPGRI), Rome, Italy, p. 83.

Dioudis, P., Filintas, A. & Koutseris, E. 2009a GPS and GIS based N-mapping of agricultural fields' spatial variability as a tool for non-polluting fertilization by drip irrigation. *Int. J. Sustainable Dev. Plann.* **4** (3), 210–225. https://doi.org/10. 2495/SDP-V5-N1-210-225.

Dioudis, S. P., Filintas, T. A. & Papadopoulos, H. A. 2009b Corn yield response to irrigation interval and the resultant savings in water and other overheads. *Irrig. Drain.* **58**, 96–104. https://doi.org/10.1002/ird.395.

Dioudis, P., Filintas, A., Papadopoulos, A. & Sakellariou-Makrantonaki, M. 2010 The influence of different drip irrigation layout designs on sugar beet yield and their contribution to environmental sustainability. *Fresenius Environ. Bull.* **19** (5), 818–831.

Environmental Sensors INC 1997 *MP-917 Soil Moisture Instrument Operational Manual*. E.S.I., Canada.

Evergetis, E. & Haroutounian, S. A. 2014 Exploitation of apiaceae family plants as valuable renewable source of essential oils containing crops for the production of fine chemicals. *Ind. Crops. Prod.* **54**, 70–77. http://dx.doi.org/10.1016/j.indcrop. 2014.01.009\.

Falloon, P., Jones, C. D., Ades, M. & Paul, K. 2011 Direct soil moisture controls of future global soil carbon changes: an important source of uncertainty. *Global Biogeochem. Cycles* **25** (3), GB3010. https://doi.org/10.1029/ 2010GB003938.

FAO 2018 *New Quality Criteria to be Developed for Booming Spice and Herb Sector*. Food and Agriculture Organization of the United Nations. Available from: http://www.fao.org/news/ story/en/item/213612/icode/ (accessed 5 February 2021).

Ferrara, G. & Flore, J. A. 2003 Comparison between different methods for measuring transpiration in potted apple trees. *Biol. Plant.* **46** (1), 41–47. https://doi.org/10.1023/ A:1022301931508.

Filintas, A. 2008 *Study and Mapping of Biomass Yield with the use of Spatial Statistics and Geoinformation*. MSc Thesis, Dept. of Natural Resources Management and Agricultural Engineering, Agricultural University of Athens, Athens, Greece (In Greek with English abstract).

Filintas, A. 2011 *Land Use Evaluation and Environmental Management of Biowastes, for Irrigation with Processed Wastewaters and Application of Bio-Sludge with Agricultural*

Machinery, for Improvement-Fertilization of Soils and Cultures, with the use of GIS-Remote Sensing, Precision Agriculture and Multicriteria Analysis. PhD Thesis, Dept. of Environment, University of the Aegean, Mitilini, Greece.

Filintas, A., Dioudis, P. & Prochaska, C. 2010 GIS modeling of the impact of drip irrigation, of water quality and of soil's available water capacity on Zea mays L, biomass yield and its biofuel potential. *Desalin. Water Treat.* **13** (1–3), 303–319. https://doi.org/10.5004/dwt.2010.1038.

Filintas, A., Gougoulias, N. & Papachatzis, A. 2019 Soil organic matter modeling and digital mapping of a Triticum turgidum cropfield using as auxiliary variables the plant available water, texture, field measurements, soil laboratory analyses, GIS and geostatistical models. *Annals of the University of Craiova, Seria Biologie, Horticultură, Tehnologia Prelucrării Produselor Agricole, Ingineria Mediului* **XXIV** (LX), 537–544.

Filintas, A., Gougoulias, N. & Papachatzis, A. 2020 Soil's plant available water and nitrogen inorganic modelling and digital GIS mapping of winter wheat, utilizing precision agriculture, geostatistical models, soil's pH, water holding capacity, sand, clay and silt. *Annals of the University of Craiova, Seria Biologie, Horticultură, Tehnologia Prelucrării Produselor Agricole, Ingineria Mediului* **XXV** (LXI), 368–375.

Garten Jr., C. T., Classen, A. T. & Norby, R. J. 2009 Soil moisture surpasses elevated CO2 and temperature as a control on soil carbon dynamics in a multi-factor climate change experiment. *Plant. Soil* **319**, 85–94. https://doi.org/10.1007/s11104-008-9851-6.

Ghazanfari, N., Mortazavi, S. A., Yazdi, F. T. & Mohammadi, M. 2020 Microwave-assisted hydrodistillation extraction of essential oil from coriander seeds and evaluation of their composition, antioxidant and antimicrobial activity. *Heliyon* **6** (9), e04893. 1-9, https://doi.org/10.1016/j.heliyon.2020.e04893.

Gil, A., De-La-Fuente, E. B., Lenardis, A. E., Pereira, M. L., Suarez, S. A., Arnaldo, B., Holsten, A., Vetter, T., Vohland, K. & Krysanova, V. 2009 Impact of climate change on soil moisture dynamics in Brandenburg with a focus on nature conservation areas. *Ecol. Modell.* **220**, 2076–2087.

Jeliazkova, E. A., Craker, L. E. & Zheljazkov, V. D. 1997 Irradiation of seeds and productivity of coriander, *Coriandrum sativum L. J. Herbs Spices Med. Plants.* **5**, 73–79.

Khodadadi, M., Dehghani, H., Jalali-Javaran, M. & Christopher, T. J. 2016 Fruit yield, fatty and essential oils content genetics in coriander. *Ind. Crops Prod.* **94** (2016), 72–81.

Koutseris, E, Filintas, A & Dioudis, P. 2010 Antiflooding prevention, protection, strategic environmental planning of aquatic resources and water purification: the case of Thessalian basin, in Greece. *Desalin. J.* **250** (1), 318–322.

Laribi, B., Kouki, K., M'Hamdi, M. & Bettaieb, T. 2015 Coriander (*Coriandrum sativum* L.) and its bioactive constituents. *Fitoterapia* **103**, 9–26.

Lenardis, A. D., De-la-Fuente, E., Gil, A. & Tubia, A. 2000 Response of coriander (*Coriandrum sativum L.*) to nitrogen availability. *J. Herbs Spices Med Plants* **7**, 47–58.

Loague, K. & Green, R. E. 1991 Statistical and graphical methods for evaluating solute transport models: overview and application. *J. Contam. Hydrol.* **7**, 51–73.

Lu, G. Y. & Wong, D. W. 2008 An adaptive inverse-distance weighting spatial interpolation technique. *Comput. Geosci.* **34** (9), 1044–1055.

Mandal, S. & Mandal, M. 2015 Coriander (*Coriandrum sativum L.*) essential oil: chemistry and biological activity. *Asian Pacific Journal of Tropical Biomedicine (Asian Pac. J Trop. Biomed.)* **5** (6), 421–428.

Meena, S. S., Singh, B., Singh, D., Ranjan, J. K. & Meena, R. D. 2013 Pre and post harvest factors effecting yield and quality of seed spices: a review. *Int. J. Seed Spices* **3**, 1–11.

Momin, A. H., Acharya, S. S. & Gajjar, A. V. 2012 Coriandrum sativum-review of advances in phytopharmacology. *Int. J. Pharm. Sci. Res.* **3**, 1233–1239.

Msaada, K., Hosni, K., Taarit, M. B., Chahed, T., Kchouk, M. E. & Marzouk, B. 2007 Changes on essential oil composition of coriander (*Coriandrum sativum L.*) fruits during three stages of maturity. *Food Chem.* **102**, 1131–1134. http://dx.doi.org/10.1016/j.foodchem.2006.06.046.

Msaada, K., Taarit, M. B., Hosni, K., Hammami, M. & Marzouk, B. 2009 Regional and maturational effects on essential oils yields and composition of coriander (*Coriandrum sativum L.*) fruits. *Sci. Hortic.* **122**, 116–124. http://dx.doi.org/10.1016/j.scienta.2009.04.008.

Muñoz-Carpena, R. 2012 Field devices for monitoring soil water content. Agricultural and Biological Engineering Department, University of Florida. BUL343. http://edis.ifas.ufl.edu.

Nadeem, M., Anjum, F. M., Khan, M. I., Tehseen, S., El-Ghorab, A. & Sultan, J. I. 2013 Nutritional and medicinal aspects of coriander (*Coriandrum sativum L.*) a review. *Brit. Food J.* **115** (5), 743–755.

Neffati, M., Sriti, J., Hamdaoui, G., Kchouk, M. E. & Marzouk, B. 2011 Salinity impact on fruit yield, essential oil composition and antioxidant activities of Coriandrum sativum fruit extracts. *Food Chem.* **124**, 221–225. http://dx.doi.org/10.1016/j.foodchem.2010.06.022.

Norusis, M. J. 2011 *IBM SPSS Statistics 19 Advanced Statistical Procedures Companion*. Pearson, London, UK.

Page, A. L., Miller, R. H. & Keeney, D. R. 1982 *Methods of Soil Analysis Part 2: Chemical and Microbiological Properties*. Agronomy, ASA and SSSA, Madison, WI, USA.

Rodriguez-Iturbe, I. 2000 Ecohydrology: a hydrologic perspective of climate–soil–vegetation dynamics. *Water Resour. Res.* **36** (1), 3–9.

Smallfield, B. M., Van-Klink, J. W., Perry, N. B. & Dodds, K. G. 2001 Coriander spice oil: effects of fruit crushing and distillation time on yield and composition. *J. Agric. Food. Chem.* **49**, 118–123.

Soil Survey Staff 1975 *Soil Taxonomy: A Basic System of Soil Classification for Making and Interpreting Soil Surveys.* USDA Natural Resources Conservation Service, Washington, DC, USA.

Stamatis, G., Parpodis, K., Filintas, A & Zagana, E. 2011 Groundwater quality, nitrate pollution and irrigation environmental management in the Neogene sediments of an agricultural region in central Thessaly (Greece). *Environ. Earth Sci.* **64**, 1081–1105. https://doi.org/10.1007/s12665-011-0926-y.

Steel, R. G. D. & Torrie, J. H. 1982 *Principles and Procedures of Statistics. A Biometrical Approach*, 2nd edn. McGraw-Hill, Inc., New York, NY, USA, p. 633.

Thomsen, A., Schelde, K., Droscher, P. & Steffensen, F. 2007 Mobile TDR for georeferenced measurement of soil water content and electrical conductivity. *Precis. Agric.* **8**, 213–223.

Topp, G. C. & Davis, J. L. 1985 Measurement of soil water content using time-domain reflectometry (TDR): a field evaluation. *Soil Sci. Soc. Am. J.* **49**, 19–24.

USDA-SCS 1970 *Irrigation Water Requirements. Technical Release No. 21.* USDA Soil Conservation Service, Washington, DC, USA.

Wan, S., Norby, R. J., Ledford, J. & Weltzin, J. F. 2007 Responses of soil respiration to elevated CO2, air warming, and changing soil water availability in a model oldfield grassland. *Global Change Biol.* **13**, 1–14.

Wei, J. N., Liu, Z. H., Zhao, Y. P., Zhao, L. L., Xue, T. K. & Lan, Q. K. 2019 Phytochemical and bioactives profile of *Cordiandrum sativum L. Food Chem.* **286**, 260–267.

Worsching, H., Becker, R., Schlaeger, S., Bieberstein, A. & Kudella, P. 2006 Spatial-TDR moisture measurement in a large scale Levee Model made of loamy soil material. In: Proc. TDR 2006, Purdue University, West Lafayette, IN, USA, September 2006, Paper ID 33, 15p. Available from: http://engineering.purdue.edu/TDR/Papers (accessed 22 April 2020).

Yu, X. & Drnevich, V. P. 2004 Soil water content and dry density by time domain reflectometry. *J. Geotech. Geoenviron. Eng.* **130** (9), 922–934.

First received 31 January 2021; accepted in revised form 1 April 2021. Available online 12 April 2021

Predicting the infiltration characteristics for semi-arid regions using regression trees

Parveen Sihag, Munish Kumar and Saad Sh. Sammen

ABSTRACT

The study of the infiltration process is considered essential and necessary for all hydrology studies. Therefore, accurate predictions of infiltration characteristics are required to understand the behavior of the subsurface flow of water through the soil surface. The aim of the current study is to simulate and improve the prediction accuracy of the infiltration rate and cumulative infiltration of soil using regression tree methods. Experimental data recorded with a double ring infiltrometer for 17 different sites are used in this study. Three regression tree methods: random tree, random forest (RF) and M5 tree, are employed to model the infiltration characteristics using basic soil characteristics. The performance of the modelling approaches is compared in predicting the infiltration rate as well as cumulative infiltration, and the obtained results suggest that the performance of the RF model is better than the other applied models with coefficient of determination (R^2) = 0.97 and 0.97, root mean square error (RMSE) = 8.10 and 6.96 and mean absolute error (MAE) = 5.74 and 4.44 for infiltration rate and cumulative infiltration respectively. The RF model is used to represent the infiltration characteristics of the study area. Moreover, parametric sensitivity is adopted to study the significance of each input parameter in estimating the infiltration process. The results suggest that time (t) is the most influencing parameter in predicting the infiltration process using this data set.

Key words | cumulative infiltration, infiltration rate, M5 tree, random forest, random tree

Parveen Sihag
Shoolini University,
Solan, Himachal Pradesh 173229,
India

Munish Kumar
National Institute of Technology,
Kurukshetra 136119,
India

Saad Sh. Sammen (corresponding author)
Department of Civil Engineering, College of Engineering,
University of Diyala,
Diyala Governorate,
Iraq
E-mail: *saad123engineer@yahoo.com*; *Dr_Saad@engineering.uodiyala.edu.iq*

HIGHLIGHTS

- Infiltration rate and cumulative infiltration characteristics in semi-arid region were modelled using regression tree.
- Random forest outperforms random tree as well as M5 tree to predict the infiltration characteristics.
- Time is observed as the most important parameter in affecting the prediction performance of both the infiltration rate and the cumulative infiltration of soil.

INTRODUCTION

Infiltration process influences the hydrological cycle, stream flow, groundwater recharge, irrigation, soil erosion and drainage system (Singh *et al.* 2018). Infiltration can be defined as irrigation or precipitation water moving into the soil (Sihag *et al.* 2018a, 2018b). The actual rate at which water moves into the soil is determined as infiltration rate (Haghighi *et al.* 2011). Infiltration process plays a significant role in the selection of crop type and agricultural practices. Prediction of infiltration is also necessary for the efficient design of irrigation systems (Hillel 1998). The sustainability of the

doi: 10.2166/ws.2021.047

groundwater system also depends on this process (Chen & Young 2006).

The quantity and quality of recharged water is the most important parameter for water resources management especially in regions where water scarcity occurs. It divides water into the two most important hydrological segments, groundwater flow and surface flow. It is the main factor in watershed modelling for the prediction of surface runoff. Accurate prediction of infiltration rate is compulsory for reliable prediction of surface runoff (Diamond & Shanley 2003). At catchment level, infiltration characteristics are one of the main factors in determining the flooding condition (Bhave & Sreeja 2013). The significance of the infiltration process has required soil and water researchers to generate several models (e.g. Green & Ampt 1911; Kostiakov 1932; Horton 1941; Philip 1957; Holtan 1961; Soil Conservation Service 1972; Smith & Parlange 1978; Swartzendruber 1987; Singh & Yu 1990; Sihag *et al.* 2017a). However, only a few developed models have been effectively used with field data. Parameter calculation is the main criterion for selection of any model over other models. These infiltration models can be divided into three groups: physical models, semi-empirical models and empirical models.

The main problem for modelling the infiltration process is the variation of soil type and texture (Machiwal *et al.* 2006). Variation in the infiltration process in any watershed causes difficulty in management of water for agricultural practices. The process of infiltration is influenced by several aspects; including soil type, texture, soil hydraulic properties, soil moisture profile, and precipitation and climate properties. The spatial and temporal prediction of the infiltration process in the real environment cannot be at present deduced by direct or indirect methods. Infiltration models are based on the law of mass conservation and Darcy's law (physical models), a few models rely on field and laboratory measurements (empirical models) and some models are based on simple hypotheses about the infiltration rate and cumulative infiltration relation quantities (semi-empirical models). These infiltration models are developed for vertically homogeneous soils with constant initial moisture content in the soil, constant density and over horizontal planes. A very few researchers have made attempts to use soft computing in the prediction of soil infiltration (Sy 2006; Sihag *et al.* 2018a, 2018b).

During the last few years, several researchers have used soft computing techniques such as adaptive neuro fuzzy inference system (ANFIS), artificial neural network (ANN), support vector machine (SVM), M5 tree, Gaussian process (GP), random forest (RF) and gene expression programming (GEP) in the prediction of hydraulic properties and infiltration process (Schaap & Leij 1998; Erzin *et al.* 2009; Anari *et al.* 2011; Das *et al.* 2011; Arshad *et al.* 2013; Al-Sulaiman 2015; Elbisy 2015; Esmaeelnejad *et al.* 2015; Al-Sulaiman & Aboukarima 2016; Rahmati 2017; Singh *et al.* 2017; Tiwari *et al.* 2017; Sihag *et al.* 2017b, 2017c; Sihag 2018) and found it to work equally well or better than empirical relations. A very few studies have been carried out on the spatial variability of soil hydraulic properties in arid regions. However, to the best of our information, no study has been carried out for describing the spatial variability of the soil water infiltration process using soft computing techniques in semi-arid regions of India. Thus, the main objective of this research is to develop infiltration models using regression-tree-based computing approaches and to determine the spatial variability at the same scale in some locations of Haryana (India). Both the cumulative infiltration as well as infiltration rate are predicted using random tree, random forest and M5 tree-based models. In the last few decades conventional infiltration models have been widely used for the calculation or estimation of infiltration rate and cumulative infiltration such as Kostiakov, Philip, Novel, Horton, Holton etc. These models are point- and location-specific. The aim of this investigation is to develop a general model for a complete study area using tree-based models.

STUDY AREA AND DATA COLLECTION

Four districts of Haryana State (Hisar, Kurukshetra, Kaithal, and Karnal) are selected for the infiltration measurements (Figure 1). The study area comprises a total of 17 sites represented by the numbers on the district maps (Figure 1).

The infiltration observations have been recorded in the field with the help of a double ring infiltrometer. The double ring infiltrometer has two concentric rings having diameter 30 and 60 cm for the inner and outer ring, respectively, with 30 cm depth. Both rings are driven about 10 cm deep into the soil by using a hammer without undue

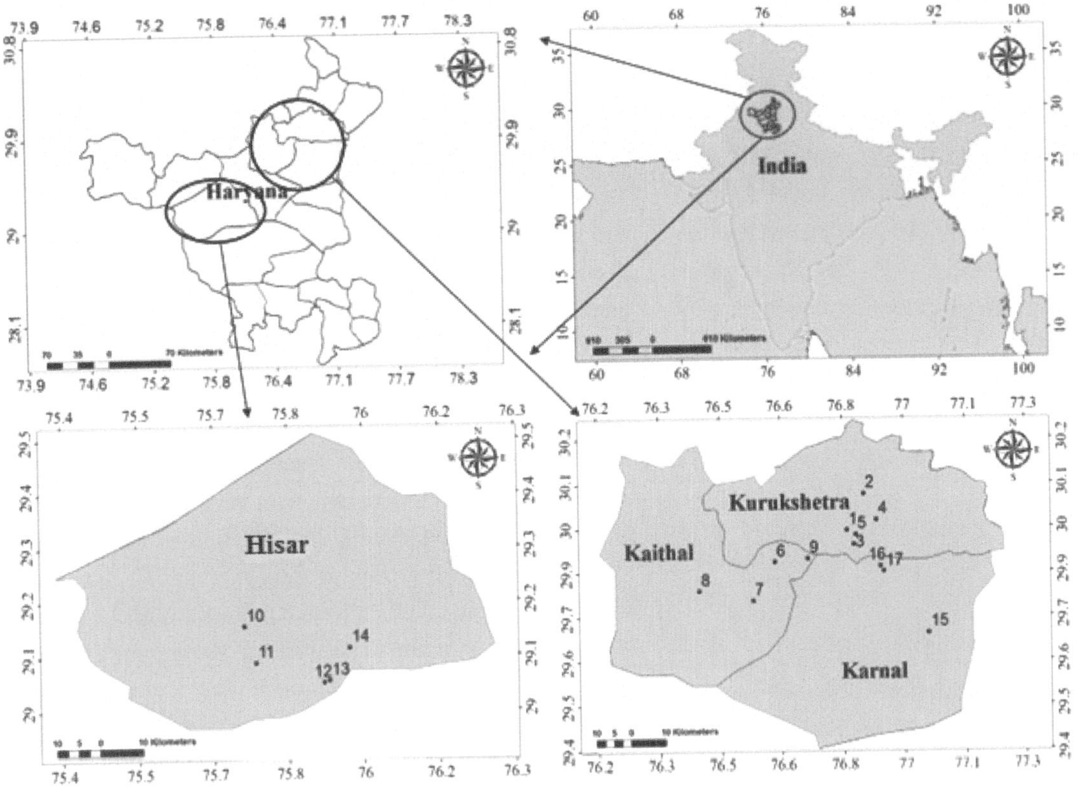

Figure 1 | Location map of the study area showing the measuring points.

disturbance to the soil surface. Both rings are filled with water to equal depth and the initial reading of the water level is noted. The rate of water level drop against time in the inner ring of the infiltrometer is noted at regular intervals until a constant rate is attained, which is called the steady-state infiltration rate. The soil samples for calculating the soil properties are collected from their respective locations selected for the measurement of infiltration rate. The sample is withdrawn along the side of the infiltrometer and represents the soil properties of that location. The properties of the soil samples are measured in the laboratory. The properties of soils respective to each location are listed in Table 1.

DATA SET

The data set used for the modelling in this study is a collection of infiltration observations yielded from the field experiments from 17 sites. In this study, infiltration rate (*I*) and cumulative infiltration (*Cu*) are considered as the

Table 1 | Basic properties of soil

Site no.	Sand (%)	Clay (%)	Silt (%)	Dry density (gm/cm^3)	Moisture content (%)
1	47.50	25.20	27.30	1.64	12.74
2	50.71	23.19	26.10	1.66	19.85
3	39.84	52.34	7.82	1.58	19.74
4	42.85	24.00	33.15	1.58	18.39
5	48.74	29.38	21.88	1.67	8.77
6	59.58	30.72	9.70	1.60	14.21
7	26.63	41.82	31.55	1.54	18.55
8	46.70	16.56	36.74	1.65	5.28
9	24.81	39.62	35.57	1.24	19.06
10	79.73	6.45	13.83	1.60	8.47
11	84.14	7.83	8.03	1.75	7.47
12	66.63	19.62	13.75	1.70	12.71
13	44.27	23.12	32.61	1.65	11.56
14	45.67	39.16	15.17	1.62	8.33
15	26.12	66.14	7.74	1.61	18.60
16	19.73	62.41	17.86	1.61	15.37
17	32.71	54.21	13.08	1.75	7.64

output variables whereas time (*t*), dry density (*ρ*), moisture content (*M*), and percentages of sand (*Sa*), silt (*Si*), and clay (*C*) are the input variables. The basic soil properties respective to each site are measured from the laboratory experiments conducted on the field soil samples. The total data set is split in two groups: one used for model development (training data) and the other used for model validation (testing data). Training data involves 70% data chosen randomly from the complete data set, and the remaining 30% data forms the testing data. The features of the data set used for modelling the infiltration characteristics of the soil are presented in Table 2.

MODELLING METHODS

Random tree (RT)

Random trees are decision/regression-tree-based models used for classification/regression problems. Random model trees are essentially the combination of two existing algorithms: single model trees are combined with random forest ideas. Every leaf in model trees holds a linear model which is optimized for the local subspace described by that leaf (Pfahringer 2010). The collection of tree predictors in random trees is called a forest. With random features at each node, a random tree is a tree drawn randomly from a set of possible trees. The distribution of trees is uniform and each tree in the set of trees has an equal chance of being sampled. Random trees can be generated efficiently and the ensemble of random trees generally leads to accurate models (Zhao & Zhang 2008).

Random forest (RF)

The random forest proposed by Breiman (2001) is a classification and regression method. The performance of single decision trees is shown to be improved considerably by adding randomness to the training data using a random forest. The algorithm combines the concept of bagging (Breiman 1996) and the random selection feature (Amit & Geman 1997). In bagging, the training data is sampled with replacement for each tree. Therefore, a new training set (bootstrap samples) is drawn with replacement from the original training set. During the selection of the bootstrap samples some of the training data may be left out of the sample and some may be repeated in the sample (Adusumilli *et al.* 2013). To generate a tree, instead of computing the best split among attributes for each node, only a random subset of all attributes is chosen at every node, and the best split for that subset is computed.

The procedure for the random forest regression model is as follows:

1. Draw bootstrap samples from the data set.
2. Generate a regression tree with the following modifications: at each node, choose the best split from the randomly sampled *k*-tree predictors.
3. The models are tuned for the user-defined parameters: number of trees (*k*) grown and number of variables used at each node (*m*) to construct a tree.

Table 2 | Features of data set used in this study

	Training data				Testing data			
Parameter	**Min.**	**Max.**	**Mean**	**Standard deviation**	**Min.**	**Max.**	**Mean**	**Standard deviation**
t (minutes)	5.00	180.00	52.08	53.21	5.00	180.00	46.50	47.10
Sa (%)	19.73	84.14	47.08	17.81	19.73	84.14	44.53	17.66
C (%)	6.45	66.14	32.45	17.29	6.45	66.14	33.98	17.45
Si (%)	7.74	36.74	20.47	10.28	7.74	36.74	21.49	10.20
ρ (gm/cm^3)	1.24	1.75	1.62	0.11	1.24	1.75	1.61	0.11
M (%)	5.28	19.85	13.20	4.93	5.28	19.85	13.47	5.00
I (mm/hr)	0.25	252.00	57.49	59.32	0.90	204.00	47.29	45.92
Cu (mm)	1.50	383.00	46.99	64.62	3.00	244.00	38.54	41.61

M5 tree (M5)

Quinlan (1992) proposed this high-accuracy regression algorithm for dealing with continuous-class problems with high dimensionality. M5 generates a tree-based model which has multivariate linear regression models at the leaves. The first step is to build a tree, which is similar to inducing a regular decision tree. The difference is that the splitting criterion is based on maximizing the expected reduction in error instead of maximizing the information gain at each interior node. The standard deviation reduction (SDR) can be calculated by:

$$SDR = sd(K) - \sum \frac{|K_i|}{|K|} sd\,(K_i) \tag{1}$$

where K represents number of instances that reach the node; K_i represents the number of instances that have the ith outcome of the potential set; and sd represents the standard deviation of the target values of the instances in the data set K.

The second step is to construct a multivariate linear regression function using the instances at each non-terminal node and then prune the built tree back from the leaves. Instead of using all attributes, this multivariate linear regression function only uses the attributes that are referenced by tests or linear functions somewhere in the subtree at this node.

The last step is to improve the prediction accuracy by applying some smoothing techniques. The smoothed predicted values of M5 are backed up from the leaf node to the root node (Li & Li 2011).

Performance assessment parameters

The performance of various modelling techniques has been carried out using various statistical performance assessment parameters, which are coefficient of determination (R^2), root mean square error (RMSE) and mean absolute error (MAE) to check the statistical difference between the actual and the predicted values of the implemented soft computing approaches.

Coefficient of determination (R^2): The coefficient of determination is used to measure of success of numeric prediction. The coefficient of determination (R^2) is computed as follows (Adnan *et al.* 2019; Malik *et al.* 2019):

$$R^2 = \left(\frac{n\sum_{i=1}^{n} A_{\text{obs}_i} A_{\text{pred}_i} - \left(\sum_{i=1}^{n} A_{\text{obs}_i}\right)\left(\sum_{i=1}^{n} A_{\text{pred}_i}\right)}{\sqrt{n\left(\sum_{i=1}^{n} A^2_{\text{obs}_i}\right) - \left(\sum_{i=1}^{n} A_{\text{obs}_i}\right)^2}\sqrt{n\left(\sum_{i=1}^{n} A^2_{\text{pred}_i}\right) - \left(\sum_{i=1}^{n} A_{\text{pred}_i}\right)^2}} \right)^2 \tag{2}$$

where A_{obs_i} is the observed values; A_{pred_i} is the predicted values; and n is the number of observations. The correlation coefficient is a statistical measure of the strength of the relationship between the relative movements of observed (A_{obs}) and predicted values (A_{pred}). The range of R^2 is 0 to 1.0. A correlation of 1.0 indicates that correlation is perfect among observed (A_{obs}) and predicted (A_{pred}) values, while a correlation of 0.0 indicates that there is no linear relationship among the observed (A_{obs}) and predicted (A_{pred}) values.

Mean absolute error (MAE)

The mean absolute the error is used to measure the success of numeric estimation. The mean absolute error (MAE) is computed as follows (Sammen *et al.* 2017; Pham *et al.* 2021):

$$MAE = \frac{1}{n}\sum_{i=1}^{n} |A_{\text{obs}_i} - A_{\text{pred}_i}| \tag{3}$$

MAE can range from 0 to ∞. An efficiency of 0 (MAE = 0) corresponds to a perfect match of modelled A to the observed data.

Root mean square error (RMSE)

Mean-squared error (MSE) is the most commonly statistical evaluator for performance evaluation, and root mean square error is the square root of the mean-squared error after giving it the same dimensions as the predicted values themselves. This method exaggerates the prediction error minus the difference between prediction value and actual value. The root mean square error (RMSE) can be computed from the following equation (Sammen *et al.* 2020):

$$RMSE = \sqrt{\frac{1}{n}\sum_{i=1}^{n}\left(A_{\text{obs}_i} - A_{\text{pred}_i}\right)^2} \quad (4)$$

RMSE can range from 0 to ∞. An efficiency of 0 (RMSE = 0) corresponds to a perfect ideal model which predicts exact values as observed values.

EVALUATION OF MODELS

The performance measures adopted for model assessment are coefficient of determination (R^2), root mean square error (RMSE) and mean absolute error (MAE). The value of R^2 determines the closeness of the regression line to the observed data. $R^2 = 1$ indicates perfect fitting of the regression line to the observed data. RMSE and MAE are the error values that measure the accuracy of the models in approximating the actual observations. Lower values of RMSE and MAE indicate better prediction of the actual data. Two output values: infiltration rate and cumulative infiltration, are predicted relative to the input variables (Table 2) using the regression-tree-based models. In this study, WEKA 3.9 software is used for all applied regression-tree-based model development and validation. The models are developed using the training data set after setting the values of tuning parameters (user-defined) respective to each regression tree model. The performance of the constructed model is tested on the unseen testing data set using the statistical performance measures (R^2, RMSE and MAE). Several models are constructed and tested by assigning the values of the user-defined parameters to the modelling techniques and the best one is chosen based on optimal fitting of the model on the testing data set. The procedure is repeated for all the regression tree models using both outputs and the tuning parameters selected for performance analysis are shown in Table 3.

Table 3 | Tuning parameters of regression models

	Tuning parameters	
Regression model	**Infiltration rate prediction**	**Cumulative infiltration prediction**
Random model tree	Number of randomly chosen attributes = 4	Number of randomly chosen attributes = 3
Random forest	$k = 200$, $m = 1$	$k = 78$, $m = 2$
M5 tree	Number of instances allowed at a leaf node = 5	Number of instances allowed at a leaf node = 5

RESULTS AND DISCUSSION

Modelling of infiltration rate

The models are formed and checked for accuracy after setting the tuning parameters to near-optimal values (Table 3) in order to predict the values of soil infiltration rate. The simulation results of training and testing data are presented in Figures 2 and 3, respectively, which show the scattering of the data points for the prediction of three regression-tree-based models viz. random tree (RT), random forest (RF) and M5 tree. To investigate the superiority of the modelling techniques, the statistical performance measures acquired by the models with the observed infiltration rate data are elucidated in Table 4. As indicated from Figure 2, the scattering of training data predicted by the M5 tree with both pruned as well as unpruned model is higher relative to random tree and random forest. The higher values of infiltration rate

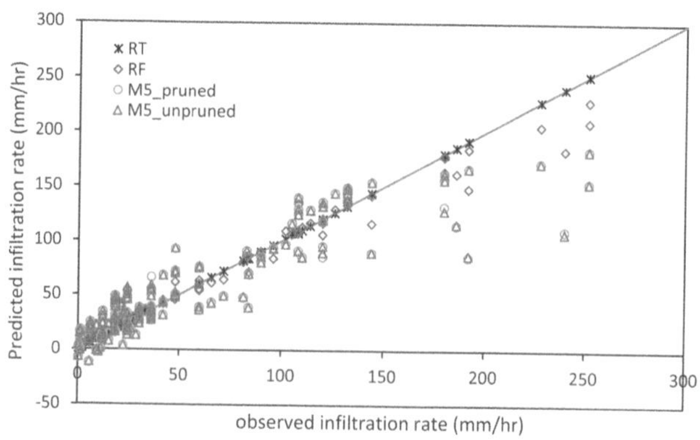

Figure 2 | Scatter plot of infiltration rate prediction by the regression models using training data.

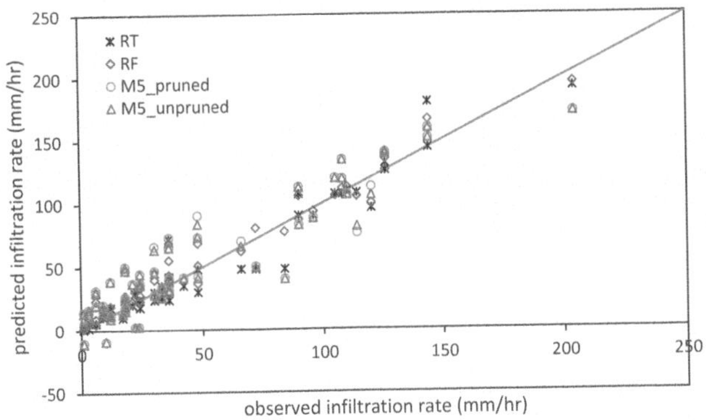

Figure 3 | Scatter plot of infiltration rate prediction by the regression models using testing data.

predicted by the M5 tree model scatter relatively more than the lower values and degrade the modelling performance. Statistical measures from the M5 tree confirm the poor performance of model development with the training data as the RMSE and MAE values are substantially higher. The performance of the model is improved with the testing data set, but the testing performance is inferior to the other regression tree models. The trained model based on the random tree is best ($R^2 = 0.99$) as the data prediction is close to the line of perfect agreement in the training stage but the model is not capable of giving accurate predictions with the testing data as compared with the random forest model. The performance of the model developed with RT drops at the validation stage having RMSE and MAE values of 10.82 and 5.98, respectively (Table 4). Closer fitting of predicted points to the perfect agreement line using the RF model in the testing stage reveals its suitability in the generalization of the infiltration rate data as the error values (RMSE = 8.10 and MAE = 5.74) are minimum (Table 4). Figure 4 shows the residual error values obtained from the experimental observations and predictions of infiltration rate with regression tree models. The residuals are minimum with the RT model during the training stage, but the deviation of the model is higher from the zero error line during the testing stage relative to the RF model. M5 tree-based models have larger residuals during model training but the quality of the models improve during testing with smaller error values, but still, the models are ranked last after the RT model. Therefore, modelling results based on this study indicate superior performance of the RF model in simulating the actual infiltration rate of soil followed by the RT and M5 tree models, respectively.

Modelling of cumulative infiltration

In this section, the cumulative infiltration of soil is simulated using random tree (RT), random forest (RF) and M5 tree (M5) models with the basic soil properties. The scattering plots of the regression-tree based models for the training and testing data sets are depicted in Figures 5 and 6, respectively. Table 5 provides the information about the statistical measures obtained from the observed training and testing cumulative infiltration data with their corresponding predicted values by the models. The scattering of the training data predicted by the M5 pruned and unpruned models is higher, particularly for the larger values, as compared with the RT and RF models (Figure 5). The predicted data of the RT and RF model fits closely to the perfect agreement line

Table 4 | Performance of regression models in simulating the infiltration rate

	Training data			Testing data		
Regression model	R^2	**RMSE (mm/hr)**	**MAE (mm/hr)**	R^2	**RMSE (mm/hr)**	**MAE (mm/hr)**
Random model tree	1.00	0.62	0.27	0.95	10.82	5.98
Random forest	0.98	9.93	4.73	0.97	8.10	5.74
M5 tree pruned	0.81	26.81	18.02	0.85	19.08	15.28
M5 tree unpruned	0.81	26.81	17.96	0.85	18.49	14.10

Table 5 | Performance of regression models in simulating the cumulative infiltration

	Training data			Testing data		
Regression model	R^2	**RMSE (mm)**	**MAE (mm)**	R^2	**RMSE (mm)**	**MAE (mm)**
Random model tree	0.99	0.92	0.54	0.94	10.76	7.87
Random forest	0.99	10.33	4.52	0.97	6.96	4.44
M5 tree pruned	0.81	28.92	16.09	0.81	18.43	11.21
M5 tree unpruned	0.82	28.68	15.91	0.81	18.11	10.97

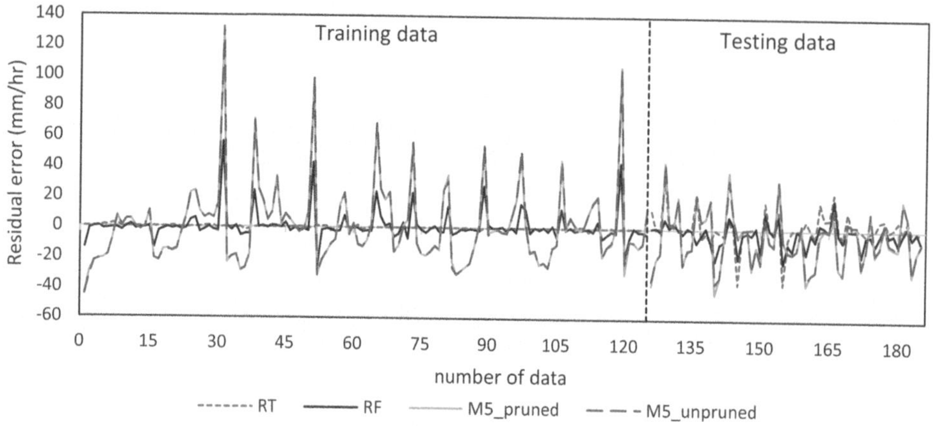

Figure 4 | Residual error values from observed and predicted infiltration rate for training and testing data sets.

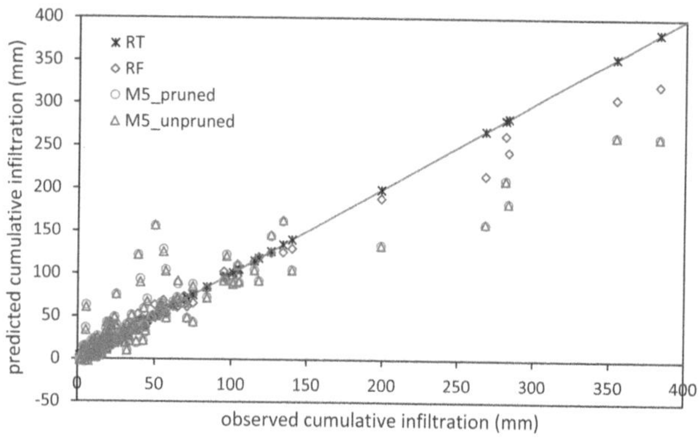

Figure 5 | Scatter plot of cumulative infiltration prediction by the regression models using training data.

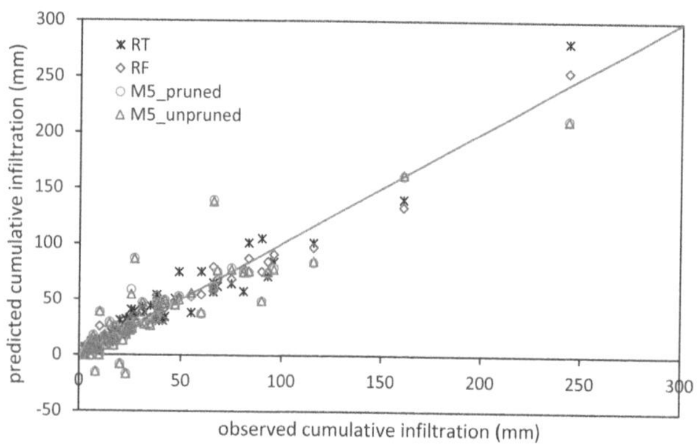

Figure 6 | Scatter plot of cumulative infiltration prediction by the regression models using testing data.

having lower values of RMSE and MAE than the M5 tree in the model development stage. The RT model perfectly predicts the training data with a maximum value of $R^2 = 0.99$ and minimum values of RMSE $= 0.92$ and MAE $= 0.54$ as most of the data points reside on the perfect agreement line, but the prediction capability of the model degrades during validation with the testing data. Analysing the testing data scatter plot (Figure 6), the prediction tendency of the RF model is superior as compared with the other regression models as the predicted points are comparatively closer to the perfect agreement line than the other models. The lowest values of RMSE $= 6.96$ and MAE $= 4.44$ obtained from the RF model at the testing stage affirm the stronger capability of the model in the generalization of the cumulative infiltration data (Table 5). So, statistical performance measures acquired from the testing data set suggest better performance by the RF model as compared with the RT and M5 tree models in approximating the cumulative infiltration of soil. Figure 7 shows the residual errors observed with the current data set using different regression models. The residual error values peak in the training stage, mostly predicted by the M5 tree model, while the RT model has smaller residual values. The residuals are minimum with the RF model relative to the RT model in the testing stage, which assures its higher predictability.

SIMULATION OF INFILTRATION CHARACTERISTICS OF SOIL BY RANDOM FOREST MODEL

Based on the modelling results discussed above, it is clear that the RF model has higher generalization performance in simulating the infiltration rate as well as cumulative

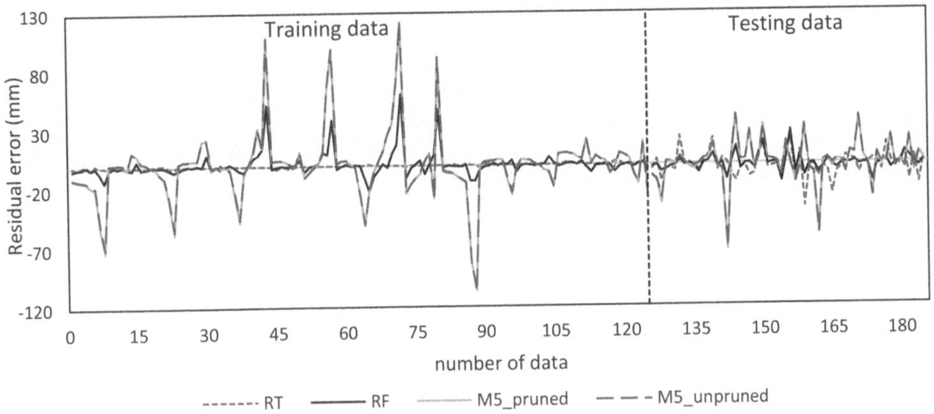

Figure 7 | Residual error values from observed and predicted cumulative infiltration for training and testing data sets.

infiltration of soil as compared with the other applied models. In this section, infiltration characteristics experimentally observed from their respective sites are compared with the predicted values of the RF model. In order to accomplish this task, the RF model developed with the training data set is tested on the complete data set of the locations. Both outputs, infiltration rate and cumulative infiltration, observed from the field are presented with the RF model-based predictions for their respective sites (Figure 8). The model trained for both outputs using the tuning parameters (Table 3) is implemented to their respective data sets for prediction. For comparative analysis, the results are plotted for each site (specified by the location number). The infiltration rate/cumulative infiltration values are plotted against time to show the variation of the model in predicting the observed data. The results reveal that the prediction of points by the RF model agrees reasonably well with the observed infiltration values, particularly in predicting the infiltration rate, and the model has a decent tendency to follow the actual observations.

PARAMETRIC SENSITIVITY

The sensitivity of each input variable in predicting the output variables (infiltration rate and cumulative infiltration of soil) is tested. The random forest (RF) model developed with the training data set is used to study the parametric importance in prediction. Models are developed using various combinations of input parameters by removing each parameter one by one and the sensitivity is judged by means of R^2 and RMSE. The fluctuation in statistical measures with various input combinations accounts for the change in performance of each model in the prediction of both outputs. Analysing the combination from Table 6, it is revealed that the removal of time (t) from the input parameters causes higher fluctuation in both R^2 as well as RMSE. The combination drastically increases the RMSE value, which makes time (t) the most important variable among the other input variables in affecting the prediction of infiltration rate as well as cumulative infiltration. The absence of other input parameters in the RF model yields a small change in R^2 and RMSE values. Removal of the parameters sand, clay and silt, one by one, marginally decrease the RMSE and hence slightly improve the efficiency of the model in predicting the infiltration rate, while on the other hand, neglecting these parameters has a negative effect in cumulative infiltration prediction with slight increment in RMSE values (Table 6).

Infiltration is characterised as the flow of water into the subsurface from above ground. The topic of infiltration has received a great deal of attention because of its importance to topics as widely ranging as irrigation, contaminant transport, groundwater recharge, and ecosystem viability. In the last few decades conventional infiltration models have been widely used for the calculation or estimation of infiltration rate and cumulative infiltration, such as Kostiakov, Philip, Novel, Horton, Holton etc. Machiwal *et al.* (2006) observed that the infiltration process was well described by the Philip model in a wasteland of Kharagpur, India. However, soil management practices influencing the final infiltration rate are a major factor in deciding the applicability of these models.

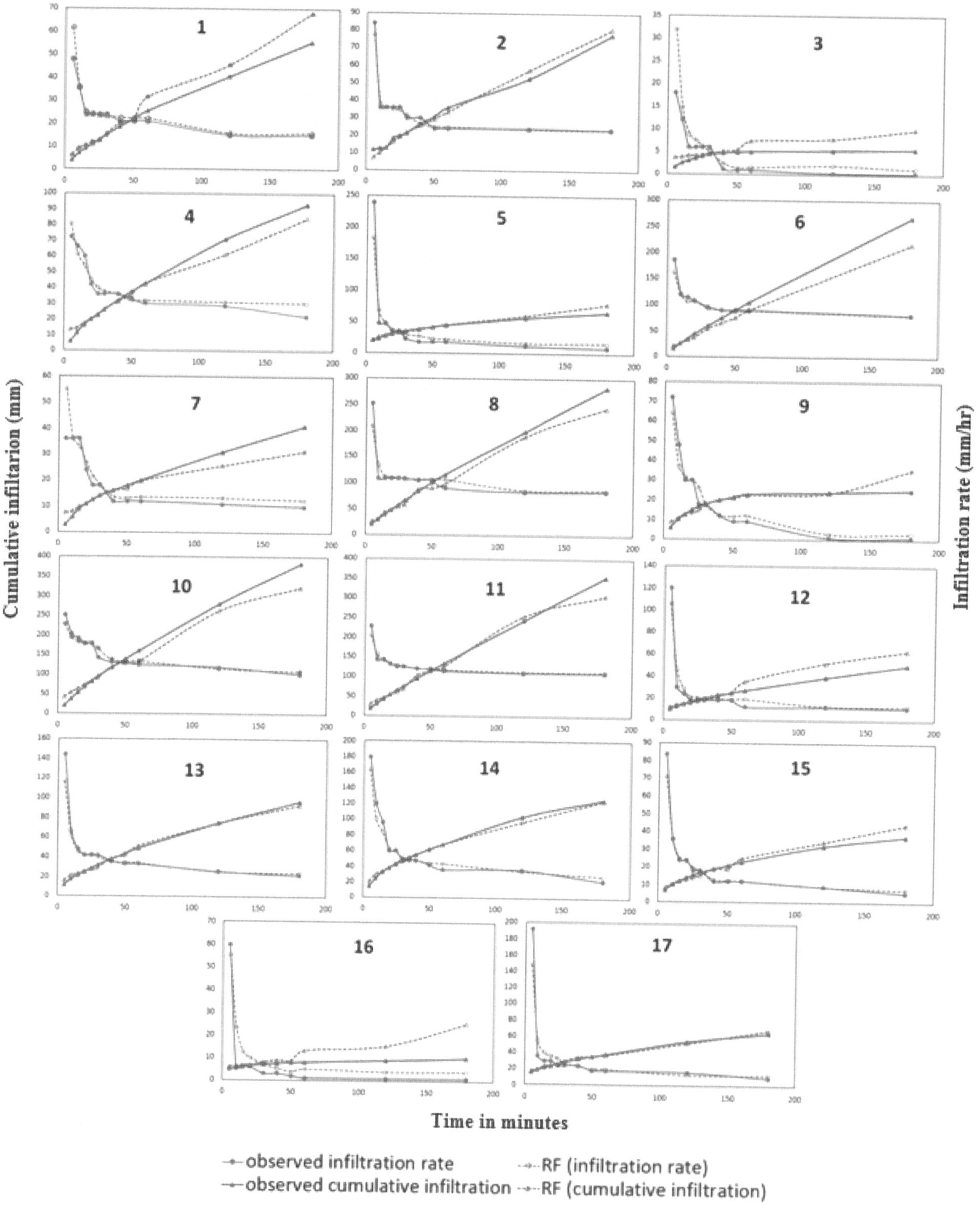

Figure 8 | Comparison of observed and RF model predicted infiltration characteristics of soil respective to each site.

Table 6 | Sensitivity analysis of parameters using random forest (RF) model

		Infiltration rate prediction		Cumulative infiltration prediction	
Combination	**Parameter removed**	R^2	**RMSE**	R^2	**RMSE**
t, ρ, M, Sa, Si, C	–	0.98	9.93	0.99	10.34
ρ, M, Sa, Si, C	*t*	0.61	37.08	0.39	50.40
t, ρ, M, Si, C	*Sa*	0.98	9.53	0.99	10.46
t, ρ, M, Sa, Si	*C*	0.98	9.41	0.99	11.49
t, ρ, M, Sa, C	*Si*	0.98	9.42	0.99	10.65
t, M, Sa, Si, C	*ρ*	0.98	9.45	0.99	10.55
t, ρ, Sa, Si, C	*M*	0.98	10.49	0.99	10.65

Thus, the variability of soil infiltration characteristics and goodness of fit of the infiltration models for different soils should be given due consideration in infiltration modelling studies for predicting the constant infiltration rate. The applicability of these models for estimating the infiltration rate with different soil management has been examined by researchers. Gifford (1976) observed that amongst the Horton, Kostiakov and Philip models, the Horton model was the best model to fit the infiltration data in mostly semi-arid regions from Australia, but under specific conditions only. Mishra *et al.* (2003) compared the performance of 14 different conventional infiltration models using 243 sets of infiltration data collected from field and laboratory tests conducted in India and the USA on soils ranging from coarse sand to fine clay. These models are point- or location-specific, and until now no general model has been developed. The results of this study suggest that the RF-based model is suitable for predicting the infiltration characteristics of the study area. The RF model is a general model for the study area and gives closer values to the observed values. The RF model is used to represent the infiltration characteristics of the study area and it saves time and effort in comparison with experimentation and conventional models.

CONCLUSIONS

Prediction of infiltration characteristics is essential for several hydrologic studies associated with surface and subsurface flow. In this study, the effectiveness of regression-tree-based models in simulating the infiltration characteristics of soil is evaluated. In estimating the infiltration rate and cumulative infiltration of soil, random forest outperforms random tree as well as M5 tree with $R^2 =$ 0.97 and 0.97, RMSE = 8.10 and 6.96 and MAE = 5.74 and 4.44, respectively. The training and testing results of the random forest encourage the utility of this method relative to the other tested methods and show a good potential in representing the infiltration characteristics of the study area. Based on the sensitivity results, time is observed as the most important parameter in affecting the prediction performance of both the infiltration rate and the cumulative infiltration of soil.

Finally, it is worth noting that the proposed approach uses only the six fundamental parameters of *t, ρ, M, Sa, Si* and *C* for prediction of the infiltration rate and cumulative infiltration of soil. This is of importance for situations where there is no possibility of measuring other parameters like porosity and lower depth soil type. Considering the ability of the RF model in handling missing values in any data sets as well as its capacity in bypassing noisy data, the implemented soft computing methodology showed the potential to be accurately used in approximation of the infiltration process. The RF model's ability needs to be compared with hybrid techniques to try to develop a more accurate model. More experimental data is also required for better and general results.

DATA AVAILABILITY STATEMENT

Data cannot be made publicly available; readers should contact the corresponding author for details.

REFERENCES

Adnan, R. M., Malik, A., Kumar, A., Parmar, K. S. & Kisi, O. 2019 Pan evaporation modeling by three different neuro-fuzzy intelligent systems using climatic inputs. *Arabian Journal of Geosciences* **12** (19), 606.

Adusumilli, S., Bhatt, D., Wang, H., Bhattacharya, P. & Devabhaktuni, V. 2013 A low-cost INS/GPS integration methodology based on random forest regression. *Expert Systems with Applications* **40** (11), 4653–4659.

Al-Sulaiman, M. A. 2015 Applying of an adaptive neuro fuzzy inference system for prediction of unsaturated soil hydraulic conductivity. *Biosciences Biotechnology Research Asia* **12** (3), 2261–2272.

Al-Sulaiman, M. A. & Aboukarima, A. M. 2016 Distribution of natural radionuclides in the surface soil in some areas of agriculture and grazing located in west of Riyadh, Saudi Arabia. *Journal of Applied Life Sciences International* **7** (2), 1–12.

Amit, Y. & Geman, D. 1997 Shape quantization and recognition with randomized trees. *Neural Computation* **9** (7), 1545–1588.

Anari, P. L., Darani, H. S. & Nafarzadegan, A. R. 2011 Application of ANN and ANFIS models for estimating total infiltration rate in an arid rangeland ecosystem. *Research Journal of Environmental Sciences* **5** (3), 236–247.

Arshad, R. R., Sayyad, G., Mosaddeghi, M. & Gharabaghi, B. 2013 Predicting saturated hydraulic conductivity by artificial intelligence and regression models. *ISRN Soil Science* **2013**, 308159.

Bhave, S. & Sreeja, P. 2013 Influence of initial soil condition on infiltration characteristics determined using a disk infiltrometer. *ISH Journal of Hydraulic Engineering* **19** (3), 291–296.

Breiman, L. 1996 Bagging predictors. *Machine Learning* **24** (2), 123–140.

Breiman, L. 2001 Random forests. *Machine Learning* **45** (1), 5–32.

Chen, L. & Young, M. H. 2006 Green–Ampt infiltration model for sloping surfaces. *Water Resources Research* **42** (7), W07420.

Das, T., Dettinger, M. D., Cayan, D. R. & Hidalgo, H. G. 2011 Potential increase in floods in California's Sierra Nevada under future climate projections. *Climatic Change* **109** (1), 71–94.

Diamond, J. & Shanley, T. 2003 Infiltration rate assessment of some major soils. *Irish Geography* **36** (1), 32–46.

Elbisy, M. S. 2015 Support Vector Machine and regression analysis to predict the field hydraulic conductivity of sandy soil. *KSCE Journal of Civil Engineering* **19** (7), 2307–2316.

Erzin, Y., Gumaste, S. D., Gupta, A. K. & Singh, D. N. 2009 Artificial neural network (ANN) models for determining hydraulic conductivity of compacted fine-grained soils. *Canadian Geotechnical Journal* **46** (8), 955–968.

Esmaeelnejad, L., Ramezanpour, H., Seyedmohammadi, J. & Shabanpour, M. 2015 Selection of a suitable model for the prediction of soil water content in north of Iran. *Spanish Journal of Agricultural Research* **13** (1), e12-002.

Gifford, G. F. 1976 Applicability of some infiltration formulae to rangeland infiltrometer data. *Journal of Hydrology* **28** (1), 1–11.

Green, W. H. & Ampt, G. A. 1911 Studies on soil physics. *The Journal of Agricultural Science* **4** (1), 1–24.

Haghighi, F., Kheirkhah, M. & Saghafian, B. 2011 Evaluation of soil hydraulic parameters in soils and land use change. In: *Earth and Environmental Sciences*. (I. A. Dar & M. A. Dar, eds), InTech, Rijeka, Croatia, pp. 457–464.

Hillel, D. 1998 *Environmental Soil Physics: Fundamentals, Applications, and Environmental Considerations*. Elsevier, London, UK.

Holtan, H. N. 1961 *A Concept for Infiltration Estimates in Watershed Engineering*. US Department of Agriculture, Washington, DC, USA.

Horton, R. E. 1941 An approach toward a physical interpretation of infiltration-capacity. *Soil Science Society of America Journal* **5** (C), 399–417.

Kostiakov, A. N. 1932 On the dynamics of the coefficient of water percolation in soils and the necessity of studying it from the dynamic point of view for the purposes of amelioration. *Trans. Sixth Comm. Int. Soc. Soil Sci.* **1**, 7–21.

Li, C. & Li, H. 2011 Learning random model trees for regression. *International Journal of Computers and Applications* **33** (3), 258–265.

Machiwal, D., Jha, M. K. & Mal, B. C. 2006 Modelling infiltration and quantifying spatial soil variability in a wasteland of Kharagpur, India. *Biosystems Engineering* **95** (4), 569–582.

Malik, A., Kumar, A. & Singh, R. P. 2019 Application of heuristic approaches for prediction of hydrological drought using multi-scalar streamflow drought index. *Water Resources Management* **33** (11), 3985–4006.

Mishra, S. K., Tyagi, J. V. & Singh, V. P. 2003 Comparison of infiltration models. *Hydrological Processes* **17** (13), 2629–2652.

Pfahringer, B. 2010 *Random Model Trees: An Effective and Scalable Regression Method*, Working Paper 03/2010. Department of Computer Science, University of Waikato, Hamilton, New Zealand.

Pham, Q. B., Mohammadpour, R., Linh, N. T. T., Mohajane, M., Pourjasem, A., Sammen, S. S., Anh, D. T. & Nam, V. T. 2021 Application of soft computing to predict water quality in wetland. *Environmental Science and Pollution Research* **28**, 185–200. https://doi.org/10.1007/s11356-020-10344-8.

Philip, J. R. 1957 The theory of infiltration: 1. The infiltration equation and its solution. *Soil Sci.* **83** (5), 345–357.

Quinlan, J. R. 1992 Learning with continuous classes. In: *AI '92: Proceedings of the 5th Australian Joint Conference on Artificial Intelligence* (A. Adams & L. Sterling, eds), World Scientific, Singapore, pp. 343–348.

Rahmati, M. 2017 Reliable and accurate point-based prediction of cumulative infiltration using soil readily available characteristics: a comparison between GMDH, ANN, and MLR. *Journal of Hydrology* **551**, 81–91.

Sammen, S. S., Mohamed, T. A., Ghazali, A. H., El-Shafie, A. H. & Sidek, L. M. 2017 Generalized regression neural network for prediction of peak outflow from dam breach. *Water Resources Management* **31**, 549–562. https://doi.org/10.1007/s11269-016-1547-8.

Sammen, S. S., Ghorbani, M. A., Malik, A., Tikhamarine, Y., AmirRahmani, M., Al-Ansari, N. & Chau, K.-W. 2020 Enhanced artificial neural network with harris hawks optimization for predicting scour depth downstream of ski-jump spillway. *Applied Sciences* **10**, 5160. https://doi.org/10.3390/app10155160.

Schaap, M. G. & Leij, F. J. 1998 Using neural networks to predict soil water retention and soil hydraulic conductivity. *Soil and Tillage Research* **47** (1–2), 37–42.

Sihag, P., Tiwari, N. K. & Ranjan, S. 2017a Estimation and inter-comparison of infiltration models. *Water Science* **31** (1), 34–43.

Sihag, P., Tiwari, N. K. & Ranjan, S. 2017b Modelling of infiltration of sandy soil using gaussian process regression. *Modeling Earth Systems and Environment* **3** (3), 1091–1100.

Sihag, P., Tiwari, N. K. & Ranjan, S. 2017c Prediction of unsaturated hydraulic conductivity using adaptive neuro-fuzzy inference system (ANFIS). *ISH Journal of Hydraulic Engineering* **25** (2), 132–142. doi:10.1080/09715010.2017.1381861.

Sihag, P. 2018 Prediction of unsaturated hydraulic conductivity using fuzzy logic and artificial neural network. *Modeling Earth Systems and Environment* **4** (1), 189–198.

Sihag, P., Tiwari, N. K. & Ranjan, S. 2018a Support vector regression-based modeling of cumulative infiltration of sandy soil. *ISH Journal of Hydraulic Engineering* **26** (1), 44–50. doi:10.1080/09715010.2018.1439776.

Sihag, P., Singh, B., Sepah Vand, A. & Mehdipour, V. 2018b Modeling the infiltration process with soft computing techniques. *ISH Journal of Hydraulic Engineering* **26** (2), 138–152. doi:10.1080/09715010.2018.1464408.

Singh, V. P. & Yu, F. X. 1990 Derivation of infiltration equation using systems approach. *Journal of Irrigation and Drainage Engineering* **116** (6), 837–858.

Singh, B., Sihag, P. & Singh, K. 2017 Modelling of impact of water quality on infiltration rate of soil by random forest regression. *Modeling Earth Systems and Environment* **3** (3), 999–1004.

Singh, B., Sihag, P. & Singh, K. 2018 Comparison of infiltration models in NIT Kurukshetra campus. *Applied Water Science* **8** (2), 63.

Smith, R. E. & Parlange, J.-Y. 1978 A parameter-efficient hydrologic infiltration model. *Water Resources Research* **14** (3), 533–538.

Soil Conservation Service 1972 *SCS National Engineering Handbook, Section 4: Hydrology*. US Department of Agriculture, Washington, DC, USA.

Swartzendruber, D. 1987 A quasi-solution of Richards' Equation for the downward infiltration of water into soil. *Water Resources Research* **23** (5), 809–817.

Sy, N. L. 2006 Modelling the infiltration process with a multi-layer perceptron artificial neural network. *Hydrological Sciences Journal* **51** (1), 3–20.

Tiwari, N. K., Sihag, P. & Ranjan, S. 2017 Modeling of infiltration of soil using adaptive neuro-fuzzy inference system (ANFIS). *Journal of Engineering & Technology Education* **11** (1), 13–21.

Zhao, Y. & Zhang, Y. 2008 Comparison of decision tree methods for finding active objects. *Advances in Space Research* **41** (12), 1955–1959.

First received 23 November 2020; accepted in revised form 29 January 2021. Available online 18 February 2021

Prediction of groundwater trends for irrigation in Northern Bangladesh

Mohammad Abid Hossain Mridha and Syed Hafizur Rahman

ABSTRACT

Groundwater trends affect the domestic, agricultural, and industrial prospects of a region. The study area is Bogura, a northern region of Bangladesh, located on the Pleistocene terrace of the Bengal Basin. The aquifer consists of medium-to-coarse sand, located at a depth of 4.66–42.68 m; groundwater is scarce during dry seasons. The water table (WT) time-series data for 2007–2019 were used for forecasting and characterizing present and future groundwater conditions using existing numerical simulations. The annual groundwater budget for discharge and storage was 2,772 and 2,442 Mm^3, respectively. Thus, the annual scarcity of groundwater was 330.4 Mm^3 (13.5%), excluding the surface water contribution of 10 Mm^3 (0.4%). The present spacing of deep tube wells (DTWs) and shallow tube wells (STWs) was 744 and 372 m, respectively. Currently, the DTW spacing ranged 744–800 m; however, the STW spacing of 250–372 m is higher than the set distance. Hence, further installations of STWs were strictly disallowed for irrigation. WT declined by 1.0 m in the last 13 years, i.e., 0.07 m or 1.2% decline rate per annum, causing water scarcity in the region during the peak period in the dry season (June–February), thus affecting irrigation and limiting agricultural production.

Key words | discharge, groundwater monitoring, groundwater storage, recharge, water table fluctuation, well spacing

Mohammad Abid Hossain Mridha
(corresponding author)
Rural Development Academy (RDA),
Sherpur, Bogura 5842,
Bangladesh
E-mail: *abidmridharda@gmail.com*

Syed Hafizur Rahman
Department of Environmental Sciences,
Jahangirnagar University,
Savar, Dhaka 1342,
Bangladesh

HIGHLIGHTS

- In the study, highest priority was given on groundwater use for irrigation.
- The study emphasized the recharge-discharge and storage phenomenon as hydrological components.
- The study forecasts and predicts the present and future threats of groundwater resources.

INTRODUCTION

Groundwater is one of the leading sources of water in Bangladesh. Owing to the inadequate availability of surface water during dry seasons, it functions as a vital, innocuous, and reliable source of water for irrigation, domestic, and industrial purposes. Groundwater is attained using various types of lifting device, such as hand tube wells (HTWs), shallow tube wells (STWs), deep tube well (DTWs), deep-set shallow tube wells (DSSTWs), and force mode tube wells (FMTWs). Moreover, the main sources of surface water in the study area are canals and the rivers of Nagar, Bengali, Ichamoti, Karatoa, Bhadrabati, Jamuna, and Tulshiganga. In the study area, a part of the surface water is obtained from low lift pumps (LLPs); however, most of the canal and rivers cannot bring water during the rainy season because of the lack of mining not only in the study area but all over the country. Furthermore, the static water

doi: 10.2166/ws.2021.076

tables (WTs) of subsurface water are decreasing in shallow aquifers in Asian mega deltas such as the Ganges–Brahmaputra–Meghna (GBM) basin owing to the large seasonal variation in the monsoon rainfall in Bangladesh. The penetration rates of the high Barind soil were $0.0015\ md^{-1}$ for wet land and $0.0075\ md^{-1}$ for dry land (MacDonald & Partners 1983). The WT is as much as 15 m below the ground surface, and groundwater fluctuation differs from 1.25 to 16.15 m (Asaduzzaman 1987). In 1990, the total amount of groundwater withdrawn was 8,806 Mm^3 against a reserve of 21,088 Mm^3 for several agricultural inputs in Bangladesh (MPO 1991). The irrigation coverage was 1.52×10^6 ha during 1982–1983, and it increased to 3.79×10^6 ha in 1996–1997 (BBS 1998). Groundwater also contributes 70% of the water used for irrigation and 90% of the drinking water in the country (Matin *et al.* 2000). The current available geological subsurface information indicates that most of the aquifers occur between depths of 30 to 130 m, and the country has heterogeneous aquifer systems (Zahid & Ahmed 2006). Further, alluvial aquifers are the proven sources of groundwater in Bangladesh, and the groundwater is annually refilled via rainfall and flooding and through the systematic recycling of water nationwide, except in Dhaka Megacity, where there is an imbalance between the asymmetrical recharge–discharge. Approximately 80% of the agricultural land is irrigated using groundwater after the start of the crop season in Bangladesh, and groundwater scarcity arose 42% each year during the operation of STWs in the dry season in northwest Bangladesh (Shahid & Hazarika 2010). In Barind Tract of Bangladesh, prospective groundwater recharge decreased by 85%, whereas the rest was moderate. In this region, only 8.6% of the total average annual rainfall (1,685 mm) infiltrates into the subsurface and ultimately contributes to the recharge of groundwater (Adham *et al.* 2010). Net recharge was increased approximately 0.005–0.015 m/year in many region of Bangladesh during 1985–2007 for increased groundwater withdrawal for irrigation as well as consumptive use, but net recharge marginally decreased from −0.0005 to −0.001 m/year, where groundwater irrigation was low; that is, <30% of total irrigation at the same period (Shamsudduha *et al.* 2011). Since the commencement of groundwater irrigation in the mid-1980s, a positive impact was observed in cropping patterns, cropping intensity, and national intensity, crop selection, and cultivation yields practices. Besides, the WT has decreased continuously at rates of 0.4 and 0.22 m/year in the wet and dry seasons, respectively (Rahman & Mahbub 2012). Generally, the rain season starts in May and ends in September, and there is low or no rainfall during the rest of the year; thus, the maximum rainfall is observed in June–August. The maximum WT is observed in July–September because of rainwater infiltration into the subsurface, and the minimum WT is observed in March–May at the time of irrigation. The WT declines continuously owing to overdrawing of groundwater for irrigation in Chapai–Nababganj in northwest Bangladesh (Hasan *et al.* 2013). The annual groundwater recharge and discharge were estimated to be 106.41–244 Mm^3 and 93.77–291 Mm^3, respectively, in Naogaon region in northwest Bangladesh; these data reveal that there existed a balance between the annual recharge and withdrawal up to 1993; however, since 1993, the discharge exceeded the recharge. In all, 23.99–42.08 Mm^3 of groundwater was discharged using STWs and DTWs, and the rest of the groundwater is discharged by natural seepage (Reza *et al.* 2013). The groundwater depletion rate ranges from 0.00 to 0.03 m/year based on the mean depth of phreatic surfaces; further, the maximum depth of the phreatic surface varies from 0.012 to 0.15 m/year (Abdullah 2014). The depth of WT and rainfall penetration are slowly declining in most wells, and the present trends of decline for the WT in Bogura district suggest that the depth of WT may double by 2060 in some regions (Hasanuzzaman *et al.* 2017). Barind, the drought prone area in Bangladesh, has a typical dry climate and high temperature except for the winter season, beginning from mid-June to October, whereas the rainfall and temperature ranges are 1,500–2,000 mm and 4 °C–44 °C, respectively, in the northern region of Bangladesh (www.bmda.gov.bd).

However, groundwater hydrology is a very complex process and is not limited to merely understanding the groundwater situation, zoning maps, temperature, humidity, rainfall, weather, climate, and aquifer properties because of the difficulties in characterizing an area based on only a few studies. Therefore, groundwater hydrology is a continuous process of providing solutions when studying a problem for a specific region. The aquifer in the Bogura region is in a poor condition – during the dry season, water is insufficient to meet the demands of both irrigation and human consumption; hence, efforts are being made to improve the availability of groundwater resources. Although research on groundwater in northern Bogura of Bangladesh is scanty, most focus on groundwater assessments (Abdullah 2014; Hasanuzzaman *et al.* 2017).

The main objective of the study is of interest to the hydrological community, particularly the Water Users Association (WUA) involved with groundwater resources, especially for irrigation. First, the aquifer is mapped using lithological data of boreholes. Second, the balance between groundwater recharge–discharge and storage is estimated. Third, the safe well spacing for installing the boreholes is determined.

The result can provide insight into the need for digging or abstaining from digging new wells, selection of the device for extracting groundwater, and operability of different modes of the water-lifting equipment. Furthermore, the results will be a handy tool for decision-makers and planners working not only in the study area but also on national and global scales.

METHODS

In this study, the present and future groundwater tendencies were predicted using some recognized empirical equations. The focus was to obtain a clear understanding of groundwater resources, including the balance between recharge–discharge and storage. The methodological steps are described in the following flowchart:

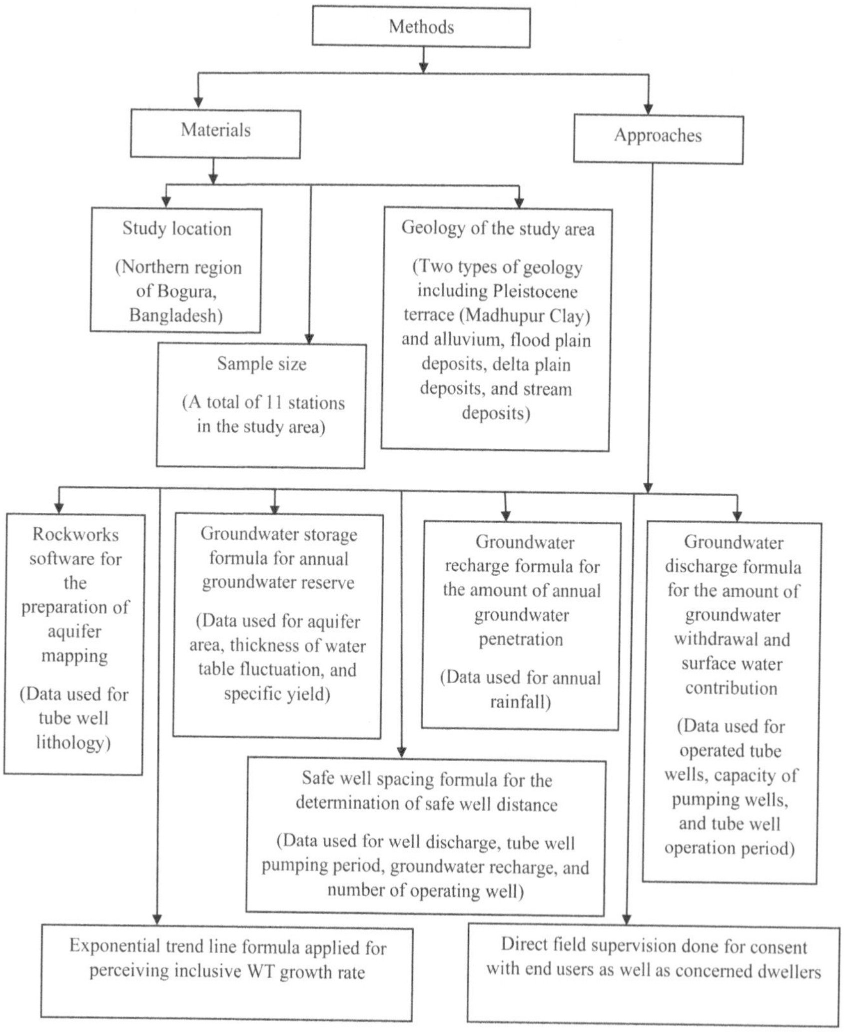

This section provides a broad stepwise description based on appropriate simulation to rationalize the projected scenario of groundwater resources.

Study location

The target area of this study was Bogura in northern Bangladesh (Figure 1). Geographically, the study area lies between longitudes 89°, 21 min, and 0 s east and from 24° and latitudes 46 min to 48° north. It is called the gateway to north Bangladesh. It is approximately 200 km from the north of Dhaka, the capital of the nation.

Geology of the study area

Seven types of geological group formations are exposed across Bangladesh (Figure 1). Of these, two types of geological formations were found in the study area. These formations include the Pleistocene terrace (Madhupur Clay) and alluvium, flood plain deposits, delta plain deposits, and stream deposits. The mappable body of rock of the Pleistocene terrace is part of the Dihing formation. This enormous formation has aquifer characteristics. The formation contains clays of various colors, such as medium-grain molted clay and yellow and gray clays, and occasional pebbly sandstone. The rock formations are partly associated in most of the locations, and similarly, alluviums are loosened materials deposited on the flood plains, delta basins, river beds, lakes, or estuaries. Alluviums comprise sand, silt, gravel, cobbles, and boulders along with organic materials that are converted into fertile soil. Generally, the term 'alluvium' is constrained to size-sorted fine sediments (sand, silt, and clay) and unlithified riverine deposits. In addition, coarse-grained and very coarse-grained sand (brown and gray) is found most of the alluvium in the study area. The color of sand is transformed from yellowish and brownish to gray.

Sample size

Lithologies for 145 boreholes were analyzed for mapping aquifers and to prepare a groundwater lithological inventory to identify the soil formation types. Precisely, 11 stations were selected for monitoring the wells (Figure 2) (one from each station or subunit), and the time-series primary data of WT were collected for the last 13 years (2007–2019). The location of the selected monitoring wells is listed in Table 1.

Rockworks software

Rockworks allow computer modeling based on the end-user specifications. The basic strategies of the software are the creation of a borehole database using the analytical results for various physical and chemical properties as a function of the depth. The database also generates images, such as block diagrams, cross sections, and fence diagrams, to check the legitimacy and geological reasonability of the modeling. It uses two-dimensional (2-D) and three-dimensional (3-D) images with three feature levels – basic, standard, and advanced. It is often difficult to determine the capabilities, operational features, and limitations of a particular groundwater modeling code from the documentation, or even without accurately running the code for situations relevant to the region for which a code is to be selected owing to the incompleteness, poor organization, or incorrectness of code documentation. Thus, this software was used in the study to map the aquifer using lithological data.

Estimation of annual groundwater storage

Groundwater refers to the water present in the pore spaces of soil. That part of the rock or unconsolidated deposit formation is called an aquifer when it can harness an effective quantity of water. Groundwater storage is the product of the area of the aquifer, depth or thickness of the WT fluctuation, and specific yield (Raghunath 1987):

$$\Delta S = A \times \varnothing H \times S_y, \tag{1}$$

where ΔS is the groundwater storage, A is the area of aquifer in the study area (m^2), $\varnothing H$ is the depth or thickness of the WT fluctuation (m), and S_y is the specific yield (%). The depth of the WT fluctuation is the difference between its maximum and minimum levels in the WT throughout the year in the given region. In this study, groundwater

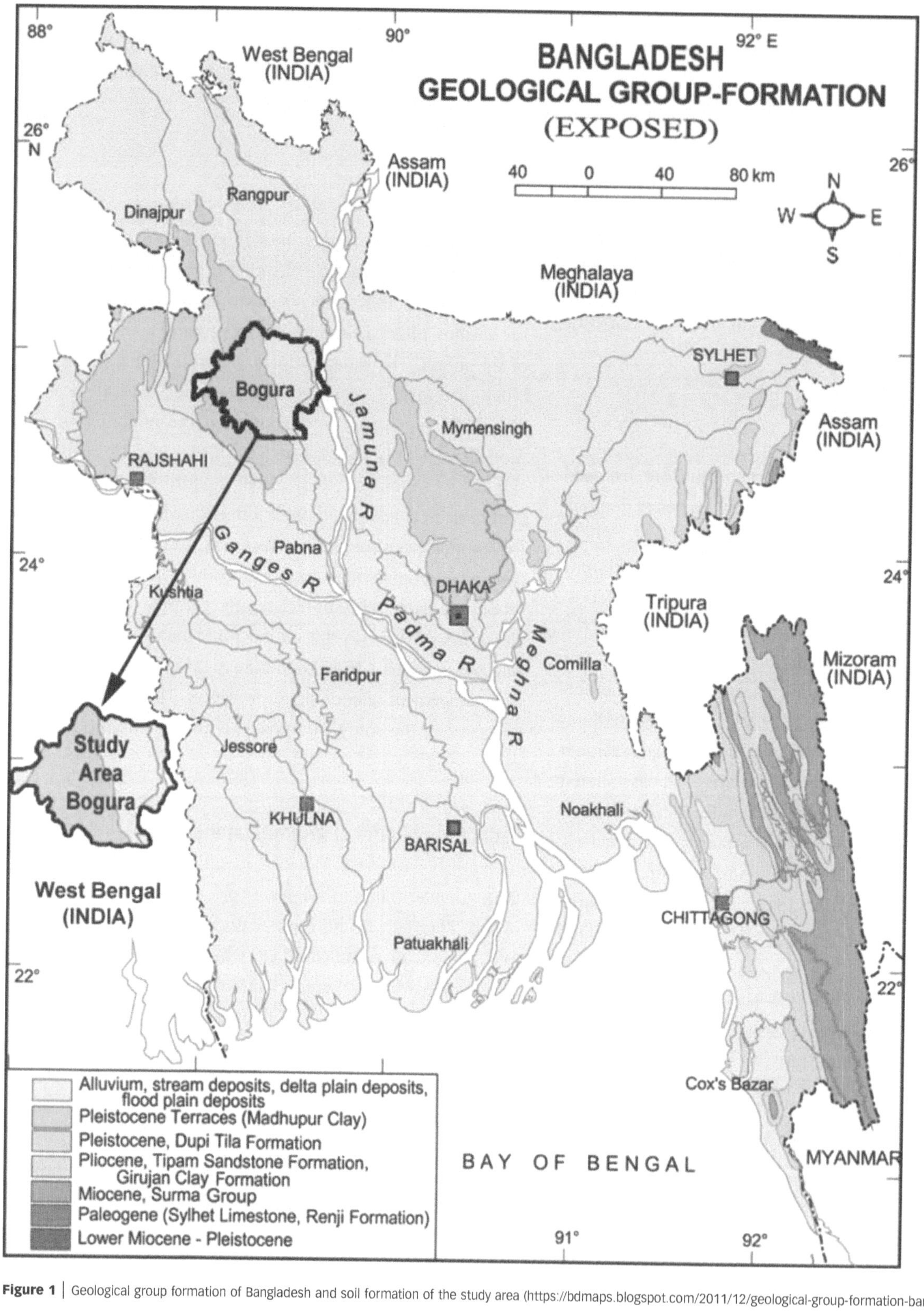

Figure 1 | Geological group formation of Bangladesh and soil formation of the study area (https://bdmaps.blogspot.com/2011/12/geological-group-formation-bangladesh.html).

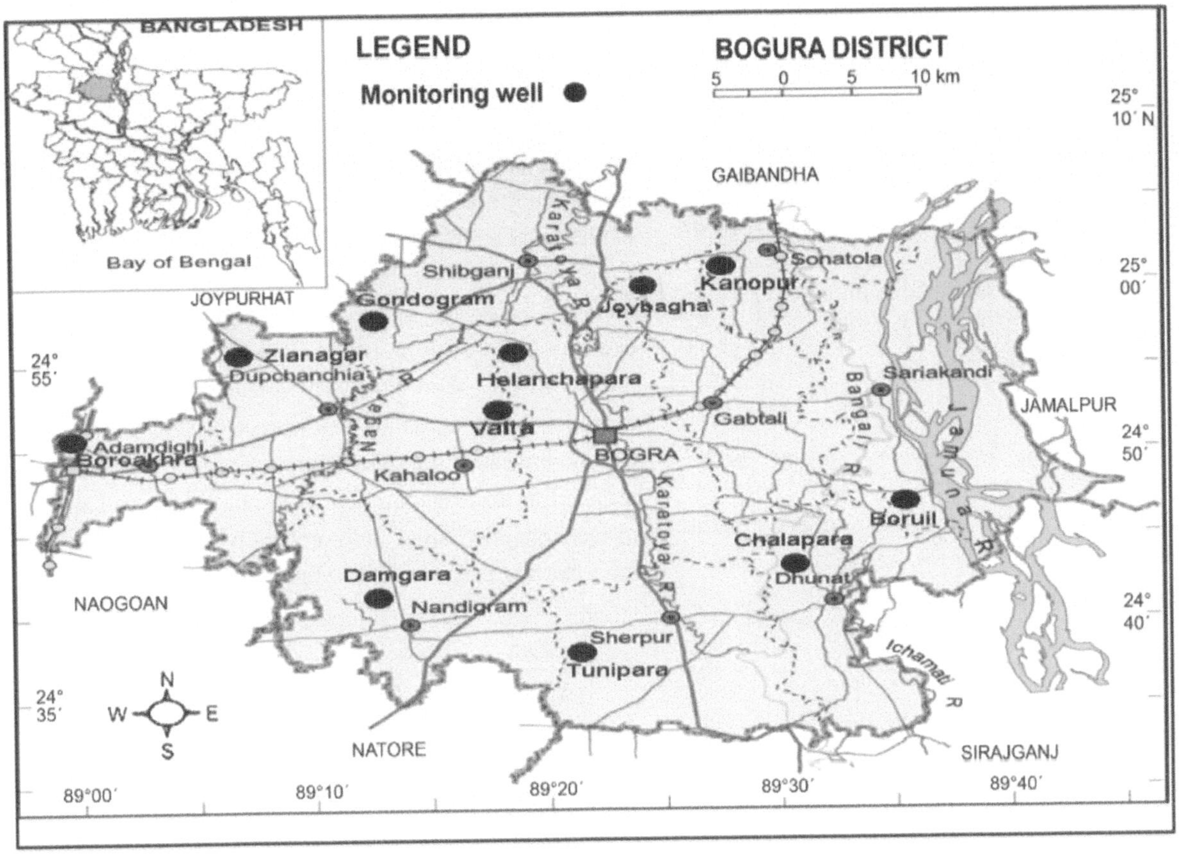

Figure 2 | Groundwater monitoring well locations.

Table 1 | Locations of the groundwater monitoring wells

Station no.	District/ Region	Station/ Subunit	Union	Mouza/Well location
1.	Bogura	Sadar	Madla	Helancha para
2.	Bogura	Sherpur	Kusumbi	Tunipara
3.	Bogura	Sariakandi	Kamalpur	Boruil
4.	Bogura	Dhunot	Dhunot	Chalapara
5.	Bogura	Gabtoli	Kagoil	Joybagha
6.	Bogura	Nondigram	Nondigram	Damgara
7.	Bogura	Kahaloo	Moral	Valta
8.	Bogura	Shibgonj	Pirab	Gondogram
9.	Bogura	Sonatola	Balua	Kanopur
10.	Bogura	Dupchachia	Zianagar	Zianagar
11.	Bogura	Adamdighi	Chhatianagram	Boroakhra

fluctuations were measured at 11 monitoring wells. They were not constant at every location, and hence, a vertical aquifer was expected. Further, the specific yield (S_y) is the ratio of the thickness of an equivalent layer of water of infinite spatial extent that is required to produce the measured change in gravity (in other words, it is the change in groundwater volume per unit area) by the change in WT elevation. Thc Equation can casily assess groundwater discharge by comparing storage, shortage, or steadiness of groundwater.

Groundwater recharge

Groundwater recharge is a hydrological process where water passes downward from the surface deeper into the ground. It is the amount of water that enters the saturated zone; that is, the enduring WT in an aquifer. The main source of recharge is rainfall, which may enter the soil directly to reach the groundwater zone. Aquifer recharge refers to the amount of water that may be available in the long term for extraction. Groundwater recharge, also referred to as deep drainage or deep percolation, is calculated for areas that receive abundant rainfall. The data of groundwater recharge is then used for calculating the well spacing in the study. Annual groundwater recharge estimated using the following formula:

(a) Chaturvedi's formula (Chaturvedi 1973):

$$R = 13.93\,(P - 381)^{0.4} \tag{2}$$

where R is the groundwater recharge (mm) and P is the annual rainfall (mm).

(b) Sehgal's formula (Sehgal 1973):

$$R = 12.6\,(P - 406.4)^{0.5} \tag{3}$$

where R is the groundwater recharge (mm) and P is the annual rainfall (mm).

(c) Datta's formula (Datta *et al.* 1980):

$$R = 0.11\,(P - 41.8), \tag{4}$$

where R is the groundwater recharge (cm) and P is the annual rainfall (cm).

Groundwater discharge and surface water contribution

Groundwater discharge is the term used to describe the movement of groundwater from the subsurface to the surface. Again, surface water contribution is the sense used to the water coming from the surface (canals or rivers). Annual discharge and surface water contribution can be easily determined to ascertain the number of wells (DTWs, STWs, and LLPs), average pumping capacity, and time period of operation. The annual discharge and surface water contribution of groundwater for irrigation in the study area is estimated using the following equation (Asaduzzaman 1987):

$$V_a = N \times Q \times t \tag{5}$$

where V_a is the annual groundwater discharge (Mm^3), N is the number of tube wells operated in the area, Q is the pumping capacity of a well ($m^3\,h^{-1}$), and t is the operation period (h/year).

Safe well spacing

This is generally defined as the maximum area of the resource basin that can be drained efficiently and economically by a well. The prevailing spacing between wells was determined for the study area by considering a uniform areal distribution of wells and entire utilization of the safe yield of the groundwater basin on the basis of the available groundwater recharge. Safe yield is defined as the amount of water that can be withdrawn from the groundwater basin without producing an undesired effect. The annual recharge must not exceed the total groundwater extracted. The well spacing clearly indicates the distribution of water-lifting devices installed in an irrigation field. The relation of well spacing to recharge, discharge, and pumping period is described by the following equation (Chowdhury & Wardlaw 1978):

$$S = 60\sqrt{\frac{QtN}{R_g}} \tag{6}$$

where S is the well spacing (m), Q is the well discharge in (L/S), t is the pumping period in (days/year), R_g is the groundwater recharge in (mm/year), and N is the number of operating well (h/day).

Growth rate and WT trends

For computing the WT growth rates, the following exponential trend line formula is used.

$$Y = ae^{bt} \tag{7}$$

$$\ln Y = \ln a + bt \tag{8}$$

where Y is the dependent variable (WT; m); t is the independent variable (month–year); a is the intercept, and b is the absolute growth rate. Thus, b is the growth with a ratio scale multiplied by 100 and expressed as a percentage of decrease. To verify the significance of the estimated regression, the following equation for the F-test, where null hypothesis H_0: $b = 0$ was used (Gujarati 1995):

$$F = \frac{R^2}{(1 - R^2)} \tag{9}$$

where R^2 is the explained sum of squares and $(1 - R^2)$ is the unexplained sum of squares.

Direct field supervision

Some data were collected through direct field supervision with consent from end users, beneficiaries, and managers handling groundwater resources in the study area.

RESULTS AND DISCUSSION

The study results are clearly systematically stated and described in this section considering the applied methods.

Aquifer mapping

A total of 145 boreholes from various locations in 11 stations of the Bogura region were considered along with the lithologies and latitudinal–longitudinal coordinates. Those borehole data were analyzed using Rockworks software. A location map of boreholes was plotted (Figure 3), and panel diagrams of groundwater resources in the north–south (N–S) direction (Figure 4(a)) and west–east (W–E) direction (Figure 4(b)) were obtained. The location map shows the borehole sites in the study area, and the panel diagrams show the aquifer formation. In addition, the panel diagrams show 25 types of soil formations using various colors for the easy identification of the present groundwater inventory of water logs. For example, subsurface layers are classified as sandy, clay,

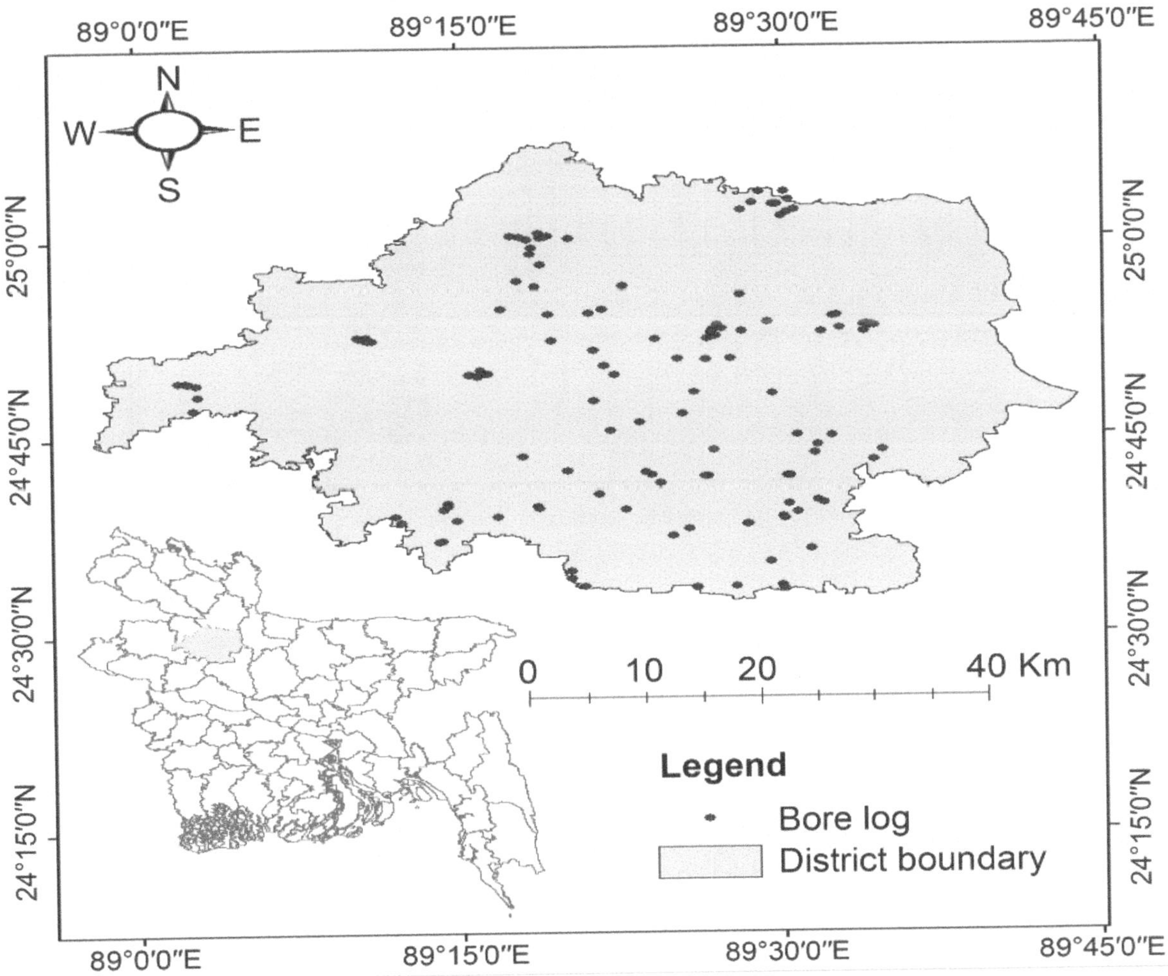

Figure 3 | Borehole locations.

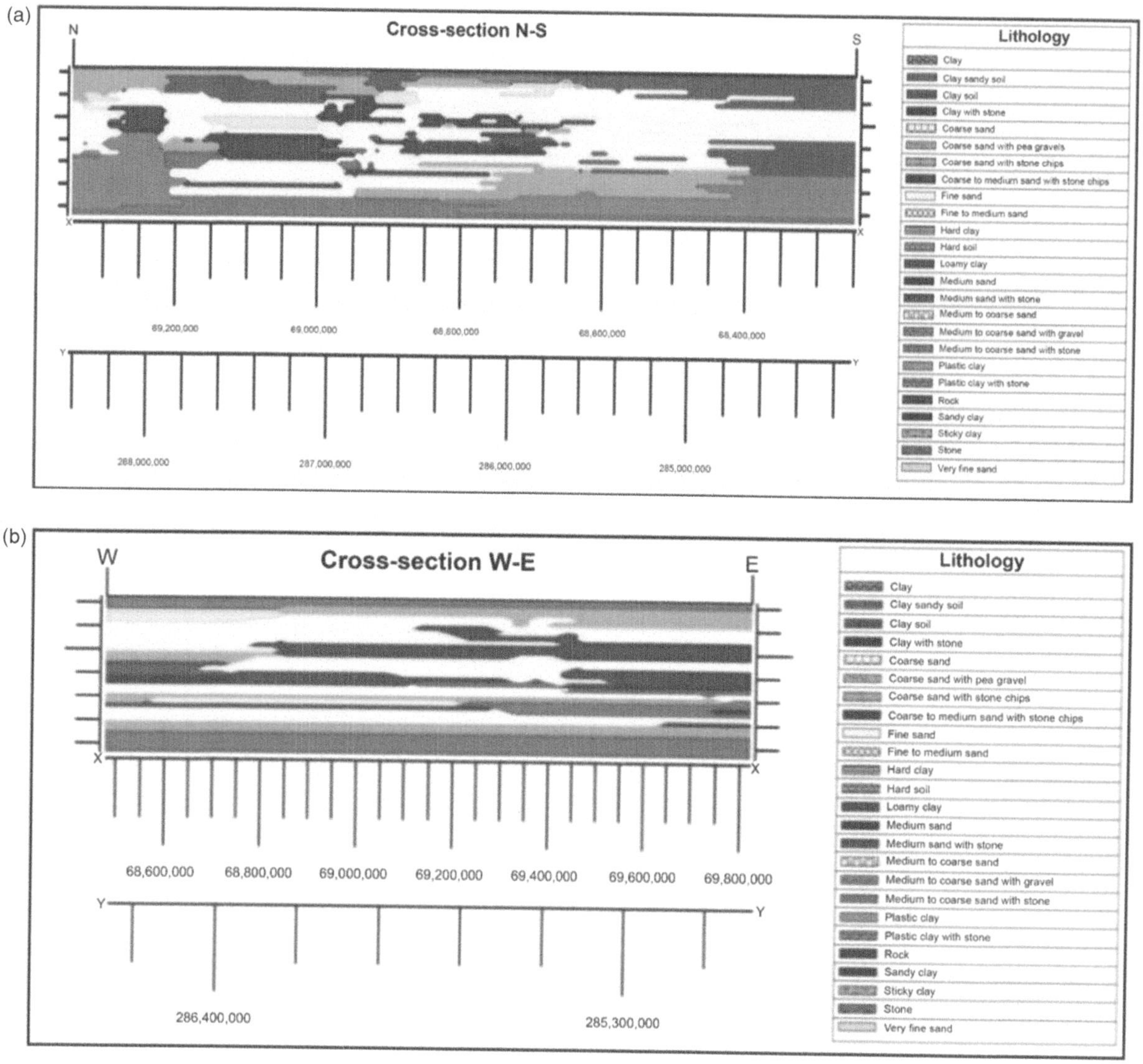

Figure 4 | (a) Panel diagram of groundwater in the North-South direction. (b) Panel diagram of groundwater in the west–east direction.

sticky, rock, plastic, stone, etc. Mostly, sandy soil remains in the water in the subsurface withdrawn using various lifting devices, such as HTWs, STWs, and DTWs, for drinking, irrigation, and industrial purposes.

Moreover, screend or strainers were settled on wells, where the sandy layer was located along the depth or thickness, using the water logs. This result was compared with the findings of earlier studies. The comparison showed that the thick layer of sticky and plastic clay hampered the recharge in WT; further, land collapse risk is low, but this layer acts as an aquitard that hampers groundwater recharging. The increased surface runoff has led to a severe decrease in the groundwater level. This area faces the risk of several adverse effects such as land subsidence and biodiversity loss. In addition, the aquifer was oriented west–east in the north-west region of Bangladesh (Rahman & Mahbub 2012).

Lithology

Borehole inventory is plotted in Figure 5 using both the borehole and aquifer thickness of 145 lithologies. From this figure, the maximum, minimum, and average borehole thickness were found as 72, 27, and 44.54 m, respectively. In addition, the maximum, minimum, and average aquifer

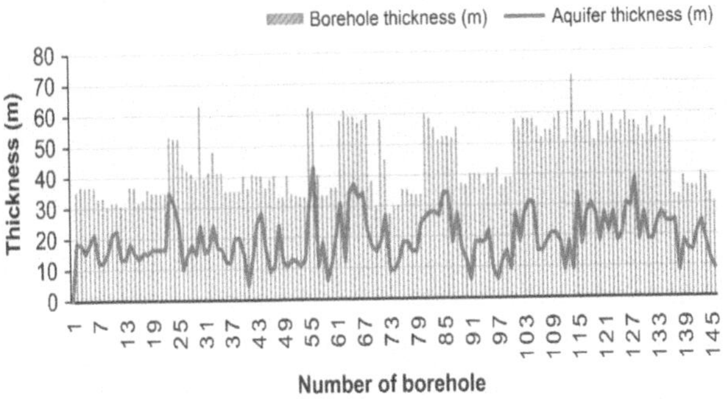

Figure 5 | Borehole inventory of the study area.

thicknesses were 42.68, 4.66, and 19.48 m, respectively. The aquifer in the study area was composed of medium sand (MS) to coarse sand (CS). This result is in agreement with the result of a recent study reporting that the highest thickness of the clay layer is 27.43 m and the lowest value is 6.09 m and that the thickness gradually increases from the east to the west; finally, the aquifer depth was 6.09–27.43 m from the surface (Rahman & Mahbub 2012).

Hydrograph

A hydrograph is a graph showing the rate of flow past a specific point such as a river, channel, or conduit. In this study, the hydrographs show the monthly rainfall and WT data from 2007 to 2019 (Figure 6). The x-axis denotes the month–year, and the y-axis represents monthly rainfall and groundwater table reflected in the hydrographs as well as in the two recognized limbs, particularly seen in the one-year water curve of Bogura Sadar upazila. The two limbs of the hydrograph show groundwater level fluctuations; one, the rising limb or crest or peaks representing the recharge phenomena, and the other, the recession limb or trough representing the discharge phenomena. The hydrographs are compared and classified with groups of identical classes. The form of the annual hydrograph in a specific well remains nearly unchanged over consequent years in spite of the deviations in its magnitude with the variation of the annual recharge or discharge to the aquifer.

Groundwater condition

The groundwater conditions shown in Figure 7 are plotted using monthly groundwater table (GWT) data for the last 13 years (2007–2019). From these figures, the maximum, minimum, and average GWT, groundwater fluctuation, and standard deviation values are obtained as listed in Table 2.

From the data in Table 2, the maximum, minimum, and average WT are 8.95, 2.95, and 6.40 m, respectively; the maximum, minimum, and average groundwater fluctuations are 4.73, 2.22, and 3.37 m, respectively; the maximum, minimum, and average standard deviation are 1.44, 0.70, and 1.03 m, respectively. Finally, the average WT and groundwater fluctuation are 6.40 and 3.37 m, respectively, in the study area. These findings were associated with earlier research on cropping patterns, cropping intensity, national intensity, crop selection, and yields practices of cultivation after groundwater irrigation commenced in the mid-1980s. WT has decreased continuously at rates of 0.4 m and 0.22 m/year in the wet and dry seasons, respectively (Rahman & Mahbub 2012).

Groundwater pressure

Generally, the suction mode of centrifugal pumps (a shallow aquifer) is employed in the case of STWs, and the force mode of submergible pumps (a deep aquifer) is employed in the case of DTWs. Using the suction mode, water can be lifted at not more than 10.33 m by exerting a pressure of 1.03 (kg cm^{-2}). The maximum WT in Sherpur, Dupchachia, and Adamdhigi stations (Table 2) are 10.95, 11.52, and 10.47 m; these values exceed the limit of 10.33 m, beyond which the suction mode is inoperable. In other words, the centrifugal pumps of STWs failed in the dry seasons. The remaining eight stations (Bogura Sadar, Sariakandi, Dhunat, Gabtali, Nandigram, Kahalo, Shibganj, and Sonatola) are safe zones for pumping up groundwater throughout the year because the WTs in these stations are within 10.33 m.

Annual groundwater storage

The analysis of the lithology and aquifer type in Bogura district revealed the presence of MS-to-CS. The specific yield (S_y) for MS and CS was 15–30% and 20–35%, respectively, for this soil formation (Walton 1996). The aquifer area of Bogura, A, is 2,898.25 km^2; average depth of the WT

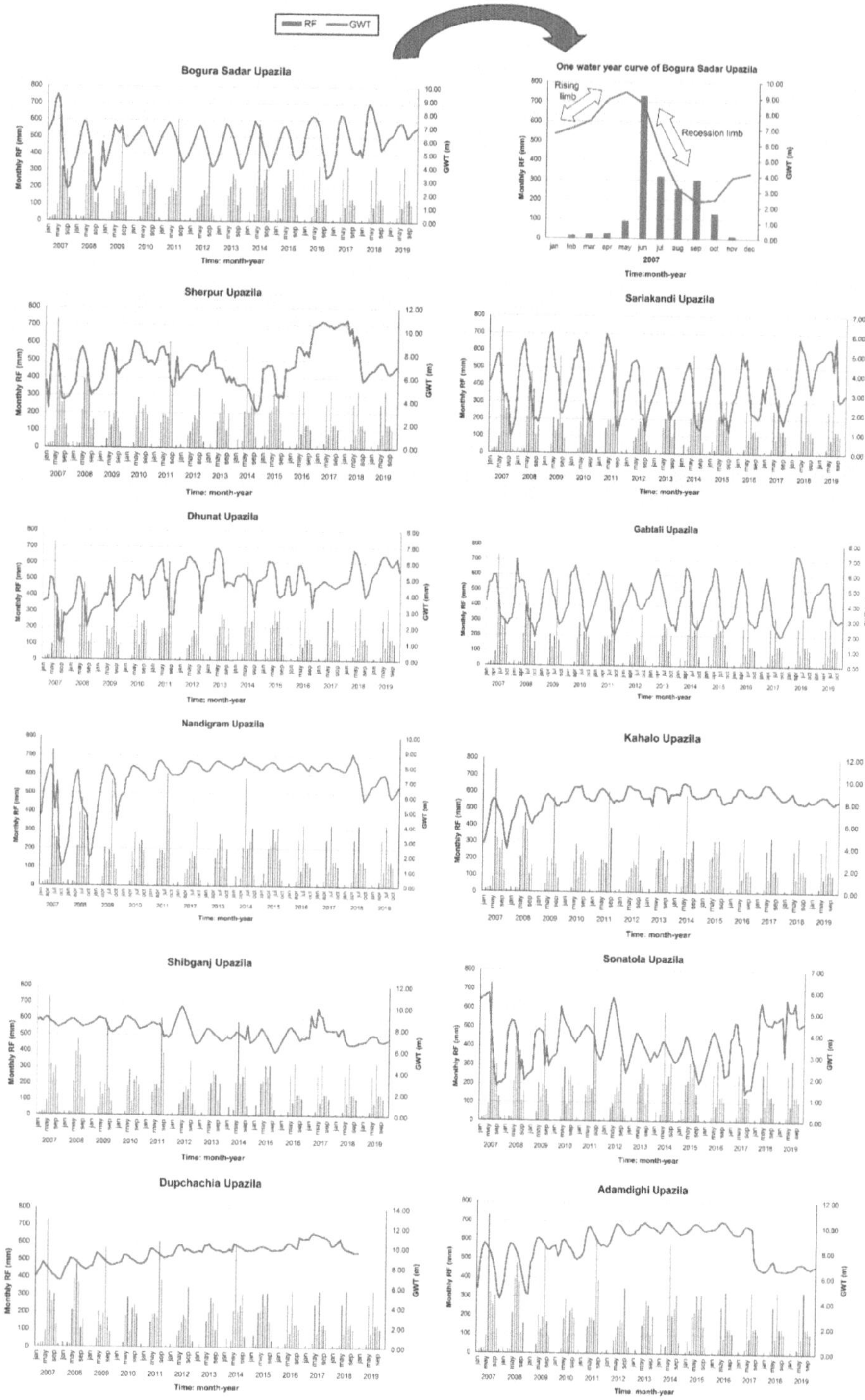

Figure 6 | Typical hydrograph of monthly rainfall and water table data along with time-series data during 2007–2019 in wells from 11 stations in the study area.

fluctuation, ØH, is 3.37 m (from Table 2). These values were substituted in Equation (1). Then, for a specific yield, S_y, of 22.5% (average for MS), the groundwater storage was estimated to be 2,898.25 $km^2 \times 3.37\ m \times 0.225$ or 2,197.59 Mm^3. Now, for a specific yield of 27.5% (average for CS), the same equation yielded an estimate of

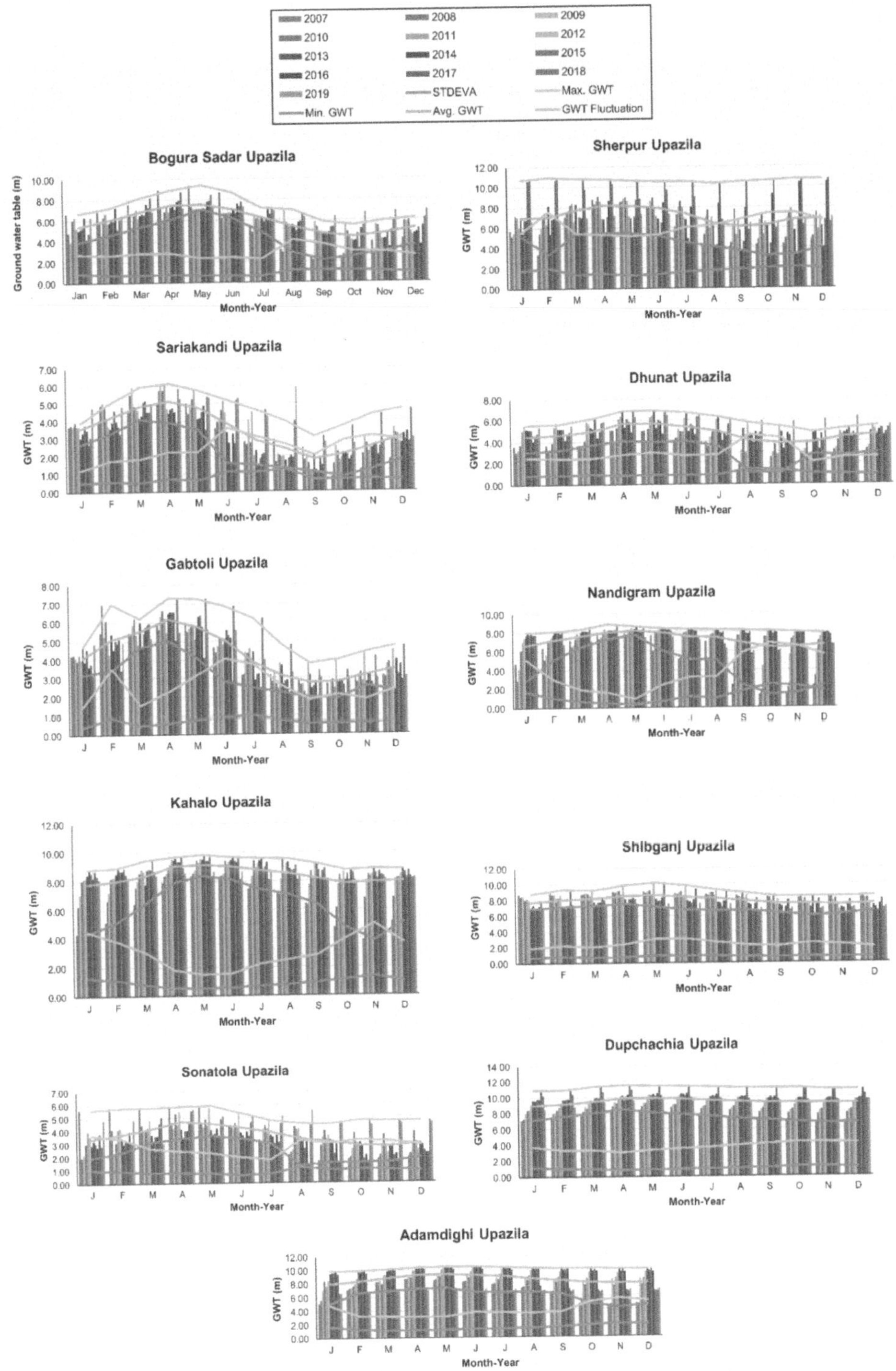

Figure 7 | Maximum, minimum, and average values of the water table, water table fluctuations, and standard deviation for the 11 stations in the study area.

$2,898.25 \text{ km}^2 \times 3.37 \text{ m} \times 0.275$ or $2,685.95 \text{ Mm}^3$ as the groundwater storage. The annual groundwater storage was taken as the average of these two values; that is, $2,441.77 \text{ Mm}^3$. Corresponding study findings reacted with these results that the annual groundwater recharge and discharge were estimated to be $106.41–244 \text{ Mm}^3$ and

Table 2 | Maximum, minimum, and average of the water table, groundwater fluctuations, and standard deviation for the period 2007–2019

		WT (m)			WT fluctuation (m)			Standard deviation (m)		
Station no.	**Name of Upazila/Subunit**	**Max.**	**Min.**	**Avg.**	**Max.**	**Min.**	**Avg.**	**Max.**	**Min.**	**Avg.**
1	Bogura Sadar	9.38	2.15	5.76	4.1	2.3	2.94	1.19	0.63	0.89
2	Sherpur	10.95	3.2	7.12	7.56	5.15	6.24	2.1	1.29	1.72
3	Sariakandi	6.15	0.83	3.44	3.6	1.23	2.46	1.24	0.51	0.77
4	Dhunot	6.91	1.07	4.7	4.61	2.07	3.02	1.42	0.83	1.01
5	Gabtoli	7.30	1.89	4.25	3.73	1.53	2.53	1.02	0.4	0.69
6	Nondigram	9.82	1.33	7.26	6.83	0.9	3.95	2.4	0.32	1.27
7	Kahaloo	9.83	3.83	8.36	4.98	1.48	3.04	1.32	0.52	0.88
8	Shibganj	10.15	5.9	7.92	3.05	1.89	2.35	0.92	0.66	0.76
9	Sonatola	5.92	1.29	3.63	3.66	1.75	2.78	1.16	0.6	0.88
10	Dupchachia	11.52	6.67	9.4	4.27	3.03	3.66	1.19	0.81	0.99
11	Adamdighi	10.47	4.27	8.53	5.62	3.12	4.05	1.89	1.14	1.44
Average		**8.95**	**2.95**	**6.40**	**4.73**	**2.22**	**3.37**	**1.44**	**0.70**	**1.03**

93.77–291 Mm^3, respectively, in Naogaon in northwest Bangladesh. The annual recharge and withdrawal remained balanced up to 1993, but the discharge has exceeded the recharge since then. In addition, 23.99–42.08 Mm^3 of groundwater was discharged using STWs and DTWs, and the rest of groundwater was discharged by natural seepage (Reza *et al.* 2013).

Groundwater recharge

Groundwater recharge was calculated using Equations (2)–(4), as listed in Table 3.

The average groundwater recharge in Table 3 is 261.82 mm. This is close to the values of 234.97 mm calculated using Chaturvedi's formulas (Chaturvedi 1973). Hence, in this study, the groundwater recharge is considered to be 261.82 mm. These results were compared with the outcomes of an alternative study, and the prospective groundwater recharge decreased by 85% and rest (15%) was moderate in the Barind Tract of Bangladesh, where only 8.6% of the total average annual rainfall (1,685 mm) infiltrates into the subsurface and ultimately contributes to groundwater recharge (Adham *et al.* 2010).

Groundwater discharge and surface water contribution

The annual discharge and surface water contribution are calculated from the total number of DTWs, STWs, and LLPs with their average pumping capacity and operating period. The study area has 2,700 DTWs, 65,607 STWs, and 273 LLPs with discharge capacities of 56 Ls^{-1} (or 201.6 m^3/h), 14 Ls^{-1} (or 50.4 m^3/h), and 14 Ls^{-1} (or 50.4 m^3/h), respectively (BADC 2019). Considering pump design, the actual pumping efficiency of each DTWs, STWs, and LLPs was regarded as approximately 25% less than the discharge capacity. Presently, the average irrigation period was considered as 960 h/year for DTWs, STWs, and LLPs (entire irrigation depends on the Rabi season, but

Table 3 | Annual groundwater recharge

Sl. no.	Equation name	Rainfall unit	Average rainfall during 2007–2019 (mm)	Annual recharge, Rg (mm)
1.	$R = 13.93\,(P - 381)^{0.4}$ Chaturvedi (1973)	$P =$ mm	1,549.62	234.97
2.	$R = 12.6\,(P - 406.4)^{0.5}$ Sehgal (1973)	$P =$ mm		426.02
3.	$R = 0.11\,(P - 41.8)$ Datta *et al.* (1980)	$P =$ cm		124.47
Average				**261.82**

supplementary irrigation is performed for the seasons of Aus and Aman cultivation; that is, July–August and December–January, respectively, if needed).

From Equation (5), the number of wells, average pumping capacity, time period of operation, annual discharge, and surface water contribution are obtained as listed in Table 4.

The annual groundwater storage and annual groundwater discharge were 2,441.77 Mm^3 and 2,772.16 Mm^3, respectively. The annual shortage of groundwater was 330.39 Mm^3 (13.53%) of the total storage in the northwest region of Bogura. Therefore, the groundwater shortage occurs primarily in the dry season and affects the amount of water available for irrigation. This result agrees with the results of an earlier study that the annual groundwater recharge and discharge were 106.41–244 Mm^3 and 93.77–291 Mm^3, respectively, in northwest Bangladesh; these values revealed the existing balance between the annual recharge and withdrawal up to 1993; thereafter, the discharge has exceeded the recharge. Further, 23.99–42.08 Mm^3 of groundwater was discharged using STWs and DTWs, and the rest of the groundwater was discharged by natural seepage (Reza *et al.* 2013). Additional, the main sources of surface water are the canal and the rivers of Nagar, Bengali, Ichamoti, Karatoa, Bhadrabati, Jamuna, Tulshiganga etc. in the study area. From Table 4, a total of 273 LLPs are used for contributing surface water with an amount of 9.90 Mm^3 (0.41%).

Safe well spacing

To calculate the well spacing, the following parameters were considered: the pumping capacity of DTWs, Q, was 151.2 m^3/h or 42 Ls^{-1}; pumping period, t, was 60 d/year; operating hours, N, was 16 h/days; and groundwater recharge, Rg, was 261.82 mm/year (Table 3). Then, using Equation (6), the well spacing of the existing DTWs was calculated as 744 m. Further, for the pumping capacity of STWs, Q, 37.8 m^3/h or 10.5 Ls^{-1} and all other parameters identical to those for the DTWs, the well spacing of existing STWs was calculated as 372 m using the same equation. However, the recommended spacing for DTWs (a well with a discharge capacity of 201.60 m^3/h or 2.01 cusec) and STWs (a well with a discharge capacity of 50.4 m^3/h or 0.50 cusec) are 800 and 250 m, respectively, in the study area in Bangladesh (MoA 2019). Thus, the results showed that although the spacing of DTWs is appropriate, that for STWs is inappropriate. Hence, further installation of STWs should be strictly disallowed in the district, and all STWs should be operated very carefully in the dry season to ensure they do not over lift groundwater. Further, the installation of DTWs is preferable for the proper management of groundwater in future. Besides, the annual shortage of groundwater due to the over-exploitation of groundwater for excess installation of STWs was 330.39 Mm^3 (13.53%); this shortage occurs in the dry period, when the WT declines slowly.

Period toward groundwater fluctuations

The list in Table 5 shows that the average WT fell by 1.02 m over the last 13 years in the period 2007–2019 in the study area. Meanwhile, the WT increased in March–May but fell during the nine months of June–February. This result is in agreement with the results of a previous study that the WT

Table 4 | Annual groundwater discharge and surface water contribution

Sl. no.	Name of devices	Total numbers of operation, nos.	Pumping capacity, Q ($m^3 h^{-1}$)	Irrigation period (16 h day^{-1} × 60 day $year^{-1}$), T (h $year^{-1}$)	Annual discharge, $V_a = N \times Q \times T$ (Mm^3)
1	2	3	4	5	6 = (3 × 4 × 5)
Annual groundwater discharge					
1.	DTW	2,700	151.2	960	391.91
2.	STW	65,607	37.8	960	2,380.74
Total					**2,772.16**
Surface water contribution					
1.	LLP	273	37.8	960	9.90

Table 5 | Monthly rise (–) or fall (+) in the groundwater table during the period 2007 to 2019 (in meters)

Station name	Jan	Feb	Mar	Apr	May	Jun	Jul	Aug	Sep	Oct	Nov	Dec	Avg.
Bogura Sadar	−0.37	−0.63	−0.75	−1.70	−2.04	−1.34	1.39	3.09	4.01	4.18	2.93	2.80	0.96
Sherpur	0.90	3.27	−0.11	−1.55	−1.21	−0.64	0.95	2.25	2.32	2.25	2.38	2.20	1.08
Sariakandi	1.10	0.77	0.5 – 2	0.25	0.29	1.72	1.44	2.95	0.33	1.86	1.56	1.20	1.16
Dhunat	1.80	1.81	2.18	1.28	1.51	2.46	2.29	4.79	4.78	3.20	3.59	2.61	2.69
Gabtali	0.65	−0.52	−0.31	−0.49	−0.36	0.93	0.84	0.16	0.11	0.21	−0.07	−0.05	0.09
Nandigram	2.00	0.42	−0.05	−0.38	−0.60	−0.19	2.02	−0.75	3.57	4.78	4.85	4.38	1.67
Kahalo	3.78	3.07	1.93	0.65	0.28	0.46	1.21	1.16	1.71	3.18	4.30	3.09	2.07
Shibganj	−1.92	−1.97	−1.51	−1.62	−1.59	−1.36	−1.20	−1.59	−1.39	−1.31	−1.30	−1.35	−1.51
Sonatola	−0.80	−0.88	−2.75	−0.22	−0.80	0.81	1.79	3.40	2.83	2.61	2.70	2.66	0.95
Dupchachia	1.35	1.16	1.38	1.06	1.69	1.95	2.15	2.27	2.40	2.49	2.81	2.82	1.96
Adamdighi	1.54	−0.47	−1.42	−1.63	−1.42	−1.12	−0.75	−0.39	0.51	1.87	2.59	2.14	0.12
Average	**0.91**	**0.55**	**− 0.08**	**− 0.40**	**− 0.39**	**0.33**	**1.10**	**1.57**	**1.92**	**2.30**	**2.39**	**2.05**	**1.02**

suffers a yearly fall of 1.20 m in some northern parts of Bangladesh because of excessive withdrawal through tube wells along with low recharge, poor management, and land use changes (Khan & Islam 2015). Moreover, a previous study on Chapai-Nawabgonj in northwest Bangladesh in 2007–2011 on the effect of rainfall and WT fluctuation showed that the rainy season usually started in May and ended in September and that low or no rainfall occurred during the rest of the year. It also showed that the maximum rainfall occurred in June–August and the maximum WT is observed in July–September owing to rainwater infiltration. The minimum WT was observed in March–May during the irrigation period of the area, and the WT declined daily because of the over withdrawal of groundwater for irrigation (Hasan *et al.* 2013).

WT trends

The annual exponential value was computed using the trend line for the 11 stations of the northern region of Bogura, Bangladesh, as listed in Table 6. The x-axis denotes the number of years, and the y-axis represents year; the WT is reflected in the exponential value *y* and regression R^2. The trend lines shown in Figure 8 were computed using

Table 6 | Growth rate of the water table

Upazila/sub unit	$Y = ae^{bt}$	$\ln Y = \ln a + bt$	Growth rate of WT = b × 100 (% per annum)	R^2	$F = \frac{R^2}{(1 - R^2)}$
BoguraSadar	$Y = 5.3459e^{0.0139x}$	1.6763 + 0.0139x	1.39	0.456	0.26352
Sherpur	Y = 6.1289e0.0187x	1.8130 + 0.0187x	1.87	0.1275	0.01752
Sariakandi	Y = 3.2934e0.0047x	1.19 + 0.0047x	0.47	0.0163	0.00037
Dhunat	Y = 3.578e0.0367x	1.2748 + 0.0367x	3.67	0.606	0.58047
Gabtoli	Y = 4.2011e0.0013x	1.4353 + 0.0013x	0.13	0.0035	0.00001
Nandigram	Y = 5.904e0.0276x	1.7756 + 0.0276x	2.76	0.3675	0.15614
Kahalo	Y = 7.4321e0.0162x	2.0058 + 0.0162x	1.62	0.3547	0.14392
Shibganj	Y = 8.8074e0.016x	2.1755 + (−) 0.016x	−1.60	0.5474	0.42785
Sonatola	Y = 3.4654e0.0051x	1.2428 + 0.0051x	0.51	0.0153	0.00023
Dupchachia	Y = 7.9958e0.0224x	2.0789 + 0.0224x	2.24	0.6788	0.85449
Adamdighi	Y = 8.1311e0.0052x	2.0956 + 0.0052x	0.52	0.0158	0.00025
Average				**1.22**	

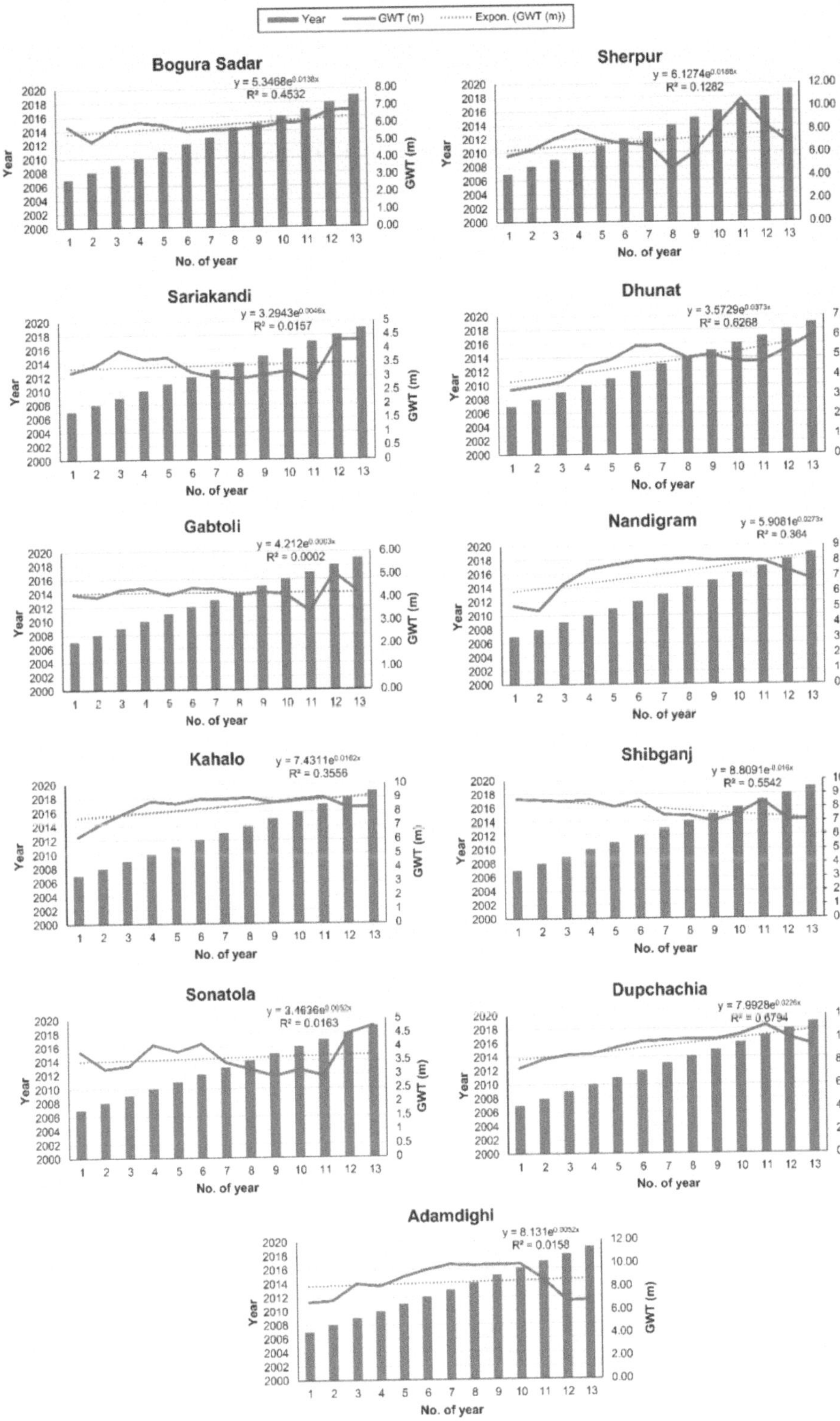

Figure 8 | Trend lines of the annual water table in the 11 stations of the study area.

Equation (7) or Equations (8) and (9), which are presented in Table 6.

The data in Table 6 show that the WT fell by 1.22% annually in the northwest region of Bogura. The WT for 10 stations, excluding Shibganj station (−1.60%), declined. The decline in the annual WT varies from 0.13 to 3.67%, indicating a wide variation in the rate of decrease in different locations because of excessive withdrawal through tube wells along with low groundwater recharge and topographical changes in the study area. This result, along with related findings, indicates that the decline increased by approximately 4% of the country's total during 1998–2002; however, it drastically increased to 11% in 2008 and 14% in 2012, and the WTs are declining gradually in the northwestern area but more slowly from 0.1 to 0.5 m $year^{-1}$ because of the intensive use of STWs in the Boro season in March–May (Qureshi *et al.* 2014).

CONCLUSIONS

In this study, groundwater trends were studied by intensively focusing on hydrological properties such as borehole lithologies, groundwater storage, the groundwater discharge–recharge phenomenon, existing well spacing (DTWs and STWs), and WT fluctuations for driving advanced groundwater enrichment plans. The borehole and aquifer thickness were 27–72 m and 4.66–42.68 m, respectively; using 145 borehole lithologies, the most helpful ones for digging of water-lifting devices containing their appropriate design. Soil formation was demarcated as MS − to − CS, which can be easily grasped for calculating the volume of water in a borehole to covering a command area. The annual groundwater budget for discharge and storage were predicted to be 2,772.2 and 2,441.8 Mm^3, respectively, and the annual scarcity of groundwater was 330.4 Mm^3; that is, 13.5%. These results can support simple perception of the groundwater shortage, steadiness, or excess condition not only in the study area but on national or global scales. The existing spacing between DTWs and between STWs was 744 and 372 m, respectively; the results can convey a message for boring water-lifting devices in the future. Furthermore, the WT increased in March–May but fell during the nine months of June–February. Monitoring the change in WT in these periods can provide the alarm for assessing the relation of the pumping head with the discharge throughout the year. WT declined by 1.0 m in the last 13 years; that is, a decline of 0.07 m or 1.2% rate per year. This result can be considered in future groundwater user plans.

The study results can be significant for decision-makers, planners, and researchers involved in groundwater management at nationwide and worldwide scales.

The recommendations made on the basis of the results of this study are as follows. (i) If WT fluctuations are identified in a region, any borehole can be easily designed for ensuring sustainability; (ii) the crop planting seasons, most notably for rice, can be alternated considering the groundwater shortage (13.53%) during the dry season in the study area; (iii) water-saving technologies for rice production, such as the use of alternating wetting and drying (AWD) methods and the raised bed crop (RBC) technology approach, can be employed; (iv) the current well spacing for DTWs is appropriate (744–800 m); however, the spacing for STWs is inappropriate (250–372 m). Hence, further installations of STWs should be strictly disallowed in northern Bogura, Bangladesh; (v) future groundwater plans should be established to ensure a balance between groundwater recharge–discharge with a reserve; and (vi) the mathematical, physical, and analogue modeling of groundwater resources in this area will require further research.

ACKNOWLEDGEMENTS

This study was legitimately supported by Rural Development Academy (RDA), Bogura, Bangladesh for PhD research work at Jahangirnagar University, Dhaka, Bangladesh. The authors thank Enago for their helpful and productive support in English language review that greatly contributed to improve the quality of our paper.

CONFLICT OF INTEREST

The authors declare no conflicts of interest.

FUNDING

This research did not receive any specific grant from funding agencies in the public, commercial, or not-for-profit sectors.

DATA AVAILABILITY STATEMENT

Data cannot be made publicly available; readers should contact the corresponding author for details.

REFERENCES

Abdullah, M. 2014 *Assessment of Groundwater Resources in Bogra District Using Groundwater Model*. A thesis for Master of Science in Department of Civil Engineering, Bangladesh University of Engineering and Technology (BUET), Bangladesh, October, 2014. Available from: https://www.researchgate.net/publication/342655878_assessment_of_groundwater_resources_in_Bogra_district_using_groundwater_model.

Adham, M. I., Jahan, C. S. J., Mazumder, Q. H., Hossain, M. M. A. & Haque, A. M. 2010 Study on groundwater recharge potentiality of Barind Tract, Rajshahi district, Bangladesh using GIS and remote sensing technique. *Journal of the Geological Society of India* **75** (2), 432–438. Available from: http://www.geosocindia.org/index.php/jgsi/article/view/57660.

Asaduzzaman, M. 1987 Development and utilization of water resources in Barind area. *The Sangbad*, June 8, Daily Newspaper, Bangladesh.

BADC (Bangladesh Agricultural Development Corporation) 2019 *District Wise Irrigation Equipment used and Total Area Irrigated. Unpublished Report, Financial Year 2018–19*, BADC Bogura region, Bangladesh.

BBS (Bangladesh Bureau of Statistics) 1998 *Yearbook of Agricultural Statistics of Bangladesh*. Bangladesh Bureau of Statistics, Dhaka, Bangladesh.

Chaturvedi, R. S. 1973 *A Note on the Investigation of Ground Water Resources in Western Districts of Uttar Pradesh, Annual Report*, UP Irrigation Research Institute, pp. 86–122.

Chowdhury, S. I. & Wardlaw, I. F. 1978 The effect of temperature on kernel development in cereals. *Australian Journal of Agricultural Research* **29** (2), 205–223. doi:10.1071/AR9780205.

Datta, A. K., Colby, B. M., Shaw, J. E. & Pagano, J. S. 1980 Acyclovir inhibition of Epstein-Barr virus replication. *Proceedings of the National Academy of Sciences* **77** (9), 5163–5166. Available from: https://www.ncbi.nlm.nih.gov/pmc/articles/PMC350017/. doi:10.1073/pnas.77.9.5163.

Gujarati, D. N. 1995 *Basic Econometrics*, 3rd edn. McGraw-Hill Inc., New York, NY, USA. ISBN; 0070252149, pp. 540–583. Available from: https://www.worldcat.org/title/basic-econometrics/oclc/31075018.

Hasan, M. R., Mostafa, M. G. & Matin, I. 2013 Effect of rainfall on groundwater level fluctuations in Chapai-Nawabgonj district. *International Journal of Engineering Research & Technology (IJERT)* **2** (4). Available from: https://www.ijert.org/effect-of-rainfall-on-groundwater-level-fluctuations-in-chapai-nawabgonj-district.

Hasanuzzaman, M., Song, X., Han, D., Zhang, Y. & Shakir, H. S. 2017 Prediction of groundwater dynamics for sustainable water resource management in Bogra district, northwest Bangladesh. *Water* **9** (4), 238. Available from: https://www.mdpi.com/2073-4441/9/4/238. doi:10.3390/w9040238.

Khan, S. S. & Islam, A. R. M. T. 2015 Anthropogenic impact on morphology of Teesta River in northern Bangladesh: an exploratory study. *Journal of Geosciences and Geomatics* **3** (3), 50–55. Available from: https://www.researchgate.net/publication/276941795_Anthropogenic_Impact_on_Morphology_of_Teesta_River_in_Northern_ Bangladesh_An_Exploratory_Study.

Matin, M. A., Khan, M. H. & Khan, N. I. 2000 *RDA Developed Low-Cost DTW Technology for Multipurpose uses (Performance Evaluation)*. Director General, Rural Development Academy (RDA), Bogra, Bangladesh.

MacDonald, M. & Partners 1983 *Water Balance Studies Bangladesh. Final Report, Report II Groundwater*, Demter House, Station Road, Cambridge, Cambridge, p. 68.

MoA (Ministry of Agriculture) 2019 *Notification of Groundwater Management Act for Agriculture in Bangladesh, 2018. S.R.O. no. 128-law/2019*. Ministry of Agriculture, Government of Bangladesh (GoB), Dhaka.

MPO (Master Plan Organization) 1991 *National Water Plan Phase II. Vol I & II*. Ministry of Water Resources, Government of Bangladesh (GoB), Dhaka.

Qureshi, A. S., Ahmed, Z. & Krupnik, T. J. 2014 *Groundwater Management in Bangladesh: an Analysis of Problems and Opportunities. Research Report No. 2, Cereal Systems Initiatives for South Asia-Mechanization and Irrigation (CSISA-MI)*. Available from: https://csisa.org/wp-content/uploads/sites/2/2014/01/Groundwater-management-in-Bangladesh-An-analysis-of-problems-and-opportunities.pdf.

Raghunath, H. M. 1987 *Groundwater*, 2nd edn. Wiley Eastern Limited, New Delhi, India, p. 6.

Rahman, M. M. & Mahbub, A. Q. M. 2012 Groundwater depletion with expansion of irrigation in Barind Tract: a case study of Tanore Upazila. *Journal of Water Resource and Protection* **4**, 567–575. http://dx.doi.org/10.4236/jwarp.2012.48066.

Reza, A. H. M., Mazumder, Q. H. & Ahmed, M. 2013 Groundwater balance study in the high Barind, Bangladesh. *Rajshahi University Journal of Science* **39** (2011), 11–26. https://doi.org/10.3329/rujs.v39i0.16539.

Sehgal, S. R. 1973 *Groundwater Resources of Punjab State*. Third annual session, Central Board of Irrigation and Power, New Delhi, India.

Shahid, S. & Hazarika, M. K. 2010 Groundwater droughts in the north-western districts of Bangladesh. *Water Resources Management* **24** (10), 1989–2006. Available from: https://link.springer.com/article/10.1007/s11269-009-9534-y.

Shamsudduha, M., Taylor, R. G., Ahmed, K. M. & Zahid, A. 2011 The impact of intensive groundwater abstraction on recharge to a shallow regional aquifer system: evidence from Bangladesh. *Hydrogeology Journal* **19**, 901–916. Available from: https://link.springer.com/article/10.1007/s10040-011-0723-4.

Walton, W. C. 1996 *Aquifer Test Analysis with Windows™ Software*, 1st edn. Lewis Publishers, USA.

Zahid, A. & Ahmed, S. R. U. 2006 *Groundwater Resources Development in Bangladesh: Contribution to Irrigation for Food Security and Constraints to Sustainability. No H039306, IWMI Books, Reports from International Water Management Institute.* Available from: https://econpapers.repec.org/scripts/redir.pf?u=https%3A%2F%2Fpublications.iwmi.org%2Fpdf%2FH039306.pdf;h=repec:iwt:bosers:h039306.

First received 20 December 2020; accepted in revised form 9 March 2021. Available online 19 March 2021

Two-dimensional numerical simulations of soil-water and heat flow in a rainfed soybean field under plastic mulching

Mohammad Abdul Kader, Kimihito Nakamura, Masateru Senge and Mohammad Abdul Mojid

ABSTRACT

Numerical simulation can help understanding water- and heat-flow systems through plastic-mulched soils. An effective simulation approach is crucial to know the role of plastic mulch in a soil ecosystem, which can save water in agriculture. A field experiment was conducted at Gifu University in a rainfed soybean cultivation under plastic mulch and bare soil treatments to clarify the soil water and heat flow mechanism. Furthermore, the two-dimensional numerical software HYDRSUS-2D model with different boundaries at the soil surface was used to simulate water and heat flows. Firstly, soil hydraulic parameters were estimated by inverse solution using laboratory-measured data and then coupled soil-water and heat flows were simulated by optimizing soil thermal parameters by inverse solution. The HYDRUS-2D model simulated water and heat flow through the root zone depths satisfactorily. The root-mean square error (*RMSE*) was 0.015–0.030, and 0.046–0.055 $cm^3\ cm^{-3}$ for the plastic mulch, and bare soil, respectively, in estimating soil moisture and 0.66–1.28, and 0.70–1.54 °C, respectively in estimating soil temperature. Water infiltration was 61% lower in the plastic-mulched soil, which reduced soil evaporation as well as soil-moisture storage changes compared to bare soil. This study can be applied to design and manage different plastic mulching patterns in rain-fed crop cultivation.

Key words | film mulching, heat transport, HYDRUS-2D, soil hydraulic function

Mohammad Abdul Kader (corresponding author)
Rural Development Academy (RDA),
Bogura 5842,
Bangladesh
E-mail: *kader@rda.gov.bd*

Kimihito Nakamura
Graduate School of Agriculture,
Kyoto University, Kitashirakawa Oiwake-Cho,
Sakyo Ward,
Kyoto 606-8502,
Japan

Masateru Senge
Faculty of Applied Biological Sciences,
Gifu University,
1-1 Yanagido,
Gifu City 501-1112,
Japan

Mohammad Abdul Mojid
Department of Irrigation and Water Management,
Bangladesh Agricultural University,
Mymensingh 2202,
Bangladesh

HIGHLIGHTS

- HYDRUS-2D simulated coupled flow of water and heat through plastic-mulched soil.
- Plastic mulch reduced soil evaporation compared to bare soil.
- Plastic mulch decreased soil moisture storage compared to bare soil.

INTRODUCTION

Global warming and irregular rainfall patterns have been limiting water resources in arid and semi-arid regions, which need effective utilization of water for enhancing crop production (Bradford *et al.* 2017). Plastic film mulching, considered as water-saving technology, is a well-known farming practice in dryland areas (Qin *et al.* 2018) for regulating soil temperature and conserving soil moisture by reducing soil evaporation (Kader *et al.* 2017a). So far, China has become the largest user of plastic film mulching (Daryanto *et al.* 2017), which is mainly popular in the Loess Plateau area where frequent water shortage occurs due to splash rainfall (Wang *et al.* 2015; He *et al.* 2016). Plastic mulch improves soil-moisture retention and availability, and exhibits soil-warming effects by regulating radiation and thermal

doi: 10.2166/ws.2021.095

properties (Tian *et al.* 2003; Kader *et al.* 2020). The use of plastic film increases the resistance to water vapour transport from the soil to atmosphere; this resistance is the major controller of evaporation (Davarzani *et al.* 2014). In addition, plastic mulch, by reducing soil evaporation and altering soil temperature in crop fields, has demonstrated great potential for improving water-use efficiency as well as crop yield (Li *et al.* 2018; Zhang *et al.* 2018). The application methods of plastic mulching in the field have been practiced by ridge furrow (Zhao *et al.* 2014), raised bed (Alliaume *et al.* 2017; Zhang *et al.* 2018), partial ridge and furrows (Chen *et al.* 2018), double layer mulching (straw incorporated with plastic) (Yin *et al.* 2016) and traditional flat film mulching (Zhang *et al.* 2015) with various thicknesses (10–20 μm) and colours. In all these application methods, the farm land is covered with plastic film, which limits the entry of rainwater and air into the soil surface. Only a little amount of water can enter into the soil through planting holes and infiltration at the edge of the mulched-bed (film-side infiltration) (Chen *et al.* 2018). Therefore, new application patterns of plastic mulching need to be developed in humid subtropical climates like the central Japan area for increasing water use efficiency in order to ensure water sustainability in irrigated farming.

Knowledge of soil thermal and optical properties is essential when assessing heat transport in mulching soil. Planting holes in plastic mulching support soil aeration for plant emergence and rainwater distribution into the soil that would increase water availability in the root zone. It is important to manage the critical parameters of plastic film (e.g. diameter of the hole and spacing of the holes) for effective conservation of water with mulches (Li *et al.* 2017). The percent open area on the plastic film may have a significant impact on soil evaporation and infiltration into soil; a greater hole ratio provides a smaller vapour transfer resistance so that water vapour can easily cross through plastic mulch to the atmosphere (Qi *et al.* 2018). Until now, soil-water and heat movement and their distributions by plastic mulch is not completely understood due to close interactions of microclimate, soil environment and plant growth (Yang *et al.* 2015; Steinmetz *et al.* 2016). Therefore, quantification of the effects of plastic mulching by numerical simulations of water and heat flows along with root-water uptake may increase the effectiveness and popularity of the system.

Numerical modelling is an effective way to explore water- and heat-flow distributions into soil ecosystems (Qi *et al.* 2018). HYDRUS is a popular software that can simulate water, heat and nutrient dynamics in variably-saturated porous media (Šimůnek *et al.* 2016). The HYDRUS-2D version has superiority over its 1D version in choosing various boundaries and providing water-flow distribution into soils. Simulations with HYDRUS-2D provides quick and accurate estimation of the hydro-thermal environment of field soils (Chen *et al.* 2018; Rai *et al.* 2019). However, selection of accurate soil water- and heat-flow boundaries in HYDRUS-2D software under plastic mulch cover is critical. In previous researches (Saglam *et al.* 2017; Chen *et al.* 2018), a 'no flux' water-flow boundary was considered in the plastic-mulched area. Moreover, several studies (e.g. Saglam *et al.* 2017; He *et al.* 2018; Qi *et al.* 2018), simulating only water flow using HYDRUS-2D under drip emitters with plastic mulching, did not discuss heat-flow distribution, which is influenced by mulch. Previous researches (e.g. Zhang *et al.* 2018; Zhao *et al.* 2018) considered a no-flux boundary for both water and heat flow for the plastic-covered area, assuming that film mulch protects water flow as well as heat flow into the soil. But, those studies did not quantify the effects of water fluxes on different boundaries and the soil-water balance. In the case of coupled flow of water and heat, the no-flux heat-flow boundary is inappropriate for plastic mulching since heat flow to the soil profile occurs through the planting holes. In actual conditions, the plastic covers prevent water flow but can influence (increase or reduce) heat flow. It is therefore important to select the correct boundary conditions in HYDRUS-2D for plastic mulching. A variable flux boundary in a mulching area where water flux is to be zero and heat flux is to be governed by air and soil surface temperatures may be the correct boundary for HYDRUS-2D simulation with plastic mulching. In this study, the crop planting holes and of the plastic mulching have been taken into the modeling approach by considering the atmospheric boundary, and the plastic-mulched area has been taken as a variable flux boundary where water flow is restricted to 0 cm d^{-1} and heat flow occurs from the plastic film to soil. The study thus simulates fully coupled water and heat flows in a rainfed soybean field.

Simulations with HYDRUS-2D under drip irrigation with ridge-furrow/raised-bed mulching were done in several regions of the world (Saglam *et al.* 2017; He *et al.* 2018; Parihar *et al.* 2019). However, in order to understand the effects of film mulching comprehensively, both plastic mulching having a potential of considerable open-hole area in plastic mulching and bare soil need to be modeled. Therefore, a comprehensive research must investigate the complete coupling of water- and heat-flow regimes with plastic mulching. With this view, the coupled-flow regimes of water and heat as well as water- and heat-distribution patterns under plastic mulching was simulated by the HYDRUS-2D model and the results were compared to that of bare soil in a rain-fed soybean field of central Japan. Furthermore, the effects of plastic mulching on soil-water balance were also quantified by using the established model.

MATERIALS AND METHODS

Description of field experiment

A field experiment was done at Gifu University farm in Japan (35° 27′N latitude and 136° 44′E longitude, 12 m above sea level) by growing a rainfed soybean cultivar (*Glycine max cv. Meguro*) under two treatments of plastic mulching, and no mulching (bare soil). The mean air temperature at the study site is 16.1 °C and annual rainfall is 1,849 mm. Figure 1 illustrates the layout of the treatments and setting of various devices in the experimental field (Kader *et al.* 2017b). The research field was divided into two plots of size 12.5 m^2 (each 5 m long and 2.5 m wide) with a buffer zone of 0.5 m surrounding each plot. A raised bed was prepared in the plots and each treatment was employed in one plot. A silver-colour plastic film (one layer of 20 μm thickness) was selected for the plastic mulching and spread manually over the entire raised bed. Soybean was grown in four rows in each plot, with row to row and plant to plant spacing of 40 and 30 cm, respectively. The diameter of the planting hole for each soybean plant was 3 cm (Figure 1). The experiment was conducted under natural rainfall (rainfed) without any irrigation; a total of 874 mm rainfall occurred during the soybean-growing period. The average pan evaporation was 3.9 mm d^{-1} (except the rainy days) during the study period.

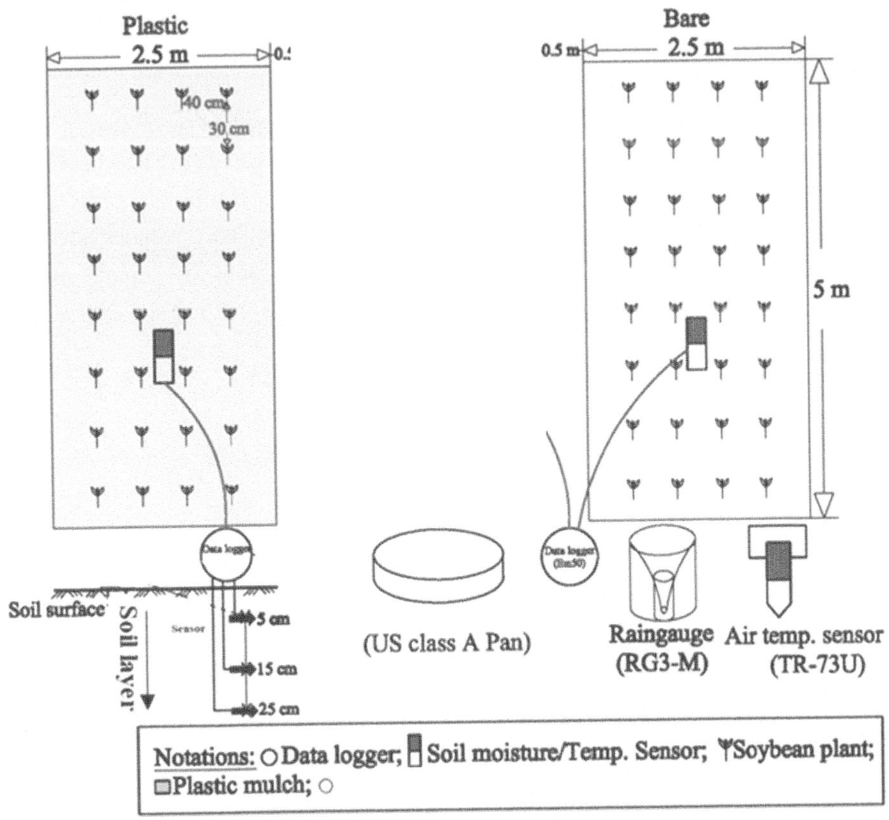

Figure 1 | Field layout of the experimental plots and set up of various devices (Kader *et al.* 2017b).

The distribution of rainfall over the crop period (Figure 2) was used as input information for defining the boundary condition in the simulation.

Measurements

Relevant meteorological data like air temperature at 2 m height, rainfall, solar radiation, relative humidity and wind speed were measured at the experimental field during the soybean cultivation period. The hourly soil moisture and temperature were measured at 5, 15 and 25 cm depths from each treatment throughout the experimental period. Necessary sensors (5TM of Decagon Devices, Inc., USA) were installed horizontally at the centre, between the rows and between the columns of each treatment, and the entire raised bed of each plot was covered with a plastic film by ensuring maximum contact of the plastic film with the soil surface. Each depth had one sensor and the sensors were connected to Em50 data loggers (Decagon Devices, Inc., USA). Soil-surface temperature (at 0 cm depth) was calculated by correlating air temperature in 2015 and regression co-efficient (R^2) of soil-surface temperature from another experiment done in 2016 (Kader *et al.* 2019). Soil samples were collected after the experiment from 0–10, 10–20 and 20–30 cm soil layers, with three replications for each treatment by using core samplers of 100 cm^3. Texture of the 0–30 cm soil profile was loamy sand (53% sand, 31% silt, 16% clay); the classifications of the soil were described by Kader *et al.* (2017b). Soil-water retention functions were determined for 0–10, 10–20 and 20–30 cm soil layers with three replications in the laboratory by centrifugation method (Russel & Richards 1938) with a Kokusan H-2000B centrifuge machine at equilibrium matric potentials of –10, –32, –100, –316 and –1,500 kPa. The saturated hydraulic conductivity (K_s) of the field soil at 5, 15 and 25 cm depths with three replications was measured in the laboratory by the falling head method (Klute & Dirksen 1986).

Numerical simulations

Soil-water flow

The Windows-based modeling software HYDRUS-2D version 2.05.0270 was used to solve the governing equation for variably-saturated water flow (Šimůnek *et al.* 2018). Neglecting vapour flow, the two-dimensional Richards' equation for water flow with root-uptake functions is expressed by (Richards 1931)

$$\frac{\partial\theta(h)}{\partial t} = \frac{\partial}{\partial x}\left[K(h)\frac{\partial h}{\partial x}\right] + \frac{\partial}{\partial z}\left[K(h)\frac{\partial h}{\partial x} + K(h)\right] - S(h) \quad (1)$$

where θ is volumetric moisture content ($cm^3\ cm^{-3}$), h is pressure head (cm), $K(h)$ is unsaturated hydraulic conductivity of the liquid phase ($cm\ d^{-1}$), t is time (d), x and z are horizontal and vertical coordinates (cm), respectively and $S(h)$ is a sink term referring to root-water uptake (d^{-1}). The unsaturated soil hydraulic properties were modeled

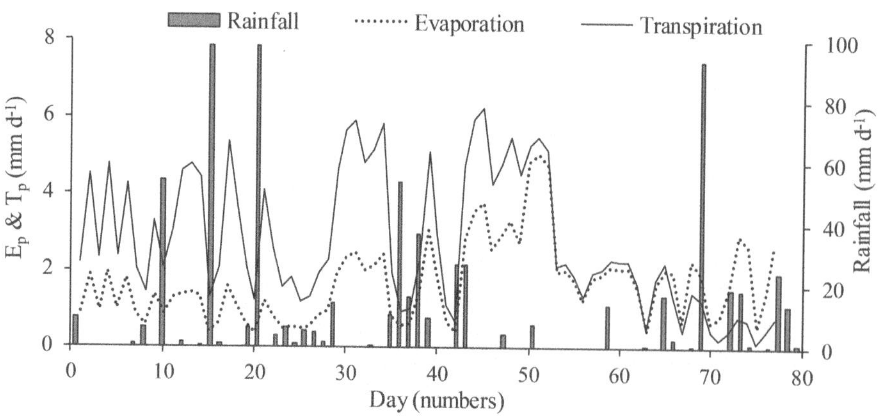

Figure 2 | Daily potential evaporation (E_p), plant transpiration (T_p) and rainfall in the experimental site during soybean-growing period.

for the entire flow domain by using the van Genuchten-Mualem model (van Genuchten 1980) expressed by

$$\theta(h) = \theta_r + \frac{\theta_s - \theta_r}{[1 + |\alpha h|^n]^m} \quad (2)$$

$$K(h) = K_s S_e^l \left[1 - (1 - S_e^{\frac{1}{m}})^m\right]^2 \quad (3)$$

and

$$S_e = \frac{\theta - \theta_r}{\theta_s - \theta_r} \quad (4)$$

where θ_s is soil-moisture content at saturation (cm^3 cm^{-3}), θ_r is residual soil-moisture content (cm^3 cm^{-3}), K_s is saturated hydraulic conductivity (cm d^{-1}), α (cm^{-1}) and n (–) are shape parameters of the soil-water retention curve, S_e (–) is effective saturation, and l (–) is a pore connectivity parameter that is assumed to be 0.50 for all cases (Mualem 1976). Here, m is given by (1–1/n). Temperature dependence of soil hydraulic parameters was taken into account in the study.

Root-uptake functions

The root zone of soybean is usually shallow; the maximum depth with roots (50 cm) was used for the transport domain in the simulation following Fan *et al.* (2016). Moreover, soybean root was found concentrated within 0–30 cm soil depth in the field observation in our experiment. The sink term, S, in Equation (1) expresses the volume of water removed per unit of time from a unit volume of soil due to plant-water uptake. It was computed according to the Feddes model (Feddes *et al.* 1978) expressed by

$$S(h, x, z) = \alpha(h, x, z) b(x, z) T_p L_t \quad (5)$$

where $\alpha(h, x, z)$ is water-stress response function (-) due to root-water uptake that is expressed as a function of soil-water pressure head, $b(x, z)$ is water-uptake distribution function (cm^{-2}), T_p is potential transpiration rate (cm d^{-1}) and L_t is surface length associated with transpiration (cm). The root-water uptake parameters were explained based on the Feddes model for bean (soybean crop) from HYDRUS internal database (Šimůnek *et al.* 2018) and are given in Table 1.

Table 1 | Root-water uptake parameters for the water-stress response function used in HYDRUS-2D simulations (Feddes *et al.* 1978)

	P_o	P_{opt}	P_{2H}	P_{2L}	P_3	r_{2H}	r_{2L}
Crops	**(cm)**					**(cm d^{-1})**	
Soybean	–10	–25	–750	–2,000	–8,000	0.5	0.1

P_o: Pressure head below which roots start to extract water from the soil; P_{Opt}: Pressure head at which maximum water uptake occurs; P_{2H}: Limiting pressure head below which roots can no longer extract water at the maximum rate; P_{2L}: Pressure head having same meaning of P_{2H}, but for a potential transpiration rate of r_{2L}; P_3: Pressure head below which root-water uptake ceases; r_{2H} and r_{2L}: High and low potential transpiration rates for an optimal range of pressure heads.

Soil-heat flow

Neglecting the effects of water-vapour diffusion, two-dimensional heat-transport function is given by (Sophocleous 1979)

$$C_p(\theta)\frac{\partial T}{\partial t} = \frac{\partial}{\partial x}\left(\lambda(\theta)\frac{\partial T}{\partial x}\right) - C_w q \frac{\partial T}{\partial x} - C_w S T \quad (6)$$

where T is soil temperature (°C), q is water flux (cm d^{-1}), $\lambda(\theta)$ is apparent thermal conductivity of the soil (W m^{-1} °C^{-1}), and C_p and C_w are volumetric heat capacities of moist soil and liquid water (J m^{-3} °C^{-1}), respectively. The thermal conductivity as a function of soil-moisture content was expressed by (Chung & Horton 1987)

$$\lambda(\theta) = b_1 + b_2\theta + b_3\theta^{0.5} \quad (7)$$

where b_1, b_2 and b_3 are empirical parameters (W m^{-1} °C^{-1}). In Equation (6), the volumetric heat capacity, originally given by de Vries (1963), was further modified by Šimůnek *et al.* (2008) as

$$C_p(\theta) = C_n\theta_n + C_o\theta_o + C_w\theta \quad (8)$$

where subscripts n, o and w represent solid phase, organic matter and liquid phase, respectively in the soil.

Model parameterizations

Geometry of HYDRUS-2D domain

HYDRUS-2D with general type of geometry uses the Galerkin finite element method in the two-dimensional vertical x-z plane. The transport domain was 250 cm deep and 250 cm wide where the length of plastic mulching was 200 cm (Figure 3(a)). Due to the deep groundwater table (much below the simulation domain) at the study site, a free drainage boundary was considered at the bottom (at 250 cm depth). A finite element mesh with radical grid

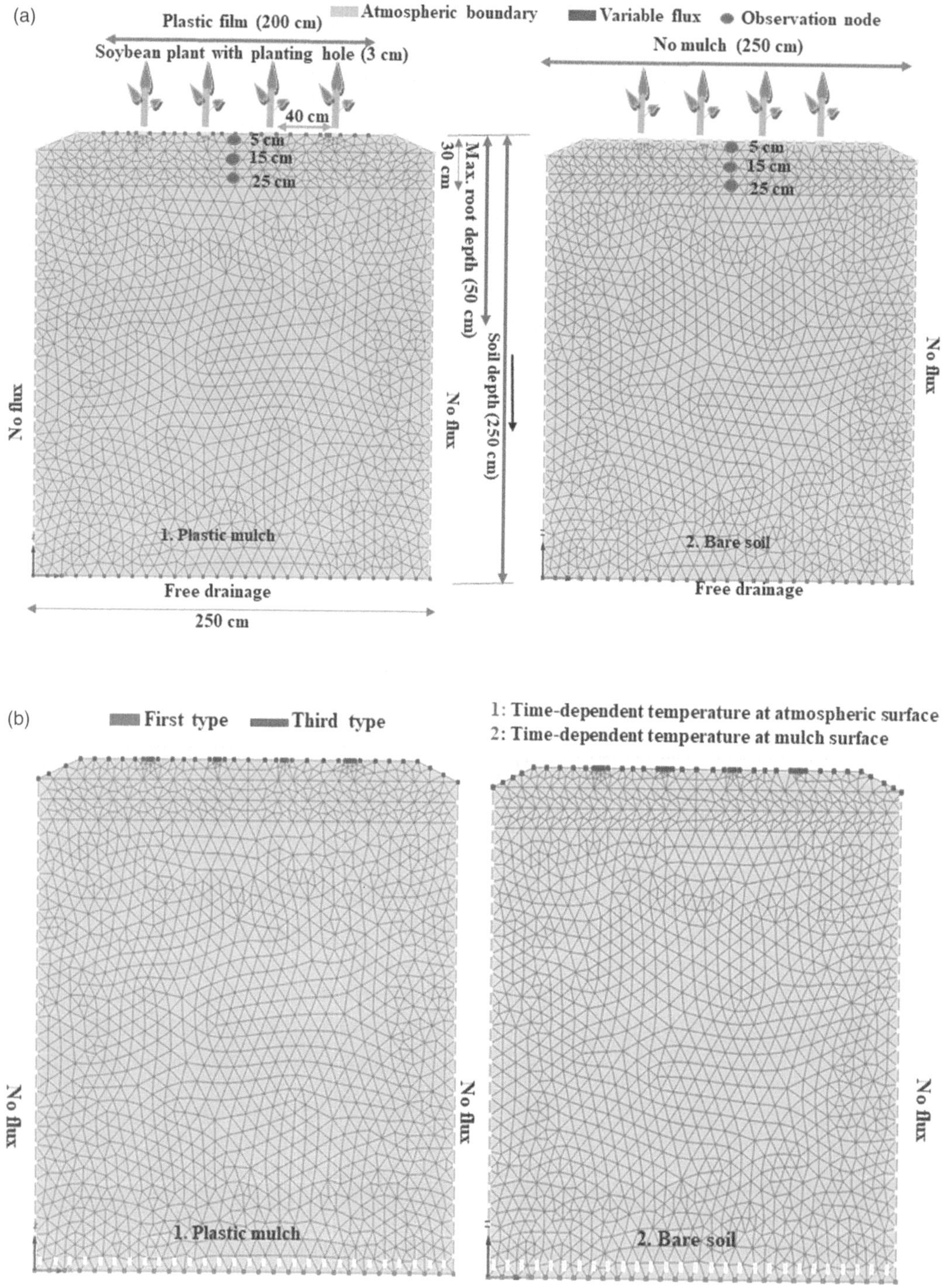

Figure 3 | (a) Geometric domain of HYDRUS-2D modeling with observation nodes, and boundary conditions for soil-water flow under two treatments: 1. plastic mulch, and 2. bare soil. (b) Geometric domain of heat-transport boundary under two treatments: 1. plastic mulch, and 2. bare soil. The full colour version of this figure is available in the online version of this paper, at http://dx.doi.org/10.2166/ws.2021.095.

spacing of 2 cm was generated by HYDRUS-2D, with a finer mesh size close to the soil surface.

Setting of boundary

The boundary and initial conditions used in the HYDRUS-2D model for simulating soil-water and heat flow under the two treatments are illustrated in Figure 3(a) and 3(b), respectively. The plastic mulching and bare soil areas of the transport domain were discriminated by choosing different boundary conditions at the soil surface (Figure 3(a) and 3(b)). The time-invariable flux ($0\ \mathrm{cm\ d^{-1}}$) boundary conditions were used for the plastic-mulched area, while bare soil planting, and planting-hole areas on the soil surface in the transport domain were represented by using an atmospheric boundary condition, with rainfall, evaporation and transpiration occurring from the soil. So, the area of raised-bed (except the plastic mulch) was considered as the atmospheric boundary. For all cases, both lateral sides (walls) of the soil profile were provided with a 'no-flux' boundary and the bottom boundary was imposed as 'free drainage' at 250 cm depth. For variable atmospheric condition, 'day' was set as the time unit. Rainfall entered bare soil area through atmospheric boundary at the soil surface, and time-invariable flux boundary through the plastic-mulched area was restricted to $0\ \mathrm{cm\ d^{-1}}$ of water flow. HYDRUS-2D model requires daily rainfall, daily potential evaporation (E_p) and potential transpiration (T_p) as inputs for defining the upper atmospheric boundary condition. Reference crop evapotranspiration (ET_o) for the experimental period was calculated from meteorological data by using Penman-Monteith equation (Allen *et al.* 1998). The crop evapotranspiration (ET_c) was calculated by

$$ET_c = K_c \times ET_0 \tag{9}$$

where the crop coefficient (K_c) for initial, development and maturity stages of soybean crop was 0.40, 1.15 and 0.50, respectively (Allen *et al.* 1998). The soybean-growing period (sowing to harvest), 10 May to 27 August (110 days) of 2015, was divided into initial stage (0–25 day after sowing (DAS)), development stage (26–85 DAS) and maturity stage (86–110 DAS) following Allen *et al.* (1998). The crop evapotranspiration, ET_c was partitioned into E_p and T_p following Ritchie (1972) as

$$E_p = ET_c \exp(-k \times LAI) \tag{10}$$

$$T_p = ET_c - E_p \tag{11}$$

where k is a constant related to the leaf extinction coefficient of soybean, which was assumed to be 0.49 (Adeboye *et al.* 2017) and LAI is Leaf Area Index, which varied linearly from 0 to 3 during the development stages of soybean (Ahmad *et al.* 2018). The variation of daily E_p and T_p over the soybean-growing period is illustrated in Figure 2. Initial conditions at the beginning of the simulation period were set in terms of field-measured soil moisture and temperature at 5, 15 and 25 cm depths for each treatment. Three observation nodes located at 5, 15 and 25 cm depths represented soil moisture and temperature sensors (Figure 3(a)).

Heat-transport boundary

First-type (Dirichlet-type) boundary, specified by temporal changes in temperature of each treatment (Figure 3(b)) at the upper surface, was adopted for heat-transport simulation against the atmospheric boundary in water-transport simulation. Heat flows from the bare soil, and planting areas occurred to the atmosphere. So, the time-dependent temperature boundaries at the upper surface were the first-type, with a pointer vector boundary of 1 and 2 (Figure 3(b)). The condition of point vector 1 was defined by air temperature, which was specified for bare soil, and planting areas. The point vector 2 was defined by soil-surface temperature just below the mulch that was specified for the plastic mulch-covered area (Figure 3(b)). The soil surface temperature of the silver colour plastic mulched soil was estimated from another study reported by Kader *et al.* (2019). Both vertical sides of the flow domain were 'no heat flux' boundary and the bottom boundary was third-type (Cauchy-type), which is of zero gradient soil temperature (Figure 3(b)).

Parameter optimization

Numerical simulations of soil-water and heat flow under the two treatments were done for the period from 12 June to 27 August (77 days) in 2015. Firstly, the soil hydraulic parameters were estimated from the laboratory-measured soil-moisture retention data of each treatment using RETC software (van Genuchten *et al.* 1991). The estimated parameters were used as initial values in inverse solution for estimating the most appropriate soil hydraulic parameters (Table 2). Then, water flow of each treatment was simulated by optimizing van Genuchten's parameters using the Levenberg-Marquardt non-linear minimization method (Šimůnek *et al.* 2018). Soil hydraulic parameters from the laboratory-measured data were used as initial soil-moisture retention functions and then θ_r, θ_s, α, n and K_s (Equations (2) and (3)) were optimized to obtain good fitting of soil-moisture contents. The optimized soil hydraulic parameters are listed in Table 2. Afterward, the coupled water and heat flow simulations were run for each treatment, where the water-flow parameters were fixed and heat-flow parameters were estimated by inverse solutions. The thermal parameters in Equation (6) for the relationship between soil thermal conductivity and volumetric moisture content for each soil layer were finally optimized by inverse solutions (Table 3). The longitudinal dispersivity (D_L) of the soil was optimized between 4 and 10 cm by inverse simulations under condition of fixed transverse dispersivity (D_T) of 1 cm for each soil layer.

Model performance

Coefficient of determination (R^2), as a statistical indicator, was used to reflect the degree of correlation between the simulated and observed values of soil moisture and soil temperature of each treatment. The performance of the model was evaluated by root-mean-square error, *RMSE*, as expressed by

$$RMSE = \left[\frac{1}{N}\sum_i (P_i - O_i)^2\right]^{\frac{1}{2}} \tag{12}$$

where N is number of observations, and P_i and O_i are the simulated and observed values, respectively, of soil moisture or soil temperature.

RESULTS

Soil-moisture retention curves

Soil-moisture retention curves from the inverse optimization and laboratory measurements at 0–10, 10–20 and 20–30 cm

Table 2 | Optimized van Genuchten's soil hydraulic parameters used in two treatments for HYDRUS-2D simulations

Treatments	Depth (cm)	θ_r (cm³ cm⁻³)	θ_s	α (cm⁻¹)	n (–)	K_s (cm d⁻¹)
Plastic	0–10	0.1230	0.5084	0.0717	1.1987	235.17
	10–20	0.0021	0.4063	0.0032	1.2649	340.97
	20–100	0.1171	0.4408	0.0082	1.3299	40.19
Bare soil	0–10	0.0080	0.5450	0.0450	1.3321	170.64
	10–20	0.0250	0.4250	0.0210	1.3020	320.00
	20–100	0.0840	0.4600	0.0180	1.3221	70.00

Table 3 | Volumetric fraction of soil solid phase (θ_n), volumetric fraction of organic matter (θ_o), longitudinal and transverse dispersivity (D_L and D_T), and soil thermal parameters (b_1, b_2 and b_3: empirical parameters of thermal conductivity; C_n, C_o and C_W: volumetric heat capacity of solid phase, organic matter and liquid water phase, respectively) used for heat flow simulations in HYDRUS-2D model

Depth (cm)	θ_n (cm³ cm⁻³)	θ_o	D_L (cm)	D_T	b_1 (W m⁻¹ °C⁻¹)	b_2	b_3	C_n (MJ m⁻³ °C⁻¹)	C_o	C_w
0–10	0.49 ± 0.046	0	4.0 ± 1.0	1	0.16 ± 0.07	0.26 ± 0.10	1.30 ± 0.17	2.48 ± 0.07	2.51	4.18
10–20	0.49 ± 0.047	0	4.0 ± 1.0	1	0.38 ± 0.10	0.41 ± 0.01	1.59 ± 0.04	2.26 ± 0.30	2.51	4.18
20–100	0.52 ± 0.06	0	10.0 ± 0	1	0.49 ± 0.18	0.47 ± 0.06	1.65 ± 0.11	2.25 ± 0.30	2.51	4.18

Values shown as mean ± SE (Standard Error). θ_n, b_1, b_2, b_3, D_L and C_n were optimized by inverse solution.

depths are depicted in Figure 4. The inverse-optimized hydraulic parameters varied slightly from those obtained from the laboratory-measured data, while the overall trend of the soil-moisture retention curves was similar for all three depths (Figure 4). It is noted that the soil hydraulic parameters obtained in laboratory tests would not be appropriate since the soil was heterogeneous over the year due to cultivation and the parameters might vary over time. The optimized soil hydraulic parameters were largely different among the three treatments to reproduce the observed soil-moisture contents. The optimized soil-moisture content at saturation, θ_s (Equation (2)), showed a decreasing pattern with increasing depth; θ_s for 0–10 cm soil layer was the highest in the plastic mulch treatment and lowest in the bare soil. The water-retention characteristics of the soils were important since they influenced water flow during the simulation process. The laboratory-measured soil hydraulic parameters (θ_r, θ_s, α, n; Equation (2)) are not always used for correct fitting in simulation study; they need to be updated in most studies (Zhao *et al.* 2018).

Soil moisture

The observed and simulated soil-moisture contents at 5, 15 and 25 cm depths in two treatments are compared in Figure 5(a) and 5(b), respectively. The fluctuation of soil moisture was the largest in bare soil followed by plastic mulch treatment. The plastic mulch showed moderate fluctuation of soil moisture. Rainfall was the main factor in altering soil moisture in all treatments. The bare soil, due to its whole open surface, received more rainwater than other treatments and showed the greatest fluctuations in soil moisture. The plastic mulch protected rainwater to a large extent and showed moderate fluctuations of soil moisture over the soybean-growing season. In plastic mulch treatment, the fluctuation of soil-moisture content at 5 cm depth was smaller than at 15 and 25 cm depths (Figure 5(a)). The variation of soil-moisture content in the bare soil was obviously larger near the soil surface (at 5 cm depth) since the effects of rainfall on evaporation and root-water uptake was greater at the soil surface than at the deeper layers.

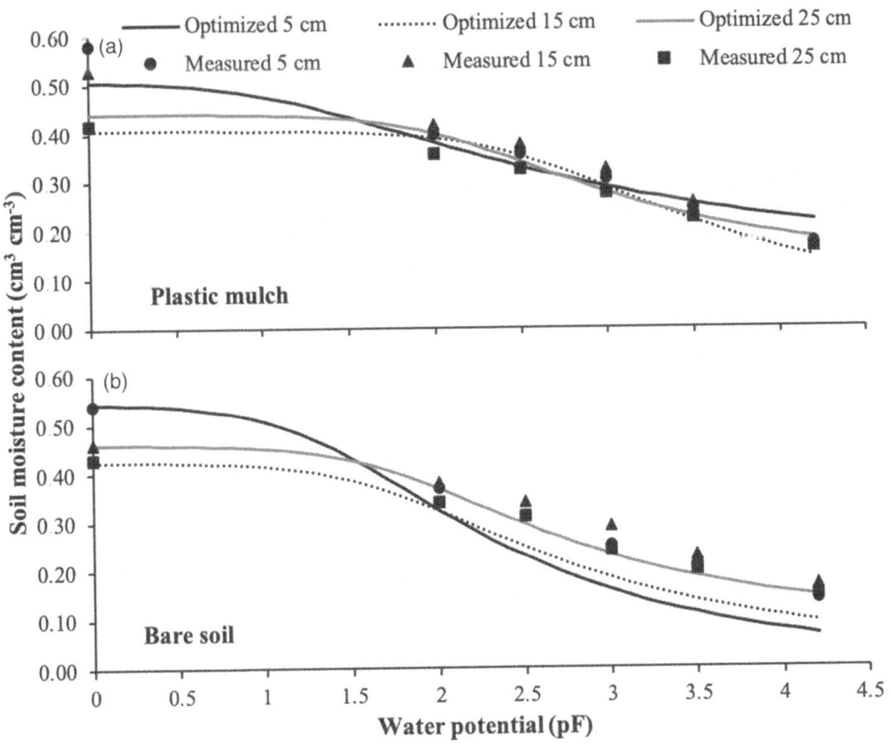

Figure 4 | Laboratory measured and HYDRUS-2D model-optimized soil-water retention curves at 5, 15 and 25 cm soil depths under the treatments of (a) plastic mulch and (b) bare soil.

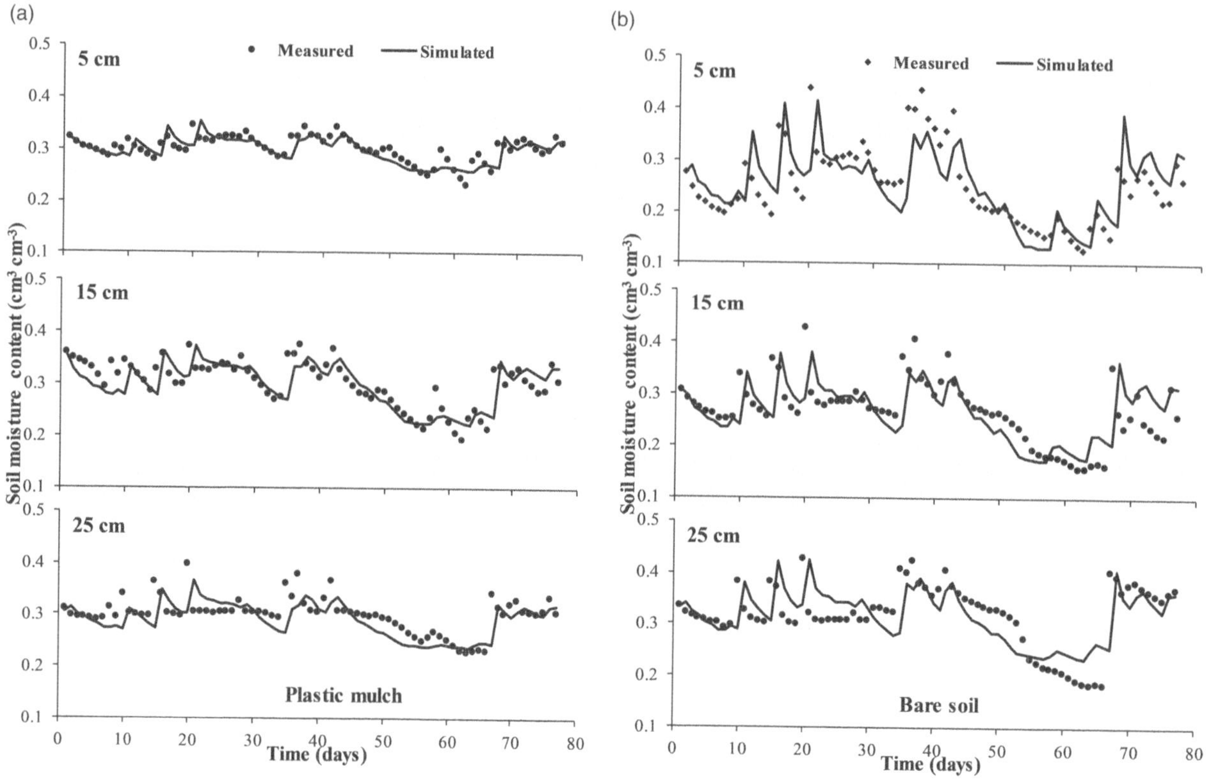

Figure 5 | Comparison of simulated and measured soil-moisture contents at 5, 15 and 25 cm depths under (a) plastic mulch and (b) bare soil treatments during soybean-growing season.

The statistical performance of HYDRUS-2D model for water-flow simulation is summarized in Table 4 in terms of root-mean square error, *RMSE*, and coefficient of determination, R^2. The model provided fairly good estimate of soil-moisture contents, with *RMSE* between the measured and simulated soil-moisture contents of 0.015–0.030, and 0.046–0.055 $cm^3\ cm^{-3}$ for the plastic mulch and bare soil treatment, respectively (Table 4). Moreover, the best fitting between the measured and simulated soil-moisture contents was obtained for the plastic mulching ($R^2 = 0.62$) followed by bare soil ($R^2 = 0.47$). There were little differences between the observed and simulated values, possibly due to close proximity of the sensors' positions in the field and observation nodes in the simulation model. Our simulation model performed well, particularly when considering the complexity of the conditions like heterogeneous soil properties, plastic mulch-covered soil, several rainfall events and non-homogeneous root distribution of soybean plants to which the model was applied. Finally, we confirmed that HYDRUS-2D model reflected reliability in simulating soil-moisture contents at the soil root zone (0–30 cm) depth.

Table 4 | Performance of HYDRUS-2D for simulating daily soil moisture and temperature in terms of root-mean square error (*RMSE*) and coefficient of determination (R^2) in two treatments during the soybean cultivation periods

		Soil moisture		**Soil temperature**	
Treatments	**Depth (cm)**	***RMSE*** **($cm^3\ cm^{-3}$)**	R^2	***RMSE*** **(°C)**	R^2
Plastic	5	0.015	0.607	1.32	0.83
	15	0.027	0.620	0.93	0.92
	25	0.030	0.345	0.90	0.95
Bare soil	5	0.055	0.471	1.54	0.76
	15	0.047	0.401	0.98	0.90
	25	0.046	0.437	0.70	0.95

Soil temperature

Figure 6(a) and 6(b) illustrate the measured and simulated soil temperatures at 5, 15 and 25 cm depths under mulching and bare soil treatments over the soybean-growing period. With the occurrence of rainfall, soil temperature decreased from top to bottom of the soil profile. The differences in

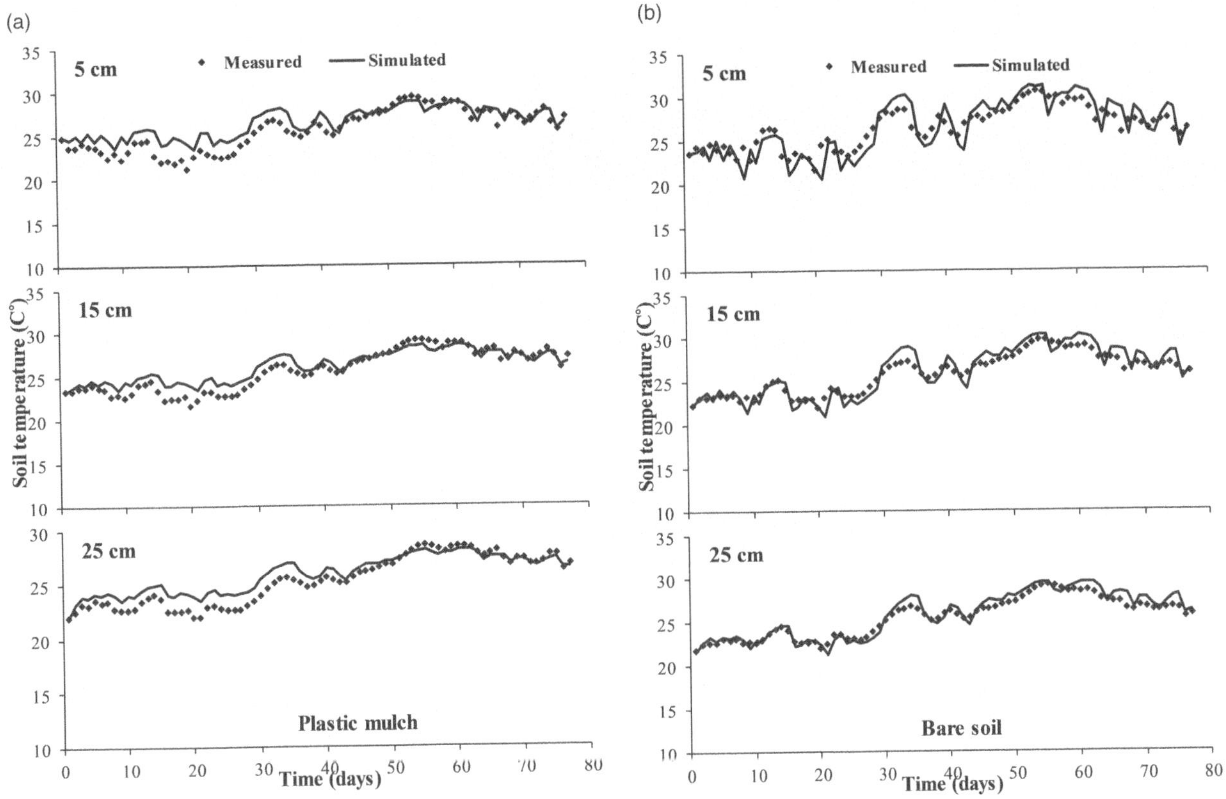

Figure 6 | Comparison of simulated and measured daily soil temperatures at 5, 15 and 25 cm depths under (a) plastic mulch and (b) bare soil treatments during soybean-growing season.

soil temperature between the plastic treatment was minimal during the crop period. The simulation of soil temperatures with thermal parameters, which were estimated by heat-transport inverse model, was reasonably good (Table 4, Figure 6). The model performed fairly well, with *RMSE* of 0.66–1.54 °C and R^2 of 0.70–0.95 for the two treatments (Table 4). The plastic mulching provided greater modeling efficiency, with *RMSE* varying from 0.90 to 1.32 °C at 5, 15 and 25 cm soil depths compared to *RMSE* 0.70–1.54 °C for bare soil (Table 4). Bare soil received rainfall and sunlight directly on the surface of the soil and part of the sunlight was blocked by the leaves, which causes higher degrees of fluctuation of soil moisture and temperature compared to plastic treatment. So, it is difficult to accurately capture the trend of fluctuations of the observed soil moisture and temperature of bare soil by HYDRUS-2D simulations. For this reason, the simulation model showed inferior performance in simulating soil moisture and temperature in bare soil (Table 4) compared to the plastic mulch treatment. Due to the lack of directly measured data of thermal conductivity and volumetric heat capacity of the soil, we optimized a minimum number of heat transport parameters in our simulations. Only the least number of heat-transport parameters (D_L, b_1, b_2, b_3 and C_n in Equations (7) and (8)) were optimized by inverse simulations for each treatment (Table 3). Moreover, the differences between the measured and simulated soil temperatures could also have occurred due to the effect of specified surface and bottom heat transport boundaries. Overall, the predicted results from the HYDRUS-2D model were fairly close to the measurements, and the model was able to accurately simulate the dynamics of soil temperature in the soybean field.

Spatial distributions of soil-moisture and temperature

Simulated distributions of soil moisture and temperature in the vertical sections of the soil profile for the two treatments are illustrated in Figures 7 and 8, respectively, for rainfall and no-rainfall (drying) days. Soil-moisture distribution greatly

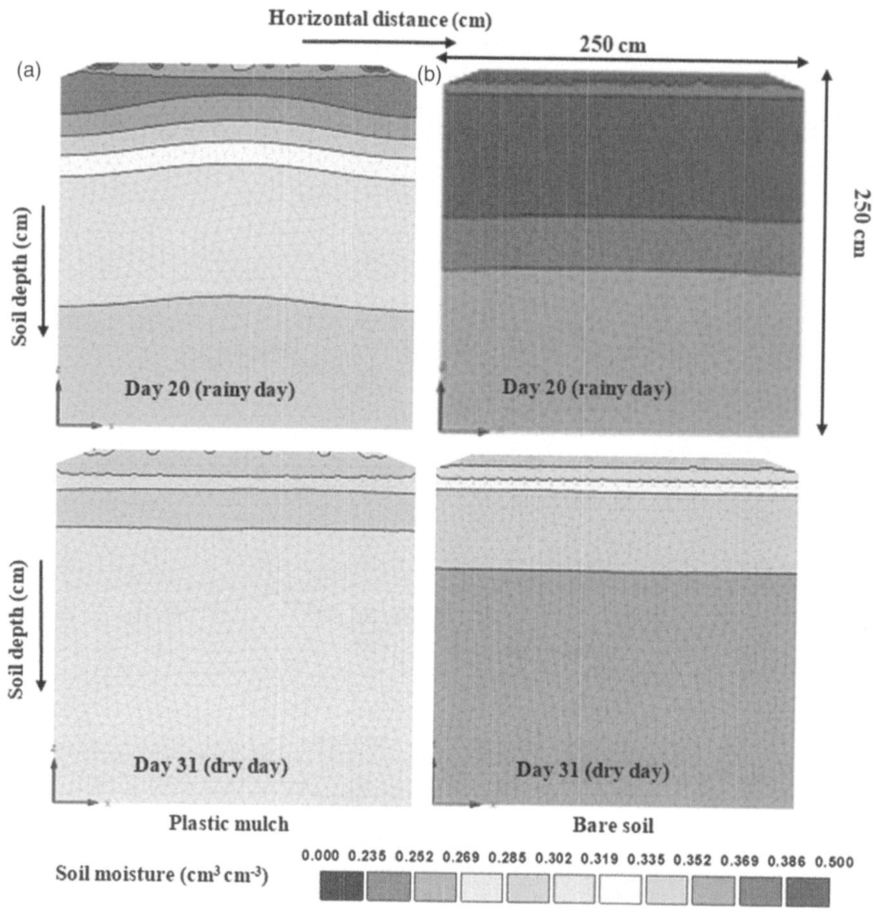

Figure 7 | Water-flow distribution on a rainy day of 20 days after simulation (DAS) and a drying day of 31 DAS under (a) plastic mulch and (b) bare soil. The full colour version of this figure is available in the online version of this paper, at http://dx.doi.org/10.2166/ws.2021.095.

varied when rainwater entered the soil profile. After a rainfall event, soil moisture moved in the vertical (from surface to deeper soil layers) and horizontal directions. In the dry days, soil moisture constantly moved, mainly in the vertical direction (Figure 7). The bare soil treatment provided higher soil moisture (wettest) at the surface during rainfall (20th day of simulation) followed by the plastic mulch, which provided the lowest soil moisture (driest) (Figure 7). On the 31st day of simulation (dry period), lower soil moisture was obtained for all three treatments compared to the wetting days (Figure 7). During the dry period, soil moisture evaporated from the bare soil and also the soil moisture moved downward due to free drainage condition, while the plastic mulch restricted evaporation from the soil.

With increasing rainfall, both the soil moisture and temperature changed significantly at the soil surface and moved in the vertical and horizontal directions from the planting and mulching holes. Rainwater could not infiltrate directly into soil through the plastic mulching as it did through the bare soil. Our simulation revealed that, with 97.5 mm rainfall on 30 June (20th day of simulation), the bare soil was completely saturated; plastic mulching partially wetted the soil, especially the planting-hole area. On the other hand, the drying day on 25 July (31st day of simulation) showed reduced soil moisture at the rooting zone under all three treatments in the order: plastic mulch > bare soil.

Figure 8 focuses on the distribution of soil temperature on day 49 (rainfall) and 56 (no rainfall, dry day) of simulations. On day 49, the variation of soil temperature was very small, especially in bare soil compared to the mulching treatments. Large variation in soil temperature was observed on the drying day 56. The distribution of soil temperature was similar for the plastic mulching, while the bare soil

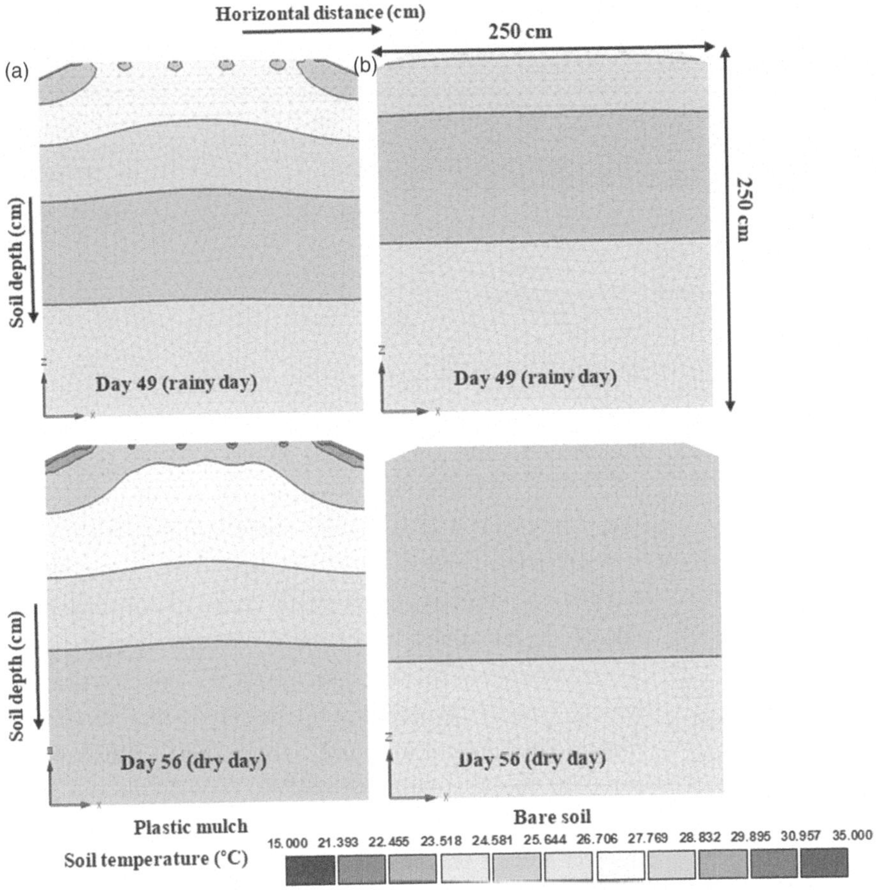

Figure 8 | Heat-flow distribution on a rainy day 49 days after simulation (DAS) and a drying day of 56 DAS under (a) plastic mulch and (b) bare soil. The full colour version of this figure is available in the online version of this paper, at http://dx.doi.org/10.2166/ws.2021.095.

showed steady change in temperature within the soil rooting zone (Figure 8). The surface mulch cover had a greater effect on the soil environment near the surface compared to the deeper soil layers, with the development of strong horizontal gradients in temperature across different mulched-soil boundaries (Zhao *et al.* 2018). Therefore, the spatial distributions of soil moisture and heat flow are influenced by the shadow between two plants, the presence of plastic film mulch and the root density of soybean plants.

Soil-water balance

Water balance in the modeling with cumulative water fluxes in and out of the simulated flow domain is described in Table 5. The water balance components include water fluxes like cumulative infiltration, evaporation and transpiration, change in water storage and deep drainage from the nodes at the bottom of the soil profile during simulation. The negative values of soil-moisture storage indicated water loss from the soil domain during the simulation period. Rainwater infiltration was 61% higher in bare soil than in plastic mulching. Similarly, cumulative evaporation from the bare soil was 62% higher compared to plastic mulching. The amount of cumulative evaporation was in the order: bare soil (126 mm) > plastic (48 mm). The bare soil caused greater infiltration of water and showed higher evaporation and bottom fluxes, indicating downward movement of water below the root zone. The change in soil-moisture storage was also the greatest in bare soil followed by plastic mulching. The water balance analysis revealed that the bare soil increased rainwater infiltration compared to plastic mulching but it reduced soil evaporation compared to the bare soil during the soybean-growing period. The plastic mulch provided direct pathways between soil and air, from which

Table 5 | Water balance components during simulation periods under plastic and bare soil treatments

	Infiltration	Evaporation	vRoot	vBottom	
	Cumulative value (mm)				ΔS (mm)
Treatments	(1)	(2)	(3)	(4)	(1) − (2) − (3) − (4)
Plastic	277	48	204	40	−14
Bare soil	709	126	209	354	20

vRoot: actual transpiration rate; vBottom: actual flux at the bottom of soil profile; ΔS: change in soil-moisture storage.

soil-moisture exited during evaporation; thus, size and density of holes might control the evaporation that occurred from the soil surface.

DISCUSSION

Modeling implications

The two-dimensional numerical simulations of soil-water and heat flows provided better understanding of infiltration and evaporation processes under plastic mulching. Although difficult to choose a correct boundary for plastic-mulched soil in the HYDRUS-2D model (Chen *et al.* 2018; Zhao *et al.* 2018), our study accurately selected the boundary conditions for the plastic mulched area, including planting holes for simulating coupled flow of water and heat under two different treatments. Therefore, the HYDRUS-2D model provided an efficient way to simulate coupled distribution of water and heat under a plastic-mulched field in a rainfed system. In our HYDRUS-2D modeling, water-flow boundary was followed by heat-flow boundary, and the model performed better in heat-flow simulation than water-flow simulation (Table 4). However, similar to the finding of Xi *et al.* (2016), infiltration of rainfall and soil evaporation considerably influenced soil-moisture contents in the upper soil layers compared to the deeper soil layers and, consequently, resulted in some deviations in the simulated soil-moisture contents. The simulation results can, however, be substantially improved by calibrating the input parameters (effective soil hydraulic parameters) against a dynamic flow experiment (Kandelous *et al.* 2011; Rai *et al.* 2019). Previous researches (Han *et al.* 2015; Saglam *et al.* 2017) simulated plastic mulch with drip irrigation, in which soil-water dynamics were controlled by both irrigation and rainfall. This study reveals that the performance of the HYDRUS-2D model strongly depended on rainfall and selection of various surface boundaries in the plastic mulch area; in this study, canopy distribution was not considered in the modeling. Moreover, the simulation of soil-surface temperature under plastic mulch is affected by latent heat, which has received limited attention in the present model. This is because it is a complex system, which involves the optical properties of the film, the thickness of the film and incoming solar radiation (Zhao *et al.* 2018). In the heat-transport models of previous research (Zhang *et al.* 2018), soil evaporation was neglected since the soil surface was fully covered with plastic film mulch. In our research, we analyzed the effects of soil evaporation from plastic mulch by quantifying the soil-water balance.

Water flow

Generally, plastic film mulching restricts water flow into the soil profile. The two-dimensional HYDRUS model quantified the effect of plastic mulching on rainwater infiltration into soil by describing water flow. The temporal variations of cumulative total seasonal infiltration and evaporation of the two treatments are illustrated in Figure 9. The cumulative water infiltration into the bare soil was greater than that into the plastic mulch. In the plastic mulch treatment, rainwater infiltrated through the planting holes. The bare soil caused more infiltration than the plastic mulch treatments since rainwater fell directly on the soil surface (Kader *et al.* 2017b). Cumulative evaporation was also greater in bare soil treatment than in plastic mulch treatment (Figure 9). In Figure 9, the fluctuations of soil-moisture storage (ΔS) are higher in bare soil than in plastic

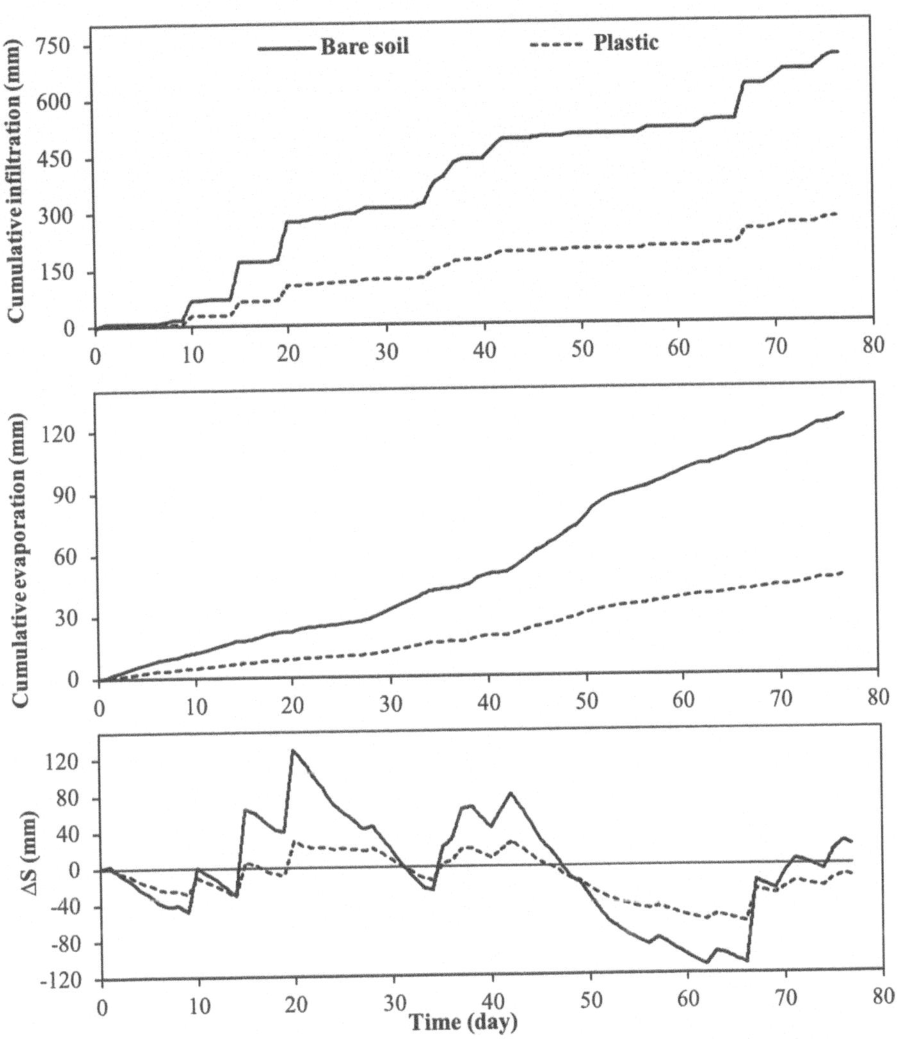

Figure 9 | Simulated cumulative total infiltration, evaporation and change in soil-moisture storage (ΔS) under plastic mulch, and bare soil treatments during soybean-growing season.

mulching. The smaller changes of soil-moisture storage by plastic mulching indicate minimized soil-moisture fluctuations, which is important for crop growth. Larger soil-moisture fluctuations may be induced frequently under soil wetting and drying conditions that have a high risk of getting good crop growth and yield (Parihar *et al.* 2019). In our rainfed crop cultivation system, rainfall was the only input of water balance where water loss included evaporation from soil, transpiration from the rooting zone, deep percolation below the root zone and lateral water movement due to surface runoff. A large part of infiltrated water in the bare soil moved to the deeper soil layers by free drainage (Table 5). Thus, the plastic mulch greatly reduced evaporation from the soil surface compared to bare soil.

CONCLUSIONS

The two-dimensional HYDRUS model first time successfully simulated soil-water dynamics and heat transport under plastic mulching in a rainfed soybean field in central Japan. In this study, water and heat distribution characteristics in soils were evaluated by HYDRUS-2D simulations under plastic and bare soil treatments. The results suggest that the measured and simulated soil moistures and temperatures are in good agreement and able to accurately capture the dynamic changes in water and heat at 5, 15 and 25 cm soil depths. The results of the soil-moisture balance of each treatment show that the bare soil can greatly increase rainwater infiltration compared to plastic mulching. This study confirmed that plastic mulch treatments reduced the changes in soil-moisture storage and soil evaporation compared to bare soil treatment, indicating greater crop-water availability by mulching. Therefore, the HYDRUS-2D model may be useful to optimize the best density of holes in the plastic mulching by scenario analysis for enhancing soil-moisture conservation under different climatic conditions; this topic can be investigated in future studies.

ACKNOWLEDGEMENTS

The authors gratefully acknowledge the Gifu University of Japan for providing agricultural land to conduct the field experiment. The first author thanks the Japanese Government for providing financial support during his doctoral study through the MEXT scholarship.

DECLARATIONS OF INTEREST

None.

DATA AVAILABILITY STATEMENT

All relevant data are included in the paper or its Supplementary Information.

REFERENCES

Adeboye, O. B., Schultz, B., Adekalu, K. O. & Prasad, K. 2017 Soil water storage, yield, water productivity and transpiration efficiency of soybeans (*Glyxine max L. Merr*) as affected by soil surface management in Ile-Ife, Nigeria. *International Journal of Soil and Water Conservation Research* **5**, 141–150. doi:10.1016/j.iswcr.2017.04.006.

Ahmad, M., Chakraborty, D., Aggarwal, P. & Bhattacharyya, R. 2018 Modelling soil water dynamics and crop water use in a soybean-wheat rotation under chisel tillage in a sandy clay loam soil. *Geoderma* **327**, 13–24. doi:10.1016/j.geoderma. 2018.04.014.

Allen, R., Pereira, L. S., Raes, D. & Smith, M. 1998 Crop evapotranspiration: Guidelines for computing crop requirements. *Irrigation and drainage paper* 56, FAO 300. doi:10.1016/j.eja.2010.12.001.

Alliaume, F., Rossing, W. A. H., Tittonell, P. & Dogliotti, S. 2017 Modelling soil tillage and mulching effects on soil water dynamics in raised-bed vegetable rotations. *European Journal Agronomy* **82**, 268–281. doi:10.1016/j.eja.2016.08.011.

Bradford, J. B., Schlaepfer, D. R., Lauenroth, W. K., Yackulic, C. B., Duniway, M., Hall, S., Jia, G., Jamiyansharav, K., Munson, S. M., Wilson, S. D. & Tietjen, B. 2017 Future soil moisture and temperature extremes imply expanding suitability for rainfed agriculture in temperate drylands. *Scientific Reports* **7**, 12923. doi:10.1038/s41598-017-13165-x.

Chen, B., Liu, E., Mei, X., Yan, C. & Garré, S. 2018 Modelling soil water dynamic in rain-fed spring maize field with plastic mulching. *Agricultural Water Management* **198**, 19–27. doi:10.1016/j.agwat.2017.12.007.

Chung, S.-O. & Horton, R. 1987 Soil heat and water flow with a partial surface mulch. *Water Resources Research* **23** (12), 2175–2186.

Daryanto, S., Wang, L. & Jacinthe, P.-A. 2017 Can ridge-furrow plastic mulching replace irrigation in dryland wheat and maize cropping systems? *Agricultural Water Management* **190**, 1–5. doi:10.1016/j.agwat.2017.05.005.

Davarzani, H., Smits, K., Tolene, R. M. & Illangasekare, T. 2014 Study of the effect of wind speed on evaporation from soil through integrated modeling of the atmospheric boundary layer and shallow subsurface. *Water Resources Research* **50**, 661–680.

de Vries, D. A. 1963 The thermal properties of soils. In: *Physics of Plant Environment* (R. W. van Wijk, ed.). John Wiley & Sons, North Holland, Amsterdam, pp. 210–235.

Fan, J., McConkey, B., Wang, H. & Janzen, H. 2016 Root distribution by depth for temperate agricultural crops. *Field Crops Research* **189**, 68–74. doi:10.1016/j.fcr. 2016.02.013.

Feddes, R. A., Kowalik, P. J. & Zaradny, H. 1978 *Simulation of Field Water Use and Crop Yield*. John Wiley & Sons, New York, NY.

Han, M., Zhao, C., Feng, G., Yan, Y. & Sheng, Y. 2015 Evaluating the effects of mulch and irrigation amount on soil water distribution and root zone water balance using HYDRUS-2D. *Water (Switzerland)* **7**, 2622–2640. doi:10.3390/w7062622.

He, J., Du, Y.-L., Wang, T., Turner, N. C., Yang, R.-P., Jin, Y., Xi, Y., Zhang, C., Cui, T., Fang, X.-W. & Li, F.-M. 2016 Conserved water use improves the yield performance of soybean (*Glycine max (L.) Merr.*) under drought. *Agricultural Water Management*. doi:10.1016/j.agwat.2016.07.008.

He, Q., Li, S., Kang, S., Yang, H. & Qin, S. 2018 Simulation of water balance in a maize field under film-mulching drip irrigation. *Agricultural Water Management* **210**, 252–260. doi:10.1016/j.agwat.2018.08.005.

Kader, M. A., Senge, M., Mojid, M. A. & Ito, K. 2017a Recent advances in mulching materials and methods for modifying soil environment. *Soil and Tillage Research* **168**, 155–166. doi:10.1016/j.still.2017.01.001.

Kader, M. A., Senge, M., Mojid, M. A., Onishi, T. & Ito, K. 2017b Effects of plastic-hole mulching on effective rainfall and readily available soil moisture under soybean (*Glycine max*) cultivation. *Paddy and Water Environment* **15** (3), 659–668. doi:10.1007/s10333-017-0585-z.

Kader, M. A., Nakamura, K., Senge, M., Mojid, M. A. & Kawashima, S. 2019 Numerical simulation of water- and heat-flow regimes of mulched soil in rain-fed soybean field in central Japan. *Soil and Tillage Research* **191**, 142–155. doi:10.1016/j.still.2019.04.006.

Kader, M. A., Nakamura, K., Senge, M., Mojid, M. A. & Kawashima, S. 2020 Effects of coloured plastic mulch on soil hydrothermal characteristics, growth and water productivity of rainfed soybean. *Irrigation and Drainage* **69**, 483–494.

Kandelous, M. M., Šimůnek, J., van Genuchten, M. T. & Malek, K. 2011 Soil water content distributions between two emitters of a subsurface drip irrigation system. *Soil Science Society of America Journal* **75**, 488. https://doi.org/10.2136/sssaj2010.0181.

Klute, A. & Dirksen, C. 1986 Hydraulic conductivity and diffusivity: laboratory methods. *Methods of Soil Analysis: Part 1—Physical and Mineralogical Methods, (methodsofsoilan1)* **5**, 687–734.

Li, Y.-B., Fan, Y.-W., Liu, Y. & Ma, X.-Y. 2017 Influencing factors and simplified model of film hole irrigation. *Water* **9**, 543. doi:10.3390/w9070543.

Li, Q., Li, H., Zhang, L., Zhang, S. & Chen, Y. 2018 Mulching improves yield and water-use efficiency of potato cropping in China: a meta-analysis. *Field Crop Research* **221**, 50–60. doi:10.1016/j.fcr.2018.02.017.

Mualem, Y. 1976 A new model for predicting the hydraulic conductivity of unsaturated porous media. *Water Resources Research* **12** (3), 513–522.

Parihar, C. M., Nayak, H. S., Rai, V. K., Jat, S. L., Parihar, N., Aggarwal, P. & Mishra, A. K. 2019 Soil water dynamics, water productivity and radiation use efficiency of maize under multi-year conservation agriculture during contrasting rainfall events. *Field Crops Research* **241**, 107570.

Qi, Z., Feng, H., Zhao, Y., Zhang, T., Yang, A. & Zhang, Z. 2018 Spatial distribution and simulation of soil moisture and salinity under mulched drip irrigation combined with tillage in an arid saline irrigation district, northwest China. *Agricultural Water Management* **201**, 219–231. doi:10.1016/j.agwat.2017.12.032.

Qin, S., Li, S., Yang, K. & Hu, K. 2018 Can plastic mulch save water at night in irrigated croplands? *Journal of Hydrology* **564**, 667–681. doi:10.1016/j.jhydrol.2018.07.050.

Rai, V., Pramanik, P., Das, T. K., Aggarwal, P., Bhattacharyya, R., Krishnan, P. & Sehgal, V. K. 2019 Modelling soil hydrothermal regimes in pigeon pea under conservation agriculture using Hydrus-2D. *Soil and Tillage Research* **190**, 92–108.

Richards, L. A. 1931 Capillary conduction of liquids through porous mediums. *Journal of Applied Physics* **1** (5), 318–333.

Ritchie, J. T. 1972 Model for predicting evaporation from a row crop with incomplete cover. *Water Resources Research* **8**, 1204–1213.

Russel, M. B. & Richards, L. A. 1938 The determination of soil moisture energy relations by centrifugation. *Soil Science Society of America Proceedings* **3**, 65–69.

Saglam, M., Sintim, H. Y., Bary, A. I., Miles, C. A., Ghimire, S., Inglis, D. A. & Flury, M. 2017 Modeling the effect of biodegradable paper and plastic mulch on soil moisture dynamics. *Agricultural Water Management* **193**, 240–250. doi:10.1016/j.agwat.2017.08.011.

Šimůnek, J., van Genuchten, M. T. & Šejna, M. 2008 Development and applications of the HYDRUS and STANMOD software packages and related codes. *Vadose Zone Journal* **7** (2), 587–600.

Šimůnek, J., van Genuchten, M. T. & Šejna, M. 2016 Recent developments and applications of the HYDRUS computer software packages. *Vadose Zone Journal* **15**. doi:10.2136/vzj2016.04.0033.

Šimůnek, J., van Genuchten, M. T. & Šejna, M. 2018 The HYDRUS software package for simulating two- and three-dimensional movement of water, heat, and multiple solutes in variably-saturated porous media. *Technical Manual Version 3*. PC Progress, Prague, Czech Republic, pp. 274.

Sophocleous, M. 1979 Analysis of water and heat flow in unsaturated-saturated porous media. *Water Resources Research* **15** (5), 1195–1206.

Steinmetz, Z., Wollmann, C., Schaefer, M., Buchmann, C., David, J., Tröger, J., Muñoz, K., Frör, O. & Schaumann, G. E. 2016 Plastic mulching in agriculture. Trading short-term agronomic benefits for long-term soil degradation? *Science of the Total Environment* **550**, 690–705. doi:10.1016/j.scitotenv.2016.01.153.

Tian, Y., Su, D. R., Li, F. M. & Li, X. L. 2003 Effect of rainwater harvesting with ridge and furrow on yield of potato in semiarid areas. *Field Crops Research* **84**, 385–391.

van Genuchten, M. T. 1980 A closed-form equation for predicting the hydraulic conductivity of unsaturated soils. *Soil Science Society of America Journal* **44** (5), 892–898.

van Genuchten, M. T., Leij, F. J. & Yates, S. R. 1991 *The RETC Code for Quantifying the Hydraulic Functions of Unsaturated Soils. Ver. 1.0 EPA Rep. 600/2-91/065*. U.S. Salinity Lab., USDA-ARS, Riverside, CA.

Wang, Z., Zhao, X., Wu, P. & Chen, X. 2015 Effects of water limitation on yield advantage and water use in wheat (*Triticum aestivum L.*)/maize (*Zea mays* L .) strip intercropping. *European Journal Agronomy* **71**, 149–159. doi:10.1016/j.eja.2015.09.007.

Xi, B., Bloomberg, M., Watt, M. S., Wang, Y. & Jia, L. 2016 Modeling growth response to soil water availability simulated by HYDRUS for a mature triploid *Populus tomentosa* plantation located on the North China Plain. *Agricultural Water Management* **176**, 243–254.

Yang, N., Sun, Z., Feng, L., Zheng, M. & Chi, D. 2015 Plastic film mulching for water-efficient agricultural applications and degradable films materials development research. *Materials and Manufacturing Processes* 37–41. doi:10.1080/10426914.2014.930958.

Yin, W., Feng, F., Zhao, C., Yu, A., Hu, F., Chai, Q., Gan, Y. & Guo, Y. 2016 Integrated double mulching practices optimizes soil temperature and improves soil water utilization in arid environments. *International Journal of Biometeorology* **60** (9), 1423–1437. doi:10.1007/s00484-016-1134-y.

Zhang, G. S., Hu, X. B., Zhang, X. X. & Li, J. 2015 Effects of plastic mulch and crop rotation on soil physical properties in rain-fed vegetable production in the mid-Yunnan plateau, China. *Soil and Tillage Research* **145**, 111–117. doi:10.1016/j.still.2014.09.010.

Zhang, Y.-L., Feng, S.-Y., Wang, F.-X. & Binley, A. 2018 Simulation of soil water flow and heat transport in drip irrigated potato field with raised beds and full plastic-film mulch in a semiarid area. *Agricultural Water Management* **209**, 178–187. doi:10.1016/j.agwat.2018.07.021.

Zhao, H., Wang, R. Y., Ma, B. L., Xiong, Y. C., Qiang, S. C., Wang, C. L., Liu, C. A. & Li, F. M. 2014 Ridge-furrow with full plastic film mulching improves water use efficiency and tuber yields of potato in a semiarid rainfed ecosystem. *Field Crop Research* **161**, 137–148. doi:10.1016/j.fcr.2014.02.013.

Zhao, Y., Zhai, X., Wang, Z., Li, H., Jiang, R., Lee Hill, R., Si, B. & Hao, F. 2018 Simulation of soil water and heat flow in ridge cultivation with plastic film mulching system on the Chinese Loess Plateau. *Agricultural Water Management* **202**, 99–112. doi:10.1016/j.agwat.2018.02.017.

First received 29 January 2021; accepted in revised form 19 March 2021. Available online 30 March 2021

The use of the pilot points method on groundwater modelling for a degraded aquifer with limited field data: the case of Lake Karla aquifer

G. Tziatzios, P. Sidiropoulos, L. Vasiliades, A. Lyra, N. Mylopoulos and A. Loukas

ABSTRACT

Groundwater depletion poses a major threat to global groundwater resources with increasing trends due to natural and anthropogenic activities. This study presents a surface-groundwater framework for water resources modelling of ill-posed problems in hydrogeologically data-scarce areas. The proposed framework is based on the application of a conceptual water balance model and composed of surface hydrological (UTHBAL) and groundwater flow simulation with the integration of a Newton formulation of the MODFLOW-2005 code (MODFLOW-NWT) and PEST suite modules. The groundwater simulation includes a preprocessor tool for automated calibration and a post-processor tool for automated validation. The methodology was applied to a rural region of Central Greece, Lake Karla Basin, which is degraded due to groundwater resources overexploitation to cover irrigation water demands. The aquifer is modelled focusing on a precise simulation–validation procedure of the conceptual model. The groundwater model was calibrated with the calibration preprocessor tool for spatially distributed hydraulic conductivity with the pilot points method. The calibration process achieved satisfactory results as validated by the post-process analysis of observed and simulated water levels. The findings for the groundwater budget indicate that the groundwater system is still under intense pressure even though farming activity in recent years has turned to less water-intensive crops.

Key words | groundwater modelling, groundwater recharge, pilot points

G. Tziatzios (corresponding author)
P. Sidiropoulos
L. Vasiliades
A. Lyra
N. Mylopoulos
Laboratory of Hydrology and Aquatic Systems Analysis, Department of Civil Engineering, School of Engineering,
University of Thessaly,
Volos 38334,
Greece
E-mail: *getziatz@civ.uth.gr*

A. Loukas
Department of Rural and Surveying Engineering,
Aristotle University of Thessaloniki,
Thessaloniki 54124,
Greece

HIGHLIGHT

- My research deals with the pilot points method on groundwater modelling in an area with scarce hydrogeologic data.

doi: 10.2166/ws.2021.133

GRAPHICAL ABSTRACT

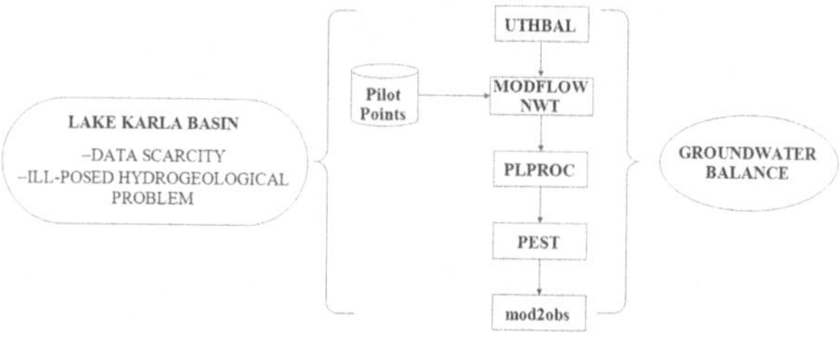

INTRODUCTION

Groundwater resources are vital in providing water for domestic, agricultural, and industrial use globally. Over two billion people rely on groundwater as their primary water source, while half or more of the irrigation water for agriculture is supplied from underground sources (Pisinaras *et al.* 2013; Famiglietti 2014). However, despite its critical importance to global water supplies, inadequate management practices are applied for monitoring groundwater resources. Furthermore, it is crucial to estimate the pressure on surface and groundwater systems due to groundwater depletion. Climate change and related changes on the components of the hydrological cycle complicate the task of sustaining groundwater supplies for the future. Hence, about 20% of the world's aquifers systems are estimated to be under over exploitation status and with increasing trends due to the climate change phenomenon and population growth (Upton *et al.* 2020).

Understanding why groundwater depletion has occurred may be the first step towards sustainable groundwater resources management. Numerical simulation models are effective and valuable tools for computing water budgets of regional groundwater systems and the planning and policy making for groundwater resources management (Konikow & Kendy 2005). The analysis of groundwater system through simulation is the main process before a researcher/modeller deals with the management of groundwater resources (Doherty & Hunt 2010). The simulation problems in groundwater hydrology are part of the complex environmental issues that are difficult to solve. These problems in groundwater modelling become extremely difficult when in addition to the lack of field data problems also meet a lack of quality data (Singh 2014).

One of the most important issues for water resources management is developing strategies for groundwater modelling that are adaptable to data scarcity and field measurements and so begins the difficulty of the inverse problem. According to Zhou and her associates (2014) many approaches have been proposed to resolve the inverse problem in hydrogeological modelling. Inverse methods are categorized as direct and indirect ones. The inverse problem in hydrogeological fields is difficult to solve because it is ill posed and the calibration parameters are more than the observations (Anderson *et al.* 2015). Common indirect methods used to solve the inverse problem are the geostatistical approach, the maximum likelihood method where zonation is employed, and pilot points. The most advanced methods where the stochastic approach is introduced are the self-calibrated method, the Markov chain Monte Carlo method, the gradual deformation method, the probability perturbation method, and the ensemble Kalman filter. The geostatistical approach has several advantages, however it is not appropriate for use in the inverse model where the hydraulic conductivity variable poses spatial distributions with significant heterogeneity, forcing the transition from linearized to non-linearized approaches (Zhou *et al.* 2014).

Zonation assumes an abrupt spatial change in parameter values, whereas the pilot points method produces smoothly distributed parameters and flexibility compared to the zonation approach (Anderson *et al.* 2015; Rajabi *et al.* 2018). These strategies are particularly important in arid and semi-arid areas where access to data is poor and data collection is difficult, such as the Lake Karla aquifer,

Greece. This study analyzes the groundwater resources status of the Lake Karla aquifer focusing on a precise simulation–validation procedure of the groundwater model. A surface hydrological simulation tool was applied to calculate the groundwater recharge while the MODFLOW with a Newton formulation was employed to simulate groundwater flow. The groundwater model was calibrated for spatial distributed hydraulic conductivity based on the pilot points method, since there are limited data for this parameter.

This paper presents a new framework for calibrating and validating the groundwater flow simulation applying two automated tools with the use of pilot points. The first relates to calibration post-process analysis and is called PLPROC and the second one deals with the post-process analysis of the results and is called mod2obs (Doherty 2016, 2018). Both tools are part of the PEST: Model-Independent Parameter Estimation and Uncertainty Analysis, an open-source, public-domain software suite (Doherty 2020a, 2020b). The PLPROC tool has been applied in the Precipice Sandstone aquifer system of the Surat Basin, Great Artesian Basin, in Australia, by Hayes *et al.* (2020). The overall purpose of this paper is to provide a useful framework in surface and groundwater modelling dealing with scarce hydrogeological information in rural areas.

Study area

The study area of about 500 km^2 is located at Eastern Thessaly, Greece and occupies the plain zone of Lake Karla watershed (Figure 1). The mean annual values of precipitation, temperature and potential evapotranspiration of the region are 560 mm, 14 °C and 775 mm respectively (Mylopoulos & Sidiropoulos 2016). The grainy, phreatic aquifer consists of alluvial deposits (Quaternary formations). The thickness varies in places reaching its maximum value of 200 m at the centre of study area. The presence of Mavrovounio Mountain at the east, where low permeable bedrocks prevail at the borderline with the lowland creates a no flow boundary. At the west, the permeable formations of Chalkodonio Mountain (south-west) and of the plain (west north) allow a hydraulic connection of adjacent aquifers with Lake Karla aquifer. The main crops are wheat, cotton, alfalfa, maize, vegetables, and trees. Lake Karla's aquifer is a characteristic example of an agricultural area in Greek territory, where surface water systems are limited and have relatively small capacity, in contrast to the great irrigation water demands imposed by the intensive agricultural activity of the area. As a result, the largest part of the region's water needs for irrigation is covered by groundwater with the use of pumping wells, most of them being private wells (Mylopoulos & Sidiropoulos 2016).

These agricultural practices and the absence of collective irrigation networks result in pumping non-renewable groundwater resources leading to groundwater depletion. Agriculture is the primary income source for most of the local population and for decades agricultural production in the region was based on subsidies from the Common Agriculture Policy (CAP) of the European Union. A typical example is the cultivation of cotton, a high-water-demanding crop which, for years, is based on the provided subsidies without considering sustainable management of groundwater resources (Alamanos *et al.* 2020). Previous studies have shown that the aquifer is in over exploitation condition and effective restoration practices are needed for sustainable water resources management (Sidiropoulos *et al.* 2015, 2019).

METHODS AND APPLICATION

The methodology is described in Figure 2 with the use of a flowchart. The first step is running the UTHBAL surface hydrology tool (Loukas *et al.* 2007). One of the outcomes of UTHBAL is groundwater recharge that is used as input parameter in the MODFLOW_NWT code (Niswonger *et al.* 2011). The preprocessor PLPROC was used during the parameterization while the optimization process follows the PEST methodology. Because the inverse problem is an ill-posed problem the regularization mode is used to prevent overfitting. Finally, the objective function results are examined. If the objective function values show a clear decrease, then post-process analysis follows (mod2obs). When there is no observed decrease, we can conclude that the groundwater hydrology model is unstable and needs to be re-evaluated.

The most common statistical criteria for the performance of a simulation model/tool are the Mean Absolute Error (MAE), Root Mean Square Error (RMSE) and for transient conditions is the Nash–Sutcliffe Efficiency (E_f). An

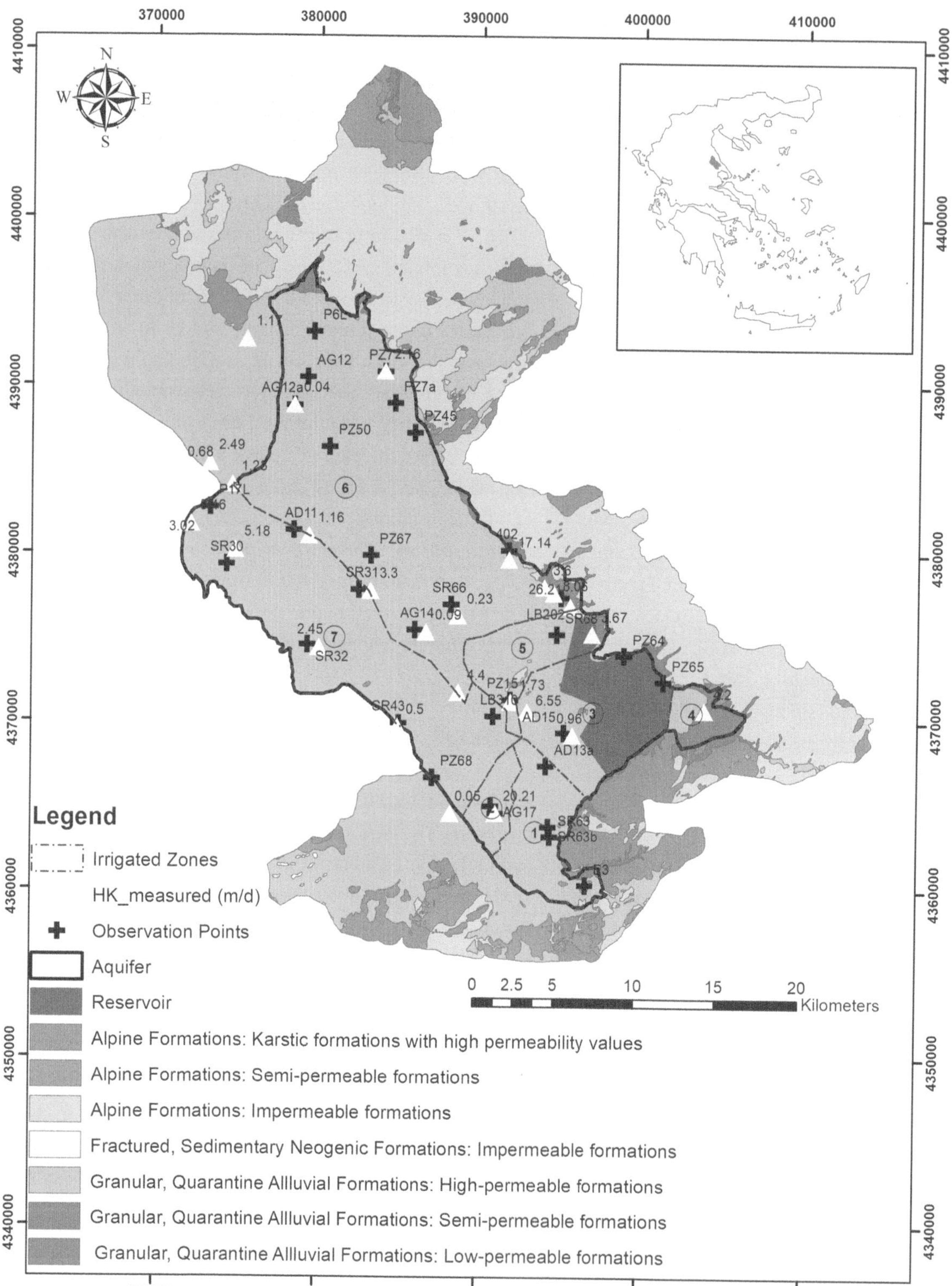

Figure 1 | The study area map indicating the geological formation, the irrigation zones, the hydraulic heads observation points and the hydraulic conductivity sample values locations.

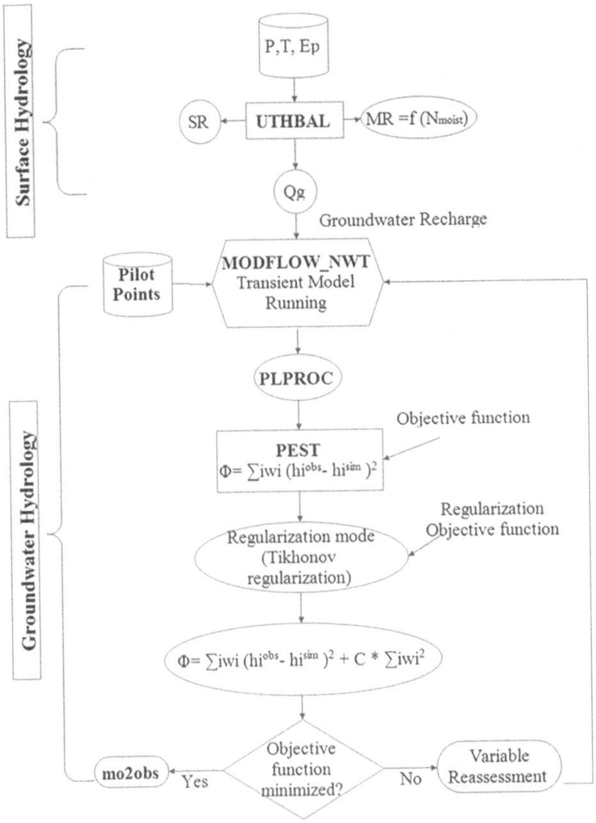

Figure 2 | Flow chart of the methodology.

acceptable criterion for the successful completion of calibration is a RMSE deviance of 10% of the water levels' total range. MAE is more robust in estimating the effects of outlier values of the spatial distribution than the RMSE, and therefore it is usually smaller than RMSE. However, if the calibration procedure should be continued, depends on score of the Nash–Sutcliffe Efficiency (E_f) (Anderson *et al.* 2015).

Estimation of groundwater recharge and irrigation demands covered by the groundwater system

The surface hydrological model (UTHBAL), is a surface hydrological simulation tool that gives a representation of the rainfall–runoff interactions of a catchment area. UTHBAL uses the components of the hydrological cycle, monthly time series of precipitation (P), mean temperature (T) and potential evapotranspiration (Ep) as inputs. The surface hydrological simulation tool disaggregates the total precipitation into rainfall and snowfall. UTHBAL distinguishes the total watershed runoff (Qsim), between the surface runoff (SR), the interflow (MR), and the baseflow (Qg) utilizing a soil moisture mechanism (Nmoist) and finally estimates the groundwater recharge to the subsurface systems (Loukas *et al.* 2007). In this study, UTHBAL has been applied in semi-distributed mode because the aquifer system covers the lower zone of the Karla watershed (e.g. Sidiropoulos *et al.* 2019). The groundwater recharge rate varied for each stress period.

The annual crop data from 2001 to 2016 come from the Integrated Administration and Control System (IACS), which is maintained by the Greek Payment Authority of Common Agricultural Policy (CAP) Aid Schemes (OPEKEPE, online). The crop pattern changes annually and the irrigation demands were estimated for every month from 2001 to 2016 with Near Irrigation Requirement (NIR). Hence, the pumping rates varied for the dry/pumping period.

Table 1 presents the estimation of irrigation water demands covered by groundwater extraction for the period 2001–2016. Table 2 shows the pumping rates for each zone and the respective irrigation area. It is worth noting that although the reservoir exists at present, it does not participate in the current study. The former natural Lake Karla had been drained in the 1960s, causing serious environmental problems in the area. A reservoir for the restoration of Lake Karla has been recently constructed to reverse the bad environmental status. The Lake Karla reservoir project construction has been completed in 2018 (Sidiropoulos *et al.* 2021). Thus, its contribution to the groundwater system is not considered during the simulation.

Groundwater simulation

Groundwater flow simulation is achieved with the use of the MODFLOW_NWT code through the commercial GMS graphic user interface (Aquaveo 2019). MODFLOW_NWT code is a Newton formulation for MODFLOW-2005 (Niswonger *et al.* 2011). A no flow boundary was used in the east part to simulate the low permeable conditions of the aquifer. While in the west, general-head boundary was used to simulate the hydraulic connection of the aquifer with the adjacent ones. Conductance was estimated by the comparison of the observation wells on either side of the

Table 1 | Irrigation demands of each zone covered by the groundwater system

	Zones (hm^3)						
Years	**1**	**2**	**3**	**4**	**5**	**6**	**7**
2001	25.23	10.19	16.36	2.40	19.41	173.17	120.79
2002	18.81	7.46	11.63	2.50	14.16	99.42	81.44
2003	19.31	7.94	12.48	3.55	14.46	105.36	81.79
2004	19.63	8.02	13.09	6.55	14.64	99.14	79.55
2005	18.67	7.68	14.36	7.12	16.04	109.54	87.32
2006	20.59	9.12	12.11	5.14	13.75	123.95	84.92
2007	21.54	9.53	14.26	6.76	16.41	72.40	103.28
2008	13.40	10.03	13.96	5.01	16.99	149.72	105.03
2009	18.58	8.86	13.11	4.73	16.67	128.99	89.61
2010	21.99	10.26	15.28	6.09	18.92	150.59	105.66
2011	19.68	9.35	12.58	4.76	18.63	127.68	90.31
2012	18.23	9.44	14.07	5.04	6.17	117.62	97.35
2013	17.93	8.49	13.57	4.07	13.51	151.78	77.50
2014	20.15	7.12	13.65	4.56	15.89	134.00	93.16
2015	17.10	6.62	14.32	3.81	13.59	113.81	78.34
2016	19.93	8.49	17.05	4.55	16.31	136.55	92.87

Table 2 | Pumping rates of each zone and the respective irrigation area in hectares

Zones	Area (ha)	Q (m^3/h)
1	3,600	60–120
2	1,490	35–100
3	4,900	70–207
4	1,400	105–250
5	3,000	55–160
6	21,000	40–120
7	14,400	35–190

border ranging from 15 to 40 m^2/d. The aquifer is represented as a single layer and the resulting grid consists of 12,500 active cells with 200 m side each. Grid refinement is observed at the wells' locations. This study examines the groundwater resources status extending the time series to November 2018 and applying the MODFLOW_NWT code instead of MODFLOW 2000. A transient groundwater flow model was developed and simulated from 1987 to 2018 with monthly changing irrigation needs from 2001 to 2016 also taking under consideration the annually different crop pattern application. Additionally, the groundwater flow was simulated with varying timesteps in accordance with the exact dates of water level observations. MODFLOW_NWT was chosen because the implemented method does not deactivate dry cells, but instead uses an upstream weighting approach to limit flow from each dry cell to an adjacent, partially saturated cell. It also uses additional smoothing for conductance and storage functions which allows the Newton numerical method application.

Investigation of the hydrogeology environment and application of a subsurface flow model

The development and application of a simulation model relies heavily on knowledge of the dominant hydrogeology conditions in the study area, the initial piezometry, the boundary conditions, the grid geometry, the trends in the groundwater levels and the assessment of hydraulic properties (Sundell *et al.* 2019). Once a groundwater model is developed and well calibrated using a technically sound hydrogeological structure for pre-process and post-process, its outcome estimates the groundwater budget on a regional scale and can be used to analyze future water management scenarios (Konikow & Kendy 2005).

Calibration and validation process

The calibration process attempts to bring the model response as close to the observed ones as possible. In the first stage, a trial-and-error method is employed whereas, in the next stage, the pilot points approach was chosen. The pilot point approach is followed due to the lack of field measurements of hydraulic conductivity. Figure 1 indicates that there are only 32 measurements of hydraulic conductivity in the study area which is not enough to correctly express the heterogeneity of such a complex aquifer system. Pilot points are used to express the heterogeneity of hydraulic variables in groundwater systems naturally without the problem of numerical instability in fields with limited measurements data (Doherty & Hunt 2010; Sundell *et al.* 2019).

As a general approach in groundwater modelling pilot points are located between the observation boreholes and the pumping wells in the direction of groundwater flow (Doherty & Hunt 2010; Baalousha *et al.* 2019). In this paper, spatial variations of hydraulic conductivity parameter were estimated using pilot points. The pilot points placement can be seen in detail in Figure 3(a) whereas Figure 3(b) is an illustration of groundwater flow. In our simulations 254 pilot points were not randomly chosen but manually set between the observation boreholes and pumping wells. However, it is worth noting that increasing the number of pilot points in the model lengthens the simulation duration. Specific yield value was set as 0.001 corresponding to the average value between fine sand and clay, according to (Johnson 1967). Hydraulic head values were calculated by following the inverse calibration process based on hydraulic conductivity variations.

The parameter estimation software suite (PEST) (Doherty 2020a, 2020b) is well known and used extensively in calibration of complex environmental systems analysis such as groundwater systems. Simulation tools in PEST have been developed to estimate appropriate values for pilot points for the model calibration using interpolation to the model grid. One such tool is the PLPROC preprocessor that interpolates pilot points to a MODFLOW grid and exports them to MODFLOW arrays. PLPROC runs as part of the 'model domain' and alters pilot point values during the calibration process using the Kriging interpolation method (Doherty 2016). In this study, the ordinary Kriging option was followed with no anisotropy, no nugget, and a sill of 1 log cycle. The main advantage of the PLPROC pre-processor choice is that it requires no additional implementation. However, the tool requires

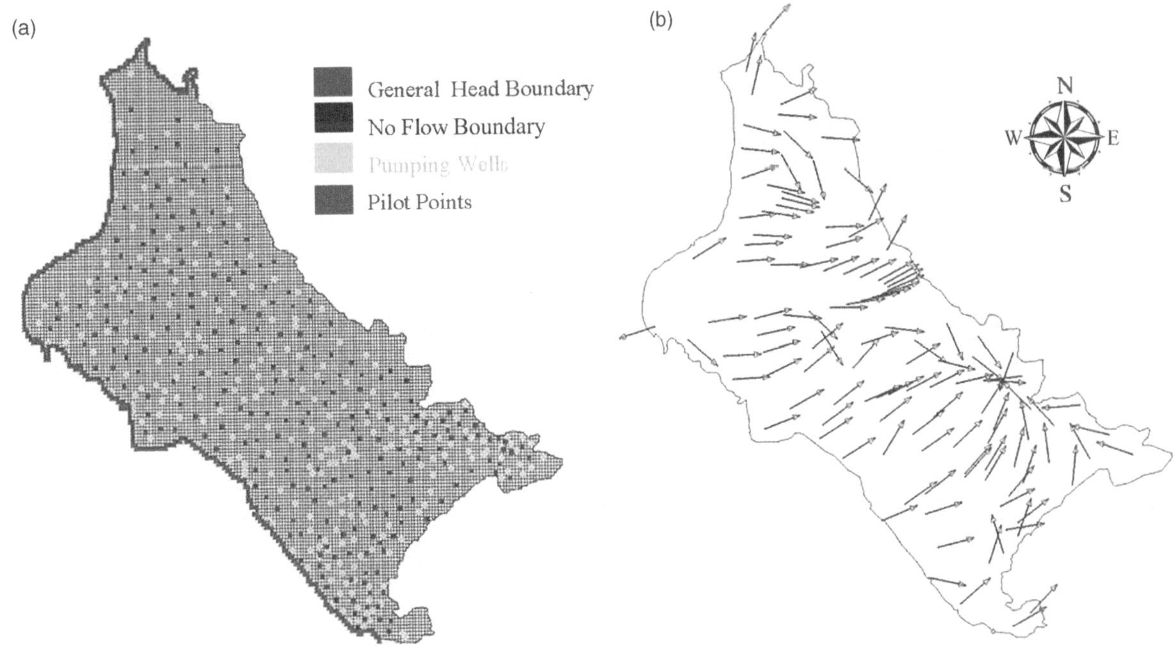

Figure 3 | Conceptual model and the location of pilot points (a) and the groundwater flow (b).

input files to be in a predefined format (Doherty 2016). Upon completion of the calibration preprocess, the template files of hydraulic conductivity are created automatically and then used during the PEST run. PEST tool is used in the model optimization process in the regularization mode. The general form of the measurement objective function (Equation (1)) is shown in Figure 2

$$\Phi = \sum_i w_i (h_i^{\text{obs}} - h_i^{\text{sim}})^2 \tag{1}$$

where (h^{obs} indicates the observation hydraulic head, h^{sim} is the simulated hydraulic head, and w the weight that has been applied to the measurement. In this study, we followed the same process as Baalousha *et al.* (2019), where an equal weight of 1 have been assigned for all the observation points as all have the same importance. As the authors mention regularization mode should be used when the inverse problem is ill posed (Doherty 2020a, 2020b).

In the presented experiments the ill-posed inverse problem was solved with the application of Tikhonov regularization method. The study uses 32 observation wells monitoring the groundwater status during 1987 (Initial Conditions), while the calibration and the validation process was conducted using datasets from 10 observation wells that monitored the groundwater status during 2015 and 2018, respectively.

Post-processing analysis

Post-processing data were analyzed using mod2obs (Doherty 2018), a PEST individual module that compares the simulated hydraulic head with the calibration target. Mod2obs converts the raw model outputs to the model generated observations. Mod2obs works like most programs provided by the PEST suite. The user needs to have prepared the dimension files of the model grid and the coordinates file of the observation points. Then mod2obs outputs a bore sample file of the model generated heads over time, interpolated to the sites and times in the borehole sample file (Doherty 2018).

RESULTS

The model calibration was carried out based on hydraulic head data for the period between 2/11/1987 and 12/11/2015. Data were validated for the period between

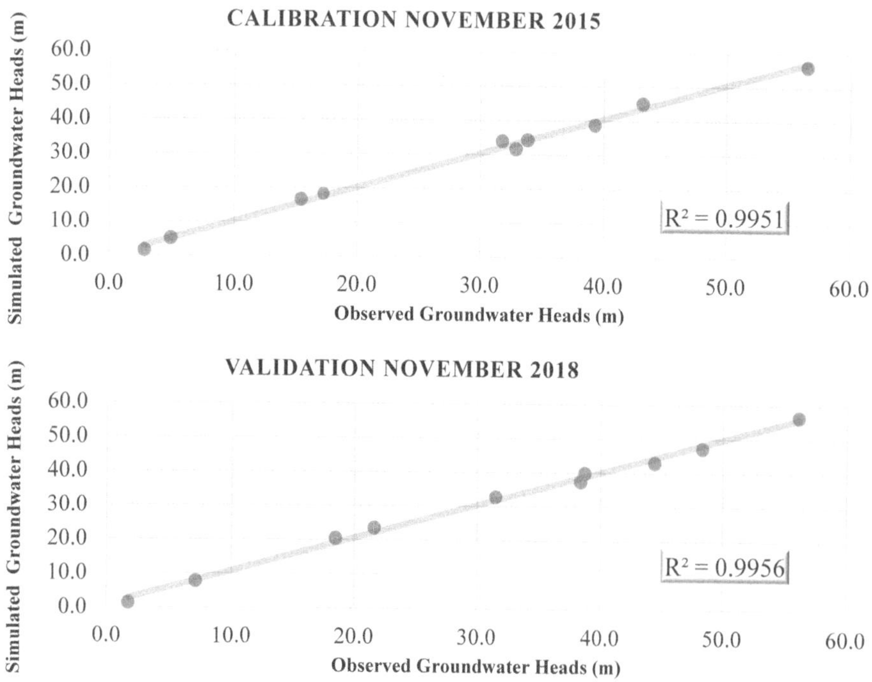

Figure 4 | Scatter plot of observed versus simulated groundwater levels in the study area for the calibration (a) and the validation process (b).

12/11/2015 and 11/22/2018. The calibration target was 12/11/2015 and the validation target was 11/22/2018. Scatter plot of observed versus simulated groundwater levels in the study area for the calibration process and the validation process are presented in Figure 4(a) and 4(b), respectively. The parameters which show the goodness-of-fit parameters are the Nash–Sutcliffe Model Efficiency (E_f) (Equation (2)), the RMSE (Equation (3)), and the Coefficient of Determination (R^2). The statistical results are presented in Table 3:

$$E_f = 1 - \frac{\sum_{t=1}^{T} (Y_m^t - Y_o^t)^2}{\sum_{t=1}^{T} (Y_0^t - \overline{Y_O})^2} \quad (2)$$

where, $\overline{Y_O}$ is the mean observed value, Y_m is the simulated value and Y_o is the observed one at time t:

$$RMSE = \sqrt{\frac{\sum_{i=1}^{n} (Y_{mi} - Y_{oi})^2}{n}} \quad (3)$$

where, Y_m is the simulated value, Y_o is the observed one, i is the code number of observation points, and n is the number of observation points.

Figures 5–7 demonstrate two types of results. The calibration and the validation using the pilot points

Table 3 | Statistical results for the calibration and the validation period

Statistical parameters	Calibration period	Validation period
MAE [m]	1.14	1.11
RMSE [m]	1.21	1.22
E_f [1]	0.99	0.99

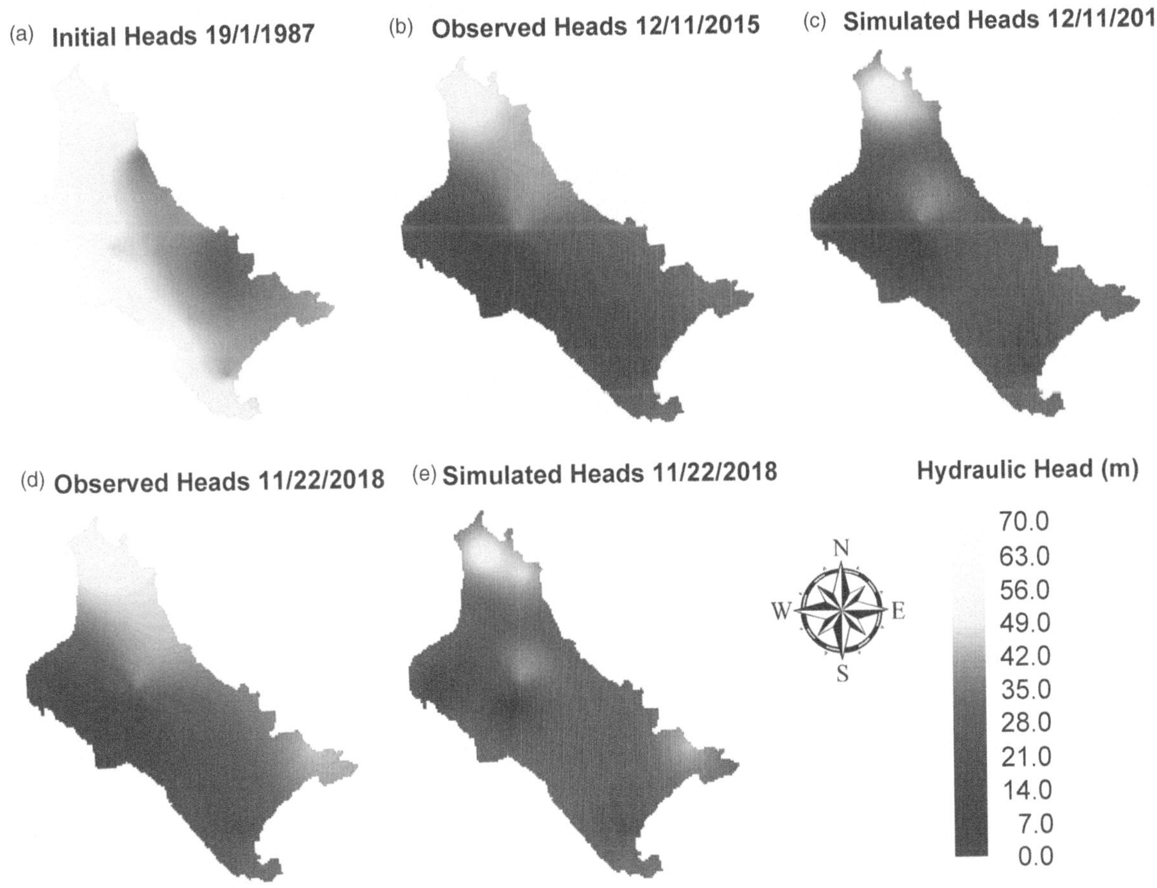

Figure 5 | The groundwater trends during the simulation period.

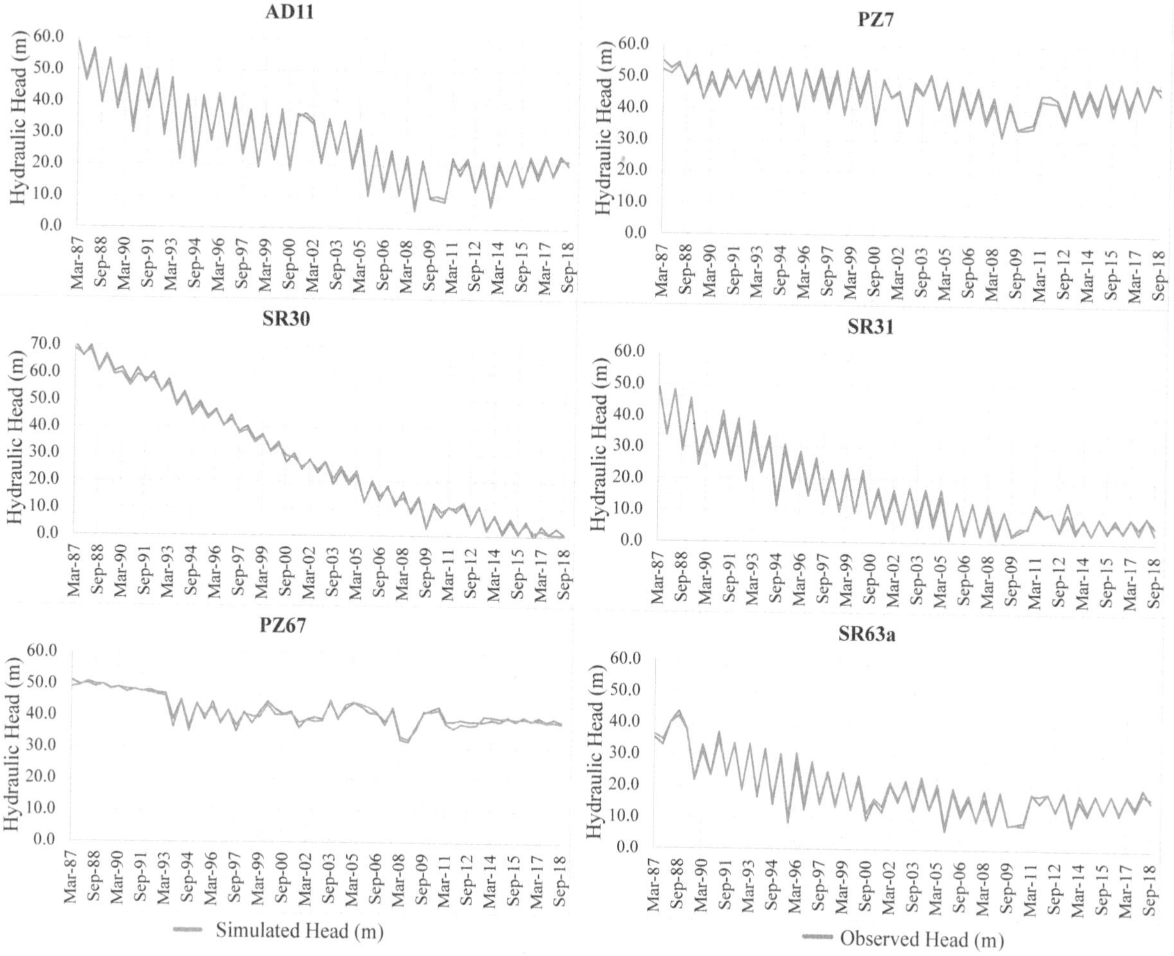

Figure 6 | Groundwater trends during the end of wet and the end of the dry period.

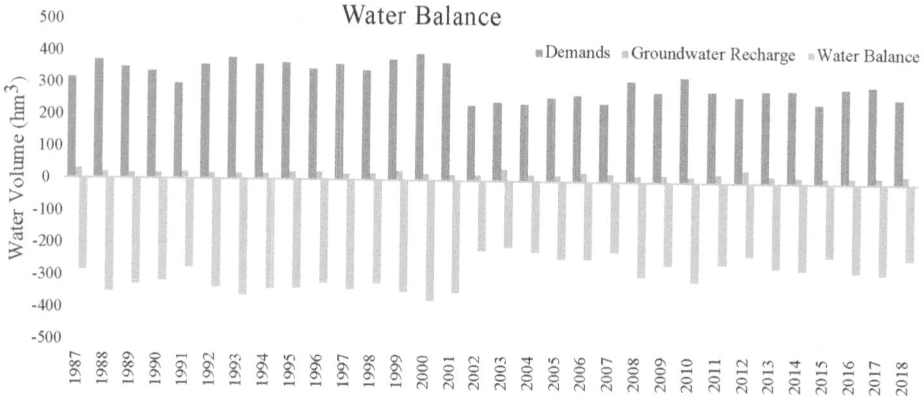

Figure 7 | The overall groundwater balance for the simulation period.

methodology that show satisfactorily results whereas a high drawdown in hydraulic heads is observed. The above findings strongly suggest and verify the presence of groundwater depletion. The phenomenon can also be observed when examining individually boreholes AD11, SR30 and SR31 which are in the central part of the study area individually including SR63a which is located in the southwest part of the study area (Figure 1). The exact drop figures for each borehole can be seen in the Table 4.

Figure 6 shows the results achieved through calibration and validation process where the observed heads converge with the simulated ones before the pumping period and at the end of the pumping period. Large groundwater fluctuations in hydraulic heads are observed between the wet and dry periods. Statistical values for each borehole are shown in the Table 5.

Figure 7 shows the groundwater balance of the area for the period 1987–2018. The groundwater recharge variable is the sum of deep infiltration variable simulated by the UTHBAL model plus the returned irrigation flow was approximately accounted as 10% of the irrigation requirements. The groundwater balance is negative throughout the years, indicating the intensive groundwater abstraction over the region's capacity to replenish the outflows with adequate recharge to minimize the deficit water balance.

DISCUSSION

The simulation of the subsurface system of Lake Karla watershed has been presented in the past by Sidiropoulos *et al.* (2015, 2019). Sidiropoulos *et al.* (2015, 2019) used the MODFLOW 2000 code and 15 observation points for hydraulic conductivity parameter applying the geostatistical approach and the model ran from 1987 to 2009. In the current study the model was expanded forward to simulate the groundwater flow from 1987 to 2018 using an updated code, the MODFLOW_NWT mode, which resolves the problem of dry cells, uses 254 pilot points, and updates the groundwater pumping data and observation during the simulation, and utilizes the automated calibration–validation tools PLPROC and mod2obs. The methodology of pilot points deals with the data scarcity, the MODFLOW_NWT does not deactivate the model cells with low water level resulting in the simulation of groundwater flow, without the numerical error that is inserted with the assumption of dry cells by the MODFLOW 2000 code. In addition to the simulation of the groundwater flow and the evolution of all simulation and analysis tools that occur over the years, it is important to mention the difference in the groundwater levels. The simulated water levels of the historical period of 1987–2009 agree well with the simulation results of the previous studies. Furthermore, the groundwater balance indicated an improving trend during the period 2009–2018 but it remains under pressure and intensive and extensive groundwater abstractions. The varying timesteps of the simulation according to the varying dates of groundwater measurements depict clearly the actual and real conditions of the groundwater with increased timely and spatial accuracy.

Table 4 | Level drop for each borehole

Boreholes	Level drop
AD11	40 m
SR30	70 m
SR31	50 m
SR63a	20 m

Table 5 | Statistical results for each borehole

Boreholes	RMSE	MAE
PZ7	1.4 m	1.22 m
PZ67	1.24 m	1.07 m
SR30	1.24 m	1.07 m
SR31	1.39 m	1.12 m
SR63	1.21 m	0.99 m

CONCLUSIONS

This study analyzes the groundwater resources status of the Lake Karla aquifer focusing on a precise simulation–validation procedure of groundwater model based upon new and innovative methodologies. The simulated annual decrease in groundwater levels verify the groundwater depletion and shows that the groundwater system is still under intense pressure even though farming activity in recent years has turned to less water-intensive crops. Moreover, the enduring water balance deficit indicates the necessity for the construction of collective irrigation networks that will eliminate the extravagant number of private own wells, and the need for sustainable water resources management. Unless appropriate measures and policies are implemented, the Lake Karla's aquifer system may face the problem of water scarcity.

This article presents a method that uses a configurable number of pilot points to calibrate and subsequently analyze the groundwater flow system. The use of the Newton smoothing technique in solving the inverse problem with the use of pilot points ensures stability in hydraulic conductivity parameter values and the other hydrogeological variables for Lake Karla's aquifer. Furthermore, the use of tools like PLPROC and mod2obs contribute significantly to the creation of a realistic image of the hydrogeological environment of the Karla aquifer. The pilot points approach is particularly important in data sparse areas of hydrogeological information. Hence, the proposed approach could be used in conjunction with stochastic field generation techniques for assessing model uncertainty and especially in contaminant fate and transport modelling studies.

DATA AVAILABILITY STATEMENT

All relevant data are included in the paper or its Supplementary Information.

REFERENCES

Alamanos, A., Latinopoulos, D., Loukas, A. & Mylopoulos, N. 2020 Comparing two hydro-economic approaches for multi-objective agricultural water resources planning. *Water Resour. Manage.* **34** (14), 4511–4526. http://doi.org/10.1007/s11269-020-02690-6.

Anderson, M. P., Woessner, W. W. & Hunt, R. J. 2015 *Applied Groundwater Modeling: Simulation of Flow and Advective Transport*, 2nd edn. Academic Press, London, UK; San Diego, CA, USA.

Aquaveo 2019 *Groundwater Modeling System Version 10.4 (Version 10.4)*. Aquaveo, Provo, UT. Available from: http://www.aquaveo.com/software.

Baalousha, H. M., Fahs, M., Ramasomanana, F. & Younes, A. 2019 Effect of pilot-points location on model calibration: application to the northern karst aquifer of Qatar. *Water* **11** (4), 679. https://doi.org/10.3390/w11040679.

Doherty, J. E. 2016 *PLPROC: A Parameter List Processor*. Watermark, Brisbane, Australia, p. 284.

Doherty, J. E. 2018 *Groundwater Data Utilities – Part B, Program Descriptions*. Watermark Numerical Computing, Brisbane, Australia, p. 387.

Doherty, J. E. 2020a *Model-Independent Parameter Estimation User Manual Part II: PEST Utility Support Software*. Watermark, Brisbane, Australia, p. 268.

Doherty, J. E. 2020b *PEST: Model-Independent Parameter Estimation and Uncertainty Analysis–User Manual Part I*. Watermark, Brisbane, Australia, p. 394.

Doherty, J. E. & Hunt, R. J. 2010 *Approaches to Highly Parameterized Inversion – A Guide to Using PEST for Groundwater-Model Calibration: U.S. Geological Survey Scientific Investigations Report 2010–5169*, p. 59. https://doi.org/10.3133/sir20105169.

Famiglietti, J. S. 2014 The global groundwater crisis. *Nat. Clim. Change* **4** (11), 945–948. https://doi.org/10.1038/nclimate2425.

Hayes, P., Nicol, C., La Croix, A. D., Pearce, J., Gonzalez, S., Wang, J., Ahmed, H., He, J., Moser, A., Helm, L., Morris, R. & Gornall, D. 2020 Enhancing geological and hydrogeological understanding of the Precipice Sandstone aquifer of the Surat Basin, Great Artesian Basin, Australia, through model inversion of managed aquifer recharge datasets. *Hydrogeol. J.* **28**, 175–192. https://doi.org/10.1007/s10040-019-02079-9.

Johnson, A. I. 1967 *Specific Yield – Compilation of Specific Yields for Various Materials*. U.S. Geological Survey, Water Supply Paper 1662-D, Washington, DC. https://doi.org/10.3133/wsp1662D.

Konikow, L. F. & Kendy, E. 2005 Groundwater depletion: a global problem. *Hydrogeol. J.* **13**, 317–320. https://doi.org/10.1007/s10040-004-0411-8.

Loukas, A., Mylopoulos, N. & Vasiliades, L. 2007 A modeling system for the evaluation of water resources management strategies in Thessaly, Greece. *Water Resour. Manage.* **21**, 1673–1702. https://doi.org/10.1007/s11269-006-9120-5.

Mylopoulos, N. & Sidiropoulos, P. 2016 A stochastic optimization framework for the restoration of an over-exploited aquifer. *Hydrol. Sci. J.* **61** (9), 1691–1706. https://doi.org/10.1080/02626667.2014.993646.

Niswonger, R. G., Panday, S. & Ibaraki, M. 2011 *MODFLOW-NWT, A Newton Formulation for MODFLOW-2005: U.S. Geological Survey Techniques and Methods 6–A37*, p. 44.

OPEKEPE. Integrated Administration and Control System (IACS). *Crop Data 2001–2016*. Ministry of Rural Development and Food, Athens, Greece. Available from http://www.opekepe.gr.

Pisinaras, V., Petalas, C., Tsihrintzis, V. A. & Karatzas, G. P. 2013 Integrated modeling as a decision-aiding tool for groundwater management in a Mediterranean agricultural watershed. *Hydrol. Processes* **27** (14), 1973–1987. https://doi.org/10.1002/hyp.9331.

Rajabi, M. M., Ataie-Ashtiani, B. & Simmons, C. T. 2018 Model-data interaction in groundwater studies: review of methods, applications, and future directions. *J. Hydrol.* **567**, 457–477. https://doi.org/10.1016/j.jhydrol.2018.09.053.

Sidiropoulos, P., Mylopoulos, N. & Loukas, A. 2015 Stochastic simulation and management of an over-exploited aquifer using an integrated modeling system. *Water Resour. Manage.* **29**, 929–943. https://doi.org10.1007/s11269-014-0852-3.

Sidiropoulos, P., Tziatzios, G., Vasiliades, L., Mylopoulos, N. & Loukas, A. 2019 Groundwater nitrate contamination integrated modeling for climate and water resources scenarios: the case of Lake Karla over-exploited aquifer. *Water* **11** (6), 1201. https://doi.org/10.3390/w11061201.

Sidiropoulos, P., Dalezios, N. R., Loukas, A., Mylopoulos, N., Spiliotopoulos, M., Faraslis, I. N., Alpanakis, N. & Sakellariou, S. 2021 Quantitative classification of desertification severity for degraded aquifer based on remotely sensed drought assessment. *Hydrology* **8** (1), 47. https://doi.org/10.3390/hydrology8010047.

Singh, A. 2014 Groundwater resources management through the applications of simulation modeling: a review. *Sci. Total Environ.* **499**, 414–423. https://doi.org/10.1016/j.scitotenv.2014.05.048.

Sundell, J., Norberg, T., Haaf, E. & Rósen, L. 2019 Economic valuation of hydrogeological information when managing groundwater drawdown. *Hydrogeol. J.* **27**, 1111–1130. https://doi.org/10.1007/s10040-018-1906-z.

Upton, K. A., Jackson, C. R., Butler, A. P. & Jones, M. A. 2020 An integrated modelling approach for assessing the effect of multiscale complexity on groundwater source yields. *J. Hydrol.* **588**, 125113. https://doi.org/10.1016/j.jhydrol.2020.125113.

Zhou, H., Gómez-Hernández, J. & Li, L. 2014 Inverse methods in hydrogeology: evolution and recent trends. *Adv. Water Resour.* **63**, 22–37. https://doi.org/10.1016/j.advwatres.2013.10.014.

First received 31 January 2021; accepted in revised form 25 April 2021. Available online 6 May 2021

doi: 10.2166/ws.2021.217

Water and soil environmental vulnerability of artificial oases in arid areas and its temporal and spatial differentiation and evolution

Cundong Xu[a,b,c], Zijin Liu[a,b,c,*], Xinglin Zhu[a,b,c], Junjiao Tian[a,b,c], Fengyou Gu[a,b,c] and Yan Wang[a,b,c]
[a] School of Water Resources, North China University of Water Resources and Electric Power, Zhengzhou 450046, China
[b] Key Laboratory for Technology in Rural Water Management of Zhejiang Province, Hangzhou 310018, China
[c] Henan Provincial Hydraulic Structure Safety Engineering Research Center, Zhengzhou 450046, China
*Corresponding author. E-mail: 1556135694@qq.com

ABSTRACT

The water and soil environmental vulnerability in arid areas is mainly affected by vegetation, hydrology, terrain, and climate. To accurately evaluate the water and soil environmental vulnerability and its evolution in the Jingtaichuan Pumping Irrigation District in China, this paper, taking 1994, 2006, and 2018 as typical years, selects 13 index factors that directly or indirectly drive the water and soil environmental evolution in this area, and adopts the cloud theory and analytic hierarchy process to determine the weight of each index factor. Spatial analysis technique and supervised classification were used to obtain and standardize the spatial distribution raster maps of each index factor. The multi-source data fusion was performed according to the index weight to analyze the evolution characteristics of soil and water environmental vulnerability. The results showed that soil salinity, vegetation coverage, and land use type have a significant impact on the water and soil environmental vulnerability in this irrigation area, and most of the area is at mild risk. High risk mainly occurs in closed hydrological units in the eastern part of the irrigation area. From 1994 to 2018, the evolution process can be divided into two stages, namely the 'environmental deterioration stage' and the 'environmental restoration stage'. The results showed that the water and soil environment of the irrigation area is developing in a healthy way.

Key words: cloud theory, Jingdian Irrigation District, remote sensing, vulnerability, water and soil environment

HIGHLIGHTS

- It fused spatial data with long series monitoring data to analyze the vulnerability of soil and water environment in typical dryland irrigation areas.
- A hierarchical analysis method improved by introducing cloud theory determined the weights of each index factor.
- Soil salinity, vegetation cover and land use type are the key factors affecting the vulnerability of soil and water environment in arid irrigation areas.

1. INTRODUCTION

Water and soil environment is an indispensable carrier for organisms to carry out life activities. It is an environmental foundation necessary for the development of agriculture and economy in a region. The water and soil environmental vulnerability is an important indicator to measure the degradation of the water and soil environment and the level of the ecological restoration technology (Braimoh and Vlek 2004; Wang *et al.* 2012; Biazin and Sterk 2013). Previous studies revealed that the vulnerability of the soil and water environment in some areas has undergone fundamental changes (Setegn *et al.* 2009; Angelovicová & Fazekašová 2014; Lamsal *et al.* 2017; Felix *et al.* 2019; Xu *et al.* 2019). These changes are mainly caused by vegetation, hydrology, terrain, and climate, with obvious manifestations. However, the internal evolution and change often take place slowly. Therefore, it is necessary to reveal the influencing process and driving mechanism of various factors on the water and soil environmental vulnerability at the spatial and temporal scale through field experiments, long-term monitoring data, and space remote sensing data, which can contribute to the study of the evolution of water and soil environment in arid areas.

The research on the vulnerability of regional water and soil environment at home and abroad has achieved many important results. For example, Rosa *et al.* (2000) analyzed the vulnerability of soil erosion in Western Europe and its hidden impact on crop productivity based on the ImpelERO model; Qing *et al.* (2016) took the Qiandao Lake as the study area, integrated the fuzzy analytic hierarchy process (FAHP) and geographic information system (GIS), and analyzed the changes in environmental vulnerability of the Qiandao Lake area from 1991 to 2010; Stampoulis *et al.* (2016) analyzed the water ecological vulnerability of different hydrogeological units in East Africa under extreme hydrological conditions using the WindSat microwave radiation measuring technique, to explore the impact of the dynamic hydrologic cycle on the ecological environment of water-deficient areas; Xu *et al.* (2011) made a systematic evaluation on the environmental vulnerability of the Pearl River Delta region using the analytic hierarchy process (AHP) and GIS based on remote sensing data. These results lay a foundation for the study of regional water and soil environment, but there are some deficiencies. In terms of research content: on the one hand, the cost of investigation poses a certain limitation to the selection of evaluation index factors, and the single or subjective method used to determine the weight of each index may have a great impact on the objectivity and accuracy of the environmental vulnerability evaluation results, which makes it difficult to accurately assess the water and soil environmental vulnerability under the combined effect of multiple factors. On the other hand, different study areas have different characteristics. The current research on environmental vulnerability evaluation mainly uses administrative regions, such as provinces, cities, and counties as evaluation units, and fewer studies are about the long-term water and soil environmental vulnerability with the study area spatially divided into grid units. In terms of research methods: on the one hand, it cannot describe the randomness and fuzziness of different environmental factors in driving the evolution of the water and soil environment in the irrigation area. On the other hand, determining the weight of each evaluation index for the water and soil environmental vulnerability is easily affected by personal subjectivity and empiricism, decreasing the accuracy and reliability of the evaluation results. The cloud theory was proposed by academician Li Deyi of the Chinese Academy of Engineering in the 1990s to deal with fuzziness and uncertainty (Li *et al.* 2004). It is more advantageous in describing the randomness and uncertainty of fuzzy evaluation systems. Previous research results have shown that the cloud model can successfully realize the conversion between qualitative concepts and quantitative values based on the principle of cloud generator. It has been widely used in many research fields and achieved good evaluation results (Zhou *et al.* 2014; Xu *et al.* 2017).

This study uses the cloud theory to improve the conventional AHP, avoiding experts' empirical and subjective determination of the index weight. This paper combines the long-term monitoring data, spatial data, and geographic information data to ensure the comprehensiveness and reliability of the evaluation data in the space-air-ground dimensions. Based on the above methods and data, a water and soil environmental vulnerability evaluation model was constructed and used to evaluate the water and soil environmental vulnerability of China's Jingdian Irrigation District. The purpose of this study was: (1) to study the dynamic evolution process and characteristics of various factors affecting the vulnerability of the water and soil environment in the irrigation area; (2) to evaluate the vulnerability of the water and soil environment in the entire area and analyze its evolution from 1994 to 2018; (3) to guide the water and soil environmental governance and sustainable development in the irrigation area.

2. MATERIALS AND METHODS

2.1. Study area

The Jingdian Irrigation District in northwestern China's Gansu Province sits between 103°20′–104°04′E and 37°26′–38°41′N (Figure 1). Affected by the geological structure and hydrological and topographical features, this irrigation area is divided into closed hydrogeological units – Caowotan and Luyang basins and open hydrogeological units – Yanghuzitan and Daduntan basins from east to west. With the total irrigated area of about 6.13×10^4 hm^2, this irrigation area is 1,470–2,368 m above sea level, and extends about 40 km from north to south and about 120 km from east to west. It borders the Tengger Desert to the north, Changling Mountain to the south, and the Yellow River to the east. In the zone of typical temperate continental climate, this irrigation area is dry with little rainfall, and features large inter-annual and day–night temperature differences. The average annual sunshine duration is up to 2,714 hours, the average annual temperature is 8.77 °C, the average annual rainfall is 185.6 mm, and the average annual evaporation is 2,433.8 mm. Long-term monitoring data show that, since the pumping irrigation in 1994, the groundwater level in part of the irrigation area has gradually increased, and the chemical features of groundwater have constantly changed, due to the irrigation mode without drainage; coupled with high evaporation, low rainfall, and unique landform, a number of water and soil environmental problems have gradually appeared in the irrigation area, posing challenges to its water and soil environmental vulnerability (Xu 2010). The above problems have severely hindered the local social, economic and ecological development. In this regard, this paper selects the representative Jingdian Irrigation District as the study area to analyze the evolution of water and soil environmental vulnerability, which is of high research value.

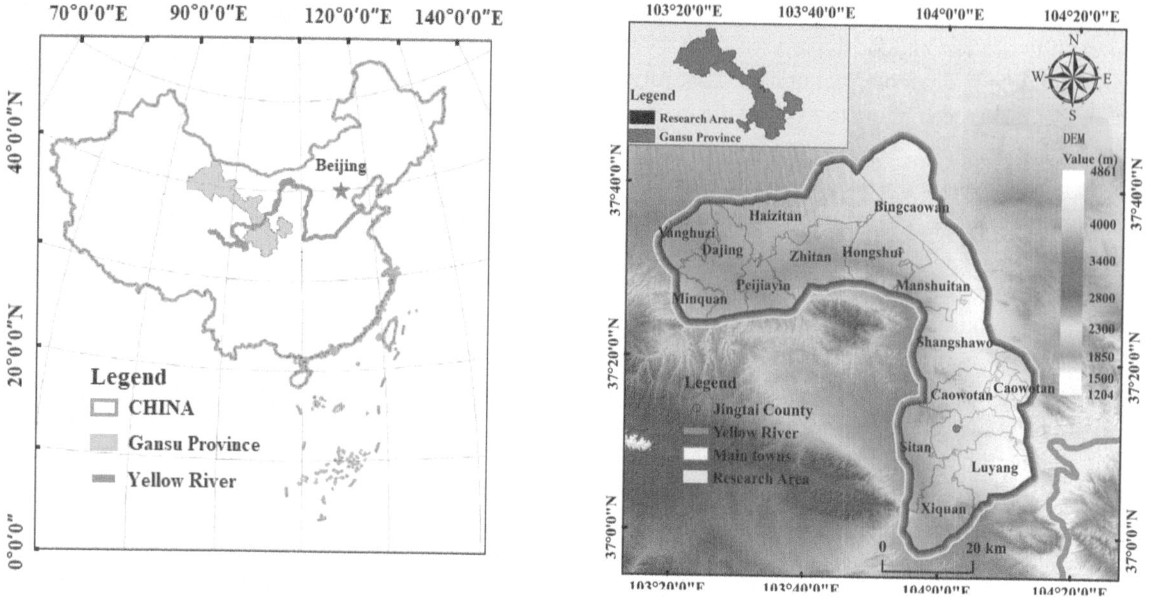

Figure 1 | Geographical location map of study area.

2.2. Index selection and data sources

To further reveal the temporal and spatial differentiation of the water and soil environmental vulnerability in the irrigation area, this paper, based on the actual situation of the study area and under the principles of scientificity, completeness and availability of data, selects the water and soil environmental vulnerability and its temporal and spatial differentiation characteristics as the target hierarchy; the soil vegetation, water resources, terrain, hydrology, and climate that significantly affect the water and soil environmental vulnerability as the criterion hierarchy; the soil salinity, vegetation coverage, land use type, and soil electrical conductivity that can reflect the water and salt content, soil erosion, deductive process of land resource utilization, and soil physical and chemical properties as the index factors of soil vegetation; the surface irrigation volume, groundwater depth, and groundwater salinity that reflect the movement and chemical properties of surface and ground water as the index factors of water resources; the elevation, slope, and aspect that directly participate in the geological structure of the water and soil environmental system as the index factors of terrain; the average annual precipitation, average annual evaporation, and average annual temperature as the index factors of climate that reflect hydrological and meteorological conditions in the area. On this basis, an evaluation index system for the water and soil environmental vulnerability of the study area was established. To make the analysis results more feasible and authentic, we selected three time nodes, namely 1994, 2006, and 2018 because: (1) these three years are highly representative. In 1994, the irrigation area was initially built, and the total irrigated area reached 40,100 hm^2, with an annual water-lifting volume of 366 million m^3, mainly distributed in the eastern part; in 2006, the rehabilitation of the irrigation area was basically completed, and the total irrigated area reached 58,000 hm^2, with an annual water-lifting volume of 551 million m^3; in 2018, the irrigation area was basically the same as its current situation, and the total irrigated area reached 68,500 hm^2, with an annual water-lifting volume of 660 million m^3. (2) The data of these three years are complete enough for the research. (3) The intervals of these three representative years are both 12 years, meeting the periodicity of the data, which can make the evaluation results closer to the actual situation.

The basic data and index factor data in this study were derived from the following sources. (1) Landsat 5 remote sensing images in 1994 and 2006 and Landsat 8 remote sensing images in 2018 downloaded from the Geospatial Data Cloud website (http://www.gscloud.cn/). After geometric correction, image registration, image mosaic and cropping, the preprocessing results of remote sensing images were obtained, as shown in Figure 2. Different gray values in the figure represent different land use conditions in the irrigation area. The preprocessing result map is mainly used for later interpretation and extraction of land use information and topographical factors. (2) Based on the *Land Survey Report of Jingtaichuan Irrigation District over the Years (1971–2018)* and the field survey of groundwater resources in Jingtaichuan, 44 salinity sampling points were selected (Figure 2(a)). Soil samples of 0–20 cm, 20–40 cm, 40–60 cm, 60–80 cm, and 80–100 cm were taken using a soil auger and then extracted in the laboratory to analyze the corresponding salinity and soil electrical conductivity. (3) Data of groundwater

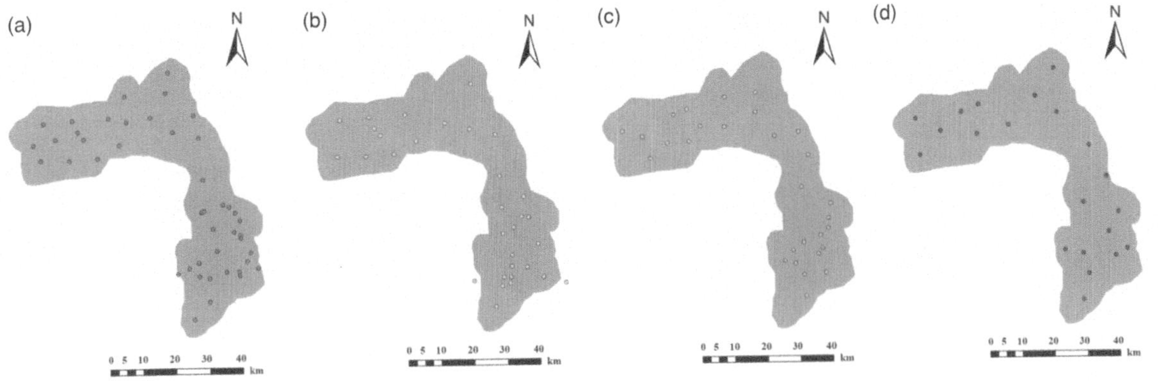

Figure 2 | Distribution of sampling points. (a) Distribution of salt index sampling points. (b) Distribution of groundwater monitoring wells. (c) Distribution of monitoring points for surface irrigation and precipitation. (d) Distribution of evaporation monitoring points.

depth and groundwater salinity were obtained from 32 monitoring wells arranged in the irrigation area (Figure 2(b)). (4) Data for surface irrigation volume and average annual precipitation were obtained from 29 water monitoring instruments and RS-QXYL-M automatic rainfall measuring stations deployed in the irrigation area (Figure 2(c)). (5) Data on average annual evaporation and average annual temperature were obtained from 21 miniature evaporation measuring meters and digital temperature measuring instruments installed in the irrigation area (Figure 2(d)). (6) The land use types were obtained from Landsat8.OLI data. In the ArcGIS10.2 Spatial Analyst module, ISO (interactive self-organization) clustering was performed first, and through image recognition and based on the actual land use conditions in the study area, the interpretation characteristics of images were divided into nine categories: cultivated land, mild saline-alkali land, moderate saline-alkali land, severe saline-alkali land, grassland, dry land, fixed sandy land, mobile sandy land, and gobi, according to the classification principle that representative pixels are in the same category. Finally, the maximum likelihood classification was used to extract the spatial distribution of each interpretation characteristics. Other supplementary materials included Report on Wasteland Resources and Their Development and Utilization in Jingtai County, Gansu Province (1971–2018), Statistics on Water Diversion and Consumption in the Irrigation Area from 1972 to 2018, and Hydrogeological Investigation Survey Report on Hexi Corridor (2018). Specific data sources and corresponding processing methods are shown in Table 1.

2.3. Interpolation accuracy validation analysis

Spatial interpolation refers to converting the limited data in the research scope into continuous surface set data using the ArcGIS software to reflect the corresponding changes of a variable in unknown parts of the study area. Commonly used spatial interpolation methods include the inverse distance weighted (IDW) method, trend surface method, spline method, and natural neighbor method based on the principle of deterministic interpolation, as well as the ordinary kriging method and universal kriging method based on the principle of geostatistical interpolation. Before interpolation analysis based on the spatial interpolation theory, data should be checked whether to meet the normal distribution. In this study, we used an Origin9.1 normal QQ graph analysis tool to test the sampled data, and found that the normal QQ test results basically present a normal distribution; thus, we performed the spatial variability analysis. In addition, we also introduced an error matrix to verify the accuracy of spatial interpolation and optimized the spatial interpolation method of each index (Kali *et al.* 2003; Li *et al.* 2016). The error matrix is a typical accuracy evaluation method in spatial interpolation, which evaluates the accuracy of spatial interpolation by calculating the comparative sequence between the measured pixel and the reference pixel. It mainly includes overall accuracy, producer accuracy, user accuracy, and Kappa coefficient. We just needed to consider the overall correct classification area and verify the accuracy of classification results in this study; thus, only the overall accuracy and Kappa coefficient were introduced (Shahid *et al.* 2018; Shahid *et al.* 2020). The calculation formulas are as follows:

$$OA = \frac{\sum_{i=1}^{r} n_{ii}}{N} \tag{1}$$

$$Kappa = \frac{P_0 - P_c}{1 - P_c} \tag{2}$$

Table 1 | Evaluation indexes and data sources of water and soil environmental vulnerability in Jingdian Irrigation District

Criterion hierarchy	Index factor	Unit	Data source	Data acquisition	Data type
Soil vegetation Y_1	Z_1 soil salinity	%	Sampling point	Laboratory analysis + spatial interpolation	Vector data
	Z_2 vegetation coverage	%	Geospatial Data Cloud	Remote sensing inversion	Raster data
	Z_3 land use type	–	Geospatial Data Cloud	ArcGIS-Spatial Analyst-ISO clustering – maximum likelihood classification	Raster data
	Z_4 soil electrical conductivity	S/m	Sampling point	Laboratory analysis + spatial interpolation	Vector data
Water Y_2	Z_5 surface irrigation volume	1×10^4 m^3	Water monitoring equipment	Monitoring data + spatial interpolation	Vector data
	Z_6 groundwater depth	m	Monitoring well	Elevation conversion + spatial interpolation	Vector data
	Z_7 groundwater salinity	g/l	Monitoring well	Laboratory analysis + spatial interpolation	Vector data
Terrain Y_3	Z_8 altitude	m	Geospatial Data Cloud	Remote sensing inversion	Raster data
	Z_9 slope	°	Geospatial Data Cloud	ArcGIS-Spatial Analyst – surface analysis	Raster data
	Z_{10} aspect	–	Geospatial Data Cloud	ArcGIS-Spatial Analyst – surface analysis	Raster data
Climate Y_4	Z_{11} average annual precipitation	mm	Rainfall measuring gauge	Monitoring data + spatial interpolation	Vector data
	Z_{12} annual evaporation	mm	Miniature evaporation measuring meter	Monitoring data + spatial interpolation	Vector data
	Z_{13} average annual temperature	°C	Geospatial Data Cloud	Monitoring data + spatial interpolation	Vector data

where, OA is the overall accuracy; N is the total number of pixels; n_{ii} is the number of correctly classified pixels; r is the number of classifications; P_0 is the overall classification accuracy, that is, the ratio of the number of correctly classified samples to the total number of samples; P_c is the expected value of the correct classification under random conditions. When the Kappa coefficient is greater than 0.75, it is generally believed that there is a high consistency between the simulated results and the measured results. When the Kappa coefficient is less than 0.4, the consistency is relatively low.

In the ArcGIS-Spatial Analyst-interpolation analysis module, the kriging method, IDW method, spline method, natural neighbor method, and trend surface method were used for spatial interpolation of each evaluation index factor. After the accuracy comparison and optimization, the final interpolation method for each index was determined (Table 2).

As seen in the above table, higher overall interpolation accuracy and Kappa coefficient were obtained by comparing the selected index factor interpolation methods, which had higher interpolation value. Since the ArcGIS-Spatial Analyst-surface analysis and ISO clustering-maximum likelihood classification inversion analysis both showed that the raster data such as elevation, slope, aspect, land use type, and vegetation coverage are highly consistent, the accuracy validation was not made separately in the study.

2.4. Evaluation factor weight cloud model

When experts use the conventional AHP to compare the influencing factors of water and soil environmental vulnerability in pairs and determine the judgment matrix and the significance of each factor, they are easily affected by their values and experience. The algebraic operation for determining weights based on the conventional principle of set-valued statistics is relatively simple, and using a single numerical scale to describe the experts' scoring it is easy to ignore the uncertainty

Table 2 | Interpolation method and accuracy of each evaluation index factor

Index	Unit	Overall accuracy (%)	Kappa
Soil salinity	%	75.00	0.7376
Soil electrical conductivity	S/m	63.00	0.6513
Average annual precipitation	mm	71.00	0.7004
Surface irrigation volume	1×10^4 m^3	73.00	0.7131
Groundwater depth	m	72.00	0.7094
Salinity of groundwater	g/L	69.00	0.6885
Average annual evaporation	mm	74.00	0.7294
Average annual temperature	%	76.00	0.7493

and discreteness generated in the scoring process. In addition, the water and soil environmental evolution and response process in arid irrigation areas is fuzzy, complex, and uncertain. It is difficult to use a single accurate value to characterize the relationship between the response index and the water and soil environmental evolution and response. The characterization of results is also full of uncertainty and does not conform to human language habits. The cloud theory can well describe the randomness and uncertainty of fuzzy systems, freely convert qualitative concepts and quantitative values, and combine the randomness and fuzziness of targets through uncertain languages, so as to overcome the shortcomings of the conventional AHP. Based on this, this paper introduces the uncertainty cloud theory into the AHP to improve the comprehensive evaluation of the water and soil environmental vulnerability in the irrigation area.

According to the cloud theory, each cloud model is characterized by its corresponding expectation (*Ex*), entropy (*En*), and hyper-entropy (*He*), namely *C (Ex, En, He)* (Guan *et al.* 2016). Expectation (*Ex*) reflects the center-of-gravity position of cloud droplets, which respectively represent the central value of the water and soil environmental vulnerability, driving process, and inducing factors' weight and membership; entropy (*En*) describes the fuzziness and randomness of cloud droplets, which respectively represent the possible value ranges of the water and soil environmental vulnerability, driving process, and influencing factors' weight and membership; hyper-entropy (*He*) describes the thickness of clouds and mainly reflects the dispersion of cloud droplets, which respectively represent the deviation from the central value of the water and soil environmental vulnerability, driving process, and influencing factors' weight and membership.

On the Satty scale of the classic AHP, experts are required to use a natural number between 1 and 9 to determine the relative importance of two factors (Nguyen and Nahavandi *et al.* 2016; Wang *et al.* 2017). In this study, we introduced the normal cloud theory and improved it by constructing a pairwise comparison judgment matrix based on influencing factors on the cloud model scale (Table 3).

The judgment matrix was constructed by the cloud model scale first, and then the square root method was used to calculate the weight of inducing factors. The weight cloud model was obtained by calculating elements of each row in the pairwise comparison judgment matrix. Calculation formulas of expectation (*Ex*), entropy (*En*), and hyper-entropy (*He*) (Zhang

Table 3 | Scale criteria of cloud model of vulnerability of water and soil environment

Importance comparison		Cloud model of scale criteria *C(Ex,En,He)*
S_i is more important than S_j	Absolutely	$C_4(9,0.33,0.01)$
	Strongly	$C_3(7,0.33,0.01)$
	Obviously	$C_2(5,0.33,0.01)$
	Slightly	$C_1(3,0.33,0.01)$
S_i and S_j are equally important		$C_0(1,0,0)$
S_i is not as important as S_j	Slightly	$C_5(1/3,0.33/9,0.01/9)$
	Obviously	$C_6(1/5,0.33/25,0.01/25)$
	Strongly	$C_7(1/7,0.33/49,0.01/49)$
	Absolutely	$C_8(1/9,0.33/81,0.01/81)$

et al. 2014) are shown as follows:

$$Ex_i = \frac{\left(\prod_{j=1}^{n} Ex_{ij}\right)^{\frac{1}{n}}}{\sum_{i=1}^{n}\left(\prod_{j=1}^{n} Ex_{ij}\right)^{\frac{1}{n}}} \tag{3}$$

$$En_i = \frac{\left(\prod_{j=1}^{n} Ex_{ij}\sqrt{\sum_{j=1}^{n}\left(\frac{En_{ij}}{Ex_{ij}}\right)^2}\right)^{\frac{1}{n}}}{\sum_{i=1}^{n}\left(\prod_{j=1}^{n} Ex_{ij}\sqrt{\sum_{j=1}^{n}\left(\frac{En_{ij}}{Ex_{ij}}\right)^2}\right)^{\frac{1}{n}}} \tag{4}$$

$$He_i = \frac{\left(\prod_{j=1}^{n} Ex_{ij}\sqrt{\sum_{j=1}^{n}\left(\frac{He_{ij}}{Ex_{ij}}\right)^2}\right)^{\frac{1}{n}}}{\sum_{i=1}^{n}\left(\prod_{j=1}^{n} Ex_{ij}\sqrt{\sum_{j=1}^{n}\left(\frac{He_{ij}}{Ex_{ij}}\right)^2}\right)^{\frac{1}{n}}} \tag{5}$$

According to the said formulas, the cloud model synthetic weight W_i (Ex_i, En_i, He_i) of 13 water and soil environmental vulnerability evaluation factors can be obtained.

2.5. Classification of water and soil environmental vulnerability

Based on the previous studies and the actual situation of this study area (Wang *et al.* 2012; Chen *et al.* 2016), the characteristic value *EVI* of environmental vulnerability was used to describe the water and soil environmental vulnerability in Jingdian Irrigation District, which was calculated by formula (6):

$$EVI = \sum_{i=1}^{n} Y_1X_1 + Y_2X_2 + Y_3X_3 + \ldots + Y_nX_n \tag{6}$$

where, *EVI* represents the characteristic value of soil and water environmental vulnerability; Y_i represents the index factor value; *Xi* represents the corresponding weight value of each index factor; and *n* represents the number of indexes.

To further enhance the comparability of the comprehensive index of water and soil environmental vulnerability in the study area in temporal and spatial distribution, the comprehensive index *EVI* was standardized according to formula (7):

$$S_i = \frac{EVI_i - EVI_{\min}}{EVI_{\max} \times 10} \tag{7}$$

where, S_i represents the standardized value of the comprehensive index of soil and water environmental vulnerability in the i-th year, which ranges from 0 and 10; EVI_i represents the actual value of the comprehensive index of soil and water environmental vulnerability in the i-th year; EVI_{max} represents the maximum value of the comprehensive index of soil and water environmental vulnerability in three periods; and EVI_{min} represents the minimum value of the comprehensive index of soil and water environmental vulnerability in three periods.

Based on the standardization of S_i and with reference to the standards of environmental vulnerability evaluation in existing studies, the water and soil environmental vulnerability of Jingdian Irrigation District was divided into five levels as per the specific characteristics (Table 4).

Table 4 | Division of soil and water environment vulnerability in Jingdian Irrigation District

Vulnerability	Grade	S_i	Ecological characteristics
Slight	I	1–2	The environmental system is perfect in structure and function, and bears little pressure. It is stable, with strong resistance to external interference, high self-resilience, and low vulnerability.
Mild	II	2–4	The environmental system is relatively perfect in structure and function, and bears little pressure. It is stable, with strong resistance to external interference and high self-resilience. However, there are possible abnormalities. The environmental vulnerability is relatively low.
Moderate	III	4–6	The environmental system structure and function can be maintained; the bearable pressure comes near the threshold value; the system is not stable, and sensitive to external interference; it is less resilient, and there have been some abnormalities; the environmental vulnerability is relatively high.
Severe	IV	6–8	The environmental system structure and function have been damaged; the pressure becomes large; the system is unstable, and highly sensitive to external interference; it is hard to restore after being damaged; there are many abnormalities; the environmental vulnerability is high.
Extremely severe	V	8–10	The environmental system structure and function are severely worsening; the pressure becomes extremely large; the system is unstable, and extremely sensitive to external interference; damage is very hard to be repaired, and may be irreversible; there are a great number of abnormalities; the environmental vulnerability is extremely high.

3. RESULTS

3.1. Weight of each index factor

We calculated the synthetic weight W_i (Ex_i, En_i, He_i) for the impact of each index factor on the water and soil environmental vulnerability using the cloud theory-based improved AHP, and ranked the evaluation factors in the index hierarchy with their expectation Ex as the first sorting element, entropy En as the second sorting element, and hyper-entropy He as the third sorting element (Xu *et al.* 2017). Accordingly, the total ranking of 13 indexes in the criterion hierarchy to the target hierarchy was obtained (Table 5).

As shown in Table 5, the factors affecting the vulnerability of the water and soil environment in the study area on regional scale are ranked as follows as per their importance: soil salinity > vegetation coverage > land use type > surface irrigation volume > groundwater depth > groundwater salinity > soil electrical conductivity > average annual evaporation > average annual precipitation > average annual temperature > elevation > slope > aspect. It can be seen that soil salinity, vegetation coverage, and land use type are the main index factors that affect the vulnerability of the water and soil environment on regional scale and its temporal and spatial differentiation process, followed by three moisture factors (surface irrigation volume, groundwater depth, and groundwater salinity), and three hydrometeorological factors (average annual evaporation, average annual precipitation, and average annual temperature). Topographical factors have the least impact on the temporal and spatial differentiation of the soil and water environmental vulnerability. The main reasons are that, in the oasis ecosystem, the soil state, vegetation coverage, and land use type are key factors that control the water and soil environmental evolution, and have the greatest impact on the temporal and spatial differentiation of the water and soil environment on the regional scale; water that transfers surface and underground solute factors directly participates in the soil–water–atmosphere continuum cycle, so it has a great impact on the regional water and soil environment; hydroclimatic factors, as representative factors of regional microclimate, participate in the regional water circulation, provide power for water and heat transfer, and play a potential driving role in the temporal and spatial differentiation of the regional water and soil environment; topographical factors, as the objective vector of the water and soil environment, only affect the temporal and spatial differentiation of the water and soil environment on a micro-scale, and has the minimal contribution to the large-scale regional water and soil environmental vulnerability.

Table 5 | Hierarchical total ranking of weights based on cloud parameters

Index	Y_1			Y_2			Y_3			Y_4			Synthetic weight	Sort
	Ex	*En*	*He*	*Ex*	*En*	*He*	*Ex*	*En*	*He*	*Ex*	*En*	*He*	W_i (Ex_i, En_i, He_i)	
X	0.468	0.471	0.471	0.279	0.275	0.275	0.075	0.095	0.095	0.178	0.109	0.109		
Z_1	0.308	0.361	0.361										(0.144,0.170,0.170)	1
Z_2	0.264	0.342	0.342										(0.124,0.161,0.161)	2
Z_3	0.256	0.194	0.194										(0.120,0.091,0.091)	3
Z_4	0.172	0.103	0.103										(0.080,0.049,0.049)	7
Z_5				0.395	0.353	0.353							(0.110,0.097,0.097)	4
Z_6				0.314	0.333	0.333							(0.088,0.092,0.092)	5
Z_7				0.291	0.314	0.314							(0.081,0.086,0.086)	6
Z_8							0.373	0.403	0.403				(0.028,0.038,0.038)	11
Z_9							0.324	0.314	0.314				(0.024,0.030,0.030)	12
Z_{10}							0.303	0.283	0.283				(0.023,0.027,0.027)	13
Z_{11}										0.342	0.519	0.519	(0.061,0.083,0.083)	9
Z_{12}										0.423	0.321	0.321	(0.075,0.051,0.051)	8
Z_{13}										0.235	0.160	0.160	(0.042,0.025,0.025)	10

3.2. Characteristics of temporal and spatial distribution of each index factor

3.2.1. Temporal and spatial distribution of soil salinity

We carried out the spatial interpolation of the measured soil salinity data in the three periods using the ArcGIS-Spatial Analyst module (Figure 3 and Table 6). According to Figure 4 and Table 6, in 1994, a great number of land resources were newly exploited due to massive land reclamation. The short irrigation time and washing salinity by irrigation made the soil salinity lower; during the period from 1994 to 2006, because of the increasing human-induced disturbance on water and soil resources in the irrigation area and the long-standing unreasonable irrigation mode, the soil salinity rose sharply, with the

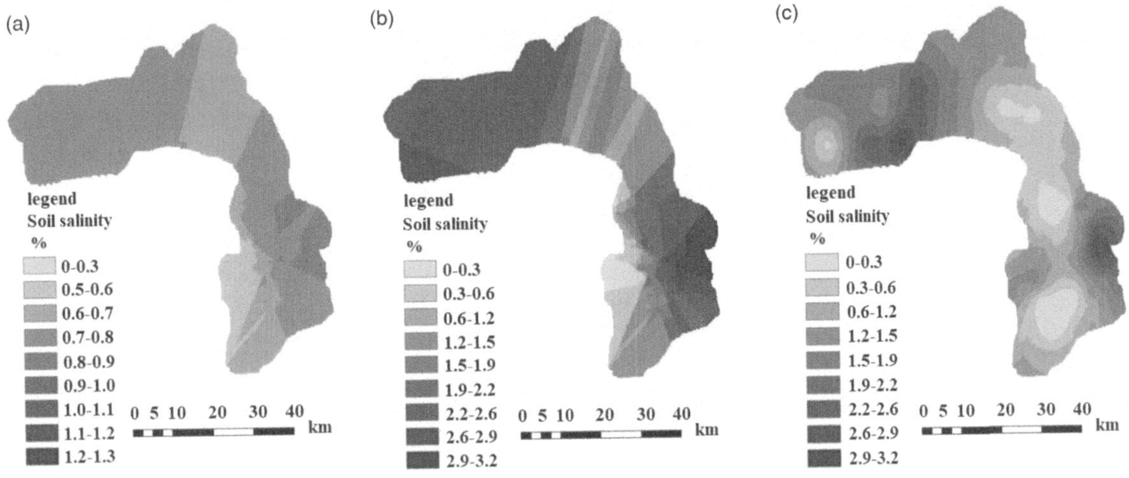

Figure 3 | Soil salinity. (a) 1994. (b) 2006. (c) 2018.

Table 6 | Characteristics of soil salinity in the three periods of Jingdian Irrigation District

Years	1994	2006	2018
Minimum value (%)	0.0324	0.0496	0.0382
Maximum value (%)	1.2948	3.1871	3.1284
Average value (%)	0.6817	1.9864	1.9685

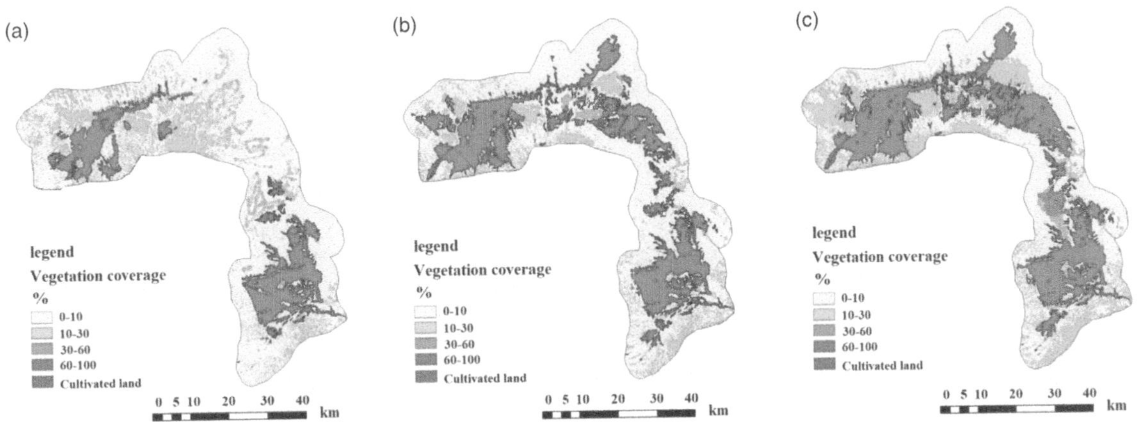

Figure 4 | Vegetation coverage. (a) 1994. (b) 2006. (c) 2018.

range expanding from 0.0324%–1.2948% in 1994 to 0.0496%–3.1871%, and the average salt salinity was 1.3047% higher than that in 1994; during the period from 2006 to 2018, due to a series of desalination measures such as washing salinity by irrigation and adding alkali drainage ditches, soil in the irrigation area continued to desalt. By 2018, except for the closed hydrogeological units in the eastern part, the soil salinity in other parts of the irrigation area significantly declined, and the characteristic value of the salt concentration decreased from 0.0496%–3.1871% to 0.0382%–3.1284%. The expansion of secondary salinized land was effectively controlled.

3.2.2. Temporal and spatial distribution of vegetation coverage

We inverted the model through ArcGIS software and calculated the area of land with different levels of vegetation coverage in 1994, 2006, and 2018 (Table 7 and Figure 4).

It can be seen from Figure 4 and Table 7 that, in Jingdian Irrigation District, the land with low vegetation coverage or below shares a relatively large area. However, from 1994 to 2018, the area of land with low vegetation coverage (including extremely low vegetation coverage) was shrinking, down from 76.6% in 1994 to 58.16% in 2018, the area of land with medium vegetation coverage increased from 0.47% in 1994 to 2.19% in 2018, and the area of cultivated land rose from 22.93% in 1994 to 39.65% in 2018, showing a significant expansion.

3.2.3. Temporal and spatial distribution of land use type

We interpreted the remote sensing images through visual interpretation and the maximum likelihood classification method, and extracted the land use information of the irrigation area in 1994, 2006, and 2018 (Figure 5).

We used ArcGIS10.2 statistical function to extract the area of each land use type (Table 8).

From the Table 8, it can be found that: in 1994, the gobi accounted for the largest proportion in the whole study area, as high as 34.13%; while the cultivated land only accounted for 20.14%. After years of irrigation by pumping and manual reclamation, the cultivated land expanded, and the gobi, grassland, and dry land shrank, forming a clear contrast. The gobi area

Table 7 | Changes of vegetation coverage in Jingdian Irrigation District in the third stage

Years		Extremely low coverage (0–10%)	Low coverage (10%–30%)	Medium coverage (30%–60%)	High coverage (60%–100%)	Cultivated land
1994	Area hm^2	183,424.68	46,908.61	1,262.92	150.34	68,949.65
	The proportion %	61	15.6	0.42	0.05	22.93
2006	Area hm^2	130,141.32	73,189.45	4,299.96	781.81	92,283.66
	The proportion %	43.28	24.34	1.43	0.26	30.69
2018	Area hm^2	83,052.29	91,832.62	5,472.67	1,112.57	119,226.04
	The proportion %	27.62	30.54	1.82	0.37	39.65

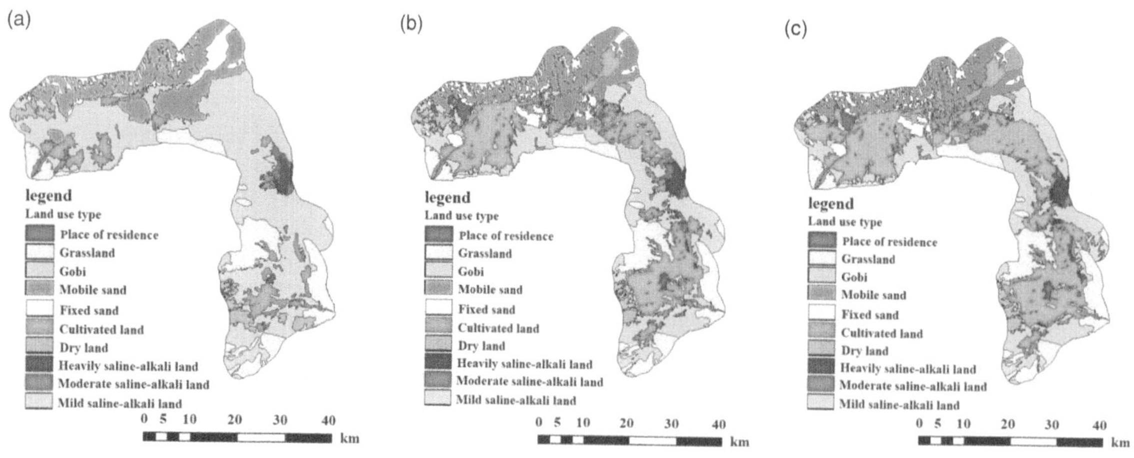

Figure 5 | Classification of land use types. (a) 1994. (b) 2006. (c) 2018.

Table 8 | Changes in land use types from 1994 to 2018

Land use types	1994		2006		2018	
	Area hm²	Proportion %	Area hm²	Proportion %	Area hm²	Proportion %
Heavily saline-alkali land	865.24	0.28	954.28	0.31	914.65	0.30
Moderate saline-alkali land	1,235.75	0.41	1,584.93	0.52	1,128.54	0.37
Mild saline-alkali land	3,215.51	1.06	3,654.89	1.21	2,924.35	0.97
Place of residence	450.98	0.15	1,385.97	0.46	1,423.68	0.47
Cultivated land	60,584.58	20.14	96,485.21	32.18	117,258.65	38.89
Grassland	72,651.85	24.16	65,579.28	21.87	61,856.47	20.52
Dry land	8,851.70	2.94	7,596.28	2.53	6,827.94	2.26
Gobi	102,645.41	34.13	75,369.48	25.13	64,382.78	21.35
Fixed sand	18,542.68	6.16	18,236.79	6.08	16,358.75	5.46
Mobile sand	31,652.51	10.52	28,967.25	9.66	28,364.82	9.40

reduced from 102,645.41 hm² in 1994 to 64,382.78 hm² in 2018, grassland area from 72,651.85 hm² in 1994 to 61,856.47 hm² in 2018, and dry land area from 8,851.70 hm² in 1994 to 6,827.94 hm² in 2018, while the cultivated land area increased from 60,584.58 hm² in 1994 to 117,258.65 hm² in 2018. Among all land types, the sand area accounted for the least proportion, and due to the restriction of local climate conditions and the large-scale afforestation by local residents, its area reduced from 50,195.19 hm² in 1994 to 44,723.57 hm² in 2018. The long-standing extensive and unreasonable irrigation mode and the high vaporization and low rainfall contributed to the concentration of deep-layer salt on the soil surface under the capillary action, enlarging the land salinization or secondary salinization area. From 1994 to 2006, the saline and alkaline land increased from 1.75% to 2.04%, mainly distributed in the connection area of Caowotan Town and Manshuitan Township. From 2006 to 2018, with the implementation of water-saving irrigation modes, such as furrow irrigation and border irrigation, the irrigation amount was controlled reasonably to realize the combination of irrigation and drainage. Targeted measures including biological, chemical, and engineering solodization were taken to slow down the expansion of the saline and alkaline land, and the proportion of saline and alkaline land decreased from 2.04% in 2006 to 1.64% in 2018.

3.2.4. Temporal and spatial distribution of soil electrical conductivity

The soil electrical conductivity can be used to objectively reflect the regional soil quality and its physiochemical properties. From Figure 6, it is seen that from 1994 to 2006, as the soil salinity concentrated, the zwitterion content in the soil increased and so did the soil electrical conductivity. From 2006 to 2018, a series of salt elimination measures were taken to decrease the

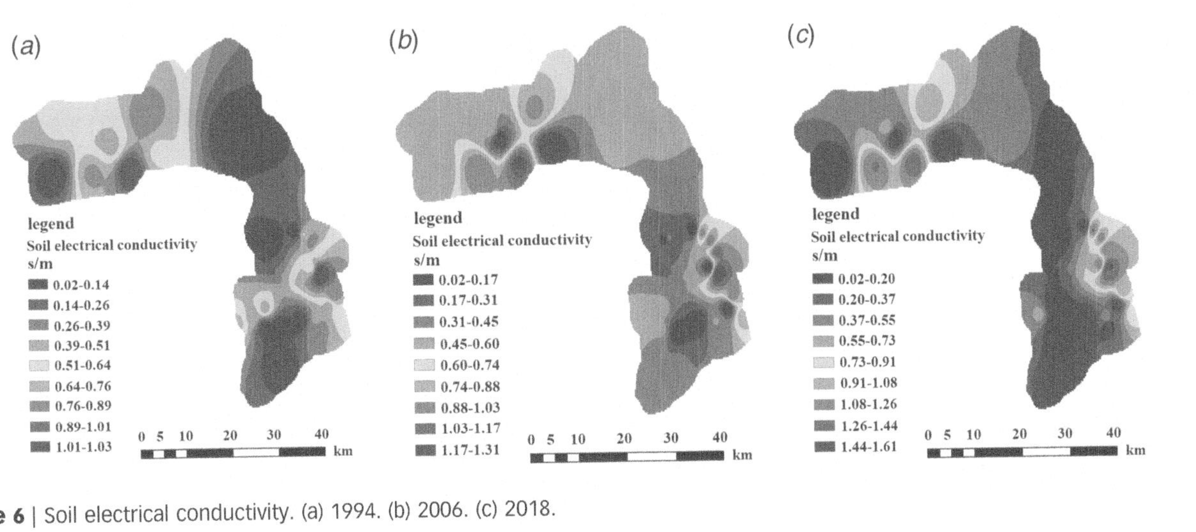

Figure 6 | Soil electrical conductivity. (a) 1994. (b) 2006. (c) 2018.

soil electrical conductivity. Combined with the soil salinity changes shown in Figure 3, the soil electrical conductivity is positively correlated to the soil salinity, which proves that the regional soil salinity is the key factor driving the change in physicochemical properties of soil.

3.2.5. Temporal and spatial distribution of surface irrigation volume

Based on the temporal and spatial distribution diagram of the surface irrigation volume (Figure 7), the surface irrigation volume witnessed a general increase from 1994 to 2018, mainly due to the continuous improvement of the ancillary irrigation equipment within the irrigation area during this period. The land resources in the irrigation area were continuously developed, leading to the expansion of cultivated land. In addition, with the implementation of some non-profit development modes and ecological breeding modes, water demand rose year after year. Concerning the overall spatial distribution of irrigation volume, the open hydrogeological units in the western part of the study area were irrigated greatly, but the irrigation volume of the closed hydrogeological units in the eastern part declined, which was mainly because the secondary salinization in the closed hydrogeological units worsened and salt-tolerant corps were planted in partial areas only. Therefore, a small amount of water was needed.

3.2.6. Temporal and spatial distribution of groundwater depth

As shown in Figure 8, the groundwater depth in the study area was generally decreasing. Among them, the change of groundwater depth was the most obvious in the Yanghuzitan Town and Daduntan Town in the open hydrogeological units, and Caowotan Town and Luyang Town in the closed hydrogeological units. The groundwater depth in Yanghuzitan Town decreased from 75 m–85 m in 1994 to 65 m–55 m in 2018, and that in Daduntan Basin from 55 m–65 m in 1994 to 45 m–55 m in 2018. It can be seen that the overall groundwater depth in the area from Yanghuzitan Town to Haizitan Town decreased by about 10 m, and that in the area from Manshuitan Town to Caowotan Town in the closed hydrogeological units declined from 65 m–80 m to 45 m–65 m. The area with the groundwater depth of 0–10 m in Luyang Town continued to expand, so that the area with the depth of 10 m–20 m in the surrounding area almost disappeared in 2018. The area with the depth of 65 m–80 m in Xiquan Town was also shrinking.

ArcGIS extraction tool was used to extract the characteristic values of groundwater depth in Jingdian Irrigation District in three periods (Table 9).

From Table 9 above, the minimum groundwater depth in the irrigation area reduced by 3.5491 m from 3.5781 m in 1994 to −0.3690 m in 2018, and the maximum groundwater depth reduced by 9.3249 m from 78.2584 m in 1994 to 68.9335 m in 2018. It can be seen that the groundwater replenishment in the irrigation area increased significantly due to unscientific irrigation methods such as long-term irrigation without drainage, unreasonable irrigation and drainage, broad irrigation, and excessive irrigation quotas. The transition from the basic balance between recharge and discharge to groundwater discharge or extraction less than recharge led to the trend of decreasing groundwater depth in some areas year by year.

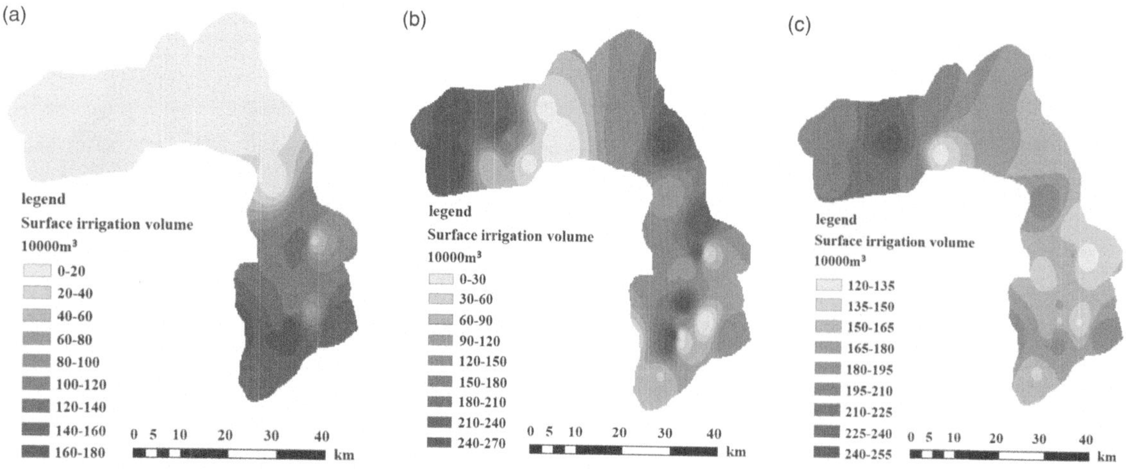

Figure 7 | Surface irrigation volume. (a) 1994. (b) 2006. (c) 2018.

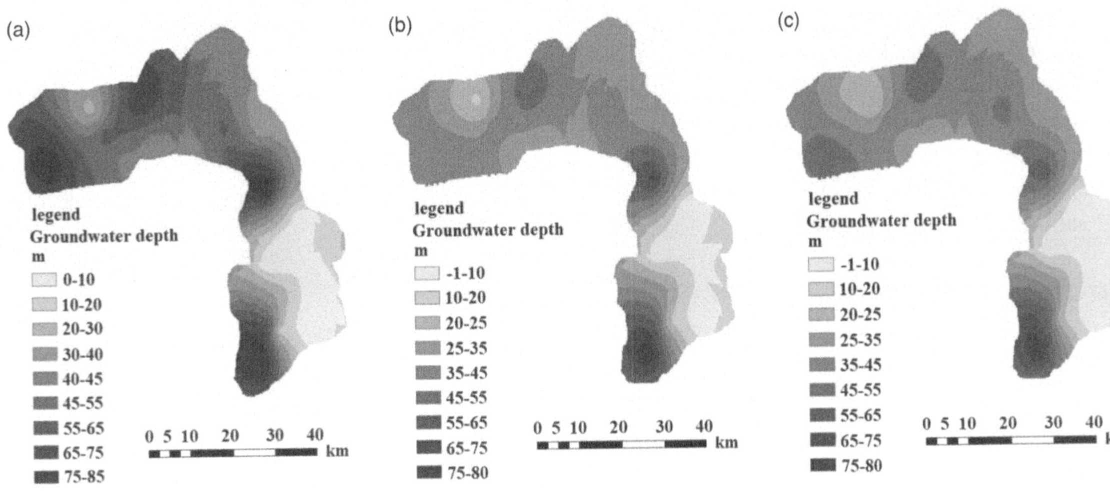

Figure 8 | Groundwater depth. (a) 1994. (b) 2006. (c) 2018.

Table 9 | Characteristic values of groundwater depth in Jingdian Irrigation District

Years	1994	2006	2018
Minimum value (m)	3.5781	−0.2744	−0.3690
Maximum value (m)	78.2584	70.2103	68.9335
Average value (m)	45.9219	41.6029	37.7474

3.2.7. Temporal and spatial distribution of groundwater salinity

As shown in Figure 9, the area with groundwater salinity ranging from 1.5 g/L to 1.9 g/L in the open hydrogeological units gradually expanded from the Yanghuzitan Town in 1994 to the Yanghuzitan Town and Haizitan Town in 2018. From 1994 to 2006, the groundwater salinity of Sitan Town in the irrigation area decreased first and then increased. In the closed hydrogeological units, the groundwater salinity from Luyang Town to Caowotan Town increased from 2.9 g/L–4.4 g/L in 1994 to 4.4 g/L–5.9 g/L in 2018. According to the overall interpolation results of groundwater salinity and inter-annual changes, the groundwater salinity in the irrigation area was increasing overall.

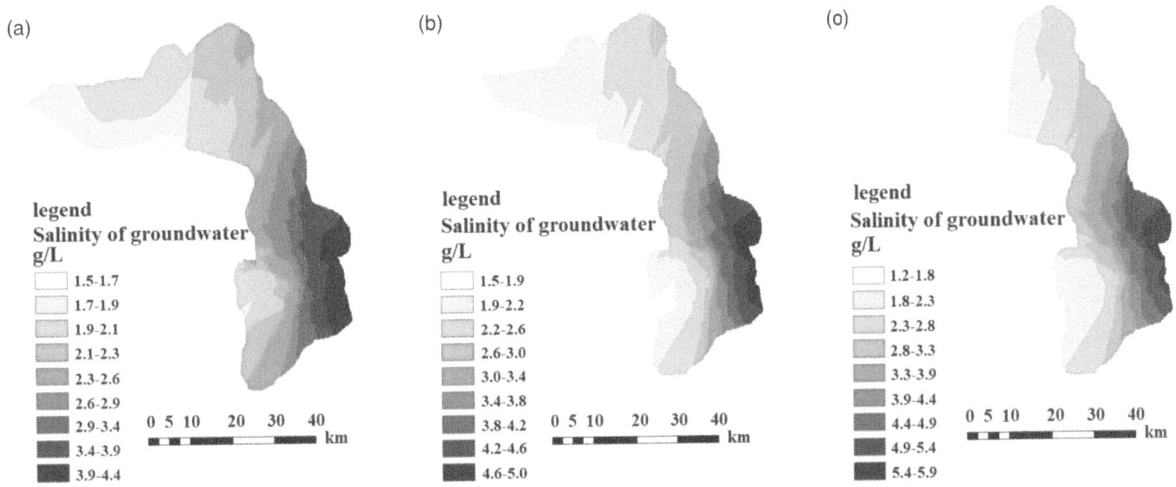

Figure 9 | Salinity of groundwater. (a) 1994. (b) 2006. (c) 2018.

The ArcGIS extraction tool was used to extract the characteristic values of groundwater salinity in the Jingdian Irrigation District in three periods (Table 10).

It can be seen from the above table that the minimum groundwater salinity increased from 1994 to 2006, which was because groundwater salinity increased and oscillated in the solute migration zone. The maximum groundwater salinity increased from 4.3285 g/L in 1994 to 5.8335 g/L in 2018, which was because that in the catchment and salt accumulation area, the dissipation of shallow groundwater caused salt migration and accumulation, leading to the gradual increase of groundwater salinity.

3.2.8. Topographical index factors

The inversion raster map of topographical factors (Figure 10) showed that the overall elevation of the study area was between 1,470 and 2,368 m, and gradually increased like an arc from east to west. The terrain sloped gently, with a slope of 0–25°, which was suitable for agricultural production and had little impact on the ecological environment.

3.2.9. Climatic index factors

As shown in Figure 11, the average annual precipitation in the study area was 162–191 mm. With the relatively uniform spatial distribution, it was typical of arid climate. The average annual evaporation was 2,274–2,358 mm. The average annual evaporation in the eastern part was higher, which was positively correlated with the high soil salinity in this part. The main reason was that soil salinity, driven by evaporation, moved to the surface through capillarity and intensified soil salinization. The average annual temperature was 7.90–9.59 °C. The high temperature mainly concentrated in the northern part of the study area, and decreased in an arc to both sides. With relatively low temperature throughout the year, the area was typical of temperate climate.

3.3. Temporal and spatial distribution of regional water and soil environmental vulnerability

According to the weights of the evaluation index factors (Table 4) and the grading standards for the water and soil environmental vulnerability in Jingdian Irrigation District (Table 5), ArcGIS software was used for weighted stack of the vulnerability

Table 10 | Characteristic values of groundwater salinity in Jingdian Irrigation District

Years	1994	2006	2018
Minimum value (g/L)	1.5169	1.5734	1.4822
Maximum value (g/L)	4.3285	4.8228	5.8335
Average value (g/L)	2.5850	3.5992	3.8952

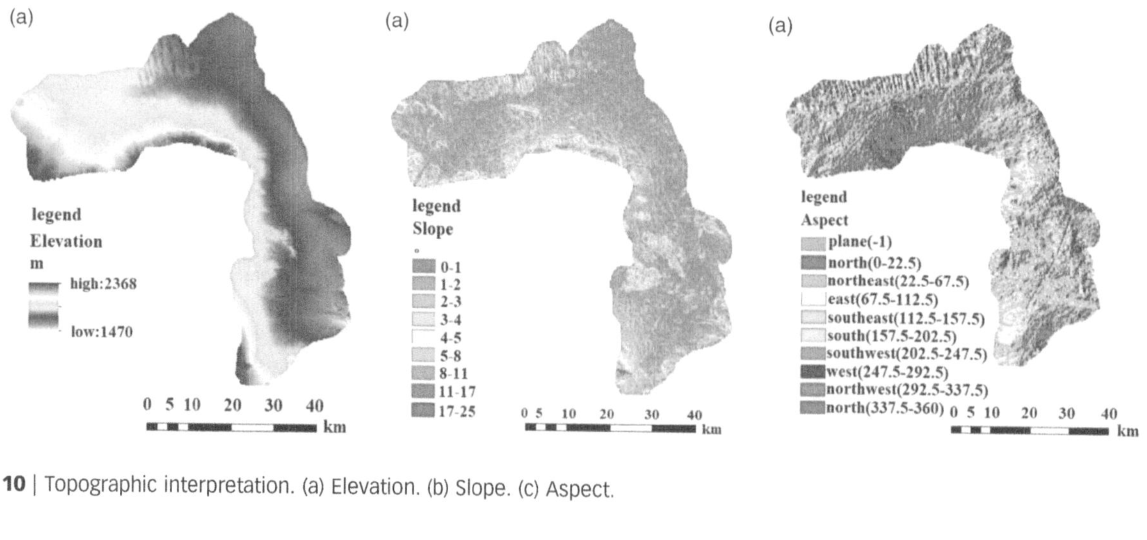

Figure 10 | Topographic interpretation. (a) Elevation. (b) Slope. (c) Aspect.

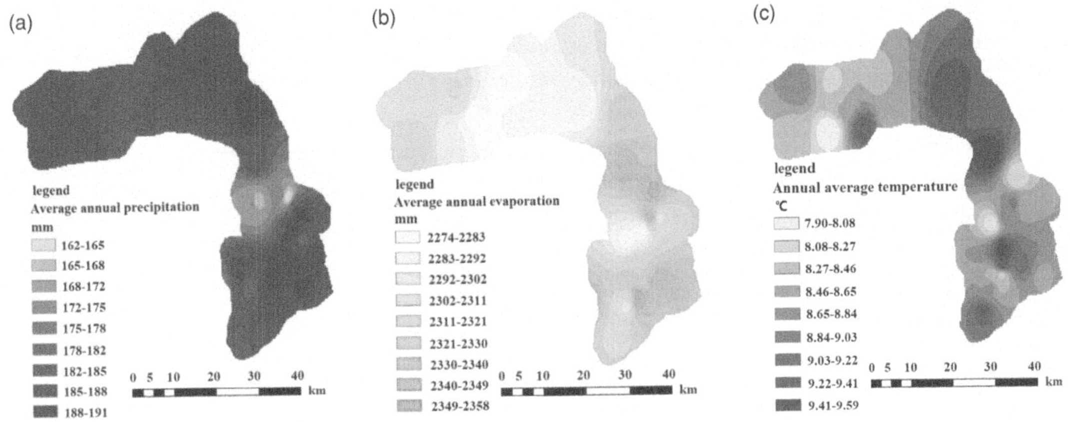

Figure 11 | Hydrological and climatic indicators. (a) Average annual precipitation. (b) Average annual evaporation. (c) Average annual temperature. (Note: as the topography and hydrological and climatic conditions of the study area are not significantly different over the years, the analysis is only for 2018.)

evaluation index factors, and finally the spatial distribution diagrams of water and soil environmental vulnerability in the irrigation area in 1994, 2006 and 2018 were obtained (Figure 12).

As can be seen from the spatial distribution of water and soil environmental vulnerability in 1994, the overall water and soil environmental vulnerability of the irrigation area was below Levels IV and V. Level II and Level III fragile areas covered about two-thirds of the study area. Level I and Level IV fragile areas covered about one-third of the study area. Level IV fragile areas were mainly distributed in the closed hydrogeological units of Luyang Town and Caowotan Town, and some of the cultivated land in these areas had been salinized. Level III fragile areas were distributed around Jingtai County and Yanghuzitan Town. Level II fragile areas were distributed in the northern part of the irrigation area, and Level I fragile areas were mainly near Xiquan Town. According to the spatial distribution of water and soil environmental vulnerability in 2006, from 1994 to 2006, the water and soil environmental vulnerability in the northern part of the irrigation area did not change much, and that in some parts of Daduntan Town and Bingcaowan Town evolved from Level II to Level I. This was because during this period, these gobi areas were gradually cultivated by local residents after pumping irrigation, that is, the land use type was changed from gobi to cultivated land. In addition, the vulnerability of local areas of Luyang Town and Caowotan Town changed greatly. Level III fragile areas had gradually evolved to Level IV, and showed a tendency towards Level V. The reason was that the groundwater level rose sharply in these areas after pumping irrigation. During the hydrothermal transport caused by strong evaporation, deep-layer soluble salt ions were transported to the soil surface along with capillary water. The rapid

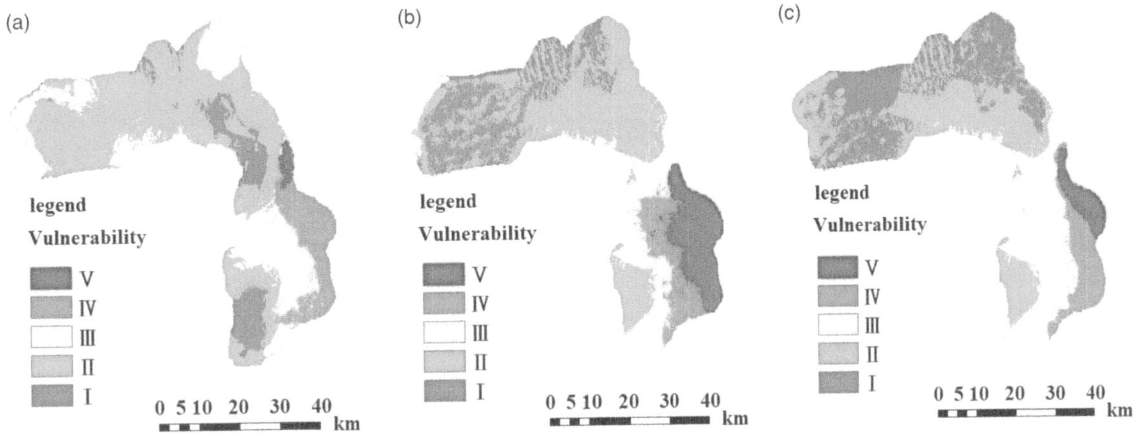

Figure 12 | Spatial distribution of soil and water environment vulnerability in the third stage of the Jingdian Irrigation District. (a) 1994. (b) 2006. (c) 2018.

increase and accumulation of surface soil salinity aggravated the evolution of secondary salinization in this area, destroyed the primitive soil and water environment, and led to the escalation of vulnerability. According to the spatial distribution of water and soil environmental vulnerability in 2018, since water-saving irrigation methods such as border irrigation and furrow irrigation were carried out to control the irrigation amount and groundwater level, and measures were taken to prevent the secondary salinization, the highly fragile areas in Caowotan Town and Manshuitan Town in 2006 were greatly reduced, with the soil and water environmental vulnerability evolving from Level V to Level IV, from Level IV to Level III. Overall, the soil and water environment tended to evolve into a safe state.

4. DISCUSSION

To objectively and accurately evaluate the spatial evolution of water and soil environmental vulnerability in the irrigation area from 1994 to 2018, we selected 13 key driving factors from four aspects, namely soil vegetation, hydrology, terrain, and climate, when constructing the index system. Moreover, there was no obvious correlation between the index systems. Judging from the ranking of index weights, soil salinity, vegetation coverage, land use type, and surface irrigation volume have the strongest driving effect on the water and soil environment. With the increase of soil salinity, the internal soluble salt content and pH increased accordingly, but the soil physical properties deteriorated, decreasing the soil fertility. This is basically consistent with the results of previous studies (Zhang *et al.* 2018). The vegetation coverage also has a great impact on the vulnerability of the water and soil environment in the irrigation area. On the one hand, as the plant species and quantity decrease, the biomass and vegetation coverage also reduce, and the community structure tends to be single, destroying the stability of the water and soil environment; on the other hand, vegetation coverage can inhibit the activity of soil organic matter through the surface temperature to a certain extent, causing decline in the soil quality, and accordingly affecting the vulnerability of water and soil environment (Huang *et al.* 2014; Liu *et al.* 2020). Under the long-term disturbance of human activities, the land use type in the irrigation area has changed its original existing form and evolution trend. Previous studies have shown that changes in the land use pattern caused by human activities are one of the important reasons for changes in the regional water and soil environmental vulnerability. In areas where the land use pattern changes frequently, the changes in water and soil environmental vulnerability are exceptionally significant (Wang *et al.* 2005; Lu *et al.* 2011). Surface irrigation volume is the most active factor affecting the evolution of water and soil environment. The Jingdian Irrigation District is an artificial oasis formed by engineering water diversion and irrigation. The diversion of such a large amount of water into the originally deserted and barren land in a short time changes its original habitat and hydrogeological conditions.

The water and soil environmental vulnerability evaluation shows comprehensive, complex, and ambiguous characteristics, while the current methods are unable to achieve a comprehensive, scientific and objective evaluation. Based on the summary of previous research results, we selected the Jingdian Irrigation District in the arid northwestern China with unique natural conditions, took 'green' and 'wet' as the main determinants and 'dry' and 'hot' as the main manifestations of the water and soil environmental vulnerability in arid areas, and made a macro, rapid, objective evaluation of the water and soil environmental vulnerability in Jingdian Irrigation District based on remote sensing and evaluation models, to reveal the evolution mechanism and process of environmental vulnerability in the study area.

5. CONCLUSIONS

To clarify the evolution trend of the water and soil environmental vulnerability in Jingdian Irrigation District and reveal the main influencing factors driving the evolution of the water and soil environment in the irrigation area, 13 key factors were selected from four aspects: soil vegetation, hydrology, terrain, and climate, the corresponding long-term monitoring data and test data were expressed using space technology, and the accuracy of the expression results was verified. The spatial processing results of all indexes are reliable and accurate, providing good conditions for the research and ensuring the credibility of the research results. The main results obtained in this study are as follows:

(1) The cloud theory-based improved AHP and multiple data fusion were applied to analyze the vulnerability of the soil and water environment in the study area from 1994 to 2018. The results showed that the soil salinity, vegetation coverage, land use type, and surface irrigation volume are the main factors influencing the evolution of water and soil environmental vulnerability in Jingdian Irrigation District.

(2) After spatial overlaying of the selected 13 key factors, we found that the overall water and soil environmental vulnerability in the study area remained below grade IV. The extremely vulnerable zone is mainly distributed in the eastern part of the study area. The environmental degradation risk in the western part is lower than that in the eastern part.
(3) The evolution of water and soil environmental vulnerability in the study area can be divided into two stages from 1994 to 2018. Phase 1 from 1994 to 2006 is the process of environmental deterioration, and Phase 2 from 2006 to 2018 is the process of environmental restoration.

This is a relatively novel research method, which can be used to effectively evaluate the evolution of the water and soil environment in the irrigation area and helps to better plan and manage the land and water resources in arid areas around the world.

ACKNOWLEDGEMENTS

This work was supported by the National Nature Science Foundation of China (Grant No. 51579102); University Scientific and Technological Innovation Talents Programme of Henan, China (Grant No. 19IRTSTHN030); Zhongyuan Science and Technology Innovation Leading Talent Support Program of Henan, China (Grant No. 204200510048). Zhejiang Province Key R&D Programme, China (2021C03019). Sincere gratitude is extended to the editor and anonymous reviewers for their professional comments, which greatly improved the presentation of the paper.

CONFLICT OF INTEREST

The authors have no conflict of interest to declare.

DATA AVAILABILITY STATEMENT

All relevant data are included in the paper or its Supplementary Information.

REFERENCES

Angelovicová, L. & Fazekašová, D. 2014 Contamination of the soil and water environment by heavy metals in the former mining area of Rudňany (Slovakia). *Soil & Water Research* **9** (1), 18–24.
Biazin, B. & Sterk, G. 2013 Drought vulnerability drives land-use and land cover changes in the Rift Valley dry lands of Ethiopia. *Agriculture Ecosystems & Environment* **164**, 100–113.
Braimoh, A. K. & Vlek, P. L. G. 2004 The impact of land-cover change on soil properties in northern Ghana. *Land Degradation and Development* **15** (1), 65–74.
Chen, J., Yang, X., Yin, S. & Wu, K. S. 2016 The vulnerability evolution and simulation of the social-ecological systems in the semi-arid area based on the VSD framework. *Acta Geographica Sinica* **71** (07), 1172–1188 (in Chinese).
Felix, O. D. H., Diekkrüger, B., Steup, G., Yira, Y., Hoffmann, T., Rode, M. & Naeschen, K. 2019 Modeling the effect of land use and climate change on water resources and soil erosion in a tropical West African catchment (Dano, Burkina Faso) using SHETRAN. *Science of the Total Environment* **653**, 431–445.
Guan, S., Kang, Z. X. & Peng, C. 2016 Analysis on cloud characteristics of wear acoustic emission signal for vehicle cutting tool. *Transactions of the Chinese Society of Agricultural Engineering* **32** (20), 63–69 (in Chinese).
Huang, J. G., Wei, X. H., Wang, X. Z., Zhou, H. Y. & Li, H. X. 2014 Effect of vegetation degradation on soil organic matte and nutrient in process of rocky desertification in typical karst area of Northern Guangdong. *Soil and Fertilizer in China* **1**, 15–18.
Kali, E. S., Leif, G. O., Nathan, J. H., Brezonik, P. L. & Bauer, M. E. 2003 Extending satellite remote sensing to local scales: land and water resource monitoring using high-resolution imagery. *Remote Sensing of Environment* **88** (1–2), 144–156.
Lamsal, P., Kumar, L., Atreya, K. & Pant, K. P. 2017 Vulnerability and impacts of climate change on forest and freshwater wetland ecosystems in Nepal: a review. *Ambio* **46** (8), 1–16.
Li, D., Liu, C., Du, Y. & Xu, H. 2004 Artificial intelligence with uncertainty. *Journal of Software* **15** (11), 1583–1594 (in Chinese).
Li, J., Li, C. & Yin, Z. 2016 ArcGIS based kriging interpolation method and its application. *Bulletin of Surveying and Mapping* **32** (20), 63–69 (in Chinese).
Liu, X., Wang, Y. & Yang, W. 2020 Study on soil quality evaluation methods under the background of vegetation degradation in Qinghai-Tibet Plateau. *Journal of Lanzhou University (Natural Sciences)* **56** (02), 143–153.
Lu, Y., Su, W. & Hua, C. 2011 Analysis of ecological vulnerability in Zuojiang river basin based on landscape pattern and ecosystem sensitivity. *Research of Soil and Water Conservation* **18** (03), 78–82 + 87.
Nguyen, T. & Nahavandi, S. 2016 Modified AHP for gene selection and cancer classification using type-2 fuzzy logic. *IEEE Transactions on Fuzzy Systems* **24** (2), 273–287.

Qing, G., Yao, Z., Ligang, M., Jiadan, L., Ke, W., Kefeng, Z., Xiaobin, Z. & Li, S. 2016 Assessment of reservoir water quality using multivariate statistical techniques: a case study of Qiandao Lake, China. *Multidisciplinary Digital Publishing Institute* **8** (3), 243.

Rosa, D. D. L., Moreno, J. A. & Mayol, F. 2000 Assessment of soil erosion vulnerability in western Europe and potential impact on crop productivity due to loss of soil depth using the ImpelERO model. *Agriculture Ecosystems & Environment* **81** (03), 179–190.

Setegn, S. G., Srinivasan, R., Dargahi, B. & Melesse, A. M. 2009 Spatial delineation of soil erosion vulnerability in the Lake Tana Basin, Ethiopia. *Hydrological Processes* **23** (26), 3738–3750.

Shahid, M., Cong, Z. & Zhang, D. 2018 Understanding the impacts of climate change and human activities on streamflow: a case study of the Soan River basin, Pakistan. *Theoretical and Applied Climatology* **134** (1–2), 205–219.

Shahid, M., Rahman, K. U., Balkhair, K. S. & Nabi, A. 2020 Impact assessment of land use and climate changes on the variation of runoff in Margalla Hills watersheds, Pakistan. *Arabian Journal of Geosciences* **13** (5), 1–14.

Stampoulis, D., Andreadis, K. M., Granger, S. L., Fisher, J. B., Turk, F. J., Behrangi, A., Ines, A. V. & Das, N. N. 2016 Assessing hydro-ecological vulnerability using microwave radiometric measurements from WindSat. *Remote Sensing of Environment* **184**, 58–72.

Wang, J., Zhao, G. & Du, C. 2005 Analysis on the regional ecological environment vulnerability based on the information of spatial structure of landscapes. *Arid Zone Research* **03**, 317–321.

Wang, R., Zhao, G., Yu, Z., Zhang, Y. & Zhang, H. 2012 Assessment of land use effects on environmental vulnerability by ecological niche suitability model. *Transactions of the Chinese Society of Agricultural Engineering* **28** (11), 218–224 (in Chinese).

Wang, M., Pei, H., Sun, J. S., Tan, X. Y. & Zhang, X. Q. 2017 Comprehensive evaluation on quality and safety of *Ganoderma lucidum* based on improved analytic hierarchy process. *Transactions of the Chinese Society of Agricultural Engineering* **33** (4), 302–308 (in Chinese).

Xu, C. 2010 *Research on the Effect of Local Water and Soil Environment Caused by Water-Salt Transportation in Jing-Dian Irrigation District. Doctoral dissertation*, Lanzhou University (in Chinese).

Xu, Q., Huang, M., Liu, P. L. & Li, R. Q. 2011 Integrated assessment of eco-environmental vulnerability in Pearl River Delta based on RS and GIS. *Chinese Journal of Applied Ecology* **22** (11), 2987–2995 (in Chinese).

Xu, C., Cheng, H., Wang, Y., Wang, R. R., Liu, L, Y. & Zhang, R. 2017 Improved multi-level fuzzy evaluation model based on cloud theory for evaluation of soil salinization degree. *Transactions of the Chinese Society of Agricultural Engineering* **33** (24), 88–95 (in Chinese).

Xu, C., Tian, J., Wang, G., Nie, J. & Zhang, H. 2019 Dynamic simulation of soil salt transport in arid irrigation areas under the HYDRUS-2D-Based rotation irrigation mode. *Water Resources Management* **33** (10), 3499–3512.

Zhang, Q., Zhang, Y. & Zhong, M. 2014 A cloud model based approach for multi-hierarchy fuzzy comprehensive evaluation of reservoir-induced seismic risk. *Journal of Hydraulic Engineering* **45** (1), 87–95 (in Chinese).

Zhang, Y., Wang, R. & Bai, Q. 2018 Development and change of soil salinization in Hetao irrigation area of Inner Mongolia. *Journal of Irrigation and Drainage* **37**, 118–122.

Zhou, Q., Zhang, X. & Wang, Z. 2014 Land use ecological risk evaluation in Three Gorges Reservoir area based on normal cloud model. *Transactions of the Chinese Society of Agricultural Engineering* **30** (23), 289–297 (in Chinese).

First received 27 February 2021; accepted in revised form 25 June 2021. Available online 6 July 2021

HYDRUS-2D simulations of nitrate nitrogen and potassium transport characteristics under fertilizer solution infiltration of furrow irrigation

Wei-Bo Nie, Kun-Kun Nie, Yi-Bo Li and Xiao-Yi Ma

ABSTRACT

Understanding the characteristics of soil solute transport is fundamental to the design and management of furrow irrigation systems. This study determined the soil hydraulic and solute transport parameters by inverse solution with HYDRUS-2D and then verified them. The experimental data were obtained from the infiltration of clay loam and sandy loam of different potassium nitrate (KNO_3) concentrations under furrow irrigation. Then, the initial soil water content (θ_0), KNO_3 concentration, and water depth (h_0) affecting the transport characteristics of nitrate nitrogen (NO_3^--N) and potassium (K^+) were analyzed. The results indicated that the soil hydraulic and solute transport parameters determined from the inversion solution with HYDRUS-2D were reliable. The soil saturated water content, saturated hydraulic conductivity, and empirical parameter *n* in the van Genuchten–Mualem model increase with the increase of KNO_3 concentrations, whereas the empirical parameter *a* shows a decreasing tendency. The distribution range of NO_3^--N increased with the increases of θ_0 and the KNO_3 concentration, which had barely any effect on the range of K^+ distribution. The horizontal distribution range of NO_3^--N and K^+ increased with the increase of h_0, but it had no obvious influence on the vertical range.

Key words | fertilizer solution infiltration, furrow irrigation, solute transport

Wei-Bo Nie (corresponding author)
Kun-Kun Nie
Yi-Bo Li
State Key Laboratory of Eco-hydraulics in Northwest Arid Region,
Xi'an University of Technology,
Xi'an 710048,
China
E-mail: *nwbo2000@163.com*

Xiao-Yi Ma
Key Laboratory of Agricultural Soil and Water Engineering in Arid and Semiarid Areas, Ministry of Education,
Northwest A&F University,
Yangling 712100,
China

HIGHLIGHTS

- Soil hydraulic and solute transport parameters were determined by inverse solution with HYDRUS-2D, and then the reliability was verified.
- θ_n, *n* and K_s (VG-M model) increased with increasing KNO_3 concentration, while *a* exhibited a decreasing trend
- The initial soil water content, KNO_3 concentration and water depth have an obvious effect on the distribution of NO_3^--N.
- K^+ was mainly distributed within 10 cm of the furrow bottom.

INTRODUCTION

Understanding the characteristics of soil solute transport is fundamental to the design and management of furrow irrigation systems. However, the study of this problem is complicated because of the nonlinear relationship between soil capillary force and gravity at different locations during furrow irrigation, which can be affected by soil physical properties and different irrigation technical factors, and the water potential gradient of fertilizer solution infiltration varies at different locations. Therefore, some researchers (Li *et al.* 2004; Ebrahimian *et al.* 2012b) have analyzed the effects

doi: 10.2166/ws.2020.266

of different factors on the transport characteristics of water content and fertilizer (such as nitrate nitrogen NO_3^--N and ammonium nitrogen $NH4^+$-N) in soil wetting patterns based on fertilizer solution infiltration, and numerous empirical models have been established, which can be used to estimate cumulative infiltration, water content of wetting patterns, and NO_3^--N and $NH4^+$-N distributions. Due to the influences of soil texture and initial condition, the distributions of different types of ions in the soil exhibit obvious differences. Silberbush & Barber (1983) indicated that potassium ions (K^+) have strong adsorption characteristics in soil. Hanson *et al.* (2006) simulated and analyzed the transport processes of K^+ under drip irrigation, and the results proved that K^+ can easily accumulate on soil surfaces. Liu *et al.* (2017) studied results that showed that the distribution of NO_3^--N was similar to that of soil water content, whereas K^+ was concentrated around the Moistube. Experimental research methods can reveal the transport characteristics of soil water and fertilizer, but experimentation is time-consuming, and most research results are limited because they are obtained under specific experimental conditions.

Researchers use various numerical simulation methods to study soil water and fertilizer transport characteristics. Zerihun *et al.* (2005a, 2005b) established a mathematical model of soil water flow and solute transport for border fertigation and verified that the model accurately simulated the soil water content and solute transport. Doltra & Muñoz (2010) simulated NO_3^--N transport processes in soil under drip fertigation, and the correlation coefficient between their simulated and measured NO_3^--N content was higher than 0.76. Tan *et al.* (2015) simulated the transport characteristics of soil water, NO_3^--N, and NH_4^+-N in paddy fields, and the results showed that the simulated values were consistent with the measured values. Li *et al.* (2019) simulated soil water and nitrogen transport processes under drip irrigation, and the relative error between the simulated and measured soil water content was within 10%; the errors for NO_3^--N and NH_4^+-N were within 20%. The aforementioned studies proved that numerical simulation can be used to study the characteristics of fertilizer solution infiltration with high accuracy. Although many studies have been done on solute transport characteristics during fertilizer solution infiltration (Deb *et al.* 2015; Iqbal *et al.* 2016; Brunetti *et al.* 2018; Ranjbar *et al.* 2019), few publications have considered the effect of the changes in fertilizer concentration on soil hydraulic parameters, and analyzed the influences of various factors (such as initial soil water content, fertilizer concentration, and water depth) on the solute transport characteristics.

Therefore, the objectives of this study were: (1) to determine the soil hydraulic and solute transport parameters by inverse solution with HYDRUS-2D from infiltration data of different potassium nitrate (KNO_3) concentrations under furrow irrigation, and then verify the reliability of the parameters; and (2) to analyze the effect of initial soil water content (θ_0), KNO_3 concentration, and water depth (h_0) on the NO_3^--N and K^+ transport characteristics by HYDRUS-2D.

MATERIALS AND METHODS

Fertilizer solution infiltration experiment design

Soil samples were taken from Yangling District (107°55′50″–108°07′50″ E and 34°14′30″–34°19′00″ N), Shaanxi Province. The terrain of Yangling District is divided into three terraces from south to north. The altitude of the first terrace is 420–430 m, and the soil texture is sandy loam. The altitudes of the second and third terraces are 450–485 m and 515–540 m, respectively, and the soil texture is clay loam. According to the topographical features of Yangling District, soil samples were collected from typical fields cultivated all year round in the first and third terraces at a depth of 0–60 cm (the depth where the roots of crops such as wheat and corn are mainly distributed). The residual water content of air-dried soil samples was measured using a weighing method. The soil texture was determined by a Mastersizer 2000 particle size analyzer. The contents of NO_3^--N and K^+ in the soil samples were determined by an ultraviolet spectrophotometer and a flame photometer, respectively. The results are listed in Table 1.

KNO_3 is a potassium–nitrogen compound fertilizer with stable physical and chemical properties. It can supplement the potassium and nitrogen required for crop growth at the same time, and it is widely used in crops and fruit trees. Therefore, KNO_3 was selected as the experimental fertilizer in this study. It was dissolved in water to carry out fertilizer solution infiltration experiments. The soil bulk density (ρ) of the first and third terraces in Yangling District varied from

Table 1 | Soil physical and chemical properties of the experimental soils

Soil samples site	Soil texture	NO_3^--N/(mg kg^{-1})	K^+/(mg kg^{-1})	θ_r/(g g^{-1})	Content of soil particle /% Clay (<0.002 mm)	Silt (0.002–0.02 mm)	Sand (0.02–2 mm)
First terrace	Sandy loam	33.4	19.9	0.05	14.76	27.12	58.12
Third terrace	Clay loam	49.6	32.7	0.03	23.63	30.25	46.12

Note: θ_r is the residual water content.

1.40 to 1.50 g cm^{-3} and 1.30 to 1.40 g cm^{-3}, respectively (Nie *et al.* 2017). Therefore, two bulk density cases were designed for sandy loam and clay loam in this study, namely 1.40 and 1.50 g cm^{-3} for sandy loam and 1.30 and 1.40 g cm^{-3} for clay loam. Based on the ranges of KNO_3 concentration set by previous studies (Zipelevish *et al.* 2000; Wang *et al.* 2010; Liu *et al.* 2017), the KNO_3 concentration was tested, in this study, at four gradients: 0, 250, 600, and 900 mg L^{-1}. Furthermore, θ_0 and h_0 were selected as the experimental design factors to carry out infiltration experiments of furrow irrigation (Table 2).

The air-dried soil sample was filtered through a 2 mm sieve, and a 5 cm layer of soil from the sieved material was placed in a soil box; the size of the soil box was 80 × 5 × 100 cm (length × width × height). A furrow was excavated along the side of the soil box, and that furrow had the trapezoidal section commonly used in northern China, with a maximum depth of 15 cm, bottom width of 20 cm and a 1:1 side-slope. Because the furrow was axisymmetric, only a half-section was used in the experiments (the range of ABCDO), as shown in Figure 1(a). The infiltration time and cumulative infiltration were recorded. After the experiments of infiltration, soil samples were collected from different locations (Figure 1(a)), and soil samples were divided into two parts. Some soil samples were measured for water content through a weighing method, and other soil samples were subjected to an ultraviolet spectrophotometer and flame photometer to determine their contents of NO_3^--N and K^+, respectively.

Model descriptions

Soil water movement model

Assuming that the soil is a homogeneous, isotropic porous medium – irrespective of the air resistance inside the soil,

Table 2 | Experimental design fertilizer solution infiltration under furrow irrigation

Clay loam No.	C /(mg L^{-1})	θ_0 /(cm^3 cm^{-3})	h_0 /(cm)	ρ /(g cm^{-3})	Sandy loam No.	C /(mg L^{-1})	θ_0 /(cm^3 cm^{-3})	h_0 /(cm)	ρ /(g cm^{-3})
T1	0	0.208	10.5	1.30	T1	0	0.168	10.5	1.40
T2	250	0.130	14.5	1.30	T2	250	0.112	14.5	1.40
T3	600	0.130	10.5	1.30	T3	600	0.112	10.5	1.40
T4	900	0.156	5.5	1.30	T4	900	0.140	5.5	1.40
T5	0	0.168	14.5	1.40	T5	0	0.150	14.5	1.50
T6	250	0.140	14.5	1.40	T6	250	0.120	14.5	1.50
T7	600	0.140	5.5	1.40	T7	600	0.120	5.5	1.50
T8	900	0.224	10.5	1.40	T8	900	0.180	10.5	1.50
V9*	0	0.208	5.5	1.30	V9*	0	0.168	5.5	1.40
V10*	250	0.156	5.5	1.30	V10*	250	0.140	5.5	1.40
V11*	600	0.208	14.5	1.30	V11*	600	0.168	14.5	1.40
V12*	900	0.156	10.5	1.30	V12*	900	0.140	10.5	1.40

Note: θ_0 is the initial soil water content; *C* is the concentration of KNO_3; h_0 is the water depth of furrow; ρ is the soil bulk density; '*' refers to the verification experiments.

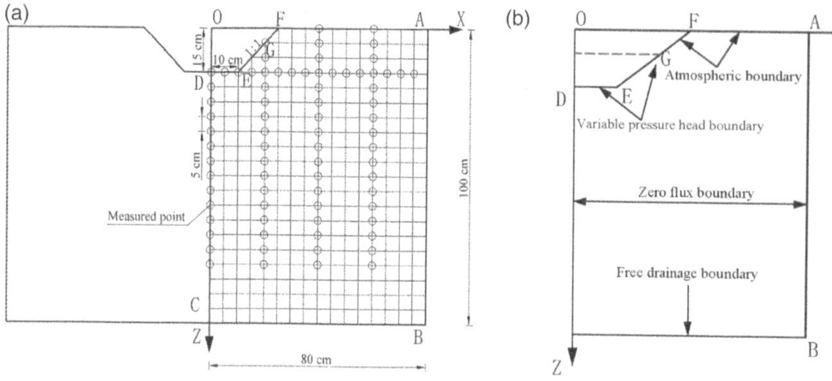

Figure 1 | Furrow cross-section and boundary conditions for model solving. (a) Furrow cross-section and measured points. (b) Boundary conditions.

temperature, and the effect of evaporation on infiltration – the equation of soil water movement in infiltration of a furrow can be expressed as (Šimůnek *et al.* 1999):

$$\frac{\partial \theta}{\partial t} = \frac{\partial}{\partial x}\left[K(h)\frac{\partial h}{\partial x}\right] + \frac{\partial}{\partial z}\left[K(h)\frac{\partial h}{\partial z}\right] - \frac{\partial K(h)}{\partial z} \tag{1}$$

where z is a vertical coordinate that is positive in a downward direction, θ is the volumetric water content ($cm^3\ cm^{-3}$), t is the time (min), h is the soil water pressure head (cm), and $K(h)$ represents the unsaturated hydraulic conductivity ($cm\ min^{-1}$). Regarding the soil water effective saturation, $K(h)$ and S_e are described using the van Genuchten–Mualem (VG-M) model (van Genuchten 1980), which is as follows:

$$K(h) = \begin{cases} K_s S_e^l [1 - (1 - S_e^{1/m})^m]^2 & h < 0 \\ K_s & h \geq 0 \end{cases} \tag{2}$$

$$S_e = \begin{cases} \dfrac{\theta - \theta_r}{\theta_s - \theta_r} = \dfrac{1}{(1 + |ah|^n)^m} & h < 0 \\ 1 & h \geq 0 \end{cases} \tag{3}$$

where θ_r is the residual soil water content ($cm^3\ cm^{-3}$); θ_s is the saturated water content ($cm^3\ cm^{-3}$); m, n, a, and l are empirical parameters of the soil–water characteristic curve, $m = 1-(1/n)$, and l is usually equal to 0.5; and K_s is the saturated hydraulic conductivity ($cm\ min^{-1}$).

Soil solute transport model

The partial differential equation governing two-dimensional solute transport during transient water flow in a variably saturated rigid porous medium is taken as (Šimůnek *et al.* 1999):

$$\begin{aligned} \frac{\partial(\theta C_k + \rho \alpha_k C_k)}{\partial t} = {} & \frac{\partial}{\partial x}\left(\theta D_{xx}\frac{\partial C_k}{\partial x} + \theta D_{xz}\frac{\partial C_k}{\partial z}\right) \\ & + \frac{\partial}{\partial z}\left(\theta D_{zz}\frac{\partial C_k}{\partial z} + \theta D_{xz}\frac{\partial C_k}{\partial x}\right) - \frac{\partial q_x C_k}{\partial x} \\ & - \frac{\partial q_z C_k}{\partial z} \end{aligned} \tag{4}$$

where C_k is the solution concentration ($g\ cm^{-3}$); $k = 1, 2$ represents NO_3^- and K^+, respectively; ρ is the soil bulk density ($g\ cm^{-3}$); α_k is the adsorption parameter, indicating the partition coefficient of the solute in the solid–liquid phase ($cm^3\ g^{-1}$); q_x and q_z are the water fluxes in the horizontal and vertical directions ($cm\ min^{-1}$), respectively; and D_{xx}, D_{xz}, and D_{zz} are components of the dispersion coefficient tensor ($cm^2\ min^{-1}$), which can be calculated by the following equation:

$$\begin{aligned} \theta D_{xx} &= D_L\frac{q_x^2}{|q|} + D_T\frac{q_z^2}{|q|} + \theta D_w \tau \\ \theta D_{zz} &= D_L\frac{q_z^2}{|q|} + D_T\frac{q_x^2}{|q|} + \theta D_w \tau \\ \theta D_{xz} &= (D_L - D_T)\frac{q_x q_z}{|q|} \end{aligned} \tag{5}$$

where $|q|$ is the absolute value of water flux ($cm\ min^{-1}$); D_L and D_T are the longitudinal and transverse dispersivities (cm), respectively; D_w is the ionic or molecular diffusion coefficient in free water ($cm^2\ min^{-1}$); and τ is the tortuosity factor, which can be calculated by the following equation:

$$\tau = \frac{\theta^{7/3}}{\theta_s^2} \quad (6)$$

Initial and boundary conditions

The initial conditions of the experiment were that the θ_0, NO_3^--N and K^+ concentrations were stable. Therefore, the initial conditions of the equations were such that all the points in the calculation area had the same matric potential:

$$h(x, z, t) = h_n \quad (0 \le x \le X, 0 \le z \le Z, t = 0) \quad (7)$$

$$C_k(x, z, t) = C_k^n \quad (0 \le x \le X, 0 \le z \le Z, t = 0) \quad (8)$$

where h_n is the pressure head corresponding to the θ_0 condition (cm); C_k^n is the initial soil solute concentration ($g\ cm^{-3}$), namely NO_3^--N and K^+ concentrations; and X and Z are the maximum horizontal length and vertical depth of the simulated domain, respectively ($X = 80$ cm and $Z = 100$ cm were adopted in this study).

Boundary conditions (Figure 1(b)): DE and EG represent the variable pressure head boundaries, and a constant water depth and solution concentration are maintained. Because of the short durations of the infiltration experiments, the scenario posits no rainfall and no plant cover on the surface, and it is considered to have no external solute supplementation or loss. Therefore, GF and FA represent the atmospheric boundaries, and evaporation can be neglected to create zero-flux boundaries. The left boundary AB is the symmetry axis, and the horizontal flux is 0. The right boundary CD is the symmetry axis of the adjacent furrow, which is regarded as a zero-flux surface. The lower boundary BC contains drainage holes. During the experiment, the wetting front does not reach the base of the soil box and does not affect the process of fertilizer solution infiltration, and the constant initial conditions are established to create a free drainage boundary.

On the basis of the initial (Equations (7) and (8)) and boundary conditions (Figure 1(b)), Equations (1) and (4) are solved by HYDRUS-2D by finite elements, which can define the computational grid automatically. The smaller finite elements are recommended in regions where rapid changes may occur of water content and pressure (such as at the infiltrating surface) and larger ones elsewhere. The resulting computational grid consisted of 2,446 nodes (not shown in Figure 1(b)).

Parameter inverse solution procedure and evaluation

A number of soil hydraulic and solute transport parameters were determined using an inverse solution procedure implementing the Levenberg–Marquardt optimization module built in HYDRUS-2D (Šimůnek *et al.* 1999), and the experimental data from the infiltration experiments of Nos. T1–T8 (Table 2). The inverse method is based on the minimization of a suitable objective function, which expresses the discrepancy between the measured and simulated values, and it can be expressed as (Ebrahimian *et al.* 2012a):

$$SSQ = \sum_{j=1}^{M} v_j [q_j*(x, z, t) - q_j(x, z, t, b)]^2 \quad (9)$$

where M is the number of the measurement set; $q_j*(x, z, t)$ is the measured values (e.g., soil water contents, NO_3^--N and K^+ concentrations) for the jth measurement set at time t (t is the experiment's end time), location x, and depth z; $q_j(x, z, t, b)$ is the corresponding model simulation obtained with the vector of optimized parameters b (i.e., soil hydraulic and solute transport parameters); and v_j is weights associated with a particular measurement set and assumed to be equal to 1 in this study. Quality in parameter estimation was assessed using two indicators: the determination coefficient (R^2) and SSQ.

This inverse solution procedure has been successfully applied by several researchers (Abbasi *et al.* 2003; Verbist *et al.* 2009; Ebrahimian *et al.* 2012a) to determine soil hydraulic and solute transport parameters. Because the θ_r has little effect on soil water movement (Wang *et al.* 2013), θ_r was set as the water content of air-dried soil in this study. Nitrogen in the KNO_3 solution is mainly in the form of NO_3^- and is negatively charged, and the adsorption of NO_3^- can be neglected. The soil wetting pattern has a high water-content in fertilizer solution infiltration, and the nitrification of nitrogen is suitable for aerobic environments; thus, the nitrification of NH_4^+-N can be neglected. Furthermore, the infiltration experiments were brief (the longest time = 6 h), and consequently the mineralization and denitrification of nitrogen in the soil can be neglected in this study. The temperature during the infiltration experiments was approximately 20–25 °C. Siyal *et al.* (2012) reported that $D_w = 0.00113$ cm^2 min^{-1} for NO_3^--N at 20–23 °C, and Ren *et al.* (2013) wrote that $D_w = 0.00395$ cm^2 min^{-1} for K^+, so the D_w of NO_3^--N and K^+ can be assigned previously published values in the solute transport parameter inversion process of this study. Therefore, the inverse solution was applied to four soil hydraulic parameters, including θ_s, n, K_s, and a, and three solute transport parameters, including α_k of K^+, D_L and D_T of NO_3^--N and K^+. The inverse optimization method simultaneously uses all measured data, i.e., soil water contents and NO_3^--N and K^+ concentrations.

In this study, the reliability of soil hydraulic and solute transport parameters was evaluated by comparing the statistical indicators of the simulated values of cumulative infiltration, soil water contents, NO_3^--N and K^+ concentrations with the measured values for the verification experiments of Nos. V9–V12 (Table 2). The indicators used were the root mean square error (*RMSE*), the mean percentage of bias error (*MPBE*) and the mean absolute percentage relative error (*MAPRE*), which were calculated as follows (Karandish & Šimůnek 2016; Nie *et al.* 2018):

$$RMSE = \sqrt{\frac{\sum_{i=1}^{N}(S_i - M_i)^2}{N}} \tag{10}$$

$$MPBE = \frac{1}{N}\sum_{i=1}^{N}\frac{S_i - M_i}{M_i} \times 100\% \tag{11}$$

$$MAPRE = \frac{1}{N}\sum_{i=1}^{N}\frac{|S_i - M_i|}{M_i} \times 100\% \tag{12}$$

where S_i and M_i are the simulated and measured data, respectively; and N is the number of observations. The *RMSE* and *MAPRE* provide an overall measure of the degree to which the data differ from the model estimations. The low *RMSE* and *MAPRE* indicate a more favorable model fit. If the *MPBE* is within ±10%, the *MPBE* is considered to be within an acceptable range (Moriasi *et al.* 2007).

RESULTS AND DISCUSSION

Parameter inversion and verification

Inversion solution of the soil hydraulic and solute transport parameters

The cumulative infiltration per unit area curve (cumulative infiltration per length divided by wetted perimeter) under different KNO_3 concentrations is shown in Figure 2. From the measured soil water contents and concentrations of NO_3^--N and K^+ at the end of the infiltration experiments of Nos. T1–T8 (Table 2 and Figure 2), the soil hydraulic and solute transport parameters were determined by inverse solution with HYDRUS-2D, and the results are listed in Table 3.

The cumulative infiltration per unit area showed an increasing trend with the increase of KNO_3 concentration. For example, for the cases where KNO_3 concentrations were 250, 600, and 900 mg L^{-1}, the cumulative infiltration per unit area of clay loam and sandy loam soil textures increased by 17.05%, 28.65%, 34.71% and 12.20%, 16.19%, 20.45% compared with 0 mg L^{-1} at the end of the infiltration experiments, respectively (Figure 2). The soil hydraulic parameter inversion results showed that the parameters of θ_s, n, and K_s increased with the increase of KNO_3 concentrations, but a showed a decreasing trend (Table 3), which is consistent with the results of Feng (2017). The reasons may be that K^+ has strong adsorption (Silberbush & Barber 1983; Hanson *et al.* 2006; Ren *et al.* 2013), which exchanges with the ions on the surface of the soil colloid, changing the properties of the soil colloid and thus affecting the pore structure of the soil. Moreover, the KNO_3 solution has a higher solute potential than the water, and the soil–water potential difference at the interface of the wetting front is larger for the KNO_3 solution than for the water, which increases the infiltration capacity

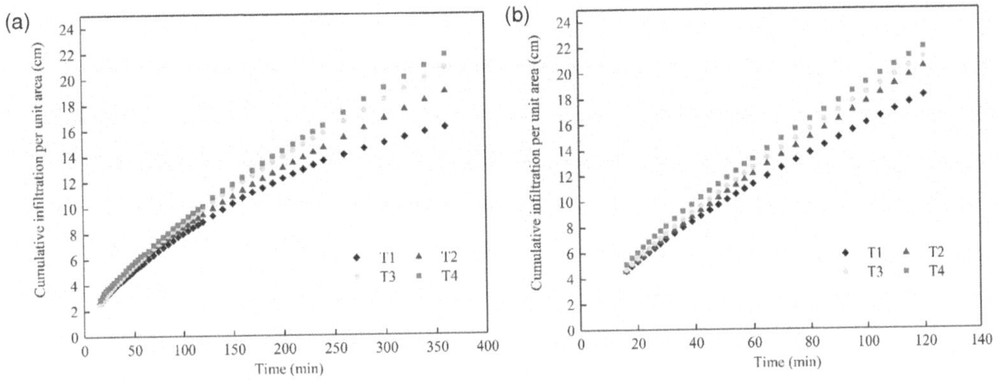

Figure 2 | Cumulative infiltration per unit area curves under different potassium nitrate concentrations. (a) Clay loam. (b) Sandy loam.

of the soil (Figure 2). This explanation would be consistent with the research conclusions of Liu *et al.* (2017). Meanwhile, it should be noted that the parameters of θ_s, n, and K_s were unlikely to increase continuously with the increase of KNO_3 concentrations, and a cannot decrease continuously. The reason may be that with the increase of KNO_3 concentrations, the pore structure of the soil is changed and the solute potential at the interface of the wetting front is increased, which makes the soil infiltration capacity gradually increase. However, with the continuous increase of KNO_3, the viscosity of water increases, that is, the fluidity of water becomes worse, which will reduce the soil infiltration capacity. Because only four concentrations of KNO_3 were considered in this study, and the concentrations are relatively small (the maximum concentration of KNO_3 is 900 mg L^{-1}), what leads to this process cannot be reflected upon, and further research is needed on this issue.

The solute dispersivities mainly depend on soil properties, and these parameters have scale characteristics. Beven *et al.* (1993) and Cote *et al.* (2003) wrote that D_L can be taken as 1/10 of the simulated length. The results of Šimůnek & van Genuchten (2008) and Siyal *et al.* (2012) proved that D_L is generally between 0.1 and 15 cm, and D_T takes 1/100–1/5 of D_L according to soil properties and can obtain a satisfactory solute transport simulation result. As isotropic soil is used in this study, and the D_L and D_T of K^+ and NO_3^--N have small differences, which were determined by inverse solution with HYDRUS-2D, so the mean values of D_L and D_T were adopted under the same bulk density of each soil texture, i.e. the ranges of D_L are 4.03–4.66 cm and 1.26–1.95 cm for clay loam and sandy loam, respectively, and the value of D_T is approximately 1/20 of D_L (Table 3). These findings are basically consistent with the aforementioned results, indicating that the inversion results of D_L and D_T of NO_3^--N and K^+ are reasonable in this study.

The soil adsorption process of K^+ can be described as a linear equation (Hanson *et al.* 2006; Grecco *et al.* 2019); that is, α_k is a constant value, which is independent of the solution concentration. In addition, the adsorption of K^+ is mainly related to soil organic matter and particle content (Long *et al.* 2001), and thus the α_k of K^+ is taken as the inversion average value under the same soil texture, namely the α_k of K^+ is 4.47 and 3.19 $cm^3\ g^{-1}$ in clay loam and sandy loam, respectively. The inversion results of α_k are close to $\alpha_k = 1.533\ cm^3\ g^{-1}$ of silt loam (Ren *et al.* 2013), but this result differs greatly from the 28.7 $cm^3\ g^{-1}$ (soil texture is loam) used in the research process of Hanson *et al.* (2006). The reasons may be that field experiments were used by Hanson *et al.* (2006), and numerous external factors are relevant. For example, salt may accumulate in the surface soil with water evaporation, and the solute that is absorbed by plant roots may affect the α_k of K^+.

Verification of the soil hydraulic and solute transport parameters

To verify the reliability of soil hydraulic and solute transport parameters determined by inversion solution, the parameters were input into HYDRUS-2D (Table 3) and the verification experiments of Nos. V9–V12 (Table 2) were simulated. The infiltration time used in the simulation

Table 3 | The soil hydraulic and solute transport parameters determined by inversion solution in HYDRUS-2D

		Soil hydraulic parameters					Solute transport parameters					
No.	$C/(\text{mg}\cdot\text{L}^{-1})$	$\theta_r/(\text{cm}^3\ \text{cm}^{-3})$	$\theta_s/(\text{cm}^3\ \text{cm}^{-3})$	a	n	$K_s/(\text{cm min}^{-1})$	D_L/cm	D_T/cm	α_K of $K^+/(\text{cm}^3\ \text{g}^{-1})$	R^2	SSQ	Soil texture
T1	0	0.042	0.472	0.030	1.392	0.021	4.66	0.245	4.47	0.9600	0.0059	Clay loam
T2	250	0.042	0.475	0.029	1.408	0.023				0.8754	0.0101	
T3	600	0.042	0.488	0.025	1.421	0.024				0.8321	0.0143	
T4	900	0.042	0.480	0.023	1.439	0.026				0.9033	0.0095	
T5	0	0.046	0.462	0.020	1.345	0.020	4.03	0.198		0.9322	0.0073	
T6	250	0.046	0.469	0.017	1.353	0.021				0.9129	0.0089	
T7	600	0.046	0.476	0.014	1.361	0.022				0.8126	0.0179	
T8	900	0.046	0.476	0.012	1.367	0.023				0.7856	0.0201	
T1	0	0.077	0.417	0.059	1.632	0.092	1.95	0.099	3.19	0.9173	0.0082	Sandy loam
T2	250	0.077	0.418	0.055	1.642	0.096				0.8649	0.0096	
T3	600	0.077	0.421	0.051	1.653	0.099				0.8433	0.0131	
T4	900	0.077	0.426	0.047	1.667	0.102				0.7981	0.0182	
T5	0	0.082	0.419	0.045	1.426	0.045	1.26	0.059		0.9024	0.0089	
T6	250	0.082	0.419	0.040	1.432	0.050				0.8632	0.0109	
T7	600	0.082	0.424	0.035	1.443	0.053				0.8459	0.0139	
T8	900	0.082	0.424	0.031	1.459	0.058				0.8522	0.0129	

Note: ρ is the soil bulk density; C is the concentration of KNO_3; θ_s is the saturated water content; K_s is the saturated hydraulic conductivity; a and n are empirical parameters; θ_r is the residual soil water content, which is determined according to the residual water content in Table 1 multiplied by the soil bulk density in Table 2; α_K is the adsorption parameter; D_L and D_T are the longitudinal and transverse dispersivities, respectively.

process was consistent with the infiltration time of the corresponding experiment treatment. The simulation results and the measured values are shown in Figures 3 and 4.

The mean values of *RMSE*, *MPBE* and *MBPRE* of cumulative infiltration and soil water content between the measured and simulated values for the verification experiments of Nos. V9–V12 were 0.66 cm, 0.24%, 3.45%, and 0.027 $\text{cm}^3\ \text{cm}^{-3}$, 1.04%, 6.04%, respectively (Table 4). The results indicated that the simulated cumulative infiltration and soil water content exhibited good consistency with measured values (Figure 3). And analysis of the causes of errors was that on the one hand that the soil hydraulic parameters were used to simulate verification experiments of Nos. V9–V12, these parameters by inverse solution from sandy loam and clay loam soil texture infiltration experiments of Nos. T1–T4, respectively (Tables 2 and 3), but some differences existed in h_0 and θ_0, resulting in errors of cumulative infiltration and soil water content between the simulated and measured experiments. On the other hand, measurement errors are inevitable during such experiments. Overall, the results indicated that the soil hydraulic parameters determined by the inversion solution are reliable and those parameters can be used to simulate the soil water movement process of furrow irrigation.

The simulated concentrations of NO_3^--N and K^+ were compared with the measured values, and they agreed (Figure 4). The mean values of *RMSE*, *MPBE* and *MBPRE* of NO_3^--N and K^+ concentrations between the measured and simulated values for the verification experiments of Nos. V9–V12 were 0.062 g cm^{-3}, −0.18%, 9.34%, and 0.066 g cm^{-3}, 0.05%, 10.54%, respectively (Table 4). The accuracy of the simulation results was slightly worse than the soil water movement simulation results, which was consistent with the research results of Ma *et al.* (2004) and Phillips (2006). The reasons may be that physicochemical reactions such as mineralization, nitrification, and denitrification of nitrogen were neglected in the process of NO_3^--N transport and parameter inversion in this study, and the effects of

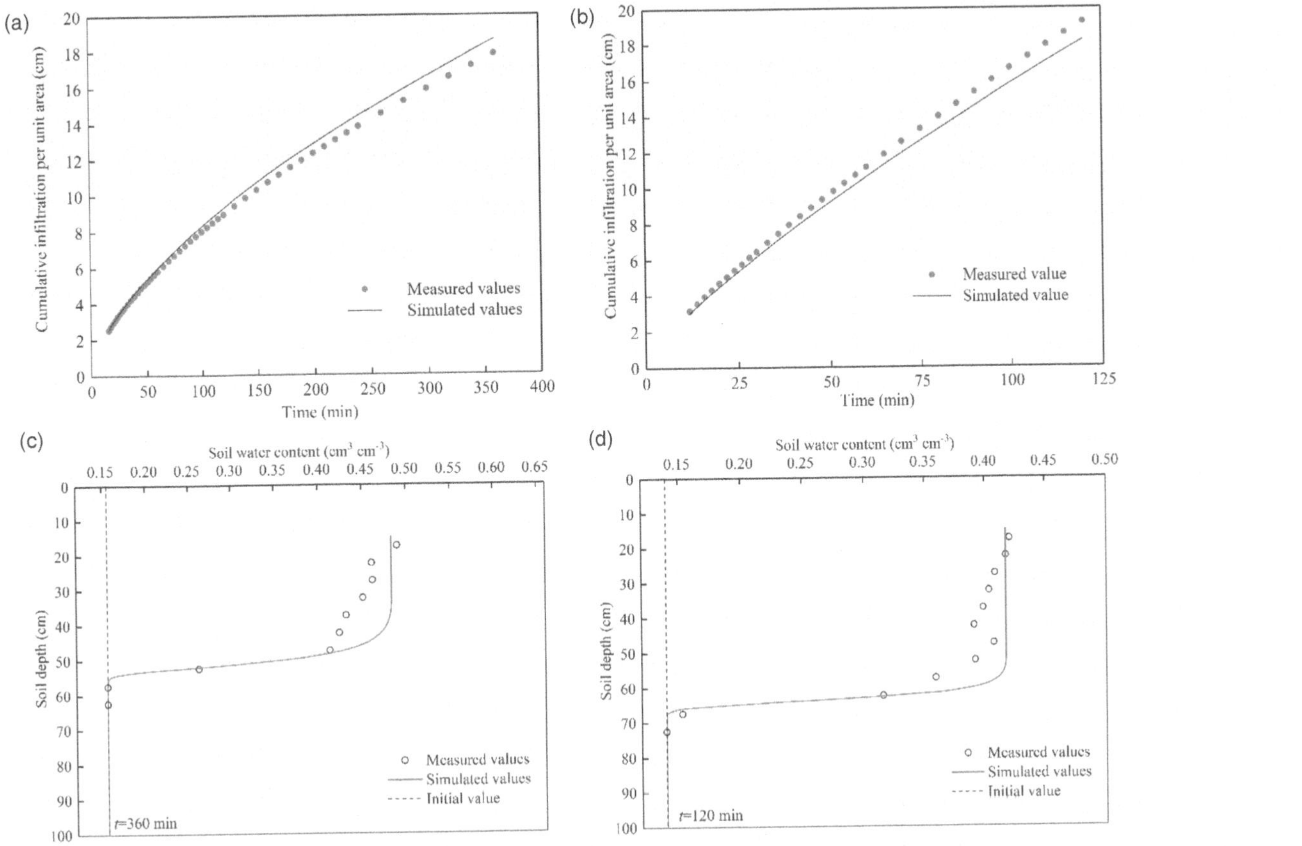

Figure 3 | Comparison between simulated and measured values of cumulative infiltration and soil water content. *Note:* vertical direction is the profile at the center of furrow. (a) Clay loam experiment No. V11. (b) Sandy loam experiment No. V10. (c) Vertical direction of clay loam experiment No. V12. (d) Vertical direction of sandy loam experiment No. V10.

soil bulk density, water content, water depth, and fertilizer concentration on α_k of K^+ were ignored. In addition, solute transport in soil is more complicated than water movement, and this study's description of the transport of convection dispersion equation was simplified. Considering the aforementioned reasons and the actual situation of the experiments, it can be concluded that the simulation accuracy of solute transport was reasonable; thus, the solute transport parameters determined in HYDRUS-2D are relatively reliable.

Solute transport analysis in fertilizer solution infiltration

Numerical simulations were used to analyze the effect of θ_0, KNO_3 concentration, and h_0 in NO_3^--N and K^+ transport characteristics by HYDRUS-2D, based on inversion solution of soil hydraulic and solute transport parameters (Table 3). Because of the similarity of the distribution characteristics of NO_3^--N and K^+ for clay loam and sandy loam, the following analysis takes clay loam as an example.

Initial soil water content

The distribution of the NO_3^--N and K^+ concentrations of the soil wetting pattern profile at the end of infiltration are shown in Figure 5, given the following values: KNO_3 concentration $C = 250$ mg L^{-1}, water depth $h_0 = 10.5$ cm, soil bulk density $\rho = 1.30$ g cm^{-3}, initial water content $\theta_0 =$ 0.130, 0.156, and 0.208 cm^3 cm^{-3} of clay loam.

The θ_0 influences the transport characteristics of NO_3^--N in the wetting pattern (Figure 5). At the end of infiltration ($t = 360$ min), for the cases where θ_0 was 0.156 and 0.208 cm^3 cm^{-3}, the vertical distribution depth of NO_3^--N increased by 5.28% and 13.53% compared with

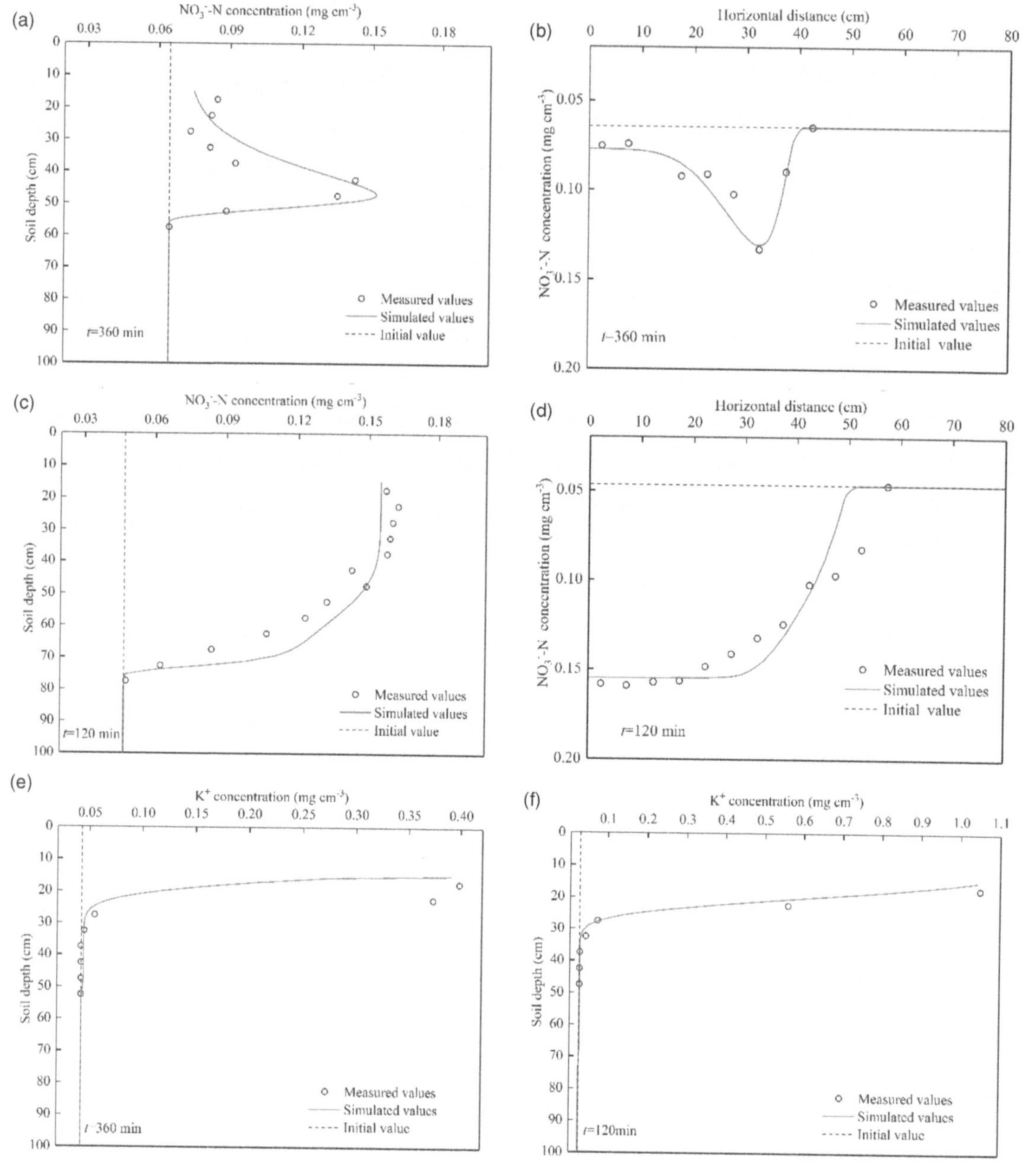

Figure 4 | Comparison between simulated and measured values of NO_3^--N and K^+ concentrations *Note:* vertical direction is the profile at the center of the furrow, and horizontal direction is the profile at the bottom of the furrow. (a) Vertical direction of clay loam experiment No. V10. (b) Horizontal direction of clay loam experiment No. V10. (c) Vertical direction of sandy loam experiment No. V11. (d) Horizontal direction of sandy loam experiment No. V11. (e) Vertical direction of clay loam experiment No. V10. (f) Vertical direction of sandy loam experiment of V11.

0.130 $cm^3\ cm^{-3}$, and the horizontal direction increased by 4.79% and 9.24%, respectively. The reasons may have been that the soil suction (mainly the matrix potential) decreased with the increase of θ_0, which may have slowed the wetting front movement. However, compared with θ_s, the water shortage ($\theta_s - \theta_0$) in the soil decreased with any increase in θ_0, which accelerated the wetting front movement. The aforementioned two reasons work together, which makes the soil wetting pattern distribution range increase with the increase of θ_0, which is consistent with the results of Lazarovitch *et al.* (2009) and Nie *et al.* (2015). Because the adsorption of NO_3^--N was not considered in this study,

Table 4 | Statistical analysis between simulated and measured values of fertilizer solution infiltration under furrow irrigation

		Cumulative infiltration per unit area			Soil water content			NO_3^--N concentration			K^+ concentration		
Soil texture	No.	*RMSE* (cm)	*MPBE* (%)	*MAPRE* (%)	*RMSE* (cm^3 cm^{-3})	*MPBE* (%)	*MAPRE* (%)	*RMSE* (g cm^{-3})	*MPBE* (%)	*MAPRE* (%)	*RMSE* (g cm^{-3})	*MPBE* (%)	*MAPRE* (%)
Clay loam	V9*	0.70	−2.37	2.37	0.023	−3.56	5.85	0.053	−6.93	6.93	0.070	−5.57	7.62
	V10*	0.85	−4.12	4.12	0.026	−3.28	6.32	0.067	−7.22	9.22	0.013	−6.72	9.77
	V11*	0.74	4.57	4.57	0.032	5.23	8.24	0.049	6.38	8.78	0.025	7.23	8.91
	V12*	0.88	3.26	3.26	0.030	7.32	7.32	0.075	8.23	10.22	0.053	9.44	10.33
Sandy loam	V9*	0.66	−3.25	3.25	0.026	−3.78	5.22	0.081	−10.02	10.02	0.082	−9.36	11.65
	V10*	0.41	−3.08	3.08	0.024	−4.49	4.49	0.057	−7.52	9.75	0.093	−12.12	14.22
	V11*	0.55	5.87	5.87	0.029	7.65	7.65	0.063	9.13	11.44	0.018	9.23	12.27
	V12*	0.49	1.06	1.06	0.022	3.22	3.22	0.047	6.53	8.32	0.011	8.29	9.56
Mean values		0.66	0.24	3.45	0.027	1.04	6.04	0.062	−0.18	9.34	0.066	0.05	10.54

Note: RMSE is the root mean square error; *MPBE* is the mean percentage of bias error; *MAPRE* is the mean absolute percentage relative error.

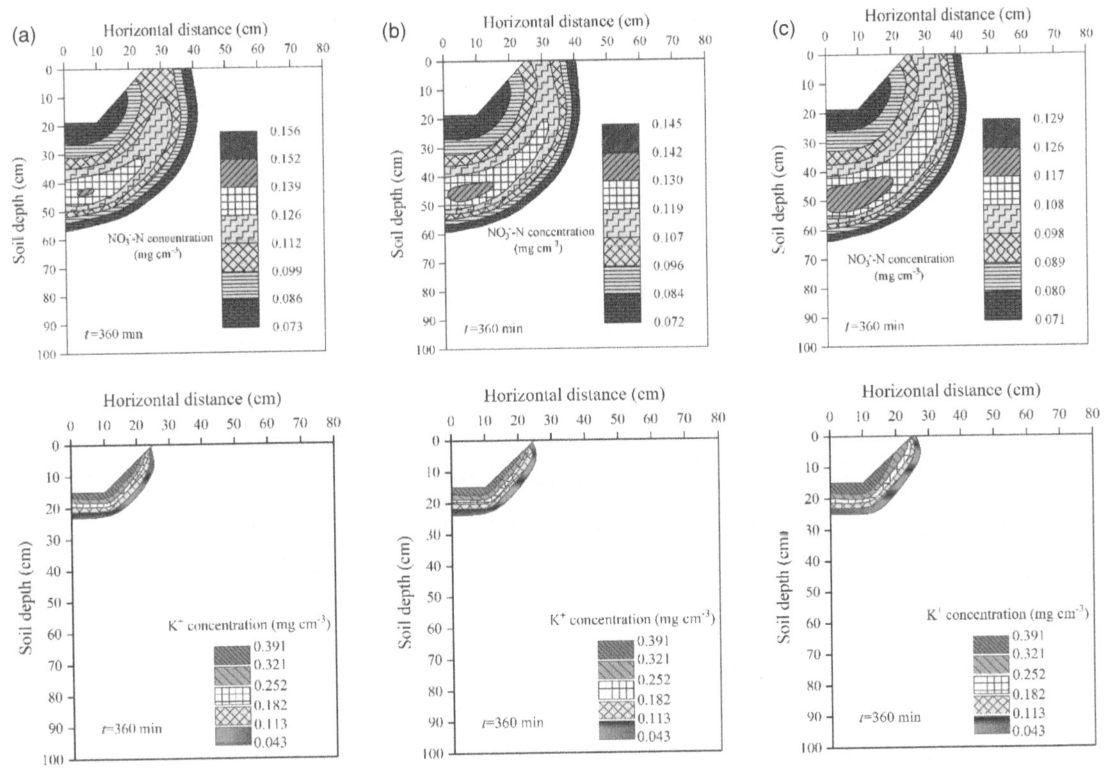

Figure 5 | NO_3^--N and K^+ concentration distributions under different initial soil water contents. (a) $\theta_0 = 0.130$ cm^3 cm^{-3}. (b) $\theta_0 = 0.156$ cm^3 cm^{-3}. (c) $\theta_0 = 0.208$ cm^3 cm^{-3}.

NO_3^--N migration was mainly affected by soil water movement; thus, the distribution range of NO_3^--N in soil was consistent with the distribution of soil water (wetting pattern), and the overall distribution range of NO_3^--N increased with the increase of θ_0. Furthermore, the results also proved that the high value region of NO_3^--N concentration had a downward trend with any increase of θ_0. For example, the high value regions of NO_3^--N concentration under θ_0 were 0.156 and 0.208 cm^3 cm^{-3}, moved downward by approximately 3.8 and 6.6 cm compared with 0.130 cm^3 cm^{-3}, respectively; this further confirmed that the characteristics of NO_3^--N easily move with soil water.

Therefore, the results suggested that in the range of soil water content suitable for crop growth, fertigation should be avoided when the soil water content is high to minimize the risk that NO_3^--N might leach deep into the soil.

K^+ mainly accumulates near the bottom of furrows (Figure 5), which indicates that θ_0 has little effect on the distribution range of K^+. In this study, the distribution depths of K^+ are mainly between 23 and 26 cm in the vertical direction, and within 25 and 27 cm in the horizontal direction under the conditions of θ_0 ranging from 0.130 to 0.208 $cm^3\,cm^{-3}$, respectively. The reason is that the soil wetting pattern movement is accelerated with the increase of θ_0, which shortens the time of colloid adsorption of K^+ in the soil, so the K^+ distribution range is slightly increased.

Potassium nitrate concentrations

The distributions of the NO_3^--N and K^+ concentrations of the soil wetting pattern profile at the end of infiltration are shown in Figure 6, on the basis of the following: initial soil water content $\theta_0 = 0.130\ cm^3\,cm^{-3}$, water depth $h_0 = 10.5$ cm, soil bulk density $\rho = 1.30\ g\,cm^{-3}$, KNO_3 concentration $C = 250$, 600, and 900 mg L^{-1} of clay loam.

The distribution range of NO_3^--N increased with the increase of KNO_3 concentration (Figure 6). At the end of infiltration ($t = 360$ min), the vertical distribution depth of NO_3^--N increased by 4.37% and 8.37% with C being 600 and 900 mg L^{-1} compared with 250 mg L^{-1}, and the horizontal direction increased by 5.02% and 9.17%, respectively. The results indicated that the soil infiltration capacity increases with the increase of KNO_3 concentration, which is consistent with the results of Liu *et al.* (2017). Furthermore, it should be noted that with the increase of KNO_3 concentration, the distribution range of NO_3^--N gradually increases, but it does not increase linearly with KNO_3 concentration and peaks may exist, which requires further study. The results also showed that when the concentrations of KNO_3 were 250 and 600 mg L^{-1}, the concentration of NO_3^--N gradually increased along the direction of the wetting front and then decreased gradually; meanwhile, the concentration of NO_3^--N gradually decreased along the direction of the wetting front when C

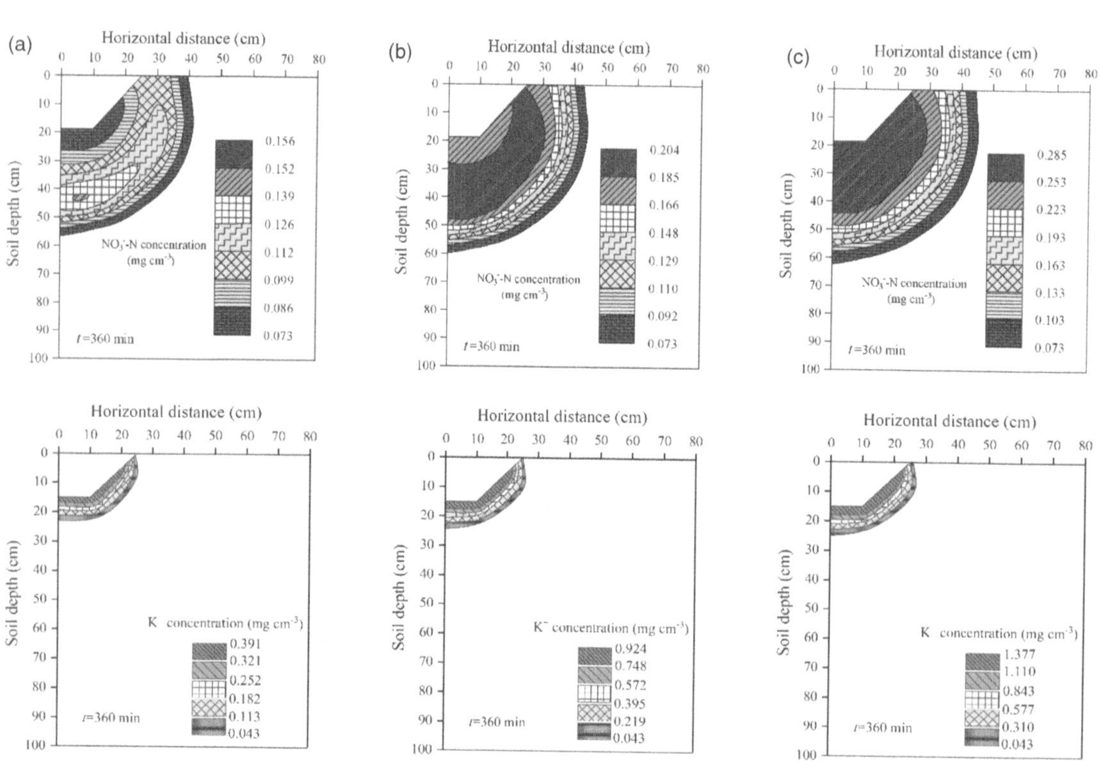

Figure 6 | NO_3^--N and K^+ concentration distributions under different KNO_3 concentrations. (a) $C = 250$ mg L^{-1}. (b) $C = 600$ mg L^{-1}. (c) $C = 900$ mg L^{-1}.

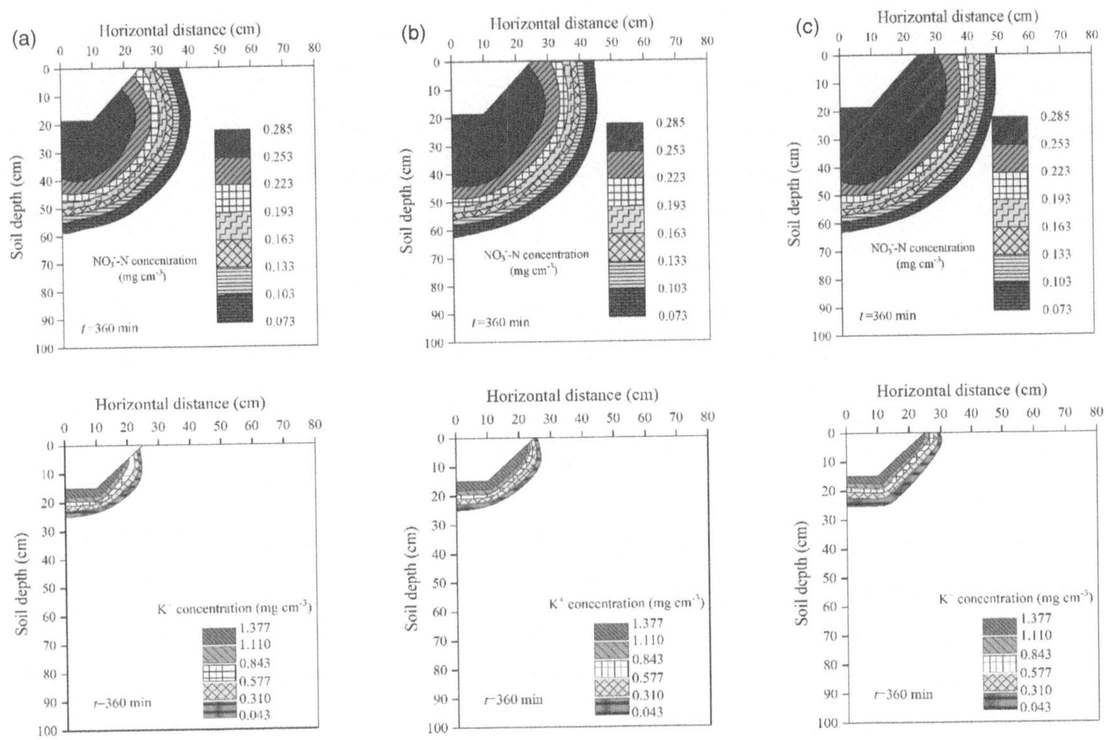

Figure 7 | NO_3^--N and K^+ concentration distributions under different water depths of furrow. (a) $h_0 = 5.5$ cm. (b) $h_0 = 10.5$ cm. (c) $h_0 = 14.5$ cm.

was 900 mg L^{-1}. The reason is that the concentration of NO_3^--N in the solution was 153 and 368 mg L^{-1} when the KNO_3 concentration was 250 and 600 mg L^{-1}, both of which were lower than the initial concentration of NO_3^--N in the soil (496 mg L^{-1}), so that the initial NO_3^--N in the soil migrated to the lower layer along with water infiltration. However, when the concentration of KNO_3 was 900 mg L^{-1}, the concentration of NO_3^--N in the solution was 552 mg L^{-1}, which is higher than the initial concentration of NO_3^--N in the soil, making the concentration distribution of NO_3^--N similar to the soil water distribution, and is consistent with the results of Cote *et al.* (2003) and Ebrahimian *et al.* (2013).

The vertical distribution depth of K^+ increased by only 1 cm when C was 900 mg L^{-1} compared with 250 mg L^{-1} (Figure 6); therefore, no obvious change exists in the distribution range of K^+, and the water was mainly concentrated near the bottom of the furrows. The results show that the change of KNO_3 concentration has little effect on the distribution range of K^+. Furthermore, the K^+ concentration near the furrow bottom increased obviously with the increase of KNO_3 concentration. For example, when KNO_3 concentration was 900 mg L^{-1}, the K^+ concentration at the bottom of the furrow increased by 248.53% compared with that at 250 mg L^{-1}. The reason may be that the K^+ concentration increases per unit volume of infiltration with the increase of KNO_3 concentration, which further indicates that K^+ has strong adsorption characteristics.

Water depths of furrow

The distributions of NO_3^--N and K^+ concentrations of soil wetting pattern profiles at the end of infiltration are shown in Figure 7, on the basis of the KNO_3 concentration $C = 900$ mg L^{-1}, initial soil water content $\theta_0 = 0.130$ cm^3 cm^{-3}, soil bulk density $\rho = 1.30$ g cm^{-3}, water depth $h_0 = 5.5$, 10.5, and 14.5 cm of clay loam.

The h_0 has little effect on the vertical distribution depth of NO_3^--N, but it has an obvious effect in the horizontal direction (Figure 7). At the end of infiltration ($t =$ 360 min), when h_0 was 14.5 and 10.5 cm compared with 5.5 cm, the distribution range of NO_3^--N increased by

1.10% and 2.71% in the vertical direction, and in the horizontal direction increased by 20.48% and 34.87%, respectively. The reason may be that h_0 was 5.5–14.5 cm, which causes the pressure potential change to be small, so that the difference in the vertical wetting front movement distance is small. For the horizontal direction, as h_0 increased, the horizontal water surface width was increased, so that the distance between the horizontal wetting front and the center of the furrow increased, which is consistent with the results of Zhang *et al.* (2012). Therefore, the results suggested that h_0 should be appropriately increased to enhance the horizontal distribution of NO_3^--N.

The influence of h_0 on the K^+ distribution was mainly in the horizontal direction (Figure 7). When h_0 was 14.5 cm, the horizontal distribution ranges of K^+ were 18.49% and 26.46% higher than for 10.5 cm and 5.5 cm, respectively. The reason is that the horizontal water surface width was increased as h_0 increased, so that the distance between the horizontal wetting front and the center of the furrow increased, thereby increasing the horizontal distribution range of K^+.

CONCLUSIONS

(1) Soil hydraulic and solute transport parameters under different potassium nitrate (KNO_3) concentration conditions were determined by inversion solution and then verified in HYDRUS-2D. The results showed that the soil saturated water content (θ_s), empirical parameter (n), and saturated hydraulic conductivity (K_s) increase with the increase of KNO_3 concentrations, but the empirical parameter (a) shows a decrease. The simulated values with HYDRUS-2D, on the basis of soil hydraulic and solute transport parameter inversion results, closely matched the measurements of cumulative infiltration, soil water content and solute transport, and the results indicated that the soil hydraulic and solute transport parameters determined from the inversion solution were reliable.

(2) The effects of initial soil water content (θ_0), potassium nitrate (KNO_3) concentration, and water depth (h_0) on the transport characteristics of nitrate nitrogen (NO_3^--N) and potassium (K^+) were analyzed The results proved that the distribution range of NO_3^--N increases with the increase of θ_0 and KNO_3 concentration. The h_0 has little effect on the vertical distribution of NO_3^--N, but it has an obvious influence on the horizontal distribution, and the horizontal distribution range increases as h_0 increases. K^+ is mainly distributed within approximately 10 cm of the bottom of the furrow, and θ_0 and the KNO_3 concentration have little effect on the K^+ distribution range, whereas h_0 has an obvious influence on the K^+ horizontal distribution, and the horizontal distribution range increases with the increase of h_0.

According to the results of this study, it is suggested that h_0 should be appropriately increased during furrow irrigation to increase the horizontal distribution of NO_3^--N. To reduce the risk of deep leaching of NO_3^--N, farmers should avoid fertigation when the soil water content is high in the range of suitable water contents for crop growth. Because K^+ easily accumulates, the concentration of KNO_3 should not be too large; if it is too high, salt can accumulate on soil surfaces. Future research will require a more suitable concentration range of KNO_3 because only four concentrations were considered in this study. In addition, nitrogen mineralization, nitrification, and denitrification were neglected, which may have influenced the results of this study.

ACKNOWLEDGEMENTS

This research was supported by grants from the National Natural Science Foundation of China (No. 51579205, 51909208), Natural Science Basic Research Program of Shaanxi (No. 2019JM-063), and Scientific Research Program by the Shaanxi Provincial Education Department (20JS099).

DATA AVAILABILITY STATEMENT

All relevant data are included in the paper or its Supplementary Information.

REFERENCES

Abbasi, F., Šimůnek, J., Feyen, J., van Genuchten, M. T. & Shouse, P. J. 2003 Simultaneous inverse estimation of the soil hydraulic and solute transport parameters from transient field experiments: homogeneous soil. *Trans. ASAE* **46** (4), 1085–1095.

Beven, K. J., Henderson, D. E. & Reeves, A. D. 1993 Dispersion parameters for undisturbed partially saturated soil. *J. Hydrol.* **143**, 19–43.

Brunetti, G., Šimůnek, J. & Bautista, E. 2018 A hybrid finite volume–finite element model for the numerical analysis of furrow irrigation and fertigation. *Comput. Electron. Agr.* **150**, 312–327.

Cote, C. M., Bristow, K. L., Charlesworth, P. B., Cook, F. J. & Thorburn, P. J. 2003 Analysis of soil wetting and solute transport in subsurface trickle irrigation. *Irrig. Sci.* **22** (3–4), 143–156.

Deb, S. K., Sharma, P., Shukla, M. K., Ashigh, J. & Šimůnek, J. 2015 Numerical evaluation of nitrate distributions in the onion root zone under conventional furrow fertigation. *J. Hydrol. Eng. ASCE* **21** (2), 05015026.

Doltra, J. & Muñoz, P. 2010 Simulation of nitrogen leaching from a fertigated crop rotation in a Mediterranean climate using the EU-Rotate_N and Hydrus-2D models. *Agric. Water Manage.* **97** (2), 277–285.

Ebrahimian, H., Liaghat, A., Parsinejad, M., Abbasi, F. & Navabian, M. 2012a Comparison of one- and two-dimensional models to simulate alternate and conventional furrow fertigation. *J. Irrig. Drain. Eng. ASCE* **138** (10), 929–938.

Ebrahimian, H., Liaghat, A., Parsinejad, M. & Playán, E. 2012b Distribution and loss of water and nitrate under alternate and conventional furrow fertigation. *Span. J. Agric. Res.* **10** (3), 849–863.

Ebrahimian, H., Liaghat, A., Parsinejad, M., Playán, E., Abbasi, F. & Navabian, M. 2013 Simulation of 1D surface and 2D subsurface water flow and nitrate transport in alternate and conventional furrow fertigation. *Irrig. Sci.* **31** (3), 301–316.

Feng, S., 2017 *Research of Soil Hydraulic Properties and Nitrogen Conversion Parameters under the Couple of Water-Temperature-Nitrogen*. Mater's thesis, Taiyuan University of Technology, Taiyuan, China.

Grecco, K. L., de Miranda, J. H., Silveira, L. K. & van Genuchten, M. T. 2019 HYDRUS-2D simulations of water and potassium movement in drip irrigated tropical soil container cultivated with sugarcane. *Agric. Water Manage.* **221**, 334–347.

Hanson, B. R., Šimůnek, J. & Hopmans, W. H. 2006 Evaluation of urea–ammonium–nitrate fertigation with drip irrigation using numerical modeling. *Agric. Water Manage.* **86** (1–2), 102–113.

Iqbal, S., Guber, A. K. & Khan, H. Z. 2016 Estimating nitrogen leaching losses after compost application in furrow irrigated soils of Pakistan using HYDRUS-2D software. *Agric. Water Manage.* **168**, 85–95.

Karandish, F. & Šimůnek, J. 2016 A field-modeling study for assessing temporal variations of soil–water–crop interactions under water-saving irrigation strategies. *Agric. Water Manage.* **178**, 291–303.

Lazarovitch, N., Warrick, A. W., Furman, A. & Zerihun, D. 2009 Subsurface water distribution from furrows described by moment analyses. *J. Irrig. Drain. Eng. ASCE* **135** (1), 7–12.

Li, J. S., Zhang, J. J. & Rao, M. J. 2004 Effect of drip irrigation system operation on water and nitrate distribution in soil. *J. Hydraul. Eng.* **35** (9), 31–37 (in Chinese).

Li, H. X., Wang, W. E. & Hu, X. T. 2019 Numerical simulation of soil water and nitrogen distribution under the integration of drip irrigation and fertilization. *Agric. Res. Arid Areas* **37** (2), 10–17 (in Chinese).

Liu, X. G., Zhu, Y. F., Yu, X. D., Li, Y. L., Tang, J. K. & Yu, L. M. 2017 Water-salinity distribution characteristics in wetted soil of moistube irrigation under different pressure heads and soil bulk densities. *T. Chin. Soc. Agric. Mach.* **48** (7), 189–197 (in Chinese).

Long, H. Y., Li, Y. Z. & Jiang, Y. Z. 2001 K+ adsorption kinetics of fluvo-aquic and cinnamon soil under different K+ concentrations. *Acta Metrol. Sin.* **38** (2), 226–234 (in Chinese).

Ma, J. H., Ren, L., Gong, Y. S. & Stahr, K. 2004 Numerical simulation of nitrogen transport during the growth of winter wheat. *J. Hydraul. Eng.* **35** (3), 103–110 (in Chinese).

Moriasi, D. N., Arnold, J. G., van Liew, M. W., Bingner, R. L., Harmel, R. D. & Veith, T. L. 2007 Model evaluation guidelines for systematic quantification of accuracy in watershed simulations. *Trans. ASABE* **50** (3), 885–900.

Nie, W. B., Fei, L. J. & Ma, X. Y. 2015 Simplified calculating method for cumulative infiltration of furrow irrigation. *Adv. Water Sci.* **26** (6), 801–810 (in Chinese).

Nie, W. B., Ma, X. Y. & Fei, L. J. 2017 Evaluation of infiltration models and variability of soil infiltration properties at multiple scales. *Irrig. Drain.* **66** (4), 589–599.

Nie, W. B., Li, Y. B., Liu, Y. & Ma, X. Y. 2018 An approximate explicit Green–Ampt infiltration model for cumulative infiltration. *Soil Sci. Soc. Am. J.* **88** (4), 919–930.

Phillips, I. R. 2006 Modelling water and chemical transport in large undisturbed soil cores using HYDRUS-2D. *Aust. J. Soil Res.* **44** (1), 27–34.

Ranjbar, A., Rahimikhoob, A., Ebrahimian, H. & Varavipour, M. 2019 Simulation of nitrogen uptake and distribution under furrows and ridges during the maize growth period using HYDRUS-2D. *Irrig. Sci.* **37** (4), 495–509.

Ren, C. J., Bai, D., Zhou, B. B., Liang, W. & Chen, Y. 2013 Using particle swarm–finite difference algorithm to identify transport parameters of water-solute in adsorptive soil. *Acta Metrol. Sin.* **50** (3), 486–491 (in Chinese).

Silberbush, M. & Barber, S. A. 1983 Prediction of phosphorus and potassium uptake by soybeans with a mechanistic mathematical model. *Soil Sci. Soc. Am. J.* **47** (2), 262–265.

Šimůnek, J., Šejna, M. & van Genuchten, M. T. 1999 *The HYDRUS-2D Software Package for Simulating Water Flow and Solute Transport in Two-Dimensional Variably*

Saturated Media (Version 2.0). US Salinity Laboratory, Riverside, CA, USA.

Šimůnek, J. & van Genuchten, M. T. 2008 Modeling nonequilibrium flow and transport processes using HYDRUS. *Vadose Zone J.* **7** (2), 782–797.

Siyal, A. A., Bristow, K. L. & Šimůnek, J. 2012 Minimizing nitrogen leaching from furrow irrigation through novel fertilizer placement and soil surface management strategies. *Agric. Water Manage.* **115**, 242–251.

Tan, X., Shao, D. G., Gu, W. Q. & Liu, H. H. 2015 Field analysis of water and nitrogen fate in lowland paddy fields under different water managements using HYDRUS-1D. *Agric. Water Manage.* **150**, 67–80.

van Genuchten, M. T. 1980 A closed-form equation for predicting the hydraulic conductivity of unsaturated soils. *Soil Sci. Soc. Am. J.* **44**, 892–898.

Verbist, K., Cornelis, W. M., Gabriels, D., Alaerts, K. & Soto, G. 2009 Using an inverse modelling approach to evaluate the water retention in a simple water harvesting technique. *Hydrol. Earth Syst. Sci.* **13** (10), 1979–1992.

Wang, C. X., Wang, Q. J., Liu, J. J., Zhuang, L. & Zhen, X. 2010 Field trial for characteristics of soil water and salt distribution in drip fertigation of kalium-using fertilizers. *Trans. CSAE* **26** (3), 25–31 (in Chinese).

Wang, Z. T., Jiao, X. Y., Han, H. L. & Yu, X. B. 2013 Sensitivity analysis of VG model parameters with vertical one-dimensional soil infiltration. *J. Hohai University (Natural Sciences)* **41** (1), 80–84 (in Chinese).

Zerihun, D., Furman, A., Warrick, A. W. & Sanchez, C. A. 2005a Coupled surface–subsurface solute transport model for irrigation borders and basins. I. Model development. *J. Irrig. Drain. Eng. ASCE* **131** (5), 396–406.

Zerihun, D., Sanchez, C. A., Furman, A. & Warrick, A. W. 2005b Coupled surface–subsurface solute transport model for irrigation borders and basins. II. Model evaluation. *J. Irrig. Drain. Eng. ASCE* **131** (5), 407–419.

Zhang, Y. Y., Wu, P. T., Zhao, X. N. & Li, P. 2012 Evaluation and modelling of furrow infiltration for uncropped ridge–furrow tillage in Loess Plateau soils. *Soil Res.* **50** (5), 360–370.

Zipelevish, E., Grinberge, A., Amar, S., Gilbo, Y. & Kafkafi, U. 2000 Eggplant dry matter composition fruit yield and quality as affected by phosphate and total salinity caused by potassium fertilizers in the irrigation solution. *J. Plant. Nutr.* **23** (4), 431–442.

First received 5 June 2020; accepted in revised form 5 October 2020. Available online 16 October 2020

Simulation soil water–salt dynamics in saline wasteland of Yongji Irrigation Area in Hetao Irrigation District of China

Chengfu Yuan

ABSTRACT

In order to explore the regional water–salt balance mechanism in Hetao Irrigation District, field experiments were conducted in 2018 and 2019 in the Heji canal study area. The SWAP model was calibrated and validated based on field experiments' observed data. The SWAP model was used to simulate soil water–salt dynamics in saline wasteland after calibration and validation. The results showed that model simulation results of soil water content and soil salt concentration agreed well with the measured values. Soil water content and soil salt concentration changed obviously under the effect of farmland irrigation in the crop growing period. Soil salt was accumulated in saline wasteland. The soil salt accumulation of each soil layer in saline wasteland was 0.164, 0.092, −0.890 and −1.261 mg/cm^3, respectively. Soil water content gradually increased and soil salt concentration gradually decreased in the autumn irrigation period. Soil salt was leached in the saline wasteland. The soil salt accumulation of each soil layer in the saline wasteland was −1.011, −1.242, −1.218 and −1.335 mg/cm^3, respectively. The saline wasteland became a drainage and salt drainage region for cultivated land. The saline wastelands had an obvious role in adjusting salt balance and maintaining salt dynamic balance in Hetao Irrigation District.

Key words | Hetao Irrigation District, saline wasteland, soil salt concentration, soil water content, SWAP model

Chengfu Yuan
College of Hydraulic Science and Engineering, Yangzhou University, Yangzhou 225009, Jiangsu, China
and
Jiangxi Water Resources Institute, Nanchang 330013, Jiangxi, China
E-mail: *cfyuan1989@163.com*

HIGHLIGHTS

- Soil water–salt transport is concentrated in saline wasteland.
- Soil water content and soil salt concentration changed obviously in crop growing period.
- Soil salt was accumulated in saline wasteland in crop growing period.
- Soil water content gradually increased and soil salt concentration gradually decreased in autumn irrigation period.
- Saline wasteland became a drainage and salt drainage region.

INTRODUCTION

Soil salinization has been one of main factors impacting agricultural production all around the world (Wang *et al.* 2007; Muyen *et al.* 2011). Global salinization land area is 9.5×10^8 hm^2, which approximately accounts for 10% of the Earth's land area, and these salinity lands are mainly distributed in arid and semi-arid areas (Ghassemi *et al.* 1995; Abbas *et al.* 2013). In China, with a total area of about 3.6×10^7 hm^2, salinity land accounts for 4.9% of the available land area. Especially, there is much salinity land in northwest China, with the arid climate. (Yang 2009; Wang *et al.* 2011). The Hetao Irrigation District (Hetao) located in the upper reach of Yellow River Basin, is a typical

doi: 10.2166/ws.2020.299

salinization irrigated district because of the dry climate and shallow groundwater. The area affected by salinization in the whole irrigated district has reached to 3.9×10^6 hm^2, accounting for 69% of the total land area (Huang *et al.* 2018). The soil salinity in the crop root zone should be maintained within the reasonable range to avoid harming crop growth. Engineering drainage and irrigation leaching were two major ways to control soil salinization (Tedeschi & Menenti 2002; Wu *et al.* 2009). From the end of the 1980s to the middle of the 1990s, a drainage-oriented irrigating supporting project was constructed for the whole of Hetao and effectively alleviated the soil salinization. A traditional engineering drainage project was difficult to carry out due to high economic cost and environmental restrictions. Dry drainage was an alternative method for effective drainage and control of salinity in the irrigation district (Wichelns & Oster 2006; Wu *et al.* 2009). Dry drainage was to set for a certain part of idle land (non-cultivated land) in the irrigated district, inputting the surplus irrigating water and salinity of cultivated land into the non-cultivated land through the movement of groundwater. The surplus water would be discharged and salinity would be stored in the non-cultivated land under soil evaporation, so as to maintain the water–salt balance of the cultivated land (Wichelns & Oster 2006). There were 5.74×10^5 hm^2 of cultivated land and 2.09×10^5 hm^2 of salt wasteland in Hetao Irrigation Area, and the salt wasteland was scattered in gaps of cultivated land, which has the conditions to realize dry drainage. Dry drainage played an important role in maintaining salt balance in Hetao irrigation area (Wang *et al.* 2019). At present, research on soil water–salt transport has mostly been concentrated on cultivated land (Tang *et al.* 2007; Feng *et al.* 2017; Yuan *et al.* 2019), while there has been little research on soil water–salt transport in saline wasteland in the irrigated district. The Soil–Water–Atmosphere–Plant (SWAP) model has been the popular tool to simulate soil water–salt transport in arid or semi-arid areas. It has been widely used in simulating soil water and salt transport in arid or semi-arid areas at home and abroad (van Dam *et al.* 2008). Singh *et al.* (2010) used the SWAP model to simulate soil water dynamics and water balance under different wheat planting conditions in Minqin County, Gansu Province. The results showed that the SWAP model could simulate soil water dynamics well and was an effective tool for analyzing soil water balance. Xu *et al.* (2013) simulated soil water and salt dynamics and spring wheat yield under different groundwater depth and irrigation water conditions in Qingtongxia Irrigation Area of Ningxia by using the SWAP model. The results showed that under the condition of maintaining the existing irrigation quota, the increase of groundwater depth would lead to a slight reduction of spring wheat yield due to soil salt stress. Yin (2013) combined one-dimensional soil water and salt transport model SWAP-EPIC and ArcGIS software to build a distributed model of Nanxiaozhao experimental area in Jiefangzha Irrigation Area of Hetao Irrigation District, and carried out distributed simulation of soil water and salt dynamics and crop yield in the experimental area. According to the dynamic law of soil water and salt in the saline wasteland of an irrigation area, the SWAP model was applied to the saline wasteland soil in Hetao Irrigation Area of Inner Mongolia, and the dynamic changes of soil water and salt in the crop growth period and autumn irrigation period were studied. The results provide a theoretical basis for water and salt management and prevention of soil salinization in Hetao Irrigation Area.

The objectives of this study were: (1) to calibrate and validate the SWAP model with two-year field observation data of water and salt in the soil profile; (2) to simulate soil water–salt dynamics of the saline wasteland in the crop growing period and autumn irrigation period using the calibrated SWAP model.

MATERIALS AND METHODS

General situation of the experiment area

The study area was selected in the Heji study area of the Yongji Irrigation Area in Hetao Irrigation District of Inner Mongolia (Figure 1), and was located in Minzhu village, Ganzhao Town, Linhe District, Bayannaoer City, Inner Mongolia Autonomous Region of northwest China. The geographical coordinates are 107°15′–107°18′ E and 40°43′–40°46′ N. The study area was located in northwest China inland, belonging to the mid-temperate semi-arid continental climate, dry and with less rain, where evaporation is strong, the annual average rainfall is 160 mm and the

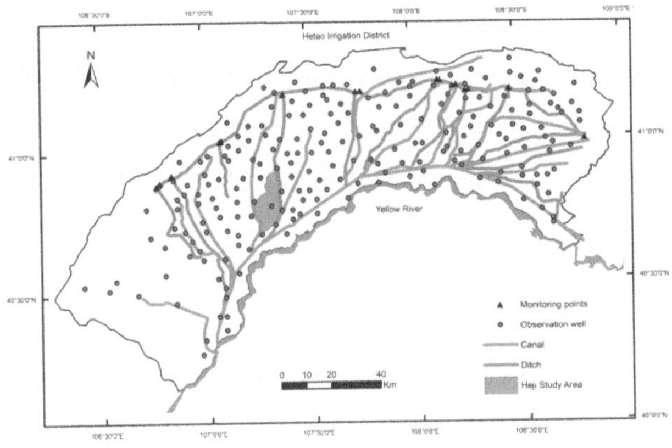

Figure 1 | Geographical location map of the study area.

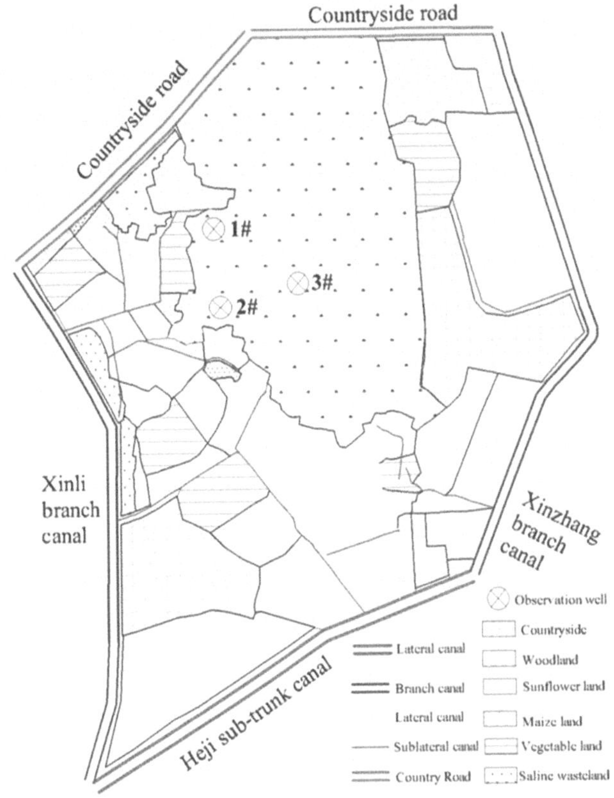

Figure 2 | The Heji study area in Hetao Irrigation District.

annual average evaporation is 2,240 mm (Ren *et al.* 2018). The average buried depth of groundwater level in the study area was 1.6–2.2 m, and the groundwater level was shallow. The terrain of the study area was relatively gentle. The terrain was high in the southeast and low in the northwest in the study area. The elevation was between 1,040 and 1,042 m, and the total area was about 507 hm^2. It was mainly controlled by two land-use types: cultivated land and salt wasteland. It was controlled by one main canal (Heji branch canal) and two branch canals (Xinli branch canal and Xinzhang branch canal). The distribution of salt wasteland in the study area was relatively concentrated, mainly concentrated in the central and northwest regions of the study area. The east, south and west sides were surrounded by cultivated land and the saline wasteland was lower than the cultivated land. Saline wasteland had the function of drainage and salt discharge, and was a typical 'dry drainage' area in Hetao Irrigation Area.

Experiment layout and observation items

Field experiments were carried out from July 2018 to October 2019. Three groundwater observation wells (Figure 2) were arranged in the saline wasteland of the study area. The depth of each observation well was 3 m and the diameter was 10 cm. The three observation wells were used to monitor the groundwater depth in the study area. Soil sampling points were arranged near the observation wells, and the distance between the soil sampling points and the observation well was 3–5 m. In the study area, soil samples were obtained from soil sampling points by soil drill before and after crop irrigation. The soil layers were divided into four layers: 0–10 cm, 10–20 cm, 20–40 cm and 40–60 cm. The soil water content was determined by the drying method, and the electrical conductivity of saturated extraction solution $EC_{1:5}$ was determined by conductivity meter (DDSJ-308A, Shanghai Leici Instrument Company), and the soil electrical conductivity was converted into soil salt content according to the formula ($S = 2.882EC_{1:5}+0.183$) (Xu *et al.* 2019). During the observation period, the groundwater depth of the observation wells was observed every seven days. The groundwater depth was measured by steel ruler water level gauge (JK 22924, Beijing Jingkaida Instrument Company). The soil particle composition of the 0–60 cm soil layer in the saline wasteland was determined by Laser Particle Sizer (Mastersizer 2000, Malvern, UK). The soil particle classification was based on the international soil texture classification standard. The soil physical parameters of the saline wasteland are shown in Table 1. According to the soil physical properties of each soil layer, the soil hydraulic characteristic parameters of each soil layer were generated

Table 1 | Soil physical properties of saline wasteland

Soil depth (cm)	Clay (<0.002 mm,%)	Silt (0.002–0.02 mm, %)	Sand (0.02–2 mm, %)	Soil bulk density (g/cm³)	International soil texture
0–10	3.47	18.61	77.93	1.56	Silty sandy loam
10–20	3.11	17.86	79.03	1.54	Silty sandy loam
20–40	2.78	14.14	83.07	1.58	Silty sandy loam
40–60	3.29	16.15	80.57	1.55	Silty sandy loam

by using the neural network soil transfer function of Rosetta Software. The initial soil hydraulic characteristic parameters of the saline wasteland are shown in Table 2. The meteorological data during the experimental observation period from 2018 to 2019 were downloaded from the meteorological station of Hetao experimental base of China Agricultural University in the study area.

The SWAP model

The SWAP model, developed by the Water Resource Group of Wageningen University, is a one-dimensional agricultural hydrological model that is used to simulate vertical soil water flow, solute and heat transport and crop growth (van Dam *et al.* 1997; van Dam *et al.* 2008). It has been widely applied to simulate irrigation, soil water–salt transport, crop evapotranspiration, crop growth, drainage, and groundwater tables in semi-arid and arid areas around the world (Vazifedoust *et al.* 2008; Noory *et al.* 2011). Soil water flow is based on the Richards' equation for vertical flow in the SWAP model, which is given by:

$$\frac{\partial\theta}{\partial t} = C(h)\frac{\partial h}{\partial t} = \frac{\partial}{\partial z}\left[K(h)(\frac{\partial h}{\partial z}+1)\right] - S(h) \tag{1}$$

where $C(\theta)$ is differential water capacity (/cm); θ is soil water content (cm^3/cm^3); h is soil water pressure head (cm); t is time (d); z is the vertical coordinate (cm, positive uptake); $K(h)$ is the hydraulic conductivity (cm/d); $S(h)$ is the soil water extraction rate by plant roots (cm^3/(cm^3/d)).

Solute transport is based on the convective–dispersive equation in the SWAP model, which is given by:

$$J = qc - \theta(D_{\text{dif}} + D_{\text{dis}})\frac{\partial c}{\partial z} \tag{2}$$

where J is total solute flux density (g/(cm^2/d)); q is vertical flow at the bottom (cm/d); c is solute concentration in the soil liquid phase (g/cm^3); D_{dif} is the diffusion coefficient (cm^2/d); D_{dis} is the dispersion coefficient (cm^2/d).

The SWAP model has two crop growth modules to simulate the crop growth process, a simple one and another detailed one based on the WOFOST model (Hijmans *et al.* 1994). Due to the first crop model requiring fewer parameters and simple crop data inputs, the simple crop growth module was applied in this study. For a more detailed introduction to the SWAP model, reference can be made to the SWAP model theory book (van Dam *et al.* 1997).

The SWAP model requires various data as inputs, which includes meteorological data, soil data, crop growth data, and initial and boundary conditions. The meteorological data was obtained from an automatic meteorological station in the study area. The meteorological data calculated by the model were the daily meteorological data observed by the weather station of Hetao experimental base of China

Table 2 | Soil hydraulic parameters for different soil layers of saline wasteland

Soil depth (cm)	Residual water content θ_r (cm³/cm³)	Saturated water content θ_s (cm³/cm³)	Saturated hydraulic conductivity K_s (cm/d¹)	Shape factor α (–)	Shape factor n (–)	Shape factor λ (–)
0–10	0.0338	0.3916	87.81	0.0451	1.6015	0.5
10–20	0.0352	0.3903	97.6	0.0449	1.6693	0.5
20–40	0.0347	0.3910	100.76	0.0455	1.6821	0.5
40–60	0.0379	0.3886	142.65	0.0439	1.9163	0.5

Agricultural University where the research area was located. Because there was no crop in the salt wasteland, it was not necessary to input irrigation data and crop growth information. Soil water content was transformed into soil water pressure head by the soil water retention curve. The soil water pressure head and soil salinity before crop sowing were regarded as the initial condition. The measured dispersion length and molecular diffusion coefficient were 12 cm and 0.5 cm^2/d, respectively. A 60 cm depth soil profile divided into five horizons in the model was represented by 33 soil compartments in the vertical plane in the cultivated land soil layer. A 60 cm depth soil profile divided into five horizons was represented by 33 soil compartments in the vertical plane in the saline wasteland soil layer. The upper boundary condition was described by the irrigation, daily rainfall and the potential ET. Groundwater level was given as the bottom boundary condition.

The root mean square error (RMSE), mean relative error (MRE) and the coefficient of determination (R^2) were used to quantify the deviation of the simulated and observed data in calibration and validation. These indicators were defined as follows:

$$RMSE = \sqrt{\frac{1}{N}\sum_{i=1}^{N}(P_i - O_i)^2} \tag{3}$$

$$MRE = \frac{1}{N}\sum_{i=1}^{N}\left|\frac{P_i - O_i}{O_i}\right| \times 100\% \tag{4}$$

where N is the total number of observations in the experiments, P_i and O_i are the ith model predicted and observed values ($i = 1, 2, \ldots, N$), respectively.

RESULTS AND DISCUSSION

Calibration and validation of SWAP model

The average groundwater depth, soil water content and soil salt concentration of three groundwater observation wells were used to calibrate and validate the parameters of the SWAP model. The observation data in 2019 were used to calibrate the model parameters, and the observation data in 2018 were used to validate the model parameters. The simulation period in 2019 was from June 7 to September 28, with a total of 113 d. The simulation period in 2018 was from July 13 to October 1, with a total of 80 days.

The comparison between simulated and measured values of soil water content in saline wasteland is shown in Figures 3 and 4 during calibration and validation. It can be seen from the figure that the simulated values of soil water content of each soil layer was basically consistent with the measured values. The calibration effect of surface soil moisture was slightly poor. This was mainly due to the influence of rainfall and evaporation on the surface soil, and the change was relatively violent, but the simulated values of soil water basically reflected the change process of the measured values. The statistical analysis of the soil water content simulation results is shown in Table 3. The RMSE value of soil water content in each soil layer was below 0.06 cm^3/cm^3, and the MRE was less than 25%, which was within the allowable error range of 25%. The simulation results were basically feasible. The results of soil hydraulic characteristic parameters after calibration of the soil water module are shown in Table 4.

The comparison between simulated and measured values of soil salt concentration in saline wasteland during calibration and validation is shown in Figures 5 and 6. It can be seen from the figure that the simulated values of soil salt concentration in each soil layer were basically consistent with the measured values. Due to the influence of rainfall and evaporation, the calibration effect of surface soil salt concentration was slightly poor, but the simulated values of soil salt concentration basically reflected the change process of the measured values. The statistical analysis of soil salt concentration simulation results is shown in Table 5. The RMSE values of soil salt concentration in each soil layer were less than 1.5 mg/cm^3. The MRE values were all lower than 25%, which was within the allowable error range of 25%. The simulation results were basically feasible. After the calibration of the soil solute transport module, the molecular diffusion coefficient was 5.0 cm^2/d, and the solute dispersion was 10 cm. The calibration and validation results show that the SWAP model can simulate the soil water–salt dynamics in saline wasteland.

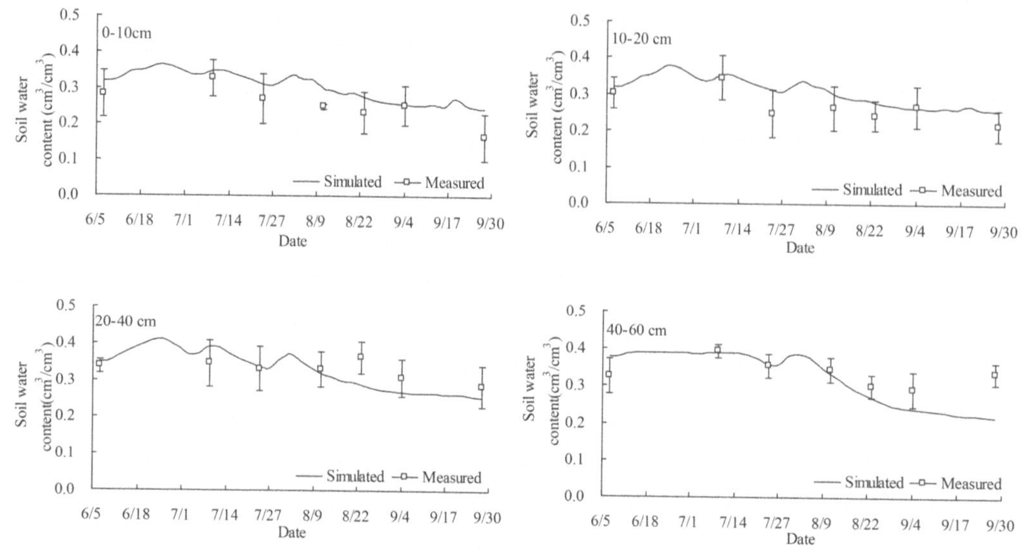

Figure 3 | Calibration for soil water content in saline wasteland (2019).

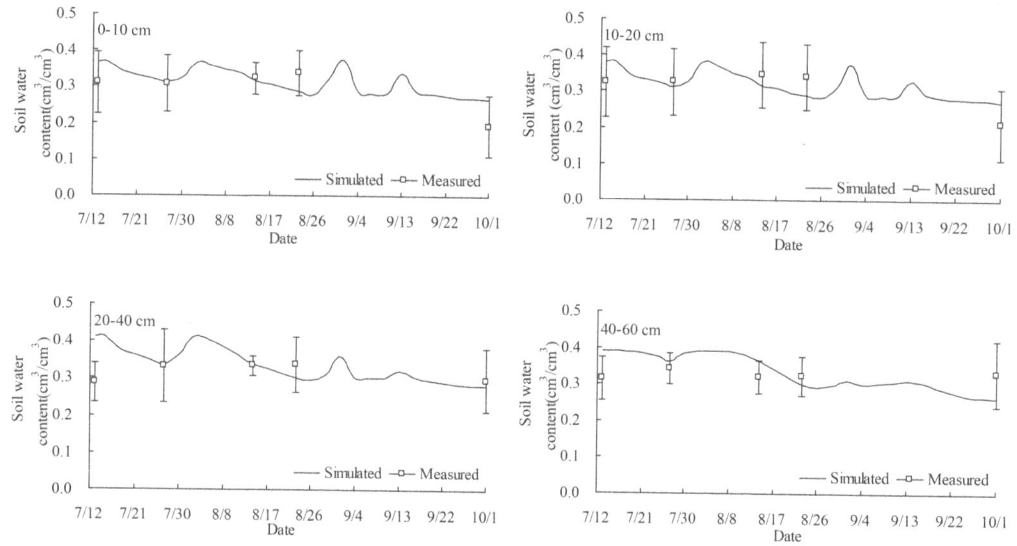

Figure 4 | Validation for soil water content in saline wasteland (2018).

Table 3 | The statistical results for soil water content in calibration and validation

	Calibration (2019)		Validation (2018)	
Soil depth (cm)	**RMSE (cm^3/cm^3)**	**MRE (%)**	**RMSE (cm^3/cm^3)**	**MRE (%)**
0–10	0.04	23.9	0.05	15.03
10–20	0.04	16.22	0.05	15.05
20–40	0.04	12.89	0.06	12.24
40–60	0.05	17.06	0.05	14.27

Simulation of soil water–salt dynamics in crop growth period

The soil water–salt dynamics of saline wasteland during April 10 to September 30 (crop growth period) and October 1 to November 20, 2019 (autumn irrigation period) were simulated by using the SWAP model after calibration of parameters.

Table 4 | Soil hydraulic parameters for different soil layers after calibration

Soil depth (cm)	Residual water content θ_r (cm³/cm³)	Saturated water content θ_s (cm³/cm³)	Saturated hydraulic conductivity K_s (cm/d)	Shape factor α (–)	Shape factor n (–)	Shape factor λ (–)
0–10	0.0312	0.3956	51.65	0.0162	1.659	0.5
10–20	0.0785	0.4195	74.49	0.0383	1.436	0.5
20–40	0.0290	0.4169	70.12	0.0349	1.423	0.5
40–60	0.0639	0.390	125.60	0.0323	1.8063	0.5

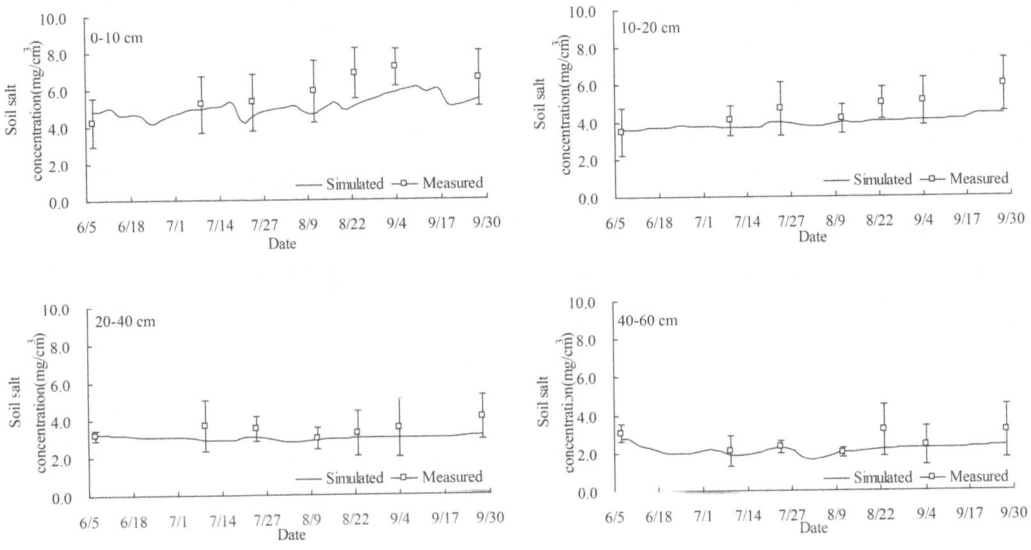

Figure 5 | Calibration for soil salt concentration in saline wasteland (2019).

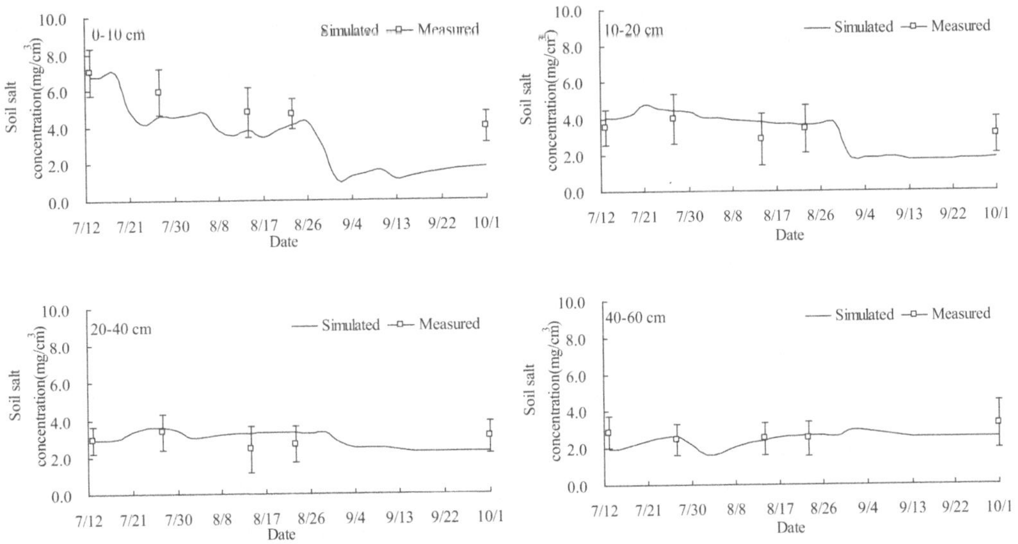

Figure 6 | Validation for soil salt concentration in saline wasteland (2018).

Table 5 | The statistical results for soil salt concentration in calibration and validation

	Calibration (2019)		Validation (2018)	
Soil depth (cm)	**RMSE (mg/cm³)**	**MRE (%)**	**RMSE (mg/cm³)**	**MRE (%)**
0–10	1.09	22.76	1.25	22.77
10–20	0.84	19.16	0.78	21.76
20–40	0.53	15.91	0.6	18.32
40–60	0.49	16.18	0.52	13.46

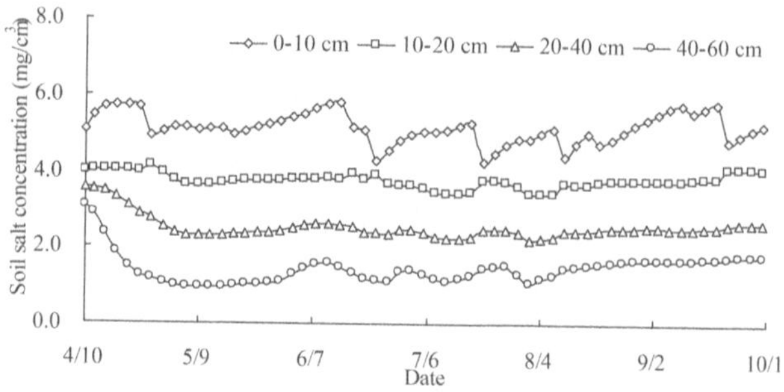

Figure 8 | Simulation of soil salt dynamics in the crop growth period.

Figure 7 shows the dynamic changes of soil water content in different soil layers of saline wasteland in the crop growth period. In the study area, the soil began to melt in the middle of April, and soil water content of each soil layer gradually increased. In early May, the spring irrigation was carried out in the study area. Affected by the spring irrigation of cultivated land, the soil water content of each soil layer in saline wasteland was higher, which was mainly because the saline wasteland was located in the low-lying area. Due to the 'dry drainage' effect, soil water content of each soil layer in the saline wasteland was relatively high. In the irrigation season, the soil water content of the saline wasteland changed obviously. After irrigation, the soil water content of the saline wasteland gradually increased, and the soil water content of the deep layer soil was the highest, and the surface water content was relatively low from May to August. From the middle of August to the end of September, the soil water content of the saline wasteland gradually decreased due to no irrigation in the cultivated land, and the soil water content of each soil layer reached the lowest values before autumn irrigation. Figure 8 shows the dynamic changes of soil salt concentration in different soil layers of saline wasteland in the crop growth period. It can be seen from the figure that the soil salt concentration of each soil layer in the saline wasteland was quite different, and soil salt concentration in the surface soil was the highest in the crop growth period. With the increase of soil depth, soil salt concentration gradually decreased, and the soil salt content of the saline wasteland mainly accumulated in the surface soil, and the change of soil salt concentration in each soil layer was obvious in the farmland irrigation season, which was mainly caused by farmland irrigation and leaching of salt, which flowed to the saline wasteland through groundwater flow. Under the action of soil evaporation, soil salt accumulated in the surface soil. At the end of the growth period, compared with the early growth stage, soil salt accumulation in each soil layer was 0.164, 0.092, −0.890 and −1.261 mg/cm³, respectively. The surface soil (0–20 cm) accumulated salt, and the lower soil (20–60 cm) desalinated. Therefore, the saline wasteland was in the state of storing salt in the crop growth period. After irrigation and leaching, soil salt flowed to the saline wasteland through groundwater flow, and salt accumulated in the surface soil of the saline wasteland. The saline wasteland had a role as drainage and salt drainage area for the cultivated land, which reduced the degree of soil salinization and created favorable conditions for crop growth in the cultivated land.

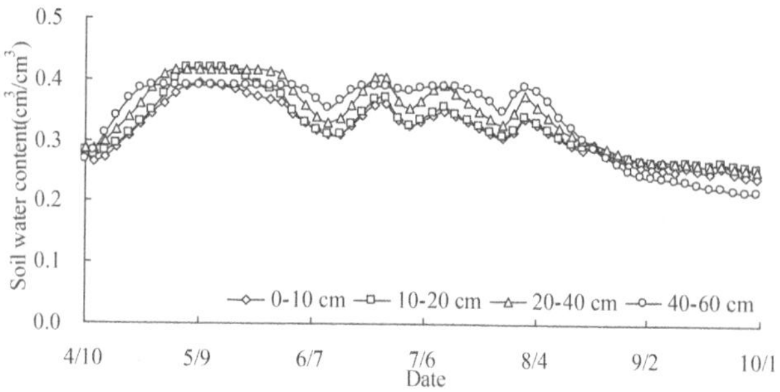

Figure 7 | Simulation of soil water dynamics in the crop growth period.

Simulation of soil water–salt dynamics in autumn irrigation period

Figure 9 shows the dynamic changes of soil water content in different soil layers of saline wasteland in the autumn irrigation period. The autumn irrigation period was from early October to mid-November in the study area. Autumn irrigation was a period of large irrigation amount in the year. The

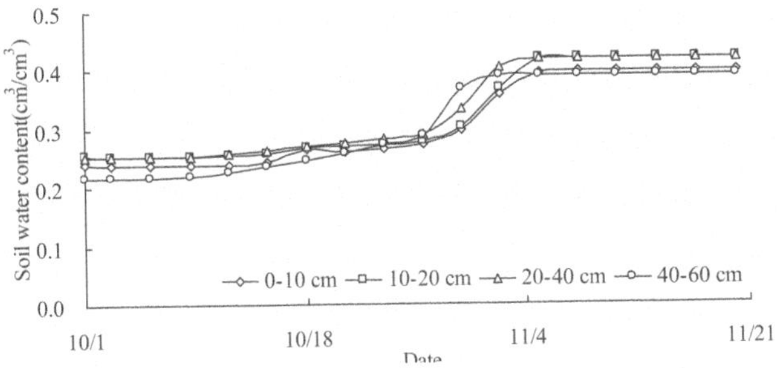

Figure 9 | Simulation of soil water dynamics in the autumn irrigation period.

main purpose of autumn irrigation was to leach soil salt in cultivated land and saline wasteland. Soil water content in saline wasteland increased gradually with the autumn irrigation of cultivated land, and soil water content of each soil layer increased gradually. In the first ten-day period of November, soil water content of each soil layer reached the maximum, and the soil reached saturation state. Figure 10 shows the dynamic changes of soil salt concentration in each soil layer of the saline wasteland in the autumn irrigation period. With the progress of autumn irrigation, soil salt concentration of each soil layer gradually decreased. After autumn irrigation, soil salt concentration of each soil layer reached the minimum values, and soil salt concentration of the saline wasteland was leached. After autumn irrigation, compared with that before autumn irrigation, the soil salt accumulation in each soil layer of saline wasteland was −1.011, −1.242, −1.218 and −1.335 mg/cm^3, respectively. Therefore, soil salt concentration of saline wasteland was less in the autumn irrigation period. The leaching soil salt entered into the groundwater outflow irrigation area and vacated the 'salt reservoir' for soil salt accumulation in the next year. The saline wasteland played an important role in maintaining the soil salt balance in the study area.

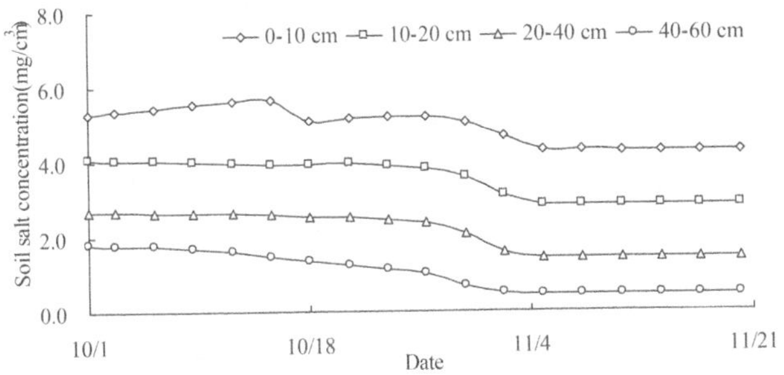

Figure 10 | Simulation of soil salt dynamics in the autumn irrigation period.

CONCLUSIONS

The SWAP model was calibrated and validated based on field experiments' observed data in 2018 and 2019 in Hetao Irrigation Area of Inner Mongolia. The SWAP model was used to simulate soil water–salt dynamics in saline wasteland in the crop growing period and autumn irrigation period after calibration and validation. The simulation of soil water–salt dynamics in saline wasteland showed that soil water content and soil salt concentration of saline wasteland were significantly affected by farmland irrigation in the crop growth period. Compared with the early growth period, soil salt accumulation of each soil layer in the saline wasteland was 0.164, 0.092, −0.890 and −1.261 mg/cm^3, respectively in the crop growth period. Saline wasteland was in the state of storing salt in the crop growth period, which was the drainage and salt drainage area for cultivated land. Soil water content gradually increased and soil salt concentration gradually decreased in the autumn irrigation period. Soil salt accumulation in each soil layer of saline wasteland was −1.011, −1.242, −1.218 and −1.335 mg/cm^3, respectively. The saline wasteland played an important role in maintaining the soil salt balance in the study area. This study will provide a theoretical basis for soil water and salt management and prevention of soil salinization in Hetao Irrigation Area of China.

ACKNOWLEDGEMENTS

This research was supported by Postgraduate Research & Practice Innovation Program of Jiangsu Province (XKYCX18_065) and the Key Popularized Programs of Water Conservancy Bureau of Jiangxi Province Science and Technology Achievement (202022TGKT07).

DATA AVAILABILITY STATEMENT

All relevant data are included in the paper or its Supplementary Information.

REFERENCES

Abbas, A., Khan, S., Hussain, N., Hanjra, M. A. & Akbar, S. 2013 Characterizing soil salinity in irrigated agriculture using a remote sensing approach. *Physics and Chemistry of the Earth* **55–57**, 43–52.

Feng, G., Zhang, Z., Wan, C., Lu, P. & Bakour, A. 2017 Effects of saline water irrigation on soil salinity and yield of summer maize (*Zea mays* L.) in subsurface drainage system. *Agricultural Water Management* **193**, 205–213.

Ghassemi, F., Jakeman, A. J. & Nix, H. A. 1995 *Salinisation of Land and Water Resources: Human Causes, Extent, Management and Case Studies*. CAB International, Wallingford, UK, pp. 31–46.

Hijmans, R. J., Guiking-Lens, I. M. & van Diepen, C. A. 1994 *User's Guide for the WOFOST 6.0 Crop Growth Simulation Model*. Technical Document 12, Winand Staring Centre, Wageningen, The Netherlands.

Huang, Q., Xu, X., Lv, L., Ren, D., Ke, J., Xiong, Y., Huo, Z. & Huang, G. 2018 Soil salinity distribution based on remote sensing and its effect on crop growth in Hetao Irrigation District. *Transactions of the Chinese Society of Agricultural Engineering* **34** (1), 102–109.

Muyen, Z., Moore, G. A. & Wrigley, R. J. 2011 Soil salinity and sodicity effects of waste water irrigation in South East Australia. *Agricultural Water Management* **99**, 33–41.

Noory, H., van der Zee, S. E. A. T. M., Liaghat, A.-M., Parsinejad, M. & van Dam, J. C. 2011 Distributed agro-hydrological modeling with SWAP to improve water and salt management of the Voshmgir irrigation and drainage network in Northern Iran. *Agricultural Water Management* **98**, 1062–1070.

Ren, D. Y., Xu, X., Engel, B. & Huang, G. H. 2018 Growth responses of crops and natural vegetation to irrigation and water table changes in an agro-ecosystem of Hetao, upper Yellow River basin: scenario analysis on maize, sunflower, watermelon and tamarisk. *Agricultural Water Management* **199**, 93–104.

Singh, U. K., Ren, L. & Kang, S. 2010 Simulation of soil water in space and time using an agro-hydrological model and remote sensing techniques. *Agricultural Water Management* **97** (8), 1210–1220.

Tang, Q., Hu, H., Oki, T. & Tian, F. 2007 Water balance within intensively cultivated alluvial plain in an arid environment. *Water Resources Management* **21**, 1703–1715.

Tedeschi, A. & Menenti, M. 2002 Simulation studies of long term saline water use: model validation and evaluation of schedules. *Agricultural Water Management* **54**, 123–157.

van Dam, J. C., Huygen, J., Wesseling, J. G., Feddes, R. A., Kabat, P., van Walsum, P. E. V., Groenendijk, P. & van Diepen, C. A. 1997 *Theory of SWAP Version 2.0: Simulation of Water Flow, Solute Transport and Plant Growth in the Soil–Water–Atmosphere–Plant Environment*. Technical Document 45, DLO Winand Staring Centre, Wageningen, The Netherlands, pp. 19–114.

van Dam, J. C., Groenendijk, P., Hendriks, R. F. A. & Kroes, J. G. 2008 Advances of modeling water flow in variably saturated soils with SWAP. *Vadose Zone Journal* **7** (2), 640–653.

Vazifedoust, M., van Dam, J. C., Feddes, R. A. & Feizi, M. 2008 Increasing water productivity of irrigated crops under limited water supply at field scale. *Agricultural Water Management* **95**, 89–102.

Wang, Y., Xiao, D., Li, Y. & Li, X. 2007 Response of salt accumulation in soil to groundwater changes at the oasis in the lower reaches of Sangong River, Xinjiang. *Acta Ecologica Sinica* **27** (10), 4036–4044.

Wang, J., Huang, X., Zhong, T. & Chen, Z. 2011 Review on sustainable utilization of salt-affected land. *Acta Geographica Sinica* **66** (5), 673–684.

Wang, G., Shi, H., Li, X., Zheng, Q., Guo, J. & Wang, W. 2019 Analysis of water and salt transportation and balance during cultivated land, waste land and lake system in Hetao Irrigation Area. *Journal of Hydraulic Engineering* **50** (12), 1518–1528.

Wichelns, D. & Oster, J. D. 2006 Sustainable irrigation is necessary and achievable, but direct costs and environmental impacts can be substantial. *Agricultural Water Management* **86**, 114–127.

Wu, J., Zhao, L., Huang, J., Yang, J., Vincent, B., Bouarfa, S. & Vidal, A. 2009 On the effectiveness of dry drainage in soil salinity control. *Science in China Series E: Technological Sciences* **52** (11), 3328–3334.

Xu, X., Huang, G., Sun, C., Pereira, L. S., Ramos, T. B., Huang, Q. & Hao, Y. 2013 Assessing the effects of water table depth on water use, soil salinity and wheat yield: searching for a target depth for irrigated areas in the upper Yellow River basin. *Agricultural Water Management* **125**, 46–60.

Xu, Y., Ge, Z., Wang, J., Li, W. & Feng, S. 2019 Study on relationship between soil salinization and groundwater table depth based on indicator Kriging. *Transactions of the Chinese Society of Agricultural Engineering* **35** (1), 123–130.

Yang, J. 2009 Development and prospect of the research on salt-affected soils in China. *Acta Pedologica Sinica* **45** (5), 837–845.

Yin, R. 2013 *Distributed Simulation of Soil Water and Salinity and Crop Water Productivity in Hetao Irrigation District*. China Agricultural University, Beijing, China, pp. 19–46.

Yuan, C., Feng, S., Huo, Z. & Ji, Q. 2019 Effects of deficit irrigation with saline water on soil water–salt distribution and water use efficiency of maize for seed production in arid northwest China. *Agricultural Water Management* **212**, 424–432.

First received 19 September 2020; accepted in revised form 22 October 2020. Available online 6 November 2020

Understanding nitrogen transport in the unsaturated zone with fluctuations in groundwater depth

Jianmin Bian, Qian Wang, Siyu Nie, Hanli Wan and Juanjuan Wu

ABSTRACT

Fluctuations in groundwater depth play an important role and are often overlooked when considering the transport of nitrogen in the unsaturated zone. To evaluate directly the variation of nitrogen transport due to fluctuations in groundwater depth, the prediction model of groundwater depth and nitrogen transport were combined and applied by least squares support vector machine and Hydrus-1D in the western irrigation area of Jilin in China. The calibration and testing results showed the prediction models were reliable. Considering different groundwater depth, the concentration of nitrogen was affected significantly with a groundwater depth of 3.42–1.71 m, while it was not affected with groundwater depth of 5.48–6.47 m. The total leaching loss of nitrogen gradually increased with the continuous decrease of groundwater depth. Furthermore, the limited groundwater depth of 1.7 m was found to reduce the risk of nitrogen pollution. This paper systematically analyzes the relationship between groundwater depth and nitrogen transport to form appropriate agriculture strategies.

Key words | ecological critical groundwater depth, fluctuations in groundwater depth, LS-SVM

Jianmin Bian
Qian Wang (corresponding author)
Hanli Wan
Juanjuan Wu
Key Laboratory of Groundwater Resources and Environment (Ministry of Education),
Jilin University,
Changchun, Jilin 130021,
China;
Jilin Provincial Key Laboratory of Water Resources and Environment,
Jilin University,
Changchun, Jilin 130021,
China;
and
College of New Energy and Environment,
Jilin University,
Changchun, Jilin 130021,
China
E-mail: *wangq0309@126.com*

Siyu Nie
College of Water Conservancy & Environment Engineering,
Changchun Institute of Technology,
Changchun, Jilin 130012,
China

HIGHLIGHTS

- The total leaching loss of nitrogen increases with the decrease of groundwater depth.
- The limited groundwater depth is proposed for determining appropriate agriculture strategies.

doi: 10.2166/ws.2021.066

GRAPHICAL ABSTRACT

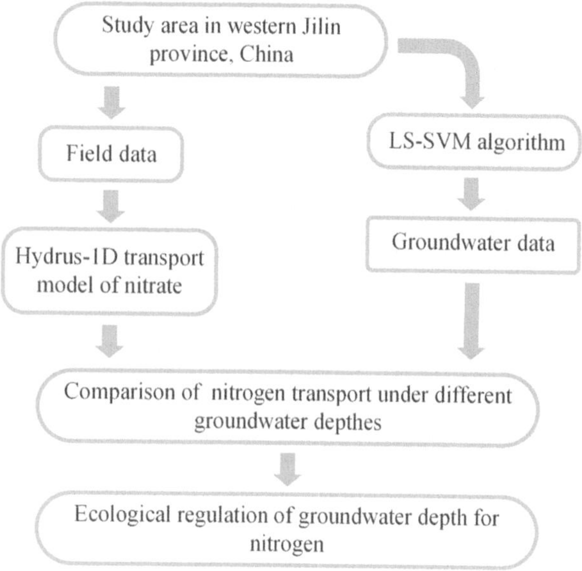

INTRODUCTION

Nitrogen fertilizer is widely used on farmland to increase crop productivity. However, the utilization rate of nitrogen fertilizer for crops (wheat, rice, and corn) is only 28–41% (Zhu & Chen 2002; China 2016). Excessive nitrogen on the soil surface infiltrates into groundwater through the unsaturated zone with precipitation or irrigation, and becomes one of the main sources of groundwater pollution (Zhang *et al.* 2018a, 2018b). Fluctuations in groundwater depth are a significant factor in changing the physical, chemical, and biological properties of the underground environment, thus leading to variations in the transport of nitrogen by leaching (Li *et al.* 2015; Akbariyeh *et al.* 2019).

In recent decades, many studies have been conducted on fluctuations in groundwater depth, mainly focusing on water and salt content (Seeboonruang 2012; Talebnejad & Sepaskhah 2015; Chen *et al.* 2018; Yang *et al.* 2019), crop growth processes (Fu & Burgher 2015; Han *et al.* 2015; Xia *et al.* 2016; Qiu *et al.* 2019), and organic matter transport and redistribution (Teramoto & Chang 2017; Gatsios *et al.* 2018; Yimsiri *et al.* 2018; Atteia *et al.* 2019).

Additionally, the correlation between groundwater depth and nitrogen content is of interest. For example, Zhiye *et al.* (2013) compared the conditions between stable and changing groundwater depth and indicated that elevated groundwater depth increased the content of nitrate nitrogen and decreased the content of ammonium nitrogen in the soil column. Li *et al.* (2015) showed that the content of nitrate nitrogen was closely related to groundwater depth through field monitoring data. Kawagoshi *et al.* (2019) and Akbariyeh *et al.* (2019) indicated that fluctuations in groundwater depth due to rainfall had an effect on the concentration of nitrogen.

The western irrigation area of Jilin is a severely saline region in China. The Chinese government launched a project to divert the Nen River to Baicheng City and developed approximately 2,000 hectares of farmland. The groundwater depth at a local location in this area changed with environmental factors (irrigation and meteorology) (Qi 2019) and had an effect on the transport of nitrogen. Considering this change, it is essential to establish an effective model to predict the groundwater depth. Empirical models, such as artificial neural network (ANN), support vector machine (SVM) and the least squares support vector machine (LS-SVM) (Ghose *et al.* 2010; Mohanty *et al.* 2015), were widely applied because they are more cost-effective at obtaining useful results with less data. LS-SVM uses different loss functions where inequality constraints are replaced by equality constraints and overcomes a huge computing burden (Wang & Hu 2005). So far, LS-SVM has been applied to estimate evapotranspiration and precipitation (Kundu *et al.* 2017), predict groundwater depth (Tang *et al.* 2018), and analytical chemistry (Balabin & Lomakina 2011). Therefore, LS-SVM was adopted to predict the groundwater depth.

However, the above literature rarely discussed the transport of nitrogen quantitatively for different groundwater levels or proposed a reliable groundwater depth limit for groundwater-ecological risk assessments. In this study, based on the field-measured and collected data, LS-SVM and Hydrus-1D were applied to evaluate the groundwater depth and nitrogen transport, respectively. Additionally, the transport of nitrogen in the unsaturated zone at varying groundwater depths was simulated. Finally, the ecological threshold for groundwater depth was determined to provide a sustainable limit for contaminated sites.

MATERIALS AND METHODS

Field site

The study area is located at Liangjiazi Town, Da'an City, west of Jilin (123° 52′ 48″ E–45° 18′ 45″ N). The annual average temperature is 5.02 °C and the annual average precipitation is 422 mm. The annual average evaporation is 1,749 mm, which is 4.14 times the precipitation. The location is shown in Figure 1 and the area is approximately 600 m^2. In the field site, the fertilize points are distributed in an 'S' type (see the right panel in Figure 1) to avoid farming errors.

The research area is a paddy field modified by saline soil and the planted rice is Jijing 88. Shallow and deep irrigation are used in the farmland during different periods of crop growth. From regreening to tillering, shallow irrigation was applied; from booting to heading and flowering, deep irrigation was applied; and then shallow irrigation was applied during the ripening period. The irrigation was stopped until

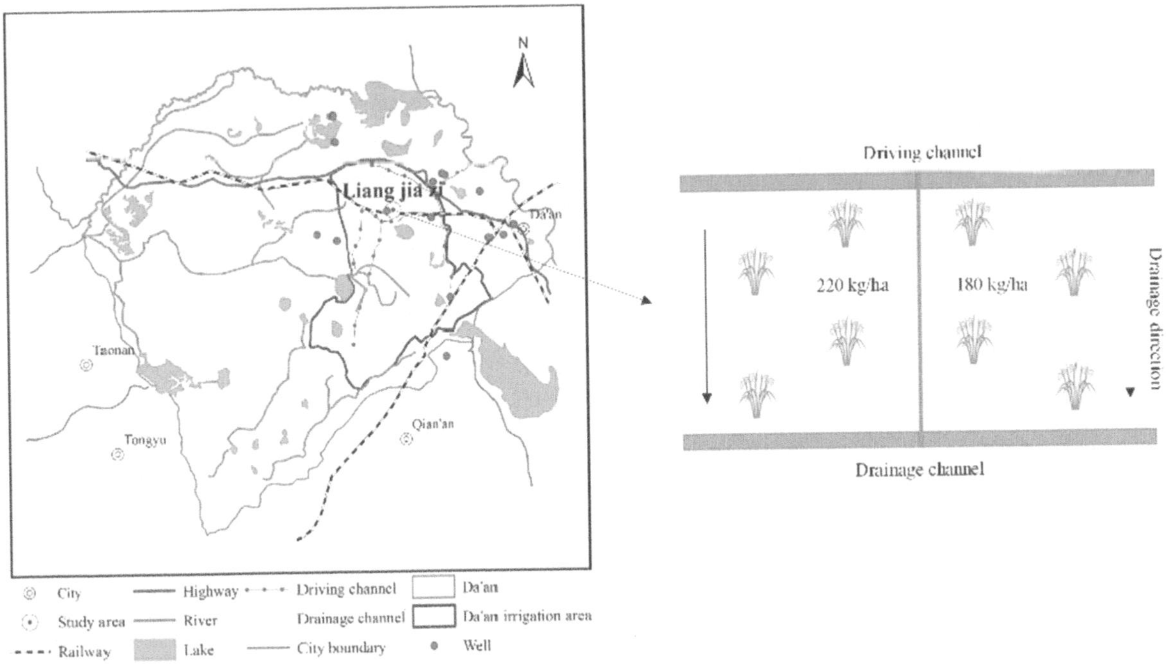

Figure 1 | The location and diagram of the field site.

the rice was harvested. According to the usual application mass of nitrogen fertilizer for paddy fields in the research area (150 kg/hm^2–230 kg/hm^2), the trial plot was fertilized in two patterns: A (180 kg/hm^2) and B (220 kg/hm^2), which were used for calibration and testing, respectively. The proportion of base fertilizer, tillering fertilizer, booting fertilizer, and granule fertilizer was 3:4:2:1. Potassium and phosphate fertilizers of 90 kg/hm^2 and 70 kg/hm^2 were applied once in the base fertilizer. The field experiments were conducted for 150 days. The rice was raised on May 15, transplanted on June 19, harvested on October 11.

During the period of regreening, tillering, booting, ripening, and harvesting, the soil samples were collected at a depth of 0–20 cm, 20–50 cm, 50–80 cm, and 80–100 cm. The soil samples were taken to the laboratory and tested for concentration of nitrate nitrogen, ammonium nitrogen, and moisture content.

Groundwater depth modeling

Groundwater depth data in the study area are estimated by LS-SVM (Levenberg 1944; Marquardt 1963). In the modeling process, the groundwater depth is simulated by mapping the input space to the high-dimensional space nonlinearly. Linear regression is then performed and the regression function by a vector form is represented as $f(x) = w\Phi(x) + b$, where w is the plane weight vector, b is the threshold value, $\Phi(x)$ is the nonlinear transfer function which maps the input vectors to the high-dimensional space. In the LS-SVM algorithm, the corresponding optimized problem is shown in Equation (1):

$$\min J(w, \xi) = \frac{1}{2}\|w\|^2 + \frac{c}{2}\sum_{i=1}^{l}\xi_i^2 \tag{1}$$

$$\text{s.t}\begin{cases} y_i\,[\Phi(x_i)w + b] = 1 - \xi_i \\ i = 1, 2, \cdots, l \end{cases} \tag{2}$$

where $J(w, \xi)$ is the objective function; c is the equilibrium constant; and ξ_i is the relaxation variable. The Lagrange function is shown as follows:

$$L(w, b, \xi, \delta) = \frac{1}{2}\|w\|^2 + \frac{c}{2}\sum_{i=1}^{l}\xi_i^2 - \sum_{i=1}^{l}\delta_i\{y_i\,[\Phi(x_i)w + b] - 1 + \xi_i\} \tag{3}$$

where δ_i is the Lagrangian operator, and the differentiation of w, b, ξ, δ with the Lagrange function is described as follows:

$$\begin{cases} \dfrac{\partial L}{\partial w} = 0 \rightarrow w = \sum_{i=1}^{l}\delta_i\Phi(x_i) \\ \dfrac{\partial L}{\partial b} = 0 \rightarrow w = \sum_{i=1}^{l}\delta_i = 0 \\ \dfrac{\partial L}{\partial \xi_i} = 0 \rightarrow \delta_i = c\xi_i,\ i = 1, 2, \cdots, l \\ \dfrac{\partial L}{\partial \delta_i} = 0 \rightarrow y_i\,[\Phi(x_i)w + b] = 1 - \xi_i \end{cases} \tag{4}$$

The kernel function is defined as $K(x_i, x_j) = \Phi(x_j)\Phi(x_i)^T$ and the regression function is described in Equation (5):

$$f(x) = sgn\left[\sum_{i=1}^{l}\delta_i K(x_i, x_j) + b\right] \tag{5}$$

In the LS-SVM algorithm, the radical basis function (RBF) is taken as the kernel function, which is pivotal for determining the mapping function and feature space.

$$K(x_i, x_j) = \exp\left(-\frac{\|x_i - x_j\|^2}{2\sigma^2}\right) \tag{6}$$

Groundwater depth data from 2003 to 2014 in the field site was collected by Da'an Water Resources Office. The distribution of groundwater wells is shown in Figure 1. In addition, monthly meteorological data for temperature, precipitation, and evaporation from 2003 to 2015 were downloaded and collated from, the China Meteorological Science Data Sharing Service Website (http://www.cma.gov.cn/2011qxfw/2011qsjgx/). For the reliability and integrity of the groundwater depth data, as well as the correlation between groundwater depth and the meteorological data, the current month's meteorological data of temperature, precipitation, and evaporation, and last month's groundwater depth, were input into the LS-SVM algorithm. The current month's groundwater depth was the output result. During the simulation process, data for 2003–2012 was set as training samples and data for 2013–2014 was set as testing samples. Finally, meteorological data for 2015 was input and the groundwater depth in 2015 was predicted by the LS-SVM algorithm.

Hydrus-1D model

Hydrus-1D can be used to simulate the transport of flow, solute, heat, and other processes in saturated and unsaturated porous media under various conditions (Dagois et al. 2017; Jha et al. 2017; Shahrokhnia & Sepaskhah 2018). More complex situations can also be modeled by Hydrus-1D, with additional modules such as UnsatChem, HP1, and Wetland (Mallants et al. 2017; Simunek et al. 2018). Hydrus-1D is widely used for research on actual problems in agriculture and groundwater pollution (Li 2020; Li et al. 2020), and it can be downloaded for free from www.hydrus2d.com (or www.pc-progress.cz). This paper established a nitrogen transport model with the modules of flow, solute, and root uptake in Hydrus-1D. The simulated depth was 0–100 cm from the soil surface. The observation points (10, 40, 70, and 100 cm) in the modeling domain were the same as the sampling points in the field experiments. The simulated period was 144 days from May 5 to October 5, 2015 (the period from transplanting to harvesting of the rice). A variable time step was used and the initial time step was 1 day. The time step was adjusted with the number of iterations and ranged from 0.01 days to 7 days. The iteration accuracy of moisture content and pressure head was 0.0001 and 0.01, respectively.

First, data for 180 kg/hm^2 were used to calibrate the parameters in the transport model of nitrogen. Then, data for 220 kg/hm^2 were applied to test the calibrated transport model. Finally, consistency between modeling and observation values was assessed using the evaluation indexes in calibration and testing processes.

Flow transport

The process of flow in the unsaturated zone is described by the modified Richard equation, which varies with soil matrix potential, as shown in Equation (7) (Simunek et al. 2013):

$$\frac{\partial \theta}{\partial t} = \frac{\partial}{\partial x_i}\left[k\left(k_{ij}^A \frac{\partial h}{\partial x_j} + k_{ij}^A\right)\right] - S \tag{7}$$

where θ is the volume moisture content; t is time; h is the pressure head; k is the unsaturated hydraulic conductivity; k_{ij}^A is the dimensionless expression of anisotropic tensor; x_i is the space coordinate; and S is the sources and sinks of root uptake. The soil moisture characteristics parameters are obtained using the van Genuchten model. The expressions are as follows:

$$\theta(h) = \begin{cases} \theta_r + \dfrac{\theta_s - \theta_r}{(1 + |\alpha h|^n)^m} & h < 0 \\ \theta_s & h \geq 0 \end{cases} \tag{8}$$

$$k(h) = k_s S_e^l \left[1 - \left(1 - S_e^{\frac{1}{m}}\right)^m\right]^2 \tag{9}$$

$$S_e = \frac{\theta - \theta_r}{\theta_s - \theta_r} \tag{10}$$

where θ_s and θ_r are the saturated and residual water content; n, m, and l are the empirical fitting parameters, $m = 1 - 1/n$; α is the reciprocal of intake pressure, $\alpha = 1/h_b$; and k_s is the saturated hydraulic conductivity.

Soil hydraulic parameters are identified by the size of soil particles. The soil moisture characteristic curves in this study are obtained using the soil conversion function in RETC Software (Genuchten et al. 1992) based on physical and chemical characteristics of collected soil samples in 2015. The simulated soil moisture data were compared with measured moisture data and fitted parameters were obtained. The related parameters including the saturated water content (θ_s), the residual water content (θ_r), the empirical fitting parameters (α, n, l) and the saturated hydraulic conductivity (k_s) are shown in Table 1.

Solute transport

During the leaching process of nitrogen in the unsaturated zone, excessive organic nitrogen undergoes biological transformation processes of mineralization, biological

Table 1 | The soil moisture characteristics parameters

Soil depth (cm)	θ_s (cm³/cm³)	θ_r (cm³/cm³)	α (cm)	n	l	k_s (cm/d)
0–20	0.4051	0.0726	0.0181	1.3174	0.5	12.160
20–50	0.4373	0.0789	0.0214	1.3482	0.5	21.468
50–80	0.4248	0.0673	0.0225	1.3837	0.5	31.016
80–100	0.39	0.0641	0.031	1.4166	0.5	14.368

immobilization, nitrification, and denitrification, and produces inorganic nitrogen of ammonium nitrogen and nitrate nitrogen. In addition, chemical processes covering volatilization, adsorption–desorption, fixation and release of nitrogen occur in leaching (Tillotson *et al.* 1980). The transformation process is described by zero-order and first-order dynamic equations in Hydrus-1D and is shown as follows:

$$\frac{\partial \theta c_1}{\partial t} = \frac{\partial}{\partial x_i}\left(\theta D_{ij,1}^{W}\frac{\partial c_1}{\partial x_j}\right) - \frac{\partial q_i c_1}{\partial x_i} - k_1\theta c_1 \quad (11)$$

$$\frac{\partial \theta c_2}{\partial t} + \frac{\partial \rho s}{\partial t} = \frac{\partial}{\partial x_i}\left(\theta D_{ij,2}^{W}\frac{\partial c_2}{\partial x_j}\right) - \frac{\partial q_i c_2}{\partial x_i} + k_0\rho c_N + k_1\theta c_1 - k_{2W}\theta c_2 - k_{2s}\rho s - Sc_2 \quad (12)$$

$$\frac{\partial \theta c_3}{\partial t} = \frac{\partial}{\partial x_i}\left(\theta D_{ij,3}^{W}\frac{\partial c_3}{\partial x_j}\right) - \frac{\partial q_i c_3}{\partial x_i} + k_{2W}\theta c_2 + k_{2s}\rho s - (k_3 + k_4)\theta c_3 - Sc_3 \quad (13)$$

where $c_1, c_2,$ and c_3 are the concentration of urea, ammonium nitrogen, and nitrate nitrogen, respectively; q_i is the soil moisture flux in i direction; s is the mass concentration of ammonium nitrogen in soils (adsorption mass), $s = k_d c_2$; k_d is the adsorption coefficient of ammonium nitrogen; c_N is the organic matter content of soils; k_0, k_1, k_2, k_3, and k_4 are the mineralization rate of organic matter, hydrolysis rate of urea, nitrification rate (k_{2W} and k_{2s} are the nitrification rates in solid and liquid phases), denitrification rate, and fixation rate of biology, respectively; ρ is the bulk density of soils; S is the term of root water uptake; and D_{ij}^{W} is hydrodynamic dispersion tensor in all directions.

Considering the concentration of nitrate nitrogen is much greater than that of ammonium nitrogen in the unsaturated zone (Li *et al.* 2007), nitrate nitrogen was regarded as the only product in the mineralization process. With the increase of soil depth, the content of soil organic matter, breathability of soil and mineralization of nitrogen decreases. Therefore, the mineralization and fixation were ignored in the deep soil and assumed to occur in the shallow soil. The denitrification was assumed to occur in the integrating domain. The volatilization of nitrogen was also ignored due to the irrigation of the paddy field (Tafteh & Sepaskhah 2012). Organic nitrogen fertilizer and nitrate nitrogen are non-adsorptive, and ammonium nitrogen is adsorbed easily (Ravikumar *et al.* 2011). The adsorption coefficients (k_d) of ammonium nitrogen are determined by batch adsorption experiments. It was difficult to obtain the remaining parameters of the nitrogen transport model so they were identified by calibration based on actual measured data. The calibrated parameters of solute transport are described in Table 2.

Root uptake

The driving force of root uptake is mainly derived from crop transpiration. The mass of root uptake is closely related to the distribution of crop root. In this research, the root uptake model (Feddes model) (Reicosky & Ritchie 1976) and the root growth model (Simunek *et al.* 2013), which are calculated in Hydrus-1D, are used and shown as follows:

$$S(x, h) = a(x, h)b(x_1)L_r T_P \quad (14)$$

where $a(x, h)$ refers to the function of soil matrix potential or suction head; $b(x_1)$ refers to the distribution of the function of relative root density; L_r refers to the root depth; and T_P refers to the potential transpiration rate.

The function of $a(x, h)$ is determined by the negative pressure head in crop root. The rice root parameters in the Feddes model used the values in Tan *et al.* (2015) and are shown in Table 3. The corresponding function of $b(x_1)$ was provided by the database in Hydrus-1D as follows:

$$b(x_1) = \begin{cases} \left(1 - \frac{x_1}{x_{1m}}\right)e^{-\left(\frac{p_x}{x_{1m}}|x_1^* - x_1|\right)} & 0 \le x_1 \le 70 \\ 0 & 70 \le x_1 \le 150 \end{cases} \quad (15)$$

Table 2 | The related parameters of solute transport

Soil depth (cm)	α_L	k_d	k_0	k_3	k_4
0–20	0.3	0.08	0.070	0.002	0.0045
20–50	0.3	0.08	0.037	0.002	0.0045
50–80	0.3	0.15	0.001	0.002	0.0045
80–100	0.3	0.15	0.001	0.002	0.0045

Table 3 | The simulated parameters of the Feddes model

Pressure head	h_1	h_2	h_{3high}	h_{3low}	h_4
cm	– 10	– 55	– 160	– 250	– 15,000

where x_{1m} refers to the vertical maximum distance of the root distribution; x_1^* represents the horizontal coordinate of the vertical maximum root density; P_x refers to the empirical parameter indicating the asymmetry of the root in the longitudinal direction. The root nitrogen uptake is the product of the water uptake mass and nitrogen concentration of the calculation point.

In addition, the parameter L_r in the calculation process of root uptake was determined by the root growth module in Hydrus-1D, which considered the actual growth situation. The model assumes that the root depth is the function of the maximum root depth and the root growth coefficient (Simunek *et al.* 2013) as follows:

$$L_r(t) = L_m \times f_r(t) \tag{16}$$

$$f_r(t) = \frac{L_0}{L_0 + (L_m - L_0)e^{-rt}} \tag{17}$$

where $L_r(t)$ represents the root depth which changes with time; L_m refers to the maximum root depth; $f_r(t)$ is the root growth coefficient; and L_0 and r are the initial root depth and the root growth rate, respectively. The rice crop in this study had a certain depth of root in transplanting. The initial root depth was 15 cm and the maximum root growth depth was 50 cm.

Boundary and initial conditions

In the simulation process, the initial moisture content and the concentration of nitrate nitrogen and ammonium nitrogen were evenly distributed in the horizontal direction. In the field site, basin irrigation was used and the depth of irrigation was 10 cm. The moisture content of soil surface was saturated after the rice transplanting and the moisture content at other points was defined by linear interpolation. The variable upper boundary condition was set up with the atmospheric boundary and the depth of surface water layer changed with rice growth (the maximal depth of the surface water layer was 20 cm). The bottom of the soil profile was set up with the free drainage boundary. The left and right positions were no-flux boundaries.

Model evaluation

Four statistical criteria – the correlation coefficient (R), the root mean square error ($RMSE$), the mean absolute error (MAE), and the Nash–Sutcliffe efficiency coefficient (NS) – were used to evaluate the model performance. R measures the correlation degree linearly between two variables. $RMSE$ and MAE provide different types of information about the model prediction capabilities. NS evaluates the reliability of the model results.

$$R = \frac{\sum_{i=1}^{n}(O_i - \bar{O})(P_i - \bar{P})}{\sqrt{\sum_{i=1}^{n}(O_i - \bar{O})^2 \sum_{i=1}^{n}(P_i - \bar{P})^2}} \tag{18}$$

$$RMSE = \sqrt{\frac{1}{n}\sum_{i=1}^{n}(P_i - O_i)^2} \tag{19}$$

$$MAE = \frac{1}{n}\sum_{i=1}^{n}|P_i - O_i| \tag{20}$$

$$NS = 1 - \frac{\sum_{i=1}^{n}(P_i - O_i)^2}{\sum_{i=1}^{n}(O_i - \bar{O})^2} \tag{21}$$

where n represents the number of input parameters; O_i and P_i refer to the observation and simulated values of groundwater depth at time t, respectively; and $\bar{O}$ and $\bar{P}$ are the average observation and simulated values of groundwater depth.

RESULT AND DISCUSSION

Groundwater depth evaluation

A model for predicting groundwater depth was established using the LS-SVM algorithm with MATLAB. The calibration and testing results are shown in Figure 2, and the evaluation results of the four statistical criteria are shown in Table 4. The values of R and NS reach 0.8 or higher and the values of $RMSE$ and MAE are closer to 0. These results indicate

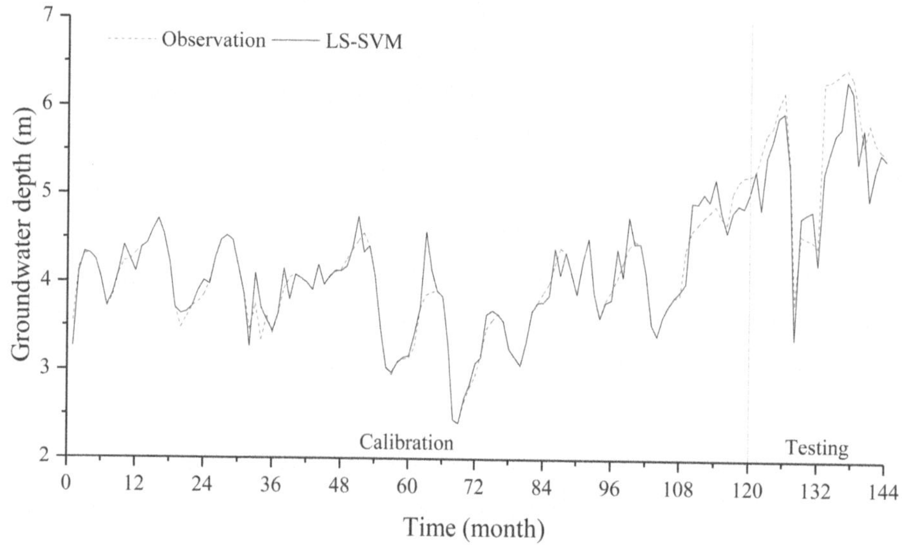

Figure 2 | The calibration and testing results of groundwater depth using the LS-SVM.

Table 4 | The evaluation results of the calibration and testing periods of groundwater depth

	LS-SVM	
	Calibration period	**Testing period**
R	0.964	0.919
NS	0.928	0.849
RMSE (cm)	0.148	0.297
MAE	0.082	0.208

that the LS-SVM method is suitable for predicting the groundwater depth.

Based on the successfully calibrated and tested LS-SVM model, the monthly mean average values of groundwater depth in 2015 were obtained by inputting the corresponding meteorological data. The results are presented in Figure 3 and Table 5. It is observed that the groundwater depth ranges from 5.48 to 6.47 m. The modeled groundwater depth manifests the actual seasonal fluctuation observed in 2015.

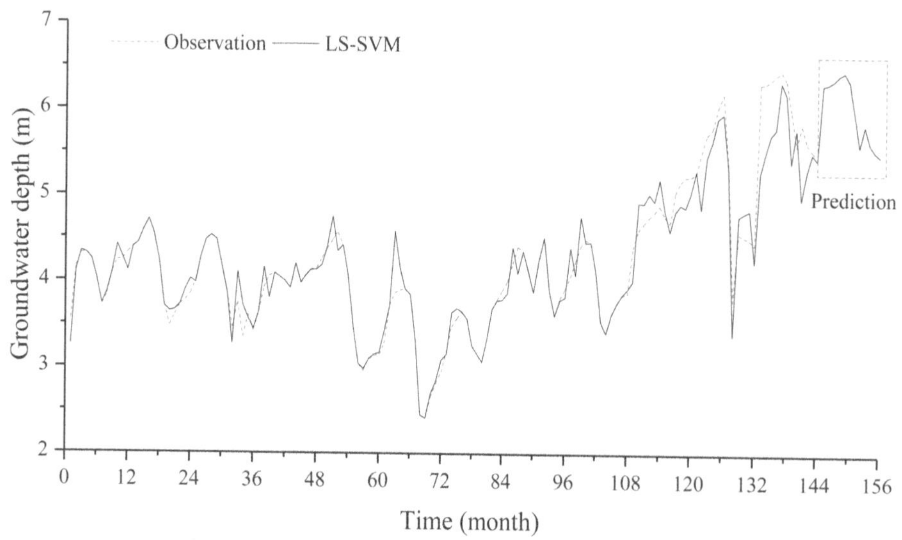

Figure 3 | The monthly mean values of groundwater depth in the field sites in 2015 using LS-SVM.

Table 5 | The monthly mean data of groundwater depth in 2015

Month	Jan	Feb	Mar	Apr	May	Jun
Groundwater depth (m)	6.31	6.32	6.36	6.42	6.47	6.35
Month	**Jul**	**Aug**	**Sept**	**Oct**	**Nov**	**Dec**
Groundwater depth (m)	5.99	5.59	5.84	5.63	5.54	5.48

Calibration and testing in Hydrus-1D

According to the defined parameters, the model of flow and solute were calibrated with a fertilization rate of 180 kg/ha and tested with a load of 220 kg/ha. The transport results of moisture content and nitrogen are shown in Figures 5–10 and the distribution of precipitation and

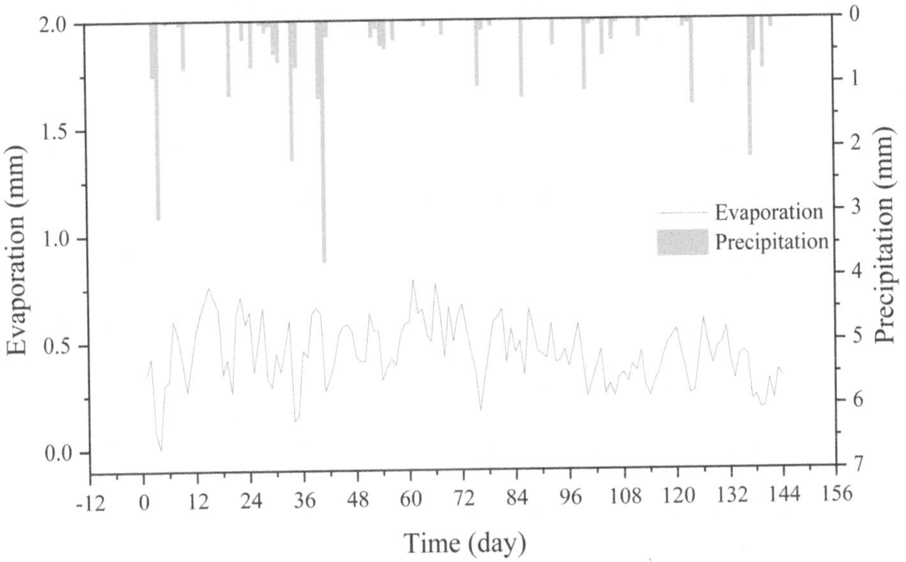

Figure 4 | The distribution of precipitation and evaporation in the simulation period.

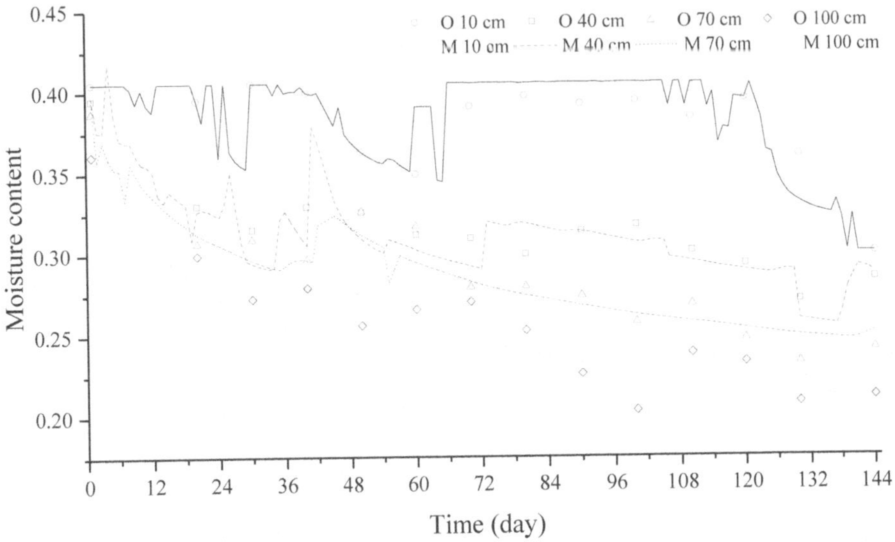

Figure 5 | The calibrated results of moisture content of 180 kg/hm^2 (O and M represent the observed and simulated data).

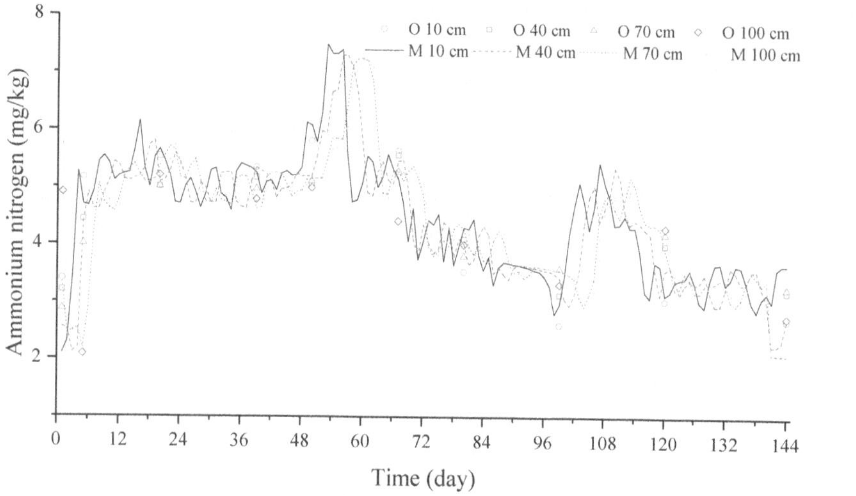

Figure 6 | The calibrated results of ammonium nitrogen of 180 kg/hm^2 (O and M represent the observed and simulated data).

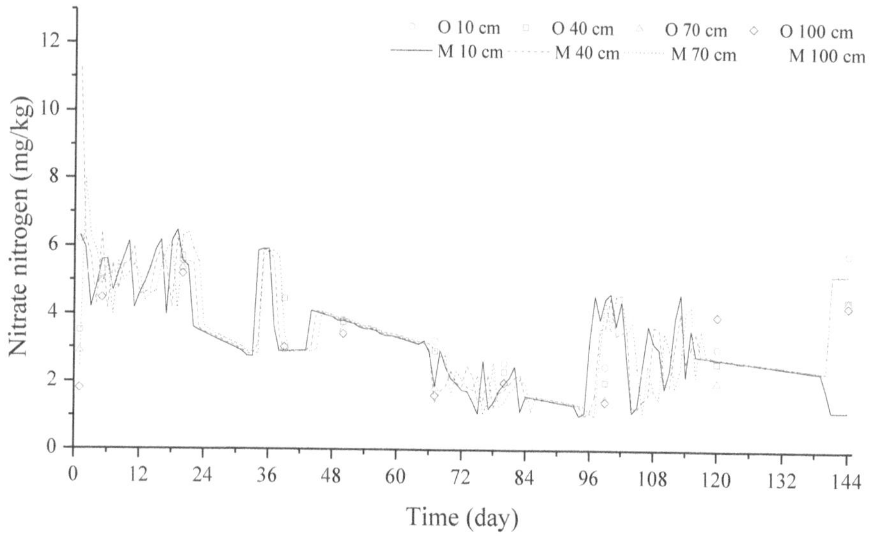

Figure 7 | The calibrated results of nitrate nitrogen of 180 kg/hm^2 (O and M represent the observed and simulated data).

evaporation during the simulation period are shown in Figure 4.

It can be seen in Figures 5–10 that moisture content, ammonium nitrogen, and nitrate nitrogen exhibit similar trends in variation at different layers. The moisture content decreases with time due to crop uptake, and fluctuates locally due to precipitation and irrigation. As the topsoil is easily recharged by precipitation and irrigation, moisture content is higher in the shallower layers than in the deeper layers.

Due to the effects of precipitation, evaporation (Figure 4), and fertilization, the concentration of ammonium nitrogen at different layers fluctuates with time, and its maximum value appears during June and July (the period of tillering and booting of crops). In general, the fluctuation range of ammonium nitrogen concentration, which first increases and then decreases, is related to the amount of fertilization, precipitation and irrigation. In addition, the concentration of ammonium nitrogen is greater in shallower layers than in deeper layers because of the adsorption behavior.

For nitrate nitrogen, the concentration in soil reduces rapidly and it fluctuates slightly until it reaches a stable state in September. Initially, nitrogen fertilizer on the soil moves downward with the flow of precipitation and

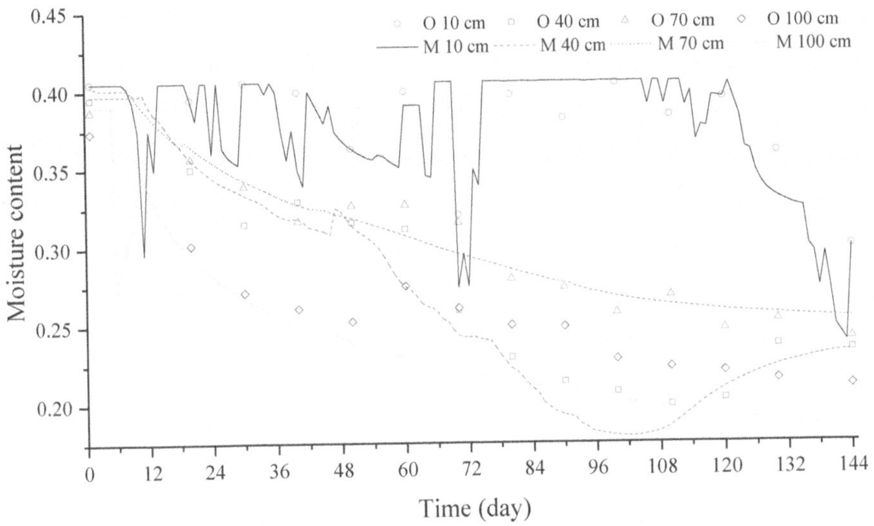

Figure 8 | The tested results of moisture content of 220 kg/hm^2 (O and M represent the observed and simulated data).

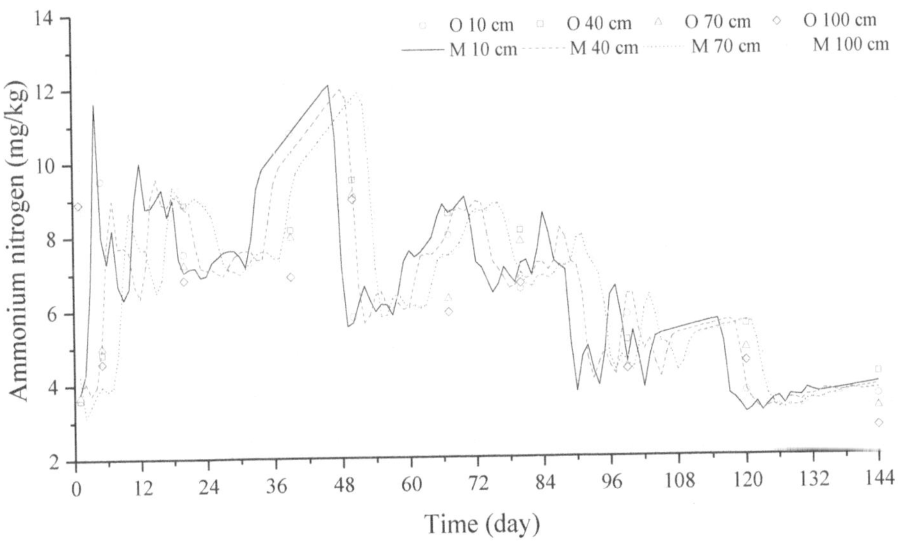

Figure 9 | The tested results of ammonium nitrogen of 220 kg/hm^2 (O and M represent the observed and simulated data).

irrigation, and the concentration of nitrate nitrogen in the unsaturated zone reaches the maximum value. During the crop growing season, nitrate nitrogen is absorbed by the crop and is affected by precipitation and irrigation, resulting in the concentration of nitrate nitrogen decreasing with time. At harvest time, the concentration of nitrate nitrogen reaches a minimum after this long-term decrease. Nitrate nitrogen also leaches into groundwater more easily and the concentration in shallower layers is lower than in deeper layers.

The values of *RMSE* are lower than 1.5 mg/kg and the values of *R* are greater than 0.8, as shown in Table 6. These results indicate that the Hydrus-1D model is viable for exploring the migration and transformation of nitrogen. In different layers, the structure of soil pores, connectivity, and physicochemical properties of salinity, pH, and moisture affect the transport and transformation of nitrogen. These related parameters change with crop growth and could not be modified in Hydrus-1D. The fitting accuracies therefore vary with different layers, as shown in Table 6.

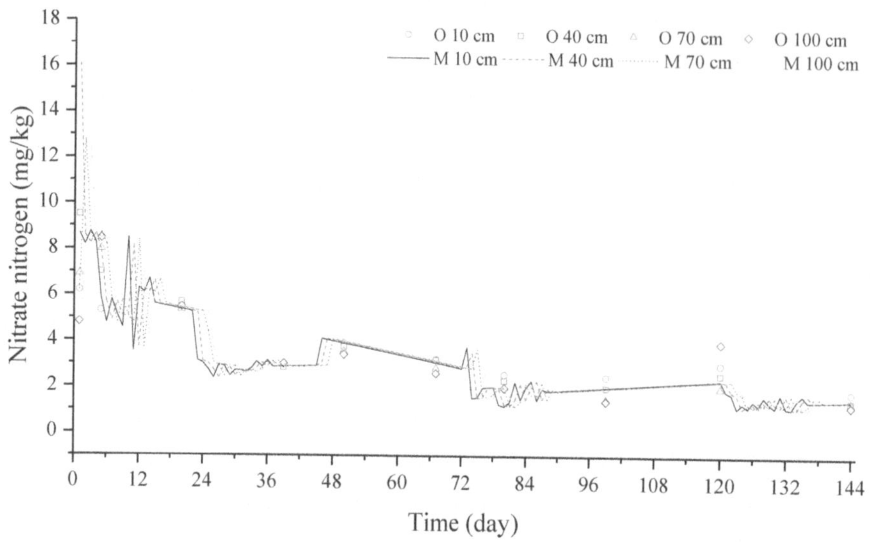

Figure 10 | The tested results of nitrate nitrogen of 220 kg/hm^2 (O and M represent the observed and simulated data).

Table 6 | The evaluation results of calibration and testing in Hydrus-1D

		Moisture content		Nitrate nitrogen		Ammonium nitrogen	
Fertilizing amount	Depth (cm)	*RMSE* (mg/kg)	*R*	*RMSE* (mg/kg)	*R*	*RMSE* (mg/kg)	*R*
180 kg/hm^2	10	0.009	0.97	0.47	0.91	0.42	0.83
	40	0.013	0.96	1.29	0.81	0.26	0.91
	70	0.014	0.95	0.64	0.83	0.68	0.84
	100	0.015	0.92	0.70	0.80	0.48	0.86
220 kg/hm^2	10	0.023	0.97	0.87	0.85	0.72	0.85
	40	0.021	0.96	1.18	0.94	1.20	0.94
	70	0.017	0.97	0.43	0.96	1.09	0.96
	100	0.016	0.94	0.81	0.88	1.90	0.86

Transport of nitrogen at 100 cm of soil under groundwater fluctuation

Considering the uncertainty of the data and parameters, the prediction interval of *t*-distribution is applied to improve the accuracies of the simulation results. The calculation result of 95% confidence interval is obtained as in Nie *et al.* (2017) and is shown in Table 7.

Based on the dynamic range of groundwater depth shown in Table 7, Plan 1 (the upper limit of the predictive interval) and Plan 2 (the lower limit of the predictive

Table 7 | The dynamic range of groundwater depth under 95% confidence interval

Month	Jan	Feb	Mar	Apr	May	Jun
Upper limit (m)	6.39	6.49	6.69	6.71	6.75	6.81
Groundwater depth (m)	6.31	6.32	6.36	6.42	6.47	6.35
Lower limit (m)	5.16	5.21	5.92	5.94	5.97	6.03
Month	**Jul**	**Aug**	**Sept**	**Oct**	**Nov**	**Dec**
Upper limit (m)	6.85	6.74	6.38	5.98	6.22	6.01
Groundwater depth (m)	5.99	5.59	5.84	5.63	5.54	5.48
Lower limit (m)	6.08	5.26	5.59	5.20	5.45	5.24

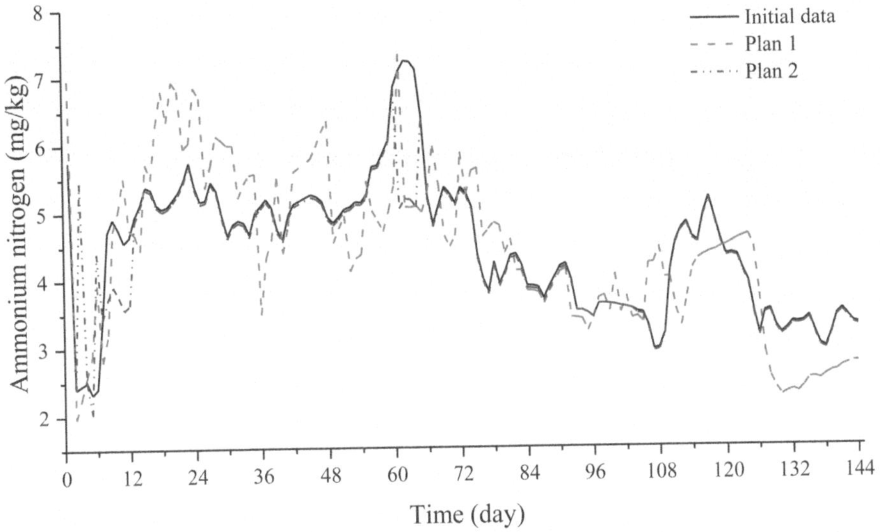

Figure 11 | The dynamic range of ammonium nitrogen in 100 cm from the surface soil.

interval) were designated as the lower boundary (100 cm from the surface soil) in Hydru-1D model. The results of the concentration of ammonium nitrogen and nitrate nitrogen for Plans 1 and 2 were then compared with those of the initial data (180 kg/hm^2).

The comparative results of the concentration of ammonium nitrogen for the two plans are shown in Figure 11. A similar trend is observed for both plans, except that the value and time of peak ammonium nitrogen are different. Furthermore, the concentration ranges of ammonium nitrogen for Plans 1 and 2 are 1.96–7.32 and 2.03–6.80 mg/kg, respectively. The difference between Plans 1 and 2 is small. Thus, the concentration of ammonium nitrogen at 100 cm is slightly affected by the groundwater depth of 5.48–6.47 m.

The comparative results of the concentration of nitrate nitrogen for different plans are shown in Figure 12. A similar trend, which shows a significant decrease with time and a different amplitude of variation in the short term, was observed for both plans. The concentration ranges of nitrate

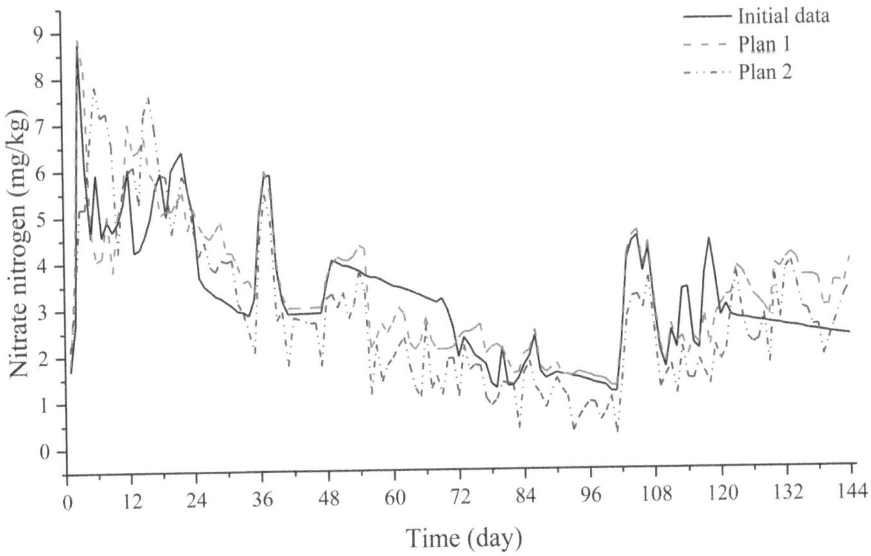

Figure 12 | The dynamic range of nitrate nitrogen in 100 cm from the surface soil.

nitrogen in Plans 1 and 2 are 1.27–8.86 and 0.24–7.82 mg/kg, respectively. The concentration of nitrate nitrogen in Plan 1 is slightly higher than that in Plan 2. It can be seen that the concentration of nitrate nitrogen at 100 cm is slightly affected by the groundwater depth of 5.48–6.47 m.

The cumulative leaching loss of nitrogen was calculated from the modeling results in Table 8. For ammonium nitrogen, the cumulative leaching loss in the initial data is 24.63 kg/ha and increased to 24.86 and 24.76 kg/ha for Plans 1 and 2, with an increase in amplitude of 0.9% and 0.5%, respectively. For nitrate nitrogen, the cumulative leaching loss in the initial data is 12.41 kg/ha, which increased to 12.46 and 12.52 kg/ha for Plans 1 and 2 with an increase in amplitude of 0.4% and 0.8%, respectively. The average increase in amplitude of nitrogen is approximately 0.6%, indicating that the mass of ammonium nitrogen and nitrate nitrogen varies little when the groundwater depth ranges from 5.48 to 6.47 m. The result was affected by the physicochemical properties of silty clay in saline areas (such as poor permeability).

Table 8 | The cumulative leaching loss of nitrogen in 0 –100 cm from the surface soil

Index	Cumulative leaching loss			Increase amplitude (%)	
	Initial data	Plan 1	Plan 2	Plan 1	Plan 2
Ammonium nitrogen (kg/hm^2)	24.63	24.86	24.76	0.9	0.5
Nitrate nitrogen (kg/hm^2)	12.41	12.46	12.52	0.4	0.8

Ecological regulation of groundwater depth

Enhanced loss of nitrogen through leaching poses a threat to the ecological environment of groundwater. Nitrogen pollution in some parts of the study area has exceeded the Level III standard of groundwater quality in China. According to the result that the loss of nitrogen by leaching is affected by fluctuations in groundwater depth, it is necessary to quantitively evaluate the ecological threshold of groundwater depth in the vadose zone in combination with crop cultivation and solute transport. In this section, the 2015 groundwater depth was continuously decreased by 50%. Two scenarios are defined: B1 (groundwater depth of 2.99–3.42 m) and B2 (groundwater depth of 1.49–1.71 m). To compare these scenarios with groundwater depth in 2015, the transport and leaching loss of nitrogen was analyzed under different conditions, as shown in Figures 13 and 14.

Figure 13 shows that the trend lines of the three situations vary similarly for ammonium nitrogen. However, there are significant differences in the concentration of ammonium nitrogen in soil in the three scenarios. In

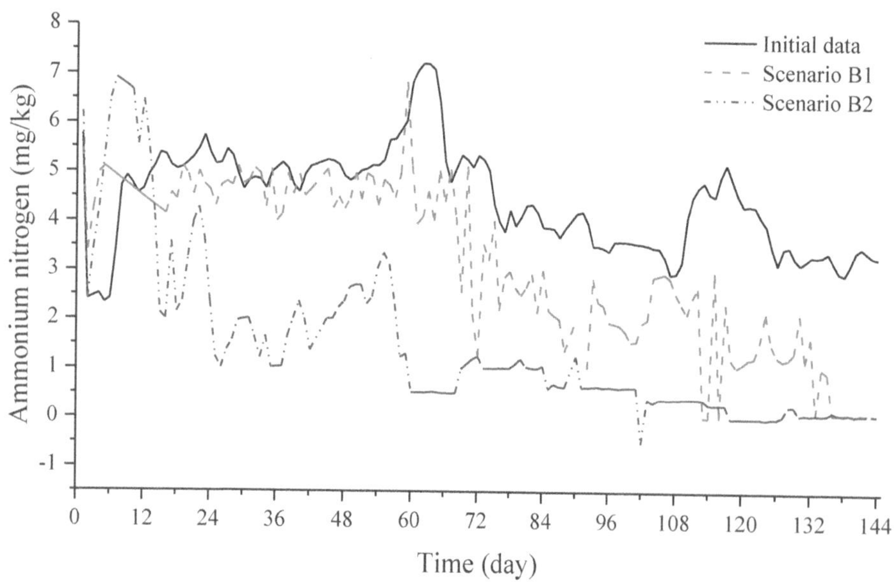

Figure 13 | The dynamic range of ammonium nitrogen in 100 cm from the surface soil.

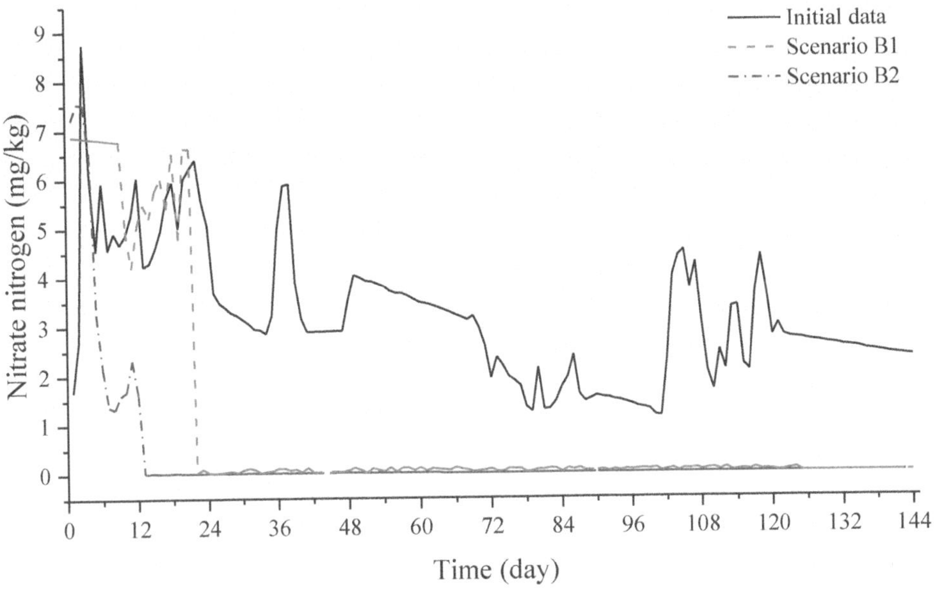

Figure 14 | The dynamic range of nitrate nitrogen in 100 cm from the surface soil.

scenario B1, the concentration of ammonium nitrogen in soil is 0.09–6.84 mg/kg and changes significantly after 72 days of crop planting compared to the initial data. In scenario B2, the concentration of ammonium nitrogen in soil is 0.007–6.91 mg/kg and changes more significantly compared with scenario B1 during the period of crop growth. Therefore, the comparative results indicate that the concentration of ammonium nitrogen in soil decreases with the decrease of groundwater depth. When groundwater depth decreases, the moisture content and root uptake of nitrogen in soil increase, which leads to a decrease in the concentration of ammonium nitrogen.

The concentration of nitrate nitrogen in soil for both scenario B1 and B2 decreases sharply with time, and the rate of reduction is different, as shown in Figure 14. In scenario B1, the initial concentration of nitrate nitrogen in soil is 7.221 mg/kg and reduces rapidly to zero on the 24th day after crop planting (the early stage of tillering). IN scenario B2, the initial concentration of nitrate nitrogen in soil is 7.543 mg/kg and reduces rapidly to zero on the 14th day after crop planting (the early stage of regreening). The results indicate that the reduction rate of nitrate nitrogen in scenario B1 is faster than that in scenario B2. The concentration of nitrate nitrogen in the initial data also varies more than in scenario B1 and B2. The reason for this is related to the lower groundwater depth, which can cause capillary moisture to be closer to the lower boundary, allowing nitrate nitrogen to migrate down more easily. Therefore, the concentration of nitrate nitrogen in soil decreases with a decrease in groundwater depth.

As shown in Table 9, for ammonium nitrogen, the cumulative leaching loss in the initial data is 24.63 kg/hm^2. This increases to 25.28 and 27.65 kg/hm^2 in scenario B1 and B2, respectively, with an increase in amplitude of 2.64% and 12.26%. For nitrate nitrogen, the cumulative leaching loss in the initial data is 12.41 kg/ha, and this increases to 15.11 and 20.01 kg/hm^2 in scenario B1 and B2, respectively, with an increase in amplitude of 21.75% and 61.24%. Combined with Figures 13 and 14, these results indicate that

Table 9 | The cumulative leaching loss of nitrogen in 0–100 cm from the surface soil

	Cumulative leaching loss			Increase amplitude (%)	
Index	**Initial data**	**Scenario B1**	**Scenario B2**	**Scenario B1**	**Scenario B2**
Ammonium nitrogen (kg/hm^2)	24.63	25.28	27.65	2.64	12.26
Nitrate nitrogen (kg/hm^2)	12.41	15.11	20.01	21.75	61.24

Table 10 | The leaching risk indicators of nitrate nitrogen in soil

Leaching risk indicators						
Level	**I**	**II**	**III**	**IV**	**V**	**VI**
Cumulative leaching loss (kg/hm^2)	0–20	20–40	40–60	60–80	80–100	100–120

cumulative leaching losses of nitrate increase with the gradual decrease in groundwater depth. In particular, for nitrate nitrogen in scenario B2, the cumulative leaching loss is 20.01 kg/hm^2 and the increase in amplitude exceeds 50% compared to the initial data. There are significant differences among the three scenarios, and the leaching risk indicators for nitrate nitrogen can be classified into six levels (Table 10) according to the study of Hasegawa & Denison (2005).

When the groundwater depth decreases to 1.37–1.62 m, the leaching risk indicator of nitrate nitrogen rises to Level II and the risk to the ecological environment of groundwater is under threat. Therefore, the groundwater depth should be limited to 1.7 m or more in the study area, and the irrigation and fertilization should remain unchanged to reduce the risk of nitrate nitrogen pollution.

CONCLUSION

This study used the LS-SVM algorithm and Hydrus-1D software to analyze the transport process of nitrogen at different groundwater depths. The groundwater depth in the western irrigation area of Jilin in China ranged from 5.48 to 6.47 m, as predicted by LS-SVM. Hydrus-1D was then used to explore the movement of nitrogen in this region. Modeling results reveal that the concentration at different layers fluctuates with time. The fluctuation range first increases and then decreases for ammonium nitrogen, while for nitrate nitrogen, the concentration in soil decreases rapidly and fluctuates slightly until reaching a stable state. In relation to predicting the effects of different groundwater depth on nitrogen transport, a groundwater depth of 5.48–6.47 m has little effect on the mass of nitrogen in the study area, but a depth of 3.42–1.71 m changes the concentration of nitrogen significantly. The cumulative leaching loss of nitrogen increases with a decrease in groundwater depth. When the groundwater depth decreases to 1.49–1.71 m, the leaching loss risk of nitrate nitrogen reaches Level II. The groundwater depth should therefore be limited to 1.7 m or more in the study area to reduce the risk of nitrate nitrogen pollution.

This work highlighted the quantitative relationship between fluctuations in groundwater depth and nitrogen mass in a case study and tried to provide a guide for the nitrate accumulation in a similar agricultural area. However, the relationship is affected by a number of key conditions, such as climate change, fertilizer application, irrigation, growth process of crops, and so on. Further efforts on developing modeling coupled with complex conditions should be considered.

ACKNOWLEDGEMENT

The research reported here was funded by the National Key R&D Program of China, grant number 2018YFC1800400.

DATA AVAILABILITY STATEMENT

All relevant data are included in the paper or its Supplementary Information.

REFERENCES

Akbariyeh, S., Pena, C. A. G., Wang, T., Mohebbi, A., Bartelt-Hunt, S., Zhang, J. & Li, Y. 2019 Prediction of nitrate accumulation and leaching beneath groundwater irrigated corn fields in the Upper Platte basin under a future climate scenario. *Science of the Total Environment* **685**, 514–526.

Atteia, O., Palmier, C. & Schafer, G. 2019 On the influence of groundwater table fluctuations on oil thickness in a well related to an LNAPL contaminated aquifer. *Journal of Contaminant Hydrology* **223**, 103476.1–103476.10.

Balabin, R. M. & Lomakina, E. I. 2011 Support vector machine regression (SVR/LS-SVM) – an alternative to neural networks (ANN) for analytical chemistry? Comparison of nonlinear methods on near infrared (NIR) spectroscopy data. *Analyst* **136** (8), 1703–1712.

Chen, W., Jin, M., Ferré, T. P. A., Liu, Y., Xian, Y., Shan, T. & Ping, X. 2018 Spatial distribution of soil moisture, soil salinity, and root density beneath a cotton field under mulched drip irrigation with brackish and fresh water. *Field Crops Research* **215**, 207–221.

Dagois, R., Faure, P., Bataillard, P., Bouzouidja, R., Coussy, S., Leguédois, S., Enjelvin, N. & Schwartz, C. 2017 From atmospheric- to pedo-climate modeling in technosols: a global scale approach. *Geoderma* **301**, 47–59.

Fu, B. & Burgher, I. 2015 Riparian vegetation NDVI dynamics and its relationship with climate, surface water and groundwater. *Journal of Arid Environments* **113**, 59–68.

Gatsios, E., Garcia-Rincon, J., Rayner, J. L., McLaughlan, R. G. & Davis, G. B. 2018 LNAPL transmissivity as a remediation metric in complex sites under water table fluctuations. *Journal of Environmental Management* **215**, 40–48.

Genuchten, M. T. V., Leij, F. J., Yates, S. R. & Williams, J. R. 1992 *The RETC Code for Quantifying the Hydraulic Functions of Unsaturated Soils* Robert S. Kerr Environmental Research Laboratory, Office of Research and Development, U.S. Environmental Protection Agency.

Ghose, D. K., Panda, S. S. & Swain, P. C. 2010 Prediction of water table depth in western region, orissa using BPNN and RBFN neural networks. *Journal of Hydrology* **394** (3–4), 296–304.

Han, M., Simunek, J., Zhao, C. & Feng, G. 2015 Evaluating the impact of groundwater on cotton growth and root zone water balance using Hydrus-1D coupled with a crop growth model. *Agricultural Water Management* **160**, 64–75.

Hasegawa, H. & Denison, R. F. 2005 Model predictions of winter rainfall effects on N dynamics of winter wheat rotation following legume cover crop or fallow. *Field Crops Research* **91** (2), 251–261.

Jha, R. K., Sahoo, B. & Panda, R. K. 2017 Modeling the water and nitrogen transports in a soil–paddy–atmosphere system using HYDRUS-1D and lysimeter experiment. *Paddy and Water Environment* **15** (4), 831–846.

Kawagoshi, Y., Suenaga, Y., Chi, N. L., Hama, T., Ito, H. & Duc, L. V. 2019 Understanding nitrate contamination based on the relationship between changes in groundwater levels and changes in water quality with precipitation fluctuations. *Science of the Total Environment* **657**, 146–153.

Kundu, S., Khare, D. & Mondal, A. 2017 Individual and combined impacts of future climate and land use changes on the water balance. *Ecological Engineering* **105**, 42–57.

Levenberg, K. 1944 A method for the solution of certain non-linear problems in least squares. *The Quarterly of Applied Mathematics* **2**, 164–168.

Li, D. F. 2020 Quantifying water use and groundwater recharge under flood irrigation in an arid oasis of northwestern China. *Agricultural Water Management* **240**, 13.

Li, X., Hu, C., Delgado, J. A., Zhang, Y. & Ouyang, Z. 2007 Increased nitrogen use efficiencies as a key mitigation alternative to reduce nitrate leaching in North China plain. *Agricultural Water Management* **89** (1–2), 137–147.

Li, X., Li, J., Xi, B., Yuan, Z., Zhu, X. & Zhang, X. 2015 Effects of groundwater level variations on the nitrate content of groundwater: a case study in Luoyang area, China. *Environmental Earth Sciences* **74**, 3969–3983.

Li, L., Song, X., Xia, L., Fu, N., Feng, D., Li, H. & Li, Y. 2020 Modelling the effects of climate change on transpiration and evaporation in natural and constructed grasslands in the semi-arid Loess Plateau, China. *Agriculture Ecosystems & Environment* **302**, 107077.1–107077.12.

Mallants, D., Simunek, J., van Genuchten, M. T. & Jacques, D. 2017 Simulating the fate and transport of coal seam gas chemicals in variably-saturated soils using HYDRUS. *Water* **9** (6), 385.

Marquardt, D. W. 1963 An algorithm for least-squares estimation of nonlinear parameters. *Journal of the Society for Industrial and Applied Mathematics* **11** (2), 431–441.

Mohanty, S., Jha, M. K., Raul, S. K., Panda, R. K. & Sudheer, K. P. 2015 Using artificial neural network approach for simultaneous forecasting of weekly groundwater levels at multiple sites. *Water Resources Management* **29** (15), 5521–5532.

Nie, S. Y., Bian, J. M., Wan, H. L., Sun, X. Q. & Zhang, B. J. 2017 Simulation and uncertainty analysis for groundwater levels using radial basis function neural network and support vector machine models. *Journal of Water Supply Research and Technology-Aqua* **66** (1), 15–24.

Qi, Q. Q. 2019 Application of a groundwater modelling system in groundwater environmental impact assessment of river and lake connection in Western Jilin Region. *Applied Ecology and Environmental Research* **17** (2), 5059–5066.

Qiu, J. X., Zipper, S. C., Motew, M., Booth, E. G., Kucharik, C. J. & Loheides, S. P. 2019 Nonlinear groundwater influence on biophysical indicators of ecosystem services. *Nature Sustainability* **2** (6), 475–483.

Ravikumar, V., Vijayakumar, G., Simunek, J., Chellamuthu, S., Santhi, R. & Appavu, K. 2011 Evaluation of fertigation scheduling for sugarcane using a vadose zone flow and transport model. *Agricultural Water Management* **98** (9), 1431–1440.

Reicosky, D. C. & Ritchie, J. T. 1976 Relative importance of soil resistance and plant resistance in root water absorption. *Soil Science Society of America Journal* **40** (2), 293–297.

Seeboonruang, U. 2012 Relationship between groundwater properties and soil salinity at the Lower Nam Kam River Basin in Thailand. *Environmental Earth Sciences* **69** (6), 1803–1812.

Shahrokhnia, M. H. & Sepaskhah, A. R. 2018 Water and nitrate dynamics in safflower field lysimeters under different irrigation strategies, planting methods, and nitrogen fertilization and application of HYDRUS-1D model. *Environmental Science and Pollution Research* **25** (9), 8563–8580.

Simunek, J., Sejna, M., Saito, H., Sakai, M. & van Genuchten, M. T. 2013 *The Hydrus-1D Software Package for Simulating the Movement of Water, Heat, and Multiple Solutes in Variably Saturated Media*. Department of Environmental Sciences, University of California Riverside, Riverside, California, USA.

Simunek, J., Sejna, M. & van Genuchten, M. T. 2018 New features of version 3 of the HYDRUS (2D/3D) computer software package. *Journal of Hydrology and Hydromechanics* **66** (2), 133–142.

Tafteh, A. & Sepaskhah, A. R. 2012 Application of HYDRUS-1D model for simulating water and nitrate leaching from continuous and alternate furrow irrigated rapeseed and maize fields. *Agricultural Water Management* **113** (10), 19–29.

Talebnejad, R. & Sepaskhah, A. R. 2015 Effect of different saline groundwater depths and irrigation water salinities on yield and water use of quinoa in lysimeter. *Agricultural Water Management* **148**, 177–188.

Tan, X., Shao, D., Gu, W. & Liu, H. 2015 Field analysis of water and nitrogen fate in lowland paddy fields under different water managements using HYDRUS-1D. *Agricultural Water Management* **150**, 67–80.

Tang, Y., Zang, C., Wei, Y. & Jiang, M. 2018 Data-driven modeling of groundwater level with least-square support vector machine and spatial–temporal analysis. *Geotechnical and Geological Engineering* **37** (3), 1661–1670.

Teramoto, E. H. & Chang, H. K. 2017 Field data and numerical simulation of btex concentration trends under water table fluctuations: example of a jet fuel-contaminated site in Brazil. *Journal of Contaminant Hydrology* **198**, 37–47.

The Ministry of Agriculture of the People's Republic of China 2016 *Trends of Agricultural Science and Education: Key Period Report about Controlling Agricultural Non-point Source Pollution*.

Tillotson, W. R., Robbins, C. W., Wagenet, R. J. & Hanks, R. J. 1980 *Soil Water, Solute and Plant Growth Simulation*. Bulletin Utah Agricultural Experiment Station.

Wang, H. F. & Hu, D. J. 2005 *Comparison of SVM and LS-SVM for Regression*. IEEE, New York.

Xia, J., Zhang, S., Zhao, X., Liu, J. & Chen, Y. 2016 Effects of different groundwater depths on the distribution characteristics of soil-*Tamarix* water contents and salinity under saline mineralization conditions. *Catena* **142**, 166–176.

Yang, Y., Wang, Z., Xie, Y., Ataie-Ashtiani, B., Simmons, C. T., Luo, Q., Chen, G., Zhang, Q., Wu, J., Wang, J. & Wu, J. 2019 Impacts of groundwater depth on regional scale soil gleyization under changing climate in the Poyang Lake Basin, China. *Journal of Hydrology* **568**, 501–516.

Yimsiri, S., Euaapiwatch, S., Flores, G., Katsumi, T. & Likitlersuang, S. 2018 Effects of water table fluctuation on diesel fuel migration in one-dimensional laboratory study. *European Journal of Environmental and Civil Engineering* **22** (3), 359–385.

Zhang, H., Yang, R., Wang, Y. & Ye, R. 2018a The evaluation and prediction of agriculture-related nitrate contamination in groundwater in Chengdu Plain, southwestern China. *Hydrogeology Journal* **27** (2), 785–799.

Zhang, L., Yan, C., Guo, Q., Zhang, J. & Ruiz-Menjivar, J. 2018b The impact of agricultural chemical inputs on environment: global evidence from informetrics analysis and visualization. *International Journal of Low-Carbon Technologies* **13**, 338–352.

Zhiye, Y., Shunguo, B., Yonghai, J., Yueying, W. & Yang, Y. 2013 *Groundwater Level Variation Effects on Nitrogen Fate and Transport Under Simulating Rainfall*. Fourth International Conference on Digital Manufacturing & Automation, Shinan, China, pp. 645–648.

Zhu, Z. L. & Chen, D. L. 2002 Nitrogen fertilizer use in China – contributions to food production, impacts on the environment and best management strategies. *Nutrient Cycling in Agroecosystems* **63** (2–3), 117–127.

First received 27 October 2020; accepted in revised form 1 March 2021. Available online 11 March 2021

Optimal water allocation method for joint operation of a reservoir and pumping station under insufficient irrigation

Cong Wei, Jilin Cheng, Yi Gong and Zhihao Gong

ABSTRACT

To address the problem of optimal allocation of water resources in water shortage areas, a reservoir and a pumping station water resource optimal scheduling model under the condition of insufficient irrigation is proposed. The model takes the maximum relative yield of crops as the objective function, the amount of water supplied, water spilled and water replenished as decision variables, the total amount of water supplied by the system, the water right of the pumping station and the operation criteria as constraints, and uses a dynamic programming method to solve the model. The optimal water supply and water spill process of the irrigation reservoir and the optimal water replenishment process of a pumping station during the entire growth period of dry crops were obtained. Moreover, under the conditions of 50 and 75% probability of exceedance, the optimized relative yields of crops increased by 17.1% and 19.6%, respectively. The results show that ensuring the optimal operation of joint water source projects can achieve the optimal allocation of limited water resources, and improve the relative yield of crops in irrigation areas, which has important guiding significance for the planning and management of water resources in similar irrigation areas.

Key words | dynamic programming, joint operation, pump station, reservoir, water right

Cong Wei
Jilin Cheng (corresponding author)
Yi Gong
Zhihao Gong
College of Hydraulic Science and Engineering,
Yangzhou University,
Yangzhou 225009,
China
E-mail: *jlcheng@yzu.edu.cn*

Jilin Cheng
Institute of Modern Rural Water Conservancy,
Yangzhou University,
Yangzhou 225009,
China

HIGHLIGHTS

- An optimal water allocation model for the joint operation of a reservoir and pump station in arid areas is established, considering regional water rights.
- An operation rule for a reservoir is integrated into the recursive process of the dynamic programming to solve the model.
- The penalty function method is introduced into the dynamic programming to deal with the constraint of water rights.

INTRODUCTION

With the rapid depletion of freshwater resources worldwide, most irrigation areas are facing a major agricultural problem, which is the shortage of irrigation water supply. With the increasing demand for water resource for various purposes and the increasing cost of regional water resources development, the contradiction between supply and demand in agricultural irrigation is also increasing.

Reservoirs are important water source projects for the planning and utilization of water resources. To alleviate the water supply pressure of irrigation reservoirs, irrigation areas require various water diversion projects to replenish water for reservoirs or farmlands (Gong *et al.* 2019; Choi *et al.* 2020; Ma *et al.* 2020). At present, a large number of studies on the optimal allocation of water resources in

doi: 10.2166/ws.2021.099

irrigation areas only focus on the optimal allocation of existing water resources of irrigation reservoirs (Vedula & Kumar 1996; Ghahraman & Sepaskhah 2002; Georgiou *et al.* 2006; Janga Reddy & Nagesh Kumar 2007; Moradi-Jalal *et al.* 2007). However, research on the optimal allocation of water resources for the joint operation of reservoirs and water diversion projects is still lacking. Simultaneously, considering the gradual clarification of the right to use water resources (Molle 2004; Heikkila 2015), the impact of regional water rights restrictions must be considered in the water supply decision-making of irrigation systems.

The optimal operation of reservoir systems is an important method in the field of water resource management (Nandalal & Sakthivadivel 2002; Suiadee & Tingsanchali 2007; Console *et al.* 2008). The application of optimization technology to solve the problem of reservoir optimal operation is also one of the hot issues in this field. In the early stage, classical linear programming and dynamic programming methods (Bellman & Dreyfus 1964) can be used to solve a single reservoir optimal operation model. This type of traditional algorithm can be proved by strict theory to obtain a global optimal solution. However, the decision-making of reservoir irrigation water supply involves many factors, such as the type of irrigation crops, growth stage and the impact of water shortage on yield (Vedula & Mujumdar 1992; Vedula & Kumar 1996; Georgiou *et al.* 2006). When traditional methods deal with multiple-decision variable problems, the existence of the "curse of dimensionality" is its evident disadvantage.

A heuristic algorithm has significant advantages in dealing with nonlinear problems and solving multi-decision variable problems and has been widely used in different fields. Among them, intelligent algorithms represented by the genetic algorithm (Tegegne & Kim 2020), simulated annealing algorithm (Rouzegari *et al.* 2019), and particle swarm optimization algorithm (Al-Aqeeli & Mahmood 2020) have also been applied to the optimal operation of reservoirs (Jothiprakash & Shanthi 2006; Chen & Chang 2007; Yun *et al.* 2010; Scola *et al.* 2014). Considering the multiple constraints in a reservoir optimization model, a heuristic algorithm usually transforms a constrained problem into an unconstrained problem. The penalty function method (Samanipour & Jelovica 2020) is the most commonly used method to deal with constraints. When there are many complex constraints, the penalty function may affect the accuracy of the solution.

Based on regional water rights, this study considers an irrigation system composed of a reservoir and pumping station. The optimal allocation model of water resources for the joint operation of the reservoir and pumping station under the condition of insufficient irrigation is established and a dynamic programming method and genetic algorithm are used to solve the nonlinear model with multiple decision variables, to determine the optimal water supply and water spill process of the reservoir and water replenishment process of the pumping station during the entire growth period of dry crops. In addition, the applicability of the two algorithms in this type of model was verified.

MODEL AND METHOD

Optimization model

Objective function

The contradiction between water supply and water demand and the relationship between water quantity and yield were considered. The Jensen model of the water production function was adopted (Jensen 1968), and an objective function was used to maximize the relative yield of crops in the irrigation district. To facilitate the calculation of the joint operation of the reservoir and supplementary pumping station, the ratio of the actual evapotranspiration and potential evapotranspiration in the water production function was transformed into the ratio of reservoir water supply and crop irrigation water demand (Wardlaw & Barnes 1999). The model is as follows:

$$\max G = \prod_{i=1}^{N} \left(\frac{X_i}{YS_i} \right)^{k_i} \tag{1}$$

where G is the relative yield of crops, i is the number of the growth period of the crop, N is the total number of crop growth periods, k_i is the sensitivity index of yield to water stress in period i, X_i is the water supply of the reservoir under insufficient irrigation in period i, (in L^3), YS_i is the irrigation water demand of the crop in period i, (in L^3).

Constraints

(1) Total quantity restriction of the water supply:

$$\sum_{i=1}^{N} X_i \leq SK + BZ \tag{2}$$

where SK is the total amount of water available in the reservoir (L^3); BZ is the water right of the pumping station, and the maximum amount of water extracted from the pump station (L^3).

(2) Operation-rule constraints:

$$V_i^{\min} \leq V_i \leq V_i^{\max} \tag{3}$$

where V_i is determined according to the water balance equation:

$$V_i = V_{i-1} + LS_i + Y_i - X_i - EF_i - PS_i \tag{4}$$

where V_i is the reservoir storage at the end of period i, (L^3); LS_i is the incoming water in period i of the reservoir, (L^3); EF_i is the water lost in period i of the reservoir, (L^3); PS_i is the abandoned water quantity in period i of the reservoir, (L^3).

According to the water balance equation of the system, the reservoir operation criteria are determined as follows:

(i) If $V_i \leq V_i^{\min}$, the reservoir needs to be replenished through the pump station in the period i, and the replenishment amount of the pump station is:

$$Y_i = V_i^{\min} - V_i \tag{5}$$

At this point, the water spilled is:

$$PS_i = 0 \tag{6}$$

(ii) If $V_i \geq V_i^{\max}$, the amount of water replenished from the pumping station is $Y_i = 0$, then the amount of water spilled is:

$$PS_i = V_i - V_i^{\max} \tag{7}$$

(iii) If $V_i^{\min} \leq V_i \leq V_i^{\max}$, the reservoir does not need to be replenished or spilled, then:

$$Y_i = PS_i = 0 \tag{8}$$

(3) Pump station constraints:

(i) Constraints on pumping capacity of pumping station:

$$Y_i \leq Y_i^{\max} \tag{9}$$

(ii) Pumping station water rights constraints:

$$\sum_{i=1}^{N} Y_i \leq BZ \tag{10}$$

where Y_i is the water replenishment of the pumping station in period i, (L^3); $Y_i^{\max}$ is the maximum water replenishment of pumping station in period i, (L^3).

(4) Constraints on crop water demand:

$$YS_i^{\min} \leq X_i \leq YS_i \tag{11}$$

where YS_i^{min} is the minimum irrigation water demand of the crop in period i, (L^3).

Solving method

Aiming at the nonlinear mathematical model established in this study, this study uses a dynamic programming method and genetic algorithm to solve the problem, and verifies the correctness of the results and the applicability of the algorithm by comparing the results of both algorithms.

Dynamic programming

Dynamic programming (DP), a branch of operations research, is a method based on the Bellman optimization principle (Bellman & Dreyfus 1964). When DP is used to solve the above model, the global optimal solution of the objective function can be obtained, and the complex constraints involved in the model can be transformed and processed in the recursive process of DP, which is as follows:

(1) Phase $i = 1$:

$$g_1(\lambda_1) = \max\left(\frac{X_1}{YS_1}\right)^{k_1} \tag{12}$$

The state variables are discretized in the corresponding feasible domain $[0, SK + BZ]$ with step size d, and m points discretized. Considering Equation (11), the feasible region of the decision variable (reservoir water supply X_1) is $YS_i^{\min} \le X_1 \le \min\{YS_1, \lambda_1\}$, and the decision variable is discretized in the feasible region with step size d. Discrete X_1 is substituted into Equations (12) and (4), so the reservoir capacity at the end of phase 1 is $V_1 = V_0 + LS_1 - EF_1 - X_1$. At this time, pumping station water replenishment or reservoir water spill operation is taken into account, and dispatching criterion Equations (5)–(9) are adopted for testing:

(1) When the reservoir water capacity is greater than the upper limit of water storage, the resulting water spill is $PS_1 = V_1 - V_1^{\max}$, the reservoir volume is revised as $V_1' = V_1 - PS_1$.

(2) When the water storage of the reservoir is less than the lower limit of water storage, the pumping station should be used to replenish the reservoir, and the amount of supplementary water is $Y_1 = V_1^{\min} - V_1$, according to Equation (9).

① When $Y_1 \le Y_1^{\max}$, the filling capacity of the pump station is sufficient, and the storage capacity of the reservoir is revised as $V_1' = V_1 + Y_1$.

② When $Y_1 > Y_1^{\max}$, the filling capacity of the pump station is insufficient, and the water storage of the reservoir is revised as $V_1' = V_1 + Y_1$. At this time, the modified reservoir water storage is lower than the dead water level and does not meet the reservoir capacity constraint, and the corresponding function value of X_1 will be sifted out (the value is assigned to be negative).

Through steps (1)–(2), the discrete X_1 is tested by the constraint of the operation rule, combined with the value of the function for each discrete X_1, then the optimal function value $g_1(\lambda_1)$ corresponding to each discrete state variable λ_1 can be obtained, as well as the optimal function value corresponding to the optimal reservoir water supply X_1^*, water replenishment Y_1^*, and water spill PS_1^*.

(2) Phase $i = 2, \ldots, N$-1:

$$g_i(\lambda_i) = \max\left[\left(\frac{X_i}{YS_i}\right)^{k_i} \cdot g_{i-1}(\lambda_{i-1})\right] \tag{13}$$

State transfer equation:

$$\lambda_{i-1} = \lambda_i - X_i \tag{14}$$

According to the state transfer equation (Equation (14)), the function value $g_{i-1}(\lambda_{i-1})$ that meets the requirements in the i-1 phase can be found and the process of step (1) repeated for each discrete state variable λ_i, the optimal value of the function $g_i(\lambda_i)$, and the optimal function value corresponding to the optimal reservoir water supply X_i^*, water requirements Y_i^*, and water spills PS_i^*.

(3) Phase $i = N$:

$$g_N(\lambda_N) = \max\left[\left(\frac{X_N}{YS_N}\right)^{k_N} \cdot g_{N-1}(\lambda_{N-1}) + p(x)\right] \tag{15}$$

A penalty term is introduced at the last stage of the recursive process to test the water rights constraint of the pumping station (Equation 10), and the expression is as follows:

$$p(x) = \begin{cases} \mu & \text{if } \sum_{i=1}^{N} Y_i > BZ \\ 0 & \text{otherwise} \end{cases} \tag{16}$$

where μ is the penalty factor, which is a sufficiently small negative number.

The state transfer equation is given by:

$$\lambda_{N-1} = \lambda_N - X_N \tag{17}$$

So far, all the constraints involved in the model have been transformed and processed. Substituting Equation (17) into Equation (15), the process of step (2) can be repeated to obtain the global optimal function value $G^* = \max g_N(\lambda_N)$, and its corresponding optimal reservoir water supply process $(X_1, X_2, \ldots, X_N)^*$, optimal water replenishment process of

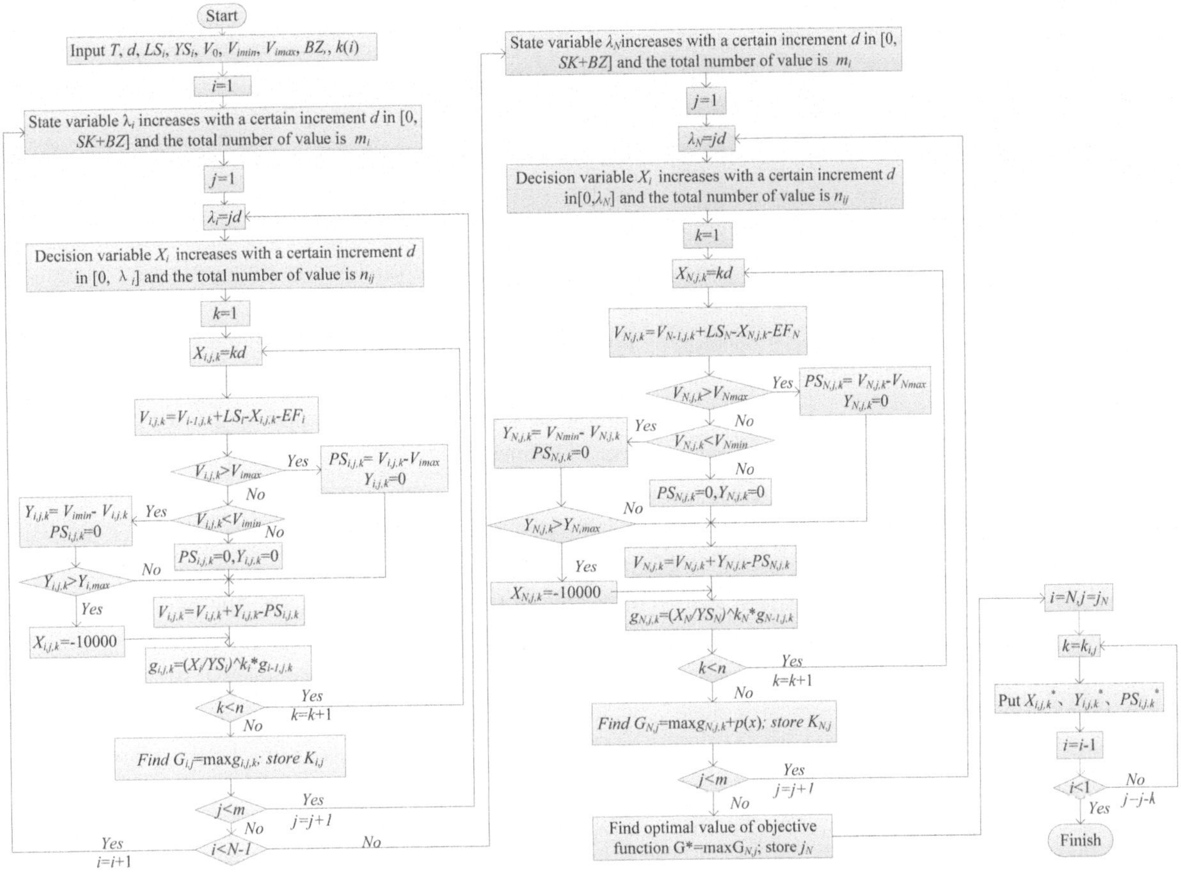

Figure 1 | System optimization flow chart.

pumping station $(Y_1, Y_2, \ldots, Y_N)^*$, and reservoir water spill process $(PS_1, PS_2, \ldots, PS_N)^*$.

The optimization flow chart of the system is shown in Figure 1.

Genetic algorithm

A genetic algorithm (GA) is an optimization algorithm developed by imitating the evolution mechanism of nature. The advantages of this algorithm are simple, general and robust, and it is suitable for parallel processing (Samanipour & Jelovica 2020); therefore, it is often used to solve the planning problem in the field of optimal reservoir operation (Jothiprakash & Shanthi 2006; Chen & Chang 2007; Yun *et al.* 2010).

In this study, the GA adopts a real-coded method, in which the reservoir water supply X_i is generated randomly in the feasible region $[YS_i^{\min}, YS_i]$, the pumping station's water replenishment Y_i is generated randomly in the feasible region $[0, Y_i^{\max}]$, and the water spill PS_i is determined according to the water balance equation. Considering that the model involves many constraints, a penalty function is constructed to deal with the constraints in the process of solving the genetic algorithm. The function expressions are as follows:

$$P_1(x) = \begin{cases} \mu_1 & \text{if } V_i > V_i^{\max} \\ 0 & \text{otherwise} \end{cases} \tag{18}$$

$$P_2(x) = \begin{cases} \mu_2 & \text{if } V_i > V_i^{\min} \\ 0 & \text{otherwise} \end{cases} \tag{19}$$

$$P_3(x) = \begin{cases} \mu_3 & \text{if } \sum_{i=1}^{N} Y_i > BZ \\ 0 & \text{otherwise} \end{cases} \tag{20}$$

$$P_4(x) = \begin{cases} \mu_4 & \text{if } V_{end} \neq V_0 \\ 0 & \text{otherwise} \end{cases} \tag{21}$$

where P_1, P_2, P_3, and P_4 are penalty functions of the upper and lower limits of storage capacity, water right of the pumping station, and boundary conditions of storage capacity, respectively; μ_1, μ_2, μ_3, and μ_4 are penalty factors respectively. The penalty factor is $-10{,}000$ in the actual solution process.

The fitness function of the GA is constructed by integrating the objective and penalty functions of the model:

$$F = G + P_1 + P_2 + P_3 + P_4 \tag{22}$$

where F is the value of the fitness function.

CASE STUDY

The Yibei (YB) irrigation district is located in the eastern mountainous area of Xinyi, Jiangsu Province, China (Figure 2). There are many low hills in the study area that afffect its topography. The main crops are winter wheat and another cash crop, and the area of the main wheat winter crop in the irrigation district is 1270 hm^2. The climate is warm and monsoonal and the climate varies markedly from year to year, with a multiyear mean rainfall of 891.5 mm and an average evaporation rate of 968.0 mm. The main source of irrigation water in the irrigated area is the Shadun (SD) Reservoir (Figure 2), and the coefficient of irrigation water use is 0.48, with a designed storage capacity of 7.5×10^6 m^3 and a lower limit of 2×10^6 m^3.

The YB irrigation district is located in a hilly area, and relying only on reservoir water storage cannot guarantee the normal agricultural water use in the irrigation district, so the surface water should be intercepted and external water should be simultaneously diverted for replenishment. Based on the local 'Xinyi Water Source Project Planning' data, and considering the division of regional water rights, the total allowable water replenishment of the pumping station is 3×10^6 m^3. In this study, winter wheat is considered as the research object, and its growth period spans from October to June of the following year. According to the division of growth stages of winter wheat, the entire water resources scheduling process was divided into six periods. The parameters of the crop experiment in the irrigation area are shown in Table 1, and the parameters of the pumping station are listed in Table 2.

With a probability of exceedance of 50 and 75%, the water inflow and irrigation water demand of each stage of the system were obtained according to the planning data of the water source project in the YB irrigation area. See Tables 3 and 4 for details.

Evaporation loss is a function of evaporation depth and average free water surface of a reservoir in each period. Evaporation depth (E_i) was obtained from evaporation data acquired by the coefficient of correction using an E_{601} evaporator (Table 5), and the data required correction

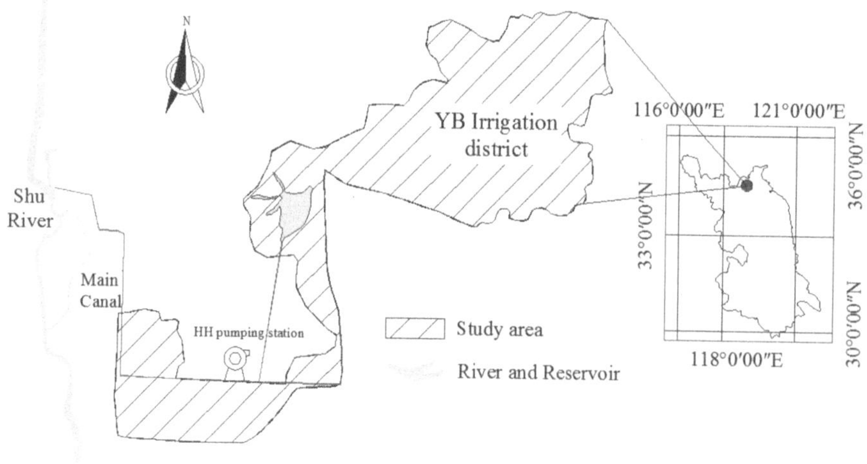

Figure 2 | Location of the study area.

Table 1 | Experimental parameters of winter wheat

Date Growth period Sensitive index		1/10 ~ 1/11 Seeding 0.2675	2/11 ~ 5/1 Returning green 0.0613	6/1 ~ 15/2 Jointing 0.3765	16/2 ~ 10/3 Heading 0.5951	11/3 ~ 25/4 Filling 0.5951	26/4 ~ 31/5 Mature 0.2981
Irrigation water demands (mm)	50%	97.3	17.3	30.4	92.1	110.7	105
	75%	107	19	33.4	101.3	121.8	115.5

Table 2 | Characteristics of pumping stations

Pumping station	Design discharge(m³/h)	Daily operation duration(h)	Water rights(10^4 m³)
HH	3,600	20	300

Table 3 | SD reservoir stage inflows (10^4 m^3)

	Period						
Probability	1	2	3	4	5	6	Total
50%	64	86	103	135	154	154	696
75%	32	60	83	126	119	143	563

Table 4 | Irrigation water demands (10^4 m^3)

	Period						
Probability	1	2	3	4	5	6	Total
50%	256	45	80	242	291	276	1,190
75%	285	50	88	266	320	304	1,313

Table 5 | E_i and w_i for each period

Period	1	2	3	4	5	6
E_i (mm)	72.8	82	24.2	31.6	87.6	74.6
w_i	1.04	1.11	1.03	0.96	0.93	0.91

by the coefficient k. The average free water surface of each reservoir in a specific period is determined by the surface-volume relationship, which was provided by the Xinan Water Authority. Finally, the reservoir evaporation for a specific period can be computed using Equation (23):

$$EF_i = 0.1 \times w_i \times E_i \times (\alpha V_i + \beta) \tag{23}$$

where EF_i (10^4 m^3) is the evaporation loss of a reservoir in period i, E_i (mm) is the evaporation depth of the E_{601} evaporator in period i, w_i is the correction coefficient for period i, V_i (10^4 m^3) is the average water storage in period i, and α and β are the reservoir coefficients. For the SD reservoir, $\alpha = 2.117 \times 10^{-3}$, $\beta = 1.863$.

RESULTS

In the conventional operation of the irrigation system, the initial boundary of the reservoir water storage is 3.18×10^6 m^3 and 2.86×10^6 m^3 when the water probability of exceedance is 50 and 75%, respectively. The optimal operation results of irrigation system are listed in Table 6.

Using the DP method to solve the model, the optimal values of the function under 50 and 75% probabilities are 0.637 and 0.373, respectively. Compared with the standard operation policy (SOP) results, the relative yield of winter wheat is increased by 17.1% and 19.6%, respectively. At the same time, using a genetic algorithm to solve the model, the optimal values of the function at 50 and 75% probabilities were 0.628 and 0.364, respectively. The comparison of the final solution results shows that the optimal function values of DP and GA are close, which indicates that the solution results of the two methods are more reliable.

COMPARISON AND DISCUSSION

Comparison of scheduling results

Figure 3(a)–3(d) show that the operation curves obtained by DP and GA are between the upper and lower limits of the reservoir capacity, indicating that the operation process of the system is reasonable. However, it can still be found that the reservoir capacity curve generated by DP is

Table 6 | Operation results of irrigation system (10^4 m^3)

Method	Probability	Water supply	Water replenishment	Water spill	Evaporation	Water shortage	Relative yield
SOP	50%	909	300	0	87	281	0.544
	75%	775	300	0	88	538	0.312
DP	50%	909	300	0	87	281	0.637
	75%	783	300	0	80	530	0.373
GA	50%	906	300	0	90	284	0.628
	75%	775	300	0	88	538	0.364

The values of crossover rate and mutation rate corresponding to GA optimal results are 0.7 and 0.15 at 50% probability and 0.8 and 0.15 at 75% probability.

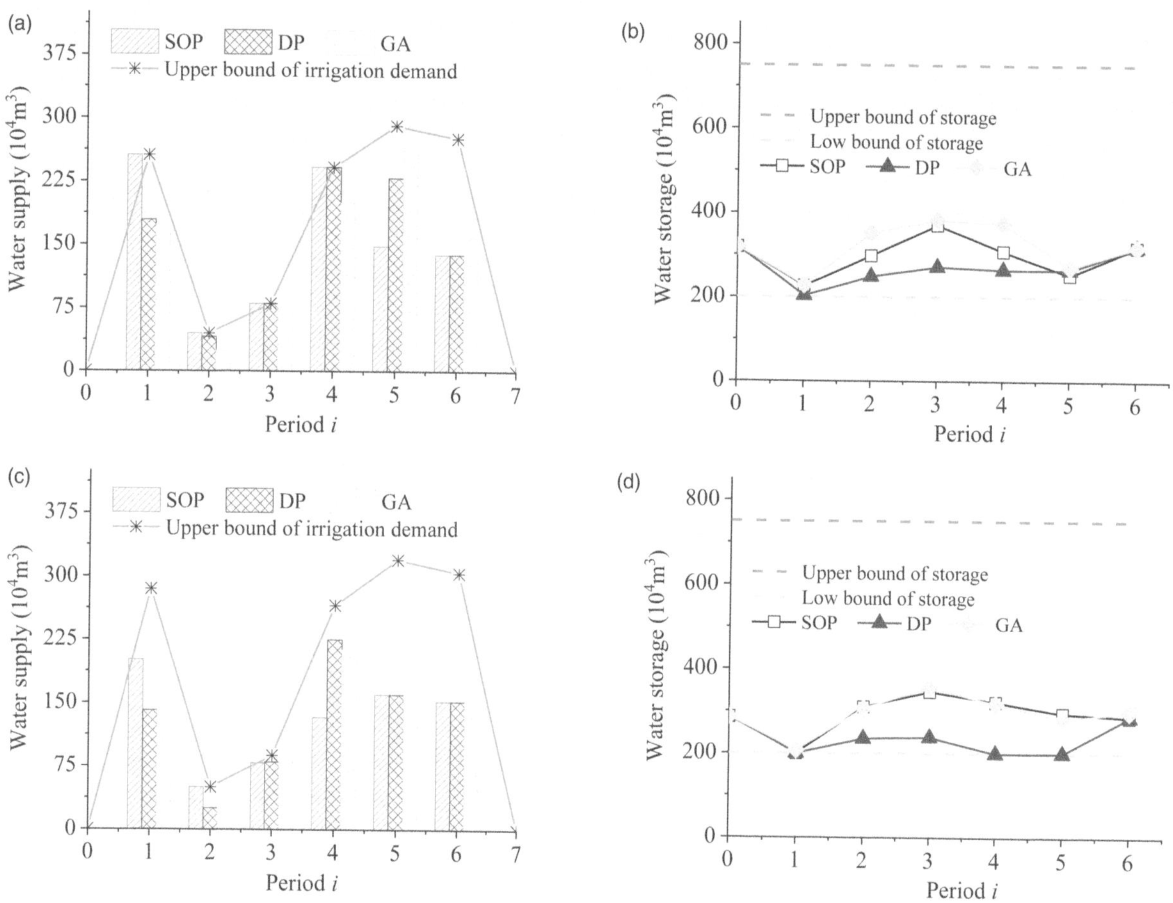

Figure 3 | (a) reservoir water supply process under 50% probability; (b) reservoir operation curve under 50% probability; (c) reservoir water supply process under 75% probability; (d) reservoir operation curve under 75% probability.

generally under the capacity curve generated by SOP and GA, which means that the operation process generated by DP maintains the average storage capacity of the reservoir at a small level and reduces the evaporation loss of the reservoir to the greatest extent. Therefore, Table 6 reveals that at 50% probability of exceedance, the system evaporation of DP is 3000 m^3 less than that of GA, and at 75% probability of exceedance, it is 8000 m^3 less than that of SOP and GA. Therefore, the optimal planning and management of local runoff resources and regional water rights can be achieved using DP to solve the optimal scheduling model of an irrigation system.

Figure 4(a) and 4(b) show that the water replenishment decision of the pumping station generated by GA is affected

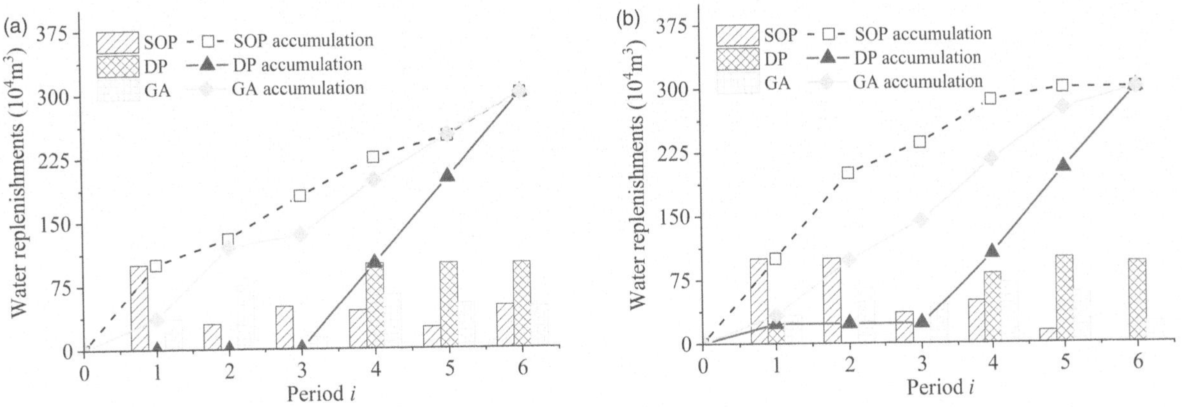

Figure 4 | (a) the water replenishment process of the pump station at 50% probability; (b) the water replenishment process of the pump station at 75% probability.

by randomness, and there are water replenishment operations in all stages of the system, which is also the main reason for the increase in average reservoir storage and evaporation loss. When DP is used to solve the problem, the joint operation criterion of the irrigation reservoir and pumping station can be coupled into the recursive process, which can clearly dictate when and how much water should be replenished. This also makes the quantity of make-up water generated by DP less than SOP and GA. In addition, the reduction in the number of water replenishments also means a reduction of the number of start-ups at the pump station, which is convenient for the staff to maintain and manage the pump unit.

Algorithm performance comparison

Considering that the solution results of GA may be affected by changes in internal parameters, the crossover rate and mutation rate of GA were analyzed first. The variation range of the crossover rate is 0.5 to 0.8, the variation range of the mutation rate is 0.05 to 0.15, and the population is 200. The optimal solution of the objective function under different GA parameter combinations is presented in Table 7.

However, as shown in Table 7, using the GA solution model, different parameter combinations in the algorithm significantly impact the final solution results. Moreover, GA is affected by the random search characteristics; even if the parameters are the same, the optimal value obtained after several runs is slightly different. When the DP method is used, the analysis of redundant parameters is not required, and the result is always unique and reliable.

CONCLUSION

In this study, the optimal allocation model of water resources for the joint operation of a reservoir and the pumping station under insufficient irrigation conditions was established considering regional water rights. The results of the joint operation of different water source

Table 7 | Parameter analysis of crossover rate and mutation rate (GA)

		Objective function				Objective function	
Crossover rate	**Mutation rate**	**50%**	**75%**	**Crossover rate**	**Mutation rate**	**50%**	**75%**
0.5	0.05	0.589	0.342	0.7	0.05	0.609	0.343
	0.1	0.621	0.355		0.1	0.565	0.347
	0.15	0.602	0.348		0.15	0.628	0.321
0.6	0.05	0.597	0.351	0.8	0.05	0.598	0.353
	0.1	0.562	0.337		0.1	0.612	0.363
	0.15	0.586	0.322		0.15	0.606	0.364

The above results are the optimal values obtained by running 10 times.

projects in the irrigation system were discussed. In addition, the optimal water supply and water spill process of the irrigation reservoir and the optimal water replenishment process of the pumping station in the whole growth period of dry crops were determined. The optimal allocation of local runoff and regional water resources was achieved which has a certain guiding significance for decision-makers in irrigation areas. In addition, the classical dynamic programming method was used to solve the mathematical model involving three decision variables, which also enriches the application of the traditional algorithm in practical problems.

ACKNOWLEDGEMENTS

This work was supported by the National Key Research and Development Program of China [grant number 2017YFC0403205] and National Natural Science Foundation of China (NSFC) [grant number 52079119].

DATA AVAILABILITY STATEMENT

All relevant data are included in the paper or its Supplementary Information.

REFERENCES

Al-Aqeeli, Y. H. & Mahmood Agha, O. M. A. 2020 Optimal operation of multi-reservoir system for hydropower production using particle swarm optimization algorithm. *Water Resources Management* **34**, 3099–3112.

Bellman, R. E. & Dreyfus, S. E. 1964 *Applied Dynamic Programming*. Princeton University Press, London.

Chen, L. & Chang, F. L. 2007 Applying a real-coded multi-Population genetic algorithm to multi-reservoir operation. *Hydrol Process* **21**, 688–698.

Choi, Y., Ahn, J., Ji, J., Lee, E. & Yi, J. 2020 Effects of inter-basin water transfer project operation for emergency water supply. *Water Resources Management* **34**, 2535–2548.

Console, S., Matarazzo, B. & Pappalardo, N. 2008 Operation rules of an irrigation purposes reservoir suing multi-objective optimization. *Water Resources Management* **22**, 551–564.

Georgiou, P. E., Papamichail, D. M. & Vougioukas, S. G. 2006 Optimal irrigation reservoir operation and simultaneous multi-crop cultivation area selection using simulated annealing. *Irrigation and Drainage* **55**, 129–144.

Ghahraman, B. & Sepaskhah, A. R. 2002 Optimal allocation of water from a single reservoir to an irrigation project with pre-determined multiple cropping patterns. *Irrigation Science* **21**, 127–137.

Gong, Z. H., Jiang, X. H., Cheng, J. L., Gong, Y. & Cheng, H. M. 2019 Optimization method for joint operation of a double reservoir and double pumping station system: a case study of Nanjing, China. *Journal of Water Supply Research and Technology-Aqua* **68** (8), 803–815.

Heikkila, T. 2015 Coordination in water resource management: the impact of water rights institutions. *Water Policy* **5** (4), 331–348.

Janga Reddy, M. & Nagesh Kumar, D. 2007 Optimal reservoir operation for irrigation of multiple crops using elitist-mutated particle swarm optimization. *Hydrological Sciences Journal* **52** (4), 686–701.

Jensen, M. E. 1968 Water consumption by agriculture plants. In: *Water Deficits and Plant Growth*, Vol. II (T. T. Kozlowski, ed.). Academic Press, New York, pp. 1–22.

Jothiprakash, V. & Shanthi, G. 2006 Single reservoir operation policies using genetic algorithm. *Water Resources Management* **20**, 917–929.

Ma, Y. S., Chang, J. X., Guo, A. J., Wu, L. Z., Yang, J. & Chen, L. 2020 Optimizing inter-basin water transfers from multiple sources among interconnected river basins. *Journal of Hydrology* **590**, 125461.

Molle, F. 2004 Defining water rights: by prescription or negotiation. *Water Policy* **6** (3), 207–227.

Moradi-Jalal, M., Bozorg Haddad, O., Karney, B. W. & Marino, M. A. 2007 Reservoir operation in assigni-ng optimal multi-crop irrigation areas. *Agricultural Water Management* **90** (1–2), 149–159.

Nandalal, K. D. W. & Sakthivadivel, R. 2002 Planning and management of a complex water reservoirs in the walawe river, Sir Lanka. *Agricultural Water Management* **57**, 207–221.

Rouzegari, N., Hassanzadeh, Y. & Sattari, M. T. 2019 Using the hybrid simulated annealing-M5 tree algorithms to extract the if-then operation rules in a single reservoir. *Water Resources Management* **33**, 3655–3672.

Samanipour, F. & Jelovica, J. 2020 Adaptive repair method for constraint handling in multi-objective genetic algorithm based on relationship between constraints and variables. *Applied Soft Computing Journal* **90**, 106143.

Scola, L. A., Tasahashi, R. H. C. & Cerqueira, S. A. A. G. 2014 Multipurpose water reservoir management: an evolutionary multiobjective optimization approach. *Mathematical Problems in Engineering* **2014**, 1–14.

Suiadee, W. & Tingsanchali, T. 2007 A combined simulation-genetic algorithm optimization model for optimal rule curves of a reservoir: a case study of the nam on irrigation project. *Thailand Hydrol Process* **21**, 3211–3225.

Tegegne, G. & Kim, Y.-O. 2020 Representing inflow uncertainty for the development of monthly reservoir operations using genetic algorithms. *Journal of Hydrology* **586**, 124876.

Vedula, S. & Kumar, N. D. 1996 An integrated model for optimal reservoir operation for irrigation of multiple crops. *Water Resources Research* **32** (4), 1101–1108.

Vedula, S. & Mujumdar, P. P. 1992 Optimal reservoir operation for irrigation multiple crops. *Water Resources Research* **28** (1), 1–9.

Wardlaw, R. & Barnes, J. 1999 Optimal allocation of irrigation water supplies in real time. *Journal of Irrigation and Drainage Engineering ASCE* **125** (6), 345–354.

Yun, R., Singh, V. & Dong, Z. 2010 Long-Term stochastic reservoir operation using a noisy genetic algorithm. *Water Resources Management* **24**, 3159–3172.

First received 29 November 2020; accepted in revised form 23 March 2021. Available online 1 April 2021

Application of the 'theory of planned behavior' to understand farmers' intentions to accept water policy options using structural equation modeling

Taghi Mahdavi

ABSTRACT

The agricultural sector is the largest sector of water consumers, and farmers are important stakeholders involved in water conservation. This research has been conducted to determine how native farmers support different policy options to reduce agricultural water consumption. Structural equations modeling was used to construct structures derived from the 'theory of planned behavior'. For each policy option, a separate model is proposed and the modeling data supports the view that attitudes, subjective norms, and perceived behavioral control have a positive and significant effect on the intention. Attitudes, and perceived behavioral control, have the strongest effect on intention. Significantly, intention also have a positive impact on farmers' behavior. According to the results of the present research, the variance explained is over 85% for intentions and the variance explained for the farmers' behavior on water policy options is above 45% which is a result that indicates the high ability of the 'theory of planned behavior' in predicting policies achievement on saving agricultural water resources. It is argued that the field of psychology, and in particular environmental psychology, can play an important role in understanding more of the drivers to reduce agricultural water consumption and contribute to the social research program for water policy.

Key words | agriculture water saving policy option, structural equation modeling, theory of planned behavior

Taghi Mahdavi
Department of Civil Engineering,
Maragheh Branch,
Islamic Azad University,
Maragheh,
Iran
E-mail: *taghi.mahdavi@iau-maragheh.ac.ir*

HIGHLIGHTS

- A socio-psychological research on agricultural water resource saving policies.
- Applies the theory of planned behavior to agricultural water resource saving policies.
- Uses Structural Equation Modeling (SEM) to investigate causal relationships between variables affecting farmers' adherence to agricultural water resource saving policies.

INTRODUCTION

One of the main reasons for food insecurity is the lack of water. Effective and integrated management of water resources has become a priority in many countries of the world. With the development of the agricultural, industrial, and service sectors, much pressure has been placed on limited water resources, while excessive use of groundwater leads to the depletion of aquifers and has significant negative consequences for natural habitats and ecosystems. According to recent research (Dalin *et al.* 2017), Iran has ranked second in the world over the past two decades in depletion of groundwater after India. In Iran, more than 90% of water use is related to the agricultural sector and

doi: 10.2166/ws.2021.138

groundwater provides 62% of the water needs of irrigated lands, which have less than 30% irrigation efficiency. As a result, the agricultural sector is a vital objective for implementing water-saving policies and increasing productivity.

Developing modern irrigation techniques and technology can significantly help to control water consumption in the agricultural sector. With water-saving in the agriculture sector, water will be released for other uses. Such techniques may also increase the quality and quantity of agricultural products whilst saving water. Therefore, many national governments, by giving incentives to farmers to improve irrigation systems, are currently seeking to reduce water use in the agricultural sector, and following water-saving measures.

Over the past few decades, Iran has had many successes in dam construction and control of surface waters, and there are many investments of this sort. Environmentalists are sometimes critical that, due to exaggerated dam construction, approximately 4 billion cubic meters (4×10^9 m^3) of wetlands water rights have been allocated for other purposes. Over the past few decades, due to a self-sufficiency policy in food products, agricultural sector development policy, and increasing employment in this sector, there is a lot of pressure on water resources from the agricultural sector. According to UNSD (2012), using more than 40% of renewable water is considered to be severe water stress, while in Iran more than 80% of renewable water is used and so there is a 15 billion cubic meters negative balance in its plains. The total amount of renewable water in Iran has decreased from a long-term average of 110×10^9 m^3 to 89×10^9 m^3 (10-year average) (Babaee & Alijani 2013).

Unfortunately, the development of water infrastructure in Iran has not been balanced and not much attention was paid to increasing transmission efficiency and field efficiency along with dam constructing. Therefore, the government has been trying to increase irrigation efficiency in recent years, and for this purpose, there are many incentives for farmers to implement modern and drip irrigation. The state subsidy grants vary from 250 USD to 1,000 USD per hectare. One of the problems is the small size of agricultural land holdings due to the law of inheritance. In the event of land accumulation and implementation of drip irrigation, the government pays 85% of the cost. The government also provides educational programs for villagers to modify the pattern of cultivation and production of economical and dehydrated products and creating other jobs. The government supports the cultivation of strategic products and guarantees their purchase. The government's environmental policies are intended to save water resources, and the present research examined the feasibility of implementing any of these policies with the 'theory of planned behavior' (TPB).

However, the question is how to increase the sensitivity of farmers to saving water and encourage them to pursue water-saving policies. Investigating factors affecting farmers' intentions and behavior can allow policymakers to target specific policies. So far, despite extensive research on this topic, empirical research on the factors affecting farmers' intentions and behaviors to adopt water policies has been very limited. Understanding farmers' incentives to adapt to water-saving policies is especially important for governments, as it increases the success of government intervention in these policies and can reduce unnecessary costs. These risks are especially important when implementing these policies in times of water scarcity. In previous research on intervention for behavior change based on De Young (2000), there are two approaches:

1. The 'antecedent approach': This approach follows strategies that change behavior by affecting determinants of behavior, such as commitment and information provision, and attitudinal changes. For example, (Kurz *et al.* 2005) concluded that the provision of information on water conservation has led to a 23% reduction in household water consumption.
2. The 'consequence approach': This approach follows strategies that change the behavior of the individuals by informing the positive or negative consequences of the behavior, such as giving a reward for specific behavior.

Other researchers such as Steg & Vlek (2009), have proposed 'information' and 'structural' approaches instead of 'antecedent' and 'consequence' approaches. The 'information' approach has the purpose of changing attitudes, beliefs, motivations, and norms. The 'structural' approach has the purpose of changing the contextual factors, such as access to products and services, regulations, or financial incentives.

In the theory of planned behavior, both approaches are considered in the components influencing the intention under the heading of attitude and behavioral control. A complete review of intervention literature goes beyond the scope of this paper. Our aim is to show how psychological research on determinants of behavior can be used to guide the development of effective interventions for implementing water policies. Our goal is to emphasize the fact that with the help of environmental psychology in the development of water policy, valuable information can be provided to inform evidence-based policy in the field of agricultural water-saving policies. Evidence-based policy (EBP) is a term often applied in multiple fields of public policy to refer to situations in which policy decisions are informed by rigorously established objective evidence.

Therefore, the present study embraced the TPB (Ajzen 1991), which is a well know framework in psychological literature, in order to assess the determinants of farmers' behavioral intentions to implement any water resource saving policy. So, the present study could be considered as a novel approach to apply the TPB to water resources policy analysis.

STUDY AREA

Our study area is adjacent to Lake Urmia, one of the most valuable aquatic ecosystems and a UNESCO biosphere reserve (Figure 1). Lake Urmia has been facing significant declines in water volume and area. Due to the interactions between Lake Urmia and the studied area, the importance of investigating the challenges of water-saving for sustainable development becomes more significant. Integrated assessment was made of the study area water resource by adopting the System of Environmental and Economic Accounts for Water (SEEA-WATER) and calculating over thirty water resource, economic and social indicators in 2006 and 2016, which indicated the study area is highly stressed in terms of water quantity and quality. In addition, the area suffers severely from unsustainability and dis-equilibrium between water resources and consumption (Mahdavi *et al.* 2019).

Table 1 presents some of the important indicators related to water issues, especially agricultural water in the study area. The indicators show that the area suffers from dehydration and the implementation of water-saving policies is important.

METHOD AND DATA SOURCES

Theoretical and applied foundations of the proposed model

The TPB attributed to Ajzen (1991) represents a refinement of earlier models of rational decision-making, like the

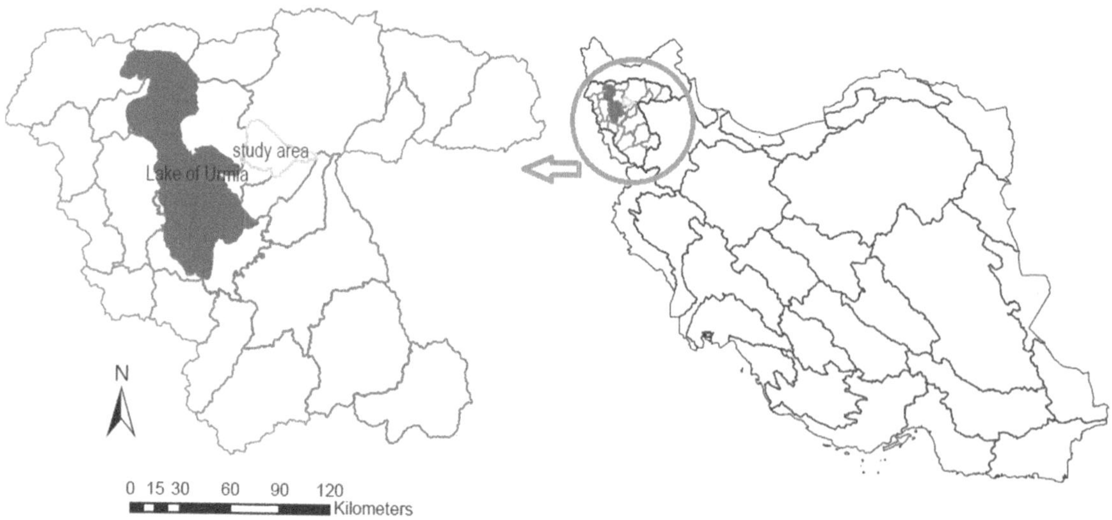

Figure 1 | Location of study area.

Table 1 | Important indicators related to water issues in the study area (Mahdavi *et al.* 2019)

Indicator	Relation	Value in 2016
Relative water stress index (RWSI)	Total consumption to natural renewable water resources	1.37
Water Use Intensity Index (WUII)	Total consumption to internal renewable water	1.883
water utilization index in agriculture sector (WUI)a	Consumption in agriculture to total consumption	94%
Economic Water Productivity Index in the agricultural sector(EcWPI)a	Agricultural gross revenue to agricultural consumption	14 (Cent/m^3)
Employment Percent in agriculture (EmP)a	Employment in agriculture to total employment	24%
Employee productivity Index in agricultural sector (EPI)a	agricultural gross revenue to agriculture employment	860 (Dollar/ employee)
Per Capita Renewable Water Resources Index (PCRWI)	Natural renewable water resources to Population	476 (m^3/person)

theory of propositional control and the theory of reasoned action (Ajzen & Fishbein 1970). The TPB (Ajzen 1991) is a socio-psychological model that assumes that the intent of a person to perform a certain behavior is the strongest predictor of that behavior. According to this theory, there are three factors that influence the intent:

1. Attitude towards a policy option (ATT): indicating the farmer's desired willingness about a policy option;
2. Subjective norm (SN): indicating the degree of social pressure of the influential persons on the farmer in following and accepting the policy option; and
3. Perceived behavioral control (PBC): indicating the level of comfort or difficulty of following the policy option.

Each of these structures arises from a set of beliefs related to the pros and cons of the intended behavior (behavioral beliefs), the support of influential individuals or government organizations (normative beliefs), and the understanding of the ease or difficulty of following and pursuing that policy option (control beliefs) (Pino *et al.* 2017). The intention is a predictor of behavior that indicates a farmer's readiness to adopt a policy option. This theory has been used in many studies to evaluate the cognitive factors influencing water conservation behavior and to evaluate behavioral interventions aimed at promoting sustainable environmental behaviors (Abrahamse *et al.* 2005). In our study, we used this theory to examine the possibility of implementing water policies, which are mainly to reduce agricultural water consumption. The TPB's ability to predict compliance with water policy options is due to the fact that, according to this theory, behavior is not only influenced by external factors such as the cost of actions and access to information, but also by conditions and decision-making motivations (Cary 2008; Dolnicar & Hurlimann 2010). Income from water policy options implementation and education, as well as the receptiveness to information on the implementation of water policy options (Trumbo & O'Keefe 2005) and concerns about future water scarcity (Clark & Finley 2007), may have a positive impact on farmers' intentions.

With regard to the intrinsic motivation forces of individuals, previous research (Lam 1999) has shown that TPB variables effectively predict people's intentions to implement water policies. Lam (1999) used the latent variable of people's perceived moral obligation to examine the behavior of water consumption reduction, but this latent variable did not have a significant effect on behavior. Recent research has attempted to determine the impact of the variable of moral commitment of individuals according to their degree of concern for their environment (so-called 'tendency to environmental values'). However, the effect of this variable on the intention to save water is contradictory. Salvaggio *et al.* (2014) found that people's attitudes toward environmental values had a positive impact on water saving policies. Also Clark & Finley (2007) found that environmental concerns were poorly significant to predict people's intent to save water. On the contrary, Chang (2013) found that this variable has no significant effect on water-saving behaviors.

Chang *et al.* (2016) found that farmers' desirable attitudes toward limiting water use could predict the adoption of water-saving policies by them. Also, Tohidyan Far &

Rezaei Moghaddam (2015) showed that attitude toward water resources management, subjective norm, and perceived behavioral control affects farmers' intentions to participate in water-saving projects. Regarding the adoption of water-saving measures, Lynne *et al.* (1995) established that farmers' attitudes toward water-saving measures, subjective norms, and perceived behavioral control are significant predictors for farmers' investment in such measures. These findings were partially confirmed by Yazdanpanah *et al.* (2014), which focused on a wide range of farmers' measures to save and protect water. They found that behavioral control perceived by Iranian farmers directly penetrated the acceptance of these behaviors, while attitudes and subjective norms indirectly affect this behavior. The TPB has been used for a wide range of protection behaviors such as environmental conservation, soil conservation, water conservation (Clark & Finley 2007; Dolnicar & Hurlimann 2010), saving water, payment of water bills, reuse of wastewater (Nancarrow *et al.* 2008), and changing the behavior of people in household water consumption (Cary 2008). However, the current understanding of the impact of the TPB variables on farmers' intentions to adopt water policy options is very limited, and this research is intended to assess the feasibility of successfully implementing water policy options.

The proposed model and hypotheses

According to the framework of TPB theory (Figure 2), four null hypotheses ($H0_A$, $H0_B$, $H0_C$ and $H0_D$) are pursued in the field of water policy options. $H0_A$ = There is a positive relationship between a person's attitude towards water policy options and the intention to implement water policy options. $H0_B$ = There is a positive relationship between the positive perceptions of influential and important individuals and organizations regarding water policy options and the farmer's intention to pursue water policy options. $H0_C$ = There is a positive relationship between perceptions of the ability to conform to water policy options and the intention to conform to water policy options. $H0_D$ = There is a positive relationship between the intention to pursue water policy options and actual compliance (behavior). All of these beliefs were assessed on a 5-point scale (1 = 'very low', 5 = 'very high').

The policy options

Since the location, type of resources, and water quality affect the acceptance of the policy option type, in this research, four regions are considered as Table 2. The policy options reviewed are:

1. Crop level reduction policy
2. Crop pattern change policy
3. Policy of increasing irrigation efficiency
4. The policy of increasing non-agricultural incomes

Structural equations for all four policy options were modeled in Smart PLS software and their results discussed.

Table 2 | The characteristics of the four regions, in terms of water resources and the number of villages

Region Number	Village Count	Surface water and Groundwater status
1	10	Surface water until mid-summer and groundwater limited to just springs
2	6	Surface water until the late spring and groundwater are decreasing
3	7	Surface water very limited and groundwater slightly salty and saline
4	7	Surface water very limited and groundwater limited to just springs

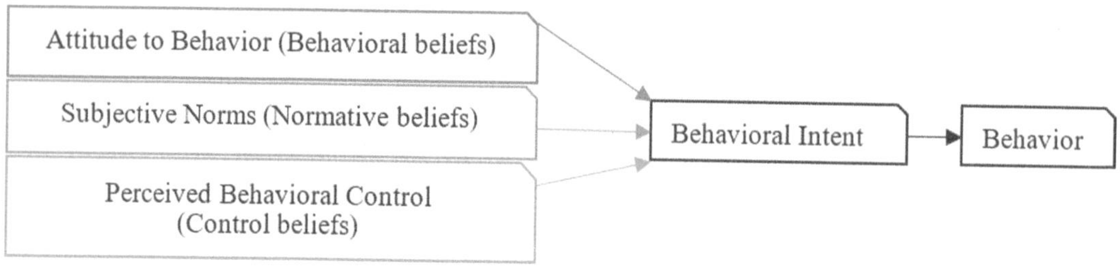

Figure 2 | The conceptual framework.

Questionnaire

The data gathering tool used was a researcher-made questionnaire designed using a literature review (Table 3). To assess the validity of the questionnaire, the content validity ratio, and content validity index were calculated using 10 expert opinions. To measure the content validity rate, experts evaluated questionnaire queries one by one on a three-point scale (1. The question is necessary; 2. The question is useful but not essential; 3. The question is not necessary) and a content validity ratio was then calculated.

To evaluate the content validity index, 10 experts were asked to specify 'relevancy', 'clarity', 'simplicity' and 'expressivity' of questionnaire queries on a 4-point ordinal scale (e.g. for 'relevancy': 1- not relevant/2- relatively relevant/3- relevant/4- completely relevant) and finally, the content validity index was calculated. In this research, after collecting expert comments, all of the questions have a content validity ratio above 0.75 and a content validity index above 0.79.

Cronbach's alpha coefficient (Cr-aco) was used to examine the reliability of the set of questions related to each variable. The mean coefficient was higher than 0.7 for

Table 3 | Sample questionnaire for policy option 'reduce cultivated area'

Intention

Q1. How much do you intend to reduce your cultivated area?
Q2. Are you going to reduce your cultivated area in the future?
Q3. How much do you intend to encourage other farmers to reduce the cultivation area?
Choose your answer: 1- Very little 2- Little 3- Medium 4- Much 5- Very much

Attitude

Q1. Should farmers be thinking about maximizing production efficiency rather than thinking about reducing their cultivation?
Q2. Do you think agricultural production more important than protecting groundwater resources?
Q3. In addition to drought conditions, is it important to reduce the area under cultivation in all conditions?
Q4. Do you believe that under the current conditions, the reduction of the area under cultivation for the conservation of water resources is unnecessary?
Q5. Do you believe that reducing the crop area is beneficial for the conservation of water resources?
Q6. Do you believe that reducing the crop area in order to protect water resources is wise?
Q7. Do you believe that reducing the cultivated areas to protect water resources is your preference?
Choose your answer: 1- Very little 2- Little 3- Medium 4- Much 5- Very much

Social norms

Q1. Most people who are important to me think that reducing the area under cultivation is considered desirable.
Q2. If I reduce my cultivated area in order to save and protect the water resources, those people who are important to me will support my actions.
Choose your answer: 1- Very little 2- Little 3- Medium 4- Much 5- Very much

Perceived behavioral control

Q1. For me, reducing the cropping areas to protect water resources is easy.
Q2. If I want to, I can easily reduce the cultivation area to protect water resources.
Q3. How much freedom do you have to reduce your cultivation area?
Q4. Reducing the cultivated area in order to preserve water resources is an option at my disposal.
Q5. How difficult is it for you to get involved in reducing the cropping areas to protect water resources?
Q6. I do not have the time and skills needed to reduce the amount of cultivation to preserve water resources.
Q7. For me, the application of tools to reduce the cultivation area to preserve water resources is costly.
Q8. Is it possible for you to reduce the cultivation area in order to preserve water resources according to your farm conditions?
Q9. It is impossible for me to carry out measures to reduce the cultivation area in order to maintain water resources.
Choose your answer: 1- Very little 2- Little 3- Medium 4- Much 5- Very much

Behavior

Q1. How much have you reduced your agricultural area?
Q2. How much have you reduced your horticultural area?
Choose your answer: 1- Very little 2- Little 3- Medium 4- Many 5- Too much

questions about the variables of attitude toward the behavior (Cr-aco = 0.71), subjective norm (Cr-aco = 0.70), perceived behavioral control (Cr-aco = 0.73), and behavioral intent (Cr-aco = 0.72).

Sample

The study area has 30 villages with a population of 4,000 farmers and only 20% (800) of these farmers own agricultural land larger than one hectare and only half of them have a reliable water source, making it possible for them to implement water resources policies. There are about 400 of these farmers, who are the statistical community of the study area and a questionnaire was completed by a statistical sample of 66 farmers (16.5% of the statistical population). The minimum sample size required can be looked up from the guidelines suggested by Marcoulides & Saunders (2006), depending on the maximum number of arrows pointing at a latent variable as specified in the structural equation model.

Since the extant research has indeed ascertained that farmers are generally unwilling to spend their time completing surveys and sharing data and/or information on themselves and their activities, in this research, sample selection from the statistical community was made purposefully and carefully until saturation state was reached, and the composition of the interview and the questionnaire used was as follows:

During the prayers of the Maghreb, when most farmers come to the mosque in order to perform a prayer, a free collective interview was held, and their suggestions and solutions for saving water were heard. In this interview, two or three farmers in each village were chosen purposefully and a questionnaire was completed by them.

RESULTS AND DISCUSSION

Structural equation modeling (SEM)

Structural equation modeling (SEM) is a second-generation multivariate data analysis method that is often used by researchers because it can test theoretically supported linear and additive causal models. With SEM, researchers can visually examine the relationships that exist among variables of interest in order to prioritize resources. The fact that unobservable, hard-to-measure latent variables can be used in SEM makes it ideal for tackling research problems such as the implementation of water-saving policies.

There are several distinct approaches to SEM: the first approach is the widely applied covariance-based SEM (CB-SEM), using software packages such as AMOS, EQS, LISREL, and MPLUS. The second approach is partial least squares (PLS), which focuses on the analysis of variance and can be carried out using PLS-Graph, VisualPLS, SmartPLS, and WarpPLS. It can also be employed using the PLS module in the 'r' statistical software package. The third approach is a component-based SEM known as generalized structured component analysis (GSCA); it is implemented through VisualGSCA or a web-based application called GeSCA. Another way to perform SEM is called nonlinear universal structural relational modeling (NEUSREL), using NEUSREL's causal analytics software (Wong 2013).

PLS is a soft modeling approach to SEM with no assumptions about data distribution (Vinzi *et al.* 2010). Thus, PLS-SEM becomes a good alternative to CB-SEM when the following situations are encountered (Wong 2013):

1. Sample size is small.
2. Applications have a weak theoretical basis.
3. Predictive accuracy is paramount.
4. Correct model specification cannot be ensured.

This paper focuses on SmartPLS because it is freely available to the research community across the globe. Attitude, subjective norm, perceived behavioral control, intention, and behavior were modeled as latent reflective variables. The crop level reduction policy conceptual model is presented in Figure 3, the crop pattern change policy conceptual model is presented in Figure 4, the policy of increasing irrigation efficiency conceptual model is presented in Figure 5, and the policy of increasing non-agricultural incomes conceptual model is presented in Figure 6.

Checking reliability and validity

To find a reliability value, the square of each of the outer loadings was verified, and all were higher than 0.70

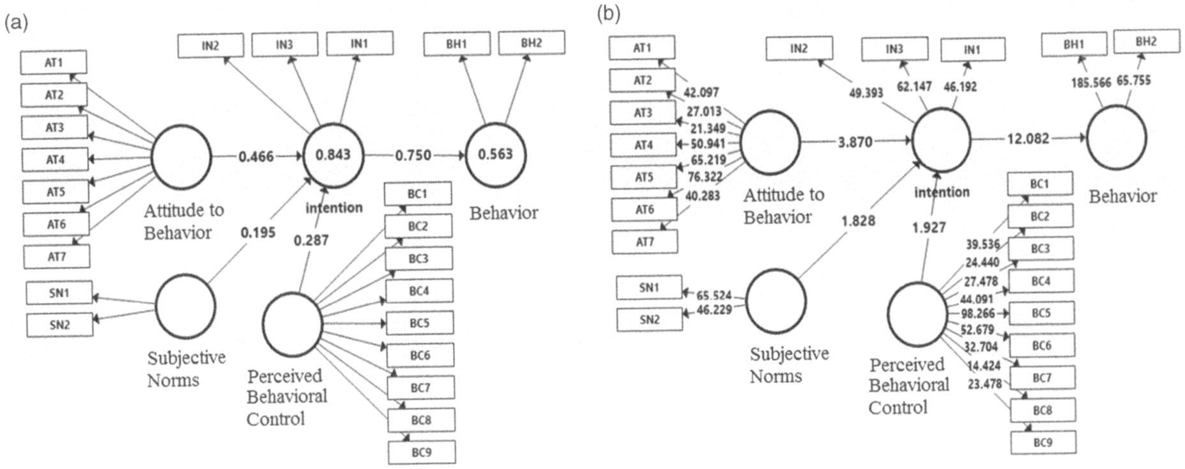

Figure 3 | Graphical representation of the conceptual crop level reduction policy model ((a) path coefficients and R^2 (b) T-values).

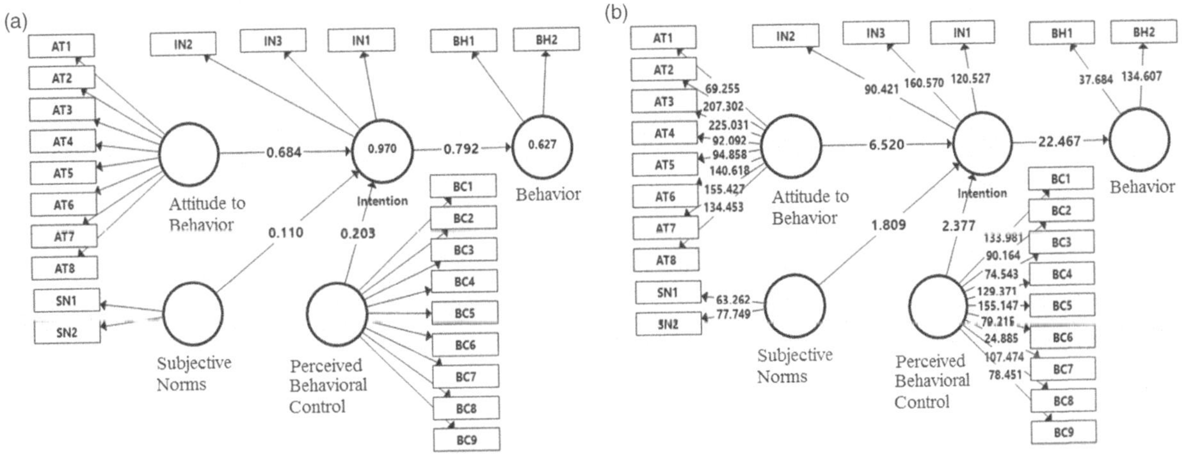

Figure 4 | Graphical representation of the conceptual crop pattern change policy model ((a) path coefficients and R^2 (b) T-values).

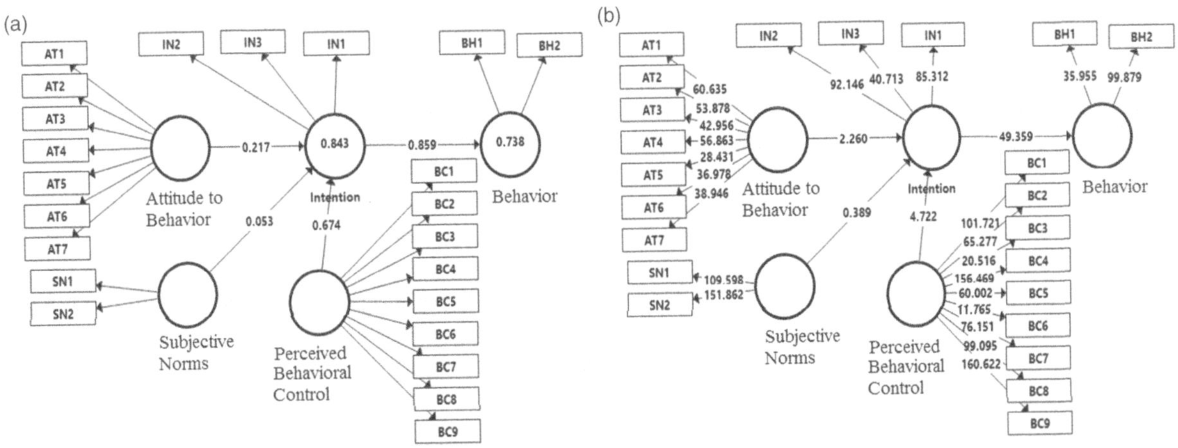

Figure 5 | Graphical representation of the conceptual policy of increasing irrigation efficiency model ((a) path coefficients and R^2 (b) T-values).

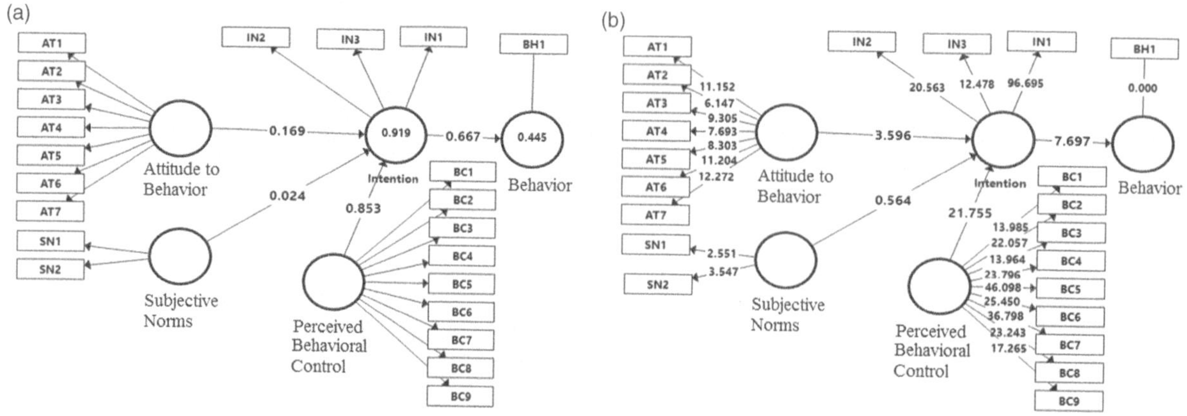

Figure 6 | Graphical representation of the conceptual policy of increasing non-agricultural incomes model ((a) path coefficients and R^2 (b) T-values).

(Hulland 1999). Traditionally, 'Cronbach's alpha coefficient' is used to measure internal consistency as a measure of scale reliability in social science research and in various research fields where questionnaires are involved, but this coefficient is used conservatively in PLS-SEM. Previous literature has suggested the use of 'composite reliability' as a replacement (Hair *et al.* 2014). In this research, 'composite reliability' values are larger than 0.6, so high levels of internal consistency reliability have been demonstrated among all five reflective latent variables. To check convergent validity, each latent variable's average variance extracted (AVE) was evaluated. It was found that all of the AVE values are greater than the acceptable threshold of 0.5, so convergent validity is confirmed. Fornell & Larcker (1981) suggest that the square root of AVE in each latent variable can be used to establish discriminant validity if this value is larger than other correlation values among the latent variables. To do this, Table 4 was created in which the square root of AVE is manually calculated and written in bold on the diagonal of Table 4. The correlations between the latent variables are copied from the 'latent variable correlation' section of the default report and appear in the lower left part of Table 4.

Checking structural path significance

SmartPLS can generate T-statistics for significance testing of both the inner and outer model, using a procedure called bootstrapping. By using a two-tailed *t*-test with a significance level of 5%, if T-statistics values are larger than 1.96, the path coefficients are significant. In our models, the results of the T-statistics values for the subjective norm to intention in Figures 3(b)–6(b), perceived behavioral control to intention in Figure 3(b), and Behavior to BH1 in Figure 6(b), mean that the path coefficients are not significant in the above cases.

Cross-validated redundancy measures

Stone-Geisser's (Q2) values (cross-validated redundancy measures) are obtained by the blindfolding procedure in SmartPLS (Table 5). Q2 values of 0.02, 0.15, and 0.35 indicate an exogenous construct has a small, medium, and large predictive relevance for an endogenous latent variable, respectively.

Cross-validated communality measures

If cross-validated communality is positive, the measurement model has a good quality. In this research all cross-validated communality for the five variables are positive.

Determining model fit

In the research literature, determining the appropriate model is one of the most important steps in the SEM (Barrett 2007; Hooper *et al.* 2008). In recent years, appropriate indicators have been under intense scrutiny, and some authors have even suggested that some of them should not be used (Barrett 2007). Some researchers find it appropriate to report a number of fit indices models because different indices reflect a different aspect of model fit. The rejection

Table 4 | Fornell-larcker criterion analysis for checking discriminant validity

Latent variable correlations	Perceived behavioral control	Behavior	Attitude toward the behavior	Intention	Subjective norm
Crop level reduction policy model					
Perceived behavioral control	0.894	0.813	0.893	0.892	0.845
Behavior	0.813	0.961	0.798	0.750	0.787
Attitude toward the behavior	0.893	0.798	0.922	0.907	0.867
Intention	0.892	0.750	0.907	0.919	0.842
Subjective norm	0.845	0.787	0.867	0.842	0.919
Crop pattern change policy model					
Perceived behavioral control	0.963	0.786	0.953	0.950	0.933
Behavior	0.786	0.943	0.792	0.822	0.650
Attitude toward the behavior	0.953	0.792	0.961	0.954	0.928
Intention	0.950	0.822	0.954	0.939	0.890
Subjective norm	0.933	0.650	0.928	0.890	0.928
Policy of increasing irrigation efficiency model					
Perceived behavioral control	0.991	0.820	0.839	0.859	0.818
Behavior	0.820	0.914	0.859	0.845	0.789
Attitude toward the behavior	0.839	0.859	0.931	0.911	0.875
Intention	0.859	0.845	0.911	0.908	0.901
Subjective norm	0.818	0.789	0.875	0.901	0.973
Policy of increasing non-agricultural incomes model					
Perceived behavioral control	0.991	0.586	0.667	0.561	−0.086
Behavior	0.586	0.821	0.677	0.600	−0.142
Attitude toward the behavior	0.667	0.677	0.876	0.866	−0.204
Intention	0.561	0.600	0.866	0.875	−0.238
Subjective norm	−0.086	−0.142	−0.204	−0.238	0.875

Table 5 | Cross-validated redundancy measures

	Crop level reduction policy model	Crop pattern change policy model	Policy of increasing irrigation efficiency model	Policy of increasing non-agricultural incomes model
Behavior	0.49	0.519	0.584	0.402
Intention	0.66	0.842	0.684	0.65

of a model shows that there are significant paths between variables that are not presented in the model.

Standardized root mean square residual (SRMR) is defined as the difference between the observed correlation and the model implied correlation matrix. Thus, it allows assessment of the average magnitude of the discrepancies between observed and expected correlations as an absolute measure of model fit criterion. A value of less than 0.10 or 0.08 is considered a good fit. Hair *et al.* (2014) introduced SRMR as a goodness of fit measure for PLS-SEM that can be used to avoid model misspecification. One of the first fit measures proposed in the SEM literature is the normed fit index (NFI) by Bentler & Bonett (1980). It computes the Chi^2 value of the proposed model and compares it against a meaningful benchmark. Since the Chi^2 value of the proposed model in itself does not provide sufficient information to judge model fit, the NFI uses the Chi^2 value from the null model, as a yardstick. The NFI is then defined as 1 minus the Chi^2 value of the proposed model divided by the Chi^2 value of the null model. Consequently, the NFI results in values between 0 and 1. The closer NFI is to 1, the better fit. NFI values above 0.9 usually represent an acceptable fit. In the 'policy of increasing non-agricultural incomes mode' model, the research data support the model weakly, and it would have been better if the data collected for this mode had been revised, which was accepted because the normed fit index is close to the acceptable threshold (Table 6).

Strength of the theoretical model to testing the research hypotheses

This research proposed a model that contained three independent variables and two dependent variables. The PLS algorithm was able to calculate an estimate R^2 for the dependent variable. The strength of the theoretical model was established by factor R^2. The R^2 was calculated using the PLS algorithm with 300 iterations. Falk & Miller (1992) suggest that the explained variance (R^2) should be greater than 0.1. All of the R^2s for the four models achieved high explained variance scores. All were above the 0.1 recommended levels. The R^2 statistics and path coefficients are shown in Figures 3(a)–6(a).

First hypothesis: there is a positive relationship between attitude towards water policy options and the intention to comply with water policy options.

Based on the results in Figures 3–6 and given T-values for variables, the null hypothesis is rejected and hypothesis 1 is confirmed. It is concluded from this discussion that there is a significant direct relationship between attitude towards water policy options and the intention to comply with water policy options.

Second hypothesis: there is a positive relationship between the positive perceptions of important others to water policy options and the intention to comply with water policy options.

Based on the results in Figures 3–6 and given T-values for variables, the null hypothesis is rejected and hypothesis 2 is confirmed. It is concluded from this discussion that there is a significant direct relationship between the positive perceptions of important others to water policy options and the intention to comply with water policy options.

Third hypothesis: there should be a positive relationship between perceptions about the ability to comply with water policy options and the intention to comply.

Based on the results in Figures 3–6 and given T-values for variables, the null hypothesis is rejected and hypothesis 3 is confirmed. It is concluded from this discussion that there is a significant direct relationship between perceptions about the ability to comply with water policy options and the intention to comply.

Fourth hypothesis: there should be a positive relationship between intention to comply with water policy options and actual compliance.

Based on the results in Figures 3–6 and given T-values for variables, the null hypothesis is rejected and hypothesis 4 is confirmed. It is concluded from this discussion that there is a significant direct relationship between intention to comply with water policy options and actual compliance.

Table 6 | Fit indices of models

Fit criterion	Crop level reduction policy mode	Crop pattern change policy model	Policy of increasing irrigation efficiency mode	Policy of increasing non-agricultural incomes mode
SRMR	0.039	0.037	0.055	0.1
NFI	0.90	0.91	0.92	0.85

Descriptive statistics of intention to comply with water policy options by region

As described above, the study area was divided into four areas according to the status of water resources. In this section, we want to compare the number of farmers' intentions regarding the four water policy options in the four regions (Figure 7).

Region '1' farmers, because of access to surface water by mid-summer, have a lower intention of implementing agricultural water-saving policies than any other group. Farmers in this area have maintained their cultivations and sometimes developed their gardens, and usually changed their cropping patterns and turned to early crop cultivation (roses, beans). Region '1' is elevated upstream and wells are rarely drilled. Region '2' has the largest number of wells that can facilitate the implementation of drip irrigation projects (irrigation efficiency improvement policy), but the small size of the land is an obstacle to this policy. Farmers in region '3' can use surface water for a short time in the event of heavy rains and floods. Due to the underground water salinity in region 3, they supply

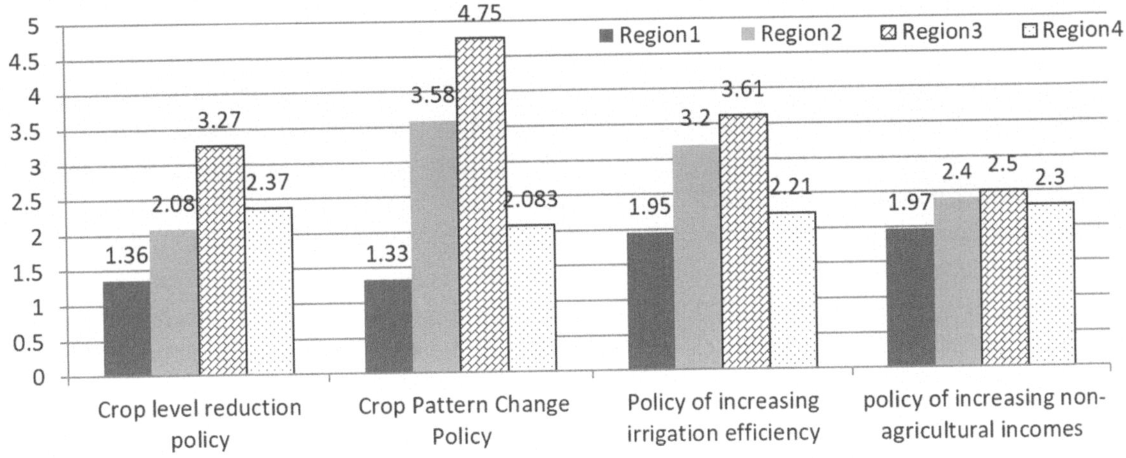

Figure 7 | Mean score of intention to comply water policy option.

their water demand by purchasing water from the neighboring area (region 2). Therefore, region 3 has the highest water stress, and in order to adapt to this stress, farmers in region 3 intend to agree on any type of agricultural water-saving policy. They have severely reduced their land cultivation. They also cultivated crops that are resistant to saline water and water stress (Alfalfa, Pistachio, and Elaeagnus Angustifolia). By lining of channels, they have increased the irrigation efficiency and now wish they also had non-agricultural incomes. Region '4' is mountainous and has a hard aquifer, and there are no wells in this area and the water resources are spring and qanats.

Because of the necessity for high initial capital to generate non-agricultural incomes, these incomes are limited and are usually agriculture-related such as livestock or processed crops, and a limited income from carpet weaving. Since implementation of crop level reduction policy and crop pattern change policy leads to a significant and modest reduction in agricultural incomes, it is noted that regions with very limited water resources are willing to comply with these policies. On the other hand, the implementation of increasing irrigation efficiency and increasing non-agricultural incomes policies requires high government investment. Therefore, considering the financial facilities of the state and the intention of the farmers, the policymaker can choose the appropriate option for each region.

Regarding the high ability of intention to predict behavior, in Table 7, based on the average intentions score, prioritizations of policies to follow in the four regions and in the whole area are given.

Table 7 | Prioritization of policies based on the average intentions score in the four regions and whole area

Policy	Region 1	Region 2	Region 3	Region 4	Whole area
Crop level reduction policy	3	4	3	1	4
Crop pattern change policy	4	1	1	4	1
Policy of increasing irrigation efficiency	2	2	2	3	2
Policy of increasing non-agricultural incomes	1	3	4	2	3

CONCLUSIONS

Uncontrolled and inefficient use of water in the study area, especially in the agricultural sector, has increased concern and reduced water security. Therefore, the attention of researchers and policymakers to the implementation of water-saving policies has been drawn. In this study, the well-known behavioral theory of Ajzen (1991) was used to identify the motivational factors that could encourage farmers to follow water-saving policies. Despite the widespread use of this theory in the literature on environmental behaviors, it has rarely been used to measure farmers' compliance with water-saving policies.

While past research has often focused on water conservation intentions this research moved toward measuring both water conservation intentions and measuring water conservation behavior. This paper drew on a social–psychological model to determine how the native farmers in a fertile region in the northwest of Iran support different agricultural policy options. Results suggest that the TPB framework is an effective tool for the feasibility of water-saving policy options. In a meta-analysis of the TPB, Armitage & Conner (2001) found that the TPB accounted for 39% of the variance in intention. In this study, the explained variance in the intention of the TPB was higher than this finding. Differences in the results of these studies may refer to the premise that water-saving behavior, as is true for many other behaviors, changes according to individuals' demographic characteristics such as age and gender, and cultural and socio-economic differences. Consistent with previous research from Chang *et al.* (2016) and Azizi Khalkheili & Zamani (2010), this study established that farmers' favorable attitudes toward water-saving policy options predict their intention to adopt them.

Therefore, one of the important duties to attract farmers' participation in water-saving policy is creating a positive attitude towards participation. To do this, solving the existing problems, providing appropriate training and publicity programs, especially by the media and service centers in agriculture, increasing farmers' knowledge, and more focus on water company staff training connected to farmers can be effective. In line with Tohidyan far & Rezaei Moghaddam (2015), the subjective norm has a very significant effect on farmers' intent to adopt water-saving policy options. Unlike Nancarrow *et al.* (2008) in Australia regarding recycled drinking water, and like Lynne *et al.* (1995), regarding soil conservation, they found that PBC is a significant predictor of intention.

Similar to research from Yazdanpanah *et al.* (2014), this study also found that perceived behavioral control had a major impact on the intentions of farmers to comply with water-saving policy options, because farmers' income is limited and usually the implementation of these policies has needed high initial investment and a sufficient supply of water with adequate quality and volume. In most of the work carried out by the TPB model in Iran, the perceived behavioral control of Iranian farmers has had a great impact on farmers' intentions.

It is concluded that farmers are more likely to comply with water-saving policy options when the government has provided financial incentives. Farmers will also follow policy options that will not reduce their income. In general, the economic interests of farmers are dominant in following water policy options. Many farmers think that drip irrigation reduces crop yields. They also generally have concerns about marketing low-water crops and have no experience growing such crops. Therefore, education and awareness by government agencies such as the agricultural organization can be effective.

Consistent with previous research (Clark & Finley 2007; Vermeir & Verbeke 2008; Yazdanpanah *et al.* 2014) the results of this study are entirely consistent with the theory of TPB because it shows that farmers' intention to adopt water policy options has a positive relationship with farmers' attitudes toward the water policy option (ATT), subjective normative (SN) and perceived behavioral control (PBC) and there is a positive relationship between intention to comply with policy option and actual compliance (behavior).

SN showed little effect on and low predictive relevance to intention in any of the four water policy option models. This shows that if farmers' attitudes to agricultural water policy are positive and that there are no barriers to doing that, they will pursue the policy without affecting social norms.

PBC had a strong influence on intentions due to the individual perception of the difficulty of implementing policy options due to financial constraints, especially in the two policy options 'Irrigation Productivity Increase Policy' and 'Non-Agricultural Income Increase Policy'. Therefore, for the implementation of these two policies, the government must provide more financial support for farmers.

ATT showed a large effect and high predictive relevance on intention in all four water policy option models. Therefore, it is recommended to design intervention programs based on promoting attitudes in order to encourage them to implement water policy.

According to the findings of this research, the explained variance over 85% for intentions and above 45% for the behavior of water policy options, suggests the high ability of the theory of planned behavior in predicting policies for saving agricultural water.

CONFLICTS OF INTEREST

The author declares no conflict of interest.

DATA AVAILABILITY STATEMENT

All relevant data are included in the paper or its Supplementary Information.

REFERENCES

Abrahamse, W., Steg, L., Vlek, C. & Rothengatter, T. 2005 A review of intervention studies aimed at household energy conservation. *Journal of Environmental Psychology* **25** (3), 273–291.

Ajzen, I. 1991 The theory of planned behavior. *Organizational Behavior and Human Decision Processes* **50** (2), 179–211.

Ajzen, I. & Fishbein, M. 1970 The prediction of behavior from attitudinal and normative variables. *Journal of Experimental Social Psychology* **6** (4), 466–487.

Armitage, C. J. & Conner, M. 2001 Efficacy of the theory of planned behaviour: a meta-analytic review. *British Journal of Social Psychology* **40** (4), 471–499.

Azizi Khalkheili, T. & Zamani, G. H. 2010 Factors affecting farmers' participation in irrigation management: the application of path analysis. *Journal of Agricultural Economics and Development* **24** (1).

Babaee, O. & Alijani, B. 2013 Spatial analysis of long duration droughts in Iran. *Physical Geography Research Quarterly* **45** (3), 1–12 (In Persian).

Barrett, P. 2007 Structural equation modelling: adjudging model fit. *Personality and Individual Differences* **42** (5), 815–824.

Bentler, P. M. & Bonett, D. G. 1980 Significance tests and goodness of fit in the analysis of covariance structures. *Psychological Bulletin* **88** (3), 588.

Cary, J. W. 2008 Influencing attitudes and changing consumers' household water consumption behaviour. *Water Science and Technology: Water Supply* **8** (3), 325–330.

Chang, G. 2013 Factors influencing water conservation behavior among urban residents in China's arid areas. *Water Policy* **15** (5), 691–704.

Chang, G., Wang, L., Meng, L. & Zhang, W. 2016 Farmers' attitudes toward mandatory water-saving policies: a case study in two basins in northwest China. *Journal of Environmental Management* **181**, 455–464.

Clark, W. A. & Finley, J. C. 2007 Determinants of water conservation intention in Blagoevgrad, Bulgaria. *Society and Natural Resources* **20** (7), 613–627.

Dalin, C., Wada, Y., Kastner, T. & Puma, M. J. 2017 Groundwater depletion embedded in international food trade. *Nature* **543** (7647), 700–704.

De Young, R. 2000 New ways to promote proenvironmental behavior: expanding and evaluating motives for environmentally responsible behavior. *Journal of Social Issues* **56** (3), 509–526.

Dolnicar, S. & Hurlimann, A. 2010 Australians' water conservation behaviours and attitudes. *Australasian Journal of Water Resources* **14** (1), 43–53.

Falk, R. F. & Miller, N. B. 1992 *A Primer for Soft Modeling*. University of Akron Press, Akron, Ohio.

Fornell, C. & Larcker, D. F. 1981 Evaluating structural equation models with unobservable variables and measurement error. *Journal of Marketing Research* **18** (1), 39–50.

Hair, J. F., Henseler, J., Dijkstra, T. K. & Sarstedt, M. 2014 Common beliefs and reality about PLS: comments on Rönkkö and Evermann. *Organizational Research Methods* **17** (2), 182–209.

Hooper, D., Coughlan, J. & Mullen, M. R. 2008 Structural equation modelling: guidelines for determining model fit. *Electron J Bus Res Methods* **2008** (6), 53–60.

Hulland, J. 1999 Use of partial least squares (PLS) in strategic management research: a review of four recent studies. *Strategic Management Journal* **20** (2), 195–204.

Kurz, T., Donaghue, N. & Walker, I. 2005 Utilizing a social-ecological framework to promote water and energy conservation: a field experiment 1. *Journal of Applied Social Psychology* **35** (6), 1281–1300.

Lam, S. P. 1999 Predicting intentions to conserve water from the theory of planned behavior, perceived moral obligation, and perceived water right 1. *Journal of Applied Social Psychology* **29** (5), 1058–1071.

Lynne, G. D., Casey, C. F., Hodges, A. & Rahmani, M. 1995 Conservation technology adoption decisions and the theory of planned behavior. *Journal of Economic Psychology* **16** (4), 581–598.

Mahdavi, T., Bagheri, A. & Hosseini, S. A. 2019 Applying the System of Environmental and Economic Accounts for Water (SEEA-Water) for integrated assessment of water security in an aquifer scale – Case study: Azarshahr aquifer, Iran. *Groundwater for Sustainable Development* **9**, 100261.

Marcoulides, G. A. & Saunders, C. 2006 Editor's comments: PLS: a silver bullet? *MIS Quarterly* **30** (2), iii–ix.

Nancarrow, B. E., Leviston, Z., Po, M., Porter, N. B. & Tucker, D. I. 2008 What drives communities' decisions and behaviours in the reuse of wastewater. *Water Science and Technology* **57** (4), 485–491.

Pino, G., Toma, P., Rizzo, C., Miglietta, P. P., Peluso, A. M. & Guido, G. 2017 Determinants of farmers' intention to adopt water saving measures: evidence from Italy. *Sustainability* **9** (1), 77.

Salvaggio, M., Futrell, R., Batson, C. D. & Brents, B. G. 2014 Water scarcity in the desert metropolis: how environmental values, knowledge and concern affect Las Vegas residents' support for water conservation policy. *Journal of Environmental Planning and Management* **57** (4), 588–611.

Steg, L. & Vlek, C. 2009 Encouraging pro-environmental behaviour: an integrative review and research agenda. *Journal of Environmental Psychology* **29** (3), 309–317.

Tohidyan Far, S. & Rezaei Moghaddam, K. 2015 Attitudes of farmers toward participation in irrigation and drainage projects: the structural equations modeling analysis. *Iran Agricultural Research* **34** (1), 80–91.

Trumbo, C. W. & O'Keefe, G. J. 2005 Intention to conserve water: environmental values, reasoned action, and information effects across time. *Society and Natural Resources* **18** (6), 573–585.

UNSD 2012 *System of Environmental-Economic Accounting for Water*. United Nations Statistics Division (UNSD), New York, NY, USA.

Vermeir, I. & Verbeke, W. 2008 Sustainable food consumption among young adults in Belgium: theory of planned behaviour and the role of confidence and values. *Ecological Economics* **64** (3), 542–553.

Vinzi, V. E., Trinchera, L. & Amato, S. 2010 PLS path modeling: from foundations to recent developments and open issues for model assessment and improvement. In: *Handbook of Partial Least Squares* (Vinzi, V. E., Chin, W. W., Henseler, J. & Wang, H., eds). Springer, Berlin, Heidelberg, pp. 47–82.

Wong, K. K. K. 2013 Partial least squares structural equation modeling (PLS-SEM) techniques using SmartPLS. *Marketing Bulletin* **24** (1), 1–32.

Yazdanpanah, M., Hayati, D., Hochrainer-Stigler, S. & Zamani, G. H. 2014 Understanding farmers' intention and behavior regarding water conservation in the Middle-East and North Africa: a case study in Iran. *Journal of Environmental Management* **135**, 63–72.

First received 27 December 2020; accepted in revised form 27 April 2021. Available online 11 May 2021

Effect of irrigation practices upon yield and fruit quality of four grapefruit (*Citrus paradisi* Mac.) cultivars

G. Morianou, V. Ziogas, N. N. Kourgialas and G. P. Karatzas

ABSTRACT

Two strategies of irrigation were applied during the phase of fruit growth in 30-year-old grapefruit trees (*Citrus paradisi* Mac.) of four varieties (cv. Marsh SRA 8, cv. Shambar SRA 22, cv. Frost Marsh, cv. Ruby). For the first strategy (T1), the trees were irrigated every week at 100% crop evapotranspiration (ETc), while for the second strategy (T2) at 60% ETc. At harvest time, tree yield was estimated along with fruit quality and water productivity. T1 strategy significantly increased the tree yield, the fruit weight and the juice content. Although an increase in water productivity (crop yield/total water use) was achieved during the water stress experiment for three out of the four varieties. The main effect of the T2 strategy was a significant increase of qualitative (commercial and nutritional) attributes like total soluble solids (TSS), citric acid, ascorbic acid (Vitamin C) and phenolic content. In addition, strategy T2 increased the maturation index and the fruits had sweeter flavor for two varieties. These results lead to the conclusion that in mature grapefruit trees, optimal irrigation practice provided better fruit yield, while water stress practices favor the accumulation of specific nutritional elements and improve specific fruit quality parameters, especially for cv. Ruby.

Key words | agronomic practices, citrus, deficit irrigation, phenols, water management

G. Morianou (corresponding author)
G. P. Karatzas
School of Environmental Engineering,
Technical University of Crete,
Polytechneioupolis, 73100 Chania,
Greece
E-mail: *morianou@elgo.iosv.gr*

G. Morianou
V. Ziogas
N. N. Kourgialas
Institute of Olive Tree, Subtropical Plants & Viticulture,
ELGO-DIMITRA,
Agrokipio, 73134, Chania,
Greece

HIGHLIGHTS

- Assess the effects on grapefruit yield, fruit quality, and water productivity under different irrigation treatments.
- Investigate differences under full irrigation and deficit irrigation among four grapefruit cultivars.

INTRODUCTION

Citrus is one of the most-economically-important crops in the word and they have great value in terms of nutrition. Grapefruits (*Citrus paradisi* L.) are an especially rich source of vitamin C, phenolic and flavonoids, carotenoids (provitamin A) and other nutritional elements. They are also famous for their antioxidant properties, which strengthen the human body against various diseases (Navarro *et al.* 2015). In Greece, citrus occupies 45,400 ha of agricultural land and the production during the year 2018 reached 1,179,555 tonnes. Grapefruit cultivation is limited in Greece, occupying an area of only 160 ha, and the production during 2018 reached 3,160 tonnes (FAO 2019). Grapefruits, like all citrus, have in general high demands for annual irrigation. The annual water requirement of citrus varies between 900 and 1,200 mm depending on the soil, climate and the age of the tree (Doorenbos & Kassam 1979; Kourgialas & Karatzas 2015).

Scarcity of water resources, due to rapidly growing cultivation demands and climate change, is one of the major factors limiting irrigated agriculture (Navarro *et al.* 2010).

doi: 10.2166/ws.2021.113

Globally, agriculture is the largest user of water, with nearly 70% of withdrawals worldwide (FAO 2019). The limited water supply worldwide and increased irrigation cost, especially in arid and semi-arid areas, demands a more efficient and optimized use of irrigation water. The aim is the maximization of water savings and the improvement of fruit yield and quality. One of the most promising irrigation strategies for accomplishing this objective might be Deficit Irrigation (DI). Deficit irrigation is based on the application of amounts of irrigation water lower than the crop evapotranspiration needs (ETc) (Romero *et al.* 2006; Fereres & Soriano 2007; García-Tejero *et al.* 2010a). DI has been examined thoroughly as a valuable and feasible way to implement sustainable agriculture in semi-arid regions (Ruiz-Sanchez *et al.* 2010). Following this trend, several authors have stated the advantages of DI strategies with regard to improving water use efficiency and fruit quality in many emblematic citrus species (Romero *et al.* 2006; García-Tejero *et al.* 2010b).

Several studies have shown that under deficit irrigation (DI), fruit quality is improved in citrus trees (Ballester *et al.* 2013; Pérez-Pérez *et al.* 2014; Navarro *et al.* 2015). The majority of the studies show that the response of citrus trees to DI treatments depends mainly on the phenological stage (cell division, cell elongation, ripening-harvest) that is applied (Ginestar & Castel 1996; García-Tejero *et al.* 2010a; Navarro *et al.* 2010, 2015). Previous studies have showed that the application of water stress during flowering, as well as during the initial stage of fruit growth, reduced fruit yield (Romero *et al.* 2006; Pérez-Pérez *et al.* 2008, 2014; García-Tejero *et al.* 2010b). Also, when water stress was applied during the stage of fruit growth, it negatively affected fruit size, yield and delayed maturation, while improving some quality parameters (Navarro *et al.* 2010, 2015; Ballester *et al.* 2013; Pérez-Pérez *et al.* 2014). Finally, applying water stress during fruit maturation mainly had a reflection on the fruit quality parameters. Citrus fruit quality characteristics such as total soluble solids and titratable acidity were positively affected, while no significant delay in maturation time or fruit yield were observed (Pérez-Pérez *et al.* 2009; García-Tejero *et al.* 2010b; Navarro *et al.* 2010, 2015). However, the citrus response to DI strategies depends not only on the phenological stage that the DI is applied, but also upon the severity of the water deficit (García-Tejero *et al.* 2010b; Aguado *et al.* 2012; Ballester *et al.* 2013; Gasque *et al.* 2016). One of the main effects of severe water stress is in the change of the maturity index. Many authors observed that DI strategies can improve water productivity (García-Tejero *et al.* 2010a; Aguado *et al.* 2012). Moderate water stress strategies are suitable for saving water and obtaining similar yield and citrus fruit quality similar to no DI strategies (García-Tejero *et al.* 2010a).

Most of the DI studies in citrus have been carried out mainly for sweet oranges and mandarin, but little work has been done upon the response of grapefruits to water stress. In order to implement an orthological DI strategy that saves water but at the same time sustains fruit quality, a comparative study is needed in order to study the effect of DI in different varieties. Pérez-Pérez *et al.* (2014) and Navarro *et al.* (2015) evaluated the sensitivity to DI of different fruit growth stages in adult 'Star Ruby' grapefruit trees. The aim of this work was to apply a Deficit Irrigation (DI) treatment versus a Full Irrigation (FI) – control treatment in order to assess the effects on yield, fruit quality, and water productivity (WP) in an experimental 30-year-old orchard with grapefruit trees (*Citrus paradisi* Mac.) of four different varieties: (i) cv. Marsh SRA 8, (ii) cv. Shambar SRA 22, (iii) cv. Frost Marsh and (iv) cv. Ruby.

MATERIAL AND METHODS

Orchard, soil and climatic characteristics

The study was carried out during 2019 at the experimental field of the Institute of Olive Tree, Subtropical plants and Viticulture (IOSV) of the Hellenic Agricultural Organization 'DIMITRA' (ELGO-DIMITRA), in Chania, Crete (south Greece). The experiment was performed on 30-year-old grapefruit trees (*Citrus paradisi* Mac.) of four varieties (cv. Marsh SRA 8, cv. Shambar SRA 22, cv. Frost Marsh, cv. Ruby) grafted onto sour orange (*C. aurantium* L.) rootstock. The soil texture was loam, with 45.28% sand, 19.34% clay and 35.38% silt, an organic matter content of 1.35%, an electrical conductivity of 0.19 dS m^{-1}, 0.32% total $CaCO_3$ and a pH of 7.3. The N-NO_3 was 8.02 mg kg^{-1} and the phosphorus was 18.28 mg P kg^{-1}Olsen. Soil water content at field capacity was 300 mm m^{-1} and the wilting point was 110 mm m^{-1}, respectively, with an available soil-water content of 190 mm in the root zone (0.6 m from the soil surface). Finally, the bulk density was 1.64 g/cm^3.

Climatic data were provided by the IOSV's meteorological station in Chania (Greece), which is located at a distance of less than 200 m from the experimental plots. The climate of the area is sub-humid Mediterranean with humid and relatively cold winters, and dry – warm summers. The annual rainfall ranges from 300 to 700 mm/year, distributed mainly during late autumn to early spring. The mean annual air temperature ranges around 18 °C.

Experimental design and description of irrigation treatments

The experimental design was two randomized complete blocks: each block had four rows with six trees per row (26 trees per block). Two irrigation treatments were applied during this experiment, a control – FI treatment (T1) and a DI treatment (T2). The DI treatments are based on the theory of non-irrigation or irrigation with less water during different phenological stages of fruit growth (Phase I: bloom and cell division; Phase II: cell elongation, rapid fruit growth period; Phase III: ripening and harvest). Phase I for grapefruits lasts from late April to early June, Phase II from early June to early October and finally Phase III lasts from October until harvest. In Crete, due the local climate conditions, the crop needs at phases I and III are mainly covered by precipitation amounts. Thus, both T1 and T2 treatments were applied from May to early October, at the fruit growth stage (during the second half of Phase I and during Phase II). T1 was applied in the first block of trees and involved irrigation every week at 100% of the crop needs (ET_C). T2, applied to the second block of trees, also involved irrigation every week at 60% of the crop needs. Irrigation was applied using 12 emitters for each tree, with a discharge rate of 4 L h^{-1}. Emitters were located in a circle 25 cm away from the tree trunk at an equal distance from each other.

Water meters were used to measure the volume of irrigation applied to each treatment. The seasonal values of ETo were determined by the FAO56 version of the Penman equation (Allen *et al.* 1998) using the climatic data recorded by the automated weather station near the orchard. In this approach, evapotranspiration of the crop (ET_C) was estimated using the following equation:

$$ET_c = k_c \cdot ET_o$$

where the k_c coefficient incorporates crop characteristics and averaged effects of evaporation from the soil. In our case, the k_c ranged between 0.5 from March to May; 0.55 from June to October and 0.5 in November and December, respectively, in accordance with BEWARE project (2005) and Kourgialas *et al.* (2019).

The fertilization applied, for both treatments, was according to soil and leaf analysis. The fertilizer applied in the experimental orchards was ammonium nitrate fertilizer – 0.5 kg per tree. All optimum agricultural practices, such as pruning, weed control and fertilizer application, were also applied for both irrigation strategies.

Fruit measurements

When fruits had reached commercial maturation, the total number of citrus fruits was recorded in order to estimate the tree yield. Then, ten fruits per variety were selected randomly from a subsample of four trees per treatment. Specifically, five fruits per side (east and west sides) were picked, at the height of the human chest. The net weight of the fruits, the equatorial diameter (mm), the fruit shape, the external fruit color, the rind thickness (mm) and the flesh firmness (Kg) and juice content (%) were measured in the laboratory. Flesh firmness was quantified by using a 10 mm cylinder Fruit Pressure Tester digital penetrometer and the fruit shape was calculated according to:

$$Fruit\ shape = \frac{polar\ diameter\ (mm)}{equatorial\ diameter\ (mm)}$$

The external fruit Color Index (CI) was measured in ten fruits per treatment, using a tristimulus color difference meter (Minolta CR300 colorimeter), at four locations around the fruit. The Hunter parameters a*, b* and L* were reported by the colorimeter, obtaining the CI using the following equation:

$$Cl = \frac{1000 \cdot a^*}{L^* \cdot b^*}$$

L* parameter gives a value of the luminance or brightness of the sample; a* parameter indicates the area of variation between red and green spectrum and b* parameter refers to the area of variation between the yellow and blue spectrum.

Subsequently, fruits were cut in the equatorial area and peel thickness was measured at four points.

Juice measurements

The fruits were squeezed with an electric citrus fruit-juicer and the juice was strained through a 1-mm-mesh sieve. The juice from three fruits were combined and considered as a single sample for further chemical analysis. Thus, there were three (3) replications per variety, for each DI strategy. The juices were analyzed for Titratable Acidity (citric acid content), pH, Total Soluble Solids, Ascorbic Acid (Vitamin C) and Total Phenolics. The Maturity Index, which is connected to the citrus fruit flavor, was also estimated.

The content of total soluble solids (TSS) in the juice was measured at 25 °C with a digital refractometer (Atago) and titratable acidity (TA) was determined by titration with NaOH and phenolphthalein indicator according to conventional methods (Kimball 2012). The maturity index (MI) was expressed as TSS × 10/TA.

Ascorbic acid analysis in the fruit juice was determined according to the Official Methods of Analysis for Ascorbic Acid in Vitamin Preparations and Juices (method number 967.21, chapter 45, AOAC). Fruit juice samples were centrifuged at 1,650 g for 5 min at 4 °C. The supernatant was used for determination of the ascorbic acid content by titration with 2,6-dichlorophenolindophenol sodium salt hydrate (DIF). The results were expressed as mg ascorbic acid 100 mL^{-1} juice.

The total phenolics were determined by the Singleton *et al.* (1999) method. According to this method, 1 mL of fruit juice was extracted with 9 mL of 80% methanol for 30 min at room temperature. After centrifugation at 5,000 rpm for 10 min, the supernatant was taken for the determination of total phenolics by the Folin–Ciocalteu method. A reagent blank using H_2O was prepared. Folin–Ciocalteu phenol reagent (0.5 mL) was added to the mixture and it was shaken vigorously. After 5 min, 5 mL of 5% Na_2CO_3 solution were added, mixing at the same time. The solution was immediately diluted to 25 mL with distilled water, mixed thoroughly and then allowed to stand for 60 min before measurement: the absorbance was measured at 750 nm versus the prepared blank. The total phenolic content of the sample (three replicates per treatment) was expressed as mg mL^{-1} of Gallic Acid Equivalent (GAE).

Water productivity

Water productivity in agriculture and landscape irrigation may be generally defined as the ratio between the actual crop yield achieved (Y_a) and the total water use (TWU = irrigation + rainfall) expressed in kg/m^3 (Pereira *et al.* 2012):

$$WP = \frac{Y_a}{TWU}$$

WP indicator express the benefit derived from the consumption of water and can be used for assessing the impact of on-farm strategies under water-scarce conditions. It provides a vision of the water that could be saved with simultaneous increase in crop yield (Singh *et al.* 2006).

Statistical analysis

Statistical analyses were performed using the SPSSv17 package (SPSS Inc., Chicago IL) with a one-way analysis of variance (ANOVA), with four treatments for each DI strategy and six replicate trees per treatment. When there was a significant difference (P-value <0.05), means were separated using Duncan's test.

A principal-component analysis (Ihaka & Gentleman 1996) was performed using SPSS v17 package (SPSS Inc., Chicago IL). Principal Component Analysis (PCA) is a factor analysis protocol that is used in order to identify variables or underline factors that better explain the correlation or covariance matrix of several attributes (Davis & Sampson 1986).

RESULTS AND DISCUSSION

Yield and water productivity

The application of water stress treatment during Phase II (T2) decreased tree fruit yield compared with the control treatment (T1), in all grapefruit varieties except for cv. Frost Marsh (Table 1). Under water stress, the overall tree yield was reduced significantly by 14, 23 and 56% in cv. Marsh SRA 8, Shambar SRA 22 and Ruby, respectively, while cv. Frost Marsh remained unaffected. This is mainly due to the reduction of fruit weight in all the grapefruit varieties, since fruit load was not significantly affected by the water stress (Table 1). Although a decrease in tree yield was observed, a significant increase in water productivity was achieved during the experiment for the three out of four varieties, namely Marsh SRA 8, Shambar SRA 22 and cv. Frost Marsh (Table 1). In the case of the Ruby cultivar, the lower water productivity during DI T2 treatment is attributed to the significant decrease of both fruit yield and fruit weight. Increasing water productivity may be a means of achieving efficient and effective water use. In agriculture, the interest is to have higher yield with less water, due to water scarcity and the necessity to use water in the most optimum way during agricultural practices (Navarro *et al.* 2015).

Table 1 | Fruit yield parameters during the experiment period for the control (T1) and the deficit irrigation (T2) treatments

Cultivar	Treatments	Yield (Kg tree^{-1})	Fruit load	Fruit weight (g)	WP (Kg m^{-3})
cv. Marsh SRA 8	T1	163 d	427 bc	381.6 cd	24.03
	T2	141 bc	409 abc	343.7 c	32.63
	ANOVA	sd	nsd	nsd	
cv. Shambar SRA 22	T1	160 cd	468 c	341.9 cd	23.62
	T2	124 b	440 c	280.7 b	28.74
	ANOVA	sd	nsd	sd	
cv. Frost Marsh	T1	160 cd	371 ab	432.2 e	23.64
	T2	143 bcd	418 abc	344.4 c	32.41
	ANOVA	nsd	nsd	sd	
cv. Ruby	T1	187 e	470 c	398.9 de	27.52
	T2	82 a	357 a	229.0 a	18.78
	ANOVA	sd	sd	sd	

Values are the mean of 3 replicates per variety. Values followed by different letter within a column are significant different (sd) and values followed by the same letter are not significantly different (nsd) at the 0.05 level of probability, according to the Duncan's test.

Field studies related to the effect of deficit irrigation upon the tree yield are contradictory, since in the work of Pérez-Pérez *et al.* (2012), the tree yield of lemon trees was not affected when they received partial root deficit irrigation treatment compared to fully irrigated ones. Furthermore, in the work of Faber & Lovatt (2012) when navel oranges received limited water in the root zone, the fruit yield and fruit size were significantly reduced. In the work of Navarro *et al.* (2010), the application of DI upon grapefruit trees, during Phase I and via the use of drip irrigation did not significantly reduce the overall tree yield. But in mature oranges, when DI was applied every 3 days during Phase II and every 17 days, during Phase III, the citrus trees maintained their yield and a limited reduction in fruit size was observed (Hutton *et al.* 2007). Furthermore, in the work of Pérez-Pérez *et al.* (2014), the application of DI water stress in grapefruit trees, during Phase II, caused a significant decrease of 25% in fruit yield, which exceeded the saving of irrigation water, compared to optimum irrigated trees, indicating that the saving of irrigation water during this stage severely affected tree yield. The different effect of DI when it is applied during Phase II upon the fruit yield of grapefruit trees, compared to other citrus species, could be due to the different pattern of vegetative-reproductive resource distribution in response to drought stress (Pérez-Pérez *et al.* 2014). The witnessed differences among these studies and our results could be due to the volumes and time of applied water stress treatment and the examined citrus species, or even cultivars put under test.

Fruit physiological and quality affect by deficit irrigation

Many authors have stated that the application of DI treatments in citrus improves fruit quality via the increase or decrease of the most important quality parameters related with taste, flavor, nutritional value and osmotic adjustment (García-Tejero *et al.* 2010b; Navarro *et al.* 2010).

In this work, fruit quality was affected by the water stress treatment during this experiment. DI significantly reduced the equatorial diameter of all the tested grapefruit cultivars (Table 2); but at the same time, the market-desired commercial equatorial diameter of near 10 cm was retained for both strategies (T1 and T2). The reduction of equatorial diameter by water stress is in line with the lower fruit

Table 2 | Fruit physiological parameters during the experiment period for the control (T1) and the water stress (T2) treatments

Cultivar	Treatments	Eq. D (mm)	Fr. Sh.	CI	R. th. (mm)	Fl. Fir (Kg)	MI
cv. Marsh SRA 8	T1	100.5 d	0.812 a	0.477 bc	4.48 a	0.91 a	5.31 a
	T2	94.9 c	0.829 a	0.295 ab	6.76 b	1.25 c	6.11 d
	ANOVA	sd	nsd	nsd	sd	sd	sd
cv. Shambar SRA 22	T1	95.9 c	0.833 a	0.243 a	4.65 a	1.05 b	5.76 c
	T2	88.0 b	0.825 a	0.343 ab	6.14 b	1.31 c	5.41 ab
	ANOVA	sd	nsd	nsd	sd	sd	sd
cv. Frost Marsh	T1	105.7 e	0.819 a	0.303 ab	5.09 a	0.86 a	5.47 abc
	T2	96.0 c	0.820 a	0.539 cd	6.48 b	1.26 c	5.76 c
	ANOVA	sd	nsd	sd	sd	sd	nsd
cv. Ruby	T1	103.9 de	0.811 a	0.586 cd	4.66 a	1.10 b	5.68 bc
	T2	81.1 a	0.866 a	0.695 d	6.50 b	1.57 d	6.11 d
	ANOVA	sd	nsd	nsd	sd	sd	sd

External parameters analyzed are: equatorial diameter (Eq. D.), fruit shape (Fr. Sh.), color index (CI), rind thickness (R. Th.) flesh firmness (Fl. Fir.), maturation index (MI).
Values are the mean of three replicates per variety. Values followed by different letters within a column are significantly different (sd) and values followed by the same letter are not significantly different (nsd) at the 0.05 level of probability, according to Duncan's test.

weight. The fruit shape, which is the ratio between the equatorial and polar diameter, was not clearly affected by the DI treatment (T2), as all grapefruit varieties maintained the same aspect ratio (Figure 1). Regarding citrus plants, it is common to observe the phenomenon of smaller mean fruit diameter under low level of water irrigation regimes (Hutton *et al.* 2007; Melgar *et al.* 2010; Hutton & Loveys 2011). In grapefruit, severe water stress during

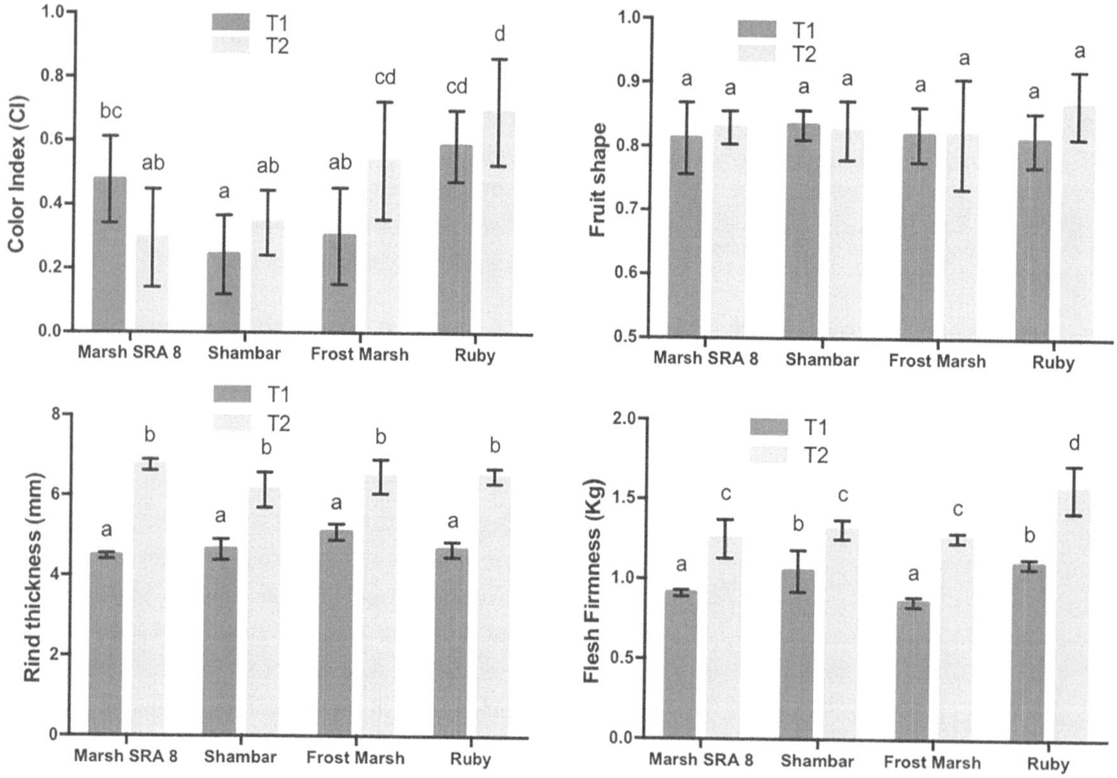

Figure 1 | External fruit Color Index (CI), Fruit shape, Rind thickness and Flesh firmness of all the studied grapefruit cultivars. Comparison between control (T1) and DI (T2) treatments. Values are the mean of three replicates, bars with the same letter are not significantly different at the 0.05 level of probability, according to Duncan's test.

Phase II significantly decreases the fruit size (Navarro *et al.* 2015). It is well documented that the reduction of fruit size in drought-stressed grapefruit trees is due to the fact that water-stressed fruits accumulate less dry matter than non-stressed citrus fruits due to active competition among the fruit tissue and other sink organs of the tree structure (Cohen & Goell 1988).

The external citrus fruit color index (CI) was also measured during this study. The obtained results did not show significant differences between irrigation treatments (T1 and T2), except for cv. Frost Marsh, which exhibited a slightly more yellowish color under DI treatment (Figure 1). This occurred due to the fact that citrus peel color changes fast during ripening period (Phase III), while the trees of both treatments received the main amount of water from precipitation. Compared to all tested grapefruit varieties, cv. Ruby exhibited, under DI and control regimes, increased citrus peel Color index. This is due to the fact that this variety is enlisted among the pink varieties. It is accepted that peel color of citrus fruit is under the tight control of environmental and nutritional factors, which control the alternation of peel chloroplasts into chromoplasts, a process that is orchestrated via the activity of various hormonal signaling molecules such as ethylene, gibberellic acid and abscisic acid (Rodrigo *et al.* 2013).

The application of DI strategy T2 significantly increased the rind thickness of all the examined cultivars (Table 2). Namely, under DI during Phase II, it increased the peel thickness of cv. Marsh SRA 8 by 51%, of cv. Shambar SRA 22 by 32%, of cv. Frost Marsh by 27% and of cv. Ruby by 39%, with respect to control. Thick peel is not a desirable feature for marketable products; some traders are interested in it for transportation and maintenance reasons. In the work of Pérez-Pérez *et al.* (2009), the implementation of DI during Phase III of orange cv. Lane Late fruit growth significantly reduced the peel thickness. Also, in the work of García-Tejero *et al.* (2010b), the peel thickness of orange fruit cv. Navellina remained unaffected when trees were under the effect of DI. While in the work of Mostert & Van Zyl (1998), peel thickness from oranges of cv. Valencia and grapefruit cv. Nelruby were significantly increased when the trees were under poor irrigation regimes.

Also, in this work, application of DI strategy during Phase II significantly increased the citrus flesh firmness of all the tested cultivars. In detail, the fruit firmness of cv. Marsh SRA 8 was increased by 37.8%, of cv. Shambar SRA 22 by 24.1%, of cv. Frost Marsh by 45.7% and of cv. Ruby by 42.7%, with respect to control. Among the tested varieties under DI regime, the flesh firmness of cv. Ruby was the one that exhibited the highest significant value. In the work of Kusakabe *et al.* (2016), the application of partial root drying in grapefruit trees during Phase II resulted in a decrease in citrus fruit firmness, but this parameter was not correlated to any other fruit or juice quality attribute.

The maturation index (MI) is considered the fine balance between the Total Soluble Solids (TSS) and the Titratable Acidity (TA) and is the most useful indicator of internal citrus fruit maturation stage (except lemons), since during ripening the TA decreases due to catabolism of citric acid and dilution phenomena and TSS (sugars) increase. In this work, under the effect of DI strategy, the MI of cv. Marsh SRA 8 and cv. Ruby exhibited an increase, while the MI of cv. Shambar SRA 22 was decreased and MI of cv. Frost Marsh was not significantly altered with respect to control. The fine balance of TSS and TA is a widely accepted method to determine citrus fruit maturation stage, and is also the most relative attribute that correlates fruit quality with consumer acceptance (Harding & Fisher 1945). In the work of Navarro *et al.* (2010), DI strategy during Phase III of citrus development increased the values of TSS and TA of fruit juices, with respect to the control fruits, and exhibited improved flavor and thus could contribute to the improvement of the commercial quality and consumer acceptance of the citrus fruit (Fellers *et al.* 1988). An increase of MI is also considered as a positive point for the commercial value of juice, and the benefit of an early harvest. Most of the previous studies on citrus showed an increase of TSS and TA values under DI strategies in citrus, although maturity index was not affected or decreased (Romero *et al.* 2006; Pérez-Pérez *et al.* 2009; Navarro *et al.* 2015). Exceptional scientific results highlight the fact that severe DI regimes implemented during citrus fruit maturity improve juice quality attributes without decreasing the maturity index (García-Tejero *et al.* 2010b). It is generally accepted that during citrus fruit ripening, external and internal ripening coincide and the peel and pulp behave in many cases as distinct organelles. Under DI strategies the delay or advance of citrus fruit maturation stage is tightly depended upon the stage of the fruit growth when the stress was applied, since it can alter the balance of TSS and TA (Pérez-Pérez *et al.* 2009; Aguado *et al.* 2012).

The T1 (full irrigation) treatment gave better results in terms of juice content. For the varieties Frost Marsh and Ruby, there is no significant difference between the two strategies, but there is a positive trend of higher juice content under the full irrigation strategy (Table 3; Figure 2). Under the effect of DI strategy, the fruit juice content of cv. Marsh SRA 8 and cv. Shambar SRA 22 were significantly reduced by 16.4% and 9.7%, respectively, with respect to control. Similar results were also reported by Navarro *et al.* (2015), where the application of DI during Phase I in grapefruit trees caused a significant reduction in fruit juice content. Also, in mandarin fruits, water stress during the Phase II of the citrus fruit developmental stage significantly decreased the juice percentage by 27% with respect to control fruits (Navarro *et al.* 2010). In the work of Pérez-Pérez *et al.* (2014), DI applied during Phase II upon grapefruit trees significantly decreased the fruit juice percentage, but considering the fact that no symptoms of water stress were observed upon the tree canopy during fruit harvest leads the authors to the fact that low juice content could be attributed to internal changes of fruit structure rather than dehydration processes.

The application of DI increased Total Soluble Solids (TSS) and Titratable Acidity (TA) – citric acid content, (Table 3; Figure 2). In detail, under DI strategy, TSS and TA of all studied varieties were increased. The highest % increase of TSS was witnessed in cv. Marsh SRA 8 and cv. Ruby, where the increase was 24.3% and 28.2%, respectively, while the highest percentage increase of TA was witnessed in cv. Shambar SRA 22 and cv. Ruby, by 22.3% and 19.2%, respectively (Table 3). It is well established that under water deficiency citrus fruit quality attributes are affected (Yakushiji *et al.* 1998) and the main effects are reflected in organoleptic parameters in juice, such as increase of Total Soluble Solids content and acidity (Chartzoulakis *et al.* 1999).

Former data indicate that the increase of TSS in juice from citrus fruits subjected to mild drought stress is due to the attempt to achieve osmotic adjustment (Yakushiji *et al.* 1998). Similar results were also reported in the work of Navarro *et al.* (2010) in clementine mandarins. The application of DI water stress, during Phase II, resulted in increased levels of TSS and TA, which were maintained for more than 3 months, resulting in passive dehydration of the juice sacs and providing significantly lower juice percentage, thus increasing the values of TSS and TA with respect to the control. In grapefruit, severe water stress during Phase II significantly increases the citrus fruit acidity (Navarro *et al.* 2015). It has been proposed that the increased levels of TA during DI regimes may be due to *de novo* biosynthesis of organic acids in an overall attempt to achieve osmotic adjustment in the fruit matrix (Navarro *et al.*

Table 3 | Fruit quality parameters during the experiment period for the control (T1) and the water stress (T2) treatments

Cultivar	Treatments	Juice (%)	TSS (°Brix)	TA (g L^{-1})	Vitamin C (mg/100 mL)	T. Phenols. (mg GAE/mL)
cv. Marsh SRA 8	T1	49.7 c	9.5 a	17.8 ab	39.0 a	1.06 a
	T2	41.5 ab	11.8 b	19.3 cd	42.6 bcd	1.24 b
	ANOVA	sd	sd	sd	sd	sd
cv. Shambar SRA 22	T1	43.0 b	10.3 a	18.0 abc	40.0 ab	1.10 a
	T2	38.8 a	11.9 b	22.0 e	45.5 d	1.35 c
	ANOVA	sd	sd	sd	sd	sd
cv. Frost Marsh	T1	41.3 ab	9.9 a	18.1 abc	39.7 ab	1.18 b
	T2	38.4 a	11.0 b	19.2 bcd	43.3 cd	1.24 b
	ANOVA	nsd	sd	nsd	sd	nsd
cv. Ruby	T1	43.4 b	9.6 a	16.9 a	41.6 abc	1.09 a
	T2	41.7 ab	12.3 b	20.1 d	49.1 e	1.40 c
	ANOVA	nsd	sd	sd	sd	sd

Parameters analyzed are: total soluble solids (TSS); titratable acidity (TA), ascorbic acid (vitamin C), phenols content (T. phenols).
Values are the mean of three replicates per variety. Values followed by different letter within a column are significantly different (sd) and values followed by the same letter are not significantly different (nsd) at the 0.05 level of probability, according to Duncan's test.

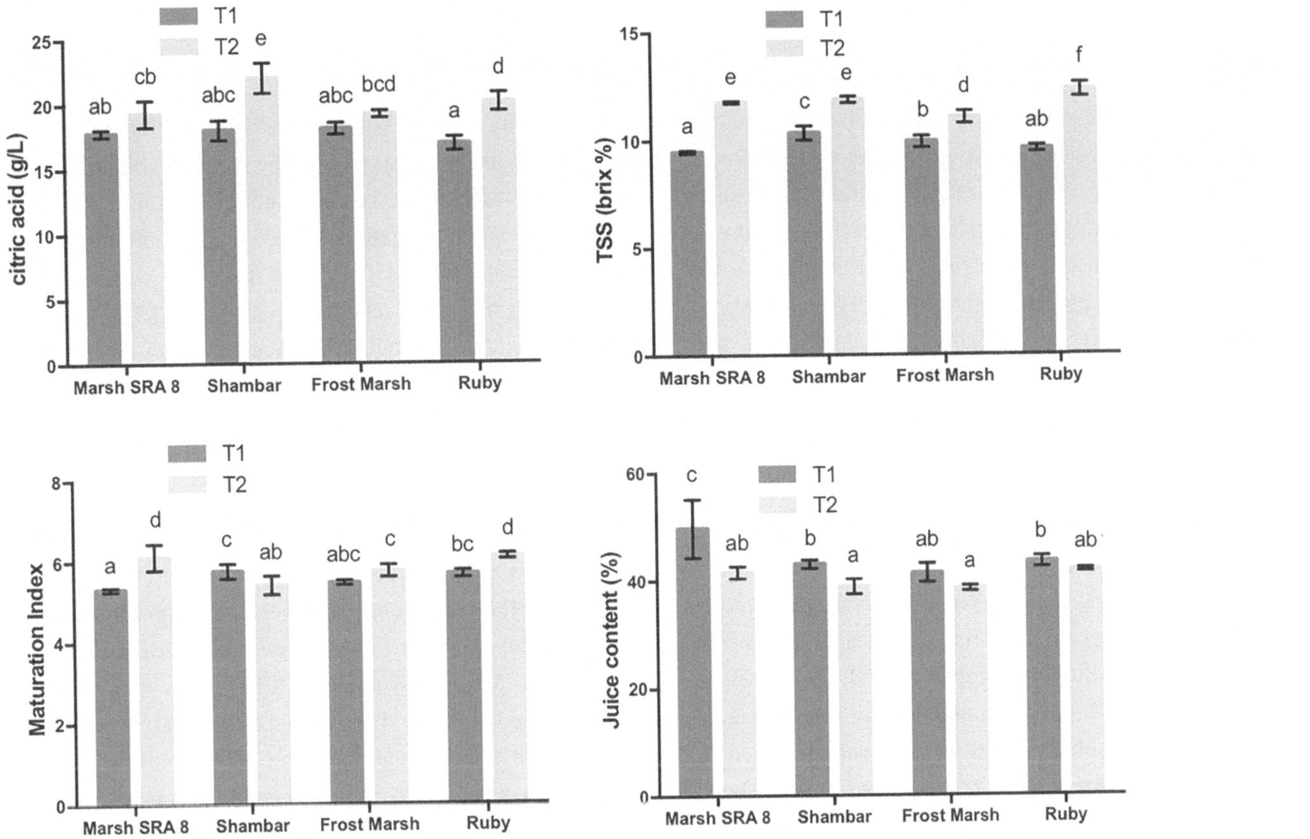

Figure 2 | Citric acid (Titratable acidity), Total Soluble Solids (TSS), Maturity Index and Juice content (%) of all the studied grapefruit cultivars. Comparison between control (T1) and DI (T2) treatment. Values are the mean of three replicates, bars with the same letter are not significantly different at the 0.05 level of probability, according to Duncan's test.

2010). Higher values of TSS and TA in juice from fruits of the DI treatment could have improved the flavor of the juice.

Vitamin C has a beneficial role in human nutrition and the juice of citrus fruits provides an important source of ascorbic acid, thus is a key parameter for the chemical analysis of juice. In the present work, the ascorbic acid content of the fruit juice of all the grapefruits varieties was also affected by the water stress treatment (Table 3; Figure 3). In detail, under the applied DI strategy, the vitamin C content (ascorbic acid) was increased by 9.1%, 13.8%, 9% and 17.8%

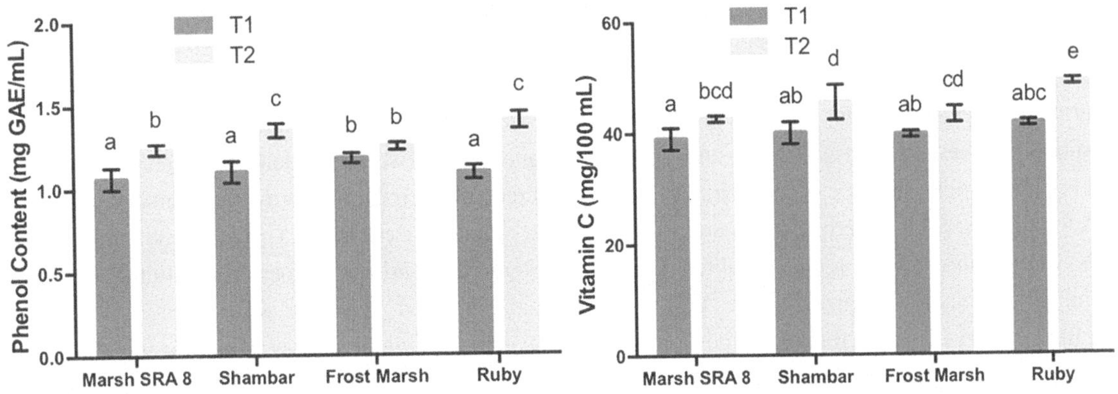

Figure 3 | Phenol Content and Vitamin C (ascorbic acid) of all the studied grapefruit cultivars. Comparison between control (T1) and DI (T2) treatment. Values are the mean of three replicates, bars with the same letter are not significantly different at the 0.05 level of probability, according to Duncan's test.

in the fruit juice of cv. Marsh SRA 8, cv. Shambar SRA 22, cv. Frost Marsh and cv. Ruby, respectively. This is in accordance with a study by Navarro *et al.* (2010), which stated that the application of DI water stress during Phase II of citrus fruit developmental stage increased the ascorbic acid content in mandarins by 15% with respect to the control juice. The witnessed effect could be related to overall attempt of the tree to combat the occurring water stress via the *de novo* synthesis of ascorbic acid (Navarro *et al.* 2010).

Several studies have highlighted the significance and importance of citrus fruit juices in dietary health (Roussos 2011), via their composition in bioactive compounds that exert beneficial effect to human health via disease prevention (Tripoli *et al.* 2007). Citrus fruits have been characterized as having a relatively high amount of phenolic content compared to other fruits and vegetables (Hunlun *et al.* 2017; Zhang *et al.* 2018). In our current work, the fruits under T2 water strategy had significantly increased levels of total phenolics compared with the control (Table 3; Figure 3). In detail, under DI the total phenolic content in the juice of cv. Marsh SRA 8 was increased by 16.3%, from cv. Shambar SRA 22 by 22.3%, from cv. Frost Marsh by 5.9% and from cv. Ruby by 28.4%. Similar results were reported by Navarro *et al.* (2010) in mandarin fruits, where the juice of clementine mandarin exhibited increased levels of total phenolics under the effect of DI treatments. The documented increase of total phenolics in fruit sunder the negative effect of drought stress could be attributed to the activation of cascades leading to new synthesis (Roby *et al.* 2004; Stefanelli *et al.* 2010) and the lower juice content (Navarro *et al.* 2010).

Principal component analysis of the overall data

Principal components analysis (PCA) was used to detect any patterns, groupings, and differences in full irrigated (T1) and deficit irrigation (T2) regimes, among the four (4) grapefruit cultivars (Marsh SRA 8, Shambar SRA 22, Frost Marsh and Ruby). PCA analysis was performed using 13 variables related to fruit physiology and quality (Figure 4, PCA A). The variance of data that was explained by the PCA model was 74.4%; in particular, PC1 explained 44.2% and PC2 30.2% of the variance. The Kaiser Meyer Olkin measure of sampling adequacy (KMO) on the traits score data was 0.768. The observed changes in PC1 scores coincided with an increasing irrigation regime, suggesting that PC1 may be associated with levels of irrigation, green color for full irrigation (T1) and red color for deficit irrigation (T2). This hypothesis was demonstrated by the fact that the juice content (JC) is closely related to PC1 and reduced sufficiently in fruits with deficit irrigation schedule (T2); on the other hand, titratable acidity (TA) was increased in the fruits under deficit irrigation schedule (red color – T2) indicating a possible way of inducing acids accumulation during fruit internal maturation (Pérez-Pérez *et al.* 2014). Moreover, other fruit quality traits such as total soluble solids (TSS), total phenols (P), and rind thickness (RT) have also participated in the structure of PC1, demonstrating a strong increase under deficit irrigation regimes. On the contrary, PC2 is more closely linked to cultivars separation mainly based on three (3) certain variables with high loadings in the PC model: these variables were Color Index (CI), maturation index (MI), and fruit shape (PED: polar to equatorial diameter ratio) (Figure 4, PCA A). In general, an obvious separation between treatments via PCA analysis on the score plot has been observed, where PC1 is mostly responsible for the observed effect (Figure 4, PCA B). A clear pattern separation between the deficit irrigation schedule (T2) upon fruits of cv. 'Ruby' over the other examined fruit cultivars was observed, while this pattern is also maintained under full irrigation, highlighting the qualitative superiority of cv. Ruby among the examined cultivars, independent of the irrigation regime (Figure 4, PCA B).

CONCLUSIONS

Lately grapefruit (*Citrus paradisi* Macf.) have been proposed as a very profitable alternative citrus cultivation compared to traditional ones that are widely cultivated. Grapefruit are characterized by increased irrigation demands for the production of quality fresh fruits. This fact drives most farmers to seek methods and implement agricultural practices that limit the applied irrigation water. Thus, there is a specific need to investigate and evaluate the impact of DI farming practices upon the most important fruit nutritional quality attributes that provide added value to the consumed fruits.

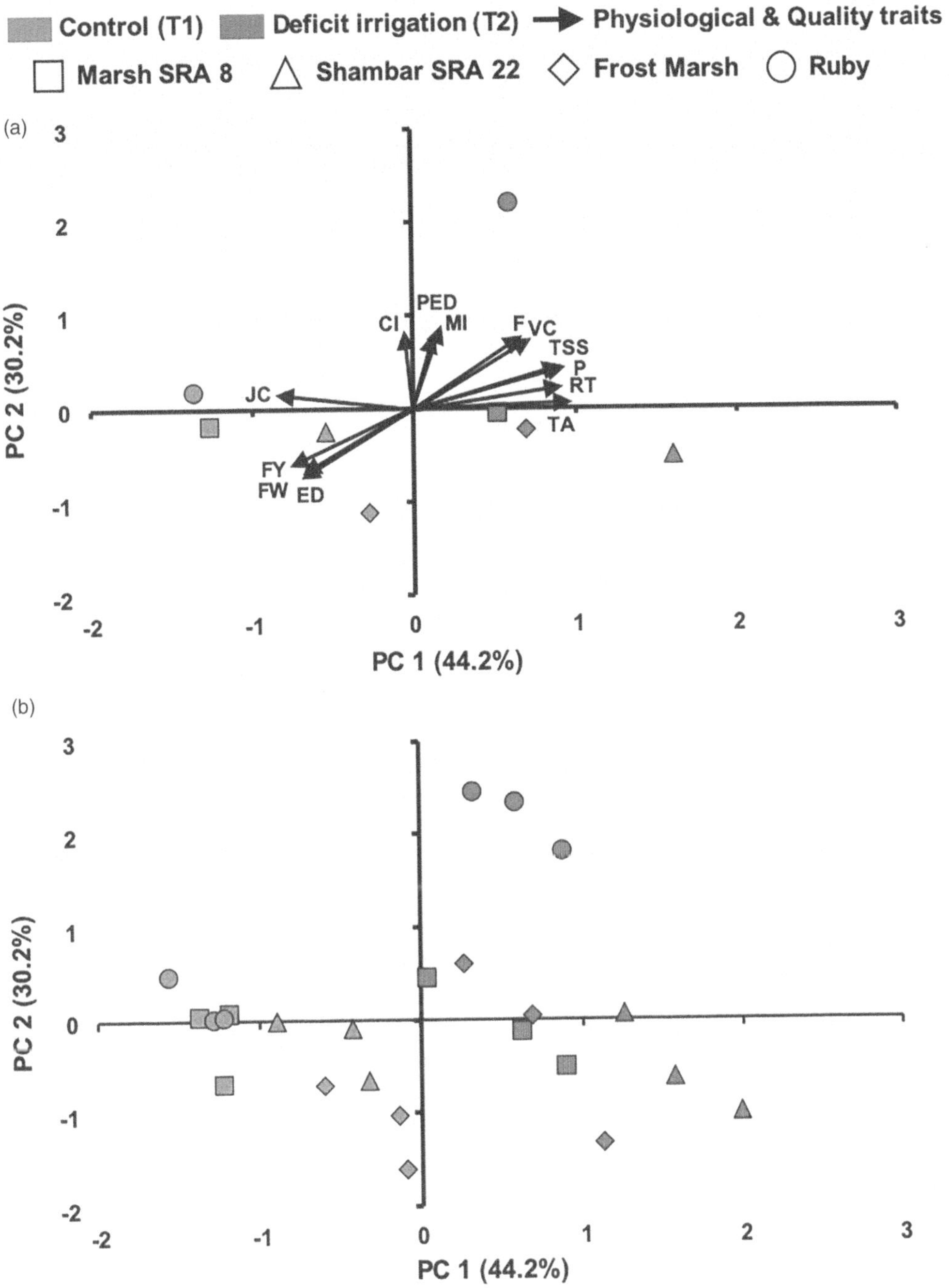

Figure 4 | Principal Component Analysis (PCA) of physiological and quality traits of four grapefruit cultivars and irrigation stress (T2) using a biplot (A) and a score plot (B) data presentation. Control – T1: green, deficit irrigation treatment T2: red, cv.'Marsh SRA 8' square, cv.'Shambar SRA 22' triangle, and cv.'Frost Marsh' rhombus, cv.'Ruby' cycle. Physiological and quality traits are represented with a black arrow: Fruit yield (FY); fruit weight (FW); equatorial diameter (ED); fruit shape (PED); color index (CI); rind thickness (RT); flesh firmness (F), maturation index (MI); juice content (JC); total soluble solids (TSS); titratable acidity (TA); ascorbic acid content (VC); total phenols content (P).

The implementation of DI during Phase II of grapefruit development strongly affects qualitative parameters in several ways. The fact that under DI cv. Ruby increased significant quality criteria like TSS, TA, ascorbic acid content and total phenolics, with respect to cv. Frost Marsh and cv. Marsh SRA 8, clearly highlights the fact that the response is tightly dependent on the cultivar genotype. Further work should be done, via the use of -omic technologies (genomic, metabolic and transcriptomic analysis) so as to investigate and pinpoint the genes and cascades that participate in the enhancement of quality parameters in grapefruit during drought stress.

ACKNOWLEDGEMENTS

The authors thank Dr Michalis Michailidis for his assistance on statistical data management and Mrs Georgia Leontopoulou for her assistance with field and laboratory measurements.

DATA AVAILABILITY STATEMENT

All relevant data are included in the paper or its Supplementary Information.

REFERENCES

Aguado, A., Frías, J., García-Tejero, I., Romero, F., Muriel, J. & Capote, N. 2012 Towards the improvement of fruit-quality parameters in citrus under deficit irrigation strategies. *International Scholarly Research Notices* **12**, 1–9.

Allen, R. G., Pereira, L. S., Raesk, D. & Smith, M. 1998 Crop evapotranspiration – guidelines for computing crop water requirements. FAO Irrigation and Drainage Paper No. 56, Rome, Italy, 300pp.

Ballester, C., Castel, J., Intrigliolo, D. S. & Castel, J. 2013 Response of Navel Lane Late citrus trees to regulated deficit irrigation: yield components and fruit composition. *Irrigation Science* **31** (3), 333–341.

BEWARE project 2005 *Final Report on 'Optimum Innovative Practices on Water Use for A Sustainable Management of Water Resources'*. Institute of Olive Tree & Subtropical Plants of Chania. Crete Innovative Region (CRINNO).

Chartzoulakis, K., Michelakis, N. & Stefanoudaki, E. 1999 Water use, growth, yield and fruit quality of 'Bonanza' oranges under different soil water regimes. *Advances in Horticultural Science* **13**, 6–11.

Cohen, A. & Goell, A. 1988 Fruit growth and dry matter accumulation in grapefruit during periods of water withholding and after reirrigation. *Functional Plant Biology* **15** (5), 633–639.

Davis, J. C. & Sampson, R. J. 1986 *Statistics and Data Analysis in Geology*. Wiley, New York, NY.

Doorenbos, J. & Kassam, A. 1979 Yield response to water. *Irrigation and Drainage Paper* **33**, 257.

Faber, B. & Lovatt, C. 2012 Effects of applying less water by partial root zone drying versus conventional irrigation on navel orange yield. *VII International Symposium on Irrigation of Horticultural Crops* **1038**, 523–530.

FAO 2019 *FAOSTAT: Statistical Reports of Plant Production*. http://www.fao.org/faostat/en/#data/QC (accessed 20 May 2020).

Fellers, P., Carter, R. & De Jager, G. 1988 Influence of the ratio of degrees Brix to percent acid on consumer acceptance of processed modified grapefruit juice. *Journal of Food Science* **53** (2), 513–515.

Fereres, E. & Soriano, M. A. 2007 Deficit irrigation for reducing agricultural water use. *Journal of Experimental Botany* **58** (2), 147–159.

García-Tejero, I., Jiménez-Bocanegra, J., Martínez, G., Romero, R., Durán-Zuazo, V. & Muriel-Fernández, J. 2010a Positive impact of regulated deficit irrigation on yield and fruit quality in a commercial citrus orchard [*Citrus sinensis* (L.) Osbeck, cv. salustiano]. *Agricultural Water Management* **97** (5), 614–622.

García-Tejero, I., Romero-Vicente, R., Jiménez-Bocanegra, J., Martínez-García, G., Durán-Zuazo, V. & Muriel-Fernández, J. 2010b Response of citrus trees to deficit irrigation during different phenological periods in relation to yield, fruit quality, and water productivity. *Agricultural Water Management* **97** (5), 689–699.

Gasque, M., Martí, P., Granero, B. & González-Altozano, P. 2016 Effects of long-term summer deficit irrigation on 'Navelina' citrus trees. *Agricultural Water Management* **169**, 140–147.

Ginestar, C. & Castel, J. 1996 Responses of young clementine citrus trees to water stress during different phenological periods. *Journal of Horticultural Science* **71** (4), 551–559.

Harding, P. L. & Fisher, D. 1945 *Seasonal Changes in Florida Grapefruit*.

Hunlun, C., De Beer, D., Sigge, G. O. & Van Wyk, J. 2017 Characterisation of the flavonoid composition and total antioxidant capacity of juice from different citrus varieties from the Western Cape region. *Journal of Food Composition and Analysis* **62**, 115–125.

Hutton, R. & Loveys, B. 2011 A partial root zone drying irrigation strategy for citrus – effects on water use efficiency and fruit characteristics. *Agricultural Water Management* **98** (10), 1485–1496.

Hutton, R., Landsberg, J. & Sutton, B. 2007 Timing irrigation to suit citrus phenology: a means of reducing water use without compromising fruit yield and quality? *Australian Journal of Experimental Agriculture* **47** (1), 71–80.

Ihaka, R. & Gentleman, R. 1996 R: a language for data analysis and graphics. *Journal of Computational and Graphical Statistics* **5** (3), 299–314.

Kimball, D. A. 2012 *Citrus Processing: Quality Control and Technology*. Springer Science & Business Media, New York, NY.

Kourgialas, N. N. & Karatzas, G. P. 2015 Groundwater contamination risk assessment in Crete, Greece, using numerical tools within a GIS framework. *Hydrological Sciences Journal* **60** (1), 111–132.

Kourgialas, N. N., Koubouris, G. C. & Dokou, Z. 2019 Optimal irrigation planning for addressing current or future water scarcity in Mediterranean tree crops. *Science of The Total Environment* **654**, 616–632.

Kusakabe, A., Contreras-Barragan, B., Simpson, C., Enciso, J., Nelson, S. & Melgar, J. 2016 Application of partial rootzone drying to improve irrigation water use efficiency in grapefruit trees. *Agricultural Water Management* **178**, 66–75.

Melgar, J., Dunlop, J. & Syvertsen, J. 2010 Growth and physiological responses of the citrus rootstock *Swingle citrumelo* seedlings to partial rootzone drying and deficit irrigation. *Journal of Agricultural Science* **148** (5), 593–602.

Mostert, P. & Van Zyl, J. 1998 Gains in citrus fruit quality through regulated irrigation. In *XXV International Horticultural Congress, Part 6: Culture Techniques with Special Emphasis on Environmental Implications 516*.

Navarro, J. M., Pérez-Pérez, J. G., Romero, P. & Botía, P. 2010 Analysis of the changes in quality in mandarin fruit, produced by deficit irrigation treatments. *Food Chemistry* **119** (4), 1591–1596.

Navarro, J. M., Botía, P. & Pérez-Pérez, J. G. 2015 Influence of deficit irrigation timing on the fruit quality of grapefruit (Citrus paradisi Mac.). *Food Chemistry* **175**, 329–336.

Pereira, L. S., Cordery, I. & Iacovides, I. 2012 Improved indicators of water use performance and productivity for sustainable water conservation and saving. *Agricultural Water Management* **108**, 39–51.

Pérez-Pérez, J. G., Romero, P., Navarro, J. M. & Botía, P. 2008 Response of sweet orange cv 'Lane late' to deficit irrigation in two rootstocks. I: water relations, leaf gas exchange and vegetative growth. *Irrigation Science* **26** (5), 415–425.

Pérez-Pérez, J., Robles, J., Tovar, J. & Botía, P. 2009 Response to drought and salt stress of lemon 'Fino 49'under field conditions: water relations, osmotic adjustment and gas exchange. *Scientia Horticulturae* **122** (1), 83–90.

Pérez-Pérez, J., Dodd, I. & Botía, P. 2012 Partial rootzone drying increases water-use efficiency of lemon Fino 49 trees independently of root-to-shoot ABA signalling. *Functional Plant Biology* **39** (5), 366–378.

Pérez-Pérez, J., Robles, J. & Botía, P. 2014 Effects of deficit irrigation in different fruit growth stages on 'Star Ruby'grapefruit trees in semi-arid conditions. *Agricultural Water Management* **133**, 44–54.

Roby, G., Harbertson, J. F., Adams, D. A. & Matthews, M. A. 2004 Berry size and vine water deficits as factors in winegrape composition: anthocyanins and tannins. *Australian Journal of Grape and Wine Research* **10** (2), 100–107.

Rodrigo, M. J., Alquézar, B., Alós, E., Lado, J. & Zacarías, L. 2013 Biochemical bases and molecular regulation of pigmentation in the peel of Citrus fruit. *Scientia Horticulturae* **163**, 46–62.

Romero, P., Navarro, J., Pérez-Pérez, J., García-Sánchez, F., Gómez-Gómez, A., Porras, I., Martinez, V. & Botía, P. 2006 Deficit irrigation and rootstock: their effects on water relations, vegetative development, yield, fruit quality and mineral nutrition of Clemenules mandarin. *Tree Physiology* **26** (12), 1537–1548.

Roussos, P. A. 2011 Phytochemicals and antioxidant capacity of orange (*Citrus sinensis* (l.) Osbeck cv. Salustiana) juice produced under organic and integrated farming system in Greece. *Scientia Horticulturae* **129** (2), 253–258.

Ruiz-Sanchez, M. C., Domingo, R. & Castel, J. R. 2010 Deficit irrigation in fruit trees and vines in Spain. *Spanish Journal of Agricultural Research* **8**, 5–20.

Singh, R., van Dam, J. C. & Feddes, R. A. 2006 Water productivity analysis of irrigated crops in Sirsa district, India. *Agricultural Water Management* **82** (3), 253–278.

Singleton, V. L., Orthofer, R. & Lamuela-Raventós, R. M. 1999 [14] Analysis of total phenols and other oxidation substrates and antioxidants by means of folin-ciocalteu reagent. *Methods in Enzymology* **299**, 152–178.

Stefanelli, D., Goodwin, I. & Jones, R. 2010 Minimal nitrogen and water use in horticulture: effects on quality and content of selected nutrients. *Food Research International* **43** (7), 1833–1843.

Tripoli, E., La Guardia, M., Giammanco, S., Di Majo, D. & Giammanco, M. 2007 Citrus flavonoids: molecular structure, biological activity and nutritional properties: a review. *Food Chemistry* **104** (2), 466–479.

Yakushiji, H., Morinaga, K. & Nonami, H. 1998 Sugar accumulation and partitioning in Satsuma mandarin tree tissues and fruit in response to drought stress. *Journal of the American Society for Horticultural Science* **123** (4), 719–726.

Zhang, H., Yang, Y.-f. & Zhou, Z.-q. 2018 Phenolic and flavonoid contents of mandarin (Citrus reticulata Blanco) fruit tissues and their antioxidant capacity as evaluated by DPPH and ABTS methods. *Journal of Integrative Agriculture* **17** (1), 256–263.

First received 31 January 2021; accepted in revised form 4 April 2021. Available online 15 April 2021

Impacts of irrigation and nitrate fertilization scenarios on groundwater resources quantity and quality of the Almyros Basin, Greece

Aikaterini Lyra, Athanasios Loukas and Pantelis Sidiropoulos

ABSTRACT

Irrigation and nitrate fertilization scenarios were combined and simulated for crop water irrigation demands and nitrogen applications based on experiments on crop cultivation practices. Two irrigation practices (deficit irrigation and rainfed conditions) were applied to the main crop types of the Almyros Basin, a coastal basin located in Thessaly, Greece. The Almyros groundwater system suffers from progressive water balance deficit, nitrate contamination and seawater intrusion due to groundwater abstractions for agricultural irrigation to cover crop water demands in the dry season. The impacts of the irrigation and nitrate fertilization scenarios on groundwater resources quantity and quality were simulated using an Integrated Modelling System consisting of models of surface hydrology (UTHBAL), groundwater hydrology (MODFLOW), crop growth/nitrate leaching (REPIC), contaminant transport (MT3DMS), and seawater intrusion (SEAWAT), for the historical period of 1991–2018. The results of the scenarios were evaluated with the indicators of Crop Water Productivity (CWP) for crop yields and irrigation water, Partial Factor Productivity (PFP) for Nitrogen Use Efficiency, and Economic Water Productivity (EWP) for the gross profits of the irrigation water.

Key words | fertilization, groundwater, integrated water resources modelling, irrigation, scenarios, water balance, water quality

Aikaterini Lyra (corresponding author)
Pantelis Sidiropoulos
Department of Civil Engineering,
School of Engineering,
University of Thessaly,
38334 Volos,
Greece
E-mail: *klyra@uth.gr*

Athanasios Loukas
Department of Rural and Surveying Engineering,
Aristotle University of Thessaloniki,
54124 Thessaloniki,
Greece

HIGHLIGHTS

- The study evaluates the impacts of irrigation and nitrate fertilization scenarios on groundwater resources.
- The Integrated Modelling System of surface hydrology (UTHBAL), groundwater hydrology (MODFLOW), crop growth/nitrate leaching (REPIC), contaminant transport (MT3DMS), and seawater intrusion (SEAWAT) has been enriched to simulate agronomic practices.
- CWP, EWP and NUE indices evaluate the performance of the scenarios.

doi: 10.2166/ws.2021.097

GRAPHICAL ABSTRACT

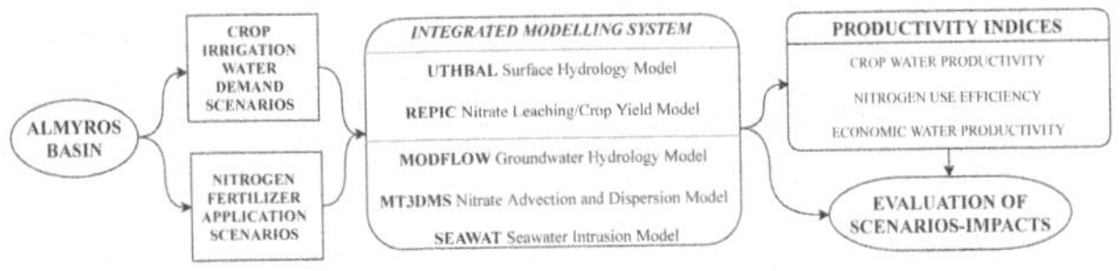

INTRODUCTION

The impacts of irrigation and fertilization practices on groundwater resources in semi-arid regions is one of the most important water supply problems of irrigated agriculture and water quality deterioration of aquifer systems (Fernández García *et al.* 2020; Lyra *et al.* 2021). In coastal water systems, over-pumping for irrigation water supply, the absence of significant surface water reservoir infrastructure (and for this reason the use of surface water), and excessive nitrogen applications for crop growth improvement, have a number of implications for groundwater, notably the water table descent, nitrate contamination and seawater intrusion (Lyra *et al.* 2021).

The continuing exacerbation of water deficit and groundwater quality poses a threat on the sustainability of irrigated agriculture in coastal regions. The water quality limit for drinking water regarding nitrates is 50 mg/L (NCESD 2018) while for chlorides it is 250 mg/L. According to the European Directives, chloride concentrations in groundwater systems can be between 24 mg/L and 12,300 mg/L depending on aquifer characteristics (EU 2010).

An integrated approach to the sustainability of crop yields and to the control of nitrate fertilization, groundwater abstractions and seawater intrusion is essential for protecting the viability of natural resources. The key points of such an approach are the understanding and assessment of water budget fluxes and safe water quality supply for irrigation and drinking water consumption (Loukas 2010).

In regions where the largest amount of water resources is consumed for agricultural production and the water balance is negative, the estimation of productivity efficiency is of paramount importance. The trend of marketable crop production is directly linked with irrigation water application, while the latter affects the availability of nitrogen in soils and nitrogen leaching to groundwater (Ullah *et al.* 2019). The indicators of Crop Water Productivity (CWP), Economic Water Productivity (EWP) and Nitrogen Use Efficiency (NUE) are often utilized for various crop irrigation and fertilization practices to define the advantages and the disadvantages of practices (Fixen *et al.* 2015; Fernández García *et al.* 2020). CWP is defined as the marketable crop yield in relation to crop evapotranspiration or irrigation water applied, and measures how much 1 m^3 of water actually contributes to crop growth in 1 ha (Ullah *et al.* 2019; Fernández *et al.* 2020). EWP is the water quantity transformed into marketable crop yield in 1ha (Fernández *et al.* 2020). The combined evaluation of EWP and CWP provides a detailed background for the assessment of impacts of irrigation practices (Fernández *et al.* 2020). Moreover, NUE may be expressed in various terms, either for short-term or long-term estimations, and is affected by irrigation water. For this reason, NUE depends on CWP and affects the marketable EWP (Ullah *et al.* 2019; Fernández *et al.* 2020). Thus, the evaluation of the impacts of irrigation and fertilization practices must be targeted at improving the agronomic, environmental and economic status and profits resulting from agricultural water supply (Ullah *et al.* 2019; Fernández *et al.* 2020; Lyra *et al.* 2021).

The scope of this study is to evaluate the impacts of irrigation and fertilization scenarios on groundwater resources, as well as their benefits compared to the current agricultural and water management practices in the Almyros Basin, a coastal agricultural basin in Thessaly, Greece. The evaluation has been performed with the application of an Integrated Water Resources Modelling System developed by the authors (Lyra *et al.* 2021) and applied to the study area for the simulation of the quantity and quality status of groundwater and crop yields for various scenarios of irrigation and nitrogen fertilization. The results of each alternative scenario were quantified using the indicators of CWP, NUE and EWP for the simulation period of 1991 to 2018.

METHODS

Study area and database

The main land use of the Almyros Basin is for agricultural fields, with a small percentage of the basin being used for urban, sub-urban and non-cultivated areas. The hydrography of the basin is characterized by ephemeral streams and absence of significant surface water storage works. The cultivated agricultural area covers most of the area of the Almyros aquifer. Areas at higher elevations are covered by natural vegetation (e.g. forests and bushland). The areas cultivated per crop were provided by the Greek Payment Authority of the Common Agricultural Policy (CAP) and grouped into nine main crop types. Cereals, olives groves, vineyards and wheat cultivars are the crops primarily cultivated in the study area for the period of 1991 to 2018. The main land uses and crops of the Almyros Basin for the year of 2010 are shown in Figure 1. The cultivated agricultural lands are totally irrigated, due to the absence of surface water storage works, with water obtained from groundwater. Groundwater also supplies water for urban use and it is taken into account in this study. However, the urban water supply volume is not as high as the water volume abstracted for irrigation. The annual agricultural water supply accounts for 29.7 hm^3, on average, while the mean annual urban water supply volume reaches 1.8 hm^3 (Lyra *et al.* 2021).

Agricultural water supply scenarios

The agricultural water demand is defined by the crop water demands for crops cultivated in the Almyros Basin. More than two thirds of the area is permanent cropland, irrigated by groundwater abstractions with private-owned wells. The number of irrigation water wells has been estimated at 2,044, indicating intensive agricultural activities (Lyra *et al.* 2021). The crop water requirements were estimated in

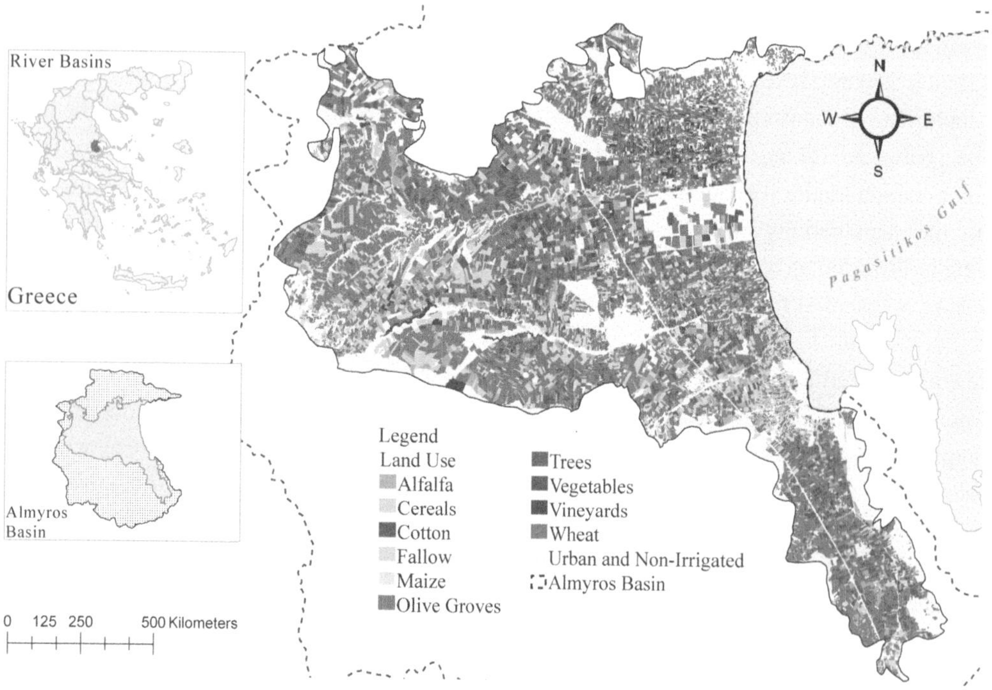

Figure 1 | Almyros Basin, its aquifer and the spatial distribution pattern of the main crops and land uses for 2010.

a previous study (Lyra *et al.* 2021) for each main crop category, based on relevant data provided by Greek Payment Authority of the CAP. The total water losses of irrigation distribution systems in the study area were estimated to 41% of the crop water requirements (Sidiropoulos *et al.* 2016).

Agronomic scenarios were combined and simulated for irrigation crop water demands and fertilizer applications based on experiments in crop cultivation practices. The main crops irrigated in the study area are alfalfa, cereals, cotton, maize, olive groves, trees, vegetables, vineyards, and wheat. Two irrigation practices were considered: (i) deficit irrigation for all crop types; and (ii) rainfed cultivation for cereals, olive groves and wheat, and deficit irrigation for alfalfa, cotton, maize, trees, vegetables, and vineyards. Additionally, a reduced nitrogen fertilization practice was implemented, based on recent field experiments in Greece and the Mediterranean basin. The irrigation and fertilization practices were coupled and produced four scenarios, as presented in the next paragraphs. The simulation results of the four scenarios were compared with the baseline scenario representing the current irrigation and nitrogen fertilization practices.

In order to define the irrigation and nitrogen fertilization scenarios, a detailed literature review of field experiments performed in Greece and in the Mediterranean area for the main crops of the study area was made. The scenarios developed are based on the reduction of irrigation water applied and nitrate fertilization. The results of field experimental studies used in this paper are outlined below.

According to field experimental research, alfalfa is characterized by high irrigation water needs and low nutrient requirements. Alfalfa presents high productivity (yield) and marketable prices as compared to other crop types and water productivity increase when applying irrigation water equal to 90% of crop water requirements, and nitrogen equal to 28 kg/ha (Djaman *et al.* 2020). The cereal cultivated in the study area is mainly barley. Barley has medium/low water demands for crop yield in irrigated conditions. The water productivity and crop yield of barley increases when applying water equal to 70 and 80% of crop water requirements (Pardo *et al.* 2020). The nitrogen fertilization promotes the crop yield of barley and increases the water productivity, with nitrogen application of 60 kg/ha and deficit irrigation practices of 75% of the crop's water requirement (Barati *et al.* 2015). Cotton is characterized by medium/high irrigation requirements, but the reduction of irrigation water to 80% of crop water requirements increases crop yield (Polychronides *et al.* 1998). The nitrogen fertilization of cotton at 109.5 kg/ha increases crop yield and maximizes the economic benefits (Polychronaki *et al.* 2012). Maize is an irrigated cultivar and the deficit irrigation at 90% of crop water demand indicated that there is no reduction in the crop yield (Fernández García *et al.* 2020). The nitrogen requirements of maize define its crop yield and the experimental low rate application of 100 kg/ha optimised dry matter yield in deficit irrigation according to Archontoulis and associates (Archontoulis *et al.* 2010). Orchards were explicitly studied for various agricultural practices, for irrigation and fertilization applications, and their impacts on water quality in the LIFE AgroClimaWater Project. Olive groves can be cultivated rainfed, but this leads to reduction of crop yield. The total irrigation amount applied in olives groves, when reduced at 89% of total crop water requirements combined with a nitrogen application rate of 91.2 kg/ha does not decrease crop yield (Arampatzis *et al.* 2018). Fruit and other orchards studied in the AgroClimaWater Project averaged values of deficit irrigation practices at 76% of crop water requirements and 73% of nitrogen fertilization for different types of orchards used in the scenarios of the study (AgroClimaWater 2016). For vegetable crop types, different crops were studied according to the literature. The irrigation water productivity of garlic and onions increases when the irrigation water is reduced at 70 and 90% of irrigation requirements, respectively (Fernández García *et al.* 2020). Tomatoes irrigated at a deficit of 75% of crop irrigation demands and fertilized at 120 kg/ha reach optimal crop yield scores (Kuscu *et al.* 2014). For these reasons an averaged amount of 80% reduction in nitrogen fertilization was considered for the vegetable crop type. Vineyards are susceptible to agronomic alterations and therefore the deficit irrigation at 60% of crop water demands combined with nitrogen application of 60 kg/ha, lower than the typical fertilization practices in the region, was considered in the study (Thomidis *et al.* 2016). Similarly to barley and cereals, the crop yield of wheat is increased, and also crop water productivity, at deficit irrigation between 60 and 85% (Ullah *et al.* 2019). Nitrogen fertilization application has been reported to reach optimum wheat yield at 102 kg/ha (Polychronaki *et al.* 2012).

The developed scenarios included combinations of irrigation and nitrogen fertilization practices. Four scenarios were developed: scenario S1 combines baseline current fertilization and deficit irrigation conditions for all crops, while scenario S2 combines baseline current fertilization and rainfed cultivation for cereals, olive grove and wheat, and deficit irrigation for alfalfa, cotton, maize, trees, vegetables, and vineyards. Scenario S3 combines reduced fertilization and deficit irrigation for all crops, while scenario S4 combines reduced fertilization and rainfed cultivation for cereals, olive grove and wheat, and deficit irrigation for alfalfa, cotton, maize, trees, vegetables, and vineyards. The consideration of a scenario with no fertilization is not applicable because the crop yields of the crop types are greatly affected by nitrogen availability (Ullah *et al.* 2019). The four developed scenarios were compared with a baseline scenario (S0) representing the current irrigation and nitrogen fertilization practices. The irrigation water and nitrogen fertilization applications are presented in Table 1.

Drinking water supply

The population of the Municipal Districts and Local Communities of Almyros were considered using the censuses of 1991, 2001 and 2011 (permanent population) by the Hellenic Statistical Authority. Drinking water supplies from groundwater covered the water needs of 21,583 people in 1991, 22,191 in 2001 and 19,405 in 2011. The urban water supply in the area of Almyros is supplied by a total of 28 wells. The possible increase in summer urban water demands is also covered by groundwater abstractions of neighbouring water systems that do not interact with the aquifer system of Almyros, and by water tanks that reserve surplus groundwater abstractions in times of low consumption. The average daily consumption per inhabitant was set at 170 L/inhabitant/day and the water losses of the distribution network were estimated to 40%. The monthly distribution of groundwater abstractions for the drinking water demands was estimated, during the calibration and validation procedures of the Integrated Modelling System, at 13% in October, 10% in November, 8% in December, 6% in January, 5% in February, 5% in March, 6% in April, 6% in May, 8% in June, 10% in July, 12% in August and 12% in September. These figures agree well with the results of a study in the nearby city of Volos (Mylopoulos *et al.* 2001). The total drinking water demands were estimated to 1.87 hm^3 for the period 1991–2000, 1.93 hm^3 for the period 2001–2010, and 1.69 hm^3 for the period 2011–2018.

Integrated modelling system

The water quantity and quality of the Almyros Basin for the alternative scenarios was simulated using an Integrated Water Resources Modelling System developed by the authors (Lyra *et al.* 2021). The modelling system consists of coupled and interrelated models of surface hydrology (UTHBAL) (Loukas *et al.* 2007), groundwater hydrology (MODFLOW) (Harbaugh *et al.* 2000), crop production/nitrate leaching (REPIC) (Lyra *et al.* 2021), nitrate transport (MT3DMS) (Zheng & Wang 1999), and seawater intrusion

Table 1 | Maximum annual irrigation water and nitrogen fertilization applications

Crop Type	Baseline Irrigation [mm]	Deficit Irrigation [mm]	Deficit Irrigation/ Rainfed [mm]	Baseline N_{Fer} [Kg/ha]	Reduced N_{Fer} [Kg/ha]
Alfalfa	893	804	804	30	28
Cereals	336	269	0	100	60
Cotton	409	327	327	140	110
Maize	389	350	350	325	100
Olives	515	407	0	125	100
Trees	515	390	390	175	128
Vegetables	271	244	244	150	120
Vineyards	297	178	178	125	60
Wheat	336	269	0	160	102

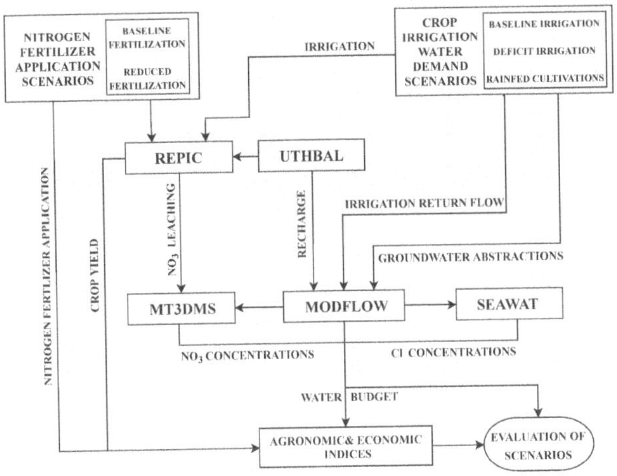

Figure 2 | Flowchart of the Integrated Modelling System for the impact assessment of the irrigation and nitrogen fertilization practices scenarios.

(SEAWAT) (Guo & Langevin 2002) (Figure 2). The modelling system was successfully calibrated and validated in Almyros Basin simulating with improved accuracy the groundwater flow, the nitrate contamination, the seawater intrusion and the crop yields (Lyra *et al.* 2021). In this study, the modelling system was used for the simulation of the irrigation and nitrogen fertilization scenarios for the period 1991–2018. The goodness-of-fit statistics for the calibration and validation of the models of the Integrated Water Resources Modelling System can be found in the Supplementary Material of the paper.

Indicators

The CWP and EWP indices regarding the irrigation water applied and the grain yield were implemented for the evaluation of the simulation results for the Almyros Basin. The CWP for irrigation water is calculated by Equation (1) (Fernández *et al.* 2020):

$$Crop\ Water\ Productivity\ (CWP_I) = \frac{Crop\ Yield\ kg\ ha^{-1}}{Irrigation\ Water\ m^3ha^{-1}} \tag{1}$$

For the calculation of the EWP indicator for irrigation water, public annual data for the crop price index from 2000–2018 with base year 2015, were obtained by the Hellenic Statistical Authority for the main crops cultivated in the Almyros Basin. The EWP is calculated by Equation (2) (Fernández *et al.* 2020):

$$Economic\ Water\ Productivity\ (EWP_I) = \frac{Profit\ ha^{-1}}{Irrigation\ Water\ m^3ha^{-1}} \tag{2}$$

The NUE is estimated in terms of Partial Factor Productivity (PFP) appropriate for long-term estimations of crop yields and nitrogen applications, by Equation (3) (Fixen *et al.* 2015):

$$Partial\ Factor\ Productivity\ (PFP) = \frac{Crop\ Yield\ (kg\ ha^{-1})}{Nitrogen\ Applied\ (kg\ ha^{-1})} \tag{3}$$

RESULTS AND DISCUSSION

UTHBAL model

The results obtained from the application of the UTHBAL model in the Almyros Basin in the earlier study (Lyra *et al.* 2021) for the period 1991–2018 were used in the simulation of the four developed scenarios. The surface hydrology model was not modified for the irrigation and fertilization scenarios, as they do not affect the physical hydrological processes in the study area.

MODFLOW model

The groundwater fluxes were simulated for the period of 1991 to 2018 for the scenarios of deficit irrigation (S1, S3) and deficit and rainfed conditions (S2, S4). The simulated groundwater heads for the four scenarios (i.e. S1, S2, S3, S4) were compared with the respective simulated results for the baseline scenario (S0) for the end of the simulation period (2018) in Figure 3. The baseline scenario (S0-continuous line) and the scenarios of deficit irrigation (S1, S3- dashed line) present a closely related variation of the water table especially in the lowland areas. The deficit irrigation and rainfed cultivation results (scenario S2, S4- dashed dot line) show slightly different hydraulic heads in the central southern parts of the aquifer where the hydraulic conductivity is low. The hydraulic heads for both irrigation scenarios (i.e. S1-S3 and S2-S4) at the end

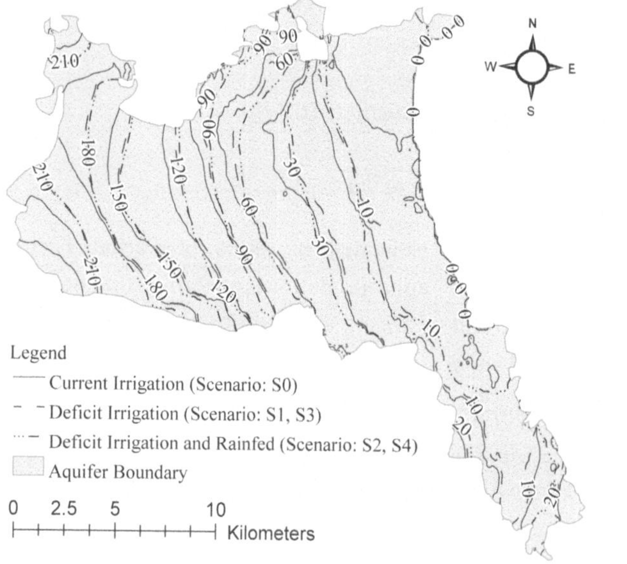

Figure 3 | Comparison of equipotential curves of groundwater heads for the developed irrigation scenarios (September 2018).

of the simulation period are higher than the baseline scenario (S0) (Lyra *et al.* 2021), indicating an increase in the elevation of the groundwater water table.

REPIC model

The crop yields of the main crops of the Almyros Basin were simulated by the REPIC model, for irrigation and fertilization scenarios for the period 1991–2018. The results of crop yields for the simulated land use periods for the baseline scenario, S0, and the alternative scenarios S1, S2, S3 and S4 are presented in Figure 4. Although, deficit irrigation and rainfed cultivation were considered, there is no significant change of the yields of the other crops for the alternative scenarios. The results indicate that the yield of vineyards, olive groves and cotton are affected by the reduction of irrigation water and nitrogen fertilization.

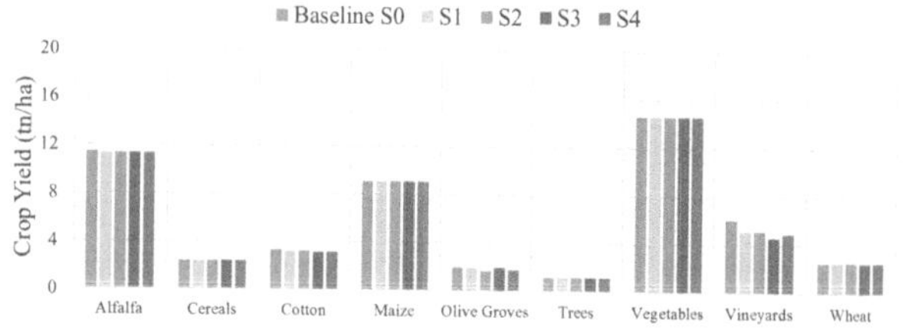

Figure 4 | Crop yields for all simulated scenarios in the period 1991–2018.

MT3DMS model

The nitrate fluxes were simulated by the MT3DMS model for the period of 1991 to 2018. The simulation was performed for all four scenarios (S1, S2, S3, S4). The differences between the simulation results for the baseline scenario, S0, and scenarios S3 and S4 are shown in Figure 5(a) and 5(b), respectively. The simulation results show that the nitrate contamination in the aquifer is mostly affected by the irrigation practices and irrigation return flow, due to the semi-arid climate and the low recharge, especially, in the southern parts of the aquifer. The nitrate concentrations of scenario S4 are spatially lower than the results for scenario S3. Although in scenario S4, a large part of the aquifer is rainfed and nitrate fertilization is reduced, the results show no reduction of the nitrate concentrations in the central coastal part, mainly due to the low recharge from precipitation and thus reduced dissolution of the contaminant. Particularly, in the low recharged and low hydraulic conductivity areas, the presence of irrigated and fertilized maize cultivars, adjacent to rainfed crops, causes the local persistence and assimilation of nitrates.

SEAWAT model

The chloride fluxes were simulated using the SEAWAT model for the period of 1991 to 2018 and the simulation was performed for all four scenarios (S1, S1, S3, S4). The differences of the simulation results of scenario S3 and scenario S4 from the baseline scenario S0 are depicted in Figure 5(c) and 5(d), respectively. The reduced irrigation scenarios have a positive influence on the delay of the evolution of seawater intrusion, as compared to the irrigation practices of the baseline scenario S0. The maximum chloride concentration reduction in all scenarios reaches a value of 18,000 mg/L in the north coastline area that is salinized in the baseline scenario S0 (Lyra *et al.* 2021). Although, scenarios S1 and S3 show a spatial average reduction of 36 mg/L, and scenarios S2 and S4 a spatial average of 26 mg/L, the latter have a larger extent of positive impact on the aquifer's water quality (Figure 5(d)).

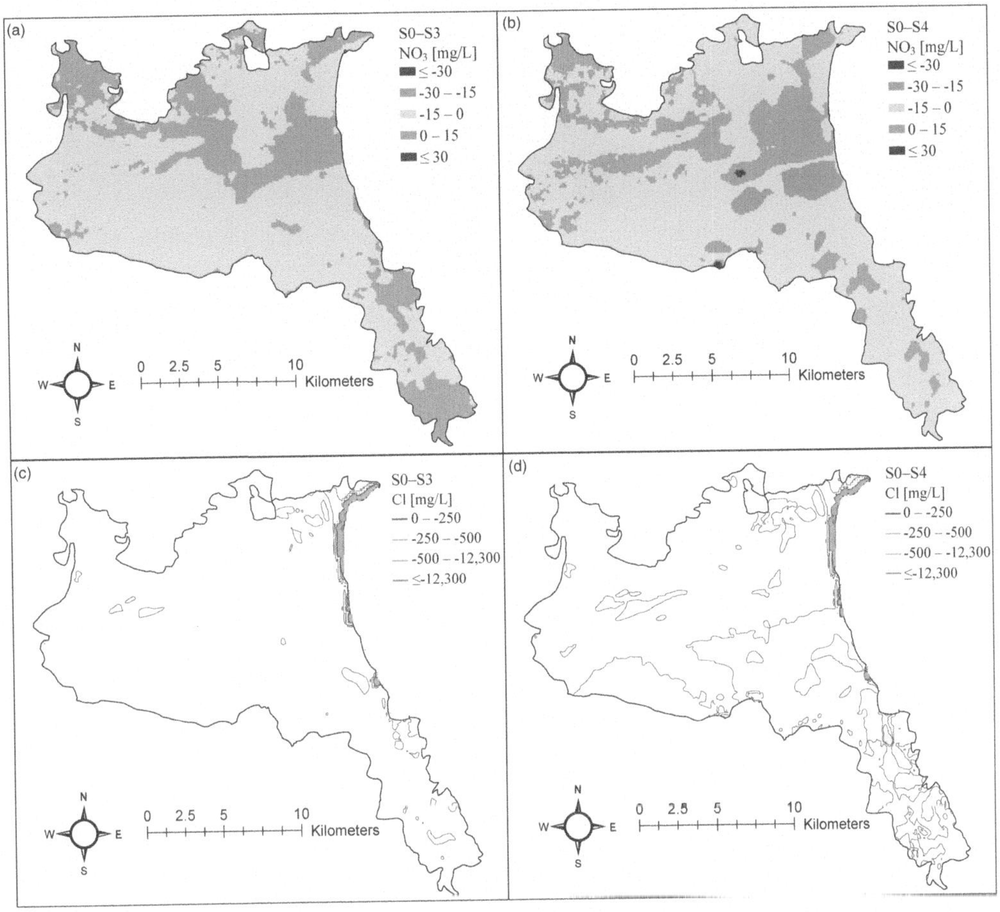

Figure 5 | Difference in the simulated nitrate concentration between scenarios, (a) S0-S3 and (b) S0-S4 and simulated chloride concentration between scenarios (c) S0-S3 and (d) S0-S4 (September 2018).

Water budget

The average monthly (Figure 6(a)) and annual (Figure 6(b)) water budget of the Almyros Basin aquifer for the deficit irrigation scenarios (S1, S3), and the deficit irrigation and rainfed cultivation scenarios (S2, S4), were calculated and compared to the current historical groundwater regime of the Almyros aquifer for the simulation years 1991–2018. Due to the large extent of the spatial distribution of cereals and wheat, the deficit irrigation scenarios (S1, S3) show a monthly higher positive inflow, which is attributed to the irrigation return flows, as compared to the scenarios of deficit irrigation and rainfed cultivation (S2, S4), when groundwater is only recharged by the infiltration of precipitation. The annual water deficit of the baseline scenario S0 is 12.02 hm^3 (Lyra *et al.* 2021), minimized in the deficit irrigation case to 2.3 hm^3 (scenarios S1 and S3), and in the deficit irrigation and rainfed cultivation case (scenarios S2 and S3) to 2.8 hm^3. The annual water deficit for the deficit irrigation and rainfed cultivation case is larger than the water deficit for the deficit irrigation case alone, due to the limited natural recharge and absence of irrigation return flow for a large extent of the aquifer area. Nevertheless, both practices have a significant positive impact on the water balance.

Crop water productivity

The CWP indicator was estimated using Equation (1) for every main crop of the Almyros Basin, based on the results of the REPIC model for crop yields, and on the irrigation water demands for the baseline scenario S0 and the scenarios S1, S2, S3 and S4. The comparative percentages of

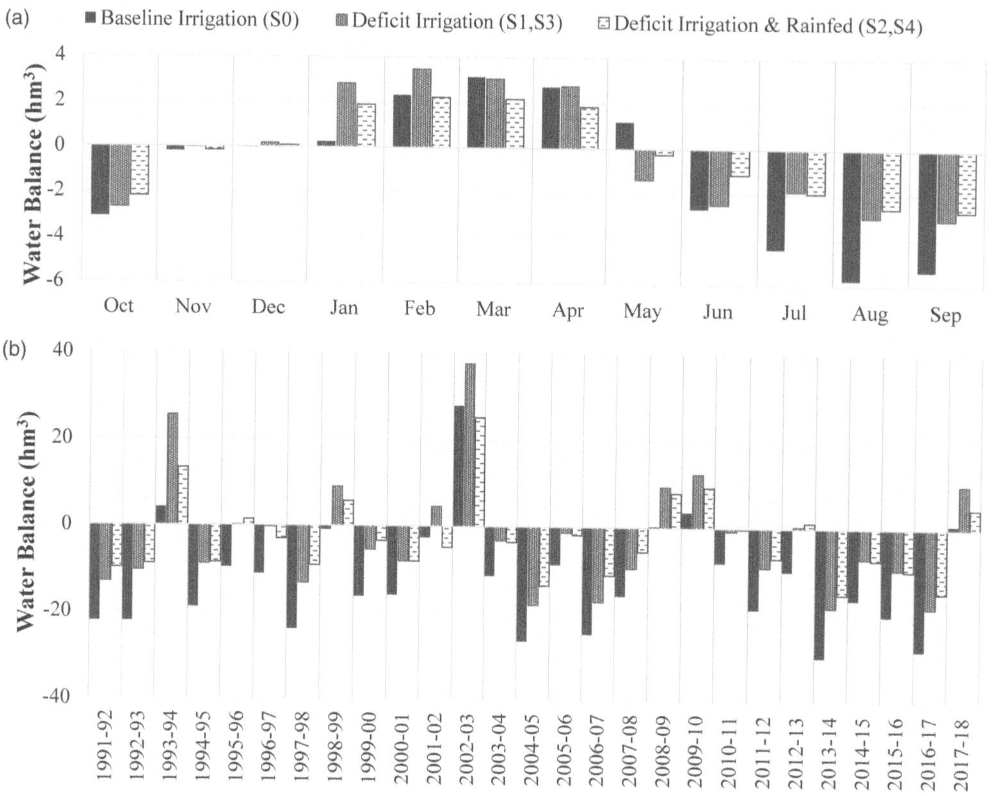

Figure 6 | Average monthly (a) and annual (b) water budget of the Almyros aquifer for the baseline scenario S0 and the scenarios of deficit irrigation (S1, S3) and deficit-rainfed irrigation (S2, S4).

CWP_I for the scenarios from the baseline scenario S0 are presented in Figure 7(a). The CWP_I is increased for the deficit irrigation conditions (i.e. S1 and S3) for all crops in the Almyros Basin.

Partial factor productivity

The PFP indicator was estimated using Equation (3) for the evaluation of the efficiency of nitrogen use on crops on the simulated scenarios for baseline fertilization (S1, S2) and reduced fertilization (S3, S4). In the reduced fertilization practices in all crop types, except for alfalfa, PFP is increased, especially for cereals, maize, vineyards, and wheat. The comparative percentages of the variations of the indicator from the baseline scenario S0 scores are presented in Figure 7(b).

Economic water productivity

The EWP indicator was estimated using Equation (2) for the irrigation water applied. The calculation was performed based on the averaged grain yield prices with base year 2015. The scores for EWP_I are presented in Table 2. The EWP_I of the crop pattern is increased in scenarios S1 and S3 while their scores indicate that the fertilization practices have minimum total impact on crop yields.

CONCLUSIONS

Irrigation water and nitrogen fertilization applications were simulated for the assessment of their impact on the groundwater quantity and quality of the Almyros Basin. The Integrated Water Resources Modelling System applied has been expanded to include the simulation of agronomic and irrigation practice alterations. All the scenarios (S1, S2, S3, S4) have a positive impact on the water balance, and the water deficit is reduced significantly by 9.7 hm^3 in scenarios S1 and S3, and 9.2 hm^3 in scenarios S2 and S4. The larger water deficit in scenarios S2 and S4 is caused by the absence of irrigation return flows in the rainfed

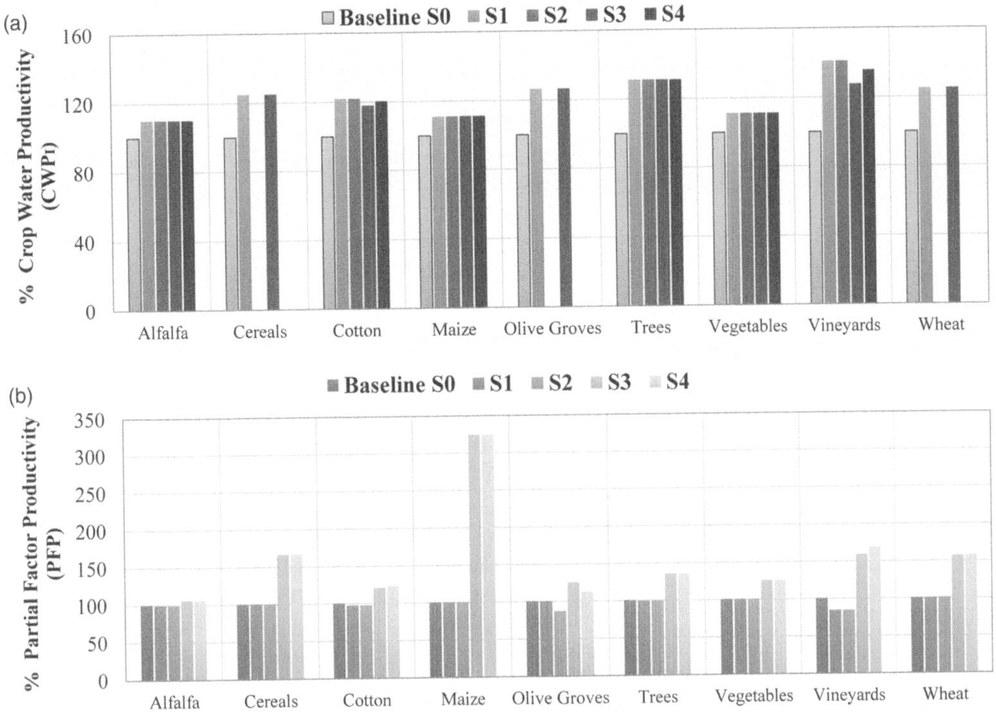

Figure 7 | Percent (%) Crop Water Productivity (a) and Partial Factor Productivity (b) for the simulated scenarios for every crop type.

Table 2 | Average of economic water productivity (EWP_I) of irrigation water applied per crop type and scenario with base year 2015

€/m³	Alfalfa	Cereals	Cotton	Maize	Olive groves	Trees	Vegetables	Vineyards	Wheat	Crop Pattern
S0	0.62	0.46	0.91	1.15	1.85	0.58	17.28	2.70	0.49	1.08
S1	0.68	0.58	1.11	1.28	2.34	0.75	19.16	3.95	0.61	1.29
S2	0.68	–	1.11	1.28	–	0.75	19.16	3.69	–	0.82[a]
S3	0.68	0.58	1.08	1.28	2.34	0.75	19.16	3.95	0.61	1.28
S4	0.68	–	1.09	1.28	–	0.75	19.16	3.69	–	0.82[a]

[a]Irrigated crops.

areas. Therefore, the latency of seawater intrusion is succeeded in the northern coastline area under all scenarios, and in scenarios S2 and S4 the aquifer's chloride concentrations are more reduced than in scenarios S1 and S3. The scenario practices show positive impacts on nitrate contamination, especially in scenarios S2 and S4. The nitrate concentrations are reduced in large extent and in all cases, but the irrigated and fertilized maize adjacent to areas of rainfed crops cause local assimilation of nitrates and have a negative impact on the aquifer's water quality. The crop yield results show that vineyards are susceptible to deficit irrigation, while cereals, maize and wheat are resilient to changes in agronomic practices. All crops had an improved CWP_I in all scenarios, as compared to baseline scenario S0, reaching maximum scores in baseline fertilization practices. Consequently, the EWP_I of the crop pattern reached an optimum score of 1.29 €/m^3 in scenario S1. However, the crops' PFP was increased when reducing nitrate fertilization in scenarios S3 and S4, especially for cereals, maize, vineyards, and wheat. In total, the results indicate that the higher positive impacts on groundwater quantity and quality are achieved in scenarios S2 and S4. The further improvement of groundwater quantity and quality may be achieved by changing the crops' cultivation patterns. The simulation of various crop pattern scenarios may reveal more about the optimal implementation of agricultural management strategies.

FUNDING

This research is co-financed by Greece and the European Union (European Social Fund–ESF) through the Operational Programme 'Human Resources Development, Education and Lifelong Learning' in the context of the project 'Strengthening Human Resources Research Potential via Doctorate Research' (MIS-5000432), implemented by the State Scholarships Foundation (IKY).

DATA AVAILABILITY STATEMENT

All relevant data are included in the paper or its Supplementary Information.

REFERENCES

AgroClimaWater 2016 *Deliverable C2: Report on Assessment of Water Efficiency of the Participant F.ORs Before LIFEAgroClimaWater.* AgroClimaWater LCG.

Arampatzis, G., Pisinaras, V., Pavlidou, E., Kourgialas, N., Psarras, G., Koubouris, G., Hatzigiannakis, E., Sismani, G. & Panagopoulos, A. 2018 Improving water use efficiency at farmers' organization level. In: *6th International Conference on Olive Tree and Olive Products*, Seville, Spain.

Archontoulis, S., Struik, P., Vos, J. & Danalatos, N. 2010 Maize and sorghum biomass and protein accumulation under adequate and limited supply of water and nitrogen in Greece. In: *Proceedings of Agro2010/XIth ESA Congress*, 29 August-3 September 2010, Montpellier, France, pp. 569–570.

Barati, V., Ghadiri, H., Zand-Parsa, S. & Karimian, N. J. 2015 Nitrogen and water use efficiencies and yield response of barley cultivars under different irrigation and nitrogen regimes in a semi-arid Mediterranean climate. *Archives of Agronomy and Soil Science* **61** (1), 15–32.

Djaman, K., Smeal, D., Koudahe, K. & Allen, S. 2020 Hay yield and water use efficiency of alfalfa under different irrigation and fungicide regimes in a semiarid climate. *Water* **12** (6), 1721.

EU 2010 *Report From the Commission in Accordance with Article 3.7 of the Groundwater Directive 2006/118/EC on the Establishment of Groundwater Threshold Values.* European Commission, Brussels.

Fernández, J. E., Alcon, F., Diaz-Espejo, A., Hernandez-Santana, V. & Cuevas, M. V. 2020 Water use indicators and economic analysis for on-farm irrigation decision: a case study of a super high density olive tree orchard. *Agricultural Water Management* **237**, 106074.

Fernández García, I., Lecina, S., Ruiz-Sánchez, M. C., Vera, J., Conejero, W., Conesa, M. R., Domínguez, A., Pardo, J. J., Léllis, B. C. & Montesinos, P. J. W. 2020 Trends and challenges in irrigation scheduling in the semi-arid area of Spain. *Water* **12** (3), 785.

Fixen, P., Brentrup, F., Bruulsema, T., Garcia, F., Norton, R. & Zingore, S. 2015 Nutrient/fertilizer use efficiency: measurement, current situation and trends. In: Drechsel, P., Heffer, P., Magen, H., Mikkelsen, R. & Wichelns, D. (eds), *Managing Water and Fertilizer for Sustainable Agricultural Intensification*, International Fertilizer Industry Association (IFA), International Water Management Institute (IWMI), International Plant Nutrition Institute (IPNI), and International Potash Institute (IPI), Paris, France.

Guo, W. & Langevin, C. D. 2002 User's Guide to SEAWAT: A Computer Program for Simulation of Three-Dimensional Variable-Density Ground-Water Flow. *Techniques of Water-Resources Investigation, 06-A7* U.S. Geological Survey, Tallahassee, Florida.

Harbaugh, A. W., Banta, E. R., Hill, M. C. & McDonald, M. G. 2000 MODFLOW-2000, The U.S. Geological Survey Modular Ground-Water Model - User Guide to Modularization Concepts and the Ground-Water Flow Process, *Report 2000-92*. Geological Survey, Reston, Virginia. doi: https://doi.org/10.3133/ofr200092.

Kuscu, H., Turhan, A., Ozmen, N., Aydinol, P. & Demir, A. O. 2014 Optimizing levels of water and nitrogen applied through drip irrigation for yield, quality, and water productivity of processing tomato (Lycopersicon esculentum mill.). *Horticulture, Environment, and Biotechnology* **55** (2), 103–114.

Loukas, A. 2010 Surface water quantity and quality assessment in Pinios River, Thessaly, Greece. *Desalination* **250** (1), 266–273.

Loukas, A., Mylopoulos, N. & Vasiliades, L. 2007 A modeling system for the evaluation of water resources management strategies in Thessaly, Greece. *Water Resources Management* **21** (10), 1673–1702.

Lyra, A., Loukas, A., Sidiropoulos, P., Tziatzios, G. & Mylopoulos, N. 2021 An integrated modeling system for the evaluation of water resources in coastal agricultural watersheds: application in Almyros Basin, Thessaly, Greece. *Water* **13** (3), 268.

Mylopoulos, N., Mentes, A. & Karamanlidou, M. 2001 Sustainable water resources management in the wider hydrological basin of Volos. In: C. A. Brebbia, P. Anagnostopoulos, K. L. Katsifarakis & A. H.-D. Cheng (eds), *Water Resources Management*. WIT Press, Southampton, pp. 99–108. doi: https://doi.org/10.2495/WRM010091.

NCESD 2018 *Greece, State of the Environment Report, Summary.* National Center of Environment and Sustainable Development, Athens. ISBN: 978-960-99033-3-2.

Pardo, J. J., Martínez-Romero, A., Léllis, B. C., Tarjuelo, J. M. & Domínguez, A. 2020 Effect of the optimized regulated deficit irrigation methodology on water use in barley under semiarid conditions. *Agricultural Water Management* **228**, 105925.

Polychronaki, E., Douma, C., Giourga, C. & Loumou, A. J. P. 2012 Assessing nitrogen fertilization strategies in winter wheat and cotton crops in northern Greece. *Pedosphere* **22** (5), 689–697.

Polychronides, M., Galanopoulou-Sendouca, S., Aggelides, S. & Danalatos, N. G. 1998 The Effect of Irrigation and Fertilization on Cotton Growth and Development under Greek Conditions. In: *Proceedings of the World Cotton Research Conference*, Athens, Greece, pp. 405–407.

Sidiropoulos, P., Loukas, A. & Georgiadou, I. 2016 Response of a degraded coastal aquifer to water resources management. *European Water* **55**, 67–77.

Thomidis, T., Zioziou, E., Koundouras, S., Karagiannidis, C., Navrozidis, I. & Nikolaou, N. 2016 Effects of nitrogen and irrigation on the quality of grapes and the susceptibility to Botrytis bunch rot. *Scientia Horticulturae* **212**, 60–68.

Ullah, H., Santiago-Arenas, R., Ferdous, Z., Attia, A. & Datta, A. 2019 Chapter Two – Improving water use efficiency, nitrogen use efficiency, and radiation use efficiency in field crops under drought stress: A review. In: Sparks, D. L. (ed.), *Advances in Agronomy*. Academic Press, Elsevier Inc., pp. 109–157.

Zheng, C. & Wang, P. P. 1999 *MT3DMS: A Modular Three-Dimensional Multi-Species Transport Model for Simulation of Advection, Dispersion and Chemical Reactions of Contaminants in Groundwater Systems, Documentation and User's Guide.* Report Contract Report SERDP-99-1, U.S. Army Engineer Research and Development Center, Vicksburg, MS.

First received 31 January 2021; accepted in revised form 23 March 2021. Available online 2 April 2021

Estimating the future hydric needs of Baja California, Mexico. Assessment of scenarios to stop being a region with water scarcity

A. Cortés-Ruiz and I. Azuz-Adeath

ABSTRACT

This paper shows the actual conditions of freshwater availability in Baja California (BC), Mexico. It aims to estimate the water needs by 2030, and propose scenarios to move out of the scarce water region classification defined by international organizations. A population of 4.1 million people was defined for year 2030 as a target to provide at least 1,000 m^3 of water *per capita*. As agriculture is the main water consumer in the region, empirical decomposition and optimization methods were used to define the trend line of the principal crops production and to establish the optimum conditions for planted surface reduction and water gain. The results show that by 2030, BC will need a total of 4,105 hm^3 of water to be classified as a non-water scarcity region; in 2018, BC had 3,045 hm^3 of renewable water per year, therefore 1,060 m^3 will be needed. The best option in economic terms to attain this goal was the reduction of croplands in Mexicali with a cost of around 82 million US dollars. Although this option is the best quantitatively, the political and social implications of it are enormous; however, the correct management of the resource in critical conditions will require difficult decisions.

Key words | agriculture water consumption, Baja California, forecast, Mexico, water scarcity

A. Cortés-Ruiz
I. Azuz-Adeath (corresponding author)
School of Engineering,
CETYS Universidad,
Camino Microondas Trinidad Km. 1 S/N, Las Palmas 3ra Sección, Moderna Oeste, c.p. 22860, Ensenada, BC,
México
E-mail: *isaac.azuz@cetys.mx*

HIGHLIGHTS

- In Baja California, 87% of the water supply is used in agricultural activities.
- In 2017, Baja California had a water availability of 849 cubic meters per capita per year, this classifies it as a region with water scarcity.
- By 2030, Baja California will need to obtain by actually unknown sources 1,060 cubic hectometres of water to be re-classified as a non-water scarcity region.
- In the short-term, the installed capacity of wastewater treatment and desalination plants in Baja California will not be enough to satisfy the future water demand.
- Technically speaking, the only feasible option to eliminate the water deficit is the reduction of 35,558 ha of croplands, but this action would have a huge political and social impact.

doi: 10.2166/ws.2020.198

GRAPHICAL ABSTRACT

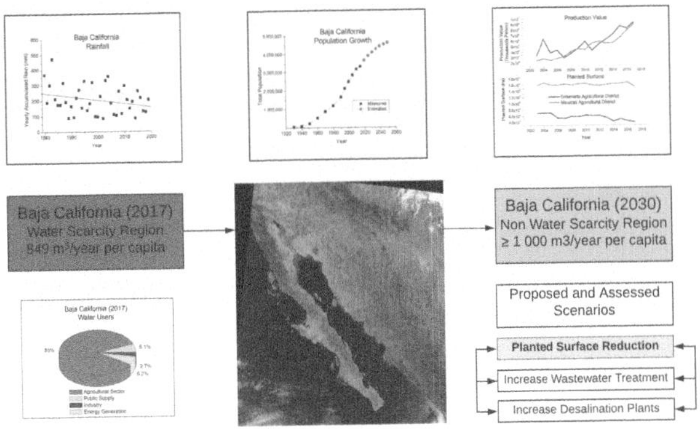

INTRODUCTION

The global population in 2019 was 7.7 billion people; in the next 30 years it is expected to increase by 2 billion, reaching 9.7 billion in 2050 (United Nations 2019). Water use has been incrementing by about 1% yearly since the 1980s, mainly due to population growth, socio-economic development and evolving consumption (UN-Water 2019). The increase of population and water consumption (overall and *per capita*) has led to the rise of water shortages. In the present, there are numerous countries that are entering a period of severe water scarcity. Water scarcity, either physical or economic, is a multi-faceted problem that not only involves water availability but also water quality and access to safe drinking water (du Plessis 2017). Also, there are several countries that have been experiencing high levels of 'water stress', defined as the proportion of water withdrawal in relation to the available water resources.

The world's average water stress level is 13% and it affects at least 2 billion people around the world (FAO & UN-Water 2018); there has been a 55% drop in globally available freshwater *per capita* since 1960; by 2030, global demand of water is expected to grow by 50%; water scarcity currently affects more than 40% of the global population; and worldwide, the total cost of water insecurity to the global economy is estimated at 500 billion US dollars (UNU-INWEH 2017). In short, there is no denying the occurrence of a global water crisis and, if not taken in to account accordingly, it will hasten the arrival of the impending 'day zero' for many countries.

At a national level, Mexico does not seem to be experiencing extreme levels of water scarcity and stress. In 2014, FAO & UN-Water (2018) determined, based on the country's total renewable freshwater, total water withdrawal and the environmental flow requirements, that Mexico has 26% of water stress. Even so, there is an alarming water crisis in Mexico which is manifesting at state and local level. The national water balance does not represent the real panorama nor the water issues that some of the federal entities have been experiencing (Otazo-Sánchez *et al.* 2020). In 2017, 11 states in the country registered less than 1,700 cubic meters *per capita* (m^3 $capita^{-1}$) and 9 out of those states registered less than 1,000 m^3 $capita^{-1}$ (EAM 2018). In accordance to the indicators of the FAO (2014), this means that from the 32 states in Mexico, 9 are identified as suffering from water scarcity and 2 from water stress (Figure 1).

In Mexico, all the matters related to water resources regulation and management belong mostly to the federal administration and are directed by the National Water Commission (CONAGUA, by its Spanish acronym), through 13 Basin Organisms (*Organismos de Cuenca* in Spanish) whose scope is the management of 13 hydrological-administrative regions (RHA in Spanish) which include all 32 states of the country (AAM 2018).

Mexico has an average of 451,585 cubic hectometres per year (hm^3 $year^{-1}$) of renewable water, 67% of this water is located in the South and Southeast regions of the country, while the remaining 33% is in the north and northeast

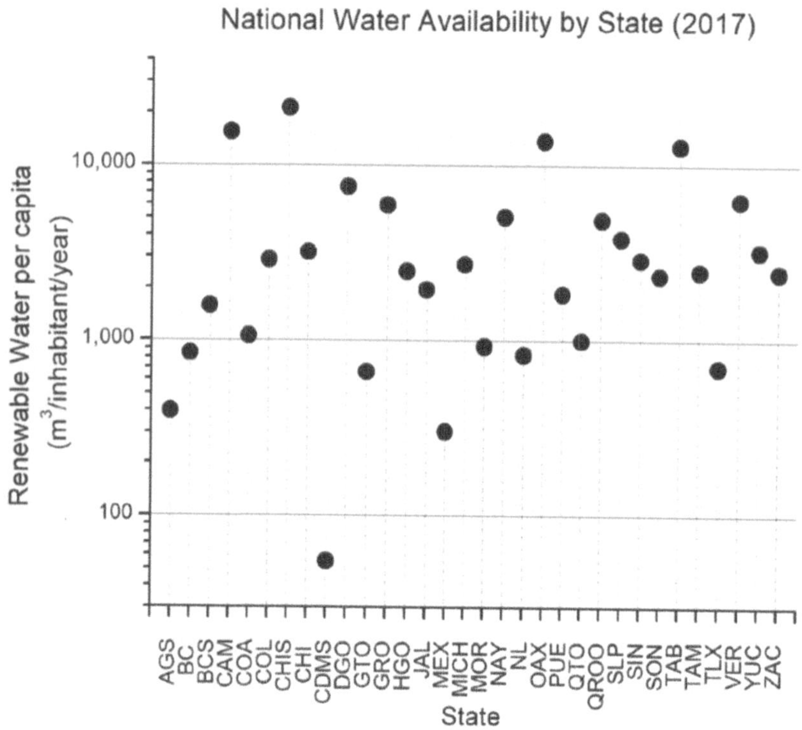

Figure 1 | Renewable water *per capita* in Mexico's states, 2017 (*data source*: EAM 2018).

sides (CONAGUA 2018). Surface and groundwater resources are the main sources of water supply in Mexico and they both play a major role in the country's hydrologic cycle. The surface water resources are divided into 1,471 river systems, 731 basins are managed for hydrological balances and are grouped in 37 hydrological units, three are shared with the United States (Tijuana, Colorado and Bravo), four with Guatemala (Grijalva-Usumacinta, Suchiate, Coatan and Candelaria), and one with Belize and Guatemala (Río Hondo) (Guerrero 2019). Regarding groundwater resources, Mexico has 653 aquifers from which 115 are over-exploited, 32 have brackish water and 18 have saline intrusion (SINA 2018).

Baja California state (BC) is located in the Northwest part of Mexico (Figure 2). It limits the North with the state of California on the Mexico-United States border. It has a territorial extension of 70,113 km^2 and occupies the Northern part of the Baja California peninsula, a piece of land surrounded by the Pacific Ocean to the West and the Sea of Cortes to the East. Baja California has 1,380 km of coastline, of which 740 km corresponds to the Pacific coast and 640 km to the Cortes Sea. The climate in BC is mostly very dry (68%), dry and semi-dry (24%), and temperate subhumid (7%). Considering the yearly information from 1980 to 2018 (SMN 2019), the average maximum atmospheric temperature was 26.8 ± 0.89 °C and the minimum 11.7 ± 1.48 °C, with an average yearly accumulated rain of 200 mm. In 2017, BC had a precipitation of 131.8 mm and a total renewable water of 3,045 hm^3 year^{-1} from which 2,093 hm^3 were surface water and 952 hm^3 were groundwater (EAM 2018). Permanent rivers are scarce and the two most important watersheds are binational basins shared with the United States (Colorado and Tijuana rivers). The major source of clean water in the state comes from the Colorado River, it has a total natural surface runoff of 1,922 hm^3 year^{-1} from which 1,850 hm^3 are delivered to Mexico due to an international treaty signed in 1944 (AAM 2018). The remaining 72 hm^3 are on Mexican territory. Water extraction from aquifers is also compromised, from 48 aquifers registered in BC, in 15 there is no water availability, 11 are overexploited, 7 have saline intrusion and in 3 the water is brackish (SINA 2018).

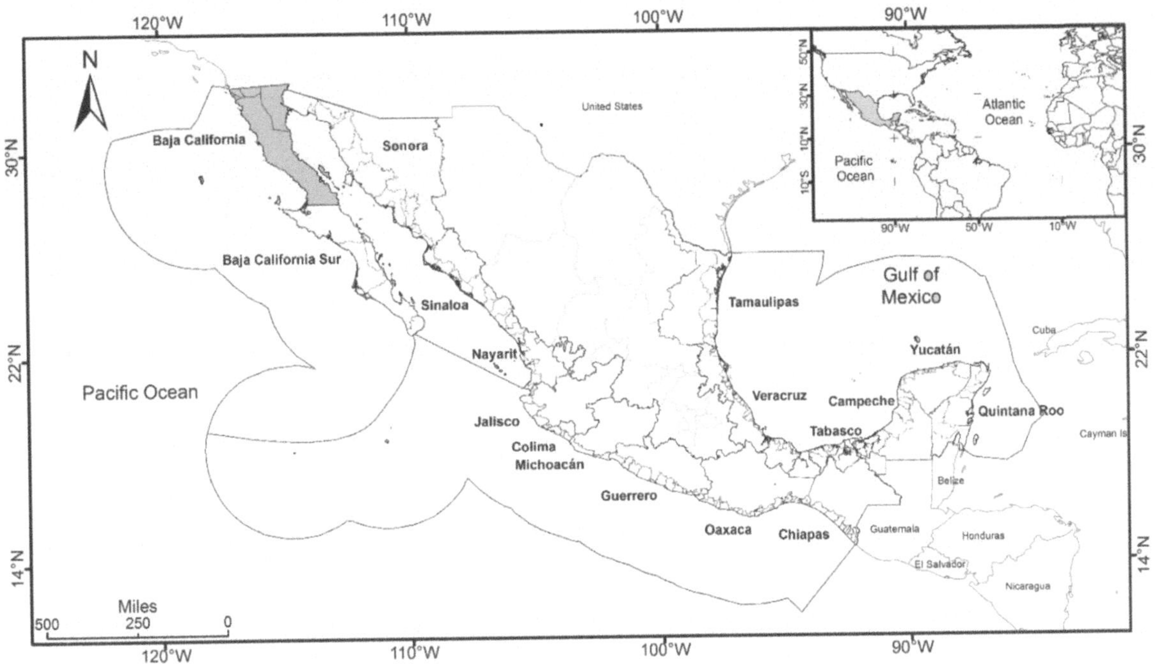

Figure 2 | Mexico in the world and Baja California state in grey colour.

Concerning hydraulic infrastructure, Baja California had a 95.43% drinking water coverage and 96.01% sewer access, both in year 2015 (SINA 2015). It has 3 main water dams: Rodriguez, El Carrizo and Emilio Lopez Zapata, which in 2018 accounted for a total of 282.11 hm^3 of water. Likewise, the state holds another source of freshwater besides rivers and aquifers, although on a smaller scale: treated water. In 2015, BC had 31 water treatment plants with an installed capacity of 12,146 litres per second ($L\,s^{-1}$) and a treated water flow of approximately 6,984.3 $L\,s^{-1}$ (CONAGUA 2015). For the same year, 114 wastewater treatment plants were registered, 71 are industrial with a combined capacity of 613 $L\,s^{-1}$ and a treated water flow of 615 $L\,s^{-1}$ (EAM 2018), and 43 are municipal with a total capacity of 7,775 $L\,s^{-1}$ and a treated water flow of 5,480 $L\,s^{-1}$ (CONAGUA 2015). The total installed capacity for wastewater treatment was 8,388 $hm^3\,year^{-1}$ and the wastewater treated amounted to 6,095 $hm^3\,year^{-1}$.

Due to its geographical location and economic development, BC is one of the states with the highest population growth rate in the country. Administratively, BC comprises five counties: Ensenada, Mexicali, Playas de Rosarito, Tecate and Tijuana. The population is mostly settled in few cities located in the northern portion of the state, outside these urban centres the population is scattered throughout its vast territory. In 2015, 3.3 million inhabitants were registered and by the year 2030, 4.1 million are expected (INEGI 2019a). Baja California has historically contributed approximately 3% of the Mexican economy, and the internal Gross Domestic Product (GDP) in Baja California depends on primary activities (3%), secondary activities (40%) and tertiary activities (57%). Of these, the manufacturing industry (electronic, automotive, aerospace and medical equipment) is the one that generates the most employment and economic resources (13% of the internal GDP).

Baja California is among the top 10 most developed states in Mexico. As Table 1 shows, BC is above the national median in several positive defined socioeconomic indicators like GDP (D1) or social security affiliations as an indicator of formal employment (D2), and below the national median in those indicators that express bad conditions like poverty (D3, D4, D10, D11, D12), education (D5 and D6) or lack of primary services and facilities (D7, D8, D9). To maintain the historical regional economic development growth rate and maintain and increase the quality of life of its inhabitants, proper water management is essential for the state.

Table 1 | General social and economic conditions of Baja California state in the context of Mexico

	D1	D2	D3	D4	D5	D6	D7	D8	D9	D10	D11	D12
Baja California State	3.00	4.34	2.88	22.75	1.96	10.46	0.26	0.47	2.82	23.03	1.15	22.85
National median	2.17	2.14	3.56	39.00	4.30	15.17	1.75	0.77	3.25	27.34	2.53	37.40
Higher state value	16.78	17.35	7.00	76.75	14.98	31.71	13.03	2.87	15.64	44.46	14.86	62.46
Lower state value	0.53	0.50	1.62	14.35	1.49	6.62	0.04	0.04	0.81	19.19	0.47	16.15

Variable/indicator description

D1 = Percentage of the National Gross Domestic Product (2017). *Source*: INEGI (2019b)

D2 = Percentage of Social Security Affiliation (2017). *Source*: INEGI (2019b)

D3 = Percentage of unemployment (2017). *Source*: INEGI (2019b)

D4 = Percentage of people in poverty (2017). *Source*: CONEVAL (2016, 2018)

D5 = Percentage of illiteracy in people under 15 years (2015). *Source*: CONAPO (2019)

D6 = Percentage of people without primary studies (2015). *Source*: CONAPO (2019)

D7 = Percentage of people living without sanitation facilities (2015). *Source*: CONAPO (2019)

D8 = Percentage of people living in houses without electricity (2015). *Source*: CONAPO (2019)

D9 = Percentage of people living in houses without clean water (2015). *Source*: CONAPO (2019)

D10 = Percentage of people living in overcrowded houses (2015). *Source*: CONAPO (2019)

D11 = Percentage of people living in houses with dirt floors (2015). *Source*: CONAPO (2019)

D12 = Percentage of employed people living with less than 4 US Dls a day (2015). *Source*: CONAPO (2019)

The main water users in BC are agriculture, public services (supply to residences and various industries), thermoelectric plants (combined cycle, carbon, turbogas, internal combustion and others) and self-supplied industries (EAM 2018). Primary activities include agriculture, which despite its low contribution to the GDP is the main water user, accounting for 85% of BC water consumption; following, with 6% each, are public services and thermoelectric plants, and self-supplied industries (secondary activities) account for the remaining 3% of the water consumption in the state (SINA 2018). In addition, water for agriculture purposes is highly subsidized in Mexico, which generates, in many cases, an inefficient use of the resource. As examples, mention may be made of the energy subsidy for pumping agricultural irrigation water (between 72% and 86% discount of the regular price) and the price paid by the farmers for the surpassed water from volumes granted to irrigation districts, which is 500 times less than the cost of the domestic urban waters (according with the Mexican Law of Rights of 2020, the cost of 100 cubic meters of irrigation waters is approximately 1 American dollar).

There has been a decrease of precipitation in the last years and an increase in temperatures, possibly associated with climate change. The condition of the aquifers is worsening both in quantity and quality along with the fact that the aquifers with water availability above 1 hm^3 are located at great distances from the main cities (273 km on average). According to SIGAGIS (2019), the Colorado River is the main source of water in BC but subject to international treaties, and to top it all there are no long-term strategies to solve or counteract this issue.

All these together create an alarming problem in BC concerning water. According with the National Water Commission of Mexico (EAM 2018), the water availability *per capita* in BC in 2017 was 849 $m^3\ year^{-1}$, which allows it to be classified, using FAO (2014) indicators, as a 'water scarcity' region, and several areas inside BC as 'extremely high water risk' (WRI-AQUEDUCT 2019).

In order to reduce the levels of water scarcity by 2030 based on the indicators of FAO (2014), the renewable water available needed, supposing some level of water stress (1,000 $hm^3\ capita^{-1}\ year^{-1}$), is 4,105 $hm^3\ year^{-1}$. Facing the best case scenario where there is no water stress (1,700 $hm^3\ capita^{-1}\ year^{-1}$ and above) the renewable water needed is 6,979 $hm^3\ year^{-1}$. Knowing that the available renewable water in 2017 in BC was 3,045 hm^3 with an agricultural consumption of 2,582 hm^3 (SINA 2018), the dimension of the problem is clear, a water deficit of at least 1,055 hm^3 can be predicted or expected for the near future.

With limited natural sources of water (rain, rivers and aquifers), a high population growth rate and two agricultural irrigation districts (Mexicali Valley and Ensenada Region) which use 85% of the water, with entirely different efficiencies (e.g. Ensenada/Mexicali ratio is 5.6 times in terms of production value in Mexican pesos per hm^3 of water), the only options to supply the future water demand in BC are: (a) reduction of water consumption in agriculture activities through the optimization of the irrigation district and reduction of cropland, this implies the reduction of food for human consumption and the income coming from this activity but an increase in water availability; (b) production of water from wastewater cleaning processes, this strategy is better implemented in addition to the strategy mentioned above since the amount of treated water generated is small in comparison, even if all the wastewater generated in the state is treated; (c) production of water through marine desalination processes, which involves an increase in water availability but also of electric power generation (sustainable practices can be implemented to reduce environmental impacts) and is also recommended to implement it in addition to the other solutions.

The main objective of this research is to estimate water needs for Baja California in the year 2030 in a scenario in which the region is not subjected to water scarcity conditions and assess the best options to satisfy those needs. In the next section will be discussed the methodology used to determine the best strategies to solve the problem addressed and the results obtained from each of them.

METHODOLOGY

The assessment of the current water availability conditions (2017 and 2018) in Baja California state was done reviewing the official information and databases from the main federal and state water-related bodies of Mexico. The estimations of total population in 2030 were done using a sigmoidal Boltzmann nonlinear model (Azuz *et al.* 2019) and the official results from the Mexican Council of Population (CONAPO is the Spanish acronym), based on decadal information since 1930 and every five years for 1995, 2005 and 2015. The agricultural information was obtained from official records of the Mexican Ministry of Agriculture (SIACON 2019) on a yearly basis for 7 crops in the irrigation district of Mexicali for the period 1980–2018, these 7 crops (alfalfa, cotton, oats, onion, asparagus, sorghum and wheat) represent 89% of the planted surface, 82% of the production and 80% of the production value. Confidence intervals, a nonlinear regression model and empirical decomposition methods were used in the estimations of agricultural production, production value and planted surface in 2030 for every individual selected crop.

To evaluate the water supply options, the following cases were analyzed (Table 2) where: (a) reduction of

Table 2 | Conditions used to theoretically solve the water deficit in Baja California

Case	Solution type	Conditions
1	'a'	(a) Reduction of agricultural lands in Mexicali: one or more crops.
2	'a' and 'b'	(a) Reduction of agricultural lands in Mexicali: one or more crops and, (b) Increase in wastewater treatment: number of facilities or efficiency increase.
3	'a' and 'c-1'	(a) Reduction of agricultural lands in Mexicali: one or more crops and, (b) Installation of desalination plants: production of water from salinized aquifers.
4	'a' and 'c-2'	(a) Reduction of agricultural lands in Mexicali: one or more crops and, (b) Installation of desalination plants: production of water from coastal waters.
5	'a', 'b' and 'c-1'	(a) Reduction of agricultural lands in Mexicali: one or more crops, (b) Increase in wastewater treatment: number of facilities or efficiency increase and, (c) Installation of desalination plants: production of water from salinized aquifers.
6	'a', 'b' and 'c-2'	(a) Reduction of agricultural lands in Mexicali: one or more crops, (b) Increase in wastewater treatment: number of facilities or efficiency increase and, (c) Installation of desalination plants: production of water from coastal waters.

planted surface for individual crops and combinations of crops; (b) increase of wastewater treatment using the same number of plants more efficiently or by the installation of new plants; (c) installation of desalination facilities that produce clean water from salinized aquifers or directly from the coast waters. Linear and nonlinear optimization processes were used to obtain the best options in terms of these variables alone or combined among cases and between cases. Economic valuation of each case was used as a selection criterion for the possible scenarios.

Improvements in irrigation technology in Mexicali Valley cannot be evaluated and incorporated into the scenarios due to the lack of local information (e.g. number of farmers by crop, losses by irrigation channel), as well the level of implementation of local government efficiency plans (CEABC 2013, 2016).

RESULTS

Possible scenarios and solutions for the year 2030

Reduction of water consumption in agricultural activities

By 2030, the state needs a total of 4,105 hm^3 of renewable water to be classified as a non-water scarcity region (1,000 m^3 *per capita* or more). According to EAM (2018), BC has 3,045 hm^3 of renewable water per year, therefore 1,060 m^3 are still needed to achieve the goal.

Before reducing the planted surface, the irrigation district of Mexicali needs to optimize its water use; as mentioned before, the Ensenada/Mexicali ratio is 5.6 times in terms of production value in Mexican pesos per hm^3 of water. To simulate the results of this optimization, the main products grown in the region were defined, and consequently, the water needed to harvest these products was determined (Table 3). Knowing the total number of planted hectares and the water needed for each product, the optimal water use for agricultural land in Mexicali in 2017 was determined to be 1,371. 55 hm^3 for the 7 products, and overall the water use in the state's capital should be approximately 1,635 hm^3 according to the State Water Plan of Baja California (PHEBC, by its Spanish acronym), which indicates that the capital is to be given said amount of water for agricultural purposes from the Colorado River basin; however, the real water volume delivered in 2017 was 2,147 hm^3. By optimizing the irrigation district the water gain is around 512 hm^3, thus decreasing the water needed to reach the goal to 548 hm^3.

To define the optimal results for cropland reduction, the issue was approached taking into account four variables: (1) number of hectares reduced from total cropland; (3) total hectares of cropland after surface reduction; (3) crop production value losses in Mexican pesos; (4) total water gain. The first step was to reduce cropland in values of 5% to the main products of the region starting from greatest to lowest number of hectares planted. Once the surface of all 7 products was reduced separately, the results were subjected to a comparative analysis to define the best option, offering the highest water gain, lowest production value losses and/or minimal reduction of cropland.

After calculating the efficiency of the production value in 2017 (57.92 thousand Mexican pesos per cultivated hectare), the analysis demonstrated that the most convenient strategy was the reduction of alfalfa, followed by cotton and then wheat (which could be an option but the monetary loss is more prominent), the rest of the products show small values of gained water, thus their surface reduction is negligible. To accurately meet the water necessities of 548 hm^3 by only using the approach of planted surface reduction, the entire surface of alfalfa would need to be removed, as well as 27% of the cotton surface. This reduction accounts for 35,558 ha of reduced surface (80% alfalfa, 20% cotton); that is, 23% of the initial hectares, a total of 118,702 ha of planted surface after the reduction, a monetary loss of 2,059,506 thousand of Mexican pesos (MXN) and a water gain of 548 hm^3. Alfalfa production in the region is intended for the feeding of livestock from private producers, and cotton is destined for industrial use for the production of clothing by transnational companies; in this sense, the reduction of its cultivated area does not affect the production of food for human consumption. Making a decision of this type raises the need to privilege the common good (water availability) over the private one, with the associated political and social consequences.

Table 3 | Main agricultural products in Mexicali, Baja California (SIACON 2019)

Product	Surface (Ha)	Production (Ton)	Production value (Thousand MXN)	Water needs per year (hm^3/ha/y)	Optimal water use (hm^3)
2017					
Alfalfa	28,558	474,793	1,245,567	0.016	457
Cotton	25,769	145,176	1,518,314	0.013	335
Oat	3,590	136,260	57,259	0.0065	23
Onion	5,430	88,975	1,642,987	0.011	60
Asparagus	2,323	23,333	1,117,125	0.014	33
Sorghum	3,646	168,026	76,835	0.0065	24
Wheat	67,744	391,224	1,463,140	0.0065	440
Sub-Total (7 products)	137,060	1,427,787	7,121,229	0.0735	1,372
TOTAL Mexicali	154,268	1,734,684	8,935,041	0.0827	1,544
2018					
Alfalfa	30,072	501,801	1,340,147	0.016	481
Cotton	28,476	179,824	1,905,364	0.013	370
Oat	5,811	213,249	91,217	0.0065	38
Onion	5,708	96,086	1,555,279	0.011	63
Asparagus	4,117	25,683	1,147,654	0.014	58
Sorghum	3,252	136,845	64,576	0.0065	21
Wheat	47,881	283,133	1,177,699	0.0065	311
Sub-Total (7 products)	125,318	1,436,621	7,281,936	0.0735	1,342
TOTAL Mexicali	144,458	1,760,725	9,210,051	0.0827	1,510

Production of water from wastewater cleaning processes

In 2017, the total wastewater generation in BC was approximately 193 hm^3; however, for this study, only the wastewater generation and plants from Tijuana and Mexicali are taken into account. According to the Hydric Program of the state of Baja California (2016) or PHEBC, by its Spanish acronym, Mexicali had a population of 998,447 inhabitants, a wastewater generation of 75.30 hm^3 and a volume of 75.42 m^3 *per capita* of wastewater. For Tijuana, the population was 1,657,289 inhabitants, 87.46 hm^3 of wastewater was generated and a total water volume of 52.77 m^3 *per capita* (CEABC 2016). Considering the approximate values of wastewater *per capita* in both municipalities and their projected population for 2030, an estimation of the wastewater generation for the same year can be made. For Mexicali, the projected population for 2030 is 1,234,931 inhabitants, therefore the wastewater generation is estimated to be 93 hm^3; continuing with Tijuana, the population will be approximately 2,023,931 with a volume of wastewater of 107 hm^3. This amounts to an estimation of 200 hm^3 of wastewater generated. Assuming that the entire volume of wastewater generated in both Mexicali and Tijuana in 2030 will be treated, the total expenses sum up to 907,729 thousand MXN, considering a treatment cost of 4.54 pesos/m^{-3}.

While the amount of water coming from wastewater treatment plants is significant, it is not enough to fulfil the goal of 548 hm^3 by 2030 on its own, thus the best way to implement this solution is by doing it in alignment with cropland reduction. The alternative is to avoid the reduction of the entire cultivated surface of alfalfa and any reduction of cotton surface by providing Mexicali with the 200 hm^3 of reusable water (while Mexicali yields 200 hm^3 of clean water coming from the Colorado River to the rest of the municipalities). Based on this, the water needed from cropland reduction is 348 hm^3, which represents the water needed for 76% of alfalfa surface; that is, 21,746 ha removed

Table 4 | Characteristics of water desalination plants in Baja California (CEABC 2016)

Municipality	Plant	Capacity (hm^3/y)	Minimum annual treatment cost (Thousand MXN/m^3)	Maximum annual treatment cost (Thousand MXN/m^3)	Phase
Ensenada	Ensenada	7.88	64,648.80	87,433.56	In operation
Ensenada	Isla de Cedros	0.15	1,197.30	1,619.27	In operation
Ensenada	Valle de San Quintín	7.88	64,648.80	87,433.56	Planning
Playas de Rosarito	Playas de Rosarito	69.38	568,909.44	769,415.33	In construction
Playas de Rosarito	La Misión	7.88	64,648.80	87,433.56	Planning
	TOTAL	**93.18**	**764,053.14**	**1,033,335.28**	

(14% of overall croplands), a total surface of 132,514 ha after reduction and a loss of 1,259,499 thousand MXN. Knowing the cost of both the reduction of 14% of cropland and the treatment of 200 hm^3, the overall expenses of implementing solution b) along with solution a) are 2,167,229 Mexican pesos (107,723 thousand MXN more expensive than implementing solution 'a' alone).

Production of water through desalination processes

Baja California has 1,380 km of coastline, of which 740 km correspond to the Pacific coast and 640 km to the Gulf of California (INAFED 2019). This opens a new window of opportunity for obtaining clean water through marine desalination methods. According to the Federal Water Commission of Baja California (2013 and 2016), or CEABC by its Spanish acronym, the state has two desalination plants in operation, both in the municipality of Ensenada. The first one, located in the main city, has an installed capacity of 7.88 hm^3 per year, while the remaining one is located in Isla de Cedros with a capacity of 0.15 hm^3 per year. Likewise, the CEABC proposed the construction of three more desalination plants, which by 2018 were either in the planning phase or starting construction.

In Table 4, the desalination projects in the state are shown; three of the plants will be located in Ensenada while the other two are in Playas de Rosarito, all five of these plants will operate or are operating with reverse osmosis, the installed capacity will be 93.18 hm^3 and all of the water is meant for public supply. Now, if this volume of water is directed to the irrigation district of Mexicali under the same conditions as in the case of wastewater reuse, this means that 93 hm^3 of freshwater coming from the Colorado River basin can be used for different purposes and be distributed between Mexicali and the other municipalities.

For 2030, the public sector will require approximately 256 hm^3 of water, meaning that the water amount coming from desalination processes is not enough to satisfy the public demand in its entirety; nevertheless, it covers 36% of the demand, which gives the possibility of using the volume of freshwater coming from basins and aquifers that was replaced by this treated water to be used for human necessities, while at the same time it contributes to the 2030 goal (4,105 hm^3) by 2.27%. The investment resulting from treating 93 hm^3 is defined by the method of extraction of water, that is through (1) marine wells or (2) direct take. For the first case, the cost of extraction per cubic meter according to CEA Sonora (2016) is 8.20 MXN, while for the second method, the cost is 11.09 pesos per m^3. This means that for marine wells the total investment is 764,053 thousand MXN, and for direct take is 1,033,335 thousand MXN.

Now, using the water treated from desalination processes for agriculture in Mexicali, means that the reduction of planted land would have to provide a volume of water of 455 hm^3, which represents 28,433 ha of alfalfa planted (99.6% alfalfa), 125,827 ha of planted surface after reduction, a loss of 1 646,839 thousand MXN and a water gain of 455 hm^3. Knowing the total loss coming from surface reduction and the cost of water desalination in both cases, the total investment will be (1) 2,410,893 thousand MXN if marine wells are used; and (2) 2,680,175 thousand MXN in the case of direct take.

Implementation of three possible solutions altogether

Considering the case in which the three solutions are implemented altogether, where all desalinated water in the state and all wastewater treated in Tijuana and Mexicali are destined for agricultural use in Mexicali, the reduction of planted surface could be minimized once more. With a total volume of water of 293 hm^3 coming from treatment processes, the amount of water needed to reach the 2030 goal would be 255 hm^3. In terms of cropland reduction, this means the reduction of 56% of alfalfa; that is, a reduction of 15,933 ha, a total planted surface of 138,327 ha after reduction, a loss of 922,839 thousand MXN and a final water gain of 255 hm^3. Now, knowing the cost of planted surface reduction, wastewater treatment and desalination treatment processes, the overall investment would be either 2,594,622 thousand MXN (marine wells) or 283,904 thousand MXN (salinized coastal aquifers).

CONCLUSION

The information of all cases addressed in this study is presented in Table 5, prepared by the authors based on information provided by the Agri-food Consultation Information System (2019) and the Federal Water Commission (2016), respectively, SIACON and CEA, by their Spanish acronyms. Solution (a) represents reduction of planted surface; solution (b) wastewater treatment; and solution (c) marine desalination processes, being c-1 marine wells and c-2 direct take from already salinized coastal aquifers.

This synthesis of information facilitates the evaluation of all cases at once and shows that financially the best option is Case 1, which is the less expensive solution by at least 107,723 thousand MXN, followed by Case 2, then Case 3 and 5, and finally Cases 4 and 6. This indicates that the use of desalination processes increases the amount of investment significantly (Cases 3–6), therefore can be concluded that this is the most expensive solution in comparison to wastewater treatment and planted surface reduction, especially with a method of water extraction of direct take (Cases 4 and 6).

Besides the economic valuation of the scenarios, it is important to note that some options considered in this study (the proposed agricultural land decrease in Mexicali), could affect the socio-economic conditions of the region, through food production reduction or unemployment increase. However, as mentioned before, the most suitable crops to be reduced (alfalfa and cotton) are not for human consumption, the first crop is for cattle feeding and the second is for industrial use (clothes). The valuation of unemployment is more difficult to assess due to the lack of specific information about agriculture-related workers; however, using a gross approximation, considering the total planted surface and the total users of water, a mean value

Table 5 | Summary of possible solutions for water scarcity in Baja California for the year 2030

Case	Implemented solution(s)	Volume of water provided (hm^3/y)	2030 goal coverage (%)	Investment by solution (thousand MXN)	Total investment (thousand MXN)
1	Solution a)	548	100	2,059,506	2,059,506
2	Solution a)	348	64	1,259,499	2,167,229
	Solution b)	200	36	907,729	
3	Solution a)	455	83	1,646,839	2,410,893
	Solution c-1	93	17	764,053	
4	Solution a)	455	83	1,646,839	2,680,175
	Solution c-2	93	17	1,033,335	
5	Solution a)	255	47	922,839	2,594,622
	Solution b)	200	36	907,729	
	Solution c-1	93	17	764,053	
6	Solution a)	255	47	922,839	2,863,904
	Solution b)	200	36	907,729	
	Solution c-2	93	17	1,033,335	

of 9.2 ha per user of water was estimated, which gives a potential unemployment number associated with the crop land reduction of 35,558 ha in Mexicali of 3,865 persons. This number represents only 0.9% of the total workers in Mexicali, and only for comparative purposes, a single maquiladora industry in Baja California that cannot operate due to lack of water could abruptly fire this number of employees. Considering the specific planted surface reduction proposed in the presented scenarios, the number of unemployed workers goes from 1,280 to 2,860, which could be easily incorporated in other economic activities in the region or subsidise by the local or federal government through – for example – by 'payment for environmental services related with water conservation' which is an already working governmental program in other Mexican zones.

Decision-making in the context of water scarcity zones is complex, but the authors consider that in order to achieve sustainable development in Baja California, in certain circumstances, the common good must be privileged over the private one, especially with a central resource like water.

DATA AVAILABILITY STATEMENT

All relevant data are included in the paper or its Supplementary Information.

REFERENCES

AAM 2018 *Atlas del Agua en México (Mexico's Water Atlas)*. Available from: https://files.conagua.gob.mx/conagua/publicaciones/Publicaciones/AAM2018.pdf (accessed 22 January 2020).

Azuz, I., Rivera, E. & Peinado, A. 2019 Current demographic conditions and future scenarios in Mexico's coastal zone. *Journal of Integrated Coastal Zone Management* **19** (2), 85–122.

CEABC 2013 *Desaladoras del municipio de Ensenada (Desalination Plants in the Municipality of Ensenada)*. Available from: http://spabc.baja.gob.mx/wp-content/uploads/2016/08/Desaladoras-Ensenada.pdf (accessed 4 March 2020).

CEABC 2016 *Programa Hídrico Estatal de Baja California (Federal Hydric Program of Baja California)*. Available from: http://www.cea.gob.mx/phebc/maqueta/index.html (accessed 4 March 2020).

CEA Sonora 2016 *Análisis Costo-Beneficio del Proyecto: "Planta desaladora para las ciudades de Guaymas y Empalme, Sonora" (Cost-Benefit Analysis of the Project: "Desalination Plant for the cities of Guaymas and Empalme, Sonora")*. Available from: http://desaladora.sonora.gob.mx/images/transparencia/dictamenes/dictamen-rentabilidad-social.pdf (accessed 9 March 2020).

CONAGUA 2015 *Inventario Nacional de Plantas Municipales de Potabilización y de Tratamiento de Aguas Residuales en Operación Diciembre 2015 (National Inventory of Municipal Purification and Waste Water Treatment Plants in Operation December 2015)*. Available from: https://www.gob.mx/cms/uploads/attachment/file/197610/Inventario_2015.pdf (accessed 5 February 2020).

CONAGUA 2018 *Numeragua México 2018*. Available from: https://files.conagua.gob.mx/conagua/publicaciones/Publicaciones/Numeragua2018.pdf (accessed 22 January 2020).

CONAPO 2019 *Índices de Marginación por Entidad Federativa 2015 (Margination Index by State 2015)*. Available from: http://www.conapo.gob.mx/es/CONAPO/Indices_de_Marginacion (accessed 27 July 2020).

CONEVAL 2016 *Medición de la Pobreza 2016 (Poverty Measurements 2016)*. Available from: https://www.coneval.org.mx/Medicion/Paginas/PobrezaInicio.aspx (accessed 28 July 2020).

CONEVAL 2018 *Medición de la Pobreza 2018 (Poverty Measurements 2018)*. Available from: https://www.coneval.org.mx/Medicion/Paginas/PobrezaInicio.aspx (accessed 28 July 2020).

du Plessis. 2017 *Global Water Scarcity and Possible Conflicts*. Available from: https://www.researchgate.net/publication/313488451_Global_Water_Scarcity_and_Possible_Conflicts (accessed 20 January 2020).

EAM 2018 *Estadísticas del Agua en México (Water Statistics in Mexico)*. Available from: https://files.conagua.gob.mx/conagua/publicaciones/Publicaciones/EAM2018.pdf (accessed 22 January 2020).

FAO 2014 *Water Stress*. Available from: http://www.fao.org/nr/water/aquastat/infographics/Stress_eng.pdf (accessed 27 January 2020)

FAO and UN-Water 2018 *Progress on Level of Water Stress – Global Baseline for SDG Indicator 6.4.2*. Available from: http://www.fao.org/3/CA1592EN/ca1592en.pdf (accessed 22 January 2020).

Guerrero, H. 2019 *Water Policy in Mexico*. https://doi.org/10.1007/978-3-319-76115-2 (accessed 22 January 2020).

INAFED 2019 *Enciclopedia de los Municipios y Delegaciones de México. Instituto Nacional para el Federalismo y el Desarrollo Municipal (Encyclopedia of Municipalities and Delegations of Mexico)*. Available from: http://siglo.inafed.gob.mx/enciclopedia/EMM02bajacalifornia/mediofisico.html (accessed 4 March 2020).

INEGI 2019a *Población Total (Total Population)*. Available from: https://www.inegi.org.mx/temas/estructura/ (accessed 5 February 2020).

INEGI 2019b *PIB y Cuentas Nacionales (GDP and National Economic Accounts)*. Available from: https://www.inegi.org.mx/datos/ (accessed 28 July 2020).

Otazo-Sánchez, E., Navarro-Frómeta, A. & Singh, V. 2020 *Water Availability and Management in Mexico*. https://doi.org/10.1007/978-3-030-24962-5 (accessed 27 January 2020).

SIACON 2019 *Sistema de Información Agroalimentaria de Consulta (Agri-Food Consultation Information System)*. Available from: https://www.gob.mx/siap/documentos/siacon-ng-161430 (accessed 13 March 2020).

SIGAGIS 2019 *Acuíferos Baja California (Aquifers in Baja California)*. Available from: https://sigagis.conagua.gob.mx/gas1/sections/Edos/BajaCalifornia/bc.html (accessed 5 February 2020).

SINA 2015 *Sistema Nacional de Información del Agua – Cobertura universal nacional (National Water Information System – Universal Coverage National)*. Available from: http://sina.conagua.gob.mx/sina/tema.php?tema=coberturaUniversal&ver=mapa&o=1&n=nacional (accessed 27 January 2020)

SINA 2018 *Sistema Nacional de Información del Agua (National Water Information System)*. Available from: http://sina.conagua.gob.mx/sina/index.phphttp://sina.conagua.gob.mx/sina/tema.php?tema=acuiferos (accessed 22 January 2020).

SMN 2019 *Servicio Meteorológico Nacional (National Weather Service)*. Available from: https://smn.conagua.gob.mx/es/climatologia/temperaturas-y-lluvias/resumenes-mensuales-de-temperaturas-y-lluvias (accessed 23 February 2020).

United Nations 2019 *World Population Prospects 2019: Highlights*. Available from: https://population.un.org/wpp/Publications/Files/WPP2019_Highlights.pdf (accessed 20 January 2020).

UNU-INWEH 2017 *Global Water Crisis: The Facts*. Available from: https://inweh.unu.edu/wp-content/uploads/2017/11/Global-Water-Crisis-The-Facts.pdf (accessed 20 January 2020).

UN-Water 2019 *The United Nations World Water Development Report 2019*. Available from: https://reliefweb.int/sites/reliefweb.int/files/resources/367306eng.pdf (accessed 20 January 2020).

WRI-AQUEDUCT 2019 *Aqueduct Water Risk Atlas*. Available from: https://www.wri.org/aqueduct (accessed 27 January 2020).

First received 21 May 2020; accepted in revised form 8 August 2020. Available online 20 August 2020

Sprinkler droplet impact angle affects shear stress distribution on soil surface – a case study of a ball-driven sprinkler

Xin Hui, Haijun Yan, Yuncheng Xu and Haibin Tan

ABSTRACT

Droplet shear stress is the main cause of soil erosion under sprinkler irrigation, and the effect of droplet impact angle on the shear stress distribution cannot be ignored. In this study, a ball-driven sprinkler was selected to investigate the radial distributions of droplet impact angles under three operating pressures (0.25, 0.30, and 0.35 MPa) and two nozzle diameters (1.9 and 2.2 mm), which are commonly used in agricultural irrigation. The effect of droplet impact angles on the distances from the sprinkler, droplet impact velocities, and shear stresses were analyzed by a 2DVD instrument. Irrespective of the nozzle diameter or operating pressure, the droplet velocities and impact angles near the sprinkler were distributed at 1.0–5.5 m s^{-1} and 70–90°, respectively, and the droplet shear stress increased with the distance from the sprinkler. Suitable operating pressure and distance from the sprinkler significantly reduced the droplet shear stress. Although the nozzle diameter had a certain effect on the maximum shear stress, the overall effect was insignificant. We developed the models for the radial distribution of droplet shear stresses, which were in good agreement with the measurement. This study proposes a new method for accurately predicating the soil erosion under sprinkler irrigation.

Key words | ball-driven sprinkler, droplet impact angle, droplet impact velocity, droplet shear stress, sprinkler irrigation

Xin Hui
Haijun Yan (corresponding author)
Yuncheng Xu
College of Water Resources and Civil Engineering,
China Agricultural University,
Beijing 100083,
China
E-mail: *yanhj@cau.edu.cn*

Haibin Tan
The Semi-arid Agriculture Engineering & Technology Research Center of P.R. China,
Shijiazhuang 050000,
China

HIGHLIGHTS

- The soil erosion risk at the end of the spray jet is highly related to droplet shear stress rather than droplet kinetic energy.
- Increasing the operating pressure or decreasing the nozzle diameter can effectively reduce the maximum droplet shear stress.
- Mathematical models are developed for predicting the radial distribution of droplet shear stress.

doi: 10.2166/ws.2021.025

GRAPHICAL ABSTRACT

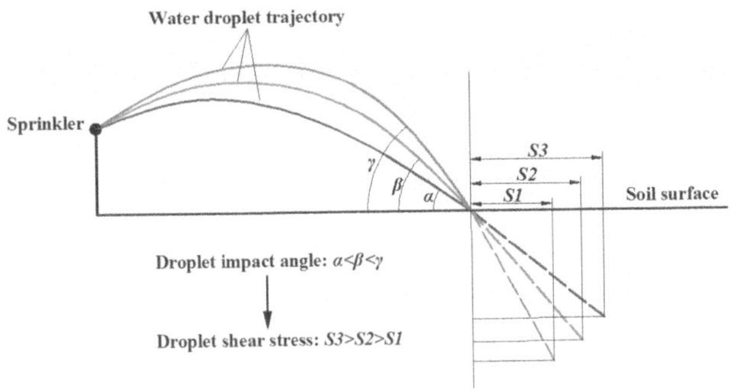

INTRODUCTION

As a high-efficiency water-saving irrigation technique, sprinkler irrigation has been developed, promoted, and applied in China since the 1950s (Yan *et al.* 2020). By the end of 2018, China's sprinkler irrigation area has reached 4.4 million hm^2, accounting for about 20% of the total area under high-efficiency water-saving irrigation. Soil surface sealing or crust formation caused by sprinkler irrigation is common (Chang & Hills 1993a). It not only reduces the water infiltration rate, but also leads to surface runoff. Previous studies have shown that soil particle detachment is the main reason for soil surface sealing (Assouline & Ben-Hur 2006) and generally related to droplet kinetic energy (Yan *et al.* 2011; Caracciolo *et al.* 2012). However, some studies (Huang *et al.* 1982; Ghadiri & Payne 2010) showed that from a mechanism perspective, the soil particle detachment from aggregates is caused by external shear stress rather than droplet kinetic energy.

Several studies have been conducted on the stress distribution due to the droplet impact on the soil surface and the soil particle decomposition. Ghadiri & Payne (1986), based on the water-hammer theory, calculated the compressive stress of falling raindrops and shear stress caused by the flow impact. They found that the lateral shear stress was several times larger than the compressive stress. In another study, Huang *et al.* (1982) used the Marker and Cell (MAC) numerical technique to simulate the impact of raindrops on a rigid surface. The results showed that the impact pressure distribution was non-uniform. The maximum pressure occurred at the contact circumference, and the rebound velocity on the rigid surface was twice the impact velocity. Ferreira *et al.* (1985) applied a modified SOLA-VOF numerical simulation scheme for simulating the raindrop's impact in a deep pool based on energy balance. Their results indicated that most of the soil aggregate fragmentation occurred during the development of the crater, mostly due to the impact pressure and shear on the bottom of the pool.

The studies mentioned above mainly focused on the vertical impact of raindrops, which was quite different from sprinkler irrigation, where droplets hit the soil at oblique angles. Chang & Hills (1993b) developed a numerical simulation model for studying the pressure and shear stress distribution on a soil surface following sprinkler droplet impact. Simulation analysis indicated that the oblique droplet impact decreased the magnitude of the impact force, but it increased the shear stress, compared with the vertical droplet impact. In another study, Chang & Hills (1993a) investigated the influence of sprinkler droplet impact angles on soil infiltration through laboratory experiments. They found that the average steady infiltration rates for all soil types increased in the following order of the impact angle: 60°, 45°, and 90°, which implied that the effects of sprinkler droplet impact angle on the shear stress distribution and soil infiltration need to be considered. However, in their research, only three droplet impact angles (90°, 45°, and 60°) were selected, which was insufficient to reflect the impact of water droplets on the ground at multiple angles. Hence, the effect of droplet impact angle on the shear stress distribution needs to be further investigated.

The sprinkler is an important component of a sprinkler irrigation system, and its performance directly affects the quality of the whole system (Zhu *et al.* 2012; Zhang *et al.* 2013). With the development of the low-pressure and energy-saving sprinkler irrigation technology, it is critical to develop sprinklers with simple structures at low operating pressure. In recent years, a ball-driven sprinkler that uses a stainless-steel ball to drive the nozzle to rotate and spray has been successfully applied in field crop irrigation (Hui *et al.* 2019). However, there are still few reports on the effects of droplet impact angle on the shear stress distribution of a ball-driven sprinkler. In this context, the main objectives of this study were (1) to investigate the radial distributions of droplet impact angles for the ball-driven sprinkler under three operating pressures (0.25, 0.30, and 0.35 MPa) and two nozzle diameters (1.9 and 2.2 mm); (2) to establish the relationships between the droplet impact angle and distance from the sprinkler, droplet velocity, and shear stresses, respectively; and (3) to further develop the models for the radial distribution of droplet shear stresses, validated by experimental data.

MATERIALS AND METHODS

Experimental setup

Figure 1 shows the flowchart of the experiment process. The experiment was performed indoors to avoid the impacts of wind. Figure 2 shows the experimental setup. We used a ball-driven sprinkler with an inlet diameter of 12.7 mm, which is not only simple in its structure and easy to clean, but also has a good application uniformity (Hui *et al.* 2019). The components of the sprinkler include a nozzle (diameter of 1.9 or 2.2 mm), a chamber (the space enclosed by the upper and lower parts of the sprinkler

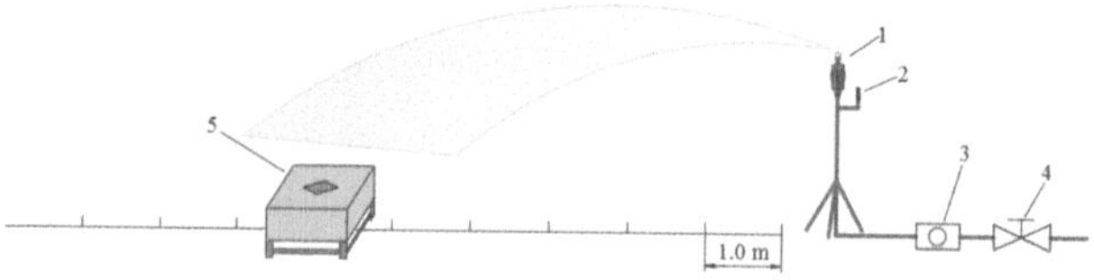

Figure 2 | Diagram of the experimental setup: 1. ball-driven sprinkler, 2. pressure transducer, 3. flowmeter, 4. pressure regulator valve, 5. 2DVD instrument.

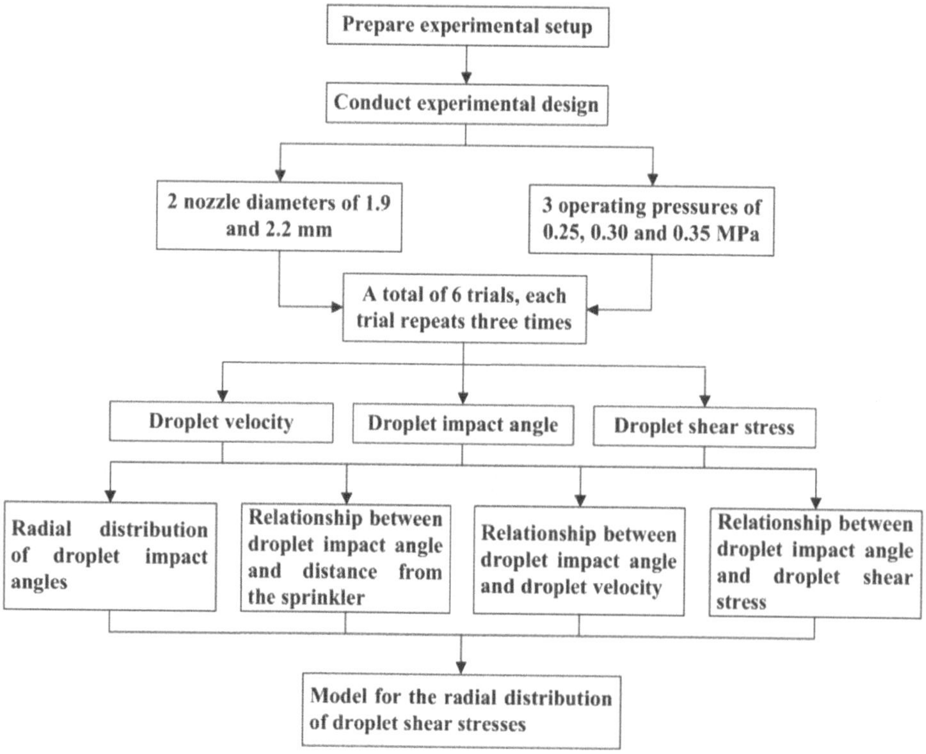

Figure 1 | Flowchart of the experimental setup and design.

body), a guide structure (with an inlet angle of 18.5°), a stainless-steel ball (diameter of 7.8 mm), and rotating components (Figure 3). During the sprinkler test, water enters the chamber via the guide structure and forms the high-speed water flow, impacting the stainless-steel ball. The ball circularly moves and continuously hits the flange connected to the nozzle, therefore the nozzle rotates and sprays.

In this experiment, the ball-driven sprinkler was mounted on a vertical riser with a height of 1 m, and its operating pressure ranged from 0.25 to 0.40 MPa. A pressure transducer (Xi'an Xinmin model CYB, accuracy of ±0.1%), with a range from 0 to 0.50 MPa, was installed at the sprinkler inlet and connected to a data logger to monitor the sprinkler pressure every 5 s. In addition, a two-dimensional video disdrometer (2DVD), developed by Joanneum Research Corp, Austria, was used to measure the droplet diameter and velocity (Figure 4). Two vertically disposed CCD cameras inside the instrument made linear scans of the droplets passing through the measurement area (100 × 100 mm^2), recording individual droplet diameter as well as the vertical and horizontal velocity components (Huang *et al.* 2010; Ge *et al.* 2018). The test accuracy of the droplet diameter was 0.19 mm.

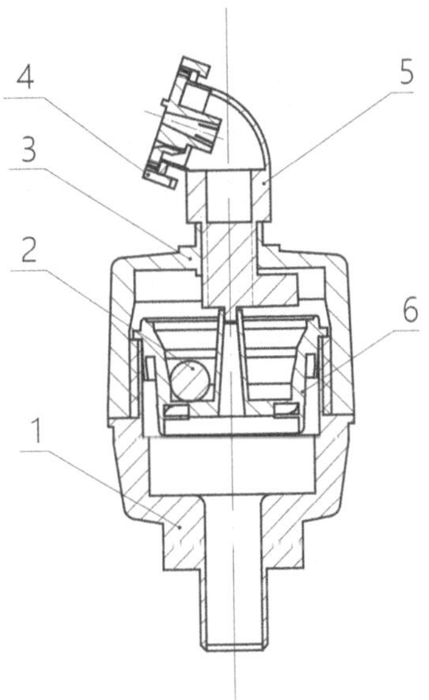

Figure 3 | Diagram of the sprinkler used in the experiment: 1. lower part of the sprinkler body, 2. stainless-steel ball, 3. upper part of the sprinkler body, 4. nozzle, 5. rotating components, 6. guide structure.

Experimental design

The operating pressure and nozzle diameter of the sprinkler were considered in the radial test of water droplets. Three operating pressures were chosen according to manufacturer's recommendation, namely 0.25, 0.30, and 0.35 MPa. Also, the two nozzle diameters of 1.9 and 2.2 mm that are commonly used in agricultural irrigation were chosen for testing. In total, six trials were performed, each with three replicates. In each trial, the measuring points were set at 1-m intervals along the radial direction, and the diameters and velocities of the sprinkler droplets falling on each measuring point were monitored by the 2DVD, as shown in Figure 2. The number of the droplets collected at each measuring point was at least 10,000. All the measurements were taken when

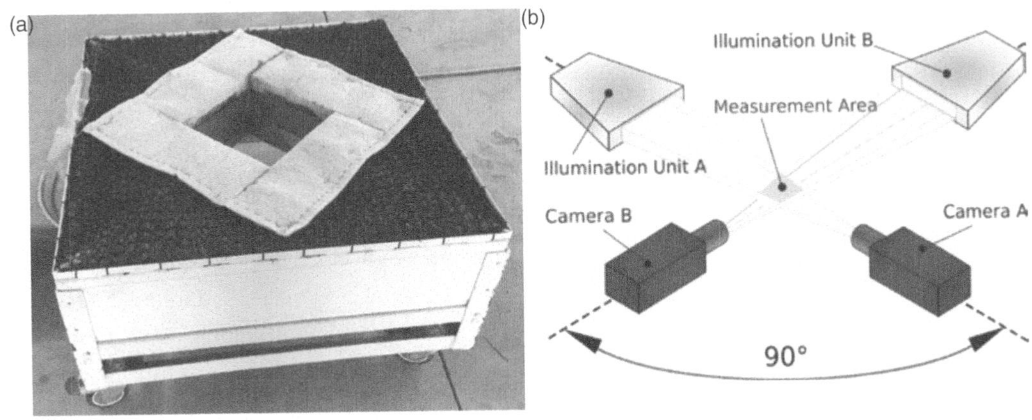

Figure 4 | Photograph of the 2DVD instrument (a) and schematic presentation of the measuring planes of 2DVD (b).

the pressure was kept stable. Indoor air and water temperature were approximately 30 and 26 °C, respectively, and relative humidity was 60%. The following design standards were adopted in the present experiment: ISO 7749-2 (ISO Standards 2004) and ISO 15886-3 (ISO Standards 2012).

Calculation of resultant droplet velocity, impact angle, and shear stress

Since the vertical (V_v) and horizontal (V_h) velocities of a water droplet could be recorded by the 2DVD, the calculation equation for the resultant velocity (V) of a water droplet was as follows:

$$V = \sqrt{V_v^2 + V_h^2} \tag{1}$$

The impact angle (θ) of a water droplet was obtained by the following equation:

$$\theta = \frac{180}{\pi} \arctan\left(\frac{V_v}{V_h}\right) \tag{2}$$

where V is the resultant droplet velocity, m·s^{-1}; θ is the droplet impact angle, °; V_v is the vertical droplet velocity, m·s^{-1}; and V_h is the horizontal droplet velocity, m·s^{-1}.

The shear stress (S) of a water droplet in this study was defined as the following formula (Ghadiri & Payne 1986):

$$S = 0.5\rho V_h^2 \tag{3}$$

where S is the droplet shear stress, N m^{-2}; ρ is the droplet density, kg m^{-3}.

Data analysis

The coefficient of variation (CV) was used to evaluate the dispersion degree of droplet impact angles at different distances from the sprinkler (Liu *et al.* 2016). A higher CV value indicated that the droplet impact angle distribution was more discrete. Regression analysis, using the Origin 8.5 software (OriginLab, Northampton, MA, USA), was employed to analyze the relationships between droplet impact angles and the distances from the sprinkler, droplet velocities, and shear stresses, respectively. The coefficient of determination (R^2) (Zapata *et al.* 2018) was used to evaluate the fitting performance of the regression models (a higher R^2 value means a higher model accuracy). The regression analysis was also applied to develop the radial distribution models of droplet shear stresses. The root mean square error (RMSE) and the normalized root mean square error (NRMSE) between the calculated and measured values were used to verify the accuracy of the distribution models (Zheng *et al.* 2018). The lower values of RMSE and NRMSE corresponded to the higher accuracies of models in predicting droplet shear stresses. Additionally, the effects of operating pressure, nozzle diameter, and distance from the sprinkler on droplet shear stresses were each subjected to multi-way analysis of variance (ANOVA). Means were separated using Fisher's Protected LSD at the 0.05 level via the software package SPSS 20.0 (IBM Corp., Armonk, NY, USA).

RESULTS AND DISCUSSION

Radial distribution of droplet impact angles

The droplet impact angle is closely related to the distribution of the droplet shear stress and the soil infiltration (Hattori & Kakuichi 2013). Figure 5 presents the radial relative frequency distributions of droplet impact angles under three operating pressures and two nozzle diameters. Within 4 m of the sprinkler, nearly 100% of the droplet impact angles were distributed in the range of 70–90°, regardless of nozzle diameters and operating pressures. This revealed that large droplet impact angles usually appeared near the sprinkler, in good agreement with the results reported by Chang & Hills (1993a). One possible explanation for this phenomenon is that many small droplets occurred near the sprinkler. For the distances of 2 and 4 m away from the sprinkler, the relative frequencies of droplets with a diameter of 0–2 mm were close to 100% under different treatments (Table 1). Smaller droplets had a larger specific surface area (surface area per unit volume), resulting in a higher air friction resistance ratio, which in turn made the horizontal droplet velocity quickly approach zero. Therefore, the horizontal distances traveled by smaller droplets

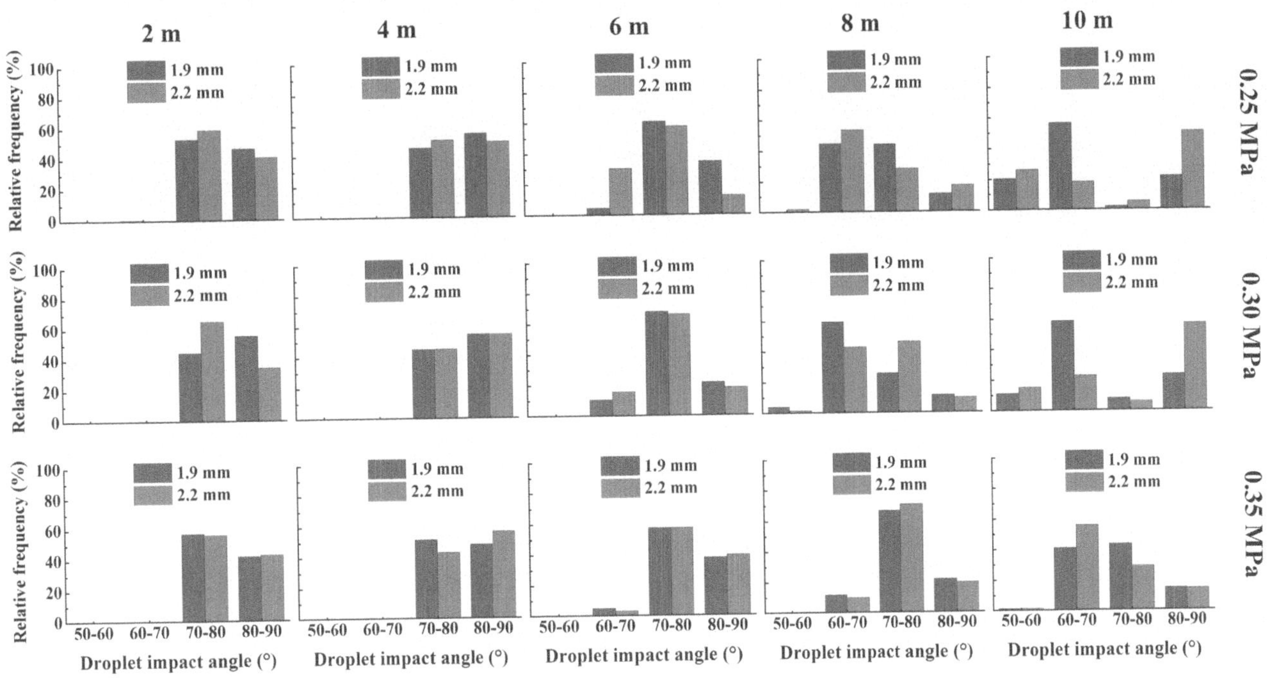

Figure 5 | Radial distributions of droplet impact angles under three operating pressures and two nozzle diameters.

Table 1 | Relative frequency distributions of droplet diameters along the spray direction under three operating pressures and two nozzle diameters

		Relative frequency (%)									
		1.9 mm					2.2 mm				
Operating pressure (MPa)	Droplet diameter (mm)	2 m	4 m	6 m	8 m	10 m	2 m	4 m	6 m	8 m	10 m
0.25	0–1	99.3	97.0	66.2	14.6	21.5	99.7	92.6	10.0	16.5	56.0
	1–2	0.7	3.0	33.7	80.3	51.8	0.3	7.4	86.7	73.8	8.0
	2–3	0	0	0.1	4.9	26.0	0	0	3.3	9.0	32.0
	3–4	0	0	0	0.2	0.7	0	0	0	0.7	4.0
0.30	0–1	98.6	95.9	29.9	12.4	24.6	99.7	95.7	19.4	11.1	61.6
	1–2	1.2	4.1	69.7	69.1	46.7	0.3	4.3	79.6	81.1	4.8
	2–3	0.2	0	0.4	18.2	27.3	0	0	1.0	7.8	28.9
	3–4	0	0	0	0.3	1.4	0	0	0	0	4.7
0.35	0–1	99.7	94.0	54.3	27.4	14.7	99.5	91.8	50.5	16.9	14.6
	1–2	0.3	6.0	45.5	70.7	73.6	0.5	8.2	48.9	80.1	60.8
	2–3	0	0	0.2	1.9	10.9	0	0	0.6	3.0	23.7
	3–4	0	0	0	0	0.8	0	0	0	0	0.9

were shorter, and their impact angles were almost perpendicular to the horizontal plane.

With increasing distance from the sprinkler, the number of droplets with an impact angle of 70–90° gradually decreased, and that of droplets with an angle below 70° increased, resulting in a dispersed distribution of droplet impact angles. For instance, when the distance from the sprinkler reached 10 m, the CV values of droplet impact angles at the operating pressure of 0.25 MPa were 17.6% and 22.4% for the nozzle diameters of 1.9 and 2.2 mm, respectively. These values increased by 10.9 and 16.4 percentage points, respectively, compared to the 2 m distance (Table 2). This implied that the end of the spray jet was probably a most vigorous area for sprinkler jet and droplet breakup and for droplets colliding

Table 2 | Coefficient of variation (CV) values of droplet impact angles along the spray direction under three operating pressures and two nozzle diameters

	CV (%)									
	1.9 mm					2.2 mm				
Operating pressure (MPa)	2 m	4 m	6 m	8 m	10 m	2 m	4 m	6 m	8 m	10 m
0.25	6.7	7.1	7.7	9.6	17.6	6.0	7.8	8.2	12.5	22.4
0.30	6.8	7.2	7.4	11.6	16.0	6.3	7.2	7.4	9.8	19.0
0.35	6.4	7.1	7.3	7.7	10.9	5.8	6.2	7.5	7.8	13.0

with each other, thereby resulting in a more dispersed distribution of droplet impact angles. Similar observations have been obtained by Zhang *et al.* (2018) and Hui *et al.* (2019) on the droplet size distribution.

Further analysis of Figure 5 indicated that when the sprinkler pressure was 0.35 MPa, the relative frequencies of two nozzles, impact angles of 70–90°, remained above 88% at the distance of 8 m from the sprinkler, far greater than for the other two pressures. Taking the nozzle diameter of 2.2 mm as an example, for 0.35 MPa, the relative frequency of droplet impact angles within 70–90° could reach 90.2% at the distance of 8 m, whereas the frequencies were only 44.7% and 56% for 0.25 and 0.30 MPa, respectively. This showed that raising the sprinkler working pressure might contribute to increasing the impact angles of water droplets at a location away from the sprinkler. This result was not surprising, since higher sprinkler operating pressures generated smaller droplets and faster initial ejecting velocities, which might carry the smaller droplets to a farther location. As shown in Table 1, at the distances of 8 and 10 m from the sprinkler, the 1.9 mm nozzle's relative frequencies of small droplets (0–2 mm) were higher by 3.2 and 15.0 percentage points under 0.35 MPa than under 0.25 MPa, respectively. The above values changed to 6.7 and 11.4 percentage points with the 2.2 mm nozzle, respectively. The observations of Hills & Gu (1989) and Zhu *et al.* (2012) also supported this finding.

The effect of nozzle diameter on the droplet impact angle was not significant ($P > 0.05$, results not presented) at the same operating pressure and measuring point, which was in good agreement with the results reported by Chang & Hills (1993a). This was partly because larger nozzles generated larger droplet sizes. Thus, the loss ratios of air resistance decreased. Simultaneously, larger nozzles also had less friction losses, which increased the initial ejecting velocities. Hence, the sum of these two effects appeared to be zero, which has also been found by Hills & Gu (1989).

Relationship between droplet impact angle and distance from the sprinkler

To further determine the relationship between droplet impact angle and distance from the sprinkler, the radial distributions of average droplet impact angles under three operating pressures and two nozzle diameters were analyzed and are depicted in Figure 6. The maximum average droplet impact angle appeared at the distance of 4 m, irrespective of nozzle size and working pressure. This could be explained by the previous findings that nearly all droplet impact angles were distributed within 70–90° at this distance. With increasing distance from the sprinkler, the average droplet impact angles under each treatment continuously tapered off and finally reached the lowest values at 10 m. This outcome suggested that if water droplets had the same diameter and velocity, the droplets located farther away from the sprinkler might generate greater shear stresses due to their lower impact angles. This result explained why the area at the end of the sprinkler jet was prone to soil erosion (King & Bjorneberg 2011; Yan *et al.* 2011).

According to regression analysis results, the average droplet impact angle and the distance from the sprinkler showed a good linear relationship (the mean of R^2 under all working conditions was 0.850, Figure 6) for three operating pressures and two nozzle diameters, which could be expressed by Equation (4):

$$\theta_a = al + b \tag{4}$$

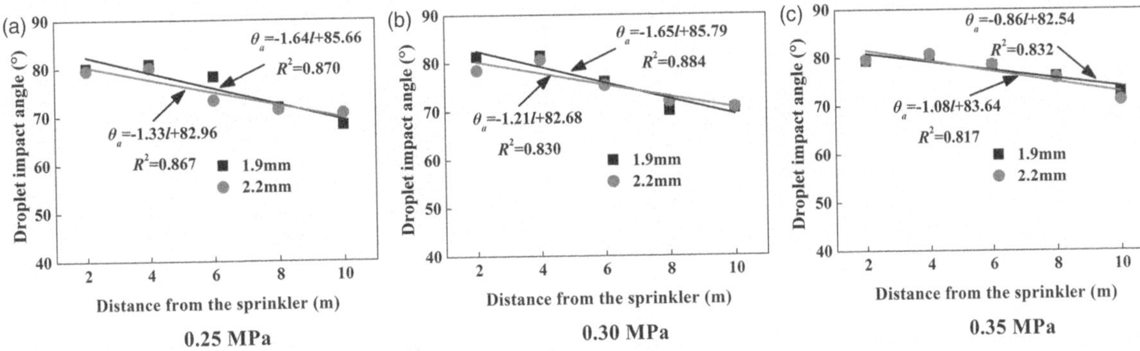

Figure 6 | Distributions of average droplet impact angles along the spray direction under three operating pressures and two nozzle diameters.

where θ_a is the average droplet impact angle, °; l is the distance from the sprinkler, m; a and b are the coefficients in the equation.

Notably, the coefficients a and b in Equation (4) for the nozzle diameters of 1.9 and 2.2 mm could be represented as the linear fitting equations of the operating pressure (p), respectively.

For the 1.9 mm nozzle diameter, the expressions were as follows:

$$a = 7.90p - 3.75 \quad (R^2 = 0.760) \tag{5}$$

$$b = -32.50p + 94.41 \quad (R^2 = 0.780) \tag{6}$$

Therefore, substituting Equations (5) and (6) into Equation (4), the average droplet impact angle could be expressed as an equation of the distance from the sprinkler and operating pressure:

$$\theta_a = 7.90pl - 3.75l - 32.50p + 94.41 \tag{7}$$

For the 2.2 mm nozzle diameter, the expressions were as follows:

$$a = 2.50p - 1.96 \quad (R^2 = 0.999) \tag{8}$$

$$b = -9.60p + 85.97 \quad (R^2 = 0.945) \tag{9}$$

Similarly, substituting Equations (8) and (9) into Equation (4), the average droplet impact angle could be expressed as an equation of the distance from the sprinkler and operating pressure:

$$\theta_a = 2.50pl - 1.96l - 9.60p + 85.97 \tag{10}$$

Relationship between droplet impact angle and droplet velocity

Droplet impact velocity is a key factor affecting the droplet shear stress, and a high droplet velocity has a higher potential to detach soil particles. Figure 7 presents the relationships between droplet impact angles and velocities for various distances to the sprinkler with three operating pressures and two nozzle diameters. Overall, droplet impact angles increased initially near the sprinkler and then decreased, with higher droplet velocities at a position far away from the sprinkler, irrespective of the nozzle pressure or diameter. When the distance from the sprinkler was 2 m, the droplet velocities and impact angles under different working conditions were mainly distributed at 1.0–5.5 m s^{-1} and 70–90°, respectively. As the distance increased to 4 m, the distribution range of droplet velocity further expanded, whereas the impact angle distribution remained unchanged. Nevertheless, the mean droplet impact angle at this distance increased slightly relative to the distance of 2 m. Taking the 1.9 mm nozzle diameter as an example, the average droplet impact angles were 0.8°, 0.1°, and 1.0° higher at 4 m than those at 2 m (80.1°, 81.4°, and 79.3°) for 0.25, 0.30, and 0.35 MPa, respectively.

With the gradual increase of the distance, the distribution range of droplet velocities expanded continually, but the impact angles of some water droplets began to

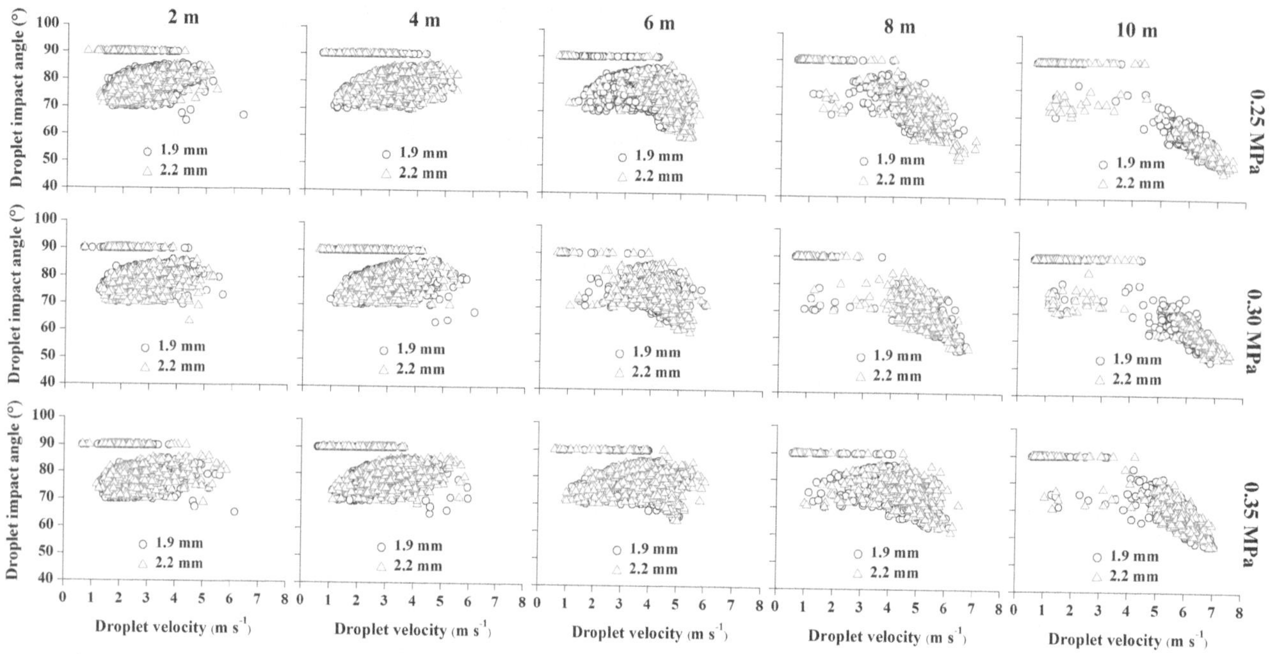

Figure 7 | Relationships between droplet impact angles and velocities for different distances from the sprinkler under three operating pressures and two nozzle diameters.

decrease. At a distance of 6 m, both nozzles' droplet velocity ranges increased to 1.0–6.2 m s^{-1} for three operating pressures, while their droplet impact angles were gradually shifted to the range of 60–70°. When the distance reached 10 m, each treatment attained the maximum on the distribution range of droplet velocities, which was 1.0–7.8 m s^{-1}. Correspondingly, the droplet impact angles under various working pressures (0.25, 0.30, and 0.35 MPa) were finally stabilized at around 68.2°, 70.5°, and 72.9° for 1.9 mm and 70.6°, 70.7°, and 71.2° for 2.2 mm. The above results implied that farther away from the sprinkler, greater droplet shear stresses occurred due to higher droplet velocities and smaller impact angles. This confirmed once again that susceptibility to soil erosion at the end of the spray jet was actually closely related to the large droplet shear stress rather than simply attributed to the large droplet kinetic energy (Ghadiri & Payne 2010).

Additionally, the distributions of average droplet impact angles with the average droplet velocities under three operating pressures and two nozzle diameters were analyzed and are depicted in Figure 8. Likewise, average droplet impact angles showed a strong linear relationship (the mean of R^2 under all working conditions was 0.880) with the average droplet velocities for different irrigation treatments, and

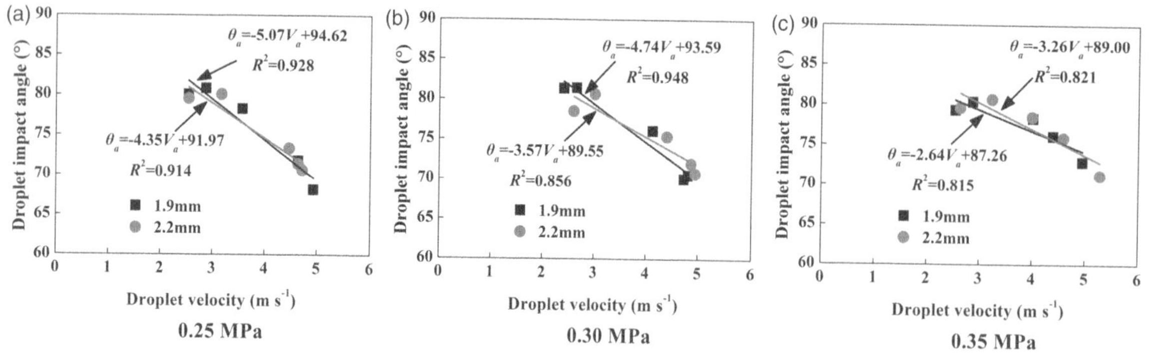

Figure 8 | Average droplet impact angle versus average droplet velocity under three operating pressures and two nozzle diameters.

their linear equations were expressed as follows:

$$\theta_a = cV_a + d \tag{11}$$

where V_a is the average droplet velocity, m s^{-1}; c and d are the coefficients in the equation.

After removal of the coefficients c and d, using a step-wise linear regression procedure, the relationships of θ_a with p and V_a for the two nozzles were determined using the following equations.

For the nozzle diameter of 1.9 mm,

$$\theta_a = 24.30pV_a - 11.44V_a - 73.60p + 113.90 \tag{12}$$

For the nozzle diameter of 2.2 mm,

$$\theta_a = 10.90pV_a - 7.00V_a - 29.70p + 99.08 \tag{13}$$

Relationship between droplet impact angle and droplet shear stress

Droplet shear stress is a major contributor to soil surface crusting (Ghadiri & Payne 1986). Figure 9 presents the relationships between droplet impact angles and shear stresses for different sprinkler working conditions. Obviously, a greater distance from the sprinkler resulted in an increased droplet shear stress, and the peak shear stresses of all irrigation treatments appeared at the end of the spray jet (10 m). For operating pressures of 0.25, 0.30, and 0.35 MPa, the maximum droplet shear stresses (the largest values of shear stresses among all water droplets measured in the test) for a nozzle diameter of 1.9 mm were 9,874.0, 8,865.0, and 7,226.5 N m^2, respectively. The corresponding values for the 2.2 mm nozzle were 10,798.4, 9,682.9, and 7,246.1 N m^2, respectively. These data revealed that the maximum shear stresses at 2.2 mm were above those at 1.9 mm, and the maximum values gradually decreased as operating pressures increased. This finding fully explained that increasing the sprinkler pressure or decreasing the nozzle size could effectively lessen the upper limit of droplet shear stresses, although the effect of nozzle size on the impact angle distribution was not significant. The main reason for this phenomenon was that the decrease of the nozzle size reduced droplet impact velocities, and therefore the shear stresses correspondingly tapered off, although the droplet impact angle was only slighlty changed (Ghadiri & Payne 1986).

Based on Figure 9, with a larger droplet impact angle, the shear stress decreased, irrespective of the nozzle diameter and the operating pressure. This implied that the droplet impact angle was closely related to the shear stress, which was in line with the results reported by Chang & Hills (1993a, 1993b). Through regression analysis, we found a strong quadratic relationship between the droplet shear stress and impact angle, with an average R^2 value of 0.916 (Figure 9). The quadratic equation could be expressed as follows:

$$S = f\theta^2 + g\theta + h \tag{14}$$

where f, g, and h are the coefficients in the equation.

Due to the good linear relationships between p and coefficients f, g, and h, the equations of S with p and θ for the two nozzles could be calculated by Equations (15) and (16), respectively.

For the 1.9 mm diameter, the equation was as follows:

$$S = -8p \cdot \theta^2 + 10.61\theta^2 + 1207p \cdot \theta - 1740.70\theta - 44700p + 71104 \tag{15}$$

For the 2.2 mm diameter, the equation was as follows:

$$S = -21.30p \cdot \theta^2 + 14.73\theta^2 + 3223p \cdot \theta - 2368.30\theta - 119210p + 94452 \tag{16}$$

Model for the radial distribution of droplet shear stresses

Scientific prediction of the radial distribution of droplet shear stresses is of great significance for optimizing sprinkler irrigation systems and reducing soil erosion risk. Figure 10 presents the distribution models of average droplet shear stresses (S_a) along the spray direction under three operating pressures and two nozzle diameters. These models could be expressed in the form of

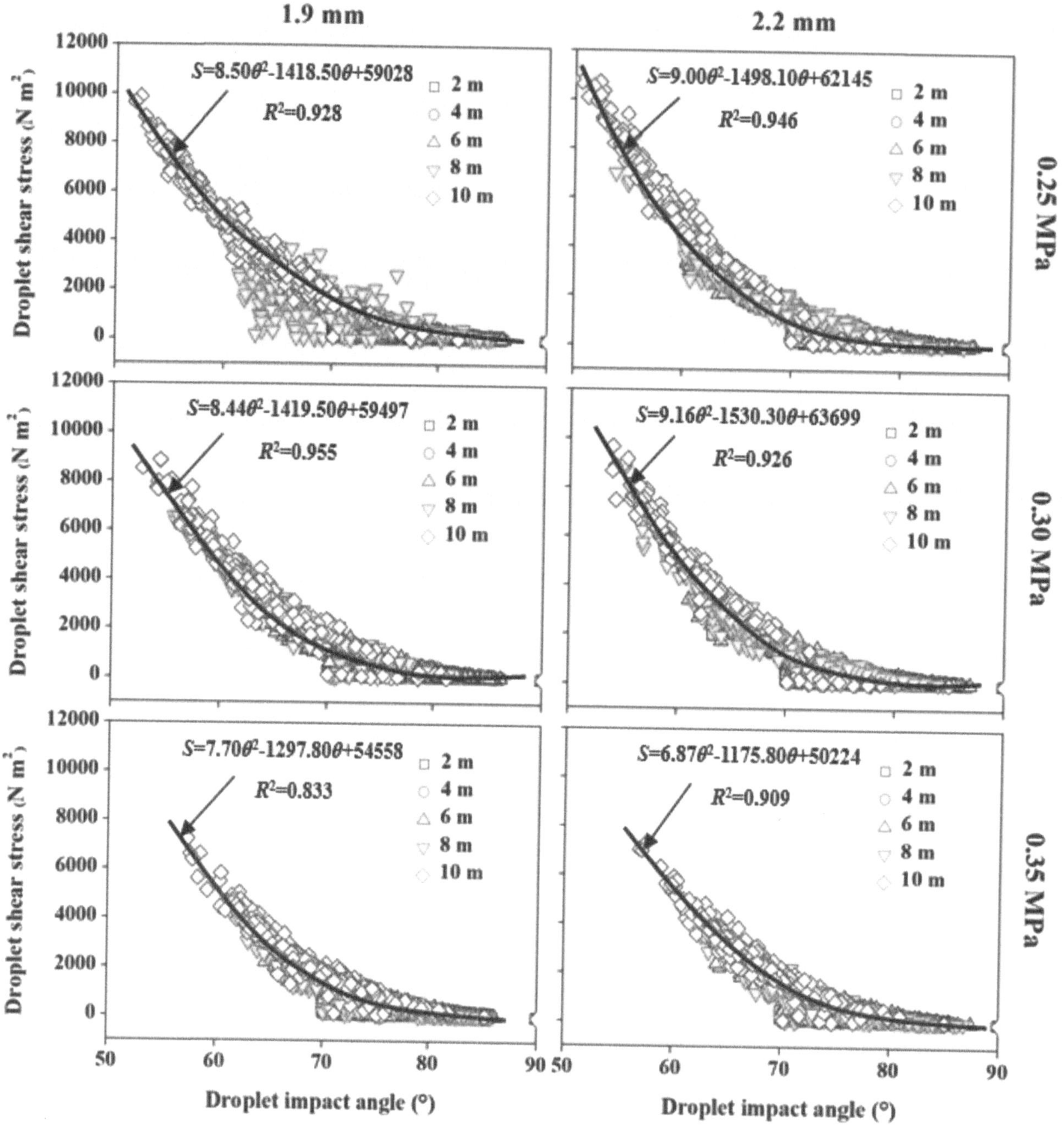

Figure 9 | Relationships between droplet impact angles and shear stresses for different distances from the sprinkler under three operating pressures and two nozzle diameters.

an exponential equation $S_a = je^{k \cdot l}$, where j and k are the coefficients. In each treatment, the average droplet shear stress increased with increasing distance from the sprinkler and decreased with decreasing operating pressure, which was consistent with the above results of the maximum shear stress.

In addition, as the operating pressure (p) had a great impact on the hydraulic performance of the sprinkler (Zhang *et al.* 2013), the following mathematical models could be attained after the coefficients j and k in the exponential equation were gradually replaced with p, according to the regression analysis.

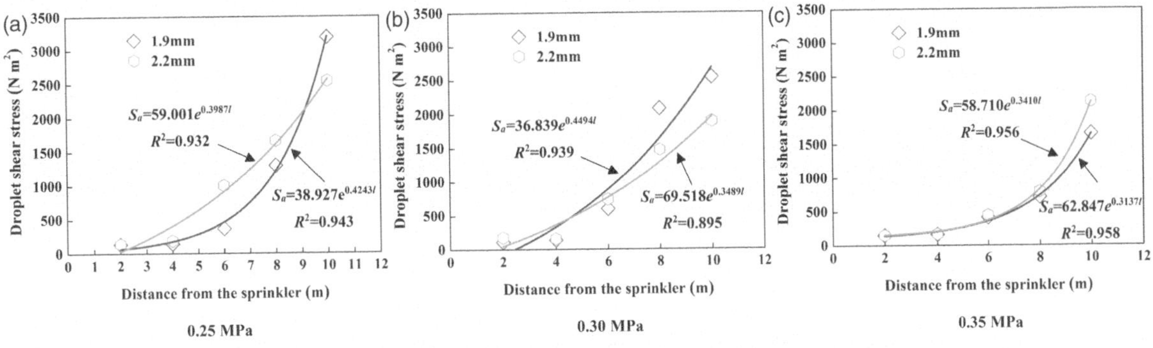

Figure 10 | Distributions of average droplet shear stresses along the spray direction under three operating pressures and two nozzle diameters.

For the 1.9 mm diameter:

$$S_a = (5619.20p^2 - 3132.30p + 470.81)e^{-32.16p^2l+18.19pl-2.1132l} \quad (17)$$

For the 2.2 mm diameter:

$$S_a = (-4265p^2 + 2556.1p - 313.46)e^{8.38p^2l-5.605pl+1.2762l} \quad (18)$$

To verify the accuracy of the models, the comparisons between the measured and calculated values of droplet shear stresses for both nozzle diameters are shown in Figure 11. The results indicated that the calculated values were in good agreement with the measured values, and their RMSE and NRMSE values were 115.5 N m^2 and 12.6% for 1.9 mm, respectively, and 134.9 N m^2 and 15.3% for 2.2 mm, respectively. Consequently, the prediction models proposed in this paper were relatively reliable and could be recommended for predicting the ball-driven sprinkler's radial distribution of droplet shear stresses under different operating pressures and nozzle sizes.

Analysis of variance of the effects of various factors on droplet shear stress

To further determine the influences of various factors (operating pressure, nozzle diameter, and distance from the sprinkler) on the droplet shear stress, we performed

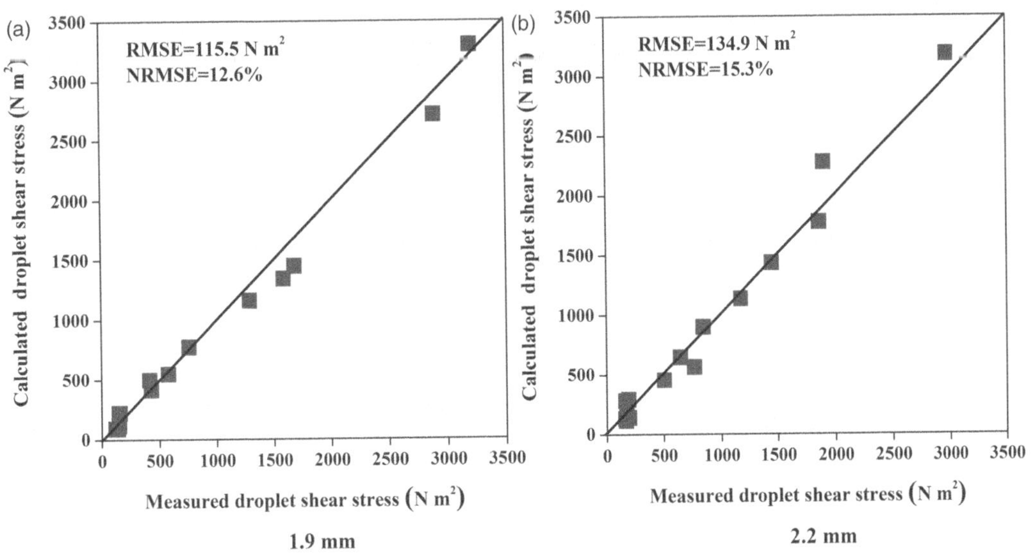

Figure 11 | Comparisons between calculated and measured results of droplet shear stresses using Equations (17) and (18).

Table 3 | Analysis of variance of the effect of three factors on the droplet shear stress

Factor	Sum of square of deviation	Degree of freedom	Mean square deviation	*F* value	*P* value
Operating pressure	4.461×10^8	2	2.230×10^8	123.047	0.000
Nozzle diameter	4.564×10^6	1	4.564×10^6	2.490	0.115
Distance from the sprinkler	1.301×10^{10}	4	3.251×10^9	2,924.770	0.000

ANOVA for the effect of each factor (Table 3). The droplet shear stress was significantly affected by the operating pressure and distance from the sprinkler ($P = 0.000$), whilst the effect of the nozzle diameter was non-significant ($P = 0.115$). These findings suggested that selecting a reasonable working pressure and sprinkler arrangement was essential to reduce the shear stress of water droplets and improve the performance of sprinkler irrigation systems. Although the nozzle diameter had a certain effect on the maximum value of the shear stress, the overall effect was not large.

CONCLUSIONS

Regardless of the nozzle diameter and operating pressure, the droplet impact angles in the range of 70–90° basically occurred within 4 m of the sprinkler. With increasing distance from the sprinkler, the higher droplet shear stress was obtained due to the increased droplet velocity and decreased impact angle. This confirmed that the soil erosion risk at the end of the spray jet was highly related to large droplet shear stresses. Moreover, the prediction models for the radial distribution of droplet shear stresses were developed using the droplet velocity and impact angle data measured by the 2DVD instrument. These models greatly improved the convenience of predicting a ball-driven sprinkler's droplet shear stress and facilitated the design optimization of the sprinkler irrigation system.

Based on the ANOVA results, the operating pressure and distance from the sprinkler significantly impacted the droplet shear stress. This finding revealed that the determination of a reasonable sprinkler working pressure and arrangement was crucial to reduce droplet shear stresses and improve the quality of sprinkler irrigation systems. Although the maximum shear stress was affected by the nozzle diameter, the overall effect was not large.

Our findings provide a new method for accurately predicating the soil erosion under sprinkler irrigation, and a basis for optimizing the working parameters of the sprinkler irrigation system. However, due to the unique structure of the ball-driven sprinkler, it may not be suitable for other typical sprinkler types, such as the impact sprinkler, fixed spray plate sprinkler and rotating spray plate sprinkler, etc., so further research is deserved. Besides, this study does not involve the relationship between the droplet shear stress and soil infiltration rate, which still needs to be determined.

ACKNOWLEDGEMENTS

This work is supported by the National Key Research and Development Program of China (Grant No. 2017YFD0201502), the National Natural Science Foundation of China (Grant No. 51939005), and the Science and Technology Planning Project of Hebei Province (Grant No. 17217002D).

DECLARATION OF CONFLICTS OF INTEREST

The authors declare that they have no conflicts of interest.

DATA AVAILABILITY STATEMENT

All relevant data are included in the paper or its Supplementary Information.

REFERENCES

Assouline, S. & Ben-Hur, M. 2006 Effects of rainfall intensity and slope gradient on the dynamics of interrill erosion during soil surface sealing. *Catena* **66** (3), 211–220. doi:10.1016/j.catena.2006.02.005.

Caracciolo, C., Napoli, M., Porcu, F., Prodi, F., Dietrich, S., Zanchi, C. & Orlandini, S. 2012 Raindrop size distribution and soil erosion. *Journal of Irrigation & Drainage Engineering* **138** (5), 461–469. doi:10.1061/(ASCE)IR.1943-4774.0000412.

Chang, W. J. & Hills, D. J. 1993a Sprinkler droplet effects on infiltration. II: laboratory study. *Journal of Irrigation and Drainage Engineering* **119** (1), 157–169. doi:10.1061/(ASCE)0733-9437(1993)119:1(157).

Chang, W. J. & Hills, D. J. 1993b Sprinkler droplet effects on infiltration. I: impact simulation. *Journal of Irrigation and Drainage Engineering* **119** (1), 142–156. doi:10.1061/(ASCE)0733-9437(1993)119:1(142).

Ferreira, A. G., Larock, B. E. & Singer, M. J. 1985 Computer simulation of water drop impact in a 9.6-mm deep pool. *Soil Science Society of America Journal* **49** (6), 1502–1507. doi:10.2136/sssaj1985.03615995004900060034x.

Ge, M., Wu, P., Zhu, D. & Zhang, L. 2018 Analysis of kinetic energy distribution of big gun sprinkler applied to continuous moving hose-drawn traveler. *Agricultural Water Management* **201**, 118–132. doi:10.1016/j.agwat.2017.12.009.

Ghadiri, H. & Payne, D. 1986 The risk of leaving the soil surface unprotected against falling rain. *Soil & Tillage Research* **8**, 119–130. doi:10.1016/0167-1987(86)90328-4.

Ghadiri, H. & Payne, D. 2010 The formation and characteristics of splash following raindrop impact on soil. *European Journal of Soil Science* **39** (4), 563–575. doi:10.1111/j.1365-2389.1988.tb01240.x.

Hattori, S. & Kakuichi, M. 2013 Effect of impact angle on liquid droplet impingement erosion. *Wear* **298–299**, 1–7. doi:10.1016/j.wear.2012.12.025.

Hills, D. J. & Gu, Y. 1989 Sprinkler volume mean droplet diameter as a function of pressure. *Transactions of the ASAE* **32** (2), 471–476. doi:10.13031/2013.31028.

Huang, C., Bradford, J. M. & Cushman, J. H. 1982 A numerical study of raindrop impact phenomena: the rigid case. *Soil Science Society of America Journal* **46** (1), 14–19. doi:10.2136/sssaj1982.03615995004600010003x.

Huang, G. J., Bringi, V. N., Cifelli, R., Hudak, D. & Petersen, W. A. 2010 A methodology to derive radar reflectivity-liquid equivalent snow rate relations using C-band radar and a 2D video disdrometer. *Journal of Atmospheric & Oceanic Technology* **27** (4), 637–651. doi:10.1175/2009JTECHA1284.1.

Hui, X., Zhu, N. J., Tan, H. B., Xie, S. H., Pang, J. R. & Yan, H. J. 2019 Hydraulic performance test and working parameters determination of ball-driven sprinkler. *Water Saving Irrigation* (4), 7–11. doi:10.3969/j.issn.1007-4929.2019.04.002.

ISO 2004 *Agricultural Irrigation Equipment-Rotating Sprinklers. 2: Uniformity of Distribution and Test Methods.* ISO 7749-2. ISO, Geneva, Switzerland.

ISO 2012 *Agricultural Irrigation Equipment-Sprinklers. 3: Characterization of Distribution and Test Methods.* ISO 15886-3. ISO, Geneva, Switzerland.

King, B. A. & Bjorneberg, D. L. 2011 Evaluation of potential runoff and erosion of four center pivot irrigation sprinklers. *Applied Engineering in Agriculture* **27** (1), 75–85. doi:10.13031/2013.36226.

Liu, J., Liu, X., Zhu, X. & Yuan, S. 2016 Droplet characterisation of a complete fluidic sprinkler with different nozzle dimensions. *Biosystems Engineering* **148**, 90–100. doi:10.1016/j.biosystemseng.2016.05.008.

Yan, H., Bai, G., He, J. & Lin, G. 2011 Influence of droplet kinetic energy flux density from fixed spray-plate sprinklers on soil infiltration, runoff and sediment yield. *Biosystems Engineering* **110** (2), 213–221. doi:10.1016/j.biosystemseng.2011.08.010.

Yan, H., Hui, X., Li, M. & Xu, Y. 2020 Development in sprinkler irrigation technology in China. *Irrigation and Drainage* **69**, 1–13. doi:10.1002/ird.2435.

Zapata, N., Robles, O., Playán, E., Paniagua, P., Romano, C., Salvador, R. & Montoya, F. 2018 Low-pressure sprinkler irrigation in maize: differences in water distribution above and below the crop canopy. *Agricultural Water Management* **203**, 353–365. doi:10.1016/j.agwat.2018.03.025.

Zhang, L., Merkley, G. P. & Pinthong, K. 2013 Assessing whole-field sprinkler irrigation application uniformity. *Irrigation Science* **31** (2), 87–105. doi:10.1007/s00271-011-0294-0.

Zhang, L., Hui, X. & Chen, J. 2018 Effect of slope on water distribution and application uniformity for sprinkler irrigation. *International Journal of Agricultural & Biological Engineering* **11** (3), 120–125. doi:10.25165/j.ijabe.20181103.2901.

Zheng, J., Fan, J., Zhang, F., Yan, S. & Xiang, Y. 2018 Rainfall partitioning into throughfall, stemflow and interception loss by maize canopy on the semi-arid Loess Plateau of China. *Agricultural Water Management* **195**, 25–36. doi:10.1016/j.agwat.2017.09.013.

Zhu, X., Yuan, S. & Liu, J. 2012 Effect of sprinkler head geometrical parameters on hydraulic performance of fluidic sprinkler. *Journal of Irrigation & Drainage Engineering* **138** (11), 1019–1026. doi:10.1061/(ASCE)IR.1943-4774.0000495.

First received 8 December 2020; accepted in revised form 12 January 2021. Available online 25 January 2021

A simplified method to improve water distribution and application uniformity for sprinkler irrigation on sloping land: adjustment of riser orientation

Xin Hui, Haijun Yan, Lin Zhang and Junying Chen

ABSTRACT

To improve the water application uniformity for sprinkler irrigation on sloping land, indoor tests were conducted on an artificial slope (slopes of 0, 0.05, 0.10 and 0.15) to evaluate the effects of two riser orientations, vertical (VO) and perpendicular (PO) to the slope, on the uniformity of sprinkler rotation, radius of throw, water distribution of an individual sprinkler and the overlapped water application uniformity (WAU). Compared with the VO, the PO could effectively improve the water distribution on sloping land and minimize the risk of soil erosion. Additionally, the PO was superior in the WAU, and a rectangular arrangement could dramatically enhance the WAU at smaller sprinkler spacing, while larger acceptable sprinkler spacing was accepted in a triangular arrangement. The riser orientation and sprinkler spacing had the most significant effect on the WAU, followed by the slope and sprinkler arrangement, suggesting that the adjustment of riser orientation or sprinkler spacing was helpful in improving the WAU. However, from the aspects of investment cost and installation convenience for irrigation projects, the method of PO was recommended. Therefore, when designing the sprinkler irrigation systems on the slope, choosing PO is the simplest and most effective way to achieve good irrigation uniformity.

Key words | overlapped application uniformity, radius of throw, riser orientation, sprinkler rotation uniformity, water distribution

Xin Hui
Haijun Yan (corresponding author)
College of Water Resources and Civil Engineering,
China Agricultural University,
Beijing 100083,
China
E-mail: *yanhj@cau.edu.cn*

Lin Zhang
Institute of Soil and Water Conservation,
Northwest A&F University,
Yangling 712100, Shaanxi,
China

Junying Chen
College of Water Resources and Architectural Engineering,
Northwest A&F University,
Yangling 712100, Shaanxi,
China

HIGHLIGHTS

- Riser perpendicular to the slope can obviously improve the water distribution on sloping land and minimize the risk of soil erosion.
- The overlapped water application uniformity is superior for riser perpendicular to the slope than for a vertical riser.
- Both the riser orientation and sprinkler spacing have the most significant effects on the overlapped water application uniformity.

doi: 10.2166/ws.2020.231

GRAPHICAL ABSTRACT

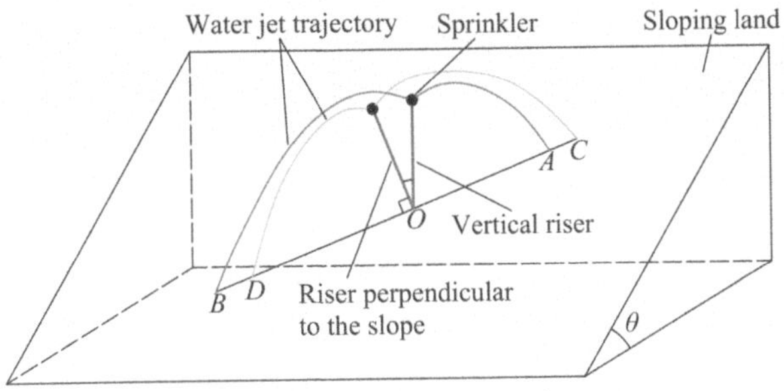

Riser perpendicular to the slope can improve the sprinkler water distribution on sloping land

INTRODUCTION

China has 135 million ha of cultivated land, of which slope cultivated land accounts for 33.33 million ha, approximately 1/4 of the total cultivated land area. Sloped farmland is the food production base for the survival and development of people in China's hilly or mountainous areas. Due to the long-term influence of terrain slope, the soil on sloping farmland is sensitive to drought, which drastically reduces crop yield and quality (Zhang *et al.* 2018). Thus, it is of great importance to determine a reasonable irrigation method for crops on sloping land. Compared with traditional irrigation methods, sprinkler irrigation on sloping land has been shown to have both higher efficiency and better adaptability to the terrain (Elwadie *et al.* 2010).

Sprinkler distribution is an important parameter in the design of irrigation engineering, and a key indicator for evaluating the water application uniformity of irrigation systems (Zhu *et al.* 2012). There are many factors that affect water distribution, including the sprinkler head, distribution system, and climatic and management factors (Mateos 1998). Because of the terrain slope, the water distribution of a single sprinkler on sloping land is quite different from that on flat land. When no wind is present, the curves of water distribution on flat ground approximate a set of concentric circles centred on the location of the sprinkler, and the water application rates at the same distance from the sprinkler are almost equal. However, when the sprinkler is used on sloping land, the water distribution patterns approximate a set of eccentric circles, and at the same distance from the sprinkler, the water application rate for uphill is greater than that for downhill, thereby resulting in poor water application uniformity (Zhang *et al.* 2018).

Extensive research has been conducted on sprinkler water distribution patterns and their application uniformity on sloping land. Montazar & Moridnejad (2008) revealed that soil moisture uniformity was more sensitive to the water application uniformity than to terrain slope. Ben-Hur *et al.* (1995) found that soil water contents were higher for downhill areas because of runoff. Cisneros Espinosa *et al.* (2007) performed an experimental assessment of the sprinkler application rate for steep sloping fields and determined the maximum application rates that caused zero runoff for slopes above 0.16 with a NAAN sprinkler. Additionally, due to the difficulty in measuring the sprinkler water distribution on sloping land, calculation models for the water distribution on sloping land were established using dynamics theory, ballistic theory and the water balance principle (Zhang *et al.* 2016).

Many previous research results have provided the basis for the design and application of sprinkler irrigation systems on sloping land (Hasebe & Mizunoe 1970; Keller & Bliesner 1990; Cisneros Espinosa *et al.* 2007). However, these results mainly aimed at determining the characteristics of water distribution and soil water infiltration, and few studies have been performed to determine how to effectively improve the water application uniformity on sloping farmland. As one of the effective methods to improve the uniformity of slope irrigation, adjusting the orientation of the sprinkler riser has been gradually applied in recent years, due to its simple operation and low investment cost. A preliminary analysis on the effects of sprinkler riser orientation on rotation uniformity and throw radius was investigated by Li (1988), but the effects on water distribution and application uniformity were not analysed. Soares *et al.* (1991) established a model to evaluate the sprinkler water distribution under different slopes and riser orientations and found that the riser perpendicular to the soil surface could maximize the uniformity of water application. However, this model was developed using water data collected on flat ground combined with ballistic theory, and there were differences from the actual water distribution of the slope. Accordingly, the effects of the riser orientation on the sprinkler water distribution and application uniformity on sloping land still requires further investigation.

In this study, an experiment on water distribution patterns was conducted under two commonly used riser orientations (vertical (VO) and perpendicular (PO) to the slope) and four slopes (0, 0.05, 0.10, and 0.15). The objectives were (1) to evaluate the effects of the two riser orientations on the sprinkler rotation uniformity, radius of throw and the water distribution of individual sprinklers on sloping land; (2) to investigate the overlapped water application uniformity (WAU) of multiple sprinklers with various riser orientations, sprinkler arrangements, spacings and slopes; and (3) to further propose a new method to improve the water application uniformity for sprinkler irrigation on sloping land. The flowchart of research methodology for this study is shown in Figure 1.

MATERIALS AND METHODS

Experimental setup

The experiment of sprinkler irrigation on sloping land was conducted indoors under no wind conditions. The experimental slope surface was artificially constructed using

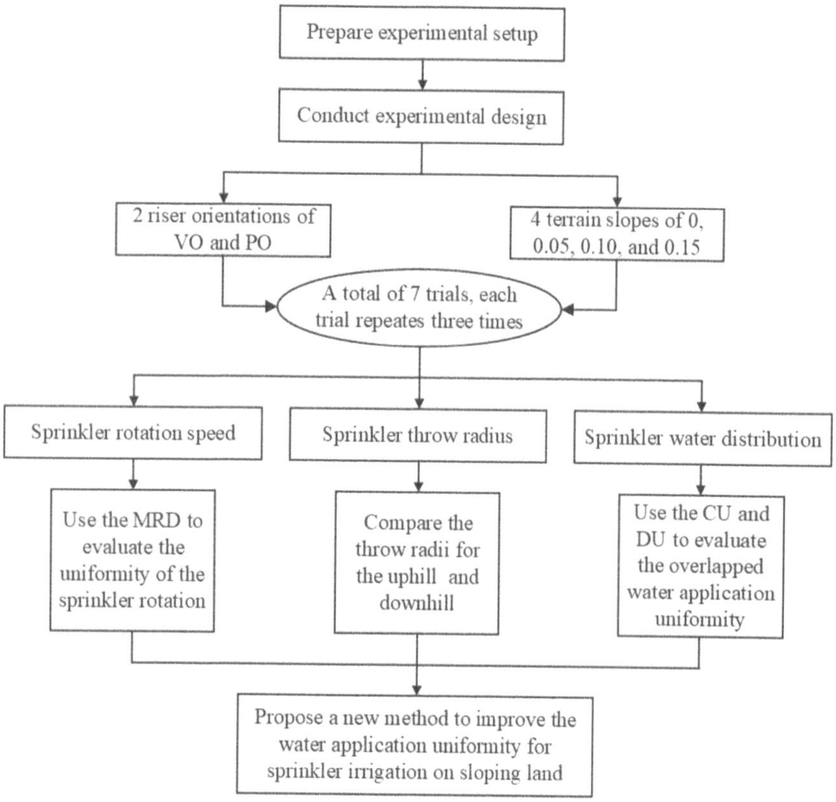

Figure 1 | Flowchart of research methodology.

height-adjustable brackets and steel channels. There were a total of 11 rows of steel channels of 12 m in length. The horizontal distance between the two adjacent rows was 1 m. The brackets were installed under the steel channels, and their heights were adjustable according to the test slope. To obtain the sprinkler water distribution on sloping land, catch cans with an opening diameter of 10.6 cm and a height of 15 cm were placed in the steel channels and arranged in a grid pattern. The grid size on the ground was 1 × 1 m. In the experiment, 11 catch cans were arranged on each row of steel channels, so there were 121 catch cans in 11 rows of steel channels (Figure 2).

An LF1200 rotating sprinkler (Rain Bird Corp., Azusa, California, USA), commonly used in agricultural irrigation, was selected for this study. The sprinkler has a 2.18 mm nozzle diameter and 17° jet angle, with a recommended operating pressure ranging from 210 to 410 kPa. In addition, a pressure transducer (Xi'an Xinmin model CYB, accuracy of ±0.1%) with a range from 0 to 500 kPa was installed at the sprinkler inlet and connected to a data logger. The pressure was recorded at 5 s intervals during each 1 h sprinkler test, and the average value was calculated for each test.

Experimental design

The sprinkler was tested with two riser orientations of VO and PO and four terrain slopes of 0, 0.05, 0.10, and 0.15. This setting was selected mainly because when the riser was PO, the riser angles (inclination of the riser from the vertical) corresponding to the various slopes (0, 0.05, 0.10 and 0.15) were 0°, 2.86°, 5.71°, and 8.53°, respectively, in accordance with the requirements of ISO 7749-1 (ISO standards 2004) that the riser angle should not exceed 10°. In total, there were seven trials in the overall experiment, and each of them were performed with three replicates in order to obtain reliable experimental data. During the sprinkler test, the working pressure of the sprinkler was stabilized at the designed pressure of 300 kPa, indoor air and water temperature were approximately 30 °C and 26 °C, respectively, and relative humidity was 60%. The test indicators included sprinkler rotation speed, radius of throw, as well as water distribution on sloping land.

Figure 2 | Experimental setup for sprinkler water distribution on sloping land: 1. pressure transducer; 2. retaining plastic; 3. sprinkler; 4. automatic pressure control system; 5. height adjustable bracket; 6. steel channel; 7. catch can.

Test of rotation speed of an individual sprinkler

Sprinkler rotation speed tests were conducted indoors for different riser orientations and slopes. The sprinkler has a full circle spray pattern and a clockwise rotation. Every 1/4 turn was regarded as a quadrant, that is, the uphill right quadrant and uphill left quadrant were defined as the first and second quadrants, respectively, and the downhill left quadrant and downhill right quadrant were defined as the third and fourth quadrants, respectively. The rotation times of four quadrants were recorded consecutively using a TF307 electronic stopwatch (Timestar Electronic Corp., Shenzhen, China) with an accuracy of 0.01 s. To comply with the ISO 7749-1 (ISO standards 2004), the rotation time for each combination of riser orientation and slope was taken as the average of five replications.

Test of water distribution and throw radius of an individual sprinkler

A full grid collector array method was used to test the sprinkler water distribution on sloping land. For convenience of the test, the sprinkler was installed at the bottom and the top of the slope surface, respectively. The height of the sprinkler riser was 30 cm, according to the manufacturer's recommendation. At the given operating pressure of 300 kPa, the sprinkler was first installed at the top of the slope, and the water distribution on the downhill was recorded. After a 1 h test, the sprinkler was installed at the bottom of the slope, and the water distribution on the uphill was recorded under exactly the same experimental conditions. The combination of the water distribution on the downhill and uphill was the water distribution on the whole slope surface. Additionally, according to the ISO 15886-3 (ISO standards 2012), the radii of throw on the slope was determined by measuring the distances to the farther points with the minimum effective water application rate of 0.26 mm h^{-1}.

Data analysis

The maximum relative sprinkler rotation deviation (MRD) (Li & Kawano 1996) was used to evaluate the uniformity of the sprinkler rotation. A higher MRD value means a lower rotation uniformity. The water distribution patterns of the individual sprinklers under different riser orientations and terrain slopes were assessed using surface maps produced with SURFER software (Golden Software Inc., Colorado, USA). Meanwhile, Catch 3D software (Merkley 2004) was used to simulate the overlapped water distribution of multiple sprinklers with various sprinkler arrangements and spacings. Additionally, the WAU was evaluated using the Christiansen uniformity coefficient (CU) (Christiansen 1942) and the distribution uniformity coefficient (DU) (Hart & Reynolds 1965). Additionally, the effects of riser orientations, sprinkler arrangements, spacings and slopes on the CU and DU values were each subjected to multi-way analysis of variance (ANOVA). Means were separated using Fisher's Protected LSD at the 0.05 level with SPSS 20.0 analytical software (IBM Corp., Armonk, NY, USA).

The MRD can be calculated by Equation (1):

$$\mathrm{MRD} = \frac{\max|T_j - \overline{T}|}{\overline{T}} \quad (j = 1, 2, 3, 4) \tag{1}$$

where T_j is the sprinkler rotation times of jth quadrant, s, and $\overline{T}$ is the average sprinkler rotation time of four quadrants, s.

The CU can be calculated by Equation (2):

$$\mathrm{CU} = \left(1 - \frac{\sum_{i=1}^{n} |h_i - \overline{h}|}{\sum_{i=1}^{n} h_i}\right) \times 100\% \tag{2}$$

where h_i is the measured water depth from an individual catch can, mm; $\overline{h}$ is the average measured water depth of all catch cans, mm; and n is the number of catch cans.

The DU can be calculated by Equation (3):

$$\mathrm{DU} = \frac{\overline{h_m}}{\overline{h}} \times 100\% \tag{3}$$

where $\overline{h_m}$ is the average of the lowest 1/4 of ranked catch can measurements, mm.

RESULTS AND DISCUSSION

Uniformity of individual sprinkler rotation

Sprinkler rotation uniformity is a key factor affecting the quality of sprinkler irrigation systems (Li & Kawano 1996). Generally, when the riser was VO, the sprinkler had uniform rotation speed, whereas the actual rotation speed varied when the riser was inclined from the vertical (Soares *et al.* 1991). The sprinkler rotation time per quadrant and MRD value versus terrain slope when the riser was PO are presented in Table 1. Overall, sprinkler rotation periods among the four slopes were basically the same, whereas a large variation was found in the MRD values. For instance, the rotation periods under the slopes of 0, 0.05, 0.10 and 0.15 all remained at approximately 18.08 s, but the corresponding MRD values were 0.96%, 3.19%, 6.81% and 9.21%, respectively (Table 1). This finding indicated that when the riser was PO, the increase in terrain slope caused the sprinkler rotation uniformity to decrease, regardless of the minimal effect on the sprinkler rotation period, which was consistent with the findings of Nderitu & Hills (1993) and Li & Kawano (1996).

Further analysis showed that the rotation times of T_1 and T_4 were always less than that of T_2 and T_3 at the slopes of 0.05, 0.10 and 0.15, which meant that when the

Table 1 | Sprinkler rotation time per quadrant, MRD value versus terrain slope when riser was PO

	Rotation time (s)					
Slope	T_1	T_2	T_3	T_4	**Period**	**MRD (%)**
0	4.42	4.48	4.40	4.45	17.75	0.96
0.05	4.50	4.69	4.57	4.42	18.18	3.19
0.10	4.32	4.82	4.61	4.30	18.05	6.81
0.15	4.22	5.01	4.94	4.18	18.35	9.21

riser was not vertical, the rotation speed slowed down when the sprinkler nozzle was moving in the uphill direction, whereas it accelerated when it was moving in the downhill direction, as previously reported by Soares *et al.* (1991). One possible explanation for this performance was that the riser inclined from the vertical led the gravity centre of the sprinkler to shift, so that the sprinkler rotating torque changed inevitably during the full circle spraying process. The sprinkler slowed down due to the reduction in the rotational torque when the nozzle turned towards the uphill direction, while the nozzle turned towards the downhill direction, a reverse result was observed.

In addition, from the analysis, the maximum MRD value appeared at the slope of 0.15, reaching 9.21%. This value was lower than the maximum allowed MRD value of 12% stipulated in Agricultural Irrigation Equipment-Rotating Sprinklers (ISO 7749-1 2004), suggesting that in this study, as the riser was perpendicular to a slope not exceeding 0.15, the uniformity of the sprinkler rotation could be guaranteed.

Throw radius of an individual sprinkler

The throw radius of a sprinkler plays an important role in the optimal selection of sprinkler spacing and lateral spacing (Ge *et al.* 2020). Table 2 presents the throw radii for the uphill (R_U) and downhill (R_D) under the two riser orientations and four slopes. It was clear that when the riser was VO, increasing the slope resulted in a gradually decreased R_U and a gradually increased R_D, thereby enlarging the difference between the R_U and R_D (ΔR), consistent with the results reported by Li (1988). Taking the slope of 0.15 as an example, the R_U decreased by 9.33% and the R_D increased by 10.10% compared to the slope of 0. Furthermore, the ΔR value reached a maximum of 2.02 m at the slope of 0.15. This occurred because, under the vertical orientation of the riser, the landing time of the water jet on sloping land was different from flat ground because of the terrain slope. When the sprinkler nozzle sprayed in the uphill direction, the water jet landed earlier than that for the flat ground, and a greater slope would result in earlier landing and a shorter throw radius. When the nozzle sprayed in the downhill direction, the result was reversed (Zhang *et al.* 2018).

Table 2 | Throw radii of sprinkler under two riser orientations and four terrain slopes

Riser orientation	Slope	R_U (m)	R_D (m)	ΔR (m)
VO	0	10.40	10.40	0.00
	0.05	10.28	10.90	0.62
	0.10	9.77	11.21	1.44
	0.15	9.43	11.45	2.02
PO	0	10.40	10.40	0.00
	0.05	10.40	10.52	0.12
	0.10	10.33	10.50	0.17
	0.15	10.30	10.34	0.04

Further analysis of Table 2 indicated that when the riser was PO, the impact of the slope on the throw radius appeared to be weakened. Both the measured R_U and R_D values at the slopes of 0.05, 0.1, and 0.15 were very close, and the ΔR values for any slope did not exceed 0.17 m. This finding was partly because this kind of riser orientation changed the initial jet angle of the nozzle, making the uphill and downhill jet trajectories similar to that of flat ground. As a result, the PO could effectively minimize the influence of terrain slope on the sprinkler throw radius and reduce the difference between the uphill and downhill throw radii, which was conducive to the design of these sprinkler irrigation projects.

Water distribution of an individual sprinkler

The riser orientation affects not only the rotation uniformity and throw radius of a sprinkler but also the trajectory of spray jet towards different directions, thereby affecting the sprinkler water distribution (Seginer *et al.* 1992). Figure 3 presents the water distribution patterns of individual sprinklers under the two riser orientations and four terrain slopes. The coordinate (0, 0) marks the location of the sprinkler. When the riser was VO, the water distribution on sloping land differed drastically from that on flat ground. From the perspective of water distribution shape, it was similar to a 'heart-shape' for the slopes of 0.05, 0.10 and 0.15, whereas it resembled a 'circle-shape' for the slope of 0 (Figure 3(a)), consistent with the results reported by Zhang *et al.* (2018). This result was primarily caused by the change in water jet landing time due to the influence of terrain slope, as in the aforementioned analysis. In contrast, when the riser was PO, the influence of the terrain slope on the water jet

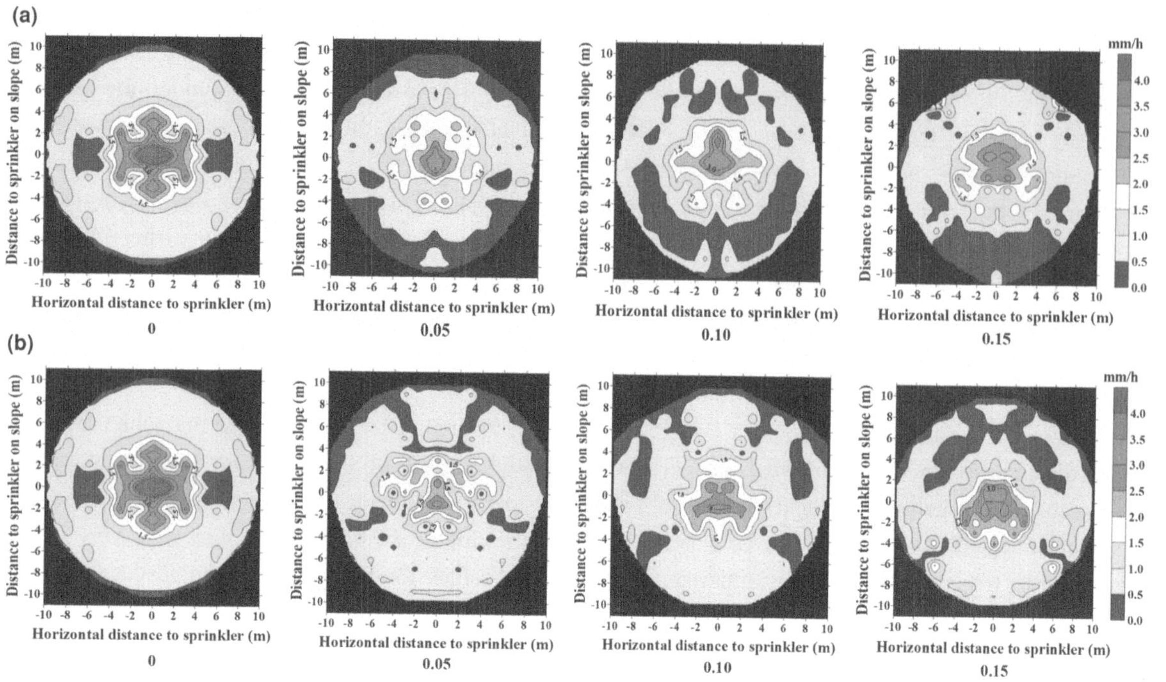

Figure 3 | Water distributions of an individual sprinkler under two riser orientations and four terrain slopes: (a) VO, (b) PO.

trajectory was greatly reduced. It was clear that as the slope increased, the water distribution curve on sloping land was gradually transformed from a 'heart-shape' to a more circular shape, and it finally approximated the water distribution curve on flat ground (Figure 3(b)).

Additionally, from the perspective of water application rate (WAR) on sloping land, the higher WAR was closer to the sprinkler and the lower WAR was further away from the sprinkler, regardless of whether the riser was VO or PO (Figure 3). Similar trends have been observed in the rotating spray plate sprinkler tests conducted by Sourell *et al.* (2003) on flat ground. Further analysis of Figure 3 indicated that, when the riser was VO, the applied water had a trend of concentrating on the uphill with the increase of the slope, while the opposite result was observed when the riser was PO. For instance, when the riser was VO, the average WAR values on the uphill were 0.02 mm h^{-1}, 0.19 mm h^{-1} and 0.22 mm h^{-1} higher than that on the downhill for the slopes of 0.05, 0.10 and 0.15, respectively. In contrast, when the riser was PO, the uphill WAR values for various slopes (0.05, 0.10 and 0.15) averaged 0.1 mm h^{-1}, 0.25 mm h^{-1} and 0.3 mm h^{-1} lower than that on the downhill, respectively (results not presented). These outcomes signified that the PO could effectively solve the problem of the concentration of irrigation water on the uphill. The greater the slope, the more pronounced was the improvement effect observed.

In an attempt to quantitatively study the water distribution of individual sprinklers on sloping land, the distribution proportions of water application rates (P_{WAR}) under the two riser orientations and four terrain slopes are presented in Figure 4. It was notable that most WAR values were below 1.5 mm h^{-1} for the different combinations of riser orientation and terrain slope, covering more than 83%. This finding demonstrated that the overall WAR values were relatively uniform, although the water distribution shape was affected by the riser orientation and slope, which might be attributed to the spray pattern of a single jet rotating sprinkler. From the aforementioned analysis, the WAR values on the slope were mainly distributed within 1.5 mm h^{-1}, but the applied water in the range of 1.5–4.0 mm h^{-1} must also be considered, because high WAR values are prone to cause soil erosion (Levy *et al.* 1992). Further investigation of Figure 4 revealed that, in terms of the WAR values within 1.5–4.0 mm h^{-1}, the P_{WAR} of the PO was lower than that of the VO. Using the three slopes (0, 0.05 and 0.15) selected in this study as an example, the P_{WAR} within 1.5–4.0 mm h^{-1} was 13.0%, 13.9% and 16.7%

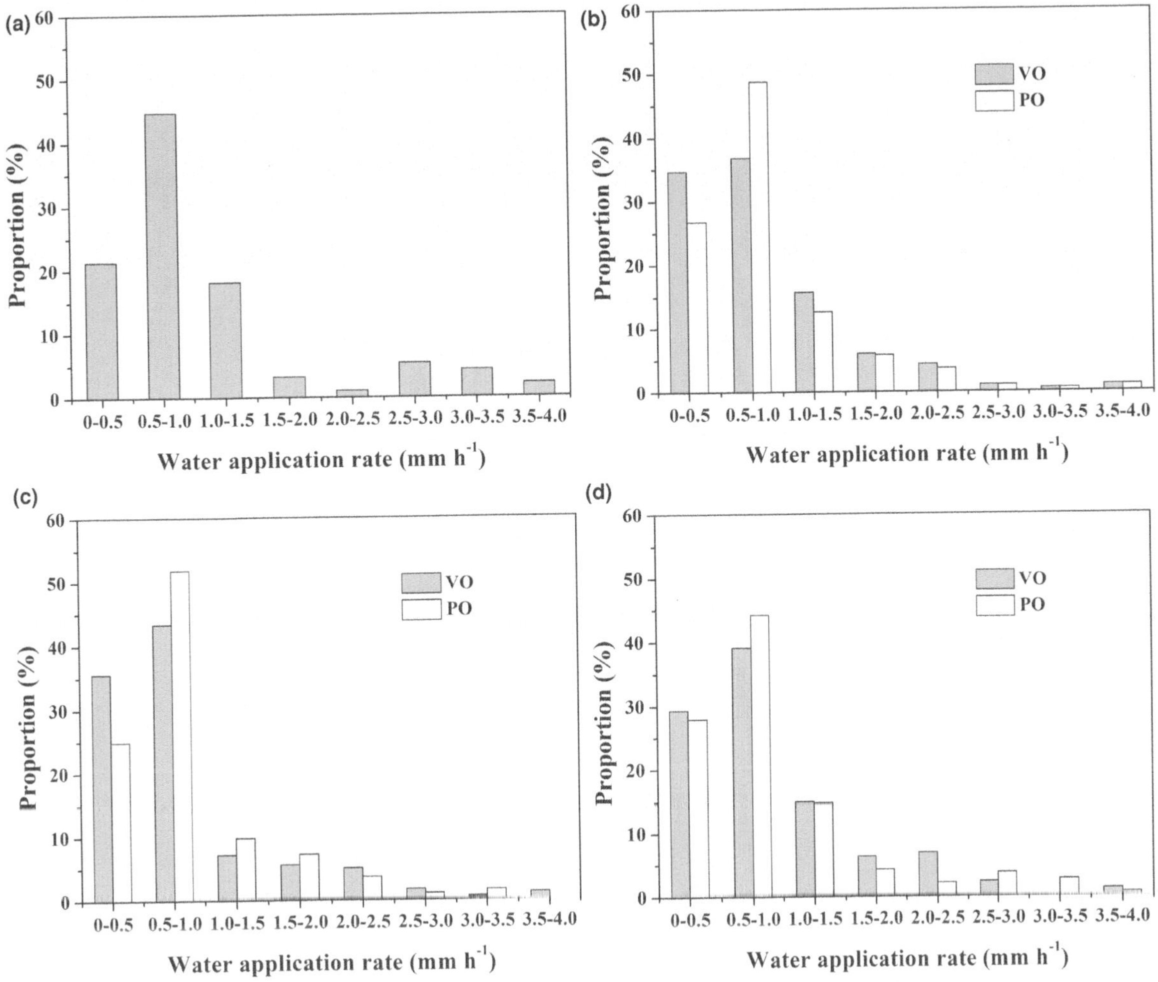

Figure 4 | Distribution proportions of water application rates under two riser orientations and four terrain slopes: (a) 0, (b) 0.05, (c) 0.10, (d) 0.15.

for the VO, respectively, whereas they were 12.0%, 13.5% and 13.2% for the PO, respectively. In summary, the PO could not only improve the water distribution of an individual sprinkler on sloping land but also minimize the risk of soil erosion, resembling the results obtained by Soares *et al.* (1991) through a ballistic trajectory model.

Overlapped water application uniformity of multiple sprinklers

Effect of riser orientations, sprinkler arrangements, spacings and slopes on overlapped water application uniformity

Statistical analysis of the WAU is usually of significant interest because, in practice, sprinklers are always overlapped (Playán *et al.* 2006). Since the radius of throw on flat ground was 10.4 m at the designed pressure of 300 kPa, seven different sprinkler spacings (8 × 8 m, 8 × 10 m, 10 × 10 m, 10 × 12 m, 12 × 12 m, 12 × 14 m, 14 × 14 m) and two sprinkler arrangements (rectangular and triangular) were considered for simulating the WAU values at the selected two riser orientations and four terrain slopes, as shown in Table 3. Overall, the VO resulted in lower WAU values compared to those with the PO, consistent with results reported by Soares *et al.* (1991). For instance, under the rectangular arrangement, the mean CU and DU values at various sprinkler spacings and slopes were 69.0% and 55.4% for the VO and 73.0% and 60.9% for the PO, respectively, whereas they were 66.7% and 51.4% for the VO, and 71.7% and 57.9% for the PO under the triangular arrangement,

Table 3 | Overlapped water application uniformities (%) for rectangular and triangular arrangements with various sprinkler spacings under two riser orientations and four terrain slopes

		8 × 8 m		8 × 10 m		10 × 10 m		10 × 12 m		12 × 12 m		12 × 14 m		14 × 14 m	
Riser orientation	**Slope**	**CU**	**DU**	**CU**	**DU**	**CU**	**DU**	**CU**	**DU**	**CU**	**DU**	**CU**	**DU**	**CU**	**DU**
Rectangular arrangement															
VO	0	82.1	77.4	77.8	66.3	74.2	61.4	69.8	54.0	66.0	65.7	67.1	56.1	64.1	49.1
	0.05	73.2	65.1	74.1	66.3	72.5	58.0	68.2	54.0	65.7	51.8	63.8	51.1	64.0	49.0
	0.10	73.0	57.4	73.6	63.1	72.3	56.7	67.2	49.8	64.5	49.5	62.5	46.2	63.4	48.9
	0.15	72.2	57.2	73.0	60.0	71.4	54.6	66.3	49.1	64.3	48.0	62.5	42.3	62.5	42.2
	Mean	75.1	64.3	74.6	63.9	72.6	57.7	67.9	51.7	65.1	53.8	64.0	48.9	63.5	47.3
PO	0	82.1	77.4	77.8	66.3	74.2	61.4	69.8	54.0	66.0	65.7	67.1	56.1	64.1	49.1
	0.05	84.6	77.2	81.8	73.9	82.1	73.1	77.9	66.5	75.3	65.1	73.6	61.9	68.2	55.1
	0.10	76.8	64.0	75.9	64.1	74.2	61.2	72.3	57.9	68.4	53.5	70.2	55.8	69.8	54.1
	0.15	77.3	61.4	76.4	60.9	74.2	58.6	71.0	56.6	64.1	49.2	64.5	55.6	65.4	48.3
	Mean	80.2	70.0	78.0	66.3	76.2	63.6	72.8	58.8	68.5	58.4	68.9	57.4	66.9	51.7
Triangular arrangement															
VO	0	74.2	59.0	68.9	51.4	74.2	63.1	75.1	66.0	70.9	58.3	68.3	56.0	56.5	44.2
	0.05	68.6	53.3	67.3	48.9	72.5	61.4	73.7	59.6	69.9	54.4	66.8	50.0	56.0	44.0
	0.10	66.9	48.9	65.4	48.2	72.3	58.0	70.6	63.4	68.8	50.6	64.4	46.2	52.9	36.8
	0.15	59.6	36.2	60.9	42.1	71.4	56.7	69.9	55.2	66.4	47.4	62.5	44.6	52.4	34.2
	Mean	67.3	49.4	65.6	47.7	72.6	59.8	72.3	61.1	69.0	52.7	65.5	49.2	54.5	39.8
PO	0	74.2	59.0	68.9	51.4	74.2	63.1	75.1	66.0	70.9	58.3	68.3	56.0	56.5	44.2
	0.05	71.3	55.8	71.3	51.2	82.1	73.1	79.1	67.2	76.3	66.0	77.2	67.9	63.9	46.8
	0.10	74.2	61.2	75.9	64.1	78.8	69.5	70.3	58.4	70.4	53.3	72.8	56.0	62.3	49.0
	0.15	74.2	58.6	76.4	60.9	80.1	72.7	71.3	57.1	67.5	50.5	66.3	50.2	56.9	32.6
	Mean	73.5	58.7	73.1	56.9	78.8	69.6	74.0	62.2	71.3	57.0	71.2	57.5	59.9	43.2

respectively. It was not difficult to find that, with the rectangular arrangement, the CU and DU values for the PO were on average 4.0 and 5.5 percentage points higher than those for the VO, while with the triangular arrangement, the values were 5.0 and 6.5 percentage points higher. One possible explanation for this performance was that when the riser was VO, the sprinkler produced a more irregular profile on the slope than that of the PO (Figure 3), and this consequently influenced the overlapped distribution pattern, as previously reported by Nderitu & Hills (1993).

Further analysis of Table 3 indicated that the WAU differed drastically between the rectangular and triangular arrangements, whether VO or PO. Among them, the CU and DU values for the rectangular arrangement mainly varied from 62.5% to 84.6% and 42.2% to 77.4%, respectively, whereas they varied from 52.4% to 82.1% and 32.6% to 73.1% for the triangular arrangement, respectively. This finding revealed that the overall WAU of the rectangular arrangement was superior to that of the triangular arrangement; despite the close upper limits of the uniformity coefficients observed under the two arrangements, their lower limits varied significantly, with differences of the lower limits of the CU and DU values attained of 10.1 and 9.6 percentage points, respectively. This finding was consistent with the results reported by Fukui *et al.* (1980) but contrasted with the results reported by Dechmi *et al.* (2004).

Additionally, the maximum values of the WAU under different slopes and riser orientations were often found to occur at 8 × 8 m for the rectangular arrangement, and the WAU values gradually decreased with the increase in sprinkler spacing (Figure 5), resembling the previous findings by Nderitu & Hills (1993) for different riser heights and sprinkler spacings. For 8 × 8 m, 10 × 12 m and 14 × 14 m spacings, the average CU and DU values of the various slopes were 75.1%, 67.9%, 63.5% and 64.3%, 51.7%, 47.3%, respectively, when the riser was VO. The corresponding values were 80.2%, 72.8%, 66.9% and 70.0%, 58.8%, 51.7%, respectively, when the riser was PO. These data suggested that for the rectangular arrangement, if high uniformity was required, reducing the sprinkler spacing might

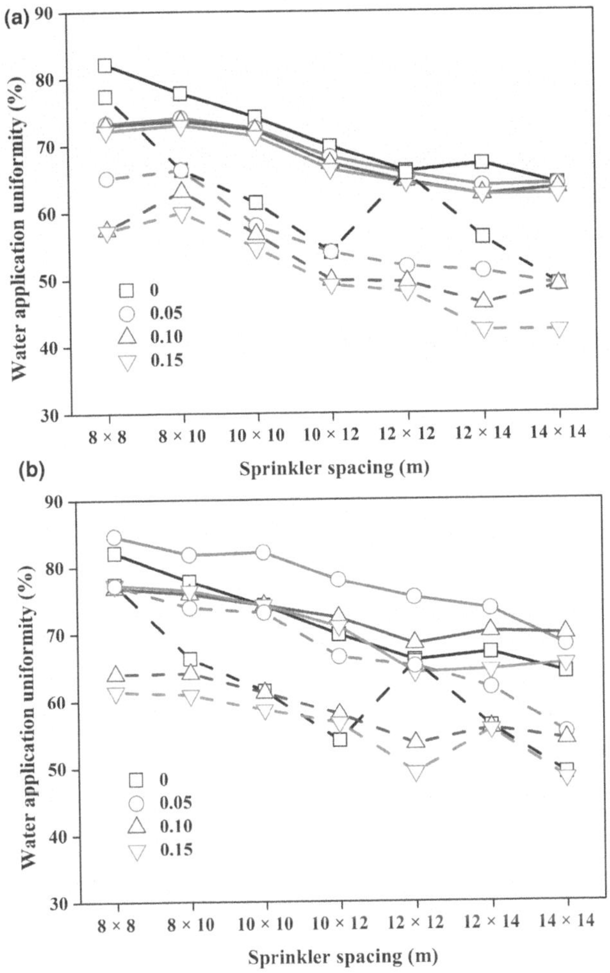

Figure 5 | CU and DU values versus sprinkler spacing for a rectangular arrangement under two riser orientations and four terrain slopes: (a) VO, (b) PO. The solid line represents CU, and the dotted line represents DU.

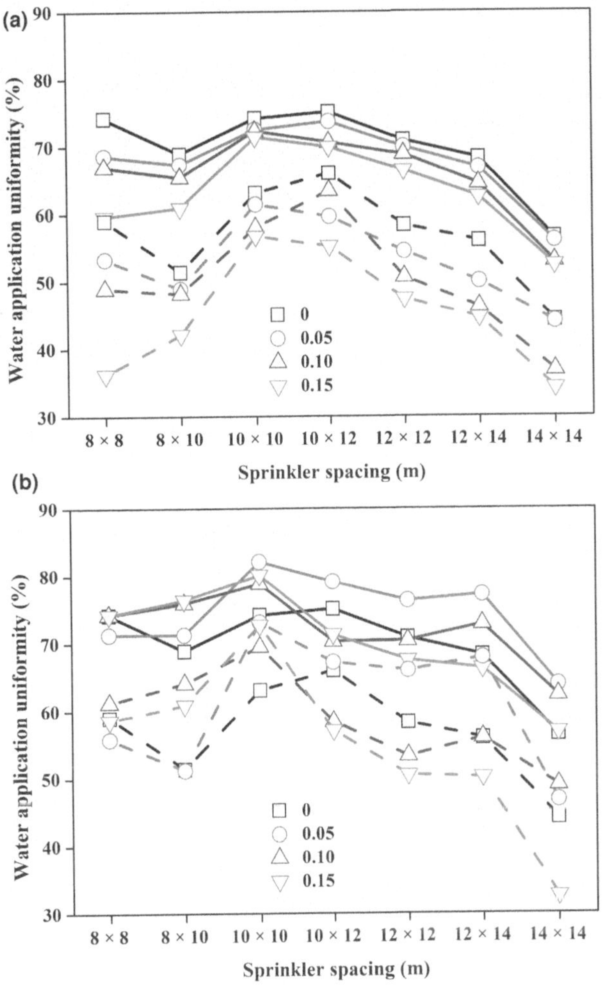

Figure 6 | CU and DU values versus sprinkler spacing for a triangular arrangement under two riser orientations and four terrain slopes: (a) VO, (b) PO. The solid line represents CU, and the dotted line represents DU.

be a good choice, in addition to the riser perpendicular to the slope. With very close spacings, however, the cost of hardware or labour would increase. The spacings, moreover, could be such that the instantaneous rates during the irrigation interval might become too high for the soil (Chen & Wallender 1984; Nderitu & Hills 1993). Hence, it was critical for determining the proper sprinkler spacing.

Compared to the rectangular arrangement, the WAU of the triangular arrangement had an initial increase and then tapered off as the sprinkler spacing increased, of which the highest WAU values at various slopes all mainly focused on the 10×10 m or 10×12 m spacings, regardless of the VO and PO (Figure 6). Moreover, it was notable that the WAU values of some triangular arrangements exceeded that of the rectangular arrangements under the three spacings of 10×12 m, 12×12 m and 12×14 m. Among these three sprinkler spacings and four different slopes, the mean CU and DU values of the triangular arrangement were 3.3% and 2.8% higher than that of the rectangular arrangement for the VO, respectively, and the above values changed to 2.1% and 0.8%, respectively, for the PO. These findings signified that the overall WAU values of the rectangular arrangement were better than that of the triangular arrangement, but the triangular arrangement seemed to improve the WAU at large sprinkler spacing in this study.

From the perspective of acceptable sprinkler spacing (ASS) (CU $\geq$ 75%, DU $\geq$ 60%) (Nderitu & Hills 1993), almost no ASS occurred in the VO in either the rectangular or triangular arrangement, suggesting that the VO indeed had poor WAU values. When the riser was PO, the maximum ASS values of the rectangular arrangement at various slopes was 8×10 m except for 12×12 m at the slope of 0.05. However, the maximum ASS values of all slopes were increased when switching from the rectangular to the triangular arrangement, especially for the slope of 0.05. The corresponding ASS values at the four slopes (0, 0.05, 0.10 and 0.15) increased to 10×12 m, 12×14 m, 10×10 m and 10×10 m, respectively. From the analysis, the rectangular and triangular arrangements had their own advantages for the PO. Among them, the rectangular arrangement played an important role in improving the uniformity under the small sprinkler spacing, while the triangular arrangement had a larger ASS value that could minimize the investment cost and risk of soil erosion. Taken together, in the design of sprinkler irrigation projects on sloping land, both of these arrangements should be used in combination; it is recommended to select the rectangular arrangement for small sprinkler spacing and the triangular arrangement for large sprinkler spacing when the riser is PO. This finding differs somewhat from those of previous studies, which suggested that either the rectangular or the triangular arrangement could be chosen in practice (Fukui *et al.* 1980; Dechmi *et al.* 2004).

Furthermore, the influence of the slope on the WAU also cannot be ignored. When the riser was VO, the WAU increased with the higher slope at the same sprinkler spacing, while for the PO, the variability of the WAU became less sensitive as the slope increased due to the reduction in the slope effect on water jet trajectory. Surprisingly, when the riser was PO, the overall WAU at the slope of 0.05 was highest compared to the other slopes, and the mean CU and DU values at different sprinkler spacings were at least 5.1 and 8.8 percentage points higher than that of the other slopes for the rectangular arrangement, as they were 2.4 and 2.3 percentage points for the triangular arrangement, respectively. Perhaps the water distribution pattern of a single sprinkler at a slope of 0.05 was more conducive to the water overlapping under the different sprinkler arrangements.

Analysis of variance of the effect of various factors on overlapped water application uniformity

To further investigate and compare the effect of different factors (riser orientation, sprinkler arrangement, spacing and slope) on the WAU, Tables 4 and 5 presented the ANOVA results of the effect of each factor on the CU and DU values, respectively. Notably, the riser orientation and sprinkler spacing had the most significant effect on the CU and DU values ($P = 0.000$) followed by the slope ($P = 0.081$ and 0.002 for CU and DU, respectively), and the sprinkler arrangement had the smallest effect ($P = 0.130$ and 0.040 for the CU and DU, respectively). This outcome suggested that, regardless of the sprinkler arrangement and slope, adjusting the riser orientation or sprinkler spacing was very helpful to improve

Table 4 | Analysis of variance of the effect of various factors on CU

Factor	Sum of square of deviation	Degree of freedom	Mean square deviation	F value	*P* value
Riser orientation	572.413	1	572.413	15.926	0.000
Sprinkler arrangement	93.623	1	93.623	2.323	0.130
Sprinkler spacing	2,235.437	6	372.573	17.079	0.000
Slope	272.262	3	90.754	2.304	0.081

Table 5 | Analysis of variance of the effect of various factors on DU

Factor	Sum of square of deviation	Degree of freedom	Mean square deviation	F value	*P* value
Riser orientation	1,006.201	1	1,006.201	13.751	0.000
Sprinkler arrangement	343.350	1	343.350	4.335	0.040
Sprinkler spacing	3,137.884	6	522.981	9.280	0.000
Slope	1,192.567	3	397.522	5.460	0.002

the WAU values on the slope as previously analysed, but from the aspects of investment cost and project convenience, changing the sprinkler spacing generally required a high level of manpower and material resources, and therefore the method of PO was recommended.

CONCLUSIONS

Compared with the vertical riser, the maximum relative sprinkler rotation deviation within the slope of 0.15 did not exceed 9.21%, indicating that the uniformity of sprinkler rotation could be guaranteed for the riser erected perpendicular to the slope. Again, the PO could effectively reduce the difference between uphill and downhill throw radii (ΔR). The ΔR values for slopes of 0.05, 0.1, and 0.15 did not exceed 0.17 m.

Despite the water distribution shape of individual sprinklers was affected by the riser orientation and slope, the overall WAR on sloping land was relatively uniform for both the VO and PO, and their WAR values below 1.5 mm h^{-1} all covered more than 83% at different combinations of riser orientation and terrain slope. Additionally, the distribution frequency of the high WAR values within 1.5–4.0 mm h^{-1} was observed to be lower for the PO than for the VO, suggesting that the PO was conducive to minimizing the risk of soil erosion.

The PO had an absolute superiority in overlapped WAU compared to the VO. Its mean CU and DU coefficient with rectangular arrangement were 4.0 and 5.5 percentage points higher than those for the VO, while with the triangular arrangement, the above values were changed to 5.0 and 6.5 percentage points. Moreover, for the PO, the rectangular arrangement could dramatically enhance the WAU at smaller sprinkler spacing, while the triangular arrangement had larger acceptable sprinkler spacing (CU ≥ 75%, DU ≥ 60%); thus, both arrangements should be used in combination for the design of sprinkler irrigation projects on sloping land.

From the ANOVA results of the effect of various factors on the WAU, the riser orientation and sprinkler spacing were found to have the most significant effect on the CU and DU values, followed by the slope and sprinkler arrangement, revealing that the adjustment of the riser orientation or sprinkler spacing was important in improving the WAU values, regardless of the sprinkler arrangement and slope. However, in view of investment cost and installation convenience for irrigation projects, the method of PO is a worthy recommendation. In summary, in order to obtain a good irrigation uniformity when designing the sprinkler irrigation system on slopes, choosing PO is undoubtedly the simplest and most effective way.

This paper presented the comparison results of the water distribution and application uniformity for the two riser orientations using a Rainbird LF1200 rotating sprinkler on an artificial slope. These results provided a basis for the design of sprinkler irrigation systems on sloping land, but there were still some limitations. In this paper, a rotating sprinkler with a unique structure was selected for experimental research. The water collected in catch cans were not weighed timely during the tests; this might lead to the measurement errors due to water evaporation. This study was conducted under indoor conditions, ignoring the influence of external factors such as wind speed and temperature, which might affect the design of sprinkler irrigation systems in the field. In addition, this study did not involve the soil water movement on sloping land, and further research is therefore required.

ACKNOWLEDGEMENT

The authors are greatly indebted to the National Key Research and Development Program of China (2017YFD0201502), National Natural Science Foundation of China (51939005, 51779246), and Key Research and Development Program of Shaanxi Province (2017NY-118).

DECLARATION OF CONFLICTS OF INTEREST

The authors declare that there are no conflicts of interest.

DATA AVAILABILITY STATEMENT

All relevant data are included in the paper or its Supplementary Information.

REFERENCES

Ben-Hur, M., Plaut, Z., Levy, G. J., Agassi, M. & Shaiberg, I. 1995 Surface runoff, uniformity of water distribution, and yield of peanut irrigated with a moving sprinkler system. *Agronomy Journal* **87** (4), 609–613. doi:10.2134/agronj1995.00021962008700040001x.

Chen, D. & Wallender, W. W. 1984 Economic sprinkler selection, spacing, and orientation. *Transactions of the ASAE* **27** (3), 737–743. doi:10.13031/2013.32863.

Christiansen, J. E. 1942 Irrigation by sprinkler. *California Agricultural Experimental Station Research Bulletin* **670**, 123.

Cisneros Espinosa, F. E., Torres, P. & Feyen, J. 2007 Experimental assessment of the sprinkler application rate for steep sloping fields. *Journal of Irrigation and Drainage Engineering* **133** (3), 276–278. doi:10.1061/(ASCE)0733-9437(2007)133:3(276).

Dechmi, F., Playán, E., Cavero, J., Martínez-Cob, A. & Faci, J. M. 2004 Coupled crop and solid set sprinkler simulation model. II: model application. *Journal of Irrigation and Drainage Engineering* **130** (6), 511–519. doi:10.1061/(asce)0733-9437(2004)130:6(511).

Elwadie, M. E. M., Mao, L. L. & Bralts, V. F. 2010 A simplified method for field evaluation of solid set sprinkler irrigation systems. *Applied Engineering in Agriculture* **26** (4), 589–597. doi:10.13031/2013.32064.

Fukui, Y., Nakanishi, K. & Okamura, S. 1980 Computer evaluation of sprinkler irrigation uniformity. *Irrigation Science* **2** (1), 23–32. doi:10.1007/BF00285427.

Ge, M., Wu, P., Zhu, D. & Zhang, L. 2020 Comparisons of spray characteristics between vertical impact and turbine drive sprinklers – a case study of the 50PYC and HY50 big gun-type sprinklers. *Agricultural Water Management* **228**, 105847. doi:10.1016/j.agwat.2019.105847.

Hart, W. E. & Reynolds, W. N. 1965 Analytical design of sprinkler systems. *Transactions of the ASAE* **8** (1), 83–0085. doi:10.13031/2013.40435.

Hasebe, J. & Mizunoe, M. 1970 Analytical methods for research into water distribution on sloping land receiving sprinkling irrigation. *Bulletin Tokai-Kinki National Agricultural Experiment Station* **19**, 150–173.

ISO 2004 *Agricultural Irrigation Equipment-Rotating Sprinklers. 1: Design and Operational Requirements.* ISO 7749-1. ISO, Geneva, Switzerland.

ISO 2012 *Agricultural Irrigation Equipment-Sprinklers. 3: Characterization of Distribution and Test Methods.* ISO 15886-3. ISO, Geneva, Switzerland.

Keller, J. & Bliesner, R. 1990 *Sprinkle and Trickle Irrigation.* Van Nostrand Reinhold, New York, NY, USA. doi:10.1007/978-1-4757-1425-8.

Levy, G. J., Levin, J., Gal, M., Ben-Hur, M. & Shainberg, I. 1992 Polymers' effects on infiltration and soil erosion during consecutive simulated sprinkler irrigations. *Soil Science Society of America Journal* **56** (3), 902–907. doi:10.2136/sssaj1992.03615995005600030037x.

Li, J. 1988 Suitable choice of sprinkler riser angle for sprinkler irrigation system in sloping land. *Journal of Hydraulic Engineering* (5), 45–48.

Li, J. & Kawano, H. 1996 The areal distribution of soil moisture under sprinkler irrigation. *Agricultural Water Management* **32** (1), 29–36. doi:10.1016/S0378-3774(96)01261-9.

Mateos, L. 1998 Assessing whole-field uniformity of stationary sprinkler irrigation systems. *Irrigation Science* **18** (2), 73–81. doi:10.1007/s002710050047.

Merkley, G. P. 2004 *Catch3D Users' Guide; Civil and Environmental Engineering Department.* Utah State University, Logan, UT, USA.

Montazar, A. & Moridnejad, M. 2008 Influence of wind and bed slope on water and soil moisture distribution in solid-set sprinkler systems. *Irrigation and Drainage: The Journal of the International Commission on Irrigation and Drainage* **57** (2), 175–185. doi:10.1002/ird.359.

Nderitu, S. M. & Hills, D. J. 1993 Sprinkler uniformity as affected by riser characteristics. *Applied Engineering in Agriculture* **9** (6), 515–521. doi:10.13031/2013.26016.

Playán, E., Zapata, N., Faci, J. M., Tolosa, D., Lacueva, J. L., Pelegrín, J. & Lafita, A. 2006 Assessing sprinkler irrigation uniformity using a ballistic simulation model. *Agricultural Water Management* **84** (1–2), 89–100. doi:10.1016/j.agwat.2006.01.006.

Seginer, I., Kantz, D., Nir, D. & Von Bernuth, R. D. 1992 Indoor measurement of single-radius sprinkler patterns. *Transactions of the ASAE* **35** (2), 523–533. doi:10.13031/2013.28630.

Soares, A. A., Willardson, L. S. & Keller, J. 1991 Surface-slope effects on sprinkler uniformity. *Journal of Irrigation and Drainage Engineering* **117** (6), 870–880. doi:10.1061/(ASCE)0733-9437(1991)117:6(870).

Sourell, H., Faci, J. M. & Playán, E. 2003 Performance of rotating spray plate sprinklers in indoor experiments. *Journal of Irrigation and Drainage Engineering* **129** (5), 376–380. doi:10.1061/(ASCE)0733-9437(2003)129:5(376).

Zhang, Y., Zhang, L., Zhu, D. & Hui, X. 2016 Simulation of water distribution and calculation on water application uniformity for sprinkler irrigation on sloping land based on ballistic theory. *Transactions of the CSAE* **32** (13), 59–66. doi:10.11975/j.issn.1002-6819.2016.13.009.

Zhang, L., Hui, X. & Chen, J. 2018 Effects of terrain slope on water distribution and application uniformity for sprinkler irrigation. *International Journal of Agricultural and Biological Engineering* **11** (3), 120–125. doi:10.25165/j.ijabe.20181103.2901.

Zhu, X., Yuan, S. & Liu, J. 2012 Effect of sprinkler head geometrical parameters on hydraulic performance of fluidic sprinkler. *Journal of Irrigation and Drainage Engineering* **138** (11), 1019–1026. doi:10.1061/(asce)ir.1943-4774.0000495.

First received 27 June 2020; accepted in revised form 8 September 2020. Available online 23 September 2020

Water reuse potential for irrigation in Brazilian hydrographic regions

Maíra Lima, Bruna Magalhães de Araujo, Sérgio Rodrigues Ayrimoraes Soares, Ana Silvia Pereira Santos and José Manuel Pereira Vieira IWA

ABSTRACT

The present paper carried out an evaluation of the reuse potential of the Wastewater Treatment Plants (WWTPs) effluents for irrigation in the 12 Brazilian Hydrographic Regions (BHRs). For this purpose, initially, the WWTPs were categorized and the effluent flow rate was estimated. Category 1 represents secondary effluent with an efficiency of organic matter removal greater than 80%; Category 2 represents effluent that underwent some disinfection step; and effluents that perform less than the other categories were called 'Uncategorized'. After that, the irrigation water demands for each BHRs were compiled, and finally, the production of water for reutilization was compared with the demand for irrigation. Thus, it was observed that all the sewage flow rates generated in Brazil classified in Categories 1 and 2 represent 9% of the total irrigation water demand in the country (1,078.71 m^3/s) and it stands out that only 7% of the flows treated in Brazil undergo a tertiary treatment step.

Key words | Brazilian Hydrographic Regions, irrigation water demand, wastewater treatment plant, water availability, water resources management, water reuse

Maíra Lima (corresponding author)
Bruna Magalhães de Araujo
Ana Silvia Pereira Santos
Engenharia Ambiental e Sanitária, Depto. de Eng. Sanitária e Meio Ambiente,
Universidade do Estado do Rio de Janeiro,
Rio de Janeiro, RJ,
Brazil
E-mail: *mairalima.90@gmail.com*

Sérgio Rodrigues Ayrimoraes Soares
Agência Nacional de Águas,
Rio de Janeiro,
Brazil

José Manuel Pereira Vieira IWA
University of Minho

HIGHLIGHTS

- In Brazil, irrigation demands 52% of water withdrawals.
- There are 12 BHRs for guiding the management of water resources.
- Flow rate of WWTP effluent with an organic matter removal efficiency greater than 80% represents 9% of the total water demand for irrigation.
- There are some BHRs with a high potential of reuse for irrigation.
- Some BHRs present high demand for irrigation, but with low sewage treatment coverage.

doi: 10.2166/ws.2020.280

GRAPHICAL ABSTRACT

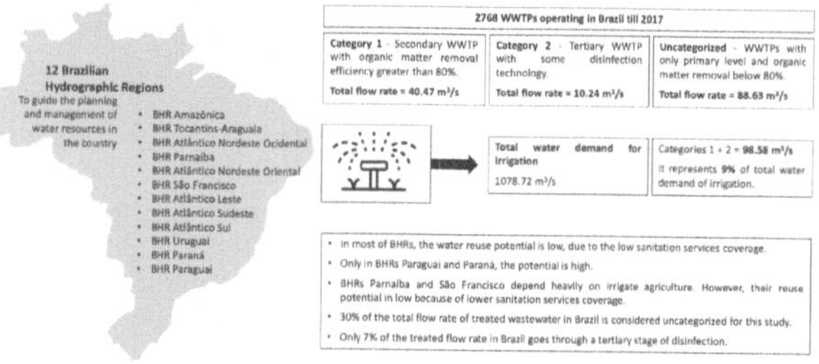

INTRODUCTION

Although Brazil has large water reserves, about 12–16% of the total amount available in the world (Ramos 2016), its water availability is not evenly distributed throughout its territory. Approximately 260 thousand m^3/s of water flows through Brazilian territory and of this amount, 80% is in the Amazon region, where there is the smallest portion of the population and the lowest water use demand (ANA 2018).

This context, which can become even more complex in a scenario of climate change and increased water use, leads to water stress in some regions of Brazil, like currently in the Southeast (ANA 2018). In the Semi-Arid Region, which covers the states of the northeast region (Alagoas, Bahia, Ceará, Paraíba, Pernambuco, Piauí, Rio Grande do Norte, and Sergipe), in addition to the north of the Minas Gerais state, the water scarcity is historic. Thus, this region has one of the lowest socioeconomic development rates in the country. The surface water availability in Brazilian territory is presented in Figure 1, with emphasis on the Semi-Arid region.

In 2003, according to Resolution No. 32 of the National Water Resources Council (Brazil 2003), Brazilian territory was divided into 12 Brazilian Hydrographic Regions (BHRs), with the objective of guiding the water resources planning and management in the country: (1) Amazônica, (2) Tocantins-Araguaia, (3) Atlântico Nordeste Ocidental, (4) Parnaíba, (5) Atlântico Nordeste Oriental, (6) São Francisco, (7) Atlântico Leste, (8) Atlântico Sudeste, (9) Paraná, (10) Uruguai, (11) Atlântico Sul, and (12) Paraguai. The 12 BHRs are represented in the map in Figure 2. The main characteristics of the 12 BHRs are shown in Table 1.

The BHRs with the most critical water levels are Atlântico Nordeste Oriental, located in the Semiarid Region; and Atlântico Sul, which has extremely high water demand for irrigation. In addition, BHRs Atlântico Leste and São Francisco have high water demands in comparison with their water availability (ANA 2018).

Throughout the world, due to water scarcity and the increase in water use conflicts, water conservation and reuse have gained prominence as water resource management tools (Angelakis *et al.* 2018). In addition, population growth, climate instability, and increased demand for food culminated in a scarcity of water of adequate quality for irrigation (Ahmadi & Merkley 2017; Maryam & Buyukgungor 2019).

The water reuse for irrigation purposes was already adopted in the world since the prehistoric period to the current, considering different aims and perspectives throughout these years (Mays *et al.* 2007; Angelakis & Spyridakis 2013; Angelakis *et al.* 2018). Nowadays, irrigated agriculture accounts for approximately 70% of total freshwater in the world, and this value is even higher in many developing countries (Peng *et al.* 2019). In Brazil, according to ANA (2019), irrigation demands about 52% of water withdrawals, followed by urban supply, the processing industry and animal supply.

Thus, the main destination of reclaimed water in the world is irrigated agriculture (Angelakis *et al.* 2018). However, it is important to highlight that this reclaimed water comes mainly from wastewater treatment plants (WWTPs). It is estimated that more than 10% of the world's population consumes agricultural products produced through wastewater irrigation (Jeong *et al.* 2016).

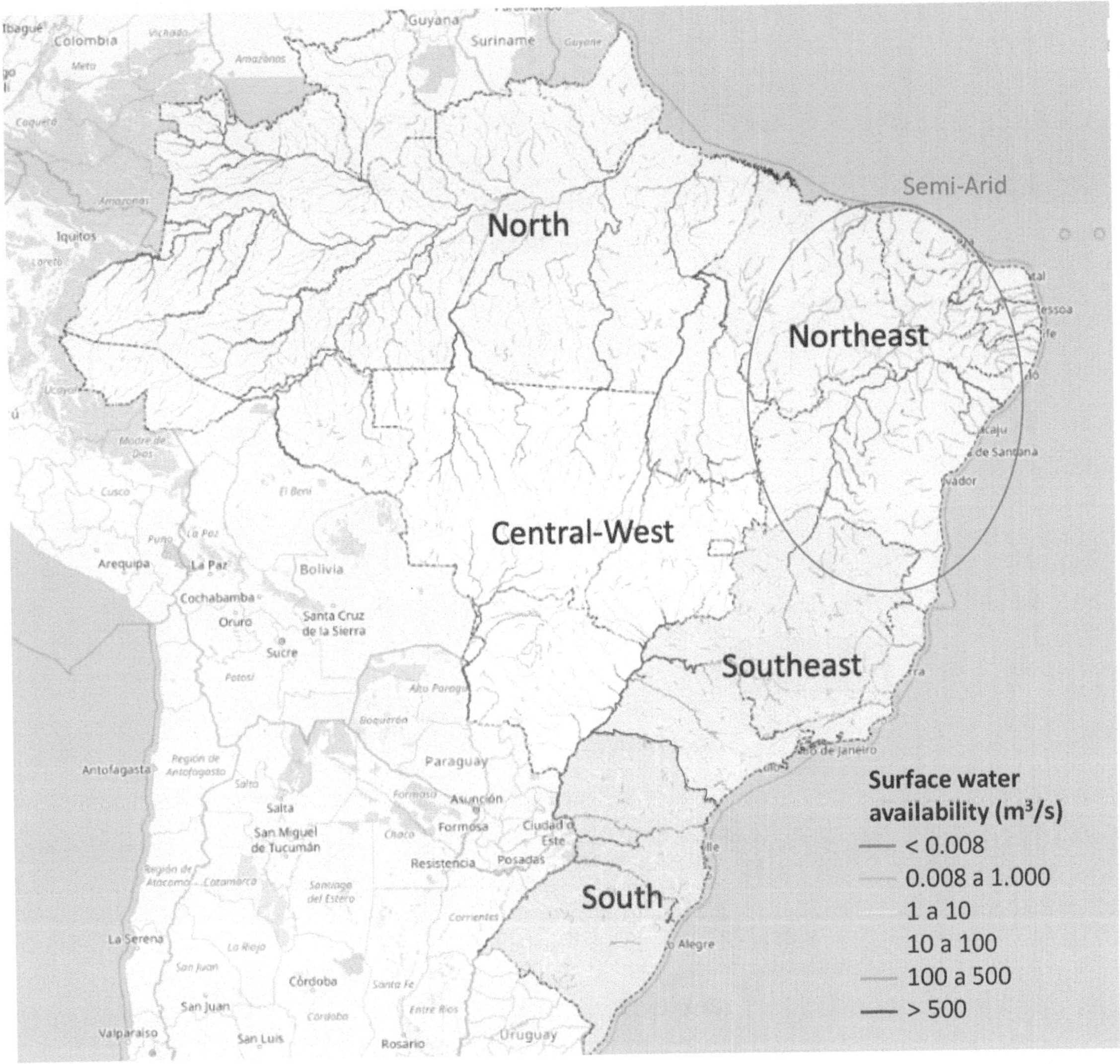

Figure 1 | Surface water availability in Brazil – highlights for the Semi Arid region modified from INDE (2020).

Regions with high rates of wastewater treatment have, theoretically, a greater potential for generating water for reutilization. This is the case in Israel, which treats 97% of the wastewater generated and reuses 80% of it in irrigation, supplying 40% of the demand for this purpose (Marin *et al.* 2017). In Brazil, only 42.6% of the wastewater generated is treated (ANA 2017), reducing its potential for reuse.

In general, water reuse in agriculture becomes an effective alternative source of water for production of different crops and also the supply of nutrients in the practice of fertigation (Maryam & Buyukgungor 2019). However, negative aspects such as the accumulation of substances that hinder plant growth, the potential damage to the soil through the transformation of its physical-chemical characteristics, and contamination by microorganisms must be evaluated (Xu *et al.* 2010).

The quality required for irrigation depends mainly on the type of consumption, cultivation, and irrigation for each crop. This quality requirement is related to the greater or lesser possibility of microbiological contamination, both from users and workers and from the soil (Beaudequin *et al.* 2015; Chhipi-Shrestha *et al.* 2017; Rock *et al.* 2018).

According to Maryam & Buyukgungor (2019), primary effluent is not recommended for reuse in agriculture;

Figure 2 | Distribution of Brazilian Hydrographic Regions and representation of the federative units – modified from ANA (2017).

secondary effluent (biological oxidation) is recommended for surface irrigation of orchards and vineyards and non-food crop irrigation; tertiary effluent (chemical coagulation, nutrient removal, filtration, and disinfection) is recommended for food crop irrigation. However, Tsagarakis *et al.* (2013) suggest that even primary treatment effluent can be used in controlled irrigation, if adequate precautionary and safety measures are taken.

Thus, the present work aims to evaluate, in a preliminary way, the potential of reuse for irrigation, of the effluents from all of wastewater treatment plants operating in Brazil (2,768 facilities) and allocated in the 12 different Brazilian Hydrographic Regions.

MATERIAL AND METHODS

The study was developed in three stages (Figure 3), as of the consolidation of public data presented in national documents (Table 2) related to all of the wastewater treatment plants in operation (2,768 facilities) and to the water demands for irrigation in each BHR.

Table 1 | Main characteristics of Brazilian hydrographic regions (Veiga & Magrini 2013)

Brazilian hydrographic regions	Area (km^2)	States[a]	Population (10^6)	Main river basins
Atlântico Sudeste	214,629	MG, ES, RJ, SP, PR	27.4	Paraíba do Sul, Doce
Paraná	879,873	SP, PR, MS, MG, GO, SC, DF	61.0	Paraná, Grande, Capivari, Jundiaí, Paranapanem, Piracicaba
São Francisco	638,576	BA, MG, PE, AL, SE, GO, DF	13.9	São Francisco
Atlântico Nordeste Oriental	286,802	PI, CE, RN, PB, PE, AL	23.4	Capibaribe, Una, Paraíba, Jaguaribe, Acaraú
Amazônica	3,869,953	AC, AM, RO, RR, AP, PA, MT	9.1	Amazonas
Tocantins-Araguaia	921,921	GO, TO, PA, MA, MT, DF	8.0	Tocantins, Araguaia
Parnaíba	333,056	PI, MA, CE	4.0	Parnaíba
Atlântico Nordeste Ocidental	274,301	MA, PA	5.8	Gurupi, Munim, Mearim, Itapecuru,
Uruguai	174,533	RS, SC	4.0	Uruguai
Paraguai	363,446	MT, MS	2.0	Paraguai
Atântico Sul	187,552	SP, PR, SC, RS	13.0	Itajai, Capivari, Taquari-Antas, Jacuí, Vacacaí, Camaquã, Lagoa Mirim, Patos
Atlântico Leste	388,160	BA, MG, SE, ES	15.1	Paraguaçu, Pardo, Contas, Salinas, Mucuri, Itapecuru, Jequitinhonha

[a]AC – Acre, AL – Alagoas, AP – Amapá, AM – Amazonas, BA – Bahia, CE – Ceará, ES – Esp + iruto Santo, GO – Goiás, MA – Maranhão, MT – Mato Grosso, MS – Mato Grosso do Sul, MG – Minas Gerais, PA – Pará, PB – Paraíba, PE – Pernambuco, PI – Piauí, RJ, Rio de Janeiro, RN – Rio Grande do norte, RS – Rio Grande do Sul, RO – Rondônia, RR – Roraima, SC – Santa Catarina, SP – São Paulo, SE – Sergipe, TO – Tocantins, DF – Distrito Federal.

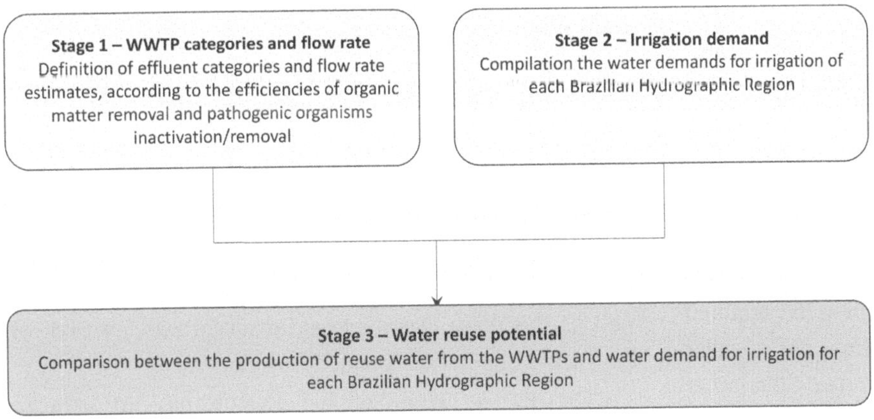

Figure 3 | Flowchart of the steps adopted in the research.

Table 2 | Main national public documents used in data generation of the research

N°	Document title (original and translation)	Description	Source
1	*Atlas esgotos: despoluição de bacias hidrográficas* (Wastewater atlas: river basins depollution)	It gathers the wastewater diagnosis of all 5,570 urban municipalities in Brazil, the impact of the effluent discharging and the planning for the wastewater treatment, with data to 2013	ANA (2017)
2	*Manual de usos consuntivos da água no Brasil* (Manual of consumptive water uses in Brazil)	It contemplates the definition of methods, the construction of a database beyond the production, storage, and availability of estimates consumptive water uses for all Brazilian municipalities	ANA (2019)

Stage 1 – definition of WWTP categories and flow rate estimative

For this stage of the research, the following data was extracted from document 1 (Wastewater atlas: water basins depollution), for each Brazilian municipality: (i) existence of WWTP; (ii) efficiency in organic matter removal; (iii) treatment technologies adopted in the WWTP; (iv) operational flow rate.

Subsequently, study categories were defined in relation to the performance of organic matter removal and pathogenic organisms' removal/inactivation. This categorization aimed to classify effluents in two situations:

- Category 1 – Secondary WWTP with organic matter removal efficiency greater than 80%: To produce effluent available to reuse in most of crops, it would be needed only to include a disinfection tertiary step.
- Category 2 – Tertiary WWTP with some disinfection technology (mostly maturation ponds): Effluent can be distributed for direct reuse for the irrigation of most crops.

The WWTPs with only primary level or only UASB (upflow anaerobic sludge blanket) reactor and organic matter removal efficiency below 80% were not categorized. They were considered for this study as 'uncategorized' because these facilities require high investments to adapt the effluent to reuse, since they would still need a secondary stage or a polishing prior to disinfection.

It is important to discuss the issue of the type of crop to be irrigated and, therefore, the quality of the water required for this purpose. In this sense, this article does not consider a discussion in relation to water quality standards, but rather evaluates possibilities for reuse depending on demand and supply in different qualities. These qualities are represented here according to the categories defined for the study and previously mentioned.

Stage 2 – irrigation water demand

For this stage of the research, the data of water demand for irrigation for each BHR were extracted and compiled from document 2 (Atlas irrigation: water use in irrigated agriculture).

The water demand for irrigation varies depending on some factors, such as rainfall regime (spatial and temporal distribution), evapotranspiration rates, seasonality, amount of water required depending on the crop to be irrigated, quality of the soil, soil capacity of water storage and the types of irrigation.

Stage 3 – water reuse potential

In this stage, a comparison was made between the flow rate data of the WWTPs, according to Categories 1 and 2, defined in Stage 1 and the water demand for irrigation in each Hydrographic Region compiled in Stage 2. Finally, it was possible to accomplish a critical analysis of the management of water resources in Brazil and the inclusion of water reuse in the national water matrix.

RESULTS AND DISCUSSION

Stage 1 – definition of WWTP categories and flow rate estimative

Initially, all the WWTPs in operation in the country were allocated in their respective Brazilian Hydrographic Regions. Furthermore, the WWTP flow rates were divided into the two categories described in the methodology.

Table 3 shows both the effluent flow rates from the WWTPs divided into the two categories (01 and 02), as well as the flows classified as 'uncategorized', in each of the 12 Brazilian Hydrographic Regions.

The Paraná Hydrographic Region has an area of approximately 10% of the national territory and covers the states of São Paulo, Paraná, Mato Grosso do Sul, Minas Gerais, Goiás, Santa Catarina, and Federal District. BHR Paraná is of great importance at national level since it is the region with the greatest economic development in the country and it also presents the greatest demand for resources (ANA 2015). In this way, it is observed that in this BHR there are the highest WWTP total flows and by category compared to the others. According to ANA (2015), its population is predominantly urban, with around 93% of the total of its inhabitants.

Table 3 | Distribution of WWTPs flow rates in Brazilian Hydrographic Regions, and the categories defined in the study

Brazilian Hydrographic Region	Total flow rate (m^3/s)			
	Uncategorized	**Category 1**	**Category 2**	**Total**
Amazônica	0.92	0.37	0.18	1.47
Atlântico Leste	0.48	0.93	0.94	2.35
Atlântico Nordeste Ocidental	0.18	0.01	0.06	0.25
Atlântico Nordeste Oriental	4.31	4.60	0.69	9.59
Atlântico Sudeste	5.07	11.69	0.86	17.62
Atlântico Sul	3.36	1.74	0.33	5.43
Paraguai	0.55	1.30	0.13	1.97
Paraná	21.32	57.88	5.50	84.70
Parnaíba	0.24	0.02	0.33	0.58
São Francisco	3.26	9.30	0.45	13.01
Tocantins-Araguaia	0.43	0.34	0.70	1.47
Uruguai	0.35	0.45	0.09	0.89

The second BHR with the highest total effluent flow rate is Atlântico Sudeste. This region has the second largest population in Brazil and a great urban concentration (92% of the total) (ANA 2015). However, in relation to Category 2, its flow rate is lower than that generated at HR Atlântico Leste. This is because, in the latter, there are 15 large WWTPs that include maturation ponds in their flowcharts.

HR Atlântico Nordeste Ocidental presents considerably reduced effluent flow rates due to the low sewage collection and treatment rates in the region: 28% of the sewage is collected and only 8% of the total generated is treated (ANA 2015). In this case, almost the entire state of Maranhão is in this region. Its most important city, the capital São Luís, has only 4% of treated wastewater, according to ANA (2017). Although the second largest municipality, the state of Maranhão, presents 100% of sewerage service covered, 95% of the population is served with septic tanks and only 5% with collection and treatment through the centralized system (ANA 2017).

In all BHRs except four (Atlântico Leste, Atlântico Nordeste Ocidental, Parnaíba, and Tocantins-Araguaia), the flow rates corresponding to Category 2 are lower to those of Category 1. This demonstrates little commitment of environmental sanitation management to adopt WWTPs with a disinfection stage in Brazil. It is also noticed that around 7% of the flow rate treated in Brazil goes through a tertiary stage of disinfection. It is noteworthy that in this country, there is a huge gap in the application of integrated wastewater management – IWM (centralized and decentralized) and integrated water and wastewater management – IWWM. According to Angelakis *et al.* (2018), IWM and IWWM are important trends in the environmental engineering and water resources field. Finally, the concept of 'one water' should merge individual water and wastewater departments into one unique department to develop more thoughtful, rational, and cost-effective solutions to meet future water needs.

Another unsatisfactory scenario is related to the BHRs that have flow rates allocated in the item 'Uncategorized', higher than the flows of Categories 1 and 2. This occurs in the BHRs Amazônica, Atlântico Nordeste Ocidental, and Atlântico Sul. This means that almost 30% of the total wastewater treated in Brazil corresponds only to the primary or advanced primary stages.

In this discussion, the importance of increasing wastewater treatment coverage rates is highlighted, not only in the sense of compliance with the national guidelines for effluent discharges, but also in relation to the production of higher volumes of water for reutilization. This action could be the target of areas with lower rates of socioeconomic development (Maryam & Buyukgungor 2019).

Stage 2 – irrigation water demand

The flow rate values corresponding to the water demands for irrigation of each BHR are shown in the schematic drawing and the graphic representation of Figure 4.

According to Figure 4, in the south of Brazil there is intense activity focused on irrigation, with high demand for this use in the BHRs Atlântico Sul, Uruguai and Paraná. It should be noted that in BHR Paraná there is the largest irrigated area in the country, corresponding to 36% of the total (ANA 2015).

In the northeast of the country, high demand for irrigation is observed only in BHR São Francisco, which has approximately half of its area in the Semi-Arid region. It is estimated that this area represents 10.9% of the total irrigated area in Brazil (ANA 2015).

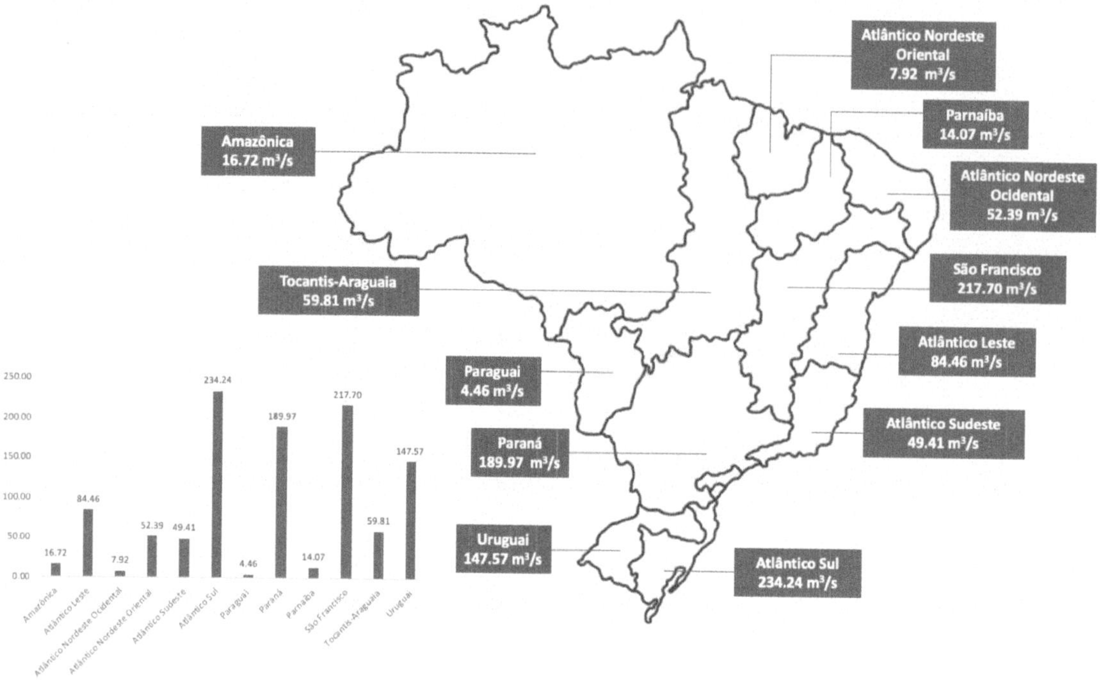

Figure 4 | Graphical representation and schematic drawing of the flow rates corresponding to the irrigation water demands for the Brazilian Hydrographic Regions.

There is lower water demand for irrigation in BHRs Paraguai, Parnaíba, and Amazônica. Much of the BHR Paraguai area is occupied by the Pantanal biome, characterized by well-defined periods of flood and drought and the main activity in this region is livestock (ANA 2015; Figueiredo *et al.* 2016). BHR Parnaíba has irrigated agriculture as its main activity. However, its bigger part is within in the Brazilian Semi-Arid region and presents characteristics of intermittent rainfall. Thus, there is a need to encourage alternative sources of water to increase the region's socioeconomic development. BHR Amazônica has a great extent of forest, with 85% of its territory occupied by native vegetation, and many of the cultivated crops do not require irrigation (ANA 2015).

For other regions, water demand for irrigation varies between 48.41 m^3/s in the BHR Atlântico Sudeste and 84.46 m^3/s in the BHR Atlântico Leste. In BHR Atlântico Leste, low water availability and intermittency of most rivers may become limiting for the expansion of agricultural activities. Thus, the availability of alternative sources of water such as water for reuse can leverage the socioeconomic development of the region. Finally, BHR Tocantins-Araguaia is important in the national context, as it is characterized by the expansion of the agricultural frontier. This expansion could be more accentuated with the inclusion of other sources of water for irrigation, since most of the current irrigated area is private (97%), fostering the potential for reuse for the development of agribusiness (ANA 2015, 2018).

Stage 3 – water reuse potential

Table 4 shows the water demand values for irrigation and effluent flow rates divided into the categories defined for the study, in each Brazilian Hydrographic Region. The values referring to uncategorized flow rates are not included in this session.

In general, in all BHRs, the effluent flow rates of Category 1 and/or Category 2 are significantly lower than the water demand for irrigation. However, it is noteworthy that for BHRs Atlântico Sudeste, Paraguai and Paraná, the effluent flows of Category 1 represent approximately 24, 29, and 30% of the water demand of irrigation, respectively. In relation to Category 2, none of effluent flow rates reaches more than 3% of demand and the one with the highest ratio is HR Paraná with 2.9%.

Table 4 | Distribution of water demands for irrigation and WWTP effluent flow rates in each category for BHRs

		Total flow rate (m^3/s)	
Brazilian Hydrographic Region	**Water demand for irrigation (m^3/s)**	**Category 1**	**Category 2**
Amazônica	16.72	0.37	0.18
Atlântico Leste	84.46	0.93	0.94
Atlântico Nordeste Ocidental	7.92	0.01	0.06
Atlântico Nordeste Oriental	52.39	4.60	0.69
Atlântico Sudeste	49.41	11.69	0.86
Atlântico Sul	234.24	1.74	0.33
Paraguai	4.46	1.30	0.13
Paraná	189.97	57.88	5.50
Parnaíba	14.07	0.02	0.33
São Francisco	217.70	9.30	0.45
Tocantins-Araguaia	59.81	0.34	0.70
Uruguai	147.57	0.45	0.09

In order to facilitate understanding and include new questions for discussion, in Figure 5, the following data are presented graphically for each Brazilian Hydrographic Region: (i) irrigation demand; (ii) effluent flow rate considered as 'uncategorized'; (iii) effluent flow rate in Category 1; (iv) effluent flow rate in Category 2; and (v) total effluent flow rate corresponding to the summation between flow rates of Categories 1 and 2.

At BHR Paraná, the total flow corresponding to the sum of Categories 1 and 2 is 63.38 m^3/s and represents 33% of the water demand for irrigation. If 'uncategorized' WWTPs were to be added to this total, with their due interventions to adjust the effluent, this value would be 44%. It should be noted that this is a considerable percentual and, therefore, the governance of this region should add this issue to the planning of water resources. Still, it is important to highlight that, because it is the most developed region in the country. The supply of an alternative source of water for irrigation could not only lead to economic growth but also the

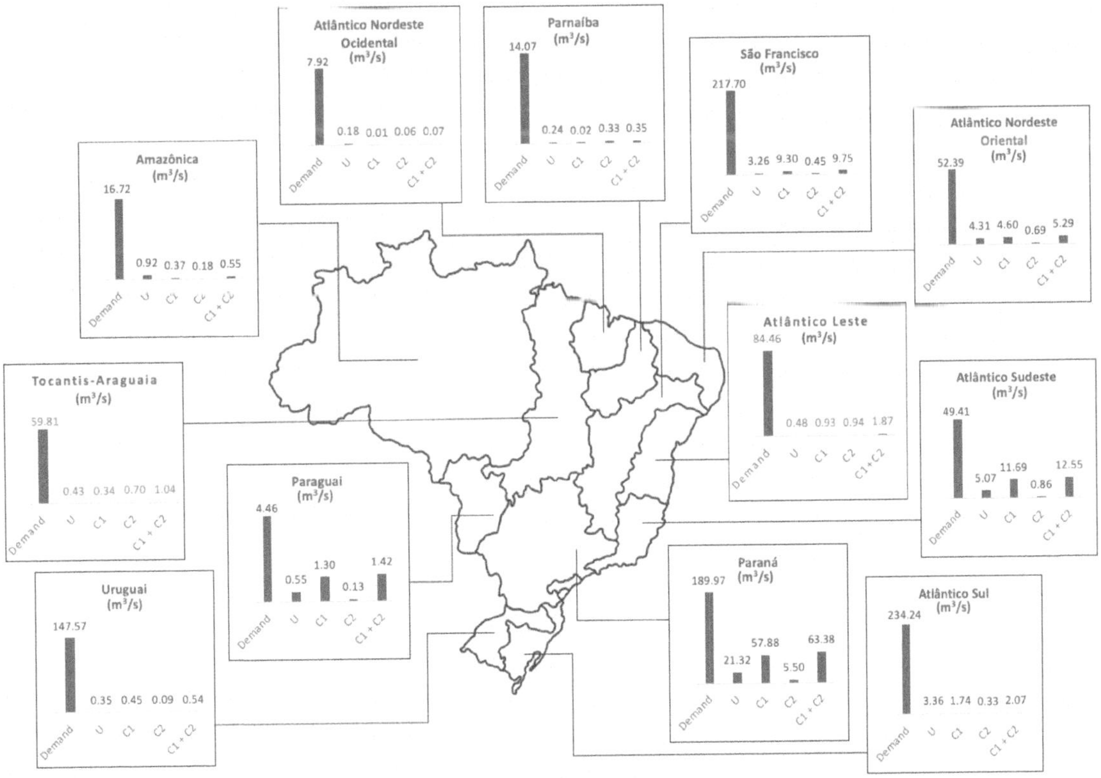

Figure 5 | Graphical and schematic representation of the flow rates corresponding to the water demands for irrigation and the effluent flow rates of WWTP for each BHR, distributed in three categories of the study. *Note:* U – Uncategorized. C1 – Category 1. C2 – Category 2.C1 + C2 – Category 1 plus Category 2.

reduction of conflicts over the use of water (Nölting & Mann 2018).

At BHR Paraguai, the uncategorized effluent (representing 12% of the water demand for irrigation) added to the effluents of Categories 1 and 2 represent 44%, similarly to HR Paraná. The water demand for irrigation in this BHR is only 4.46 m^3/s and in BHR Paraná is 189.97 m^3/s. Thus, there is a need to adapt these WWTPs (uncategorized), not only to comply with the current legislation on effluent discharges, but also to structure reuse planning in emergency situations.

The BHR Atlântico Sudeste also deserves to be highlighted for its high potential for the adoption of water reuse in irrigation. The flow rates of Categories 1 and 2 combined represent 25% of water demand for irrigation. When the 'uncategorized' flow rate is added, this ratio increases to 35%. Again, the discussion supports the need for more attention to WWTPs and their performance.

In the 12 HRs, eight of them (Amazônica, Atlântico Leste, Atlântico Nordeste Ocidental, Atlântico Sul, Parnaíba, São Francisco, Tocantins-Araguaia and Uruguai) present ratios of Categories 1 and 2 flow rates below 5% of the water demand for irrigation. This scenario demonstrates the country's fragility in relation to wastewater treatment and, consequently, the generation of water for reutilization as an alternative source for the irrigation of several crops. It is observed that the water demand in these regions is not high; actually, the wastewater treatment rates are low. This is the case, for example, of BHR Parnaíba, where the sewage collection rate in 2012 was 18%, the lowest among the 12 Brazilian Hydrographic Regions, as mentioned by ANA (2015).

Regarding the sanitary deficit faced in Brazil and the imminent need to apply the practice of water reuse to minimize the drought impacts that will be aggravated by climate changes and water use growth, the need to achieve universal sanitation is evident. This is not only to minimize the pollution of water bodies, but also to provide, in quantitative terms, effluents treated for the practice of water reuse.

In the Brazilian Semi-Arid region, the driest area in the country, 470 municipalities have intermittent or ephemeral water bodies. Thus, in addition to the need to remove BOD (Biochemical Oxygen Demand), it is important to take into account the practice of water reuse and/or prioritize treatment processes that result in high removal and inactivation of pathogenic microorganisms (ANA 2017).

The BHR Nordeste Oriental has a relative potential for water reuse in irrigation. For this, Category 1 effluent represents 9% of the water demand and if added to the 'uncategorized' effluent, this ratio becomes 17%. Apparently, this is a good percentual. However, the Category 2 rate represents less than 2%. Again, it demonstrates the fragility in terms of tertiary effluent generation, with good performance in relation to the removal and inactivation of pathogenic organisms.

The BHR Uruguai has the fourth largest water demand for irrigation in the country. However, the sum of all effluent flow rates, including 'uncategorized' ones, represents less than 1% of this demand. Thus, the need for adequate planning for the generation of water for reutilization is highlighted, with the aim of achieving universalization.

According to the results presented in Table 4, it can be observed that, hypothetically, if all the WWTPs in Categories 1 and 2 reused 100% of their effluents, it would be possible to produce almost 100,000 liters of water for reutilization per second, substantially reducing the amount of water captured from the sources. Although this statement seems inaccessible, it is worth remembering that Israel reuses 87% of all the effluent generated in the country (Marin *et al.* 2017).

Angelakis *et al.* (2018) indicate some important issues and challenges will need to be resolved to optimize water reclamation and reuse: (a) the development of more effective techniques and methods incorporating risk assessment to assess human and environmental health effects of wastewater constituents, and (b) the development of appropriate water reclamation and reuse regulations, applicable to different situations, which will both help to promote reuse as well as regulate it. However, there is still no adequate framework in Brazil for laws that regulate the practice of water reuse in the country.

CONCLUSION

Generally, in quantitative terms, the WWTPs in operation in Brazil generate 88.18 m^3/s of effluent in Category 1 and 10.24 m^3/s in Category 2. These values in total represent 9% of the total water demand for irrigation in the country (1,078.71 m^3/s).

Only in two Brazilian Hydrographic Regions (Paraná and Paraguai), is the installed potential high, around 44% of demand, considering the effluents from all categories. However, in most of them, the potential is low, due to the low attendance rates in relation to sanitation.

In situations where the installed potential is high, it is necessary to consider the operational levels of the WWTPs, to guarantee an adequate quality for reuse. In cases of low installed potential, it is necessary to assess the capacity to implement complementary units for such a practice, in addition to advances in service rates, with a view to universalization and water security.

The Brazilian Hydrographic Regions Parnaíba and São Francisco depend heavily on irrigated agriculture. Thus, water scarcity can curb socioeconomic growth of these regions, bringing water for reutilization with an important planning factor. In these two cases, the intermittent rainfall accompanied by climate change and population growth aggravate the situation. Furthermore, the sanitation services coverage is low and must be assessed for adequate water resources and sanitation management.

Only 7% of the treated wastewater flow rate in Brazil goes through a tertiary stage of disinfection, showing a fragility in relation to the quality of the effluent both for discharging and for the application of the practice of water reuse.

It is reinforced that the present work took into consideration only the amount of effluent generated in relation to the water demand for irrigation in each Brazilian Hydrographic Region. Thus, it is concluded that for the effective adoption of the practice of water reuse in planning, it is necessary to take into account the locations of both WWTPs, irrigated fields and transport, and storage logistics, in addition to the different levels of quality required for each type of culture. Still, in this work, an estimate with punctual flows was adopted. However, these flow rates that represent demands can occur on a seasonal basis in irrigation.

ACKNOWLEDGEMENT

The authors would like to acknowledge Mrs Vimbai Pachawo, from the Ministry of Lands, Agriculture, Water, Climate and Rural Resettlement (Zimbabwe), for the English review. She is a student of PhD Program in University of Minho and member of the research group.

DATA AVAILABILITY STATEMENT

All relevant data are included in the paper or its Supplementary Information.

REFERENCES

Ahmadi, L. & Merkley, G. P. 2017 Wastewater reuse potencial for irrigated agriculture. *Irrigation Science* **35** (4), 275–285. doi:10.1007/s00271-017-0539-7.

ANA 2015 *Conjuntura dos recursos hídricos no Brasil – Informe 2015 (Conjuncture of Water Resources in Brazil – Report 2015)*. Brasília.

ANA 2017 *Atlas Esgotos: Despoluição de bacias hidrográficas (Wastewater Atlas: Water Basins Depollution)*. Brasília. Available from: http://atlasesgotos.ana.gov.br/ (accessed 2019)

ANA 2018 *Conjuntura dos recursos hídricos no Brasil (Conjuncture of Water Resources in Brazil)*. Agência Nacional de Águas (ANA), Brasília. Available from: http://www.snirh.gov.br/portal/snirh/centrais-de-conteudos/conjuntura-dos-recursos-hidricos

ANA 2019 *Manual de usos consuntivos da Água no Brasil (Manual of Consumptive Water Uses in Brazil)*. Retrieved 2019.

Angelakis, A. & Spyridakis, S. 2013 Major urban water and wastewater systems in Minoan Crete, Greece. *Water Science &Technology: Water Supply* **13** (3), 564–573.

Angelakis, A., Takashi, A., Bahri, A., Jimenez, B. & Tchobanoglous, G. 2018 Water reuse: from ancient to modern times and the future. *Frontiers in Environmental Science* **6** (28). 6(26), doi: 10.3389/fenvs.2018.00026.

Beaudequin, D., Harden, F., Roiko, A., Stratton, H., Lemckert, C. & Mengersen, K. 2015 Modelling microbial health risk of wastewater reuse: a system perspective. *Environmental International* (84), 131–141. doi:10.1016/j.envint.2015.08.001.

Brazil 2003 *Resolução n° 32 do Conselho Nacional de Recursos Hídricos (Resolution n° 32 of the National Resources Council)*. Brasília, Brasil.

Chhipi-Shrestha, G., Hewage, K. & Rehan, S. 2017 Microbial quality of reclaimed water for urban reuses: probabilistic risk-based investigation and recommendations. *Science of the Total Environment* (576), 738–751. doi:10.1016/j.scitotenv.2016.10.105.

Figueiredo, D., Dores, E., Paz, A. & Souza, C. 2016 Avaiability, uses and management of water in the Brazilian Pantanal. In: *Tropical Wetland Management – The South American*

Pantanal and International Experience (A. R. Ioris, ed.). Routledge, New York, NY, p. 333.

INDE 2020 *Infraestrutura Nacional de Dados Espaciais (National Spatial Data Infrastructure)*. Available from: https://www.inde.gov.br/VisualizadorMapas (accessed 12 December 2020).

Jeong, H., Kim, H. & Jang, T. 2016 Irrigation water quality standards for indirect wastewater reuse in agriculture: a contribution toward sustainable wastewater reuse in South Korea. *Water* **8** (4). *8*(169). https://doi.org/10.3390/w8040169.

Marin, P., Tal, S., Yeres, J. & Ringskog, K. 2017 *Water Management in Israel: Key Innovations and Lessons Learned for Water Scarce Countries*. World Bank, Washington, DC.

Maryam, B. & Buyukgungor, H. 2019 Wastewater reclamation and reuse trends in Turkey: oportunities and challanges. *Journal of Water Process Engineering* **30**, 100501.

Mays, L. W., Koutsoyiannis, D. & Angelakis, A. 2007 A brief history of urban water supply in antiquity. *Water Science & Technology: Water Supply* **7** (1), 1–12.

Nölting, B. & Mann, C. 2018 Governance strategy for sustainable land management and water reuse: challenges for transdisciplinary research. *Sustainable Development* **172** (6), 691–700.

Peng, Y., Xiao, Y., Fu, Z., Dong, Y., Zheng, Y., Yan, H. & Li, X. 2019 Precision irrigation perspectives on the sustainable water-saving of field crop production in China: water demand prediction and irrigation scheme optimization. *Journal of Cleaner Production* (239), 365–367. doi:10.1016/j.jclepro.2019.04.347.

Ramos, R. 2016 *Future Directions*. Available from: http://www.futuredirections.org.au/publication/world-stage-set-brazils-olympic-games-climate-change-water-security/ (accessed 20 May 2020).

Rock, C., Brassill, N., Dery, J., Carr, D., McLain, J., Bright, K. & Gerba, C. 2018 Review of water quality criteria for water reuse and risk-based implications for irrigated produce under the FDA Food Safety Modernization Act, produce safety rule. *Environmental Research*. doi:10.1016/j.envres.2018.12.050.

Tsagarakis, K., Menegaki, A., Siarapi, K. & Zacharapoulou, F. 2013 Safety alerts reduce willingness to visit parks irrigated with recycled water. *Journal of Risk Research* **16** (2), 133–144. doi:10.1080/13669877.2012.726246.

Veiga, L. & Magrini, A. 2013 The Brazilian water resources management policy: fifteen years of success and challenges. *Water Resource Management* **7** (27), 2287–23032. doi:10.1007/s11269-013-0288-1.

Xu, J., Wu, L., Chang, A. & Zhang, Y. 2010 Impact of long-term reclaimed wastewater irrigation on agricultural soils: a preliminary assessment. *Journal of Hazardous Materials* **183** (1–3), 780–786.

First received 29 July 2020; accepted in revised form 8 October 2020. Available online 22 October 2020

Characteristics and evaluations of heavy metal contents in irrigation water sources on a suburb farmland in the lower reaches of Xijiang River in the early busy season of farming

Ren Jie, Liu Xiao-wen and Wu Ying-xin

ABSTRACT

The pollution of heavy metals in irrigation water was related to the management-control of the risk of heavy metals in farmland and the safe production of crops. There were various sources of irrigation water in the suburb farmlands of Xijiang River. We investigated the sources and distribution characteristics of heavy metals in the irrigation water sources on a riverside and inside-dyke farmland of Xijiang River. According to sample test results, it was found that nearly 100% of heavy metals that we were concerned about were detected in the irrigation water sources in the study. The box plot showed that the average content of other elements ranged from 1.65 to 43.75 $\mu g{\cdot}L^{-1}$, and the average level of cadmium was 0.02 $\mu g{\cdot}L^{-1}$ in the studied area. According to quantitative evaluation index and analysis of distribution characteristics, most heavy metals were at relatively low concentration in the irrigation water sources, while mercury content was far greater than other elements, followed by arsenic. Arsenic and mercury had become increasingly prominent problems in local irrigation water sources. River quality was more likely to be affected by river water and agricultural activities, while groundwater quality was more likely to be affected by natural causes. According to single/comprehensive water-quality evaluations, the comprehensive water quality evaluation ranged from level I to level III, but the single-factor index evaluations of mercury in most water samples and arsenic in two groundwater samples were of 'Inferior' level. Attention should be paid to the monitoring of mercury and arsenic (in the groundwater) in the studied area. The disposal methods of farmland water inflow and drainage should be increased if necessary, and reducing the influence of human activities on the natural weathering and leaching of heavy metals was an effective measure to avoid risk of agricultural production.

Key words | groundwater, heavy metals, irrigation water sources, rainwater runoff, reservoir water, river water

Ren Jie
School of Environmental and Municipal Engineering,
Lan Zhou JiaoTong University,
Lanzhou, Gansu, 730070,
China
and
Chinese Academy of Sciences,
Institute of Modern Physics,
Lanzhou, Gansu, 730000,
China

Ren Jie
Liu Xiao-wen (corresponding author)
Wu Ying-xin
South China Institute of Environmental Science,
Ministry of Ecology and Environment,
Guangzhou, Guangdong 510655,
China
E-mail: *liuxiaowen@scies.org*

HIGHLIGHTS

- Nearly 100% of heavy metals that we were concerned about were detected in the irrigation water sources in the studied area in the early busy season of farming.
- The studied area should reduce the frequency and use of groundwater irrigation.
- The monitoring of mercury and arsenic should be strengthened immediately in the studied area.
- Scientific research on the irrigation water sources in the early busy season of farming can react to the maximum anthropogenic influence on irrigation.

doi: 10.2166/ws.2020.281

GRAPHICAL ABSTRACT

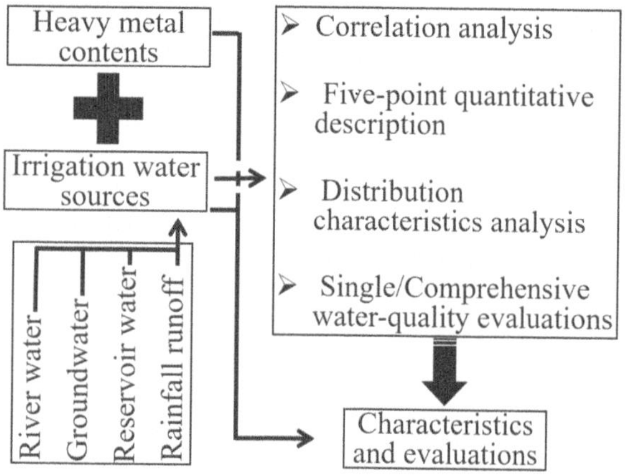

INTRODUCTION

Suburb farmlands in the lower reaches of Xijiang River were mostly irrigated by natural water. In recent years, the deterioration of water quality in rivers and lakes has become more and more serious, especially in industrial areas, and living/industrial drainage areas. Landfill leachate, industrial pollution sources and breeding industry were suspected sources of heavy metal pollution in natural water (Zhang *et al.* 2018a). Arsenic (As), mercury (Hg) and other heavy metals in the study of other natural water bodies of Xijiang River showed a greater risk. According to the recent study of surface water (large area) in the Pearl River Delta (almost all the lower reaches of Xijiang River were in the area), the content of arsenic varied dramatically from 0.2 to 722.5 $\mu g{\cdot}L^{-1}$ (Zhang *et al.* 2018a). The highest total mercury content in irrigation water in Nanhai District of Foshan City exceeded the Chinese standard of *Environmental Quality Standards for Surface Water* (China 2002a) by 1.3–6 times (Chen *et al.* 2011). In the process of agricultural production, the secondary release of chemical fertilizers and pesticides was one of the main factors of heavy metal pollution in the surrounding water system (Du *et al.* 2005). The water quality of Xijiang River was one of the important reasons for heavy metal pollution of irrigation water (mainly rivers and other natural water) in Foshan City (a big city in the lower reaches of Xijiang River), but industrial wastewater and domestic sewage were the direct causes of water pollution of Xijiang River (Chen *et al.* 2011). And the local effect of anthropogenic influence continued to deepen in the process of the deterioration of water quality in rivers and lakes (Kumarathilaka *et al.* 2018) resulting in the increasing potential risk from irrigation using natural water in the area.

Scholars have focused on the purification of sewage and reused water, irrigation risk and its transport mechanism in crops and farmland environment in recent years (Chaney *et al.* 2004; Fiol *et al.* 2006; Galletti *et al.* 2010; Ahsan *et al.* 2018; Huang *et al.* 2018; Zhang *et al.* 2018b), while there have been few trace investigations of heavy metals from irrigation sources in the Xijiang River Basin. The distribution of heavy metals in irrigation water was greatly affected by human activities during the early busy season of farming. This article took the irrigation water sources (river overlying water, groundwater, reservoir overlying water and initial rainwater runoff) on a suburb farmland in the lower reaches of Xijiang River as an example. We fully reported the pollution characteristics of heavy metals in each potential irrigation water source, and preliminarily explored the influence of human activities (other cases with relatively small impact) on the distribution of heavy metals in the irrigation water sources in the studied area. This was expected to provide help for irrigation water management and purification in the lower reaches of Xijiang River.

MATERIALS AND METHODS

Studied area

Agriculture had been a long history in Xijiang River Basin. As a large number of farmlands occupied by industrialization and urbanization, the importance of suburb farmlands along the river rose in the local agricultural supply chain. The lower reaches of Xijiang River were a typical example. The studied area was located at the intersection of rivers in the lower reaches. There were adequate surface water resources, abundant groundwater reserves and large rainfall. The rainfall was mainly concentrated from April to September, the total precipitation accounted for 80% of the annual precipitation, and the total precipitation days accounted for 74% of the annual precipitation days. Among them, the monthly average precipitation and the average monthly precipitation days were the highest in June.

The studied area was located in a town in Foshan City, Guangdong Province. The distribution of the Xijiang River Basin and the layout of the sampling points are shown in Figure 1. River water samples were named as FSJS 1–7, groundwater samples were named as FSDX 1–7, reservoir water samples were named as FSXS 1–7, and rainwater runoff samples were named as FSYS 1–5. The studied area was about 15.74 hm^2 and included river beach farmlands and inside-dyke farmlands. The irrigation forms included rainwater storage irrigation, river water direct irrigation, river water storage irrigation and groundwater storage irrigation in the studied area. River water direct irrigation and groundwater storage irrigation were the most common. The main pollution sources included: (a) Land and water transportation, (b) industrial sewage outlet near the sampling point FSJS1, (c) agricultural irrigation and drainage, (d) river water quality, (e) rainwater and rainwater runoff.

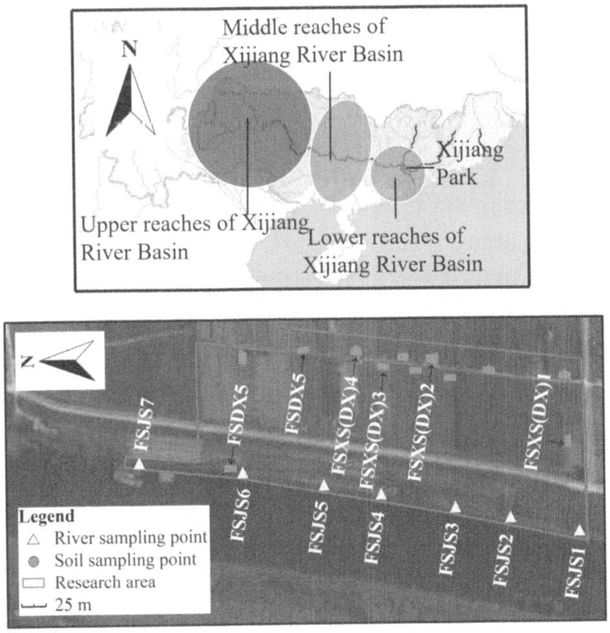

Figure 1 | Distribution of Xijiang River Basin and layout of sampling points.

Sample collection and pretreatment

The sampling period was in the early busy season of farming (summer), and the stage of preparation, sowing and topdressing as well. After long-term crop absorption, farmland self-purification and environmental purification, base fertilizer applied in the autumn of the previous year moved outward and its impact on the surrounding environment decreased significantly. During the period from May to July, frequent moderate rain and rainy weather had scoured the water environment of the irrigation system. At this time, external influences on the irrigation water environment were the smallest in this year, and this can feed back to some extent the impact of agricultural activities on the river overlying water, groundwater, reservoir overlying water and initial rainwater runoff.

The river water was taken from the overlying water with relatively stable fluctuations. Groundwater was taken from the outlet of the pump pipe (buried about 10 metres deep). The greater the water tables were (the average water table was greater than 7 m), the less was the possibility of pollutants entering, and the risk of external pollution was relatively reduced. The reservoir water samples were collected 0.5–1 h after the water inflow in the reservoir ended. Local rainfall was frequent, and rainwater runoff was a potential irrigation source. The rainwater runoff samples were collected at multiple points from a colored steel panel roof with concave and convex structure beside FSXS3. The surrounding area was open and no one had lived there for a long time, so it can be preliminarily judged that dry and wet atmospheric deposition was the main pollution source. Except for the rainwater runoff samples that were collected in July and August, the rest of the water samples were collected in August (the early busy season of farming). The water source of the inside-dyke reservoir was mainly groundwater, followed by river water, rainwater and rainwater runoff with fewer contributions. The river overlying water came from upstream migration, rainwater, runoff and farmland drainage.

Sample collection and pretreatment: (a) From July to August 2019, through the weather forecast published by the China Weather Network, auxiliary tools were reasonably selected and samples taken. (b) Positioning software assisted sampling. The layout of sampling points was completed through OvitalMap browser (mobile version), then we took samples according to the positioning information successively. The water quality status of the irrigation process was restored as much as possible. Sampling details referred to the Chinese standard of *Technical Specification Requirements for Monitoring of Surface Water and Wastewater* (China 2002b). The sampling bottles were a batch of 500 ml mineral water bottles; firstly the mouth of the bottle was blocked with fingers when collecting water, and after reaching a certain depth, it was filled with water and the cap was tightened. Finally, it was taken out and stored in a low-temperature box and transported back to the laboratory for pretreatment as soon as possible. (c) The sample testing was outdoor testing (pH value was detected on site) and indoor testing. The water samples were mixed into two parts ($n = 2$). According to different test content requirements, the pretreatment was completed as soon as possible, stored in a refrigerator at 4 °C, and the sample testing was completed within the specified time. Each water sample was set to be tested three times, and the average content (±standard deviation) was used for analysis.

Reagents and instruments

Water pH was measured with an acidity meter (STARTER 3100, Ohaus Instrument Co., Ltd). Total content of cadmium (Cd), lead (Pb), copper (Cu) and zinc (Zn) was digested using the method with HNO_3-$HClO_4$ mixed acid, and determined using an atomic absorption spectrometer (Pin AAcle 900T, PerkinElmer, USA). Total arsenic was digested using the sample pretreatment method, referring to the Chinese standard of *Monitoring and Analysis Methods for Water and Wastewater* (China 2002c), and determined using an atomic fluorescence spectrometer (AFS-9700, Beijing Haiguang Instrument Co., Ltd, China). Sample pretreatment and testing of total chromium (Cr), nickel (Ni) and mercury referred to the Chinese standard of *Monitoring and Analysis Methods for Water and Wastewater* (China 2002c).

Quality assurance

The experimental reagents were all high grade pure. The laboratory utensils were soaked in 10% nitric acid for more than 24 h before use, and washed with high-purity water before use. In the experiment, reagent blank and 20% parallel samples were made for each batch of samples. The recovery rate of heavy metal blank or sample standard addition was controlled within 10%. The relative standard error of repeated tests of each sample was less than 2%. The R^2 value of the standard curve of the detection items was higher than 0.995, and the data accuracy met the requirements.

Data processing

The statistical analysis of experimental data and the production of data graphs was made based on Microsoft Excel 2016, SPSS 22.0 and Origin 9.0.

Assessment methods

The coefficient of variation (CV) can be used to describe the dispersion degree of each sample: CV < 10% belongs to weak variation, 10% ≤ CV ≤ 100% belongs to moderate variation, and CV > 100% belongs to strong variation (Zhou *et al.* 2016). The kurtosis is the characteristic number of kurtosis at the average content of the data distribution, which is used to describe the steepness and slowness of the data distribution. The skewness is used to describe the digital characteristics of the asymmetric degree of the data distribution. If its value is less than 0, it is left skewness, and greater than 0 it is right skewness. The Kolmogorov–Smirnov test (K-S test) is used to determine whether there is a normal distribution.

Identification index method

The single/comprehensive water quality identification index method was used to evaluate the water environment quality of irrigation water sources in the studied area, and the method was suitable for surface water evaluation. The grading process referred to the Chinese standard of *Environmental Quality Standards for Surface Water* (China 2002a). For multiple functions in the same water area, the standard value corresponding to the highest function was used.

The single-factor water quality identification index method (P_i) was composed of an integer and two or three decimal places, and its structure was according to Equation (1):

$$P_i = X_1.X_2X_3 \tag{1}$$

The comprehensive water quality identification index method (P) was composed of an integer and three or four decimal places, and its structure was determined by Equation (2):

$$P = X_1.X_2X_3X_4 \tag{2}$$

In Equations (1) and (2), the symbol content and calculation method of X_1, X_2, X_3 and X_4, was based on two works in the literature (Xu 2005a, 2005b) for details.

Primary quantitative evaluation method

Recently, there have been many studies on water-saving rate, water-saving irrigation area and irrigation water utilization coefficient. It was found that appropriate grading of the existing evaluation system was of great help to further research. In order to facilitate the evaluation and quantitative description of water quality, reference was made to Wang *et al.* (2018) on the classification method of heavy metal pollution in paddy soil. Irrigation water quality limits referred to the Chinese standard of *Standards for Farmland Irrigation Water Quality* (China 2005). On referring to the Chinese standard of *Environmental Quality Standards for Surface Water* (China 2002a), a five-point quantitative description method is provided in Table 1. In this, nickel in groundwater refers to the Chinese standard of *Standards for Groundwater Quality* (China 2017). Since there are no detailed regulations for nickel in the Chinese standards of *Standards for Farmland Irrigation Water Quality* (China 2005) and *Environmental Quality Standards for Surface Water* (China 2002a), this article did not quantitatively evaluate nickel in water.

RESULTS AND DISCUSSION

pH

During the studied period, river water pH ranged from 7.12 to 7.56, groundwater pH ranged from 7.92 to 8.33, reservoir water pH ranged from 8.47 to 9.12, and rainwater runoff pH ranged from 6.11 to 6.69.

Heavy metal contents of sampling points

The heavy metal contents of each sampling point are shown in Figure 2. The samples were mixed uniformly to the

Table 1 | Evaluation method of heavy metal contents in irrigation water (μg·L^{-1})

Level	Evaluation	Cd	As	Pb	Cu	Zn	Cr	Hg
I	Very low	≤ 2	≤ 10	≤ 40	≤ 200	≤ 400	≤ 20	≤ 0.2
II	Low	2–4	10–20	40–80	200–400	400–800	20–40	0.2–0.4
III	General	4–6	20–30	80–120	400–600	800–1,200	40–60	0.4–0.6
VI	Higher	6–8	30–40	120–160	600–800	1,200–1,600	60–80	0.6–0.8
V	High	8–10	40–50	160–200	800–1,000	1,600–2,000	80–100	0.8–1
VI	Inferior	> 10	> 50	> 200	> 1,000	> 2,000	> 100	> 1

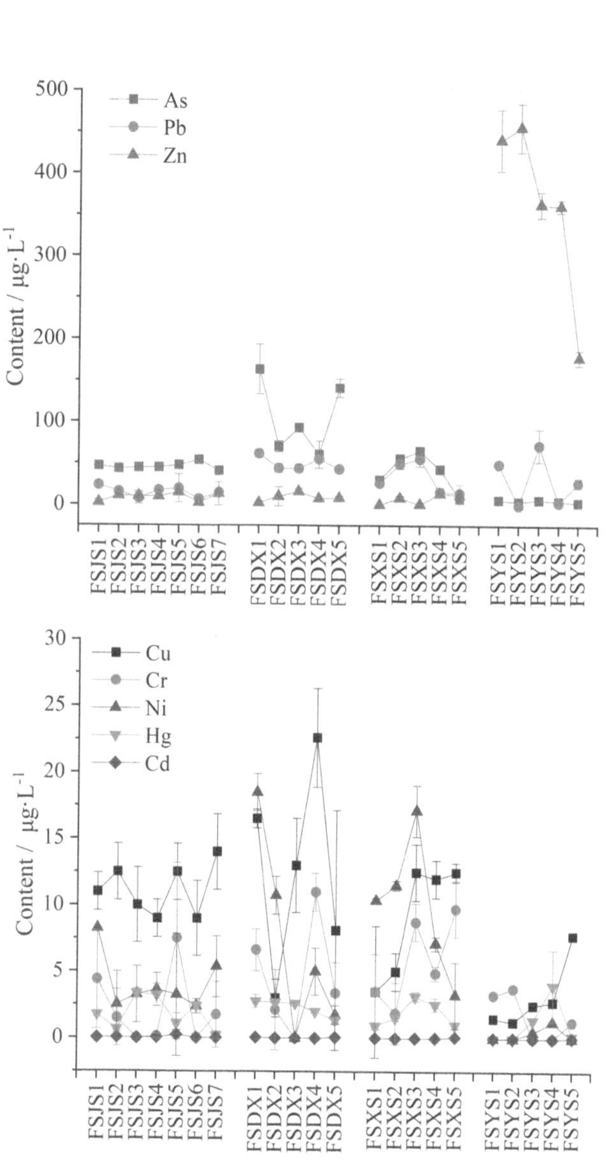

Figure 2 | Heavy metal contents of every sampling point. (*Note*: ND is less than the detection limits, and it is calculated as 0 μ·L^{-1} in data processing.)

greatest extent before analysis, and the element difference of the same sample may be related to the suspended phase (type and quantity). Nearly 100% of heavy metals that we were concerned about were detected in the irrigation water sources. Except for cadmium, the contents of the other seven elements were mostly in the range of a few to dozens of µg·L^{-1}. On referring to the Chinese standard of *Standards for Farmland Irrigation Water Quality* (China 2005), mercury exceeded the standard by 74.36% (44 samples), and the maximum content ranged from 3.1 to 4.0 µg·L^{-1}, which exceeded the mercury concentration in most natural water bodies reported at home and abroad. This may be related to farmland drainage and industrial pollutant emissions in the studied area (Yin *et al.* 2019). Arsenic exceeded the standard by 10.26% (44 samples), and the other elements did not exceed the standard (44 samples).

The box plot of heavy metal contents in the whole of the samples is shown in Figure 3. According to the test results, the median was selected as the average level of heavy metal contents in all irrigation water sources in the area. The box plot shows that the contents of cadmium, arsenic, lead, copper, zinc, chromium, nickel, mercury were 0.02 µg·L^{-1}, 43.75 µg·L^{-1}, 22.94 µg·L^{-1}, 9.50 µg·L^{-1}, 10.38 µg·L^{-1}, 2.69 µg·L^{-1}, 3.25 µg·L^{-1}, 1.65 µg·L^{-1}, respectively.

Characteristics of heavy metal contents in the river overlying water

Comparing Tables 1 and 2, cadmium, lead, copper, zinc and chromium were at 'Very low' level, arsenic was at 'High' level and mercury was at 'Inferior' level. No other heavy metals exceeded the standard except for arsenic and mercury. The standard-exceeding ratio of arsenic was 14.29% and the standard-exceeding rate of mercury was 64.29%. The standard-exceeding points were mostly concentrated in areas with more irrigation and drainage activities. During the sampling period, the sampling point of FSJS2 was river water direct irrigation. The analysis of the pipeline effluent is shown in Table 3. In the process of water intake, the heavy metal contents at FSJS2 were relatively lower than that at the other points in the same period, and the heavy

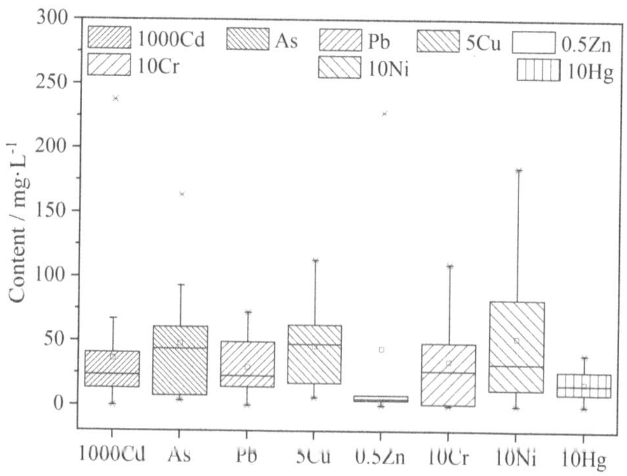

Figure 3 | Box plot of heavy metal contents in all samples.

Table 2 | Descriptive characteristics of heavy metal contents in the river overlying water ($n = 14$)

Elements	Min/(μg·L^{-1})	Max/(μg·L^{-1})	SD/(μg·L^{-1})	CV/%	Limits/(μg·L^{-1})	OSR/%	KU	SK	K-S test	
									Z	P
Cd	ND	0.480	0.123	230.01	≤ 10	0.00	12.92	3.54	0.410	0.000
As	39.46	56.92	4.53	9.89	≤ 50	14.29	2.07	1.19	0.196	0.182
Pb	2.50	31.50	7.89	55.45	≤ 200	0.00	0.60	0.62	0.098	0.200
Cu	7.00	16.00	2.51	22.50	≤ 1,000	0.00	− 0.17	0.15	0.152	0.200
Zn	ND	23.25	6.57	72.37	≤ 2,000	0.00	0.22	0.57	0.122	0.200
Cr	ND	11.50	3.41	156.66	≤ 100	0.00	3.62	1.91	0.285	0.003
Ni	ND	8.25	2.65	64.68	/	/	− 0.93	0.25	0.122	0.200
Hg	ND	3.80	1.22	67.44	≤ 1	64.29	− 1.13	0.21	0.119	0.200

Notes: ND is less than the detection limit, SD means standard deviation, CV means coefficient of variation, KU means kurtosis, SK means skewness, and similarly below.

Table 3 | Average content of heavy metals at FSJS2 and its pipeline effluent (μg·L^{-1})

Sampling points	Cd	As	Pb	Cu	Zn	Cr	Ni	Hg
FSJS2	0.04	43.05	15.63	12.50	10.63	1.50	2.50	0.65
Pipe mouth	0.03	43.18	16.00	12.00	13.88	4.88	7.13	2.55

metal contents in the pipeline outlet water were obviously higher. Water lifting irrigation reduced the mercury content of the overlying water in the local river. It can be seen that the migration of river water was not the main source of mercury in the river overlying water and it was most likely related to the release of sediments. Jia *et al.* (2018) have pointed out that mercury entering the water body would be converted into methyl mercury quickly, and would be easily enriched into sediments. He (2010) and Guo *et al.* (2018) have pointed out that sediments were the input source of mercury in overlying water, and mercury concentration in overlying water was affected by temperature and was higher in summer than in winter. It could be preliminarily judged that the excessive mercury in the local river water may be related to irrigation drainage and release of sediments at high temperature.

The contents of arsenic and mercury were relatively high. It can be seen from the CV value that arsenic belonged to weak variation. In other words, the arsenic contents did not change much in the river water. Mercury (0.2 ± 0.28 μg·L^{-1}, Figure 2) in the upstream did not exceed the standard. At the sampling points where the farmland drainage was discharged into the river overlying water, the water sample detection exceeded the standard (1.1 ± 0.14–3.35 ± 0.35 μg·L^{-1}). The mercury content (1.75 ± 0.21 μg·L^{-1}) in the sewage discharge outlet downstream exceeded the Chinese standard of *Environmental Quality Standards for Surface Water* (China 2002a).

It can be seen from the CV value that cadmium and chromium belonged to strong variation. According to the kurtosis values, the mercury distribution was relatively uniform in the river water, and standard exceeding points were not abnormally higher than the average content. Other elements had one or several abnormal distribution points in the river water, and their contents were abnormally higher than the average contents; cadmium especially was the most obvious, and it can be seen that the self-purification effect of river water failed to adjust the local distribution of elements in time. All skewnesses were greater than 0, the tailings were on the left, and it can be seen that the heavy metal contents were very low in the river water. The K-S test showed that cadmium ($Z = 0.410$, $P = 0.000 < 0.05$) and chromium were right-skewed distributions, and arsenic, lead, copper, zinc, nickel and mercury were approximately normal distributions, which showed that the river water was affected by external input, while the impact was limited, and the heavy metal contents were mostly low in the river water.

Above all, all the heavy metals in the river water were subject to a certain extent to external influences. The upstream river water was less affected by local drainage and had the lowest arsenic and mercury content. Compared with arsenic, other elements were more affected along the way. Mercury had a trend of enrichment at individual points, which was related to drainage.

Characteristics of heavy metal contents in the groundwater

Comparing Table 1 and Table 4, cadmium, copper, zinc and chromium were at 'Very low' level, lead was at 'Low' level, and arsenic and mercury were at 'Inferior' level. On referring to the Chinese standard of *Standards for Groundwater Quality* (China 2017), the nickel contents ranged from 0.00 to 19.50 $\mu g \cdot L^{-1}$, the average was 7.20 $\mu g \cdot L^{-1}$, and this belonged to level II.

The main heavy metal contaminants were arsenic and mercury in the groundwater of the studied area. Arsenic content in the groundwater of the studied area was much higher than that in healthy groundwater studied by Cheng *et al.* (2019). The mercury content of the groundwater was slightly lower than that of the reservoir.

It can be seen from the CV value that all elements belonged to moderate variation. According to the kurtosis value, the spatial distribution of each element was relatively uniform in the groundwater, other elements were closer to uniform distribution except for mercury, and it can be seen that the external influences of groundwater were the same/similar in the studied area. The skewnesses of cadmium, arsenic, lead, zinc, chromium and nickel were greater than 0, the tailings were on the left side, and the contents were relatively low in the groundwater. Copper and mercury skewnesses were less than 0, the tailings were on the right side, and the contents were relatively high in the groundwater. According to the K-S test results, cadmium, arsenic, lead, copper, zinc, chromium, nickel and mercury presented an approximately normal distribution. Considering every evaluation indicator, all the elements were relatively less affected by external influences, mainly by natural influences.

High arsenic groundwater is a global problem, and it can exceed drinking water standards by more than a hundred times due to the influences of groundwater irrigation, changes in agricultural methods and construction of dyke works (Eid *et al.* 2019). Research on shallow groundwater in the Pearl River Delta found that the total arsenic content ranged from 0 to 560 $\mu g \cdot L^{-1}$, and the average was 12.78 $\mu g \cdot L^{-1}$, which was subject to industrial pollution sources and the breeding industry (Zhang *et al.* 2018a). Research on spring water, civilian wells and drilling wells in the Pearl River Delta found that excessive copper and arsenic were from natural sources, lead was basically derived from rock weathering, and lead concentration in some shallow groundwater was abnormally increased due to anthropogenic pollution. Cadmium had no obvious distribution pattern, and the reason for exceeding the standard was likely to be the coexistence of natural and anthropogenic influences (Cheng *et al.* 2019).

Table 4 | Descriptive characteristics of heavy metal contents in the groundwater ($n = 10$)

Elements	Min/($\mu g \cdot L^{-1}$)	Max/($\mu g \cdot L^{-1}$)	SD/($\mu g \cdot L^{-1}$)	CV/%	OSR/%	KU	SK	K-S test	
								Z	*P*
Cd	ND	0.080	0.028	92.19	0.00	− 0.82	0.51	0.186	0.200
As	49.20	184.80	43.98	41.64	90.00	− 0.84	0.54	0.183	0.200
Pb	40.25	65.00	8.51	17.20	0.00	− 0.77	0.59	0.192	0.200
Cu	1.75	25.25	7.93	62.70	0.00	− 0.96	− 0.17	0.192	0.200
Zn	1.50	18.50	6.21	69.61	0.00	− 0.97	0.47	0.133	0.200
Cr	ND	12.00	4.30	93.01	0.00	− 0.94	0.47	0.159	0.200
Ni	ND	19.50	7.16	99.40	/	− 0.85	0.74	0.185	0.200
Hg	1.20	3.10	0.60	26.30	100	− 0.42	− 0.63	0.179	0.200

Considering the surrounding environmental characteristics, it could be seen that the problem of arsenic and mercury in groundwater in the studied area might be related to groundwater irrigation, agricultural deposited waste leachate and geochemical migration of the surrounding industrial pollution sources, mainly groundwater irrigation. Considering factors such as the surrounding environment of the pumping wells, installation depth and water table, it could be judged that the groundwater was affected by natural and anthropogenic influences, mainly natural influences.

Anthropogenic sources/anthropogenic activities were the main factors leading to the high regionalization of elements. The high background of the region was an important reference for risk variation, and reducing anthropogenic influences on the natural weathering and leaching of heavy metals was an effective measure to avoid risk from agricultural production in high background regions.

Characteristics of heavy metal contents in the reservoir overlying water

Comparing Table 1 and Table 5, cadmium, lead, copper, zinc and chromium were at the 'Very low' level, arsenic was at the 'High' level, and mercury was at the 'Inferior' level.

It can be seen from the CV value that cadmium belonged to strong variation, and other elements belonged to moderate variation. It can be seen from the kurtosis value that there were one or several distribution points abnormally higher than the average content for cadmium. The spatial distribution of other elements was relatively uniform, nickel distribution was close to the normal distribution, and the distribution of arsenic, lead, zinc, chromium and mercury was closer to the uniform distribution. It can be seen that the arsenic, lead, zinc, chromium and mercury were affected by external influences (groundwater and river water underwent natural influences such as standing still, precipitation and sun exposure), which were relatively close. It can be seen that the skewnesses of cadmium, lead, zinc, nickel and mercury were greater than 0, the tailings were on the left, and the contents of most heavy metals were relatively low. Research points that were abnormally higher than the average may be affected by human/agricultural activities, such as the use of reservoir water to wash vegetable roots and leaves. Some reservoirs were adjacent to the drainage ditch and vegetable stems and other agricultural waste accumulation points. The skewnesses of arsenic, copper and chromium were less than 0, the tailings were on the right side, and the contents of most heavy metals were high. According to the K-S test results, cadmium presented a right-skewed distribution, and arsenic, lead, copper, zinc, chromium, nickel and mercury presented an approximately normal distribution.

Characteristics of heavy metal contents in the initial rainwater runoff

Comparing Table 1 and Table 6, cadmium, arsenic, lead, copper, zinc and chromium were at the 'Very low' level, and mercury was at the 'Inferior' level.

Table 5 | Descriptive characteristics of heavy metal contents in the reservoir overlying water ($n = 10$)

								K-S test	
Elements	**Min/(μg·L^{-1})**	**Max/(μg·L^{-1})**	**SD/(μg·L^{-1})**	**CV/%**	**OSR/%**	**KU**	**SK**	***Z***	***P***
Cd	ND	0.210	0.062	187.68	0.00	9.01	2.95	0.388	0.000
As	4.19	66.20	21.39	52.85	40.00	− 0.81	− 0.52	0.141	0.200
Pb	7.00	62.00	18.78	57.96	0.00	− 1.37	0.29	0.187	0.200
Cu	ND	14.00	4.63	50.87	0.00	− 0.17	− 0.91	0.259	0.055
Zn	ND	16.75	6.46	89.75	0.00	− 1.34	0.31	0.189	0.200
Cr	0.25	11.25	3.52	61.28	0.00	− 0.81	− 0.01	0.128	0.200
Ni	1.50	18.50	4.99	50.54	/	0.01	0.12	0.154	0.200
Hg	0.80	3.40	0.95	51.70	80.00	− 1.32	0.52	0.178	0.200

Table 6 | Descriptive characteristics of heavy metal contents in the initial rainwater runoff ($n = 10$)

Elements	Min/(μg·L^{-1})	Max/(μg·L^{-1})	SD/(μg·L^{-1})	CV/%	OSR/%	KU	SK	K-S test	
								Z	P
Cd	ND	0.048	0.017	80.03	0.00	1.47	0.60	0.209	0.200
As	3.66	6.88	1.30	24.11	0.00	− 1.34	− 0.29	0.182	0.200
Pb	0.00	72.25	30.67	100.56	0.00	− 1.56	0.45	0.211	0.200
Cu	1.25	7.75	2.65	84.10	0.00	3.89	1.92	0.360	0.033
Zn	178.50	456.00	110.33	30.65	0.00	2.23	− 1.42	0.305	0.143
Cr	ND	3.75	1.77	107.46	0.00	− 2.83	0.32	0.224	0.200
Ni	ND	1.25	0.55	156.49	/	1.74	1.53	0.339	0.062
Hg	ND	4.00	1.71	152.88	40.00	2.57	1.69	0.304	0.146

It can be seen from the CV value that lead, chromium, nickel and mercury belonged to strong variation. It can be seen from the kurtosis value that cadmium, copper, zinc, nickel and mercury had one or more distribution points higher than the average content, the spatial distribution of arsenic, lead and chromium was relatively uniform, and the distribution of arsenic was closer to a uniform distribution. It can be seen from the skewness value that, except for arsenic and zinc, the skewnesses of other elements were greater than 0, the tailings were on the left, and the contents of most heavy metals were relatively low. The distribution points that were abnormally higher than the average contents may be affected by external influences, such as: atmospheric deposition mostly including exhaust gas, wastewater and solid waste caused by transportation, industrial production, agricultural activities, residents' daily life, nearby restaurants and tea houses, etc. According to the K-S test results, it can be seen that cadmium, arsenic, lead, zinc, chromium, nickel and mercury presented an approximately normal distribution, and copper presented a right-skewed distribution.

Correlation analysis of heavy metals in the irrigation water sources

The correlation matrix of heavy metal contents of each irrigation water source is shown in Table 7.

The correlation of many elements in the river water was significant, indicating that they might have the same source. There was a significant negative correlation between cadmium and arsenic at the $P < 0.01$ level, and a significant negative correlation between arsenic and chromium at the $P < 0.05$ level, so it was inferred that they may not have the same source. Considering the farming slack period, arsenic concentration in the river water ranged from 3.22 to 4.35 μg·L^{-1}, and it could be inferred that arsenic concentration increase in the river water was affected by irrigation and drainage, and lead and arsenic were consistent. It could also be preliminarily judged that cadmium and chromium in the river water were relatively less affected by agriculture.

All heavy metals in the groundwater were significantly correlated at the level of $P < 0.01$, suggesting the sources of heavy metals in the groundwater were stable and were affected similarly by industrial and agricultural activities. Considering the depth of groundwater, it could be inferred that the risk of leaching was not high. Its concentration was relatively higher than other environments and was related to frequent agricultural irrigation. The results showed that the elements in the groundwater tended to be natural sources.

Except for cadmium, other heavy metals in the reservoir water were all significantly correlated at the $P < 0.01$ level, and it can be seen that heavy metal sources were relatively stable. The preliminary judgment was less affected by external influences, mainly from the migration of water sources. Considering the 'original' water sources in the environment, individual elements tended to accumulate and increase.

The heavy metals in the rainwater runoff were mainly composed of air dust particles. The correlation of many elements in the rainwater runoff was significant, indicating that most elements in the rainwater runoff had the same source.

Table 7 | Correlation matrix of heavy metal contents in the irrigation water sources

Type	Elements	Cd	As	Pb	Cu	Zn	Cr	Ni	Hg
River water	Cd	1							
	As	−0.873**	1						
	Pb	−0.294	0.717**	1					
	Cu	0.680**	−0.273	0.456	1				
	Zn	0.741**	−0.340	0.406	0.970**	1			
	Cr	0.872**	−0.551*	0.146	0.835**	0.899**	1		
	Ni	0.597*	−0.141	0.579*	0.955**	0.968**	0.838**	1	
	Hg	0.613*	−0.165	0.559*	0.966**	0.975**	0.848**	0.986**	1
Groundwater	Cd	1							
	As	0.958**	1						
	Pb	0.973**	0.990**	1					
	Cu	0.955**	0.925**	0.941**	1				
	Zn	0.971**	0.956**	0.970**	0.945**	1			
	Cr	0.995**	0.966**	0.979**	0.961**	0.984**	1		
	Ni	0.968**	0.976**	0.978**	0.915**	0.983**	0.981**	1	
	Hg	0.909**	0.918**	0.908**	0.973**	0.917**	0.920**	0.883**	1
Reservoir water	Cd	1							
	As	0.575	1						
	Pb	0.685*	0.955**	1					
	Cu	0.523	0.976**	0.899**	1				
	Zn	0.673*	0.941**	0.961**	0.894**	1			
	Cr	0.688*	0.982**	0.986**	0.940**	0.962**	1		
	Ni	0.744*	0.957**	0.945**	0.934**	0.953**	0.978**	1	
	Hg	0.704*	0.926**	0.986**	0.864**	0.961**	0.963**	0.925**	1
Rainwater runoff	Cd	1							
	As	0.929*	1						
	Pb	0.927*	0.961**	1					
	Cu	0.926*	0.788	0.889*	1				
	Zn	0.881*	0.938*	0.813	0.641	1			
	Cr	0.867	0.953*	0.983**	0.806	0.801	1		
	Ni	0.911*	0.820	0.916*	0.956*	0.663	0.882*	1	
	Hg	0.920*	0.817	0.917*	0.976**	0.659	0.870	0.997**	1

Note: **is significantly correlated at level 0.01 (two-sided); and *is significantly correlated at level 0.05 (two-sided).

Evaluation of irrigation water quality in the studied area

The evaluation results of each sample are shown in Table 8. The comprehensive water quality evaluation results of each sampling point ranged from level I to level III, which met the requirements for the use of water environment function areas. According to the single-factor water quality identification index method, mercury in each irrigation water source was the main pollution factor and most sampling points had an index greater than 6, and water quality evaluation was 'Inferior' level; the arsenic indexes of two groundwater sampling points were greater than 6 and the water quality evaluations were 'Inferior' level.

Status analysis

The average contents of heavy metals are shown in Table 9 in the irrigation water sources in the studied area. The comparison of the heavy metal contents between the reservoir and the corresponding groundwater showed that there was no significant difference. The average contents of certain heavy metals in individual reservoirs had been greatly reduced compared with the corresponding groundwater. It can be seen that the method of irrigation after a period of precipitation certainly had scientific and practical effectiveness and was one of the simplest and most effective methods for removing heavy metals in irrigation water.

Table 8 | Evaluation results of single-factor water quality identification index P_i and comprehensive water quality identification index P at each sampling point

	P_i							Comprehensive evaluation		
Sample No.	**Cd**	**As**	**Pb**	**Cu**	**Zn**	**Cr**	**Hg**	***P***	**Level**	**Evaluation**
FSJS1	1.00	1.90	3.30	2.00	1.00	1.40	**6.81**	2.510	II	Up to standard
FSJS2	1.00	1.90	3.10	2.00	1.20	1.20	4.60	2.100	II	Up to standard
FSJS3	1.00	1.90	1.80	2.00	1.20	1.00	**8.43**	2.510	II	Up to standard
FSJS4	1.00	1.90	3.20	1.90	1.20	1.00	**8.23**	2.610	II	Up to standard
FSJS5	1.20	1.90	3.20	2.00	1.30	1.80	**6.11**	2.510	II	Up to standard
FSJS6	1.00	4.10	1.70	1.90	1.10	1.00	**7.42**	2.610	II	Up to standard
FSJS7	1.00	1.80	3.10	2.00	1.30	1.20	4.10	2.100	II	Up to standard
FSDX1	1.00	**6.61**	5.20	2.00	1.00	1.70	**7.82**	3.620	III	Up to standard
FSDX2	1.00	4.40	3.80	1.30	1.20	1.20	**7.72**	2.910	II	Up to standard
FSDX3	1.00	4.90	3.80	2.00	1.30	1.00	**7.62**	3.110	III	Up to standard
FSDX4	1.00	4.20	5.10	2.00	1.20	2.00	**7.02**	3.210	III	Up to standard
FSDX5	1.10	**6.41**	3.80	1.80	1.20	1.30	**6.41**	3.120	III	Up to standard
FSXS1	1.00	1.60	3.40	1.40	1.00	1.40	4.90	2.100	II	Up to standard
FSXS2	1.00	4.10	5.00	1.50	1.20	1.20	**6.61**	2.910	II	Up to standard
FSXS3	1.00	4.30	5.10	2.00	1.00	1.90	**8.23**	3.410	III	Up to standard
FSXS4	1.00	1.90	3.10	2.00	1.30	1.50	**7.62**	2.610	II	Up to standard
FSXS5	1.10	1.20	3.10	2.00	1.20	2.00	6.00	2.400	II	Up to standard
FSYS1	1.00	1.10	5.00	1.20	2.40	1.30	1.00	1.900	I	Up to standard
FSYS2	1.00	1.10	1.00	1.10	2.40	1.40	1.00	1.300	I	Up to standard
FSYS3	1.00	1.10	5.40	1.30	2.30	1.00	**6.41**	2.610	II	Up to standard
FSYS4	1.00	1.10	1.30	1.30	2.30	1.00	**9.04**	2.410	II	Up to standard
FSYS5	1.00	1.10	3.40	1.80	2.10	1.10	4.10	2.100	II	Up to standard

Table 9 | Average contents of heavy metals in the irrigation water sources ($\mu g \cdot L^{-1}$)

Content	Cd	As	Pb	Cu	Zn	Cr	Ni	Hg
River water	0.05	45.80	14.23	11.14	9.07	2.18	4.09	1.81
Groundwater	0.03	105.62	49.48	12.65	8.93	4.63	7.20	2.28
Reservoir water	0.03	40.47	32.40	9.10	7.20	5.75	9.88	1.84
Rainwater runoff	0.02	5.41	30.50	3.15	360.00	1.65	0.35	1.12

For individual elements exceeding the standard, the addition of pretreatment devices could further optimize all elements (Eid *et al.* 2019). Taking the study of low concentrations of cadmium as an example, the constructed wetland system treated irrigation influent with an average influent concentration of 0.232 μg/L, which was down between 0.054 and 0.063 μg/L after treatment (Jiang *et al.* 2019). It can be seen that there was a greater possibility of repurification of trace cadmium in irrigation water.

Taking the purification study of cadmium in the irrigation water as an example, it will bring a chain of ecological benefits: the use of the plant pond constructed wetland system effectively intercepted cadmium in the irrigation water (average content $0.232\ \mu g \cdot L^{-1}$), the removal rate reached 72.03%, compared with the unpurified irrigation water, and the cadmium content of brown rice irrigated with optimized water was down between 10.50% and 24.51% (Zhang *et al.* 2019). The plant pond constructed wetland system effectively intercepted the cadmium-excess in the irrigation water (average content 6.65 $\mu g \cdot L^{-1}$) in the typical mining area of Hunan, and the total removal rate of the system reached 87.94%. Compared with the unpurified irrigation water, the cadmium content of brown rice irrigated with optimized water was reduced 0.12 $mg \cdot kg^{-1}$ (Liu *et al.* 2019).

CONCLUSIONS

(1) Irrigation water sources may be slightly polluted by between one and three elements in the lower reaches of Xijiang River and the entire basin. Except for arsenic and mercury, other elements met the irrigation requirements and were at a relatively safe 'Very low' level in the irrigation water sources. The heavy metals were mostly in low concentrations in the irrigation water sources, and sampling points that were abnormally higher than the average contents may be affected by agriculture-related activities. Agricultural irrigation drainage and groundwater pumping were the main reasons for the relatively high mercury content of corresponding water sources.

(2) The comprehensive water quality evaluation results of each sampling point ranged from level I to level III, which met the requirements for the use of water environment functional areas. It should be noted that the mercury contents of most water samples and arsenic contents of two groundwater samples were relatively high, and the single-factor index evaluation results were 'Inferior' level. The simple standing-precipitate of the reservoir could no longer ensure that the arsenic and mercury inflow reached the standard, especially mercury.

(3) In order to ensure safe irrigation water in the agricultural production process, it is suggested to consider the following two suggestions. For one thing, the monitoring of mercury in most water sources and arsenic in the groundwater should be strengthened immediately in the irrigation influent; and it is recommended that the local area reduce the frequency and use of groundwater irrigation. For another, it is possible to appropriately add a pre-reservoir for irrigation water (such as a plant pond constructed wetland system or a rapid purification device), plant water hyacinths along the way and add irrigation drainage treatment facilities such as ecological ditch constructed wetlands.

DATA AVAILABILITY STATEMENT

All relevant data are included in the paper or its Supplementary Information.

REFERENCES

Ahsan, M., Younis, A., Jaskani, M. J., Tufail, A., Riaz, A., Schwinghamer, T., Tariq, U. & Nawaz, F. 2018 Heavy metal accumulation imparts structural differences in fragrant *Rosa* species irrigated with marginal quality water. *Ecotoxicology and Environmental Safety* **159**, 240–248.

Chaney, R. L., Reeves, P. G., Ryan, J. A., Simmons, R. W., Welch, R. M. & Angle, J. S. 2004 An improved understanding of soil Cd risk to humans and low cost methods to phytoextract Cd from contaminated soils to prevent soil Cd risks. *Biometals* **17** (5), 549–553.

Chen, B. L., Guan, S. H., Guan, H. K., Huang, X. M., Ke, Z. P., Tang, K. L., Huang, J. Z. & Li, Z. M. 2011 Heavy metal contamination of vegetables grown in Nanhai District and risk analysis on human health. *Preventive Medicine in South China* **37** (5), 3–8 (in Chinese).

Cheng, Y., Wang, Y., Zhou, Y. Z. & Gu, T. 2019 Status and sources of heavy metal/metalloid pollution in the groundwater system of Zhuhai City. *Bulletin of Mineralogy, Petrology and Geochemistry* **38** (3), 595–603 (in Chinese).

China, GAQSIQ & Standardization Authority 2005 *Standards for Farmland Irrigation Water Quality*. General Administration of Quality Supervision, Inspection and Quarantine of the People's Republic of China and Standardization Administration, Beijing, China (in Chinese).

China, GAQSIQ & Standardization Authority 2017 *Standards for Groundwater Quality*. General Administration of Quality Supervision, Inspection and Quarantine of the People's Republic of China and Standardization Administration, Beijing, China (in Chinese).

China, SEPA 2002a *Environmental Quality Standards for Surface Water*. State Environmental Protection Administration of China, Beijing, China (in Chinese).

China, SEPA 2002b *Technical Specification Requirements for Monitoring of Surface Water and Wastewater*. State Environmental Protection Administration of China, Beijing, China (in Chinese).

China, SEPA 2002c *Monitoring and Analysis Methods for Water and Wastewater*, 4th edn. State Environmental Protection Administration of China, Beijing, China (in Chinese).

Du, M., Chen, J. & Ma, Z. 2005 Impact of water price and water pollution on farmers in Weihe River Valley. *Environmental Protection* **2005** (10), 39–43 (in Chinese).

Eid, E. M., Shaltout, K. H., Moghanm, F. S., Youssef, M. S. G., El-Mohsnawy, E. & Haroun, S. A. 2019 Bioaccumulation and translocation of nine heavy metals by *Eichhornia crassipes* in Nile Delta, Egypt: perspectives for phytoremediation. *International Journal of Phytoremediation* **21** (8), 821–830.

Fiol, N., Villaescusa, I., Martínez, M., Miralles, N., Poch, J. & Serarols, J. 2006 Sorption of Pb(II), Ni(II), Cu(II) and Cd(II) from aqueous solution by olive stone waste. *Separation & Purification Technology* **50** (1), 132–140.

Galletti, A., Verlicchi, P. & Ranieri, E. 2010 Removal and accumulation of Cu, Ni and Zn in horizontal subsurface flow constructed wetlands: contribution of vegetation and filling medium. *Science of the Total Environment* **408** (21), 5097–5105.

Guo, P., Sun, T., Yang, G. & Ma, M. 2018 Migration and transformation of mercury at sediment–water interface of the Dahong Lake Reservoir in the Simian Mountains. *Environmental Science* **39** (12), 5473–5479 (in Chinese).

He, X. 2010 *Study on Migration and Enrichment of Arsenic in Groundwater under the Influence of Agricultural Irrigation in Hetao Plain*. China University of Geosciences, Wuhan, China (in Chinese).

Huang, Q., Yu, Y., Wan, Y., Wang, Q., Luo, Z., Qiao, Y., Su, D. & Li, H. 2018 Effects of continuous fertilization on bioavailability and fractionation of cadmium in soil and its uptake by rice (*Oryza sativa* L.). *Journal of Environmental Management* **215**, 13–21.

Jia, Y. N., Tian, H., He, J. H. & Feng, L. L. 2018 Progress in enrichment and separation of HCH and mercury in water system. *Chemistry* **82** (10), 878–885 (in Chinese).

Jiang, K., Deng, X., Zhou, H., Long, J., Li, X. Y., Dong, X., Liu, W. H., Hou, H. B., Peng, P. Q. & Liao, B. H. 2019 Ecological interception effect of plant pond constructed wetland system on irrigation water Cd. *Research of Agricultural Modernization* **40** (3), 518–526 (in Chinese).

Kumarathilaka, P., Seneweera, S., Meharg, A. & Bundschuh, J. 2018 Arsenic speciation dynamics in paddy rice soil–water environment: sources, physico-chemical, and biological factors – a review. *Water Research* **140**, 403–414.

Liu, S. T., He, Z. X., Xu, M., Li, D. Y., Luo, H. Y., Liu, X. L. & Tie, B. Q. 2019 Effects of plant pond and constructed wetland system on irrigation water purification and rice cadmium control. *Journal of Agro-Environment Science* **38** (5), 1060–1068 (in Chinese).

Wang, Q., Li, F. B., Huang, X. Z., Ma, Y. B., Lin, Y. S. & Liu, X. W. 2018 A classification approach of heavy metal pollution of paddy soil based on risk management. *Ecology and Environmental Sciences* **27** (12), 2321–2328 (in Chinese).

Xu, Z. X. 2005a Single factor water quality identification index for environmental quality assessment of surface water. *Journal of Tongji University* **33** (3), 321–325 (in Chinese).

Xu, Z. X. 2005b Comprehensive water quality identification index for environmental quality assessment of surface water. *Journal of Tongji University* **33** (4), 482–488 (in Chinese).

Yin, S. H., Zhang, Z. B., Xiao, Y., Zhang, Z., Yao, C. X., Deng, H. G. & Zhang, J. 2019 Distribution characteristic of mercury and arsenic in the *Potamogeton crispus*-overlying water-sediment system of Dongping Lake. *Environmental Chemistry* **38** (3), 635–643 (in Chinese).

Zhang, C. Y., He, J. T., Zhang, X. W. & Ni, Z. H. 2018a Geochemical characteristics and genesis analyses of high-arsenic groundwater in the Pearl River Delta. *Environmental Science* **39** (8), 3631–3639 (in Chinese).

Zhang, J., Li, H., Zhou, Y., Dou, L. & You, J. 2018b Bioavailability and soil-to-crop transfer of heavy metals in farmland soils: a case study in the Pearl River Delta, South China. *Environmental Pollution* **235**, 710–719.

Zhang, W. Y., Ma, L., Jilili, A., Ge, Y. X., Pu, J. & Hua, Y. S. 2019 Source analysis and pollution assessment of heavy metals in surface water of Bortala River, Northwest China. *Journal of Arid Land Resources and Environment* **33** (7), 100–106 (in Chinese).

Zhou, J., Feng, K., Pei, Z. & Lu, M. 2016 Pollution assessment and spatial variation of soil heavy metals in Lixia River Region of Eastern China. *Journal of Soils & Sediments* **16** (3), 748–755.

First received 11 August 2020; accepted in revised form 3 October 2020. Available online 19 January 2021

Effect of water compounds on photo-disinfection efficacy of TiO_2 NP-embedded cellulose acetate film in natural water

Jing Xie and Yen-Con Hung

ABSTRACT

Photocatalysis disinfection has great potential for irrigation water disinfection to improve fresh produce safety. Titanium dioxide (TiO_2) nanoparticle (NP)-embedded cellulose acetate (CA) film has shown effectiveness against *Escherichia coli* (*E. coli*) O157:H7 in water. The current study evaluated the effect of natural water compounds on the photo-disinfection efficacy of TiO_2 NP-embedded CA film. Humic acid, calcium carbonate ($CaCO_3$), and kaolin clay solutions were prepared at four concentrations, respectively. When concentration increased from 0 to 20 ml/L, inactivation of *E. coli* O157:H7 in humic acid, $CaCO_3$, and kaolin clay solutions decreased from 6 log to 5, 4, and 2 log CFU/ml, respectively after 3 h treatment. Turbidity, UVT-254, water hardness, total suspended solids (TSS), and total organic carbon (TOC) of the solutions were measured. UVT-254 and turbidity had the highest correlation with the inhibition effect of water compounds on photo-disinfection efficacy. A prediction equation was developed with UVT-254 and water hardness as independent variables to predict photo-disinfection efficacy in natural water. *E. coli* O157:H7 decreased by 1 and 2.5 log CFU/ml in unfiltered and filtered natural creek water samples after treatment. The results from this study showed promise in the use of TiO_2 NP-embedded CA film to inactivate pathogens in natural water.

Key words | *E. coli* O157:H7, natural water compound, photocatalysis, TiO_2 nanoparticle, water disinfection

Jing Xie
Yen-Con Hung (corresponding author)
Department of Food Science and Technology,
University of Georgia,
1109 Experiment Street, Griffin, GA 30223-1797,
USA
E-mail: *yhung@uga.edu*

HIGHLIGHTS

- Natural water compounds affect TiO_2 photo-disinfection efficacy in different degrees.
- UV transmittance and water hardness are important water quality parameters that can affect TiO_2 photo-disinfection efficacy.
- Prediction equation was developed to predict TiO_2 photo-disinfection efficacy in natural water.
- Filtration is an effective pre-treatment method for improving TiO_2 photo-disinfection in natural water.

INTRODUCTION

Agriculture consumes 70% of the freshwater used worldwide, and the number increases to 95% in developing countries (FAO 2008). As a crucial part of agricultural water, irrigation water is one of the main pathways by which pathogenic microorganisms can reach fresh produce (FDA 2013). Irrigating produce to be eaten raw with contaminated water may increase the risk of foodborne illness (Chalmers *et al.* 2000; Gu *et al.* 2019; Rodrigues *et al.* 2020). Evidence of produce contamination through contaminated irrigation water has been found in epidemiological investigations (Steele &

doi: 10.2166/ws.2020.271

Odumeru 2004). Irrigation water has been implicated in several *E. coli* O157:H7 outbreaks (Ackers *et al.* 1998; CDC 1999; Iwu & Okoh 2019). Several studies have demonstrated that pathogens can transfer from contaminated irrigation water to soil and fresh produce and survive for a long period (Islam *et al.* 2004, 2005; Jokinen *et al.* 2019; Kabir *et al.* 2020). Moreover, the occurrence of the internationalization of pathogens in plants through contaminated irrigation water has been reported (Erickson *et al.* 2010; Erickson *et al.* 2019). Improving irrigation water safety can effectively reduce the risk of fresh produce contamination.

Chlorination is the most commonly used disinfection method, however, disinfection byproducts (DBPs) that are potentially carcinogenic may be produced when chlorine reacts with natural organic matter (Marugán *et al.* 2008). Treatment methods such as UV and ozone are effective and have been demonstrated for drinking water treatment (Solomon *et al.* 2002; Loeb *et al.* 2012). However, these methods are costly and may not be suitable for irrigation water disinfection. The development and implementation of alternative disinfection methods are therefore needed.

Photocatalysis disinfection inactivates microorganisms by reactive oxygen species (ROS) generated when a semiconductor photocatalyst is activated by light in the presence of water and oxygen. Photocatalysis disinfection is an economic and environmentally friendly process and might be an alternative water disinfection method for irrigation water (Dasgupta *et al.* 2017). Titanium dioxide (TiO_2) has been considered the most suitable photocatalyst for water disinfection due to its strong oxidizing abilities, chemical stability, low cost, and low toxicity (Chong *et al.* 2010). TiO_2 has a bandgap energy of about 3.2 eV, which means light with a wavelength of approximately 385 nm (i.e. UV-A light) can activate TiO_2 photocatalysis (Foster *et al.* 2011). TiO_2 has three crystalline structures: anatase, rutile, and brookite. The photocatalytic effectiveness and the agglomeration behavior of TiO_2 depends on its structures (Sygouni & Chrysikopoulos 2015; Yemmireddy & Hung 2015b). Mixtures of anatase and rutile have shown higher effectiveness as photocatalysts than other structures (Ishibashi *et al.* 2000). More than 1,000 research papers have been published regarding TiO_2 photocatalytic disinfection of microorganisms (Byrne *et al.* 2015). It has been proved that TiO_2 can kill a wide range of microorganisms, including Gram-negative and Gram-positive bacteria, bacteriophage, fungi, protozoa, and viruses (Foster *et al.* 2011). More recently, the interaction of human adenoviruses and TiO_2 nanoparticles has been investigated (Syngouna *et al.* 2017).

TiO_2 NP-embedded cellulose acetate (CA) film has shown effectiveness in the inactivation of bacteria in water (Xie & Hung 2019). This film might be combined with different types of solar photocatalytic reactors for water treatment. However, when applyied in natural water, various organic and inorganic compounds may detrimentally affect photocatalysis disinfection of water. The effect of these water compounds on photocatalysis disinfection needs to be understood before this technology can be practically applied. Humic substances constitute the major fraction of natural organic matter (NOM) in natural water and 90% of total dissolved organic carbon (DOC) in surface water may come from humic acid (Alkan *et al.* 2007). It has been reported that humic acid can inhibit the photocatalysis process by acting as an ·OH scavenger (Garbin *et al.* 2007; Cheng *et al.* 2018). Calcium carbonate ($CaCO_3$) occurs in rocks, soil, and natural water world-wide. The presence of $CaCO_3$ affects water hardness and alkalinity, and it is the predominant component of scales deposited from natural water (Wang *et al.* 2009). The effect of $CaCO_3$ on photocatalysis inactivation is rarely reported. Kaolin clay is one of the common inorganic particles in natural water, and it contributes to turbidity in water and might be the worst particle for shielding (Liu & Zhang 2006). The effect of kaolin clay on photocatalysis disinfection in water has not been reported. Also, there is a lack of information on the comparison of the inhibition effect of different water compounds on photocatalysis disinfection.

The overall purpose of this study was to determine the effect of natural water compounds on the photo-disinfection efficacy of TiO_2 NP-embedded CA films. Specific objectives include:

(i) to determine the photo-disinfection efficacy of TiO_2 NP-embedded CA film in water containing humic acid, $CaCO_3$, and kaolin clay;
(ii) to evaluate the correlation between water quality parameters and the photo-disinfection efficacy of TiO_2 NP-embedded CA film in water;
(iii) to develop a prediction equation for predicting the disinfection efficacy of TiO_2 NP-embedded CA film in natural water.

MATERIALS AND METHODS

Preparation of TiO_2 NPs-embedded film

Aeroxide® P25 TiO_2 nanoparticles (anatase–rutile), acetone, cellulose acetate (average Mn ~30,000 by GPC), and triethyl citrate (TEC) (>99%) were purchased from Sigma-Aldrich (St Louis, MO, USA). TiO_2 NPs-embedded CA film was prepared using a solution casting method as described in Xie & Hung (2018). An optimum TiO_2 NPs concentration at 0.82 mg/cm^2 on the film to achieve the highest bactericidal effect had been determined in a previous study (Xie & Hung 2020 submitted). To fabricate the film, 4 g of CA and 0.4 g of TiO_2 were dissolved and suspended in 20 ml of acetone at room temperature (24 °C), separately. TEC (1.2 g) was added to TiO_2 NPs suspension as plasticizer. An ultra-sonication bath (Model FS60, Fisher Scientific, Waltham, MA, USA) was used to assist the suspension of TiO_2 NPs. After sonication, TiO_2-solvent suspension was added into the CA solution gradually using a pipette with stirring, and the solution was continuously stirred for 2 h. Five millilitres of the mixed solution was added into each glass Petri dish (88 mm in diameter, Corning®, Sigma-Aldrich (St Louis, MO, USA)) and allowed to dry with Petri dish lid covering in a fume hood at 24 °C overnight, and then stored in a vacuum desiccator for 24 h. The thickness of the prepared film was 53.5–54.7 μm.

Preparation of bacterial strains and inoculum

A five-strain cocktail of nalidixic-acid-adapted *E. coli* O157:H7 strains (E009 (beef), EO932 (cattle), O157-1 (beef), O157-4 (human), O157-5 (human)) was used in this study. The strains were stored at −70 °C in tryptic soy broth (Becton Dickson and Company, Sparks, MD, USA; TSB) containing 20% glycerol. To resuscitate the bacteria, each strain was streaked on Sorbitol MacConkey agar (Hardy Diagnostics, Santa Maria, CA, USA) supplemented with 50 μg/ml nalidixic acid (Sigma-Aldrich, St Louis, MO, USA; SMACNA) and incubated at 37 °C for 24 h.

One isolated colony of each strain was then transferred into 10 ml of tryptic soy broth supplemented with 50 μg/ml nalidixic acid (TSBNA) using a sterile inoculation loop and incubated at 37 °C overnight. Then one loopful (10 μL) of bacterial culture was transferred again into 10 ml of TSBNA and incubated overnight. After incubation, bacterial suspension was centrifuged at 4,000 × g for 12 min, and the supernatant was decanted and the cell pellet was resuspended in 9 ml of sterilized phosphate-buffered saline (Acros Organics, NJ, USA; PBS, pH 7.2). The inoculum was prepared by mixing 9 ml of each strain to obtain a five-strain cocktail with a concentration of about 9 log CFU/ml. The inoculum was further adjusted to a concentration of about 8 log CFU/ml by making ten-fold dilution.

Preparation of water samples

$CaCO_3$ was purchased from Fisher Scientific (Waltham, MA, USA). Kaolin clay and humic acid were purchased from Sigma-Aldrich (St Louis, MO, USA). Stock solutions of $CaCO_3$, kaolin clay and humic acid were prepared at 200 mg/L each and stored at 4 °C until use. Water samples containing various levels of $CaCO_3$, kaolin clay, or humic acid were prepared using the stock solutions before the experiment. The pH of these water samples was measured as about 6.5 ± 0.4. The natural water sample was obtained from a creek in Griffin, Georgia (33.2545488, −84.3114407).

Photo-disinfection of *E. coli* O157:H7 in water

Photo-disinfection of *E. coli* O157:H7 in water was performed as described in Xie & Hung (2019). In brief, the photocatalytic inactivation of *E. coli* O157:H7 in water was carried out in 150 ml glass beakers. TiO_2 NP-embedded CA films (4.5 cm in diameter) were adhered to the bottom of the beakers using ethylene-vinyl acetate (EVA, AdTech™, adhesive technologies, Hampton, NH, USA). The beakers were sterilized with UV-C light for 30 min in a bio-safety cabinet (Class II Type A/B3, Nuaire, Plymouth, MN, USA) before the experiment. The beakers were then placed on a platform shaker (Model Classic 10, New Brunswick Scientific Co., Inc., Edison, NJ, USA) inside a photocatalytic disinfection chamber (Figure 1). Four 40 W UV-A lamps (American DJ®, Model UV Panel HP™, LL-UV P40, Los Angeles, CA, USA) were fitted on the inside top of the photocatalytic disinfection chamber. The shaker was set at a speed of 100 rpm during the photo-disinfection experiment. Water samples were inoculated right before the

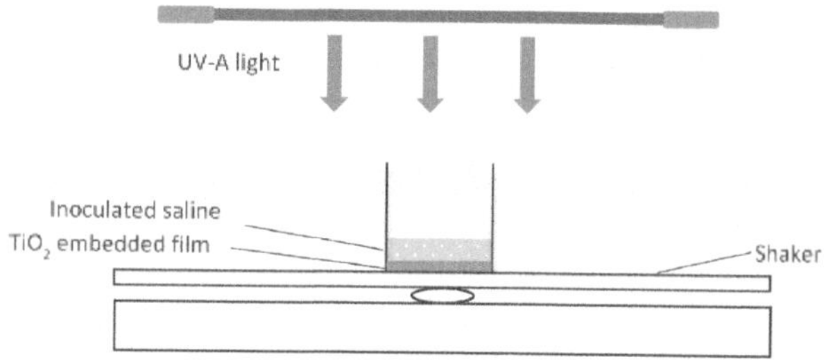

Figure 1 | Experiment setting of photo-disinfection of water using TiO_2 NP-embedded CA film.

photo-disinfection experiment. In this study, 30 ml of water sample was inoculated with 300 µl inoculum. The experiments were carried out at a light intensity of 1 mW/cm^2.

Sampling and bacterial enumeration

Water samples were taken hourly for three hours and 1 ml of sample was taken each time. Serial dilutions were made in PBS and appropriate dilutions were plated on SMACNA agar and incubated at 37 °C for 24 h. Colonies were counted and recorded as log CFU/ml.

Water quality parameter measurement

Water quality parameters including turbidity, UV transmittance at 254 nm (UVT-254), total organic carbon (TOC), total suspended solids (TSS), and total hardness of all the lab-prepared water samples and natural water samples were measured. Turbidity, TOC, and TSS were measured following Hach methods 8237, 10129, and 8006, respectively, using a DR/90 colorimeter (HACH, Loveland, CO, USA). UVT-254, which measures the percentage of light that passes through a water sample at 254 nm, was measured using a UV-vis spectrophotometer (Orion™ AquaMate 8000, Thermo Fisher Scientific, Waltham, MA, USA). Total hardness was determined following USEPA method 8226.

Statistical analysis

Experiments were replicated at least twice. Duplicate measurements were made on each sample. Pearson correlation analysis and partial correlation analysis on the water quality parameters and bacterial reduction were conducted using JMP 14 (SAS Institute, Cary, NC, USA). The regression equation was developed by least squares regression also using JMP 14. All the tests were performed at a significance level of 0.05.

RESULTS AND DISCUSSION

Effect of water compound on photo-disinfection efficacy of TiO_2 embedded CA film

Figure 2 shows the results of the photo-disinfection of *E. coli* O157:H7 using TiO_2 NP-embedded CA film in DI water and lab-prepared water samples with different natural water compounds under UV-A light illumination. In DI water, *E. coli* O157:H7 population was reduced by about 5.8 log after 3 h of photo-disinfection treatment. Increasing humic acid concentration from 2 to 20 mg/L reduced bacterial reduction from 3.9 to 0.8 log CFU/ml. Increasing $CaCO_3$ concentration from 2 to 20 mg/L reduced bacterial reduction from 4.3 to 1.9 log CFU/ml. In comparison with humic acid and $CaCO_3$, kaolin clay had the least inhibition effect on photo-disinfection. When adding 2 mg/L of kaolin clay in water, 5 log reductions were achieved after 3 h of photo-disinfection treatment. When the concentration of kaolin clay was increased to 20 mg/L, more than 4 log reductions could still be achieved after 3 h of treatment.

Inactivation of bacteria using TiO_2 immobilized systems has been demonstrated in published studies. Xiong & Hu (2013) evaluated the inactivation of *E. coli* using a UVA/

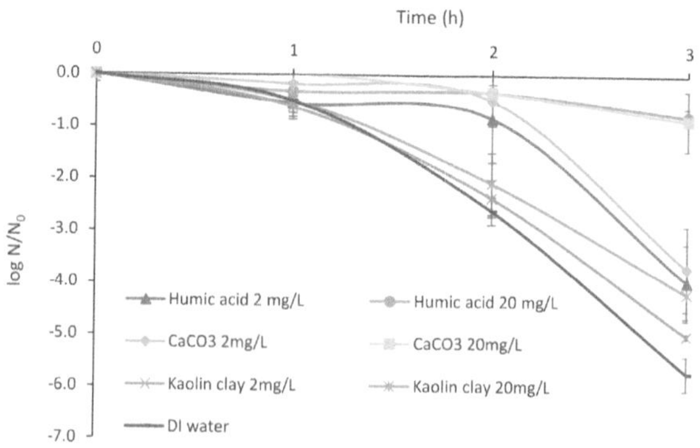

Figure 2 | Effect of different water compounds on photo-disinfection of *E. coli* O157:H7 using TiO_2 NP-embedded CA film. Vertical bars represent the standard error of three replicates.

LED system with a crystallizing dish coated with TiO_2. They reported that at a light intensity of 6 mW/cm^2, it took 145 min for 3 log inactivations of *E. coli* in 30 ml of inoculated distilled water. Rodrigues *et al.* (2007) evaluated the inactivation of *E. coli* in synthetic water and natural water using a glass reactor with immobilized TiO_2 (catalyst). They found 100% bacterial reduction in synthetic water, and 80% bacterial recution in natural water after 1 h treatment under solar light illumination. Due to the variation in factors such as experimental setting, TiO_2 immobilized system, light intensity, and water volume among different studies, the treatment efficiency may not be compared directly.

The disinfection kinetics in the current study follows a non-linear trend with a 'shoulder'. The 'shoulder' can be explained by the cumulative-damage nature of photo-disinfection treatment on the cytoplasmic membrane rather than an instantly lethal effect (Gyürék & Finch 1998). The disinfection kinetics for $CaCO_3$ and humic acid tend to have a longer 'shoulder' period. This indicates that the bacterial inactivation process was subject to interference by these two water compounds. Many empirical kinetic models have been reported in the literature for the interpretation of photo-disinfection data. The following models have been summarized as the most well-known disinfection kinetic models (Marugán *et al.* 2008; Chong *et al.* 2010; Yemmireddy & Hung 2015a).

The Chick–Watson (C-W) model (Chick 1908; Watson 1908), which is the most commonly employed disinfection model in photo-disinfection:

$$\log\frac{C}{C_0} = -kt \quad (1)$$

where C/C_0 is the reduction in bacterial concentration, k is the disinfection rate constant, and t is the treatment time.

The delayed Chick–Watson model (Cho *et al.* 2004) was developed to accommodate any initial lag time (t_0):

$$\log\frac{C}{C_0} = \begin{cases} 0 & for\ t \leq t_0 \\ -k(t-t_0) & for\ t > t_0 \end{cases} \quad (2)$$

The modified Chick–Watson model (Cho *et al.* 2003) was to fit either an initial shoulder or a tail at the end of the reaction:

$$\log\frac{C}{C_0} = k_1[1-\exp(-k_2 t)] \quad (3)$$

The Homs model (Hom 1972) was reported for predicting a curvilinear or non-linear function:

$$\log\frac{C}{C_0} = -kt^h \quad (4)$$

where h is the second parameter. If $h = 1$, the Homs model becomes a linear Chick–Watson equation and, if $h > 1$, this model can fit a line with a tail.

The modified Homs model (Cho *et al.* 2003), which expands the applicability of the Homs model for the fitting of disinfection data with an initial shoulder, log-linear reduction, and prolonged tailing behaviors:

$$\log\frac{C}{C_0} = k_1[1-\exp(-k_2 t)]^{k_3} \quad (5)$$

The data generated in the current study were fitted using these empirical models and RMSE (root mean square error) was calculated using the following equation:

$$RMSE = \sqrt{\frac{\sum_{n=1}^{N}(\hat{y}_n - y_n)^2}{N}} \quad (6)$$

where N is the number of observations, y_n is the observed value, and $\hat{y}_n$ is the predicted value. The results in Table 1 show that the delayed Chick–Watson model provided a better fit with the smallest RMSE for all water compound solutions tested. The three water compounds inhibited the photo-disinfection of CA film following the order of: humic acid > $CaCO_3$ > kaolin clay.

Effect of kaolin clay on water property and disinfection efficacy

The effects of kaolin clay on water quality parameters and bacterial reduction are shown in Table 2. Kaolin clay did not have a strong inhibition effect on photo-disinfection in the current study compared with humic acid and $CaCO_3$.

Table 1 | Comparison of kinetic models to predict photo-disinfection of TiO_2 NP-embedded CA film in different water compound solutions

Testing solution and concentration	Root mean squared error (RMSE)					Model selected
	Chick–Watson	Delayed Chick–Watson	Modified Chick–Watson	Homs	Modified Homs	
2 mg/L humic acid	0.7901	0.6603	0.7920	1.3289	0.7920	Delayed Chick–Watson
20 mg/L humic acid	0.0612	0.1732	0.0625	10.0704	0.1871	Chick–Watson
2 mg/L $CaCO_3$	0.8688	0.6229	0.8688	1.5220	0.7920	Delayed Chick–Watson
20 mg/L $CaCO_3$	0.4980	0.3873	0.4981	0.7381	/	Delayed Chick–Watson
2 mg/L kaolin clay	0.5904	0.3502	0.5909	4.8774	1.5642	Delayed Chick–Watson
20 mg/L kaolin clay	0.5009	0.2933	0.5014	1.3154	0.5014	Delayed Chick–Watson
DI water	1.0485	0.7071	1.0516	1.9343	1.0516	Delayed Chick–Watson

Table 2 | Effect of water compounds on water quality parameters and bacterial reduction

Water compounds	Concentration (mg/L)	Turbidity (FAU)	UVT-254 (%)	Total organic carbon (mg/L)	Total hardness (mg/L)	Total suspended solids (mg/L)	Bacterial reduction at 3 h (log CFU/ml)
Humic acid	2	3 ± 1	89.74 ± 2.67	1.2 ± 0.1	2 ± 0	3 ± 0	3.9 ± 0.6
	5	9 ± 4	74.45 ± 3.04	2.2 ± 0.5	2 ± 0	9 ± 1	2.1 ± 0.5
	10	16 ± 5	59.36 ± 3.41	3.1 ± 0.5	2 ± 0	12 ± 3	1.6 ± 0.6
	20	37 ± 6	40.35 ± 3.79	5.8 ± 0.6	2 ± 0	27 ± 7	0.8 ± 0.6
$CaCO_3$	2	2 ± 0	93.96 ± 2.01	0.0 ± 0.0	2 ± 2	1 ± 1	4.3 ± 0.6
	5	6 ± 2	93.49 ± 2.23	0.0 ± 0.0	5 ± 2	0 ± 0	3.4 ± 0.6
	10	8 ± 2	91.80 ± 1.23	0.0 ± 0.0	12 ± 2	1 ± 1	2.8 ± 0.7
	20	13 ± 2	89.90 ± 3.52	0.0 ± 0.0	22 ± 4	2 ± 1	1.9 ± 0.9
Kaolin clay	2	2 ± 0	91.50 ± 1.45	0.0 ± 0.0	0 ± 0	2 ± 2	5.0 ± 0.1
	5	6 ± 2	90.25 ± 2.70	0.0 ± 0.0	0 ± 0	4 ± 2	5.0 ± 0.1
	10	13 ± 3	81.04 ± 2.38	0.0 ± 0.0	0 ± 0	10 ± 3	4.6 ± 0.4
	20	24 ± 4	72.50 ± 2.08	0.0 ± 0.0	0 ± 0	23 ± 2	4.2 ± 0.5
DI water	–	0 ± 0	0.00 ± 0.00	0.0 ± 0	0 ± 0	0 ± 0	5.8 ± 0.3

Results of water property analysis showed that increasing kaolin clay concentration in water did not significantly change TOC and hardness. UVT-254 slightly decreased when kaolin clay concentration increased. However, kaolin clay concentration significantly affected the total suspended solids (TSS) and turbidity. It has been reported that kaolin clay might be the worst particle for shielding light (Liu & Zhang 2006) and hence may affect photo-disinfection efficacy. However, the current study showed that when the turbidity and TSS of kaolin clay solution were increased to 24 FAU and 24 mg/L, respectively, more than 4.2 log bacterial reductions were achieved. This indicates that turbidity and TSS might not be significant parameters affecting photo-disinfection efficacy.

Effect of $CaCO_3$ on water properties and disinfection efficacy

Results in Table 2 show that increasing $CaCO_3$ concentration did not affect TOC and TSS concentration in water. Increasing $CaCO_3$ slightly increased UVT-254 and turbidity. Water hardness significantly increased when increasing $CaCO_3$ concentration. When the hardness of the $CaCO_3$ solution increased to 22 mg/L, bacterial reduction decreased to about 1.9 log CFU/ml. Similar findings regarding the effect of $CaCO_3$ on photo-disinfection or photo-degradation have been reported in other published reports. Cohen Yaniv *et al.* (2008) studied the inactivation of *Flavobacterium* and *E. coli* in water by a continuous stirred tank reactor (CSTR) fed with suspended or glass immobilized TiO_2. They found water hardness increase reduced photocatalytic inactivation efficiency, and they suggested the treatment of water chemically before disinfection. Sreethawong *et al.* (2014) studied the degradation of Congo Red (CR) azo dye using nano-Ag/sol-gel TiO_2-In_2O_3 mixed oxide mesoporous-assembled nanocrystals. Their results showed that the water hardness reduced the photocatalytic CR dye degradation activity. In natural water, $CaCO_3$ is the major compound that causes an increase in water hardness (Sreethawong *et al.* 2014). Increased water hardness can cause a variety of problems such as reduced efficiency of chlorine treatment and decreased life of plumbing and appliances (Pangloli & Hung 2013). Results from the current study suggest that water hardness might be a significant parameter affecting photo-disinfection efficacy.

Effect of humic acid on water properties and disinfection efficacy

Results in Table 2 show that increasing humic acid concentration significantly increased TOC, turbidity, and TSS in water, whereas the UVT-254 reading decreased significantly. When humic acid concentration increased to 20 mg/L, turbidity increased to 37 FAU, TSS increased to 27 mg/L, TOC increased to 5.8 mg/L, while UVT-254 decreased to 40.35%. At this humic acid concentration, bacterial reduction decreased significantly to 0.8 log CFU/ml. Although humic acid significantly affected the readings of turbidity, UVT-254, TOC, and TSS, turbidity and TSS were found not significantly to affect microbial inactivation as reported above.

The inhibition effect of humic acid on photo-disinfection has also been reported in published reports. Cheng *et al.* (2018) studied the effect of humic acid on virus removal using a photocatalytic membrane reactor (PMR). They found that when increasing humic acid concentration from 0 to 5 and 7.5 ml/L, the photocatalytic disinfection efficiency of phage f2 decreased from 5.4 to 3.5 and 2.2 log, respectively. Marugán *et al.* (2008) studied the effect of humic acid on the photocatalytic disinfection kinetics of *E. coli* suspensions using TiO_2/SiO_2 photocatalysts. They found that low humic substance concentration could inhibit the disinfection process and when humic acid concentration increased from 0 to 2.0 mg/L, disinfection efficacy reduced from 5 log to about 2 log after 1 h treatment. According to these studies, the main reason that humic acid inhibited photo-disinfection efficacy was due to the consumption of hydroxyl radicals produced by TiO_2 photocatalysis.

Humic acid is a major fraction of natural organic matter (NOM) in natural water (Wang & Hsieh 2001). The C = C double bonds and C = O double bonds in humic acid chemical structure can absorb UV light at 254 nm and prevent light at this wavelength transmitting through water (Cheng *et al.* 2018), and hence caused low UVT-254 readings for humic acid solutions in the current study (Table 2). UVT-254 measures the transmittance of UV light at 254 nm through the water. It has been reported that the efficiency of UV disinfection will be affected by the turbidity of the target water and the resultant transmittance of UV light through the water (Cantwell & Hofmann 2011), while the current study showed that for photocatalytic disinfection, the effect of turbidity and UVT-254 on disinfection efficiency varied among the different compounds studied. For example, kaolin clay caused a significant increase in turbidity, but it did not have as much effect on bacterial reduction as other compounds such as humic acid.

Prediction equation development

As discussed above, various water compounds such as humic acid and $CaCO_3$ have strong inhibition effects on the TiO_2 NP-embedded CA film photo-disinfection effect. However, in a practical situation, it is unrealistic to monitor the level of all compounds in water. Identifying common water quality parameters that can be used as water quality indicators for photo-disinfection is a more efficient solution.

To select the proper water quality parameters that can be used for predicting TiO_2 NP-embedded CA film disinfection efficacy, the Pearson correlation of all the water quality parameters reported in Table 2 was first performed and the results are presented in Table 3. It shows that several variables are strongly correlated with another variable. For example, turbidity and TSS have a correlation coefficient of 0.899, TOC and UVT-254 have a correlation coefficient of −0.884, turbidity and UVT-254 have a correlation coefficient of −0.881, and UVT-254 and TSS have a correlation coefficient of −0.872. These suggest collinearity exists

Table 3 | Pearson correlations of water quality parameters and bacterial reduction

	Turbidity	UV-254	Total organic carbon	Total hardness	Total suspended solids	Bacterial reduction
Turbidity	1.000	– 0.881	0.674	0.027	0.899	– 0.617
UV-254		1.000	– 0.884	0.179	– 0.872	0.685
Total organic carbon			1.000	– 0.161	0.672	– 0.702
Total hardness				1.000	– 0.284	– 0.432
Suspended solids					1.000	– 0.433
Bacterial reduction						1.000

among these variables and using all the variables for model development would cause an inaccurate predictor contribution to the model. Therefore, only one variable between highly correlated pairs should be considered for regression development.

To determine which variable should be excluded from model development, a partial correlation between the predictor variables and the response variable was performed and the results are shown in Table 4. Pearson correlations reported in Table 3 demonstrated the linear correlation between two variables without controlling other variables in the model, while partial correlation measures the correlation between two variables with the effect of other controlling variables removed. Results show that turbidity, TOC, and TSS have Pearson correlation coefficients of -0.617, -0.702, and –0.433 with bacterial reduction, respectively. However, these three variables have much lower partial correlation coefficients of 0.161, –0.272, and 0.097 with bacterial reduction, respectively, after controlling other confounding variables in the model. As a result, UVT-254 and total hardness with high partial correlation coefficients with bacterial reduction (Table 4) are further considered for regression analysis. This conclusion further supports the observations reported above that changes in total hardness and UVT-254, respectively can explain the changes in bacterial reduction.

In addition, for practical applications, some water properties can be affected by more than one chemical compound. For example, UVT-254 and TOC can all be used to indicate organic matter in water. However, TOC is a more time-consuming measurement compared with UVT-254. It is also desirable to select the parameters that are easy to measure. In the current study, UVT-254 and water hardness are the two parameters that are easy to measure and can explain the changes in bacterial reduction. A linear prediction equation is then developed using the least squares method and listed in Equation (7) with the adjusted R^2 and root mean square error of 0.78 and 0.74, respectively:

$$\text{Bacterial reduction} = -2.150 + 0.075 \times \text{UVT} - 254 - 0.142 \times \text{Hardness} \tag{7}$$

Verification of prediction equation using laboratory-prepared water samples and natural water

In order to test whether the prediction equation can accurately predict bacterial reduction in a complex environment that contains different water compounds, four different simulated water samples were prepared with different combinations of the three water compounds based on the formula shown in Table 5. Table 6 shows the observed and predicted bacterial reductions using TiO_2 NP-embedded CA film in these simulated water samples.

Table 4 | Partial correlations between water quality parameters and bacterial reduction

	Turbidity	UVT-254	Total organic carbon	Total hardness	Total suspended solids
Bacterial reduction	0.161	0.461	– 0.272	– 0.726	0.097

Table 5 | Formula of simulated water samples

Water sample	Kaolin clay (mg/L)	Humic acid (mg/L)	$CaCO_3$ (mg/L)
#1	10	10	10
#2	5	5	20
#3	5	20	5
#4	20	5	5

The highest bacterial reduction was detected in sample 4 (2.1 log CFU/ml), followed by sample 2, sample 1, and sample 3 with reductions of 1.8, 1.6, and 1.5 log CFU/ml, respectively. Water property analysis shows that sample 3 had the lowest UVT-254 of 61.62%. The UVT-254 of sample 1, 4, and 2 was 73.3%, 79.47%, and 81.85%, respectively. Sample 2 had the highest hardness of 20 mg/L. The hardness of sample 1 was 14 mg/L and the hardness of both samples 3 and 4 was 6 mg/L. The prediction results show that half of the samples have a prediction percentage error of less than 15%, however, the other half have a prediction percentage error of more than 30%. The prediction equation developed in the preceding section was based on the disinfection experiment carried out in individual water compounds. The prediction equation might be improved by studying the interaction effects between different water compounds during treatment.

Table 7 shows that in natural creek water, bacteria can be reduced by 1 log CFU/ml after 3 h treatment using TiO_2 NP-embedded CA film. After filtration using a 0.45 μm filter, 1.4 log reductions were achieved. After filtration using a 0.2 μm filter, bacterial reductions increased to 2.5 log CFU/ml. As shown in Table 7, filtration also improved water quality. The hardness, turbidity, TSS, and TOC all decreased after filtration, and UVT-254 increased with filtration. Filtration can remove insoluble particles that cause the increase of these parameters and can therefore improve water quality (O'Melia 1985). Bacterial reductions were also calculated using Equation (7). The predicted bacterial reduction was −0.9 CFU/ml reduction for the water sample without filtration, and 0.9 and 1.1 CFU/ml reductions for samples filtered with 0.2 and 0.45 μm filters, respectively. The total hardness of the unfiltered water sample was over the ranges of the hardness of water samples used for the development of the prediction equation, which may affect the prediction accuracy. More types of natural water and more water property information are needed to further improve the prediction equation.

Nevertheless, the current study has demonstrated that using TiO_2 NP-embedded CA film has the potential to inactivate pathogens in natural water. Filtration has also been found as a simple and effective method to remove water compounds that may affect bacterial inactivation during the photo-disinfection process.

CONCLUSIONS

The results of this study showed that natural water compounds such as humic acid, $CaCO_3$, and kaolin clay can all affect the photo-disinfection efficacy of TiO_2 NP-embedded CA film. Humic acid had the highest inhibition effect on photo-disinfection, followed by $CaCO_3$ and kaolin clay. The effect of different water quality parameters on TiO_2 NP-embedded CA film indicates that different water compounds affect photo-disinfection efficacy through different mechanisms. It was found that UVT-254 and turbidity can be used as indicators for predicting the effect of natural

Table 6 | Water properties and disinfection efficacy of simulated water samples

			Bacterial reduction at 3 h (log CFU/ml)		
water sample	Hardness (mg/L)	UVT-254 (%)	Observed	Predicted	Prediction percentage error (%)[a]
#1	14 ± 1	73.30 ± 0.32	1.6 ± 0.3	1.4 ± 0.1	−13
#2	20 ± 1	81.95 ± 0.20	1.8 ± 0.2	1.2 ± 0.1	−33
#3	6 ± 1	61.62 ± 0.53	1.5 ± 0.7	1.7 ± 0.1	13
#4	6 ± 1	79.47 ± 0.32	2.1 ± 0.5	3.0 ± 0.1	43

[a]Prediction percentage error(%) = $\frac{\text{Predicted bacterial reduction} - \text{Observed bacterial reduction}}{\text{Observed bacterial reduction}} \times 100$.

Table 7 | Water properties and disinfection efficacy of natural creek water samples

Filter (μm)	Hardness (mg/L)	Turbidity (FAU)	Total suspended solids (mg/L)	Total organic carbon (mg/L)	UVT-254 (%)	Bacterial reduction at 3 h (log CFU/ml) actual	predicted	Prediction percentage error (%)[a]
0.2	20 ± 1	3.5 ± 2.1	0	4.9 ± 1.6	85.75 ± 2.25	2.5 ± 0.4	1.4 ± 0.2	−44
0.45	24 ± 1	7 ± 2.8	1.5 ± 0.7	5.9 ± 1.6	80.71 ± 0.95	1.4 ± 0.1	0.5 ± 0.1	−64
w/o	32 ± 1	10.5 ± 2.1	3 ± 1.4	8.7 ± 3.3	77.82 ± 1.90	1.0 ± 0.3	−0.9 ± 0.1	−190

[a]Prediction percentage error(%) = $\frac{\text{Predicted bacterial reduction} - \text{Observed bacterial reduction}}{\text{Observed bacterial reduction}} \times 100.$

water compounds on photo-disinfection efficacy. A predictive equation was developed using UVT-254 and turbidity as independent variables. Photo-disinfection using TiO_2 NP-embedded CA film reduced *E. coli* O157:H7 by 1 log CFU/ml in the inoculated natural creek water sample. After filtration using a 0.2 μm filter, about 2.5 log CFU/ml reductions were achieved with the same treatment. Hence, filtration can be used as an effective pre-treatment for photo-disinfection by TiO_2 NP-embedded CA film.

ACKNOWLEDGEMENTS

This research was supported by State and Hatch funds allocated to the University of Georgia Agriculture Experiment Station, Griffin Campus.

DATA AVAILABILITY STATEMENT

All relevant data are included in the paper or its Supplementary Information.

REFERENCES

Ackers, M. L., Mahon, B. E., Leahy, E., Goode, B., Damrow, T., Hayes, P. S., Bibb, W. F., Rice, D. H., Barrett, T. J., Hutwagner, L., Griffin, P. M. & Slutsker, L. 1998 An outbreak of *Escherichia coli* O157:H7 infections associated with leaf lettuce consumption. *Journal of Infectious Diseases* **177** (6), 1588–1593. doi:10.1086/515323.

Alkan, U., Teksoy, A., Ateşli, A. & Başkaya, H. S. 2007 Influence of humic substances on the ultraviolet disinfection of surface waters. *Water and Environment Journal* **21** (1), 61–68. doi:10.1111/j.1747-6593.2006.00047.x.

Byrne, J. A., Dunlop, P. S. M., Hamilton, J. W. J., Fernández-Ibáñez, P., Polo-López, I., Sharma, P. K. & Vennard, A. S. M. 2015 A review of heterogeneous photocatalysis for water and surface disinfection. *Molecules* **20** (4), 5574–5615. doi:10.3390/molecules20045574.

Cantwell, R. E. & Hofmann, R. 2011 Ultraviolet absorption properties of suspended particulate matter in untreated surface waters. *Water Research* **45** (3), 1322–1328. doi:10.1016/j.watres.2010.10.020.

CDC 1999 Outbreak of *Escherichia coli* O157:H7 and *Campylobacter* among attendees of the Washington County Fair – New York, 1999. *Morbidity and Mortality Weekly Report* **48**, 803–805.

Chalmers, R. M., Aird, H. & Bolton, F. J. 2000 Waterborne *Escherichia coli* O157. *Journal of Applied Microbiology* **88** (S1), 124S–132S. doi:10.1111/j.1365-2672.2000.tb05340.x.

Cheng, R., Shen, L., Wang, Q., Xiang, S., Shi, L., Zheng, X. & Lv, W. 2018 Photocatalytic membrane reactor (PMR) for virus removal in drinking water: effect of humic acid. *Catalysts* **8** (7), 284. doi:10.3390/catal8070284.

Chick, H. 1908 An investigation of the laws of disinfection. *Journal of Hygiene* **8** (1), 92–158. doi:10.1017/s0022172400006987.

Cho, M., Chung, H. & Yoon, J. 2003 Disinfection of water containing natural organic matter by using ozone-initiated radical reactions. *Applied and Environmental Microbiology* **69** (4), 2284–2291. doi:10.1128/aem.69.4.2284-2291.2003.

Cho, M., Chung, H., Choi, W. & Yoon, J. 2004 Linear correlation between inactivation of *E-coli* and OH radical concentration in TiO_2 photocatalytic disinfection. *Water Research* **38** (4), 1069–1077. doi:10.1016/j.watres.2003.10.029.

Chong, M. N., Jin, B., Chow, C. W. K. & Saint, C. 2010 Recent developments in photocatalytic water treatment technology: a review. *Water Research* **44** (10), 2997–3027. doi:10.1016/j.watres.2010.02.039.

Cohen-Yaniv, V., Narkis, N. & Armon, R. 2008 Photocatalytic inactivation of *Flavobacterium* and *E. coli* in water by a continuous stirred tank reactor (CSTR) fed with suspended/immobilised TiO_2 medium. *Water Science and Technology* **58** (1), 247–252. doi:10.2166/wst.2008.664.

Dasgupta, N., Ranjan, S. & Ramalingam, C. 2017 Applications of nanotechnology in agriculture and water quality management. *Environmental Chemistry Letters* **15** (4), 591–605. doi:10.1007/s10311-017-0648-9.

Erickson, M. C., Webb, C. C., Diaz-Perez, J. C., Phatak, S. C., Silvoy, J. J., Davey, L., Payton, A. S., Liao, J., Ma, L. & Doyle, M. P. 2010 Surface and internalized *Escherichia coli* O157:H7 on field-grown spinach and lettuce treated with spray-contaminated irrigation water. *Journal of Food Protection* **73** (6), 1023–1029. doi:10.4315/0362-028x-73.6.1023.

Erickson, M. C., Liao, J. Y., Payton, A. S., Cook, P. W. & Ortega, Y. R. 2019 Survival and internalization of *Salmonella* and *Escherichia coli* O157:H7 sprayed onto different cabbage cultivars during cultivation in growth chambers. *Journal of the Science of Food and Agriculture* **99** (7), 3530–3537. doi:10.1002/jsfa.9573.

FAO 2008 *AQUASTAT – FAO's Global Information System on Water and Agriculture*. Available from: http://www.fao.org/nr/water/aquastat/main/index.stm.

FDA 2013 *Food Safety Modernization Act: Standards for the Growing, Harvesting, Packing, and Holding of Produce for Human Consumption; Proposed Rule, 2013 (Docket No. FDA-2011-N-0921)*. Available from: https://www.fda.gov/Food/GuidanceRegulation/FSMA/ucm334114.htm.

Foster, H. A., Ditta, I. B., Varghese, S. & Steele, A. 2011 Photocatalytic disinfection using titanium dioxide: spectrum and mechanism of antimicrobial activity. *Applied Microbiology and Biotechnology* **90** (6), 1847–1868. doi:10.1007/s00253-011-3213-7.

Garbin, J. R., Milori, D. M. B. P., Simões, M. L., da Silva, W. T. L. & Neto, L. M. 2007 Influence of humic substances on the photolysis of aqueous pesticide residues. *Chemosphere* **66** (9), 1692–1698. doi:10.1016/j.chemosphere.2006.07.017.

Gu, G. Y., Strawn, L. K., Zheng, J., Reed, E. A. & Rideout, S. L. 2019 Diversity and dynamics of *Salmonella enterica* in water sources, poultry litters, and field soils amended with poultry litter in a major agricultural area of Virginia. *Frontiers in Microbiology* **10**, 2868. doi:10.3389/fmicb.2019.02868.

Gyürék, L. L. & Finch, G. R. 1998 Modeling water treatment chemical disinfection kinetics. *Journal of Environmental Engineering* **124** (9), 783–793.

Hom, L. W. 1972 Kinetics of chlorine disinfection in an ecosystem. *Journal of the Sanitary Engineering Division* **98** (1), 183–194.

Ishibashi, K., Fujishima, A., Watanabe, T. & Hashimoto, K. 2000 Generation and deactivation processes of superoxide formed on TiO_2 film illuminated by very weak UV light in air or water. *Journal of Physical Chemistry B* **104** (20), 4934–4938. doi:10.1021/jp9942670.

Islam, M., Doyle, M. P., Phatak, S. C., Millner, P. & Jiang, X. P. 2004 Persistence of enterohemorrhagic *Escherichia coli* O157:H7 in soil and on leaf lettuce and parsley grown in fields treated with contaminated manure composts or irrigation water. *Journal of Food Protection* **67** (7), 1365–1370. doi:10.4315/0362-028x-67.7.1365.

Islam, M., Doyle, M. P., Phatak, S. C., Millner, P. & Jiang, X. P. 2005 Survival of *Escherichia coli* O157:H7 in soil and on carrots and onions grown in fields treated with contaminated manure composts or irrigation water. *Food Microbiology* **22** (1), 63–70. doi:10.1016/j.fm.2004.04.007.

Iwu, C. D. & Okoh, A. I. 2019 Preharvest transmission routes of fresh produce associated bacterial pathogens with outbreak potentials: a review. *International Journal of Environmental Research and Public Health* **16** (22), 4407. doi:10.3390/ijerph16224407.

Jokinen, C. C., Hillman, E. & Tymensen, L. 2019 Sources of generic *Escherichia coli* and factors impacting guideline exceedances for food safety in an irrigation reservoir outlet and two canals. *Water Research* **156**, 148–158. doi:10.1016/j.watres.2019.03.013.

Kabir, M. N., Aras, S., Wadood, S., Chowdhury, S. & Fouladkhah, A. C. 2020 Fate and biofilm formation of wild-type and pressure-stressed pathogens of public health concern in surface water and on abiotic surfaces. *Microorganisms* **8** (3), 408. doi:10.3390/microorganisms8030408.

Liu, W. J. & Zhang, Y. J. 2006 Effects of UV intensity and water turbidity on microbial indicator inactivation. *Journal of Environmental Sciences* **18** (4), 650–653.

Loeb, B. L., Thompson, C. M., Drago, J., Takahara, H. & Baig, S. 2012 Worldwide ozone capacity for treatment of drinking water and wastewater: a review. *Ozone: Science & Engineering* **34** (1), 64–77. doi:10.1080/01919512.2012.640251.

Marugán, J., van Grieken, R., Sordo, C. & Cruz, C. 2008 Kinetics of the photocatalytic disinfection of *Escherichia coli* suspensions. *Applied Catalysis B: Environmental* **82** (1–2), 27–36. doi:10.1016/j.apcatb.2008.01.002.

O'Melia, C. R. 1985 Particles, pretreatment, and performance in water filtration. *Journal of Environmental Engineering* **111** (6), 874–890. doi:10.1061/(asce)0733-9372(1985)111:6(874).

Pangloli, P. & Hung, Y. C. 2013 Effects of water hardness and pH on efficacy of chlorine-based sanitizers for inactivating *Escherichia coli* O157:H7 and *Listeria monocytogenes*. *Food Control* **32** (2), 626–631. doi:10.1016/j.foodcont.2013.01.044.

Rodrigues, C. P., Ziolli, R. L. & Guimarães, J. R. 2007 Inactivation of *Escherichia coli* in water by TiO_2-assisted disinfection using solar light. *Journal of the Brazilian Chemical Society* **18** (1), 126–134. doi:10.1590/s0103-50532007000100014.

Rodrigues, C., da Silva, A. L. B. R. & Dunn, L. L. 2020 Factors impacting the prevalence of foodborne pathogens in agricultural water sources in the southeastern United States. *Water* **12** (1), 51. doi:10.3390/w12010051.

Solomon, E. B., Potenski, C. J. & Matthews, K. R. 2002 Effect of irrigation method on transmission to and persistence of *Escherichia coli* O157:H7 on lettuce. *Journal of Food Protection* **65** (4), 673–676. doi:10.4315/0362-028x-65.4.673.

Sreethawong, T., Ngamsinlapasathian, S. & Yoshikawa, S. 2014 Photochemically deposited nano-Ag/sol–gel TiO_2-In_2O_3 mixed oxide mesoporous-assembled nanocrystals for photocatalytic dye degradation. *Journal of Colloid and Interface Science* **421**, 191–198. doi:10.1016/j.jcis.2014.02.002.

Steele, M. & Odumeru, J. 2004 Irrigation water as source of foodborne pathogens on fruit and vegetables. *Journal of Food Protection* **67** (12), 2839–2849. doi:10.4315/0362-028x-67.12.2839.

Sygouni, V. & Chrysikopoulos, C. V. 2015 Characterization of TiO_2 nanoparticle suspensions in aqueous solutions and TiO_2 nanoparticle retention in water-saturated columns packed with glass beads. *Chemical Engineering Journal* **262**, 823–830. doi:10.1016/j.cej.2014.10.044.

Syngouna, V. I., Chrysikopoulos, C. V., Kokkinos, P., Tselepi, M. A. & Vantarakis, A. 2017 Cotransport of human adenoviruses with clay colloids and TiO_2 nanoparticles in saturated porous media: effect of flow velocity. *Science of the Total Environment* **598**, 160–167. doi:10.1016/j.scitotenv.2017.04.082.

Wang, G. S. & Hsieh, S. T. 2001 Monitoring natural organic matter in water with scanning spectrophotometer. *Environment International* **26** (4), 205–212. doi:10.1016/s0160-4120(00)00107-0.

Wang, C., Li, S. P. & Li, T. D. 2009 Calcium carbonate inhibition by a phosphonate-terminated poly(maleic-co-sulfonate) polymeric inhibitor. *Desalination* **249** (1), 1–4. doi:10.1016/j.desal.2009.06.006.

Watson, H. E. 1908 A note on the variation of the rate of disinfection with change in the concentration of the disinfectant. *Journal of Hygiene* **8** (4), 536–542. doi:10.1017/s0022172400015928.

Xie, J. & Hung, Y. C. 2018 UV-A activated TiO_2 embedded biodegradable polymer film for antimicrobial food packaging application. *LWT – Food Science and Technology* **96**, 307–314.

Xie, J. & Hung, Y. C. 2019 Methodology to evaluate the antimicrobial effectiveness of UV-activated TiO_2 nanoparticle-embedded cellulose acetate film. *Food Control* **106**, 106690.

Xie, J. & Hung, Y. C. 2020 Effect of TiO_2 loading, water depth and light intensity on photo-disinfection efficacy of Escherichia coli O157:H7 using TiO_2 NP-embedded cellulose acetate film in water. (Submitted.)

Xiong, P. & Hu, J. Y. 2013 Inactivation/reactivation of antibiotic-resistant bacteria by a novel UVA/LED/TiO_2 system. *Water Research* **47** (13), 4547–4555. doi:10.1016/j.watres.2013.04.056.

Yemmireddy, V. K. & Hung, Y. C. 2015a Effect of food processing organic matter on photocatalytic bactericidal activity of titanium dioxide (TiO_2). *International Journal of Food Microbiology* **204**, 75–80.

Yemmireddy, V. K. & Hung, Y. C. 2015b Selection of photocatalytic bactericidal titanium dioxide (TiO_2) nanoparticles for food safety applications. *LWT – Food Science and Technology* **61** (1), 1–6. doi:10.1016/j.lwt.2014.11.043.

First received 14 August 2020; accepted in revised form 19 September 2020. Available online 19 October 2020

Water–sediment separation efficiency prediction of gill-piece separation device

Hongfei Tao, Yang Zhou, Mahemujiang Aihemaiti, Qiao Li, Wenxin Yang, Youwei Jiang and Zijing Wu

ABSTRACT

As the key piece of equipment of a micro-irrigation system, the filter can prevent clogging of the emitter and ensure normal operation of the micro-irrigation system. A gill-piece separation device is used for the removal of viscous sediment, which helps to reduce the sediment concentration and decrease the burden on the filter. In this study, using the water–sediment separation efficiency as an evaluation index, a uniform orthogonal experiment was conducted to study the flow rate, sediment concentration, and gill-piece spacing using a physical model. Based on the experimental results of the physical model, multiple linear regression and projection pursuit regression were used for analysis. The results showed that the order of the factors affecting the water–sediment separation efficiency was as follows: flow rate at muddy water inlet > gill-piece spacing > sediment concentration. The correlation coefficients of the water–sediment separation efficiency models established using multiple linear regression and projection pursuit regression were 0.93 and 0.98, respectively. Both models could predict the water–sediment separation efficiency and determine the optimal working conditions of the gill-piece separation device.

Key words | filter, gill-piece separation device, model, multiple linear regression, projection pursuit regression, water–sediment separation efficiency

Hongfei Tao (corresponding author)
Yang Zhou
Mahemujiang Aihemaiti
Qiao Li
Wenxin Yang
Youwei Jiang
Zijing Wu
College of Water Conservancy and Civil Engineering,
Xinjiang Agricultural University,
Urumqi 830052,
China
and
Xinjiang Key Laboratory of Hydraulic Engineering Security and Water Disasters Prevention,
Urumqi 830052,
China
E-mail: *304276290@qq.com*

HIGHLIGHTS

- A prediction model for water–sediment separation efficiency of the gill-piece separation device was established under the conditions of sediment concentration of 2–12 kg/m^3, flow rate of 0.3–1.1 m^3/h, and gill-piece spacing of 5–11 cm.
- The order of influence of each factor on the water–sediment separation efficiency is as follows: flow rate at muddy water inlet > gill-piece spacing > sediment concentration.

INTRODUCTION

China is one of many countries with a severe sediment problem. In sediment-laden rivers, cascade treatment methods are mainly used to achieve efficient water diversion and sand removal. Common first-level sediment treatment engineering measures are sluice-type, bend-type, bottom-stockade-type, and layered water diversion projects. However, the treated water flow often cannot meet the requirements on sediment concentration or particle size. Therefore, secondary sediment treatment engineering measures must be built on the intake gate or downstream channel to further treat the sediment, such as a vortex tube sediment extractor, sand funnel, and circular-ring

doi: 10.2166/ws.2021.079

desilting and sediment ejection basin (Zhang 2015). To prevent the clogging of emitters in micro-irrigation systems with surface water as the water source, a settling basin is usually set up to allow the sediment to settle, and then a single or combined filter is installed to further purify the water. Next, the treated purified water is transported to the emitter via a pipeline system to meet the water requirements for crop growth.

Extensive studies have been carried out on filters, which have yielded fruitful results. Sand filters exhibit the best filtration performance, especially when the water source contains algae and organic impurities, and daily backwash is conducive to maintaining the good performance of the sand filter (Duran-Ros *et al.* 2009a; Elbana *et al.* 2012). Disc filters have the advantages of low costs and convenient management. With a reasonable and high-quality design, a disc filter can achieve a filtering effect similar to a sand filter (Capra & Scicolone 2005; Capra & Scicolone 2007). Screen filters have simple structures and easy operation, and they are mainly used for the removal of inorganic impurity particles (Adin & Alon 1986). Many experimental studies have been conducted on the filtration mechanisms of different types of filters and the performances of filters and drippers(Zeier & Hills 1987; Capra & Scicolone 2005; Puig-Bargués *et al.* 2005a; Capra & Scicolone 2007; Demir *et al.* 2009; Duran-Ros *et al.* 2009a, 2009b; Elbana *et al.* 2012; Bové *et al.* 2015; Wen-Yong *et al.* 2015), which provides a basis for the structural optimization of filters. The performance parameters of filters include filter head loss, treatment capacity (treatment speed and treated water quality), cost, and ease of maintenance. Researchers have adopted physical model tests, theoretical analysis, and numerical simulation to study different types of filters, tested the head loss and treatment capacity of different filters under different test conditions, given the applicable conditions for each type of filter, and pointed out the direction of structural optimization. Using theoretical analysis, they have studied the filtration mechanism of filters and the formation of filter cake. The computational flow dynamics (CFD) technique was used to study the flow field distribution of different filters under different test conditions, which laid the foundation for an in-depth understanding of the filtering process and the internal flow field of the filter (Puig-Bargués *et al.* 2005a; Yurdem *et al.* 2008; Duran-Ros *et al.* 2010; Yurdem *et al.* 2010; Elbana *et al.* 2012; Wu *et al.* 2014; Yurdem *et al.* 2015; Zong *et al.* 2015).

In Northwest China and North China, the sediment in most rivers has high concentrations and small particle sizes, which increases the difficulty of sediment treatment. As the key equipment of a micro-irrigation system, the filter can remove most of the impurities in the water. However, when the filter is used to treat muddy water with a high sediment concentration, problems such as easy clogging and frequent flushing occur. Although increasing the area of the settling basin in front of the filter can reduce the sediment concentration, it will increase the system investment and occupy more floor area, and it cannot effectively process viscous fine sediment. Studies have shown that the sediment concentration and particle size are the main factors that cause the clogging of emitters. When the particle size is less than 0.1 mm, the sediment concentration has a significant impact on clogging. When the sediment concentration is greater than 1.25 kg/m^3, the risk of clogging will increase (Niu *et al.* 2013). Therefore, determining how to effectively purify fine sediment with a high sediment concentration is very important for delaying the clogging of emitters and ensuring the safe operation of micro-irrigation systems. To meet the demands of agricultural water, high turbidity water must be purified, and the addition of flocculants plays an important role. However, inorganic salt and organic polymer flocculants have problems such as requiring large dosages, being harmful to the human body, and creating environmental pollution. Separating fine sediment, reducing the sediment concentration in water resources, and improving the separation efficiency have been the subjects of sediment studies. Qiu improved the traditional inclined-plate settling basin and developed a new water–sediment separation device called the gill-piece separation device. The gill-piece separation device could accelerate the sedimentation of sediment particles, with sedimentation velocities that were 1.9–3.7 times faster than those under static conditions. A series of subsequent studies was carried out on the gill-piece separation device under static water conditions. In terms of structural optimization, Zhu *et al.* (2008, 2009a, 2009b) carried out sediment deposition experiments on different sandy water flows and studied the effects of the gill piece type, inclination, and spacing on the efficiency of the water–sediment separation. The results showed that the gill-piece in the gill-piece separation device should be arranged in a single row of gill-pieces, and the optimal arrangement for the gill-piece inclination angle was $\alpha = 60^\circ$ and $\beta = 45^\circ$. The gill-piece spacing had a great influence on the efficiency of the water–sediment separation. The smaller the gill-piece spacing was, the higher the separation efficiency became. Yan *et al.* (2011a) studied the influence of the sediment settling channel width on water–sediment separation. Experiments showed that the critical width for silting in the sediment settling channel in the gill-piece separation device was 1 mm, and the channel width should be greater than or equal to 5 mm. In terms of the acceleration mechanism, Zhu *et al.* (2009c) pointed out that the reason the gill-piece separation device accelerated sediment settling was that gill-pieces destroyed the rigid spatial structure network formed during sediment settling. Tao (2014) summarized the separation mechanism of the gill-piece separation device by combining physical model experiments and numerical calculation results and pointed out that the structure of the gill-piece separation device was conducive to water–sediment separation. The addition of gill-pieces improved the hydraulic conditions, reduced the flow turbulence, increased the settling area, and improved the separation efficiency. The transverse density current formed between the gill-pieces and the entrainment effect of the local eddy current on the flow and sediment accelerated the process of water–sediment separation. The settling velocities on the upper surfaces of the gill-pieces and sediment channel were larger, which was conducive to the rapid separation of water and sediment. Yan *et al.* (2011b) obtained images of the sediment movement in the gill-piece separation device using high-speed camera technology, which proved that vertical density and transverse density current phenomena occurred in the gill-piece separation device. In addition, Yan *et al.* carried out a mechanical analysis on the sediment in the gill-piece separation device and described the movement trajectories of the sediment. Zhao *et al.* (2012) measured the velocity field of the gill-piece separation device using particle image velocimetry. The experimental results showed that a vertical density current appeared in the gill-piece separation device, in which the sediment moved downward along the upper surface of the gill-piece and the clean water moved upward along the lower surface of the gill-piece. The analysis results were consistent with the theoretical analysis and the observed phenomena.

In terms of flow field studies, Tao *et al.* (2013b), (2014a), (2015) carried out numerical simulations on the flow field of the gill-piece separation device with different sediment particle sizes and sediment concentration conditions and determined that the Euler model in the Fluent software could well simulate the flow field of the water–sediment two-phase flow in the gill-piece separation device. The average sediment concentration on the upper surfaces of the gill-pieces was far greater than that on the lower surface of the gill-pieces. The larger the sediment particle size was, the higher the water–sediment separation efficiency became. The smaller the sediment concentration was, the higher the water–sediment separation efficiency became. These studies laid a foundation for the practical application and structural optimization of gill-piece separation devices.

Many studies have been carried out on the gill-piece separation device using physical experiments and numerical simulations. However, the relationships between the water–sediment separation efficiency of the gill-piece separation device and the key structural (gill-piece spacing) and operating (sediment concentration and inlet flow) parameters have not yet been established. In this study, based on previous study results, a uniform orthogonal experiment was designed. The results were analyzed using multiple linear regression and projection pursuit regression (PPR) to obtain the order of the factors that affected the water–sediment separation efficiency of gill-piece separation devices and to construct a water–sediment separation efficiency prediction model.

MATERIALS AND METHODS

Device

The experiments were conducted in the hydraulic engineering laboratory of the College of Hydraulic and Civil Engineering of Xinjiang Agricultural University. The gill-piece separation device was a new water–sediment separation device, which is mainly used for the treatment of viscous fine sediment. It was composed of gill-pieces and a gill-duct. The gill-piece separation device used in this experiment was made of polymethyl methacrylate with a uniform texture and no bubbles. The circulation system is shown in Figure 1. It was composed of a water tank, mixing pump, water pump, gill-piece separation device, and gate valve, and it was provided with a clean water outlet, sediment outlet, and diversion port. Before the experiment, the prepared water and sediment were added into the water tank and fully mixed using the mixing pump. The pump was turned on to pump the well-mixed sandy water into the gill-piece separation device for the water–sediment separation experiments. After settling due to the gill-piece separation device, the sediment particles were discharged into the water tank from the bottom sediment outlet, and the clean water flowed from the upper overflow port back to the water tank. The water and sediment in the water tank were then remixed evenly, and the experiment entered the next cycle. The structure of the gill-piece separation device is shown in Figure 2. The device had a length a of 200 mm, width b of 100 mm, and height h of 1,000 mm. The gill-piece spacing was d. The gill-pieces were fixed on the two side walls in the length direction of the ordinary tube at an inclination angle of $\alpha = 60°$, forming an inclination angle of $\beta = 45°$ with the two side walls in the width direction. A clean water ascending channel was set between the gill-piece and the ordinary tube, with a width e of 10 mm, and the sediment descending channel had a

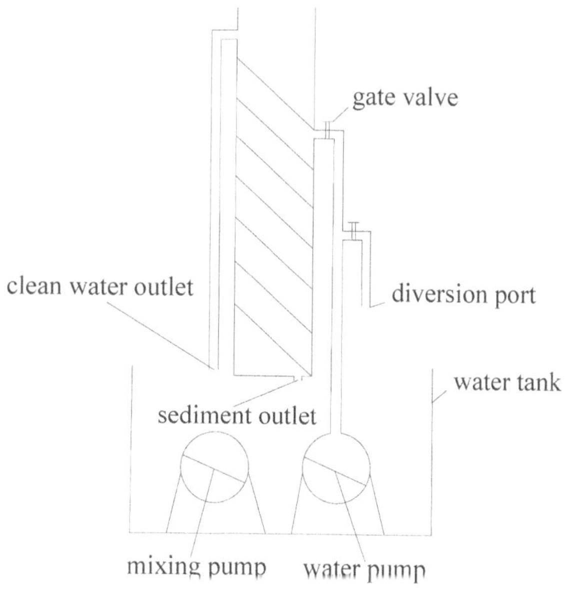

Figure 1 | Circulation system of gill-piece separation device experiment.

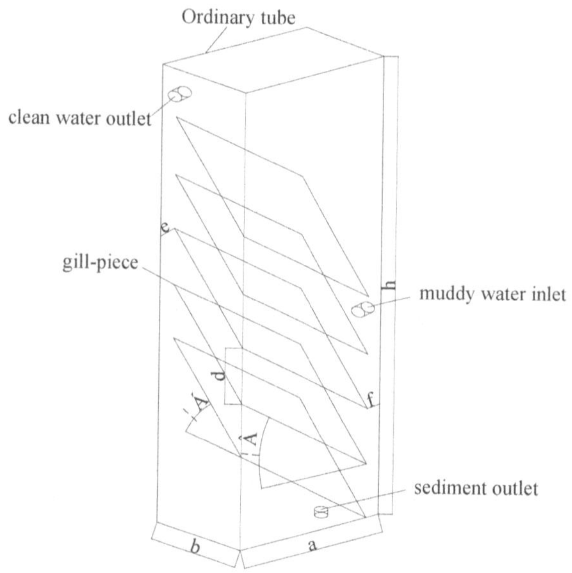

Figure 2 | Three-dimensional structure of gill-piece separation device.

width f of 10 mm. The diameter of the sediment outlet at the bottom of the gill-piece separation device was 2.5 mm. There were sandy water inlet and clean water outlet channels with diameters of 20 mm on both sides in the width direction, and the clean water outlet was 950 mm away from the bottom of the gill-piece separation device.

Working principle of gill-piece separation device

The experiments were carried out under static and dynamic water conditions. The static water conditions were as follows. The gill-piece separation device was only open at the top, with a free surface, closed periphery, and bottom, and it did not discharge sand. The dynamic water conditions were as follows. The top and bottom of the gill-piece separation device were open. Water was introduced through the water inlet, and the water and sand outlets facilitated the discharge of clean water and sand, respectively. Figure 3 shows the working principle of the gill-piece separation device. The sediment particles settled freely under the action of gravity. When the particles reached the upper surfaces of the gill-pieces, they slid from the top to the bottom along the upper surfaces of the gill-pieces due to gravity. During the sliding process, the particles encountered other particles and gathered under the action of particle contact, friction, collision, and adhesion to form sediment flow. The sediment

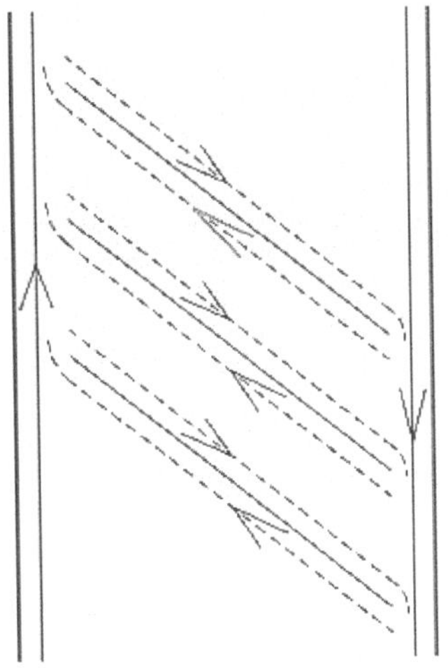

Figure 3 | Working principle of gill-piece separation device.

stream flowed vertically downward into the triangle sediment channel, while the clean water flowed from bottom to top along the lower surface of the gill-pieces and moved vertically upward in the clean water channel, thus achieving the separation of water and sediment. The movement of the sediment flow on the upper surfaces of the gill-pieces and the movement of the clean water flow on the lower surfaces of the gill-pieces formed an anticlockwise transverse density current between the gill-pieces, marked by the dotted line with an arrow in Figure 1. The sediment at the lower end of the gill-pieces slid down to the sediment channel and moved vertically downward, while the clean water at the upper end of the gill-pieces moved vertically upward when it flowed into the clean water channel, thus forming a clockwise vertical density current along the side wall of the gill-piece separation device, marked by a solid line with an arrow in Figure 3.

Sediment

The gill-piece separation device was mainly used for the treatment of viscous fine sediment. In this experiment, the natural loess from Xishan Mountain, Urumqi City, Xinjiang Uygur Autonomous Region was selected as the experimental sediment, and the sediment density was $\rho_s = 2{,}650\ \mathrm{kg/m^3}$. The particle size distribution was as follows: 100% of the particles were smaller than 0.075 mm, 80.4% were smaller than 0.048 mm, 47.8% were smaller than 0.023 mm, 26.0% were smaller than 0.01 mm, 13.5% were smaller than 0.005 mm, and 6.6% were smaller than 0.0015 mm. The median diameter D_{50} was 0.025 mm. Figure 4 shows the particle size distribution.

Steps

Before the experiment, the viscous sediment and water were mixed evenly in various proportions to prepare sediment-

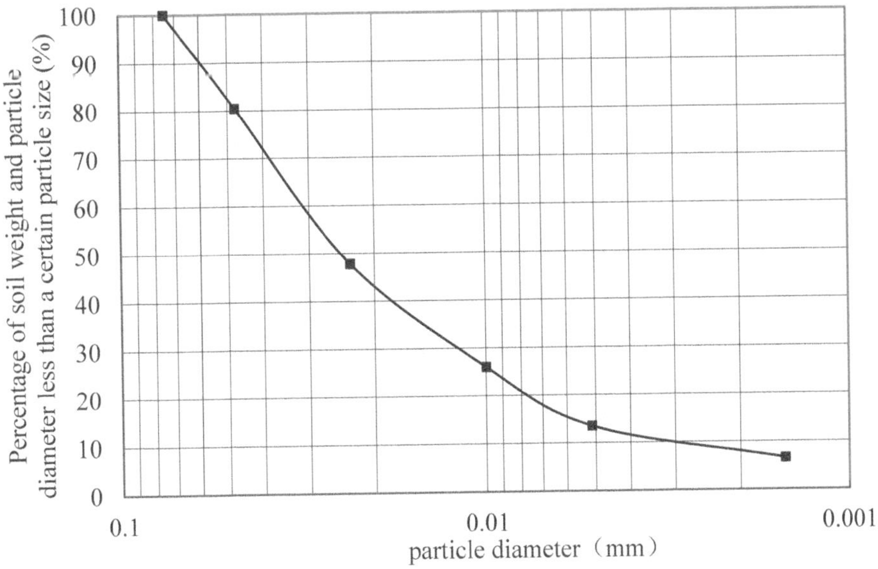

Figure 4 | Sediment particle size distribution.

laden water with specified concentrations. The experiment was started after the viscous sediment was fully mixed. During the experiment, the water samples were collected from the inlet of the muddy water, the outlet of the clean water, and the desilting outlet, and the parameters such as the flow rate and sediment concentration were measured and calculated. The experimental phenomena in the gill-piece separation device with different filtration times were observed, and the water–sediment separation efficiency of the gill-piece separation device under different conditions was analyzed.

Based on the results of the physical experiment and the numerical simulation under static and dynamic water conditions (Tao *et al.* 2013a, 2014b, 2015; Zhang *et al.* 2019), a three-factor, three-level, uniform orthogonal table UL_9 (3^4) was selected for this experiment, as shown in Table 1. The experiments were designed mainly to investigate the influence of the flow rate of muddy water at the inlet, the sediment concentration, and the gill-piece spacing on the water–sediment separation efficiency of gill-piece separation device.

Flow rate measurements

The flow rate was determined by measuring the volume. Glass beakers were used to take water samples with certain volumes at the clean water outlet and the desilting outlet. The corresponding time T was recorded with a stopwatch. The water sample volume V was measured using a graduated cylinder. Sampling was performed three times at each position, and average value was taken as the volume of the water sample. The corresponding flow rate could be calculated using the following equations:

$$Q = \frac{V}{T} \tag{1}$$

$$Q_c + Q_d = Q_m \tag{2}$$

where Q is the flow rate (m^3/h), V is the volume of the water sample (m^3), T is the sampling time (h), Q_c is the flow rate at the clean water outlet (m^3/h), Q_d is the flow rate at the desilting outlet (m^3/h), and Q_m is the flow rate at the muddy water inlet (m^3/h). The flow rate Q_m at the muddy water inlet could be calculated by substituting the flow rate Q_c at the outlet of the clean water and the flow rate Q_d at the desilting outlet into Equation (2).

Table 1 | Design of the uniform orthogonal experiments

Experiment no.	Inlet flow Q (m^3/h)	Sediment concentration S (kg/m^3)	Gill-piece spacing d (cm)
1	1 (0.3)	1 (2)	1 (5)
2	1 (0.3)	2 (7)	3 (11)
3	1 (0.3)	3 (12)	2 (8)
4	2 (0.7)	1 (2)	3 (11)
5	2 (0.7)	2 (7)	2 (8)
6	2 (0.7)	3 (12)	1 (5)
7	3 (1.1)	1 (2)	2 (8)
8	3 (1.1)	2 (7)	1 (5)
9	3 (1.1)	3 (12)	3 (11)

Measurement of sediment concentration

The principle of the replacement method was used to measure the sediment concentration quickly and accurately. The volume of the conical flask was calculated using Equation (3), where the sample was weighed using an electronic balance with an accuracy of 0.01 g, and the sediment concentration of the sample was calculated using Equation (4):

$$V_f = \frac{M_{f+c} - M_f}{\rho_w} \tag{3}$$

$$S_m = \frac{(M_{f+m} - M_f - \rho_w V_f)\rho_s}{(\rho_s - \rho_w)V_f} \tag{4}$$

where S_m is the sediment concentration of the muddy water (kg/m^3), M_{f+m} is the mass of the conical flask and the muddy water (kg), M_{f+c} is the mass of the conical flask and the clean water (kg), M_f is the mass of the conical flask (kg), and V_f is the volume of the conical flask (m^3).

Experimental index

The water–sediment separation efficiency was used to evaluate the filtration performance of the gill-piece separation device. The water–sediment separation efficiency refers to the ratio of the difference between the sediment concentration at the inlet of the muddy water and that at the outlet of the clean water to the sediment concentration of the muddy water. The equation is:

$$\eta = \frac{S_m - S_c}{S_m} \times 100\% \tag{5}$$

where η is the water–sediment separation efficiency (%), S_m is the sediment concentration at the inlet of the muddy water (kg/m^3), and S_c is the sediment concentration at the outlet of the clean water (kg/m^3).

RESULTS AND DISCUSSION

Uniform orthogonal experiment results

In the uniform orthogonal experiment, the water–sediment separation efficiency was the index, and the results are shown in Table 2. R is the range of factors, which reflects the variation range of the indices when the levels of the factors change. The larger the range is, the greater the influence of this factor on the index becomes. According to the results of the range analysis, the order of influence of each factor on water–sediment separation efficiency was as follows: flow rate at muddy water inlet > gill-piece spacing > sediment concentration. The optimum conditions were a flow rate of 0.3 m^3/h at the muddy water inlet, a gill-piece spacing of 11 cm, and a sediment concentration of 2 kg/m^3.

Table 2 | Results of uniform orthogonal experiments

Experiment no.	Flow rate at muddy water inlet A_1 (m^3/h)	Sediment concentration A_2 (kg/m^3)	Gill-piece spacing A_3 (cm)	Water–sediment separation efficiency η (%)
1	0.3	2	5	33.00
2	0.3	7	11	36.50
3	0.3	12	8	34.35
4	0.7	2	11	44.05
5	0.7	7	8	27.30
6	0.7	12	5	19.90
7	1.1	2	8	20.80
8	1.1	7	5	15.80
9	1.1	12	11	20.36
Mean 1	34.62	32.62	22.90	
Mean 2	30.42	26.53	27.48	
Mean 3	18.99	24.87	33.64	
Range R	15.63	7.75	10.74	
Order	1	3	2	

Establishment of multiple linear regression model for water–sediment separation efficiency

A multiple linear regression is mainly used to study the linear relationship between a dependent variable and multiple independent variables. Its objective is to use the values of multiple independent variables ($x_1, x_2, x_3, \cdots, x_k$) to estimate the dependent variable y.

In this study, SPSS 25.0 software was used to analyze the experimental results, and the statistical significance threshold was set to $P < 0.1$. Non-significant parameters were excluded from the model. The observed values in the study were independent of each other. The Durbin–Watson experiment value was 2.016. The regression tolerances were all greater than 0.1, and there was no multicollinearity. The chart of the regression standardized residuals shows that the data were distributed on and near the diagonal line (Figure 5). The regression model met the normality assumption and could be used for regression analysis.

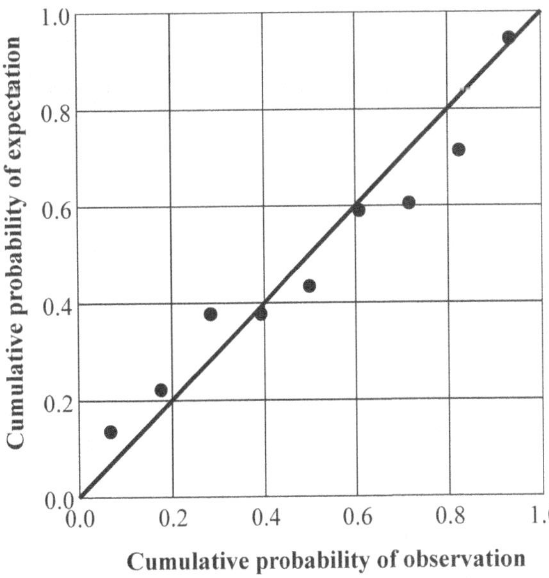

Figure 5 | Normal P–P plot of regression standardized residuals.

The regression model was statistically significant. $R^2 = 0.874$, suggesting that the fit was good. Table 3 shows that among the independent variables, the flow rate at the muddy water inlet, sediment concentration, and gill-piece spacing had a significant impact on the water–sediment separation efficiency ($P < 0.1$). According to the standardized coefficient of the regression model, the level of influence of each factor on the water–sediment separation efficiency was in the following order: flow rate at muddy water inlet > gill-piece spacing > sediment concentration, which was consistent with the results of the range analysis.

According to the results of the SPSS regression analysis, the obtained prediction model of the water–sediment separation efficiency was as follows:

$$\eta = 32.79 - 19.537Q - 0.775S + 1.789d \quad (6)$$

where η is the water–sediment separation efficiency (%), Q is the flow rate (m^3/h), S is the sediment concentration (kg/m^3), and d is the gill-piece spacing (cm).

According to Equation (6), when the minimum flow rate was 0.3 m^3/h, the minimum sediment concentration was 2 kg/m^3, and the maximum gill-piece spacing was 11 cm. The water–sediment separation efficiency of the gill-piece separation device reached a maximum of 45.06%, which was consistent with the results of the range analysis.

The water–sediment separation efficiency predicted using Equation (6) was compared with the measured value, as shown in Figure 6. Under the same conditions, the predicted water–sediment separation efficiency was close to the measured value. The mean relative error between the measured value and the predicted value was 8.44%, and the R was 0.93. Therefore, this model could be used to predict the water–sediment separation efficiency of gill-piece separation device.

Establishment of PPR water–sediment separation efficiency model

Projection pursuit regression (PPR) (Zheng *et al.* 1995) is a method for processing and analyzing high-dimensional data, and the model of PPR, a new multivariate statistical analysis technique, can be expressed as follows:

$$\hat{y} = E(y|x_1, x_2, \cdots, x_p) = \bar{y} + \sum_{i=1}^{M} \beta_i f_i \left(\sum_{n=1}^{p} \alpha_{in} x_n \right) \quad (7)$$

where f_i is the ith ridge function, $Ef_i = 0$, $Ef_i{}^2 = 1$, and $\sum_{n=1}^{p} \alpha_{in}^2 = 1$. The solution steps have been published previously (Jiang *et al.* 2019).

The PPR was used to analyze the experimental data of nine groups of physical gill-piece separation device models. The model projection parameters were: $N_m = 9$, $p = 3$, $Q = 1$, $M = 5$, $Mu = 3$, and $Span = 0.5$, where N_m is the number of modeling data, p is the number of influencing factors, Q is the number of dependent variables, Mu is the number of the optimum of ridge function, M is the number of the upper limit of the ridge function. Mu and M determine the fineness of the internal structure of the data that the model is looking for, while $Span$ is the smooth coefficient, whose value depends on the accuracy of the experimental data and determines the sensitivity of the model. The range of $Span$ is $0 < Span < 1$. The smaller the value, the more sensitive the model is.

In the process of establishing the water–sediment separation efficiency model using PPR, the calculated contribution weight coefficient β and projection direction

Table 3 | Results of multiple linear regression analysis

Dependent variable	Regression coefficient R^2	Independent variable	Non-standardized coefficient B	Non-standardized coefficient Standard error	Standardized coefficient β	Significance level P	Root mean square error RMSE
Water-sediment separation efficiency	0.874	Intercept	32.790	6.229		<0.05	4.265
		Q	−19.537	4.353	−0.713	<0.05	
		S	−0.775	0.348	−0.354	<0.1	
		d	1.789	0.580	0.490	<0.05	

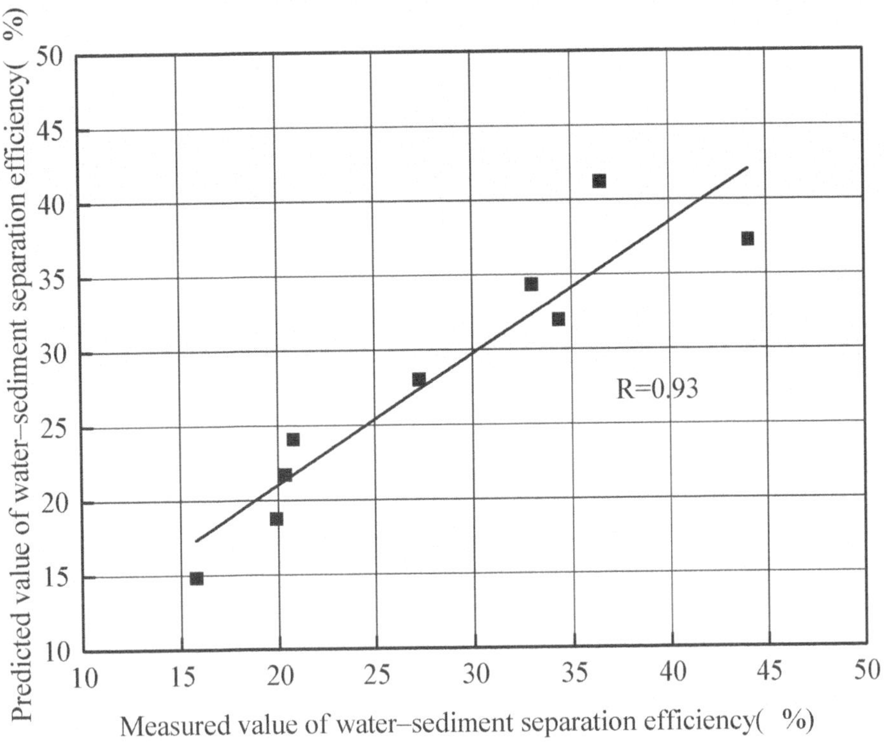

Figure 6 | Comparison of measured and predicted values of water–sediment separation efficiency of multiple linear regression model.

α of the ridge function are as follows:

$$\beta = (1.099768, 0.276116, 0.285926) \tag{8}$$

$$\begin{pmatrix} \overline{a_1} \\ \overline{a_2} \\ \overline{a_3} \end{pmatrix} = \begin{pmatrix} -0.99605 & -0.02558 & 0.085002 \\ -0.71178 & -0.53415 & 0.456119 \\ 0.991148 & -0.12661 & -0.03994 \end{pmatrix} \tag{9}$$

The weight coefficients of the influence of the flow rate at the muddy water inlet, sediment concentration, and gill-piece spacing on the water–sediment separation efficiency of the gill-piece separation device are shown in Table 4. The flow rate at the muddy water inlet had the greatest effect on the water–sediment separation efficiency, followed by the gill-piece spacing, and the sediment concentration had the least effect, which is consistent with the conclusions of the range analysis.

Table 4 | Relative weight of influencing factors

Influencing factors	Flow rate at muddy water inlet	Gill-piece spacing	Sediment concentration
Weight	1.000	0.635	0.522

Comparison of measured and predicted values of water–sediment separation efficiency model based on PPR

The above PPR model was used to predict the water–sediment separation efficiency of the gill-piece separation device under different working conditions, and the measured value was compared with the predicted value of the model, as shown in Figure 7. R reached 0.98. The mean relative error between the measured and predicted values was 5.11%. Compared with the multiple linear regression model, the mean relative error was reduced by 3.33%, indicating that the PPR model had a higher prediction accuracy.

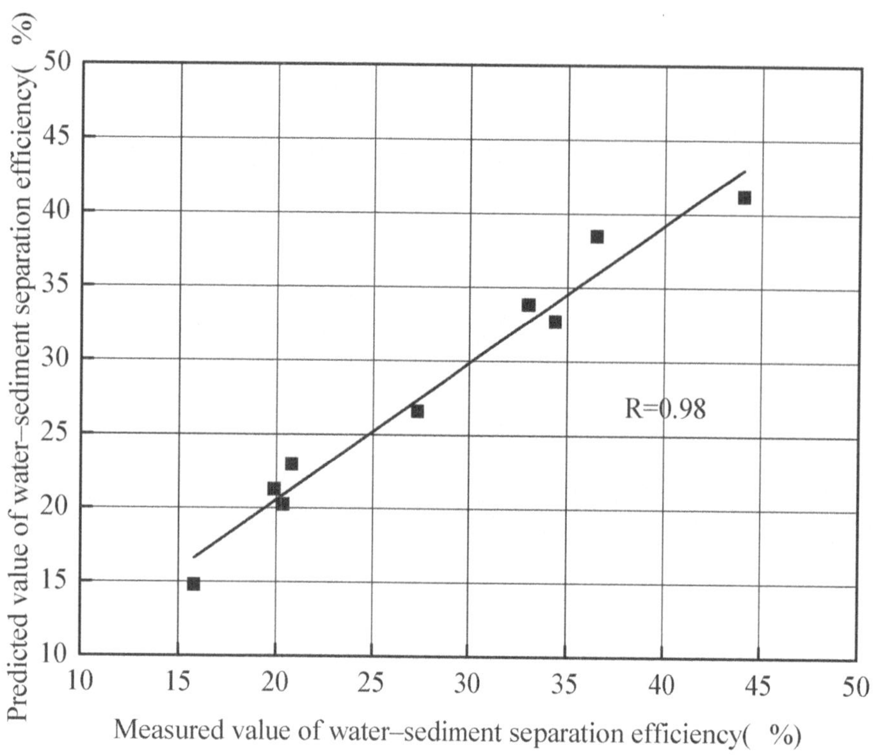

Figure 7 | Comparison of measured and predicted values of water–sediment separation efficiency of the PPR model.

Prediction of water–sediment separation efficiency based on the PPR model

The established PPR prediction model was used to predict the water–sediment separation efficiency of the gill-piece separation device when the flow rate at the muddy water inlet was 0.3–1.1 m^3/h, the gill-piece spacing was 5–11 cm, and the sediment concentration was 2–12 kg/m^3. The results showed that when the flow rate at the muddy water inlet was 0.3 m^3/h, the gill-piece spacing was 11 cm, and the sediment concentration was 2 kg/m^3, the water–sediment separation efficiency reached the maximum value. The optimal working conditions predicted using PPR model were consistent with those obtained using range analysis and multiple linear regression analysis, which validated the reliability of the PPR model prediction. Therefore, the PPR model could be used to predict the water–sediment separation efficiency of the gill-piece separation device under different working conditions.

The use of multiple linear regression analysis requires some conditions to be met, such as correlation, independence, variance homogeneity, and normality. These conditions will increase the difficulty of data processing, and human judgment will also affect the experimental results to a certain extent. To avoid the artificial assumptions and transformations of the original data and seek the real inherent dependencies of the data itself, the PPR method was applied for the material selection and concrete performance optimization, which validated the advantages of the PPR modeling method (Sheng & Wen 1997; Zhu *et al.* 2014; Jiang *et al.* 2019). In this study, a uniform orthogonal experiment was carried out on a gill-piece separation device under muddy water conditions. Multiple linear regression and the PPR method were used to establish the prediction model of the water–sediment separation efficiency. R of the water–sediment separation efficiency prediction model established using the multiple linear regression method was 0.93, and that of the PPR model was 0.98. The model established using these two methods could predict the water–sediment separation efficiency, and the prediction accuracy of the PPR model was higher, indicating that PPR method could also be used for modeling and analysis of normal data. In a practical application, the multiple linear regression and PPR model can well predict the water–sediment separation efficiency. When the experimental data are not normally distributed, the PPR method can be used to establish the prediction model. The advantage of PPR modeling is its assumption-free modeling characteristics, that is, there is no need to assume the distribution type of the experimental data, thus avoiding the unreasonable constraints of human factors on the regression model. PPR can make full use of the information and real dependencies of the data itself and solve the optimization problem by exploring the internal structure of the data, giving it a higher degree of fitting.

It was found that, although the gill-piece separation device could treat the viscous sediment, its water–sediment separation efficiency was not high. To further improve the water–sediment separation capacity of the gill-piece separation device, the structural optimization requires further study. In previous studies, the water inlets were all arranged on the side of the device, and the size of the device was relatively small. Therefore, further improvement measures can be performed. The size of the gill-piece separation device could be increased to increase its sediment treatment capacity. The position of the water inlet could be changed, e.g., the water inlet could be moved to the bottom. The material or the structural form of the gill-pieces could be changed to increase the speed and efficiency of sediment treatment. Considering the influence of drainage solid flux may make the efficiency prediction more accurate. We will study this factor in further.

CONCLUSION

In this study, the effects of the flow rate at the muddy water inlet, sediment concentration, and gill-piece spacing on the water–sediment separation efficiency were investigated using a uniform orthogonal experimental design. The order of influence of each factor on the water–sediment separation efficiency was as follows: flow rate at muddy water inlet > gill-piece spacing > sediment concentration. Multiple linear regression and the PPR method were used to establish a prediction model of the water–sediment separation efficiency. The models established using the above two methods could well predict the water–sediment separation efficiency of the gill-piece separation device. The prediction results of the PPR model were more accurate than those of the multiple linear regression, and its R reached 0.98. The established PPR model was used to predict the optimal working conditions. When the gill-piece spacing was 11 cm, the sediment concentration was 2 kg/m^3, and the flow rate at the muddy water inlet was 0.3 m^3/h, the water–sediment separation efficiency reached the maximum value. The models in this study were both established under a sediment concentration of 2–12 kg/m^3, a flow rate of 0.3–1.1 m^3/h, and a gill-piece spacing of 5–11 cm. Whether these models are applicable under other conditions should be studied in the future.

ACKNOWLEDGEMENTS

This study is financially supported by National Natural Science Foundation of China (50969009), Xinjiang Uygur Autonomous Region Innovation Environment (Talents, Bases) Construction Special Project-Natural Science Foundation Project (2017D01B18), Xinjiang Uygur Autonomous Region University Scientific Research Program General Project (XJEDU2017M011).

CONFLICTS OF INTEREST

The authors declare that they have no known competing financial interests or personal relationships that could have appeared to influence the work reported in this paper.

DATA AVAILABILITY STATEMENT

All relevant data are included in the paper or its Supplementary Information.

REFERENCES

Adin, A. & Alon, G. 1986 Mechanisms and process parameters of filter screens. *Journal of Irrigation and Drainage Engineering* **112** (4), 293–304. https://doi.org/10.1061/(ASCE)0733-9437(1986)112:4(293).

Bové, J., Arbat, G., Pujol, T., Duran-Ros, M., de Cartagena, F. R., Velayos, J. & Puig-Bargués, J. 2015 Reducing energy requirements for sand filtration in microirrigation: improving the underdrain and packing. *Biosystems Engineering* **140**, 67–78. https://doi.org/10.1016/j.biosystemseng.2015.09.008.

Capra, A. & Scicolone, B. 2005 Assessing dripper clogging and filtering performance using municipal wastewater. *Irrigation and Drainage* **54** (S1), S71–S79. https://doi.org/10.1002/ird.187.

Capra, A. & Scicolone, B. 2007 Recycling of poor quality urban wastewater by drip irrigation systems. *Journal of Cleaner Production* **15** (16), 1529–1534. https://doi.org/10.1016/j.jclepro.2006.07.032.

Demir, V., Yurdem, H., Yazgi, A. & Degirmencioglu, A. 2009 Determination of the head losses in metal body disc filters used in drip irrigation systems. *Turkish Journal of Agriculture and Forestry* **33** (3), 219–229. https://doi.org/10.3906/tar-0811-1.

Duran-Ros, M., Puig-Bargués, J., Arbat, G., Barragán, J. & Cartagena, F. R. D. 2009a Effect of filter, emitter and location on clogging when using effluents. *Agricultural Water Management* **96** (1), 67–79.

Duran-Ros, M., Puig-Bargués, J., Arbat, G., Barragán, J. & Ramírez De Cartagena, F. 2009b Performance and backwashing efficiency of disc and screen filters in microirrigation systems. *Biosystems Engineering, (ASCE)* **103** (1), 35–42. https://doi.org/10.1016/j.biosystemseng.2009.01.017.

Duran-Ros, M., Arbat, G., Barragán, J., Ramírez De Cartagena, F. & & Puig-Bargués, J. 2010 Assessment of head loss equations developed with dimensional analysis for micro irrigation filters using effluents. *Biosystems Engineering (ASCE)* **106** (4), 521–526. https://doi.org/10.1016/j.biosystemseng.2010.06.001.

Elbana, M., de Cartagena, F. R. & Puig-Bargués, J. 2012 Effectiveness of sand media filters for removing turbidity and recovering dissolved oxygen from a reclaimed effluent used for micro-irrigation. *Agricultural Water Management* **111**, 27–33. https://doi.org/10.1016/j.agwat.2012.04.010.

Jiang, C., Gong, J., Tang, X., Jiang, L. & Zheng, Z. 2019 Optimization method of comprehensive properties of low heat cement cementitious system based on projection pursuit regression. *Journal of Building Materials* **22** (03), 333–340. https://doi.org/10.3969/j.issn.1007-9629.2019.03.002.

Niu, W., Liu, L. & Chen, X. 2013 Influence of fine particle size and concentration on the clogging of labyrinth emitters. *Irrigation Science* **31** (4), 545–555. https://doi.org/10.1007/s00271-012-0328-2.

Puig-Bargués, J., Barragán, J. & de Cartagena, F. R. 2005a Development of equations for calculating the head loss in effluent filtration in microirrigation systems using dimensional analysis. *Biosystems Engineering (ASCE)* **92** (3), 383–390. https://doi.org/10.1016/j.biosystemseng.2005.07.009.

Sheng, J. & Wen, Q. 1997 Application of the projection pursuit regression technique in the rural land suitability evaluation. *Arid Zone Research (II)* 36–41. https://doi.org/10.13866/j.azr.1997.02.007.

Tao, H. 2014 *Numerical Analysis of Water-Sediment Flow Field and Experiment Study in Gill-Piece Separation Device.* Xinjiang Agricultural University, Urumqi.

Tao, H., Qiu, X., He, Q., Li, W. & Xu, L. 2013a Sediment concentration on water-sediment separation effect test of separation device. *Water Resources and Power* **31** (07), 93–95.

Tao, H., Qiu, X., Zhao, L. & Li, L. 2013b Numerical simulation of internal flow field in gill-piece separation device. *Transactions of the Chinese Society of Agricultural Engineering* **29** (17), 38–46 + 293. https://doi.org/10.3969/j.issn.1002-6819.2013.17.006.

Tao, H., Qiu, X., Li, Q., Tan, Y. & Wang, M. 2014a 3D numerical simulation of internal flow field in gill-piece separation device under different gill-piece spacing. *Transactions of the Chinese Society for Agricultural Machinery* **45** (06), 183–189. https://doi.org/10.6041/j.issn.1000-1298.2014.06.028.

Tao, H., Qiu, X., Li, Q. & Su, J. 2014b Three-dimensional numerical simulation of internal flow field in gill-piece separation device under different sediment grain size. *Advances in Science and Technology of Water Resources* **34** (5), 17–23. https://doi.org/10.3880/j.issn.1006-7647.2014.05.004.

Tao, H., Qiu, X. & Yang, H. 2015 Three-dimensional numerical simulation of internal flow in gill-piece separation device under muddy water of different sediment concentrations. *Water Resources and Power* **34** (02), 64–71.

Wen-Yong, W., Yan, H., Hong-Lu, L., Shi-Yang, Y. & Yong, N. 2015 Reclaimed water filtration efficiency and drip irrigation emitter performance with different combinations of sand and disc filters. *Irrigation and Drainage* **64** (3), 362–369. https://doi.org/10.1002/ird.1909.

Wu, W., Chen, W., Liu, H., Yin, S. & Niu, Y. 2014 A new model for head loss assessment of screen filters developed with dimensional analysis in drip irrigation systems. *Irrigation and Drainage (ASCE)* **63** (4), 523–531. https://doi.org/10.1002/ird.1846.

Yan, Y., Qiu, X., Zhang, X. & Gong, S. 2011a Experimental study on influence of sediment settling access width in two-phase flow separation device on water-sediment separation. *Journal of Xinjiang Agricultural University* **34** (06), 526–528. https://doi.org/10.3969/j.issn.1007-8614.2011.06.015.

Yan, C., Qiu, X., Zhang, X. & Gong, S. 2011b Mechanical analysis of sediment motion path and accelerated slip on two-phase flow separation device. *Advances in Science and Technology of Water Resources* **31** (05), 27–29 + 42. https://doi.org/10.3880/j.issn.1006-7647.2011 .05.007.

Yurdem, H., Demir, V. & Degirmencioglu, A. 2008 Development of a mathematical model to predict head losses from disc filters in drip irrigation systems using dimensional analysis. *Biosystems Engineering (ASCE)* **100** (1), 14–23. https://doi.org/10.1016/j.biosystemseng.2008.01.003.

Yurdem, H., Demir, V. & Degirmencioglu, A. 2010 Development of a mathematical model to predict clean water head losses in hydrocyclone filters in drip irrigation systems using dimensional analysis. *Biosystems Engineering (ASCE)* **105** (4), 495–506. https://doi.org/10.1016/j.biosystemseng.2010.02.001.

Yurdem, H., Demir, V. & Mancuhan, A. 2015 Development of a simplified model for predicting the optimum lengths of drip irrigation laterals with coextruded cylindrical in-line emitters. *Biosystems Engineering (ASCE)* **137**, 22–35. https://doi.org/10.1016/j.biosystemseng.2015.06.010.

Zeier, K. R. & Hills, D. J. 1987 Trickle irrigation screen filter performance as affected by sand size and concentration. *Transactions of the ASAE* **30** (3), 0735–0739. https://doi.org/10.13031/2013.30468.

Zhang, J. 2015 *Research on Model Experiment and Numerical Simulation of Flow Field in Circular-Ring Sediment Desilting Basin*. Xinjiang Agricultural University, Urumqi.

Zhang, J., Tao, H., Q, L. I., Yang, W., Ma, H. & Song, R. 2019 Effect of sediment concentration on water-sediment separation efficiency of gill- piece separation device under dynamic water condition. *Water Resources and Power* **37** (10), 92–96.

Zhao, L., Qiu, X., Chen, B. & Zou, Q. 2012 Test and analysis on flow field in a two-phase flow water-sediment separation device by PIV techniques. *Journal of Xinjiang Agricultural University* **35** (02), 144–148.

Zong, Q., Zheng, T., Liu, H. & Li, C. 2015 Development of head loss equations for self-cleaning screen filters in drip irrigation systems using dimensional analysis. *Biosystems Engineering (ASCE)* **133**, 116–127. https://doi.org/ 10.1016/j.biosystemseng.2015.03.001.

Zheng, Z., Ge, Y., Yang, L. & Sun, Z. 1995 Application of projection pursuit regression technique in optimization of cement formulation. *Journal of Xinjiang Agricultural University* (01), 20–24.

Zhu, C., Qiu, X. & Liu, Y. 2008 Effects of gill-piece pattern of vertical-component density flow water-sediment separation device on water sand separation. *Journal of Xinjiang Agricultural University* **31** (06), 72–75. https://doi.org/10.3969/j.issn.1007-8614.2008.06.017.

Zhu, C., Gong, S., Qiu, X., Sun, X. & Zhang, X. 2009a Effects of spacing of gill-piece in vertical-component density flow water-sediment separation device on water sand separation. *Journal of Xinjiang Agricultural University* (02), 81–84.

Zhu, C., Qiu, X. & Sun, X. 2009b Experimental study on the effects of gill angle of separating gill on water and sediment separation. *Yellow River* **31** (10), 86–87 + 89. https://doi.org/10.3969/j.issn.1000-1379.2009.10.046.

Zhu, C., Qiu, X. & Yan, Y. 2009c Analysis of mechanism of vertical-component density flow water-sediment separation device. *Advance in Science and Technology of Water Resources* **29** (05), 20–23. https://doi.org/10.3880/j.issn.1006-7647.2009.05.006.

Zhu, L., Mu, Z. & Yang, L. 2014 Experiment on uniform orthogonal design and projection pursuit regression of energy dissipator of suspension gird in stilling pool. *Journal of Water Resources Water Engineering* **25** (06), 210–214. https://doi.org/10.11705/j.issn.1672-643X.2014.06.044.

First received 2 December 2020; accepted in revised form 8 March 2021. Available online 22 March 2021

Influence of the soil water retention curve type and magnetic water treatment on lettuce irrigation management responses

Lis Tavares Ordones Lemos, Fábio Ponciano de Deus, Michael Silveira Thebaldi, Adriano Valentim Diotto, Valter Carvalho de Andrade Júnior and Rodrigo César de Almeida

ABSTRACT

This study aimed to evaluate the influence of magnetized water use on lettuce irrigation management responses, and based on the generated data, to evaluate by simulation the influence of soil water retention curve type on the lettuce irrigation management responses. This work was divided into three stages: 1 – determination of field and laboratory soil water retention curves; 2 – lettuce crop irrigation management experiment using soil water retention curve with field data, evaluating different soil water tensions to start irrigation and different water types (magnetically treated water, and non-treated water); and 3 – estimation of the irrigation management responses (simulation) using the soil water retention curve performed in the laboratory (using non-treated water), compared with the experimental results (stage 2). The use of magnetically treated water determined the soil moisture maintenance for a longer time and fewer irrigation events, leading to less water being applied and electricity consumption. The use of the soil water retention curve derived from the field data determined less water and electric energy consumption in Lucy Brown lettuce irrigation, in comparison with the simulated use of the soil water retention curve from laboratory data.

Key words | microirrigation, optimization, soil water matric potential, tensiometry

Lis Tavares Ordones Lemos
Fábio Ponciano de Deus
Michael Silveira Thebaldi (corresponding author)
Adriano Valentim Diotto
Valter Carvalho de Andrade Júnior
Rodrigo César de Almeida
Universidade Federal de Lavras Caixa,
Postal 3037, Campus Universitário,
37200-900 Lavras,
MG,
Brazil
E-mail: *michael.thebaldi@ufla.br*

HIGHLIGHTS

- Iceberg lettuce irrigation with magnetically treated water determined lower number of irrigations.
- Soil irrigated with magnetized water maintained its moisture for a longer time.
- Magnetically treated water irrigation may have hindered the capacity of water uptake by lettuce.

INTRODUCTION

Over the years, water availability has been a concern of many economic sectors. In irrigated agriculture, researchers have been currently concerned with the development of techniques and equipment for water use optimization. Irrigation with magnetically treated water has been scientifically evaluated over the years, and has brought positive improvements in water use reduction, and qualitative and productive benefits for some crop production (Maheshwari & Grewal 2009; Surendran *et al.* 2016; Yusuf & Ogunlola 2017).

doi: 10.2166/ws.2021.038

According to some authors, magnetic treatment modifies the physical structure of water molecules and their chemical composition, altering the hydrogen bonds, van der Waals forces, and sizes of the water clusters, decreasing the surface tension and increasing the viscosity (Toledo *et al.* 2008; Mostafazadeh-Fard *et al.* 2011; Khoshravesh-Miangoleh & Kiani 2014; Surendran *et al.* 2016). According to Mostafazadeh-Fard *et al.* (2011), Surendran *et al.* (2016) and Selim *et al.* (2019) these modifications in water molecules promote the maintenance of soil moisture for a longer time, which may explain the productive benefits and water savings.

Another factor that has been investigated is the effectiveness of the irrigation management methodology. The use of soil water retention curves for irrigation management, performed in the laboratory, may overestimate the values of water content, especially the field capacity, and consequently the irrigation depths (Morgan *et al.* 2001; Jabro *et al.* 2009; Brito *et al.* 2011). According to the authors, this overestimation may be due to the combination of entrapped air in the soil and the soil sample compaction in the collection, increasing the soil density and microporosity. Additionally, there are studies highlighting the occurrence of errors from laboratory methods, especially in high water potentials, associated with fine and extra-fine texture soil, leading to water content overestimation (Bittelli & Flury 2009; Solone *et al.* 2012; van Lier *et al.* 2019). A potentially more reliable alternative is the use of soil water retention curves performed in the field (Morgan *et al.* 2001; Jabro *et al.* 2009; Brito *et al.* 2011). Although there are several studies evaluating the differences between the soil water retention curves carried out in the field and in the laboratory, there are few studies that have assessed the impact on the irrigation management responses on different crops.

Based on the given information, it is expected that the use in irrigation of water subjected to magnetic treatment, associated with the use of the soil water retention curve with data from field testing for irrigation management, will determine reduction in water amount and in electricity consumption. In this sense, this study aimed to evaluate the influence of magnetized water use on lettuce irrigation management responses, and based on the generated data, to evaluate by simulation the influence of soil water retention curve type on the lettuce irrigation management responses.

METHODS

Experimental planning

Figure 1 shows a flowchart of the development stages of the experiment.

Initially, two forms of soil water retention curve construction were performed (in the field, and in the laboratory) – Stage 1. The objective was to evaluate the influence of the soil water retention curve determination method on the lettuce crop irrigation management responses. Subsequently, a lettuce crop irrigation management experiment was performed using the soil water retention curve made by the field data – Stage 2. It evaluated different combinations of soil water tensions to start irrigation and types of water (magnetically treated water, and non-treated water). Finally, using the data generated in the field experiment, an estimate (simulation) of the irrigation management responses was performed, using the soil water retention curve performed in the laboratory – Stage 3. These results were compared with the Stage 2 results.

Location of the experiment and water magnetic treatment

The field essays were conducted in a greenhouse (7×30 m in size, with transparent plastic anti-UV additives (150 μm)), in the south of Minas Gerais state, Brazil (21°14′S, 45°00′W and 910 m). According to the Köppen-Geiger climate classification, the region's climate is Cwa, with an average annual air temperature of 20.4 °C and an average annual precipitation of 1,460 mm (Alvares *et al.* 2013).

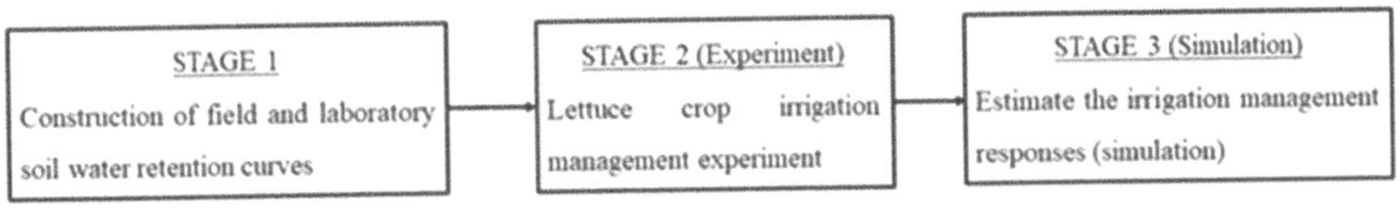

Figure 1 | Development stages of the experiment.

The measured air temperatures inside the greenhouse during the experiment were: maximum of 31.0 ± 2.4 °C, minimum of 19.0 ± 1.8 °C, and average of 25.0 ± 1.5 °C. The relative humidity values were: maximum of 91.2% ± 6.3%, minimum of 41.0% ± 9.2%, and average of 66.1% ± 5.9%.

A Sylocimol Residencial magnetizer from the Timol Indústria e Comércio de Produtos Magnéticos (Uberlândia, MG, Brazil) was used to magnetize the water. This equipment is composed of alternating magnets and covered by stainless-steel protection, and subjects the water to a magnetic field of 3,860 gauss, with the capacity to magnetize 1,000 L in one hour of exposure (TIMOL 2012). The equipment was allocated and kept inside a 500 L water tank throughout the experiment.

Determination of the soil water retention curve

The soil of the experimental area was classified as Typic Hapludox (Soil Survey Staff 1999). Table 1 shows the results of the chemical and physical soil analysis, referring to depths of 0–20 cm and 20–40 cm.

Table 1 | Chemical and physical soil analysis results from the experimental area soil, referring to the depths of 0–20 and 20–40 cm[1]

Parameter (unit)	Depth (cm) 0–20	Depth (cm) 20–40	Parameter (unit)	Depth (cm) 0–20	Depth (cm) 20–40
pH	6.60	6.40	MO (dag kg^{-1})	2.96	2.39
K (mg dm^{-3})	87,00	39.22	P-Rem (mg L^{-1})	67.10	67.70
P (mg dm^{-3})	10.03	4.89	Zn (mg dm^{-3})	2.98	1.79
Na (mg dm^{-3})	–	–	Fe (mg dm^{-3})	63.13	57.76
Ca (cmolc dm^{-3})	4.82	4.16	Mn (mg dm^{-3})	85.78	68.12
Mg (cmolc dm^{-3})	2.10	1.83	Cu (mg dm^{-3})	5.46	5.27
Al (cmolc dm^{-3})	0.04	0.04	B (mg dm^{-3})	0.28	0.27
H + Al (cmolc dm^{-3})	1.03	1.10	S (mg dm^{-3})	64.37	61.33
SB (cmolc dm^{-3})	7.14	6.09	Sand (%)	10	
t (cmolc dm^{-3})	7.18	6.13	Silt (%)	29	
T (cmolc dm^{-3})	8.17	7.19	Clay (%)	61	
V (%)	87.43	84.71	ρ_s (g cm^{-3})	1.14	
Mg (%)	0.56	0.65	–	–	

[1] K – potassium; P – phosphorus; Na – sodium; Ca – calcium; Al – aluminum; H + Al – potential acidity with SMP extractor; SB – sum of exchangeable base; t – effective cation exchange capacity; T – cation exchange capacity at pH 7.0; V – base saturation index; Mg – magnesium; MO – organic matter; P-Rem – remaining phosphorus; Zn – zinc; Mn – manganese; Cu – copper; B – boron; S – sulfur; ρ_s – soil bulk density; pH in water, KCl and CaCl2 – 1:2.5 ratio; Ca – Mg – Al – extractor: KCl – 1 mol/L; S – extractor – monocalcium phosphate in acetic acid; P – Na – K – Fe – Zn – Mn – Cu – Mehlich extractor I, H + Al – extractor: SMP, organic matter (MO) – oxidation: Na2Cr2O7 4N + H2SO4 10N; B – hot water extractor.

To determine the soil water retention curve in the field, two beds in the area (1.2 × 2.4 m) were selected: one for the magnetically treated water evaluation (MW), and one for non-treated water evaluation (OW). The beds were delimited by a PVC sheet 20 cm high above the ground surface to avoid water runoff, and 50 cm below the ground surface to prevent lateral water flow. The objective was to promote only drainage and evaporation of water during the test (Figure 2).

To evaluate the soil water retention characteristics to which plants were subjected, the soil water retention curves were made after incorporating the planting fertilization. In order to promote repetitive assessment, in each bed six tensiometers were installed at 12.5 cm (depth defined to indicate irrigation), six at 25 cm (to check the amount of water at the effective depth of the lettuce roots), and three at 40 cm (evaluation depth to ensure saturation of the soil profile).

The soil was subjected to saturation (zero reading on the tensiometer at all depths) using 30 drip emitters with a flow rate of 4 L h^{-1}. Once the saturation condition was reached, water application was interrupted, and the beds were covered with impermeable material to prevent water evaporation and allow only drainage. The beds were kept covered until the field capacity condition occurred in the soil. This condition was defined considering the first derivative of the soil moisture model as a function of time equal to −0.01 (Oliveira *et al.* 2001; Casaroli & van Lier 2008). A regression analysis ($p \leq 0.05$) was performed to define the most representative soil moisture model as a function of time. Subsequently, the fitted mathematical models were compared by the *F*-test ($p \leq 0.05$).

After reaching equilibrium (field capacity condition), the impermeable material was removed in order for the soil to dry until the recommended reading limit of the tensiometer, i.e. −80 kPa. The maintenance of the water column inside the tensiometers was performed routinely.

Tensiometer readings were performed concomitantly with the soil moisture measurement in the laboratory by the standard method (gravimetric method). The soil water retention curve was determined to a depth of 12.5 cm. Soil volumetric water content was obtained from gravimetric water content multiplying it by the soil bulk density after

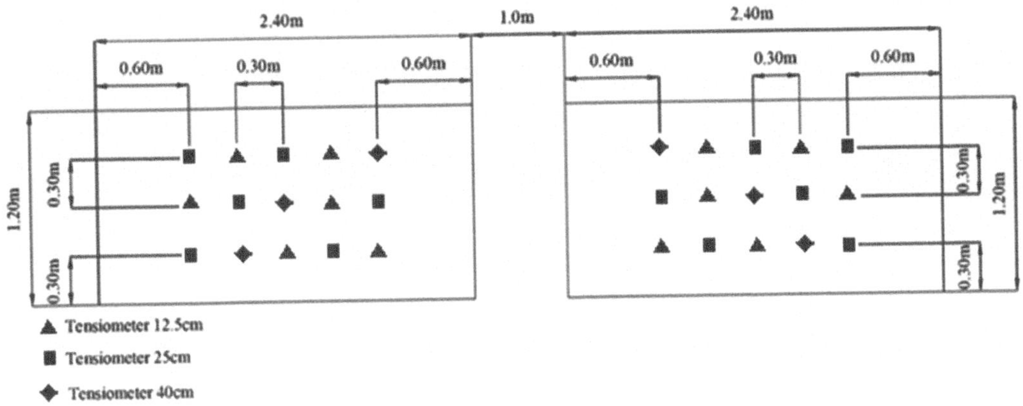

Figure 2 | Detail of the beds delimited with PVC sheet and the tensiometer distribution: tensiometers at 12.5 cm depth in gray, tensiometers at 25 cm depth in red, tensiometers at 40 cm depth in yellow. Please refer to the online version of this paper to see this figure in colour: doi: 10.2166/ws.2021.038.

bed-making (1.14 ± 0.05 g cm^{-3}). Sampling was more frequent on the first day (0, 3, 6, 12 and 24 hours after saturation), and later evaluations were daily.

A digital tensimeter (model TENSIMETER, from the company Hidrodinâmica Irrigation. URL: http://tensiometro.com.br/) was used to read the tensiometers. Equation (1) was used to determine the soil water tension or the matric potential from the reading of the tensiometer:

$$\Psi_m = L - 0.098.h \quad (1)$$

where Ψ_m is the soil water tension or matric potential (kPa), L is the tensiometer reading (kPa), and h is the water column height in the tensiometers (cm).

The F-test ($p \leq 0.05$) was performed to evaluate the water type (MW and OW) influence on the behavior of the field soil water retention curves.

A soil water retention curve was also determined in the laboratory, using soil cores made from deformed soil samples, from saturated condition to a tension of −98 kPa, being −0.98, −1.96, −3.92, −5.88, −7.84 and −9.8 kPa in the hanging-water funnel, and −98 kPa in the pressure plate apparatus. For this, we used six soil deformed samples submitted to a 2 mm sieve, from 0 to 0.20 m deep to prepare the soil cores with a bulk density of 1.14 g cm^{-3}, the same found in field conditions, in order to compare its hypothetical irrigation management responses with the results from the actual use of the soil water retention curve performed in the field (comparison only with non-treated water).

For the field (magnetically treated water – MW, and non-treated water – OW) and laboratory data (non-treated water – OW), regression analyses ($p \leq 0.05$) were performed to define the most appropriate mathematical model of the relationship between volumetric soil moisture and soil water tension. Linear models, second-degree polynomial, power and logarithmic models were evaluated.

Lucy Brown lettuce irrigation management

The irrigation management evaluation was carried out in a Lucy Brown lettuce (*Lactuca sativa* L.) cultivation, transplanted 23 days after sowing, at a spacing of 0.30×0.30 m (32 plants per bed). During the experiment, temperature and relative humidity were monitored with a digital thermo-hygrometer (model HT-600, from the company Instrutherm), installed in a weather shelter (2 m in relation to the soil surface) inside of the greenhouse.

The experimental design was completely randomized, in factorial 2×4 with three replications (R1, R2 and R3). The treatments were composed by the combination of two water types (MW and OW) and four soil water tensions to start irrigation (T1 −15 kPa, T2 −25 kPa, T3 −40 kPa and T4 −70 kPa), the irrigation process being performed to raise the soil moisture to field capacity condition. As a limit of −80 kPa is indicated for proper functioning of the tensiometers (Azevedo & Silva 1999), −70 kPa was defined as the treatment with the greatest distance from the field capacity tension. The soil water retention curve used for the crop management was the one derived from the field with non-treated water.

In one repetition of each experimental combination (type of water and soil water tension), three tensiometers at 12.5 cm was used to manage the irrigation. The readings of the tensiometers were performed daily at 9 am and 3 pm.

A drip irrigation system (distribution uniformity – DU equal to 95.4%) was used for irrigation, with self-regulating dripper ClickTif NaanDanJain (average flow of 2.14 ± 0.08 L h^{-1}), with a spacing of 30 cm (16 emitters per bed). Additionally, two irrigation lines were used per bed, spaced by 60 cm. Equation (2) was used to estimate the irrigation water depth. The irrigation time was calculated by Equation (3). There was assumed to be 95% application efficiency, and according to Yuri *et al.* (2002) an effective depth of the root system of 25 cm (250 mm).

$$LI = \frac{(\theta_{cc} - \theta_l).z}{E_a.DU} \quad (2)$$

where LI is the irrigation depth (mm), θ_{cc} is the volumetric soil moisture at field capacity (cm^3 cm^{-3}), θ_l is the volumetric soil moisture at the time of reading (cm^3 cm^{-3}), z is the effective depth of the root system (mm), E_a is the application efficiency (decimal), and DU is the uniformity of distribution of the irrigation system (decimal).

$$T = LI.\frac{A}{n_e\, q_a}.60 \quad (3)$$

where T is the irrigation time (min), A is the bed area (m^2), q_a is the average flow of the emitters (L h^{-1}), and n_e is the number of emitters per bed.

Electric energy consumption was estimated for each treatment by Equation (4), proposed by Bilibio *et al.* (2010) and adapted for this work. Electric energy consumption was estimated per unit area and crop cycles to facilitate later estimation by users.

$$EC = 2.78 \times 10^{-7} \frac{\Sigma LI \text{ Hman}}{\eta_{MB}} \quad (4)$$

where EC is the electric energy consumption (kWh m^{-2} $cycle^{-1}$), ΣLI is the total applied irrigation depth during the crop cycle (mm $cycle^{-1}$), *Hman* is the manometric height of the pumping system (kPa), and η_{MB} is the pump efficiency (decimal).

The irrigations required a manometric height of 176.4 kPa (18 mca) from the pumping system. Based on the pump performance curve, the pump efficiency was 47% (model BC-91S, 1/2 cv, Schneider Motobombas).

Irrigation management responses

Evaluation of field data

The field data were analyzed to evaluate the influence of the irrigation water magnetic treatment on the lettuce irrigation management responses. For the irrigation management, the soil water retention curve constructed in the field was used. We evaluated the following irrigation management responses: soil water tension data during the experiment for each treatment; number of irrigation events (*NI*); average interval between irrigations (*Ii*); total applied irrigation depth during the crop cycle (ΣLI); and electric energy consumption estimate (*EC*).

For the soil water tension data comparison between each treatment, an analysis of variance on ranks was used, by the Kruskal–Wallis test with 5% of probability, followed by Tukey's test, also at 5% of significance for pairwise multiple comparison.

Simulation of irrigation management responses

Using the soil water tension data measured during the experiment, simulation of irrigation management responses was estimated using the soil water retention curve made in the laboratory. The objective was to evaluate the influence of the soil water retention curve data source on the lettuce irrigation management responses. Total applied irrigation depth during the crop cycle (ΣLI) and electric energy consumption (*EC*) were estimated.

RESULTS AND DISCUSSION

Soil water retention curve

Soil water retention curve performed in the field

Table 2 presents the summary of the regression analysis ($p \leq 0.05$) of the soil moisture data (θ_l) as a function of the time (t) for the field test.

Table 2 | Summary of regression analysis ($p \leq 0.05$) in adjusting the soil moisture data ($cm^3\ cm^{-3}$) as a function of time (days), using non-treated water (OW) and water with magnetic treatment (MW), referring to the field test

	Non-treated water (OW)			Magnetized water (MW)		
Model	**Regression mean square**	**Residual error**	R^2	**Regression mean square**	**Residual error**	R^2
Linear	3.17*	0.022	82.6	3.00^{ns}	0.022	76.6
2nd degree	2.12*	0.019	87.4	2.01*	0.015	88.5
3rd degree	1.59*	0.019	88.0	1.51*	0.015	89.0
Power	3.17*	0.028	73.0	2.99*	0.026	67.2
Logarithmic	3.17*	0.025	78.5	3.00*	0.024	72.2

* – significant at 5% probability, ^{ns} – not significant at 5% probability.

All mathematical models presented significant fitting to the data for both types of water (except the linear model for magnetized water use). The third-degree polynomial model was selected for both types of water due to the higher value of the determination coefficient (greater than 85%). Figure 3 shows the behavior of the data fit and the respective model for each treatment.

Comparing the fitted mathematical models using the *F*-test ($p \leq 0.05$), significant differences were found between data generated by the models, demonstrating that the type of water significantly influenced the behavior of the soil moisture over time. The soil moisture at field capacity for magnetically treated and non-treated water was 0.356 and 0.340 $cm^3\ cm^{-3}$ respectively, reached in 3.35 and 4.98 days, respectively. The soil moisture at saturation for magnetically treated and non-treated water was 0.359 $cm^3\ cm^{-3}$ ($\pm 0.005\ cm^3\ cm^{-3}$) and 0.456 $cm^3\ cm^{-3}$ ($\pm 0.019\ cm^3\ cm^{-3}$), respectively.

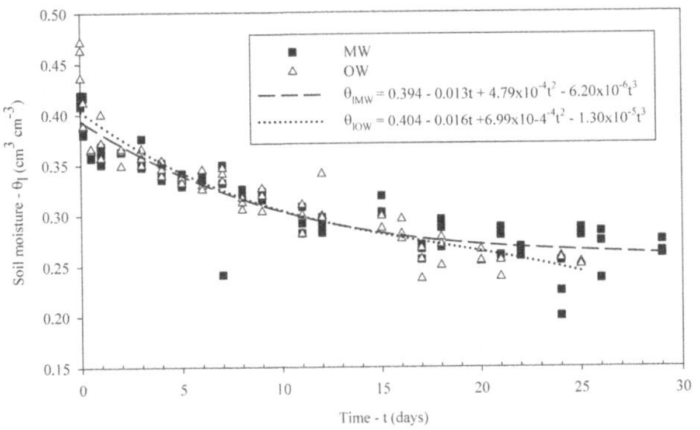

Figure 3 | Behavior of the variation in the soil moisture (θ_I) as a function of the time (days) with magnetically treated water (MW) and non-treated water (OW) with details of the model fit.

It was possible to observe that the use of magnetically treated water showed greater mobility compared with the non-treated water use, reaching the soil moisture at field capacity 1.63 days before. This result is in agreement with Khoshravesh-Miangoleh & Kiani (2014), where greater accumulated infiltration and water infiltration rate were observed in the use of magnetically treated water.

Table 3 shows the summary of the regression analysis ($p \leq 0.05$) of the volumetric soil moisture data adjustment as a function of the soil water tension, referring to the field test, in the use of non-treated water (OW) and water subjected to magnetic treatment (MW).

Almost all mathematical models represented significantly ($p \leq 0.05$) the soil moisture data as a function of the soil water tension for both types of water (except the linear model for magnetized water use). The logarithmic model was selected to represent the soil water retention curve for both types of water. Figure 4 shows the behavior of the data fitting for each treatment, as well as the fitted mathematical models.

Comparing the fitted mathematical models by the *F*-test ($p \leq 0.05$), there were not found to be any significant differences between the data generated, demonstrating that the type of water did not significantly influence the soil water retention behavior.

Considering the estimated values of soil moisture at field capacity (0.356 $cm^3\ cm^{-3}$ for magnetized water, and 0.340 $cm^3\ cm^{-3}$ for non-treated water) for the fitted mathematical models presented in Figure 3, tensions equivalent to 7.22 kPa for magnetized water and 10.54 kPa for non-treated water were observed. Although there are no significant differences between models, it is interesting to

Table 3 | Summary of the regression analysis ($p \leq 0.05$) in adjustment of the volumetric soil moisture data as a function of the soil water tension, using non-treated water (OW) and water with magnetic treatment (MW), referring to the field test

	Non-treated water (OW)			Magnetized water (MW)		
Model	**Regression mean square**	**Residual error**	**R^2**	**Regression mean square**	**Residual error**	**R^2**
Linear	3.30*	0.028	72.4	3.21^{ns}	0.033	55.9
2nd degree	2.20*	0.022	82.7	2.15*	0.023	78.4
3rd degree	1.65*	0.019	88.1	1.62*	0.021	82.4
Power	3.31*	0.013	94.4	3.23*	0.022	80.9
Logarithmic	3.31*	0.013	94.3	3.23*	0.022	79.8

* – significant at 5% probability, ^{ns} – not significant at 5% probability.

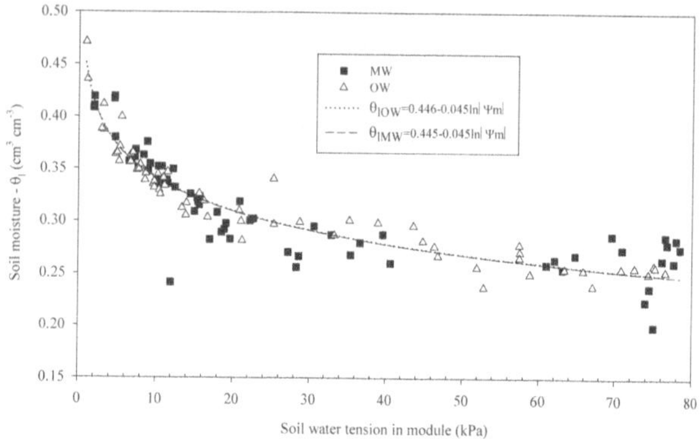

Figure 4 | Soil water retention curves performed in the field with magnetically treated water (MW) and non-treated water (OW) with details of the data adjustment model.

note that the soil moisture at field capacity using water subjected to magnetic treatment was obtained with less retention force (31.5% less). According to some authors, the cause of this effect may be associated with the surface tension decrease and water viscosity increase (Toledo *et al.* 2008; Mostafazadeh-Fard *et al.* 2011; Khoshravesh-Miangoleh & Kiani 2014; Surendran *et al.* 2016). An analogous result was found by Zlotopolski (2017) that observed lower soil water tension values using magnetic water in lettuce irrigation, the soil moisture being maintained for a longer time.

As there was no statistical difference between the curves when changing the type of water, the soil water retention curve performed in field conditions for non-treated water was selected for the irrigation management due its higher value of the determination coefficient for the curve fitting ($R^2 = 94.3\%$). Considering the soil water tension equal to 10 kPa in the module to represent the field capacity in the irrigation management, this resulted in a soil moisture of $0.342\ cm^3\ cm^{-3}$.

Soil water retention curve performed in the laboratory

Table 4 shows the summary of the regression analysis ($p \leq 0.05$) of the fitting of the volumetric moisture data as a function of the soil water tension, referring to the laboratory analysis.

Almost all mathematical models represented the data significantly ($p \leq 0.05$), the exception being the linear model. The power model was selected (Figure 5) due its highest value of the determination coefficient ($R^2 = 94.1\%$).

Applying the F-test ($p \leq 0.05$) between the fitted models of soil water retention curves performed in the field and in the laboratory, both with non-treated water, it was observed that the obtained models were statistically different.

Assuming the same soil water tension value regarding the field capacity found in the field test (10 kPa in the module), this would result in a volumetric soil moisture

Table 4 | Summary of the regression analysis ($p \leq 0.05$) in adjustment of the volumetric soil moisture data as a function of soil water tension, referring to the laboratory analysis

	Laboratory test		
Model	**Regression mean square**	**Residual error**	**R^2**
Linear	5.14^{ns}	0.070	55.7
2nd degree	3.48*	0.027	93.4
Power	5.22*	0.026	94.1
Logarithmic	5.22*	0.031	91.4

* – significant at 5% probability, ^{ns} – not significant at 5% probability.

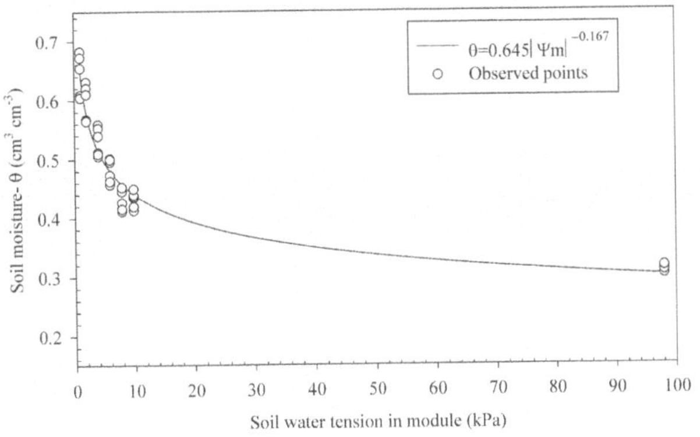

Figure 5 | Soil water retention curve obtained with data collected in laboratory with the fitted mathematical model.

equivalent to 0.439 $cm^3\ cm^{-3}$. Pinto *et al.* (2018) found similar values (around 0.4115 $cm^3\ cm^{-3}$) considering the same soil water tension (10 kPa) in the same type of soil.

Comparing the values of soil moisture at the field capacity obtained in the field (0.342 $cm^3\ cm^{-3}$) and in the laboratory (0.439 $cm^3\ cm^{-3}$) (both using non-treated water), overestimation was observed in the laboratory estimate (28.36% increase). Some authors state that the soil water retention characteristics, as well as the soil moisture at field capacity obtained in field conditions, represent the soil more adequately when compared with laboratory methods (Morgan *et al.* 2001; Jabro *et al.* 2009; Brito *et al.* 2011). According to the authors, soil sampling can modify the original soil bulk density because of sample compaction, which leads to a reduction of the porous space.

Irrigation management responses

Soil water tension data during the experiment, number of irrigations (*NI*) and average interval between irrigations (*Ii*)

Figure 6 shows the estimated soil water tensions (Ψ_m) during the experiment for the different experimental combinations.

It was possible to observe less data variability using magnetized water in almost all treatments. Table 5 shows the Tukey's test significance of pairwise multiple comparison for soil water tension data in different combinations of soil water tensions to start irrigation (T1, T2, T3 and T4) and types of water (MW and OW).

Varying the types of water for each soil water tension to start irrigation, no significant differentiation was observed ($p \leq 0.05$). In the other way, comparing the soil water tensions to start irrigation, considering fixed the type of water, it was observed: MW – T2 treatment being an exception that was statistically equal to T3, all other treatments were statistically different from each other; OW – the soil water tension variation during the experiment for T2 was statistically equal to T3; and T3 was statistically equal to T4, all other treatments being different from each other.

Table 6 shows the number of irrigations (*NI*) carried out and the average interval between irrigations (*Ii*) for each experimental combination (water type and soil water tension).

It was possible to observe a smaller number of irrigations (*NI*) in the use of magnetized water (with the exception of the T2 treatment, where the use of magnetized water determined a greater number of events). Analogous behavior was observed in several studies, where, in general, less variability of soil moisture in time was observed in irrigations using magnetized water, determining a lower frequency of irrigation (Mostafazadeh-Fard *et al.* 2011; Surendran *et al.* 2016; Al-Ogaidi *et al.* 2017). Additionally Zlotopolski (2017) observed lower values of soil water tension when using magnetized water.

According to some authors, the magnetization process makes the water more cohesive, defining smaller dimensions of water clusters, determining greater attraction to soil particles and less movement of water in soil pores (Mostafazadeh-Fard *et al.* 2011; Surendran *et al.* 2016; Al-Ogaidi *et al.* 2017). The authors explain that the water cohesion increase was the result of molecules released in reaction with ions by hydrogen bonds and van der Waals forces, leading to greater ease of water penetration in soil micropores, reducing percolation. Additionally, the authors explain that calcium carbonate ions in the water form aragonite crystals after magnetizing the water, which are deposited in the soil. This process determines the elevation of soil osmotic potential, decreasing crop evapotranspiration and maintaining soil moisture for a longer time. Surendran *et al.* (2016) pointed out that the maintenance of soil moisture for a longer time may also be associated with reduced water evaporation capacity. Kareem & Adeniran (2020) also observed a reduction in Lagos spinach

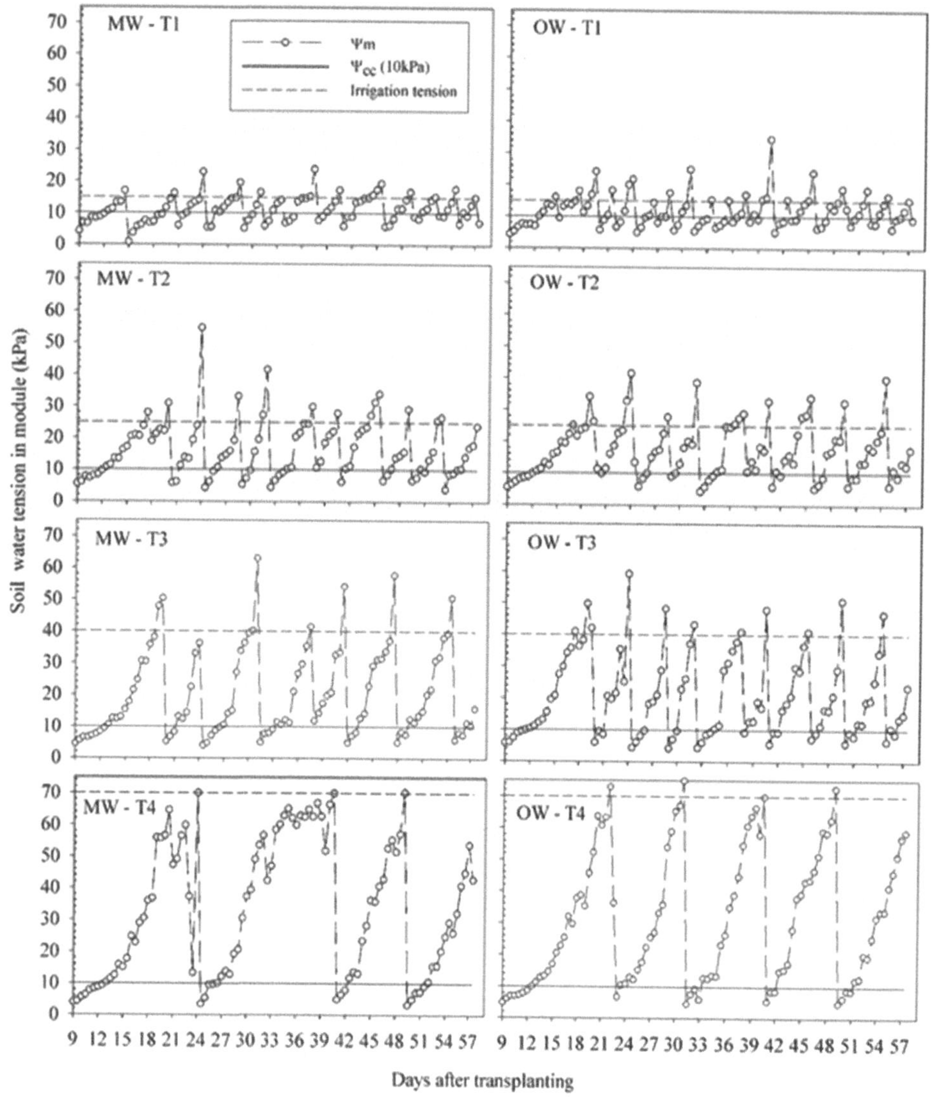

Figure 6 | Estimated soil water tension (kPa) during the experiment for different combinations of soil water tensions to start irrigation (T1, T2, T3 and T4) and types of water (MW and OW). Ψ_{cc} – soil water tension referring to field capacity (10 kPa).

Table 5 | Tukey's test significance of pairwise multiple comparison soil water tension data in different combinations of soil water tensions to start irrigation (T1, T2, T3 and T4) and types of water (MW and OW)

Treatments	MWT1	MWT2	MWT3	MWT4	OWT1	OWT2	OWT3	OWT4
MWT1	–							
MWT2	*	–						
MWT3	*	ns	–					
MWT4	*	*	*	–				
OWT1	ns	ns	*	*	–			
OWT2	*	ns	ns	*	*	–		
OWT3	*	ns	ns	*	*	ns	–	
OWT4	*	*	*	ns	*	*	ns	–

* – significant at 5% probability, ns – not significant at 5% probability.

Table 6 | Number of irrigations performed and average interval between irrigations (days) for different combinations of soil water tension to start irrigation (T1, T2, T3 and T4) and types of water (OW and MW)

	NI		*Ii* (days)	
Soil water tension	Non-treated water (OW)	Magnetized water (MW)	Non-treated water (OW)	Magnetized water (MW)
T1	18	13	2.7	3.8
T2	9	10	5.4	4.9
T3	9	7	5.4	7.0
T4	4	3	12.3	16.3

(*Celosia argentea*) evapotranspiration with the use of magnetized water.

Total applied irrigation depth during the crop cycle (Σ*LI*)

Table 7 shows the Tukey's test result ($p \leq 0.05$) to evaluate the total applied irrigation depth for different combinations of soil water tension to start irrigation and types of water, considering the soil water retention curve built in the field (actual applied values).

A significant reduction was observed in the total applied irrigation depth using magnetized water for irrigation (with T2 treatment as exception). The reduction was 34.62%, 18.62% and 26.34% for T1, T3 and T4, respectively. Selim *et al.* (2019) also observed reduction in the amount of water applied (25% reduction) for wheat crops irrigated with magnetically treated water compared with non-treated water.

Regarding the influence of soil water tension to start irrigation on total irrigation water depth, it was observed that T3 > T1 > T2 > T4 for non-treated water, and T2 > T3 > T1 > T4 for magnetized water. In both types of water, T4 treatment obtained the lowest irrigation depths. Koetz *et al.* (2006) points out that generally, longer intervals between irrigations determine lower total applied irrigation depth.

Table 7 | Total applied irrigation depth (mm) for different combinations of soil water tensions to start irrigation (T1, T2, T3 and T4) and types of water (OW and MW), for the soil water retention curve performed in the field

	Soil water tension			
Type of water	T1	T2	T3	T4
Non-treated water	138.72 aB	132.87 bC	169.71 aA	97.45 aD
Magnetized water	90.69 bC	146.11 aA	138.11 bB	71.48 bD

Different lowercase letters in the vertical differ significantly ($p \leq 0.05$) with changing water type, and different uppercase letters in the horizontal differ significantly ($p \leq 0.05$) with changing soil water tension.

Electric energy consumption (*EC*)

Table 8 shows the Tukey's test result ($p \leq 0.05$) to evaluate the electric energy consumption estimate for different combinations of soil water tensions to start irrigation and the types of water, considering the soil water retention curve from the field (actual applied values).

The use of water subjected to magnetic treatment determined lower electric energy consumption in all soil water tensions (with the exception of T2 treatment). Reductions were of 34.7%, 18.6% and 26.5% for T1, T3 and T4 respectively. For T2, the use of magnetized water increased 10% the electric energy consumption in relation to the use of non-treated water. The lowest *EC* value was reached using magnetically treated water with 70 kPa tension (0.0075 kWh m^{-2} $cycle^{-1}$), due to the lower total applied irrigation depth and longer interval between irrigations. The highest consumption was reached using non-treated water with 40 kPa for soil water tension to start irrigation (0.0177 kWh m^{-2} $cycle^{-1}$), this being the treatment that demanded a larger total applied irrigation depth.

Simulation of irrigation management responses

Comparison between applied and simulated irrigation depth during the crop cycle

Table 9 shows the Tukey's test result ($p \leq 0.05$) to evaluate the soil water retention curve data source (field – applied

Table 8 | Electric energy consumption estimate (kWh m^{-2} $cycle^{-1}$) for different combinations of soil water tensions to start irrigation (T1, T2, T3 and T4) and types of water (OW and MW), using soil water retention curve from the field

	Soil water tension			
Type of water	T1	T2	T3	T4
Non-treated water	0.0145 aB	0.0139 bC	0.0177 aA	0.010 aD
Magnetized water	0.0095 bC	0.0153 aA	0.0144 bB	0.0075 bD

Different lowercase letters in the vertical differ significantly ($p \leq 0.05$) with changing water type, and different uppercase letters in the horizontal differ significantly ($p \leq 0.05$) with changing water tension in the soil.

Table 9 | Total applied irrigation depth (mm) for different soil water tensions to start irrigation (T1, T2, T3 and T4), considering the soil water retention curves from the field (applied) and from the laboratory (simulated) for each type of water

		Soil water tensions			
Type of water	**Retention curve**	**T1**	**T2**	**T3**	**T4**
Non-treated water	Field (applied)	138.72 b	132.87 b	169.71 b	97.45 b
	Laboratory (simulated)	211.86 a	194.59 a	242.19 a	134.29 a
Magnetized water	Field (applied)	90.69 b	146.11 b	138.11 b	71.48 b
	Laboratory (simulated)	139.62 a	213.93 a	195.82 a	99.20 a

Different lowercase letters in the vertical differ significantly ($p \leq 0.05$) with changing water type.

irrigation depths, and laboratory – simulated irrigation depths) considering the total applied irrigation depth for different soil water tensions to start irrigation. The total applied irrigation depths using the laboratory soil water retention curve were simulated.

It was observed in all soil water tensions that the hypothetical use of the soil water retention curve performed in the laboratory in the lettuce irrigation management provided significantly the highest total applied irrigation depth. Increases were observed of 52.72%, 46.45%, 42.71% and 37.80% for T1, T2, T3 and T4 respectively using non-treated water, and 53.95%, 46.42%, 41.79% and 38.20% for T1, T2, T3 and T4 respectively using magnetized water. This result highlights the importance of the representativeness of the soil water retention curve for effective irrigation management.

Some authors point out that the determination of the soil water retention curve in the laboratory can be subject to errors due to inadequate procedures such as: inadequate contact between soil and porous plate; decrease in hydraulic conductivity of soil due to sample compaction; soil dispersion as a result of the sample moistening process, changing the natural flow characteristics in the soil; and reduced soil sample height determining less soil profile representativeness (Bittelli & Flury 2009; Solone *et al.* 2012; van Lier *et al.* 2019).

Electric energy consumption (*EC*)

Table 10 shows the Tukey's test result ($p \leq 0.05$) to evaluate soil water retention curve type (field – applied irrigation depths, and laboratory – simulated irrigation depths) considering the electric energy consumption estimate for different soil water tensions to start irrigation. The electric energy consumption estimates using the laboratory soil water retention curve were simulated.

In all soil water tensions to start the irrigation, a significant increase in the electric energy consumption was observed in the hypothetical use of the soil water retention curve from the laboratory in the lettuce irrigation management. The increase in consumption was of 52.7%, 46.5%, 42.7% and 37.8% for T1, T2, T3 and T4, respectively using non-treated water, and 53.9%, 46.4%, 41.8% and 38.19% for T1, T2, T3 and T4 respectively, using magnetized water.

Table 10 | Electric energy consumption for different soil water tensions to start irrigation (T1, T2, T3 and T4), considering the soil water retention curves from the field (applied) and from the laboratory (simulated) in each type of water

		Soil water tensions			
Type of water	**Retention curve**	**T1**	**T2**	**T3**	**T4**
Non-treated water	Field (applied)	0.0145 bB	0.0139 bC	0.0177 bA	0.0102 bD
	Laboratory (simulated)	0.0221 aB	0.0203 aC	0.0253 aA	0.0140 aD
Magnetized water	Field (applied)	0.0095 bC	0.0153 bA	0.0144 bB	0.0075 bD
	Laboratory (simulated)	0.0146 aA	0.0223 aA	0.0204 aB	0.0104 aD

Different lowercase letters in the vertical differ significantly ($p \leq 0.05$) with changing water type, and different uppercase letters in the horizontal differ significantly ($P \leq 0.05$) with changing soil water tension.

Practical observations

The result of this research presents, to the scientific and growers' community, a discussion that should be the focus of today's research on irrigated agriculture, which is the rational and sustainable use of water. In terms of water subjected to magnetic treatment use on different crops' irrigation, despite this experiment showing water and energy savings, this study should be extended to the productive and economic evaluation of the use of water magnetization technology.

In addition, this research demonstrated the need for further studies on the rational assessment of irrigation management methods. The use of soil water retention curves developed in the laboratory has long been preferred for irrigation management of various crops: although there is a lot of knowledge of the negative impacts of the soil sampling process, as well as the process of making the curve in the laboratory compared with the field reality, there are practically no studies that assess the practical impact, i.e., evaluations in irrigated crops. Even with this study having demonstrated this impact, it is still necessary to continue evaluations, to strengthen the debate with the scientific community and irrigating farmers regarding the improvement of irrigation management methods.

CONCLUSIONS

The use of the water subjected to magnetic treatment in the irrigation of iceberg lettuce cv. Lucy Brown determined the maintenance of soil moisture for a longer time, defining a smaller number of irrigation events, less volume of water applied, and less electric energy consumption. Despite the need to continue evaluating the impact of this technology on other combinations of crops, soils and climates, it has potential in saving water and electricity, improving the management of water resources.

The use of the soil water retention curve derived from the field data determined less water and electric energy consumption in Lucy Brown lettuce irrigation, in comparison with the simulated use of the soil water retention curve derived from laboratory data. This finding presents itself as an opportunity for the growers themselves to make their retention curves in the irrigation area, with the potential to achieve high irrigation efficiencies, providing savings in water and electricity.

ACKNOWLEDGEMENT

This study was financed in part by the Coordenação de Aperfeiçoamento de Pessoal de Nível Superior – Brasil (CAPES) – Finance Code 001.

DATA AVAILABILITY STATEMENT

All relevant data are included in the paper or its Supplementary Information.

REFERENCES

Al-Ogaidi, A. A. M., Wayayok, A., Rowshon, M. K. & Abdullah, A. F. 2017 The influence of magnetized water on soil water dynamics under drip irrigation systems. *Agricultural Water Management* **180**, 70–77. https://doi.org/10.1016/j.agwat.2016.11.001.

Alvares, C. A., Stape, J. L., Sentelhas, P. C., de Moraes Gonçalves, J. L. & Sparovek, G. 2013 Köppen's climate classification map for Brazil. *Meteorologische Zeitschrift* **22** (6), 711–728. https://doi.org/10.1127/0941-2948/2013/0507.

Azevedo, J. A. & Silva, E. M. 1999 Tensiômetro: dispositivo prático para controle da irrigação, Circular Técnica No 001. Embrapa Cerrados, Planaltina, DF, Brazil.

Bilibio, C., Carvalho, J. A., Martins, M., Rezende, F. C., de Freitas, W. A. & Gomes, L. A. A. 2010 Função de produção da berinjela irrigada em ambiente protegido (Irrigated eggplant crop function in protected environment). *Irriga* **15** (1), 10–22 (in Portuguese).

Bittelli, M. & Flury, M. 2009 Errors in water retention curves determined with pressure plates. *Soil Science Society of America Journal* **73** (5), 1453–1460. https://doi.org/10.2136/sssaj2008.0082.

Brito, A. S., Libardi, P. L., Mota, J. C. A. & Moraes, S. O. 2011 Estimativa da capacidade de campo pela curva de retenção e pela densidade de fluxo da água (Field capacity estimation based on retention curve and soil water flux density). *Revista Brasileira de Ciência do Solo* **35** (6), 1939–1948 (in Portuguese). https://doi.org/10.1590/S0100-06832011000600010.

Casaroli, D. & van Lier, Q. J. 2008 Critérios para determinação da capacidade de vaso (Criteria for pot capacity determination). *Revista Brasileira de Ciência do Solo* **32** (1), 59–66 (in

Portuguese). https://doi.org/10.1590/s0100-06832008000100007.

Jabro, J. D., Evans, R. G., Kim, Y. & Iversen, W. M. 2009 Estimating in situ soil–water retention and field water capacity in two contrasting soil textures. *Irrigation Science* **27**, 223–229. https://doi.org/10.1007/s00271-008-0137-9.

Kareem, K. Y. & Adeniran, K. A. 2020 Effects of electromagnetic irrigation water on Lagos spinach evapotranspiration using lysimetric method. *Agricultural Engineering International: CIGR Journal* **22** (2), 41–48.

Khoshravesh-Miangoleh, M. & Kiani, A.-R. 2014 Effect of magnetized water on infiltration capacity of different soil textures. *Soil Use and Management* **30** (4), 588–594. https://doi.org/10.1111/sum.12140.

Koetz, M., Coelho, G., Costa, C. C., Lima, E. P. & Souza, R. J. 2006 Efeito de doses de potássio e da freqüência de irrigação na produção da alface-americana em ambiente protegido (Effect of potassium doses and irrigation frequency in the production of the American lettuce in greenhouse condition). *Engenharia Agrícola* **26** (3), 730–737 (in Portuguese). https://doi.org/10.1590/S0100-69162006000300009.

Maheshwari, B. L. & Grewal, H. S. 2009 Magnetic treatment of irrigation water: its effects on vegetable crop yield and water productivity. *Agricultural Water Management* **96** (8), 1229–1236. https://doi.org/10.1016/j.agwat.2009.03.016.

Morgan, K. T., Parsons, L. R. & Wheaton, T. A. 2001 Comparison of laboratory- and field-derived soil water retention curves for a fine sand soil using tensiometric, resistance and capacitance methods. *Plant and Soil* **234**, 153–157. https://doi.org/10.1023/A:1017915114685.

Mostafazadeh-Fard, B., Khoshravesh, M., Mousavi, S.-F. & Kiani, A.-R. 2011 Effects of magnetized water and irrigation water salinity on soil moisture distribution in trickle irrigation. *Journal of Irrigation and Drainage Engineering* **137** (6), 398–402. https://doi.org/10.1061/(ASCE)IR.1943-4774.0000304.

Oliveira, L. F. C., Bonomo, R. & Cortês, F. C. 2001 Desenvolvimento matemático de uma equação para estimativa da capacidade de campo (Mathematical development of an equation for the estimate of field capacity). *Pesquisa Agropecuária Tropical* **31** (1), 65–69 (in Portuguese).

Pinto, L. C., Chagas, W. F. T. & Amaral, F. H. C. 2018 Qualidade física de um Latossolo Vermelho distroférrico sob diferentes usos (Water retention and S index of an Oxisol under different uses). *Revista Agrogeoambiental* **10** (3), 47–58 (in Portuguese). https://doi.org/10.18406/2316-1817v10n320181103.

Selim, D. A. H., Nassar, R. M. A., Boghdady, M. S. & Bonfill, M. 2019 Physiological and anatomical studies of two wheat cultivars irrigated with magnetic water under drought stress conditions. *Plant Physiology and Biochemistry* **135**, 480–488. https://doi.org/10.1016/j.plaphy.2018.11.012.

Soil Survey Staff 1999 *Soil Taxonomy: A Basic System of Soil Classification for Making and Interpreting Soil Surveys*, 2nd edn. Natural Resources Conservation Service, US Department of Agriculture, Washington, DC, USA.

Solone, R., Bittelli, M., Tomei, F. & Morari, F. 2012 Errors in water retention curves determined with pressure plates: effects on the soil water balance. *Journal of Hydrology* **470–471**, 65–74. https://doi.org/10.1016/j.jhydrol.2012.08.017.

Surendran, U., Sandeep, O. & Joseph, E. J. 2016 The impacts of magnetic treatment of irrigation water on plant, water and soil characteristics. *Agricultural Water Management* **178**, 21–29. https://doi.org/10.1016/j.agwat.2016.08.016.

TIMOL 2012 TIMOL GROUP: Área Rural. Available from: http://www.timolgroup.com.br/content.as p?contentid=290

Toledo, E. J. L., Ramalho, T. C. & Magriotis, Z. M. 2008 Influence of magnetic field on physical–chemical properties of the liquid water: insights from experimental and theoretical models. *Journal of Molecular Structure* **888** (1–3), 409–415. https://doi.org/10.1016/j.molstruc.2008.01.010.

van Lier, Q. J., Pinheiro, E. A. R. & Inforsato, L. 2019 Hydrostatic equilibrium between soil samples and pressure plates used in soil water retention determination: consequences of a questionable assumption. *Revista Brasileira de Ciência do Solo* **43**, e0190014. https://doi.org/10.1590/18069657rbcs20190014.

Yuri, J. E., Mota, J. H., Souza, R. J., Resende, G. M., Freitas, S. A. C. & Rodrigues Júnior, J. C. 2002 *Alface americana – Cultivo comercial (American Lettuce – Commercial Cultivation)*. UFLA, Lavras, MG, Brazil (in Portuguese).

Yusuf, K. O. & Ogunlela, A. O. 2017 Effects of deficit irrigation on the growth and yield of tomato (*Solanum lycopersicum*) irrigated with magnetised water. *Environmental Research, Engineering and Management* **73** (1), 59–68.

Zlotopolski, V. 2017 Magnetic treatment reduces water usage in irrigation without negatively impacting yield, photosynthesis and nutrient uptake in lettuce. *International Journal of Applied Agricultural Sciences* **3** (5), 117–122. https://doi.org/10.11648/j.ijaas.20170305.13.

First received 13 December 2020; accepted in revised form 28 January 2021. Available online 10 February 2021

Effects of magnetized fresh water on seed germination and seeding growth of cotton

Jihong Zhang, Kai Wei, Quanjiu Wang, Yan Sun and Weiyi Mu

ABSTRACT

Magnetized water treatment technology is usually used to improve poor quality water, and there is still a lack of study on fresh water. To understand the biological effects of different strength magnetized fresh water (MFW), seed germination and potted experiments on cotton were carried out to study the effects of MFW with different magnetic intensity (0, 100, 300, 500 mT). Results showed that the surface tension coefficient of MFW reduced by 7.3–10.5%, whilst dissolved oxygen concentrations increased by 8.8–12.7%. Germination strength indexes of cotton cultivated with MFW significantly increased, showing potential and vigor indexes of 16.8–22.4% and 47.4–78.0%, respectively. The emergence rate of cotton irrigated with MFW was faster and higher, with recorded values of 7.7–13.1%. The net photosynthetic rate (P_n) and instantaneous water use efficiency (*iWUE*) of cotton increased significantly, whereas the stomatal limit value (L_s) decreased. In all, results suggest the total biomasses of MFW irrigated cotton have significantly increased. Therefore, it is suggested that MFW may more effectively promote the utilization of water and light in cotton under magnetic field intensities of 300–500 mT. The results can provide guidance for the efficient utilization of magnetized fresh water in arid and semi-arid areas.

Key words | cotton seedling, fresh water, magnetic field intensity, photosynthesis, seed germination

Jihong Zhang
Kai Wei
Quanjiu Wang (corresponding author)
Yan Sun
Weiyi Mu
State Key Laboratory of Eco-Hydraulics in Northwest Arid Region of China,
Xi'an University of Technology,
Xi'an 710048,
China
E-mail: *wquanjiu@163.com*

HIGHLIGHTS

- Magnetized fresh water can improve the activity of water molecules and the amount of dissolved oxygen.
- The biological magnetic effect of magnetized fresh water is closely related to magnetization intensity.
- Magnetization treatment can bring into play the physiological effect of irrigation water and improve water production capacity.

INTRODUCTION

As a new type of green and pollution-free water treatment technology (Esmaeilnezhad *et al.* 2017), magnetization is simple, efficient, and low input (Luo *et al.* 2020). It has been widely used in wastewater treatment (Ji *et al.* 2010), water purification (Yao *et al.* 2019), scale reduction (Sohaili *et al.* 2016), construction (Ghorbani *et al.* 2019), materials (Pour *et al.* 2017), textiles (Lipus *et al.* 2013), and aquaculture (Hassan *et al.* 2019). The basic principle of magnetized water is that after the liquid water passes through a certain intensity magnetic field at a given flow rate, a series of changes have taken place in the physical and chemical properties of water under the influence of Lorentz force (Chibowski

doi: 10.2166/ws.2021.051

& Szcześ 2018), which appears to create the possibility of improving the production capacity of fresh water (Ambashta & Sillanpää 2010). The development of magnetized water research has achieved good results in agricultural irrigation (Selim & Selim 2019; Sultan *et al.* 2019), and study on irrigation experiments with magnetized brackish water showed reduced harm of brackish water, promotion of crop growth, and increased crop yields (Hachicha *et al.* 2018; Liu *et al.* 2020). Surendran *et al.* (2016) indicated that under field conditions, the yield of eggplant under magnetized water irrigation increased by 17.0%. Mahmoud *et al.* (2019) showed that the grain yield, straw yield, and biological yield of magnetized brackish water increased by 19.24%, 33.97%, and 26.99%, respectively. Moreover, studies on the early growth of crops found that magnetized water irrigation significantly improved germination rate, germination index, activity index, salt tolerance index, and other physiological factors of corn (*Zea mays* L.) seeds, and promoted the growth of maize seedlings under saline conditions (Aghamir *et al.* 2015). Sayed (2014) indicated that magnetized water treatment had a significant positive effect on plant growth parameters (i.e. height, leaf area, leaf, stem, root fresh weight, and dry weight) of broad bean (*Vicia Faba* L.) seedlings. Studies on the utilization of light energy of crops by magnetic water showed that the net photosynthetic rate, stomatal conductance, intercellular CO_2 concentration and water use efficiency of *Populus × euramericana 'Neva'* in magnetized brackish water irrigation were all increased, while transpiration rate and stomatal limit values decreased compared with that of unmagnetized water (Liu *et al.* 2019). Alfaidi *et al.* (2017) found that the contents of chlorophyll a, b, carotenoids, total pigment, soluble protein, and total protein in guinea grass (*panicum maximum*) leaves significantly increased after irrigating with magnetized water. Research indicates that the growth promoting effect of magnetized water appears to be related to the magnetic field strength and irrigation water quality. Massah *et al.* (2019) indicated that the intensity of the magnetic field and the type of treated water had a significant effect on the germination and growth characteristics of wheat seeds. The germination rate of seeds treated with 400 mT of distilled water was the highest (53.3%), the fresh weight of seedlings treated with 600 mT of distilled water was the highest, and the root length of wheat seeds treated with 400 mT of groundwater was the largest (155.3 mm).

In summary, remarkable achievements have been made in irrigated wheat, corn, broad bean, and eggplant with magnetized water; nevertheless, research on irrigating cotton with magnetized water is scarce. At the same time, most of the studies on magnetized water irrigation focus on the application of brackish water, while there are few studies on magnetization of fresh water. Therefore, it is of interest to explore the effects of MFW with different magnetic fields on the growth of cotton, especially on the fragile and sensitive seedling stage of cotton growth. Thus, this paper target the analysis of cotton seed germination, seedling emergence, seedling growth, photosynthetic characteristics parameters, and biomass distribution pattern, along with studying physical and chemical properties of MFW, and the effects of MFW with different magnetization intensities (0, 100, 300, 500 mT). Cotton seedling growth and photosynthetic characteristics were discussed, and suitable magnetization intensities for cotton seed germination and seedling growth determined. The results can provide theoretical basis and technical support for the efficient utilization of fresh water resources in arid and semi-arid areas.

MATERIALS AND METHODS

Overview of test area

The experimental area is in the Kunqi River alluvial plain (41 °35′N, 86 °10′E, and 901 m a.s.l.), the Bazhou irrigation experimental station of Xinjiang, on the edge of Tarim Basin at the southern foot of Tianshan Mountain. Lying in the middle of the Eurasian continent, it belongs to the typical dry continental climate in the warm zone, with scarce rainfall, intense positive and negative rainfall, large temperature difference between day and night, sufficient sunshine in the test area, averaging 3,036.2 h of annual sunshine (Chen *et al.* 2018). Average temperatures oscillate at 11.5 °C, with minimum and maximum temperatures of −30.9 °C and 42.2 °C, respectively (Yang *et al.* 2016). Annual precipitation falls between 53.3–62.7 mm, whilst annual evaporation, annual wind and maximum wind speed are recorded amomg 2,273–2,788 mm (Liang *et al.* 2019), 2.4 m s^{-1} and

22 m s^{-1} (Li *et al.* 2019), in that order. The fresh water used for the test came from Kongke River with an average salinity of 0.72 g L^{-1} and pH of 7.80.

Magnetizers and magnetized water devices

Three external sintered Rufe-B CHQ permanent magnet magnetizers (Shanghai Juncai Magnetic Materials Co., LTD., China) were used with magnetic field strengths of 100, 300, and 500 mT. The effective magnetic field area was 8 cm × 10 cm and magnetic field intensities was calibrated with a 5180-Gaussian Meter (F.W. BELL Co., USA). The magnetized water device is composed of a water box, peristaltic pump, magnetizer, water pipeline, and valves (Figure 1). Water pipelines are of PVC with a 2.5 cm diameter (cross-sectional area of 4.91 cm^2), and 10 cm of effective magnetic distance (i.e. effective pipiline length through which the magentic field was applied). As shown in Figure 1, 100 L of fresh water placed in the water box and pipelines opened, one at a time, through manually controlled valves. A peristaltic pump (0.74–12 L h^{-1}) was used to circulate the water. A section of the circulating pipeline was placed between the two poles of the magnetizer, and the magnetic induction line was cut perpendicular to the magnetic field. The flow rate was adjusted to 0.5 m s^{-1} by the peristaltic pump, and the flow was cyclically magnetized by the magnetic field. The magnetization time was 30 min, and the retention time of the magnetized water was 8 days (Coey & Cass 2000).

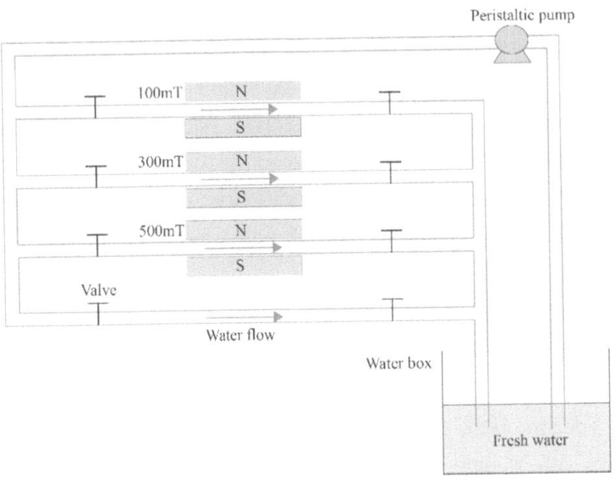

Figure 1 | Schematic diagram of magnetized water device.

Experimental design and process

The magnetized water device mentioned above was used to magnetize fresh water, whilst unmagnetized fresh water was used as a control. A total of four water clusters were formed, including unmagnetized fresh water (FM0), 100 mT magnetized fresh water (FM1), 300 mT magnetized fresh water (FM3) and 500 mT magnetized fresh water (FM5). A series of experiments including cotton seed germination, potted experiments in a cotton field, and physicochemical properties test were carried out for each cluster. The objective was to analyze the influence and mechanics of MFW under diversified magnetic field intensities on cotton seed germination and seedling growth.

Experiment 1–seed germination test

Seed germination tests were performed as described in a previous study with three repetitions (Fu *et al.* 2010). Fifty cotton seeds of the same size and full grain were selected and arranged evenly in a petri dish with tweezers. The petri dish was then marked and covered with a layer (1.7 mm) of fine filter paper (qualitative filter with 3 μm of mesh size). 20 mL of cluster water was added to the corresponding culture dish and the seeds were covered with a layer (the same as the filter paper layer described previously). After having absorbed enough water (24 h), the excess water was drained, and the seeds were placed in the culture dish in the incubator at 28 ± 1 °C, with a light intensity of 800 Lx. The numbers of germinations were recorded every day, and the corresponding cluster water (5 mL) was added every day to maintain the moisture content of the filter paper. The radicle length of cotton seeds was measured on the eighth day. The germination condition was evaluated by calculating germination potential (*GP*), germination rate (*GR*), germination index (*GI*), and vigor index (*VI*) as follows (Mosse *et al.* 2013):

$$GP = \frac{N_4}{N_T} \times 100\% \tag{1}$$

$$GR = \frac{N_8}{N_T} \times 100\% \tag{2}$$

$$GI = \sum \frac{G_t}{D_t} \tag{3}$$

$$VI = GI \times L \tag{4}$$

Table 1 | Soil physical properties in experimental cotton field

Soil texture	Depth/cm	γ_d/(g cm^{-3})	θ_s/(cm^3 cm^{-3})	FC/(cm^3 cm^{-3})	PWP/(% cm^3 cm^{-3})	ECe /(ds m^{-1})
Sandy loam	0–5	1.60	0.396	0.197	0.05	5.52
	5–10	1.62	0.394	0.195	0.05	5.49
	10–15	1.63	0.396	0.196	0.04	5.45
	15–20	1.67	0.395	0.194	0.06	5.43
	average	1.63	0.395	0.196	0.05	5.47

γ_d, θ_s, FC, PWP, and K_s represent soil bulk density, saturated volume water content, field capacity, wilting coefficient, and extract conductivity, in that order.

where, N_4 and N_8 are the numbers of seeds germinated in 4 and 8 days, respectively; N_T is the total number of seeds; G_t is the number of seeds germinated at day t (D_t); L is the average radicle length of seed on the eighth day.

Experimental 2–potted experiments in a cotton field

To simulate a cotton field growth environment, all potted plants were buried, randomly divided, and repeated three times for each cluster. Saline alkali soil from 0–20 cm depth (Table 1) from the surface of the cotton field was used in the experiment. The inner diameter and height of the plastic basin used in the test were 30 and 20 cm respectively, and eight air holes with a diameter of 10 mm were set at the bottom of the basin. Each barrel was uniformly loaded with 20 kg of soil sample and fertilizer with the following composition: carbamide (N 45.4%) 6.8 g, compound fertilizer (N 12%, P 18%, K 15%) 6.4 g, organic fertilizer (organic matter >35%) 24 g. After, the fertilizer and soil were evenly mixed, 7.065 L of cluster water was irrigated to the corresponding plastic basin. Each pot was seeded with 20 cotton seeds, with a sowing depth of 2–3 cm. As in the field, the soil surface was covered with plastic film mulch after sowing.

Fourteen days after sowing, the rate of emergence was measured and only 4 cotton plants were kept in each pot, with the remaining seedlings being removed. The length of cotton seedlings was then measured, the dust on the surface removed, and fresh weight of the seedling measured after washing and drying. For that purpose, seedlings were placed in an oven at 105 °C for 1 h and dried at 75 °C. The dry weight of the seedling was measured after cooling down. Plant height, stem diameter at cotyledon node, number of leaves per plant, and area of single leaf of potted cotton were measured. The stem diameter was measured with an electronic vernier caliper with an accuracy of 0.01 mm. The number of leaves was counted manually. Plant height, leaf width, and vein length were measured with measuring tape with an accuracy of 1 mm. The area of a single leaf was calculated with the formula length × width × 0.84 (Tan *et al.* 2017). The growth indexes of potted cotton seedlings were measured every 10 days. Thirty days after final singling, the net photosynthetic rate (P_n), stomatal conductance (G_s), intercellular CO_2 concentration (C_i), and transpiration rate (T_r) of the main functional leaves (from the top to the bottom of the fourth leaf) of cotton seedlings were measured using LC Pro SD full-automatic portable photosynthetic instrument (UK ADC) equipped with an LED artificial light source. Then, the stomatal limit $L_s = 1 - C_i/C_a$ (C_a is atmospheric CO_2 concentration) and instantaneous water use efficiency $WUE = P_n/T_r$ were calculated (Liu *et al.* 2017). When determining the photosynthesis parameters of cotton seedlings, according to the meteorological environment conditions during 10:00–12:00 am in a cotton field, the light intensity was set as 1,100 μ mol m^{-2} s^{-1}, the concentration of CO_2 was 36 μ mol mol^{-1}, and the temperature was 30 °C. The SPAD value of chlorophyll in cotton seedling leaves was measured using the SPAD-502 Plus Chlorophyll Meter (Konica Minolta, China). Forty days after final singling, the cotton seedlings were removed slowly. The cotton branches were then cut with fruit scissors, the dust was removed from the surface, the cotton branches were washed and dried, and the dry weight of each part was weighed.

Determination of physicochemical properties of magnetized water

In order to clarify the mechanism, the physicochemical properties of magnetized water, including surface tension, dissolved oxygen content, electrical conductivity (EC), and pH were measured. The surface tension was determined by a SCZL202 automatic tension tester (Zibo Shengkang Electric Co., Ltd, China), and the dissolved oxygen content was measured by HQ40 portable dissolved oxygen meter (HACH, USA). EC and pH were determined by DDS-307 electrical conductivity meter and PHSJ-4A pH meter, respectively (Shanghai INESA Scientific Instrument Co., Ltd, China).

Data processing and analysis

Data were recorded in Excel 2016 and analyzed using SPSS 22.0 software (IBM Corp. USA). The least significant difference (LSD) method ($P < 0.05$) was used for comparison of multiple values.

RESULTS AND DISCUSSION

Germination characteristics of cotton seeds

GP, *GR*, *GI*, and *VI* are important indicators to reflect seed germination strength (Kumar *et al.* 2011). *GP* and *GR* directly reflect the initiation of the metabolic process, while *GI* and *VI* reflect the seed's potential for rapid and uniform emergence. Compared with FM0, the *GP*, *GR*, *GI*, and *VI* of cotton irrigated with MFW were significantly increased ($P < 0.05$), in the ranges of 22.2–26.3%, 16.8–22.4%, 22.2–27.0%, and 47.4–78.0%, respectively (Table 2). The increased range of *GP* was greater than that of *GR*, while the amplitude of *VI* was greater than that of *GI*, indicating that fresh water magnetization could significantly improve seed strength and enhance the germination potential of cotton. This is consistent with the results of studies on the application of magnetized water to promote the germination of wheat seeds (Massah *et al.* 2019). As reported, smaller water cluster structure and higher dissolved oxygen after magnetization of fresh water were found, which provide good water and gas environments for seed germination and are conducive to improved seed germination (Surendran *et al.* 2016). Comparing cotton seeds cultivated with fresh water against those under different magnetized intensities, the results suggested that *GI*'s cotton seeds cultivated by FM3 improved the most, whereas *GP*, *GR*, and *GI* of cotton seeds treated with FM1 were slightly larger than FM5. At the same time *VI* was slightly smaller than FM5, but none of them was significant ($P > 0.05$).

Table 2 | Germination indexes of cotton seeds under different strengths of magnetized fresh water

Treatment	*GP* (%)	*GR* (%)	*GI*	*VI*
FM0	53.3 ± 1.2b	71.3 ± 1.2c	17.6 ± 0.7b	26.4 ± 0.8c
FM1	66.0 ± 2.0a	86.0 ± 2.0ab	21.6 ± 0.5a	38.9 ± 1.3b
FM3	67.3 ± 3.1a	87.3 ± 1.2a	22.4 ± 0.8a	47.0 ± 2.5a
FM5	65.3 ± 1.2a	83.3 ± 2.3b	21.5 ± 0.5a	41.6 ± 1.4b

GP, *GR*, *GI*, and *VI* represent germination potential, germination rate, germination index, and vigor index, in that order. Different letters within a column indicate significant differences among all treatments at $P < 0.05$.

Effect of MFW on the emergence rate of cotton

High seedling emergence rate and seedling uniformity are the guarantees for high and stable yield (Mo *et al.* 2017). The emergence of cotton seeds started 3 days after sowing (Figure 2). With the increase of emergence time, the emergence rate of cotton gradually increased and tended to stabilize. As shown in Figure 2, the emergence rate of cotton under MFW irrigation was faster and higher when compared to freshwater irrigated seeds. The emergence rate and full seedling formation of cotton irrigated with FM0 was 77.3% and 8 d, respectively, while the emergence

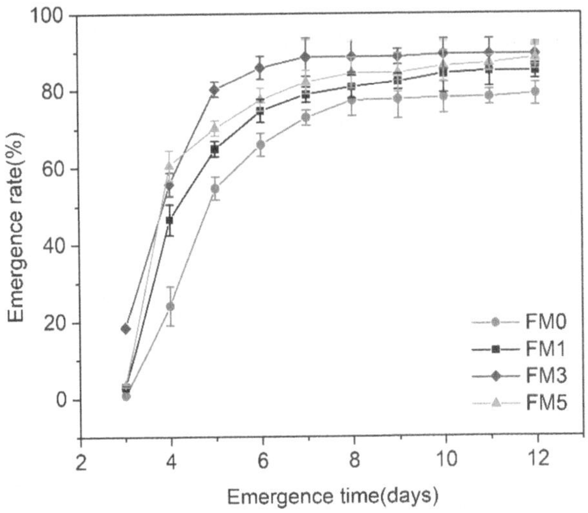

Figure 2 | The effect of different strengths of magnetized fresh water irrigation on the emergence rate of cotton.

rates of cotton irrigated with FM1, FM3 and FM5 were 80.8%, 88.6%, and 84.4% respectively, with full seedling formations of 8 d, 7 d, and 8 d, in that order. Compared with FM0, the emergence rate of cotton irrigated with MFW increased by 7.7–13.1%, and the full seedling formation of cotton irrigated with FM3 was advanced by 1 d. Similarly, Moussa (2011) applied magnetized water irrigation to improve the emergence rate of snow peas and chickpeas. In general, the most remarkable seeding emergence was by FM3, followed by FM5 and FM1.

The seedling strength was analyzed by seedling length, young root length, seedling fresh weight, seedling dry weight, and seedling water content. The cotton seeds were removed 14 days after sowing and seedling activity indexes were measured and analyzed (Table 3). Compared to FM0, the activity of cotton seedlings under MFW irrigation was significantly increased ($P < 0.05$), and the seedling length, root length, fresh weight, dry weight, and water content increased by 4.0–8.0%, 20.3–35.5%, 40.1–73.1%, 18.7–34.4%, and 2.7–4.3%, respectively. Among them, the seedling root length and seedling fresh weight increased condiderably while the water content of the seedling showed moderate intensification, indicating that MFW mainly improves seedling vitality and ensures the seedling health by promoting root growth and boosting fresh weight. The effects of MFW on the activity of cotton seedlings were significantly different ($P < 0.05$). By and large FM5 had a greater promoting effect on root length, fresh weight, and dry weight, while FM3 had the greatest promotion effect on length and water content.

Effect of MFW on cotton morphological development

The growth indexes of cotton enlarged with time, and a small difference between growth indexes of cotton irrigated with magnetized and unmagnetized fresh water was observed, 10 d after final singling. The longer the time of growth, growth indexes for both cotton irrigated with magnetized and unmagnetized fresh water gradually increased (Figure 3). In the same growth period, plant height, stem diameter, leaf number, and single leaf area of cotton irrigated with MFW were larger than those of FM0. After 10–40 d of final singling an average growth rate (slope of plant height growth curve) of 0.46 cm d^{-1} was recorded for FM0, with a peak growth rate on plant height between 10–20 d after final singling. At the same time the average growth rates of cotton plant height under FM1, FM3, and FM5 were 0.56, 0.70, and 0.63 cm d^{-1}, respectively. All peaks of plant height growth rate occurred between 30–40 d after final singling. 40 d after seedling, plant height of cotton irrigated with FM1, FM3, and FM5 increased by 21.5%, 51.1%, and 33.0%, respectively, compared to FM0 (Figure 3(a). After 10–40 d of final singling, an average growth rate (slope of stem diameter growth curve) of 0.11 mm d^{-1} was recorded for FM0, with a peak growth rate on stem diameter between 20–30 d after final singling. Meanwhile the average growth rates of cotton stem diameter under FM1, FM3 and FM5 were 0.15, 0.17, and 0.14 cm d^{-1}, respectively. All the peaks of stem diameter growth rate occurred between 30–40 d after final singling. 40 d after final singling, stem diameter of cotton irrigated with FM1, FM3, and FM5 increased by 27.8%, 47.3%, and 28.0%, respectively, compared to FM0 (Figure 3(b)). In addition, 40 d after final singling, the number of cotton leaves irrigated with FM1, FM3, and FM5 increased by 20.0%, 46.7%, and 26.7% (Figure 3(c)), respectively, while the single leaf area increased by 20.1%, 47.0%, and 33.8% (Figure 3(d)), respectively, compared to FM0. Generally speaking, FM3 had the greatest promoting effect on cotton growth, followed by FM5 and FM1.

Table 3 | Characteristics of seedling activity of cotton under magnetized and unmagnetized fresh water

Treatment	Seedling height (cm)	Seedling root length (cm)	Fresh weight (g)	Dry weight (g)	Seedling water content (%)
FM0	7.5 ± 0.1c	4.6 ± 0.2c	5.89 ± 0.11d	0.87 ± 0.02d	85.17 ± 0.04d
FM1	7.8 ± 0.1b	5.5 ± 0.3b	8.25 ± 0.15c	1.04 ± 0.02c	87.43 ± 0.04c
FM3	8.1 ± 0.1a	5.9 ± 0.2ab	9.57 ± 0.28b	1.07 ± 0.03b	88.81 ± 0.06a
FM5	7.9 ± 0.2b	6.2 ± 0.4a	10.19 ± 0.12a	1.17 ± 0.01a	88.49 ± 0.09b

Different letters within a column indicate significant differences among all treatments at $P < 0.05$.

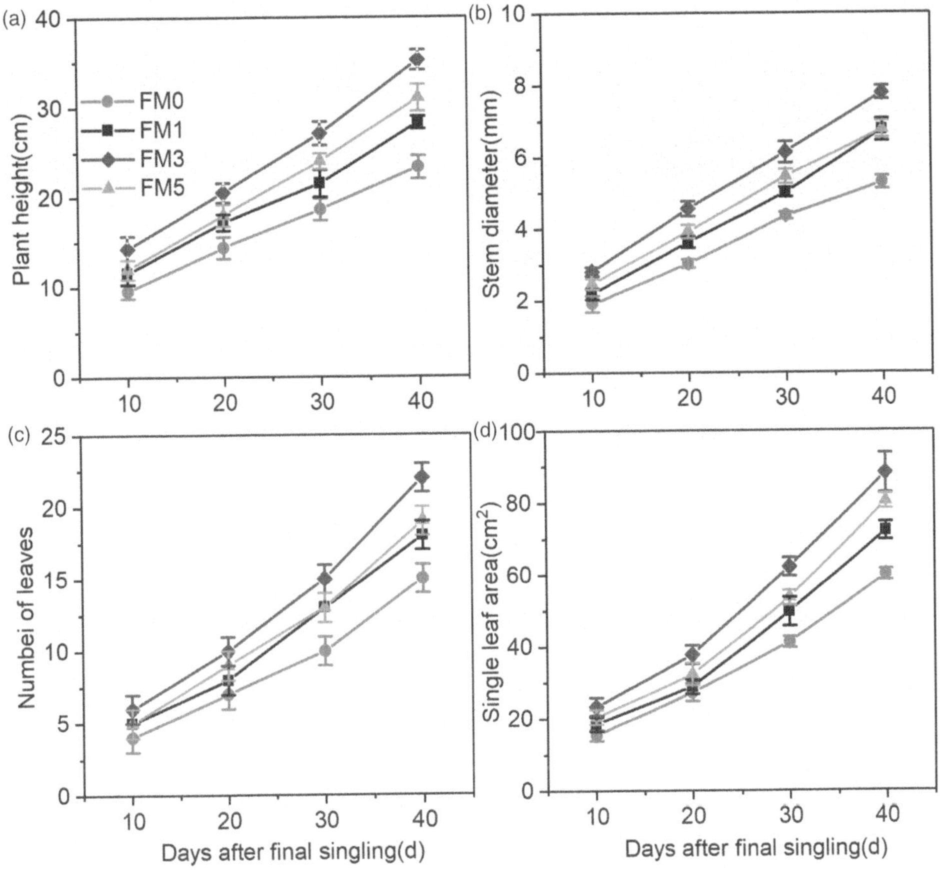

Figure 3 | Effect of magnetized and unmagnetized fresh water irrigation on cotton morphological indexes. FM0 represents unmagnetized fresh water, while FM1, FM3 and FM5 represent fresh water treated with magnetic field intensity of 100, 300 and 500 mT, respectively.

Effect of magnetized fresh water irrigation on photosynthetic parameters of cotton

The P_n of cotton irrigated with FM3 and FM5 increased by 33.9% and 32.8%, respectively, nevertheless there was no significant difference between FM1 and FM0 (Figure 4(a)). The T_r of cotton irrigated with FM1 was slightly smaller than FM0, and the T_r of cotton irrigated with FM3 and FM5 was slightly increased, but not significant (Figure 4(b)). The G_s of cotton irrigated with FM3 and FM5 increased by 69.6% and 40.2%, respectively; however, there was no significant difference between FM1 and FM0 (Figure 4(c)). The C_i of cotton irrigated with FM3 and FM5 increased by 51.3% and 37.9%, respectively, yet there was no significant difference between FM1 and FM0 (Figure 4(d)). Compared with FM0, the *iWUE* of cotton irrigated with FM3 and FM5 increased by 24.4% and 22.3% (Figure 4(e)), while L_s decreased by 21.5% and 25.9% (Figure 4(f)). In general, the most remarkable photosynthetic characteristic parameters were by FM3, followed by FM5 and FM1. The results indicated that MFW could effectively improve stomatal conductance of cotton, reduce stomatal limitation and increase the supply of CO_2, thus improving the photosynthetic carbon assimilation capacity and enhancing the utilization efficiency of light energy and water (Qiu *et al.* 2011). This is consistent with the results of Hasan *et al.* (2017). As reported, magnetization can improve the activity of related enzymes in plants and increase the metabolism rate of crops (ul Haq *et al.* 2016). Free water increased in magnetized crop cells, the photochemical activity of chlorophyll increased, the rate of photophosphorylation accelerated, and the net photosynthetic rate increased (Huuskonen *et al.* 1998).

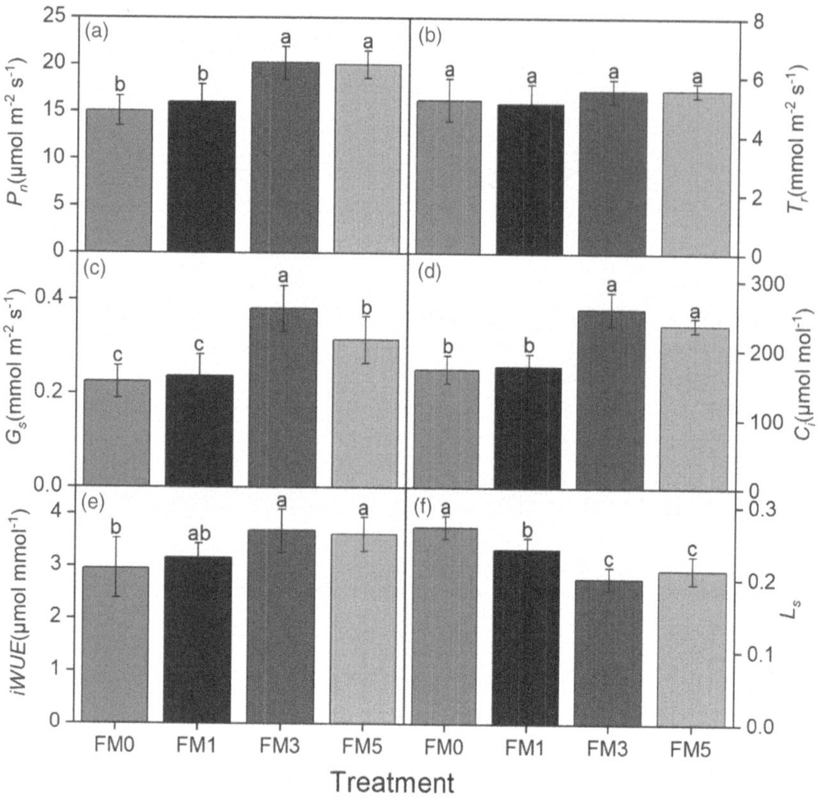

Figure 4 | Characteristic photosynthetic parameters of cotton irrigated with magnetized and unmagnetized fresh water. FM0 represents unmagnetized fresh water, while FM1, FM3 and FM5 represent fresh water treated with magnetic field intensity of 100, 300, and 500 mT, respectively. P_n, G_s, C_i, T_r, *iWUE* and L_s represent the net photosynthetic rate, stomatal conductance, intercellular CO_2 concentration, transpiration rate, instantaneous water use efficiency and stomatal limit, respectively. Different letters within a column indicate significant differences among all treatments at $P < 0.05$.

Effects of magnetized fresh water irrigation on cotton biomass and its distribution

The cotton biomass showed a significant difference between magnetized and unmagnetized fresh water, but not a significant difference among different magnetization intensities (Table 4). Compared with unmagnetized fresh water, leaves, root dry weight and total dry matter of cotton irrigated with MFW increased by 10.7–12.5%, 22.9–33.3%, and 10.3–15.1%, respectively, which conformed to the results of Yusuf & Ogunlela (2015). The stem weight of cotton irrigated with FM5 increased by 18.8% compared with FM0; however, FM1 and FM3 had no significant difference from FM0. In addition, the stem weight ratio of cotton irrigated with FM5 increased by 3.1% compared to FM0, yet FM1 and FM3 decreased slightly. The leaf weight ratio of

Table 4 | Biomass and its distribution of cotton irrigated with magnetized and unmagnetized fresh water

Treatment	Stem (g)	Leaf (g)	Root (g)	Total dry weight (g)	Stem weight ratio (%)	Leaf weight ratio (%)	Root shoot ratio (%)
FM0	5.76 ± 0.12b	11.49 ± 0.20b	1.57 ± 0.12b	18.82 ± 0.42b	30.61 ± 0.21ab	61.06 ± 0.33a	9.10 ± 0.57a
FM1	5.90 ± 0.18b	12.93 ± 1.12a	1.94 ± 0.13a	20.76 ± 1.25a	28.44 ± 0.91bc	62.21 ± 1.76a	10.33 ± 1.10a
FM3	6.11 ± 0.12b	12.83 ± 0.15a	1.93 ± 0.09a	20.87 ± 0.18a	29.28 ± 0.33b	61.47 ± 0.18a	10.2 ± 0.62a
FM5	6.84 ± 0.58a	12.72 ± 0.48a	2.09 ± 0.12a	21.66 ± 0.98a	31.55 ± 1.4a	58.76 ± 0.54b	10.73 ± 1.06a

Different letters within a column indicate significant differences among all treatments at $P < 0.05$.

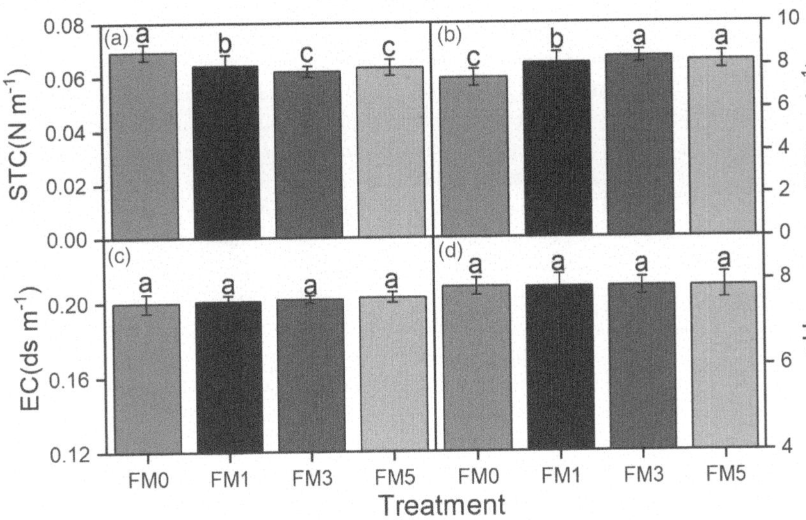

Figure 5 | Changes of physicochemical properties magnetized fresh water. FM0 represents unmagnetized fresh water, while FM1, FM3 and FM5 represent fresh water treated with magnetic field intensity of 100, 300, and 500 mT, respectively. STC, DO, and EC represent the surface tension coefficient, dissolved oxygen content, and electrical conductivity, respectively. Different letters within a column indicate significant differences among all treatments at $P < 0.05$.

cotton irrigated with FM5 reduce by 3.8% compared to FM0. For root shoot ratio, there was no significant difference between magnetized and unmagnetized irrigated cotton. Results indicated that magnetization could promote the overall development of cotton plants by promoting the growth and development of seedling roots.

Changes of physicochemical properties of magnetized fresh water

The surface tension is a macroscopic expression of liquid micro molecular structure, and its change can reflect the change of internal molecular structure (Wang *et al.* 2013; Amor *et al.* 2017), which can be described by the surface tension coefficient (STC). The STC of FM0 was 0.069 N m^{-1}, while the STCs of MFW were 0.062–0.064 N m^{-1}, which reduced by 7.3–10.5% compared to FM0 (Figure 5(a)). The reduced surface tension of MFW was similar to the results of Pang & Deng (2008), which seemed that the internal structure changed (Toledo *et al.* 2008), the average distance between water molecules increased (Wang *et al.* 2013), the hydrogen bonds weakened (Cai *et al.* 2009), the number of free monomer water molecules and dimer water molecules increased (Kney & Parsons 2006), and the chemical bond angle and the radius of water ion gel decreased (Moosavi & Gholizadeh 2014). The dissolved oxygen content (DO) of FM0 was 7.46 mg L^{-1}, while the DOs of MFW were 8.12–8.41 mg L^{-1}, which increased by 8.8–12.7% compared to FM0 (Figure 5(b)), which also verified the results of Lee *et al.* (2019). The electrical conductivity (EC) of FM0 was 0.200 ds m^{-1}, and the ECs of MFW increased slightly but not significantly (Figure 5(c)). The pH of FM0 was 7.86, while the pH of MFW decreased slightly but not significantly (Figure 5(d)). In all, MFW had a great influence on the STC and DO, and FM3 had the best magnetization effect on STC and DO, followed by FM5 and FM1.

CONCLUSIONS

The surface tension coefficient of magnetized fresh water (MFW) decreased by 7.3–10.5%, the dissolved oxygen content increased by 8.8–12.7%, and the water molecular activity was significantly improved. The emergence rate of cotton irrigated with MFW was 7.7–13.1%, and the seedling vigor indexes were significantly increased ($P < 0.05$). MFW could improve the emergence and seedling vigor of cotton by increasing *GP* and *VI* of seeds. The P_n and *iWUE* of cotton irrigated with MFW were increased by 32.8–33.9% and 22.3–24.4%, respectively, and MFW promoted morphological development and biomass by improved cotton photosynthesis and water use efficiency. The water molecular activity of MFW with 300–500 mT magnetic field intensity was best, which was beneficial to the growth of cotton seedlings. The results can provide guidance for the efficient utilization of MFW in arid and semi-arid areas. However, this study only focused on the magnetization of fresh water quality and its effect on the growth of early cotton seedlings, while the effects of magnetized water with different water quality on the physiological growth characteristics, yield, and quality of cotton at the later growth stage need to be further studied.

ACKNOWLEDGEMENTS

This study was supported by the National Natural Science Foundation of China (41830754, 51239009 and 41907010), Basic research plan of Natural Science of Shaanxi Province (2020JQ-616).

DECLARATION OF COMPETING INTEREST

The authors declared that they have no conflicts of interest to this work.

DATA AVAILABILITY STATEMENT

All relevant data are included in the paper or its Supplementary Information.

REFERENCES

Aghamir, F., Bahrami, H. A., Malakouti, M. J. & Eshghi, S. 2015 Magnetized water effects on seed germination and seedling growth of corn (*Zea mays*) under saline conditions. *American Journal of Life Science Researches* **3** (2), 184–195.

Alfaidi, M. A., Al-Toukhy, A. A., Al-Zahrani, H. S. & Howladar, M. M. 2017 Effect of irrigation by magnetized sea water on Guinea grass (*panicum maximum*) leaf content of chlorophyll a, b, carotenoids, pigments, protein & proline. *Advances in Environmental Biology* **11** (1), 73–84.

Ambashta, R. D. & Sillanpää, M. 2010 Water purification using magnetic assistance: a review. *Journal of Hazardous Materials* **180** (1–3), 38–49.

Amor, H. B., Elaoud, A., Salah, N. B. & Elmoueddeb, K. 2017 Effect of magnetic treatment on surface tension and water evaporation. *International Journal of Advance Industrial Engineering* **5**, 119–124.

Cai, R., Yang, H., He, J. & Zhu, W. 2009 The effects of magnetic fields on water molecular hydrogen bonds. *Journal of Molecular Structure* **938** (1–3), 15–19.

Chen, W., Jin, M., Ferré, T. P., Liu, Y., Xian, Y., Shan, T. & Ping, X. 2018 Spatial distribution of soil moisture, soil salinity, and root density beneath a cotton field under mulched drip irrigation with brackish and fresh water. *Field Crops Research* **215**, 207–221.

Chibowski, E. & Szcześ, A. 2018 Magnetic water treatment – a review of the latest approaches. *Chemosphere* **203**, 54–67.

Coey, J. M. D. & Cass, S. 2000 Magnetic water treatment. *Journal of Magnetism and Magnetic Materials* **209** (1–3), 71–74.

El Sayed, H. E. S. A. 2014 Impact of magnetic water irrigation for improve the growth, chemical composition and yield production of broad bean (*Vicia faba* L.) plant. *Journal of Experimental Agriculture International* **4** (4), 476–496.

Esmaeilnezhad, E., Choi, H. J., Schaffie, M., Gholizadeh, M. & Ranjbar, M. 2017 Characteristics and applications of magnetized water as a green technology. *Journal of Cleaner Production* **161**, 908–921.

Fu, Q., Liu, C., Ding, N., Lin, Y. & Guo, B. 2010 Ameliorative effects of inoculation with the plant growth-promoting rhizobacterium *Pseudomonas* sp. DW1 on growth of eggplant (*Solanum melongena* L.) seedlings under salt stress. *Agricultural Water Management* **97** (12), 1994–2000.

Ghorbani, S., Sharifi, S., De Brito, J., Ghorbani, S., Jalayer, M. A. & Tavakkolizadeh, M. 2019 Using statistical analysis and laboratory testing to evaluate the effect of magnetized water on the stability of foaming agents and foam concrete. *Construction and Building Materials* **207**, 28–40.

Hachicha, M., Kahlaoui, B., Khamassi, N., Misle, E. & Jouzdan, O. 2018 Effect of electromagnetic treatment of saline water on soil and crops. *Journal of the Saudi Society of Agricultural Sciences* **17** (2), 154–162.

Hasan, M. M., Alharby, H. F., Hajar, A. S. & Hakeem, K. R. 2017 Leaf gas exchange, Fv/Fm ratio, ion content and growth conditions of the two *Moringa* species under magnetic water treatment. *The Pakistan Journal of Botany* **49** (3), 921.

Hassan, S. M., Sulaiman, M. A., Madlul, N. S., Fadel, A. H. & Abdul Rahman, R. 2019 Influence of continuous magnetic field exposure on water properties and subsequent effects on the growth performance, plasma biochemistry, nutritive value and liver histopathology of Jade Perch Scortum barcoo in a recirculating system. *Aquaculture Research* **50** (7), 1931–1941.

Huuskonen, H., Lindbohm, M. L. & Juutilainen, J. 1998 Teratogenic and reproductive effects of low-frequency magnetic fields. *Mutation Research/Reviews in Mutation Research* **410** (2), 167–183.

Ji, Y., Wang, Y., Sun, J., Yan, T., Li, J., Zhao, T., Yin, X. & Sun, C. 2010 Enhancement of biological treatment of wastewater by magnetic field. *Bioresource Technology* **101** (22), 8535–8540.

Kney, A. D. & Parsons, S. A. 2006 A spectrophotometer-based study of magnetic water treatment: assessment of ionic vs. surface mechanisms. *Water Research* **40** (3), 517–524.

Kumar, B., Verma, S. K. & Singh, H. P. 2011 Effect of temperature on seed germination parameters in Kalmegh (*Andrographis paniculata* Wall. ex Nees. *Industrial Crops and Products* **34** (1), 1241–1244.

Lee, H. S., Yap, A. C. W., Ng, C. C., Mohd, N. S. & Loo, J. L. 2019 Increased electron density and dissolved oxygen level in water through magnetic effect. *IOP Conference Series: Earth and Environmental Science* **257** (1), 012010.

Li, M., Du, Y., Zhang, F., Bai, Y., Fan, J., Zhang, J. & Chen, S. 2019 Simulation of cotton growth and soil water content under film-mulched drip irrigation using modified CSM-CROPGRO-cotton model. *Agricultural Water Management* **218**, 124–138.

Liang, J., Shi, W., He, Z., Pang, L. & Zhang, Y. 2019 Effects of poly-γ-glutamic acid on water use efficiency, cotton yield, and fiber quality in the sandy soil of southern Xinjiang, China. *Agricultural Water Management* **218**, 48–59.

Lipus, L. C., Ačko, B. & Neral, B. 2013 Influence of magnetic water treatment on fabrics' characteristics. *Journal of Cleaner Production* **52**, 374–379.

Liu, X., Ma, F., Zhu, H., Ma, X., Guo, J., Wan, X., Wang, L., Wang, H. & Wang, Y. 2017 Effects of magnetized water treatment on growth characteristics and ion absorption, transportation, and distribution in Populus × euramericana 'Neva' under NaCl stress. *Canadian Journal of Forest Research* **47** (6), 828–838.

Liu, X., Wang, L., Cui, H., Zhu, H., Bi, S., Zhang, Z., Meng, S., Song, C., Wang, H. & Ma, F. 2019 Effects of magnetic treatment on the ascorbate–glutathione cycle and endogenous hormone levels in Populus × euramericana 'Neva' under cadmium stress. *Canadian Journal of Forest Research* **49** (9), 1147–1158.

Liu, X., Wang, L., Wei, Y., Zhang, Z., Zhu, H., Kong, L., Meng, S., Song, C., Wang, H. & Ma, F. 2020 Irrigation with magnetically treated saline water influences the growth and photosynthetic capability of *Vitis vinifera* L. seedlings. *Scientia Horticulturae* **262**, 109056.

Luo, J., He, W., Qi, S., Wu, J. & Gu, X. S. 2020 A novel phytoremediation method assisted by magnetized water to decontaminate soil Cd based on harvesting senescent and dead leaves of *Festuca arundinacea*. *Journal of Hazardous Materials* **383**, 121115.

Mahmoud, M. H., Salim, M. A., El-Monem, A., Amany, A. & El-Mahdy, A. A. 2019 Effect of magnetic brackish-water treatments on morphology, anatomy and yield productivity of wheat (*Triticum Aestivum*). Alexandria. *Science Exchange Journal* **40**, 604–617.

Massah, J., Dousti, A., Khazaei, J. & Vaezzadeh, M. 2019 Effects of water magnetic treatment on seed germination and seedling growth of wheat. *Journal of Plant Nutrition* **42** (11–12), 1283–1289.

Mo, Y., Li, G. & Wang, D. 2017 A sowing method for subsurface drip irrigation that increases the emergence rate, yield, and water use efficiency in spring corn. *Agricultural Water Management* **179**, 288–295.

Moosavi, F. & Gholizadeh, M. 2014 Magnetic effects on the solvent properties investigated by molecular dynamics simulation. *Journal of Magnetism and Magnetic Materials* **354**, 239–247.

Mosse, K. P., Verheyen, T. V., Cruickshank, A. J., Patti, A. F. & Cavagnaro, T. R. 2013 Soluble organic components of winery wastewater and implications for reuse. *Agricultural Water Management* **120**, 5–10.

Moussa, H. R. 2011 The impact of magnetic water application for improving common bean (*Phaseolus vulgaris* L.) production. *New York Science Journal* **4** (6), 15–20.

Pang, X. & Deng, B. 2008 Investigation of changes in properties of water under the action of a magnetic field. *Science in China Series G: Physics, Mechanics and Astronomy* **51** (11), 1621–1632.

Pour, A. N., Gholizadeh, M., Housaindokht, M., Moosavi, F. & Monhemi, H. 2017 A new method for preparing mono-dispersed nanoparticles using magnetized water. *Applied Physics A* **123** (4), 269.

Qiu, N., Tan, Y., Shen, X., Han, R., Lin, Y. & Ma, Z. 2011 Biological effects of magnetized water on seed germination, seedling growth and physiological characteristics of wheat. *Plant Physiology Communications* **47** (8), 803–810.

Selim, A. F. H. & Selim, D. A. 2019 Physio-biochemical behaviour, water use efficiency and productivity of wheat plants exposed to magnetic field. *Journal of Plant Production* **10** (2), 185–191.

Sohaili, J., Shi, H. S., Zardari, N. H., Ahmad, N. & Muniyandi, S. K. 2016 Removal of scale deposition on pipe walls by using magnetic field treatment and the effects of magnetic strength. *Journal of Cleaner Production* **139**, 1393–1399.

Sultan, W., Hussien, A. & Hussein, N. 2019 Using magnetic water technique in fertigation process. *Journal of Soil Sciences and Agricultural Engineering* **10** (2), 95–98.

Surendran, U., Sandeep, O. & Joseph, E. J. 2016 The impacts of magnetic treatment of irrigation water on plant, water and soil characteristics. *Agricultural Water Management* **178**, 21–29.

Tan, S., Wang, Q., Xu, D., Zhang, J. & Shan, Y. 2017 Evaluating effects of four controlling methods in bare strips on soil temperature, water, and salt accumulation under film-mulched drip irrigation. *Field Crops Research* **214**, 350–358.

Toledo, E. J., Ramalho, T. C. & Magriotis, Z. M. 2008 Influence of magnetic field on physical–chemical properties of the liquid water: insights from experimental and theoretical models. *Journal of Molecular Structure* **888** (1–3), 409–415.

ul Haq, Z., Iqbal, M., Jamil, Y., Anwar, H., Younis, A., Arif, M., Fareed, M. Z. & Hussain, F. 2016 Magnetically treated water irrigation effect on turnip seed germination, seedling growth and enzymatic activities. *Information Processing in Agriculture* **3** (2), 99–106.

Wang, Y., Zhang, B., Gong, Z., Gao, K., Ou, Y. & Zhang, J. 2013 The effect of a static magnetic field on the hydrogen bonding in water using frictional experiments. *Journal of Molecular Structure* **1052**, 102–104.

Yang, P., Hu, H., Tian, F., Zhang, Z. & Dai, C. 2016 Crop coefficient for cotton under plastic mulch and drip irrigation based on eddy covariance observation in an arid area of northwestern China. *Agricultural Water Management* **171**, 21–30.

Yao, Z., Jiao, W., Shao, F., Song, H., Zhang, H., Zhou, Q. & Li, A. 2019 Fabrication and characterization of amphiphilic magnetic water purification materials for efficient PPCPs removal. *Chemical Engineering Journal* **360**, 511–518.

Yusuf, K. O. & Ogunlela, A. O. 2015 Impact of magnetic treatment of irrigation water on the growth and yield of tomato. *Notulae Scientia Biologicae* **7** (3), 345–348.

First received 19 January 2021; accepted in revised form 8 February 2021. Available online 17 February 2021

Investigation of flow measurement mechanism and hydraulic characteristics of the NACA airfoil pillar-shaped flume with different wing lengths in a rectangular channel

Bin Sun, Lei Yang, Shun Zhu, Haijia Zhang, Chao Zhang and Jinping Zhang

ABSTRACT

In this paper, the application of NACA airfoil pillar-shaped flumes and the influence of their wing lengths on flow measurement characteristics are discussed. The standard k-ϵ three-dimensional turbulence model and the volume of fluid (VOF) method were used for numerical simulation of the flow in six NACA airfoil pillar-shaped flumes with different wing lengths. In addition, to verify the accuracy of numerical simulations, the corresponding experiments were conducted. The Buckingham Pi theorem of dimensional analysis coupled with the incomplete self-similarity theory was applied to deduce the theoretical flow calculation formula of these flumes. Moreover, Froude number, velocity distribution, backwater height, critical submergence, and energy loss of the flume were analyzed, for which the experimental and numerical results are compared and further discussed. The results show that the backwater height is directly proportional to the wing length, while the head loss and critical submergence are inversely proportional to the wing length. Based on the results, in terms of backwater height, it is recommended to use the NACA airfoil pillar-shaped flume with a smaller wing length, while, in terms of head loss and critical submergence, the NACA airfoil pillar-shaped flume with a larger wing length should be used.

Key words | critical submergence, energy loss, flow measurement, NACA airfoil pillar-shaped flume

Bin Sun
Lei Yang (corresponding author)
Shun Zhu
Haijia Zhang
Chao Zhang
Jinping Zhang
School of Water Conservancy Engineering,
Zhengzhou University,
Zhengzhou 450001,
China
E-mail: *1206938199@qq.com*

HIGHLIGHTS

- The application of NACA airfoil pillar-shaped flumes is discussed.
- The Buckingham Pi theorem of dimensional analysis coupled with the incomplete self-similarity theory is applied to deduce the theoretical flow calculation formula.
- Laboratory experiments and numerical simulations are combined to provide a theoretical reference for applications of this flume.

doi: 10.2166/ws.2021.098

INTRODUCTION

Flow measurement in open channels is the key means for optimal allocation of water resources, flood discharge forecasting, and water-saving irrigation (Dabrowski & Polak 2012; Carollo *et al.* 2016; Jesson *et al.* 2017). The measuring flume is a special facility for measuring the flow through various channels. Flumes that are designed according to critical flow theory are widely used in agricultural irrigation and sewage monitoring because of their advantages such as simple structure, large measuring range, and small head loss (Xiao *et al.* 2016). These flumes can be exemplified by long throat flumes, short throat flumes, such as the Parshall flume (Li *et al.* 2004; Thornton *et al.* 2009; Cox *et al.* 2013), no throat flumes, such as the U-shaped channel parabolic flume (Hu *et al.* 2014), and cut-throat flumes (Manekar *et al.* 2007; Torres & Merkley 2008; Das *et al.* 2017). In terms of effective factors (cost, practicality, stability, and reliability), long throat flumes are more suitable for medium-sized channel flow measurement, while the hydraulic structures, such as sluice and culvert, are more suitable for large channel flow measurement, and cylindrical flumes are more suitable for measuring a massive amount of flow in small channels. In addition, cylindrical measuring flumes have the advantages of simple structure, convenient production, and installation. In addition, these flumes can be made into a fixed type or a mobile type so that they are very suitable for small channels (Ferro 2018).

Hager (1985) proposed a portable flume for measuring flow in rectangular, trapezoidal, and U-shaped channels. In traditional venturi channels, both sides of the channel are commonly contracted to achieve a narrow cross-section, while Hager suggested placing a vertical cylinder on the channel axis to form the contraction. Samani (2017) conducted laboratory experiments and evaluated the hydraulic characteristics of several simple mobile circular flumes. Later, following the experimental results of Hager and Samani *et al.*, Ferro (2002) applied dimensional analysis and self-similarity theory to determine the stage-discharge relationship for a flume with a horizontal bottom and without submergence effects.

It can be concluded from the literature that a cylindrical measuring flume, as a portable venturi flume, is widely used for flow measurement. However, in order to measure the flow in a small channel, the water measuring device not only needs to have portability and high measurement accuracy, but also needs smaller backwater height and larger critical submergence degree. Note that streamlined flow measuring flumes can meet the above requirements at the same time. In addition, most of the above-mentioned types of measuring flumes cannot be applied to established channels. Therefore, it is of practical significance to study streamlined flow measuring flumes.

The National Advisory Committee for Aeronautics (NACA) conducted a systematic study on airfoils in the 1930s and provided a general analytical expression for airfoil thickness distribution. Based on this expression, Liu *et al.* (2014) proposed the airfoil pillar-shaped flume and carried out model experiments on this flume. The experiments proved that this airfoil has more efficient hydraulic characteristics and can be used for flow measurement in rectangular channels. In addition, it is known that computational fluid dynamics has the advantages of low cost, short time, and easy access to flow field data. Thus, it has become one of the most indispensable methods for hydraulics analyses such as study of the hydraulic characteristics of open channels (Nazari-Sharabian *et al.* 2020a, 2020b), measuring flumes (Ramamurthy & Tadayon 2008; Qi *et al.* 2014; Li *et al.* 2020), and weirs (Aydin 2012; Ziaei *et al.* 2019; Saadatnejadgharahassanlou *et al.* 2020). Therefore, in the present work, NACA airfoil pillar-shaped flumes with different wing lengths are further studied by combining laboratory experiments and numerical simulations to provide a theoretical reference for applications of this flume in rectangular channels.

MATERIALS AND METHODS

Structure of NACA airfoil pillar-shaped flume

The NACA airfoil pillar-shaped flume can be considered as a special venturi measuring flume because it can be installed or built as an airfoil-shaped cylinder in the center of the channel to make the channel section contracted. It is known that when the water flow sufficiently increases in the flume section the critical flow can be generated, which is not affected by the downstream flow conditions. Therefore, by measuring the upstream water depth, the flow rate can be calculated using a single stable relationship between water level and flow rate.

The contour curve of the NACA airfoil pillar-shaped flume is illustrated in Figure 1, where the structural design of the flume is shown in Figure 1(b). Note that the contour curve of the NACA airfoil is controlled by the NACA airfoil equation as presented in Equation (1) (Wang & Wen 1990):

$$y = 10P\left[0.2969\sqrt{\frac{x}{C}} - 0.1260\frac{x}{C} - 0.3516\left(\frac{x}{C}\right)^2 + 0.2843\left(\frac{x}{C}\right)^3 - 0.1015\left(\frac{x}{C}\right)^4\right] \tag{1}$$

where P is the maximum thickness of flume (m), C represents its length (m), and x and y represent the coordinate of the equation on the X-axis and Y-axis.

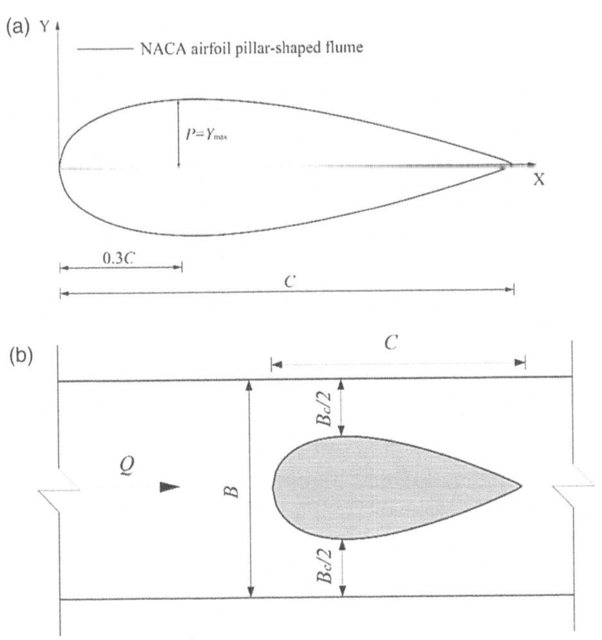

Figure 1 | The curve of NACA airfoil equation (a) and the shape of the NACA airfoil-shaped flume (b).

The ratio of the wing length (C) to the wing height (P) is defined as r:

$$r = \frac{C}{P} \tag{2}$$

In this work, six NACA airfoil-shaped flumes (Figure 2) with $r=2$, 3, 4, 5, 6, and 7 were studied, for which the specific dimensions are provided in Table 1.

Experimental setup

Laboratory experiments were conducted to investigate the effect of different wing lengths ($C=15.0$, 22.5, 30.0, 37.5, 45.0, and 52.5 cm) and flow rates (from 7 to 27 L/s) on hydraulic characteristics and measurement accuracy of the NACA airfoil-shaped flume. The working conditions of experiments are presented in Table 2. The experiments were carried out in a plexiglass channel located at the hydraulic laboratory of the Water Engineering Department in Zhengzhou University, Zhengzhou, China. The experimental setup of the NACA airfoil-shaped flume is shown in Figure 3.

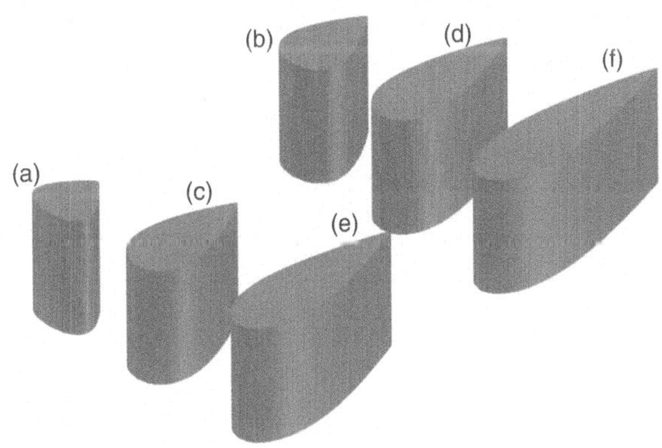

Figure 2 | The NACA airfoil-shaped flumes with different wing lengths.

Table 1 | The parameters of NACA airfoil-shaped flumes

r	B_c (cm)	C (cm)		r	B_c (cm)	C (cm)	
2	15.0	15.0	Figure 2(a)	5	15.0	37.5	Figure 2(d)
3		22.5	Figure 2(b)	6		45.0	Figure 2(e)
4		30.0	Figure 2(c)	7		52.5	Figure 2(f)

Table 2 | Experimental working conditions

No.	r	B_c (cm)	C (cm)	Q (L·s^{-1})	No.	r	B_c (cm)	C (cm)	Q (L·s^{-1})
1	2	15	15.0	7	19	5	15	37.5	7
2				11	20				11
3				15	21				15
4				19	22				19
5				23	23				23
6				27	24				27
7	3	15	22.5	7	25	6	15	45.0	7
8				11	26				11
9				15	27				15
10				19	28				19
11				23	29				23
12				27	30				27
13	4	15	30.0	7	31	7	15	52.5	7
14				11	32				11
15				15	33				15
16				19	34				19
17				23	35				23
18				27	36				27

All dimensions of NACA airfoil-shaped flumes are fitted with channel centerlines. Measuring points are set every 0.5 m in the upstream section and the canal section after the hydraulic jump. In the contraction section and the hydraulic jump section, measuring points are set every 0.1 m. Finally, five typical sections were selected for analysis: the upstream canal section, the inlet section of the flume, the throat section, the outlet section of the flume, and the downstream canal section. To ensure that the upstream water flow of the flume is stable, the inlet section of the flume is set to 3.0 m away from the inlet of the channel. A centrifugal pump is used to supply water from the downstream reservoir to the upstream reservoir and an electromagnetic flowmeter is installed between the pump and the valve for real-time monitoring of the flow through the flume. For a given discharge value, the water depth at all sections of the rectangular channel is measured.

Discharge equation

In this paper, a dimensional analysis (P. W. 1922) was used to predict the discharge equation, by which the free flow hydraulics of the NACA airfoil-shaped flume can be expressed using the following functional relationship:

$$F(Q, h, B, B_c, C, g, \mu) = 0 \tag{3}$$

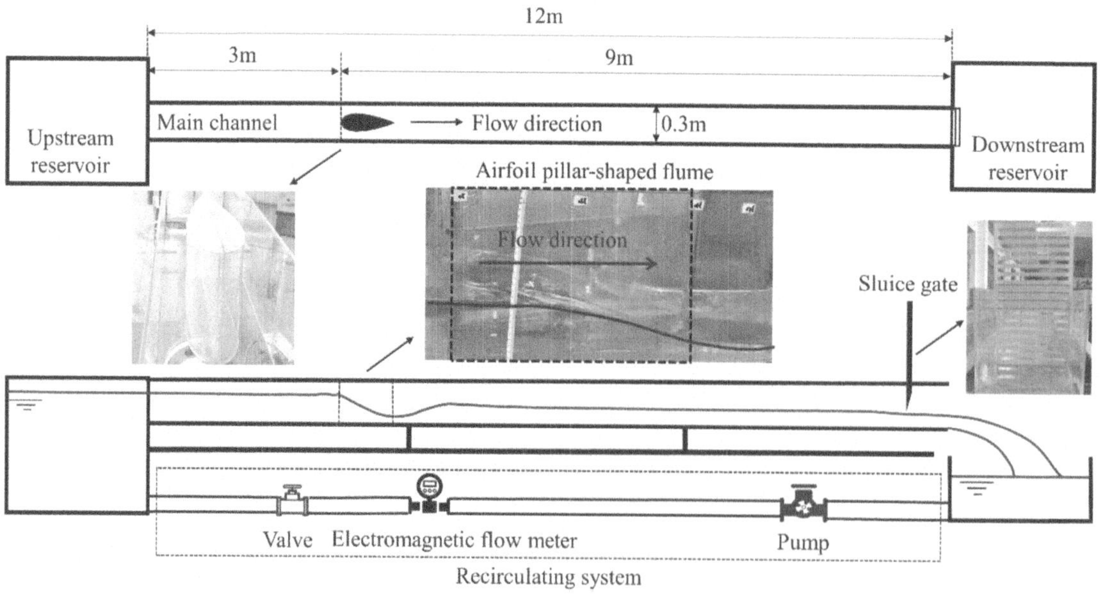

Figure 3 | The experimental setup of NACA airfoil-shaped flume.

where F is a functional symbol, Q is the discharge (m^3/s), h is the upstream flow depth (m), B is the channel width (m), B_c is the throat width (m), g is the gravitational acceleration (m/s^2), and μ is the water viscosity (Ns/m^2).

Considering B_c, μ, and g as reference variables, the following dimensionless groups could be obtained based on the Buckingham Pi theorem:

$$\Pi_1 = \frac{Q}{B_c^{5/2} g^{1/2}} = \left(\frac{K}{B_c}\right)^{3/2} \tag{4}$$

$$\Pi_2 = \frac{h}{B_c} \tag{5}$$

$$\Pi_3 = \frac{C}{B_c} = r \tag{6}$$

$$\Pi_4 = \frac{B}{B_c} \tag{7}$$

where Π_1, Π_2, Π_3 and Π_4 are dimensionless groups and $K(m)$ is the critical water depth. Equations (4)–(7) are substituted into Equation (3) as follows:

$$F\left(\frac{K}{B_c}, \frac{h}{B_c}, r, \frac{B}{B_c}\right) = 0 \tag{8}$$

Therefore, for a given contraction ratio (B/B_c), and for a given r, the stage-discharge relationship is expressed by:

$$\frac{K}{B_c} = \varphi\left(\frac{h}{B_c}\right) \tag{9}$$

where φ is a functional symbol.

As shown in Figure 4, correlation analysis of K/B_c and h/B_c of six NACA airfoil-shaped flumes with different r clearly shows that K/B_c and h/B_c have an excellent correlation.

Consider r as a variable, the stage-discharge relationship can be expressed:

$$\frac{Q}{B_c^{5/2} g^{1/2}} = \varphi\left(\frac{h}{B_c}, r\right) \tag{10}$$

Note that Equation (10) can also be obtained through self-similarity theory (Bridgman 1922). For a given dimensionless group Π_n, the functional relationship $\Pi_1 = \phi$ (Π_2, Π_3, ……, Π_n) is independent of Π_n. When Π_n tends to zero or infinity and the corresponding limits of the function ϕ are equal to zero or infinity, incomplete self-similarity would occur, which, in this work, is expressed by the following functional relationship:

$$\Pi_1 = \Pi_n^{\varepsilon} f(\Pi_2, \Pi_3, \ldots\ldots, \Pi_{n-1}) \tag{11}$$

where f is a functional symbol and ε is a numerical constant

According to incomplete self similarity, Equation (12) can be rewritten as follows:

$$\frac{Q}{B_c^{5/2} g^{1/2}} = a\left(\frac{h}{B_c}\right)^m r^n \tag{12}$$

where, a, m, n are undetermined numbers that can be obtained by experimental data.

Numerical simulation

Governing equations and turbulence model

The continuity and Navier-Stokes equations for incompressible viscous flow are given by Equations (13) and (14):

$$\text{Continuity equation} \quad \frac{\partial \rho}{\partial t} + \frac{\partial(\rho u)}{\partial x} + \frac{\partial(\rho v)}{\partial y} + \frac{\partial(\rho w)}{\partial z} = 0 \tag{13}$$

Navier-Stokes equation

$$\begin{cases} \dfrac{\partial(\rho u)}{\partial t} + \text{div} \quad (\rho u \mathbf{u}) = \dfrac{\partial p}{\partial x} + \text{div} \quad (\mu \,\mathbf{grad} u) + F_x \\ \dfrac{\partial(\rho v)}{\partial t} + \text{div} \quad (\rho u \mathbf{v}) = \dfrac{\partial p}{\partial y} + \text{div} \quad (\mu \,\mathbf{grad} v) + F_y \\ \dfrac{\partial(\rho w)}{\partial t} + \text{div} \quad (\rho u \mathbf{w}) = \dfrac{\partial p}{\partial z} + \text{div} \quad (\mu \,\mathbf{grad} w) + F_z \end{cases} \tag{14}$$

The VOF method (Hirt & Nichols 1981) is employed to describe the free-surface motion and the standard k-ε three-dimensional turbulence model (Saadatnejadgharahassanlou *et al.* 2020) is used to calculate the turbulence variables and to analyze the internal flow field because this turbulence model can calculate sufficiently accurate

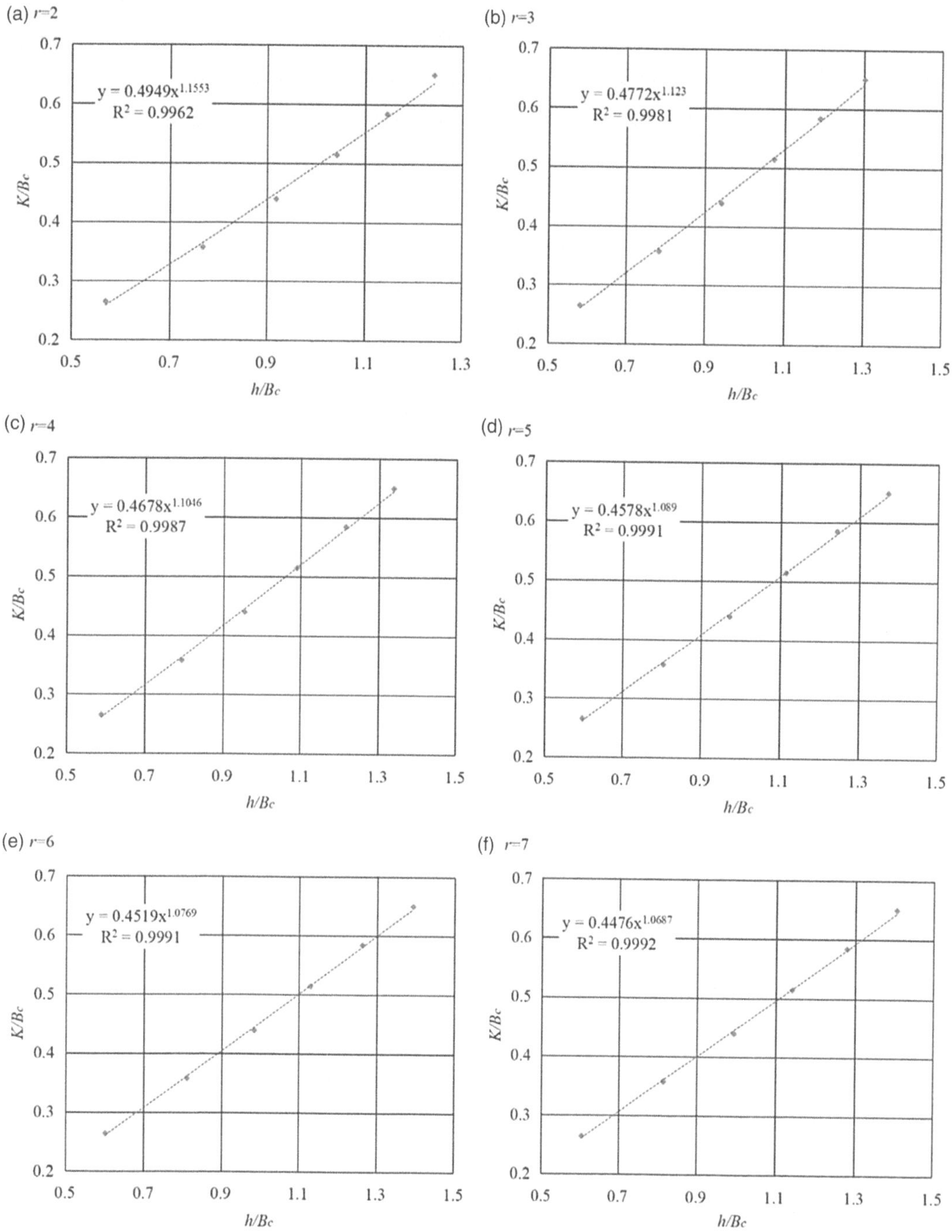

Figure 4 | The functional relationship of K/B_c and h/B_c with different r.

results for open channel flow problems. This model calculates the turbulent kinetic energy and turbulent dissipation rate by solving partial differential equations, which are provided in Equations (15) and (16), respectively.

Turbulent kinetic energy equation:

$$\rho \frac{dk}{dt} = \frac{\partial}{\partial x_i}\left[\left(\mu + \frac{\mu_t}{\sigma_k}\right)\frac{\partial_k}{\partial x_i}\right] + G_k + G_b - \rho\alpha \tag{15}$$

Turbulent dissipation rate equation:

$$\rho\frac{d\varepsilon}{dt} = \left[\left(\mu + \frac{\mu_t}{\sigma_\varepsilon}\right)\frac{\partial\varepsilon}{\partial x_i}\right] + C_{1\varepsilon}\frac{\varepsilon}{k}(G_k + C_{3\varepsilon}G_b) - C_{2\varepsilon}\rho\frac{\varepsilon^2}{k} \quad (16)$$

where k is the turbulent kinetic energy (m^2/s^2), α is the dissipation rate of turbulent kinetic energy ($kg{\cdot}m^2/s^3$), G_k is the turbulent energy caused by the average velocity gradient, G_b is the turbulent energy caused by buoyancy, and μ_t is the turbulent viscosity of the fluid ($N{\cdot}s/m^2$). In this paper, σ_k and σ_α were set as 1.39, $G_{1\epsilon}$ was set as 1.42, $G_{2\epsilon}$ was set as 1.68, and $G_{3\epsilon}$ was set as 0.09.

Geometric models

In order to completely and accurately simulate the real internal flow field, similar to the actual size of the NACA airfoil pillar-shaped flumes and the experimental setup, three-dimensional geometrical models were developed using the DesignModeler software. The center point of the bottom of the flume inlet section is taken as the origin of the coordinate system, the water flow direction is assumed in the positive direction of the X-axis and the gravity direction is set to the negative direction of the Z-axis. Five typical sections are selected for in-depth analyses of the velocity distribution, as shown in Figure 5(a), where the geometric model is illustrated as well.

Mesh generation and boundary conditions

The grid division and its quality directly affect the calculation accuracy and efficiency of the numerical simulation. The purpose of this study is to more accurately simulate the flow state of the channel and reduce the calculation time as much as possible. Using five different sizes of grids to simulate the same working condition, as shown in Table 3, the hexahedral grid was used to discretize the entire fluid domain and the grid size was set to $0.01 \times 0.01 \times 0.01$ m (M4). For the boundary conditions, the water inlet was set as the velocity inlet and the air inlet was set as the pressure inlet. In addition, the outlet was set as the pressure outlet, the wall was set to no-slip condition and the roughness height was set to 0.000011 m. The gravity was set to 9.81 m/s^2 and the air pressure was 101,325 Pa. The mesh generation and boundary conditions of the NACA airfoil pillar-shaped flume are shown in Figure 5.

RESULTS AND DISCUSSION

Verification of the numerical model

After constructing the measuring flume, the variation of the water surface height along the channel can be directly affected by the water surface line. By measuring the water depth at 17 sections of six NACA airfoil pillar-shaped flumes in the rectangular channel, and under different working conditions, the variation of the water surface line through the channel can be obtained. In order to verify the reliability of numerical simulation for further analyses, the numerical solutions of the water surface line are compared with the experimental data. Figure 6 shows the comparison of the measured and simulated results of the free water surface height in the channel centerline of the NACA airfoil pillar-shaped flumes. Figure 7 shows the comparison of water depth of the numerical results and the experimental data ($r = 4$). The maximum error between the experimental data and the numerical results is 7.86%, the minimum error is 0.46%, and the average error is 4.83%. These errors indicate that the results of this numerical simulation can be used for in-depth analyses of the internal flow field.

Flow pattern analysis

The flow pattern can directly reflect the movement state of the water flowing through the flume. Note that the Fluent software can clearly simulate the flow pattern diagram of the flume such as the flow diagram, Figure 8(a), the streamline diagram, Figure 8(b), and the longitudinal section diagram, Figure 9.

In the NACA airfoil pillar-shaped flume, the upstream water flow is relatively smooth and the flow direction is basically parallel. After entering the contraction section, the water surface declines slowly, the streamline is bent, caused by the column, and the upstream water flow is divided into two parts. In addition, the throat cross-section shrinks

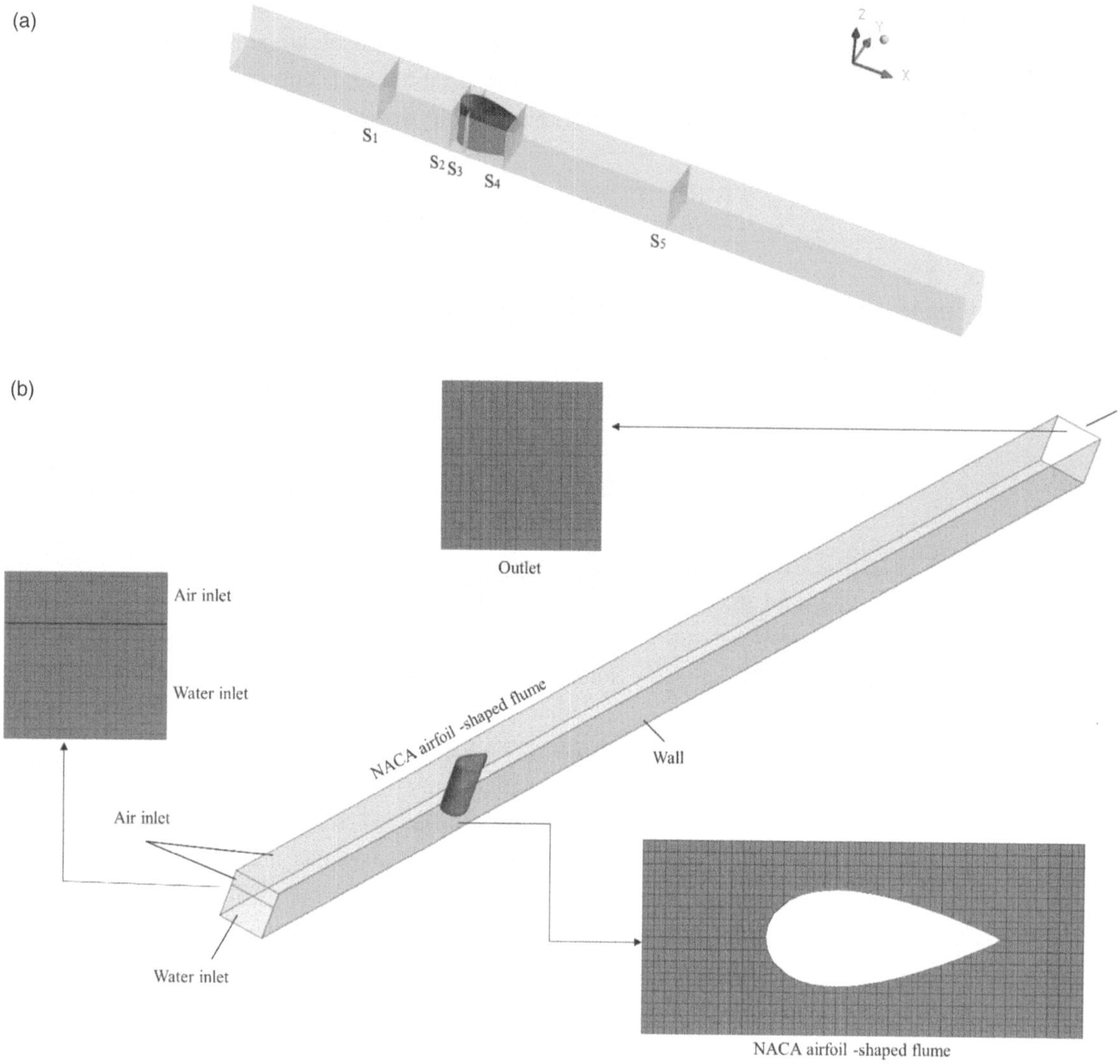

Figure 5 | Geometry and boundary conditions.

Table 3 | Grid-independence tests

Mesh	Number of elements	Water depth of S1 (m)	Error (%)
M1	25,032	15.22	
M2	66,962	14.49	4.80%
M3	164,350	14.41	0.69%
M4	198,722	14.34	0.35%
M5	1,292,270	14.33	0.07%

significantly leading to a sharp decrease in water surface level and a sharp fluctuation of the water flow. Due to inertial force, the water flow continuously diffuses to both sides, where the water depth is higher near both walls than near the center. Subsequently, the water surface after the diffusion section of the measuring flume slowly rises and smoothly connects with the downstream water surface and the horizontal distribution of the water surface gradually recovers and becomes balanced. It can be seen in Figure 9 that the water flow patterns of the six NACA airfoil pillar-shaped flumes under the condition of free flow have produced two hydraulic phenomena, hydraulic drop and hydraulic jump. Under the same working conditions, as the value of r increases, the water surface drops more slowly and the distance required for the hydraulic jump is longer.

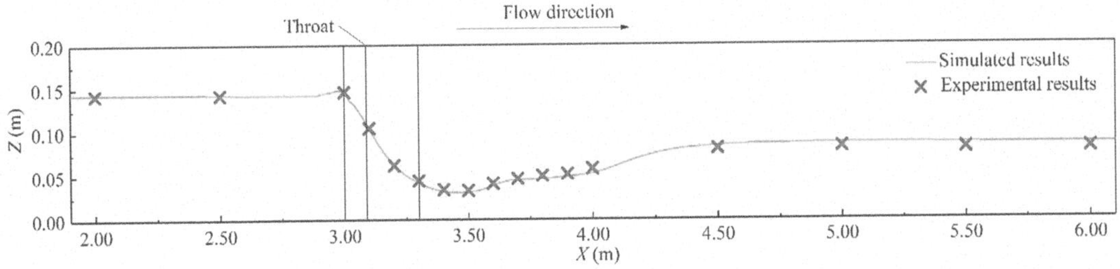

Figure 6 | Comparison of the measured and simulated value of the free water surface height in the channel centerline of NACA airfoil pillar-shaped flumes ($r = 4$, $Q = 15$ L/s).

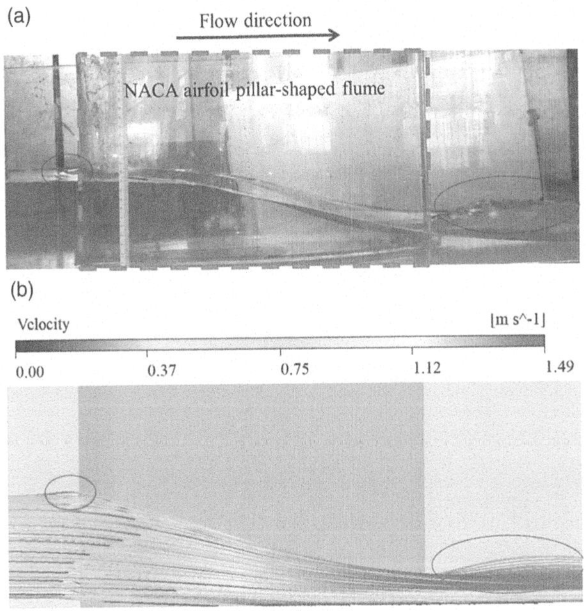

Figure 7 | Comparison of water depth between simulation results and experimental data ($r = 6$).

Flow measurement accuracy

Using the data fitting software to customize Equation (12), and using the values of Q, h, and r, the unknown parameters a, m, and n can be calculated as 0.7304, 1.7030, −0.1412, respectively, and $R^2 = 0.9981$. Consequently, the stage-discharge formula can be expressed by the following function:

$$Q = 0.1610g^{0.5}r^{-0.1412}h^{1.7030} \quad (17)$$

Equation (17) can be used for calculating the discharge when r is in the range of 2–7. In addition, Table 4 compares the measured discharge (Q_1) and the calculated discharge (Q_2) for all working conditions. The error distribution between the calculated flow and the measured flow is shown in Figure 10. As the average error is 1.72%, evidently the flow discharge calculated by Equation (17) matches the measured flow well. Thus, it can be used to accurately calculate the flow of the NACA airfoil pillar-shaped flume.

Froude number

Froude number (Fr) is a dimensionless number that can be used as a criterion to judge the open channel flow pattern (Karakouzian *et al.* 2019). Physically, it compares the inertial and gravity forces of water flows. When $Fr = 1$, it means that the magnitude of the inertial force is equal to the gravity force and the flow is called critical flow. When $Fr > 1$, it means that the inertial force is larger than the gravity force and the water flow is in the state of a jet stream. Also, $Fr < 1$ indicates that the inertial force is less than the gravity force and the water flow is in a slow flow state. Note that the flume measurement is based on the critical flow principle. In the measuring process, the water flow changes from slow flow to jet stream state and then to slow flow state. Figure 11 shows the variation of Fr along the course of the six NACA airfoil pillar-shaped flumes. As can be seen, the six NACA airfoil pillar-shaped flumes have realized the flow pattern conversion between slow flow and rapid flow states and when r increases, a larger distance is required to complete the conversion. Furthermore, Fr in a certain area in front of the airfoil pillar approaches zero and as r increases, the area of this part increases.

Generally, the flow upstream of the measuring flume is required to have Fr less than 0.5 to ensure the water surface is stable and to facilitate the accurate measurement of water level. Figure 11 illustrates the variation trend of Fr at the upstream sections of the six measuring flumes under

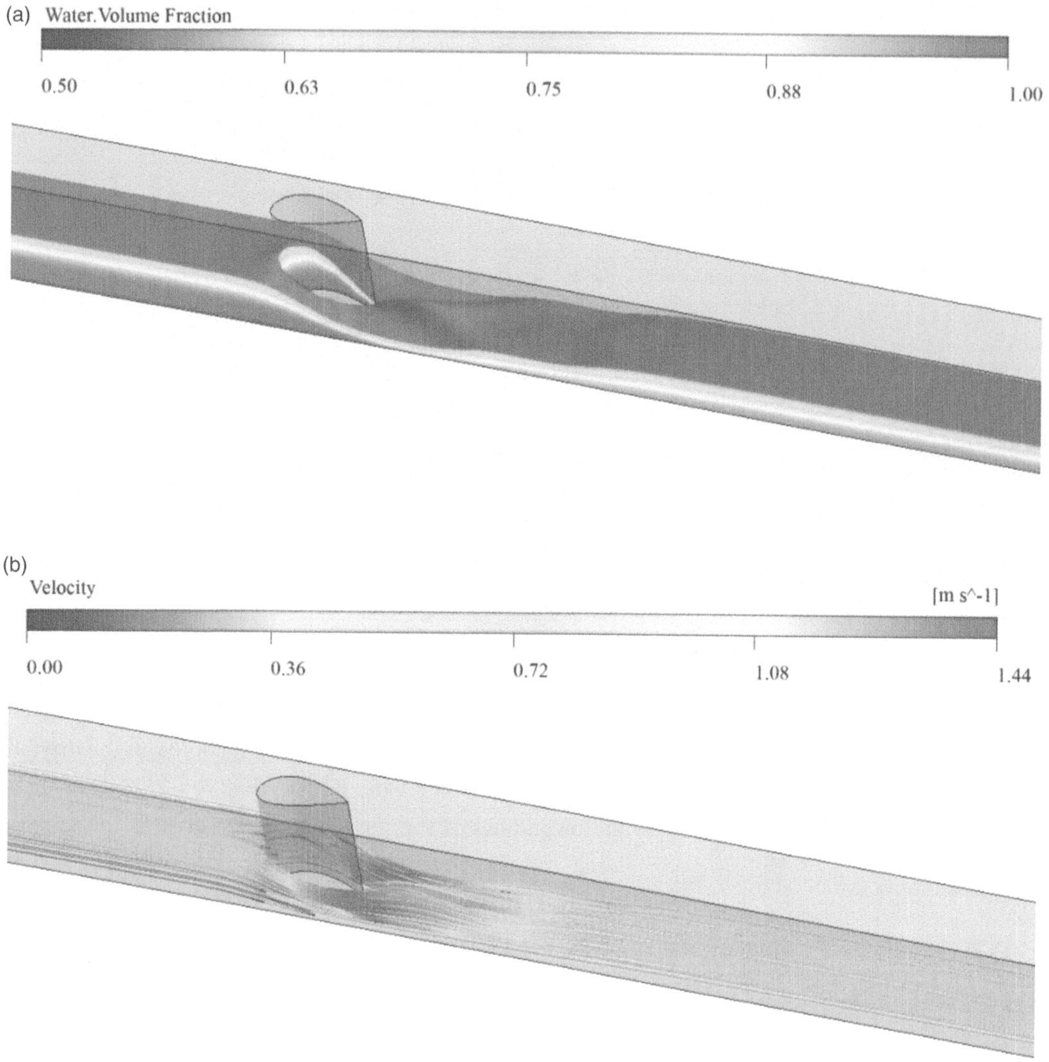

Figure 8 | Simulated flow pattern (a) and streamline diagram (b) when $r = 4$ and $Q = 15$ L/s.

different working conditions. It can be seen from Figure 12 that, for the same flow rate, *Fr* continuously decreases with an increase in *r*, while for the same *r*, *Fr* tends to increase with an increase in flow rate.

Velocity distribution

The velocity distribution can reflect the basic law of water flow movement in the NACA airfoil pillar-shaped flume. The flow velocity distribution in the inlet section (a), throat section (b), and outlet section (c) of six NACA airfoil pillar-shaped flumes with different *r* for the flow rate of 15 L/s are shown in Figure 13. As can be seen, the lateral flow velocity of the upstream section, from the center to the sidewall of the channel, is a symmetrical distribution, which is first increasing and then decreasing. Due to the influence of the airfoil pillar, the flow velocity in the center area of the channel is close to zero, and the area increases as *r* decreases. In addition, Figure 13 illustrates that as r increases, the upstream water flow rate of NACA airfoil pillar-shaped flumes decreases continuously, which is caused by different degrees of cross-section contraction. On the throat section, due to the reduction of the cross-section, the flow velocity further increases compared to

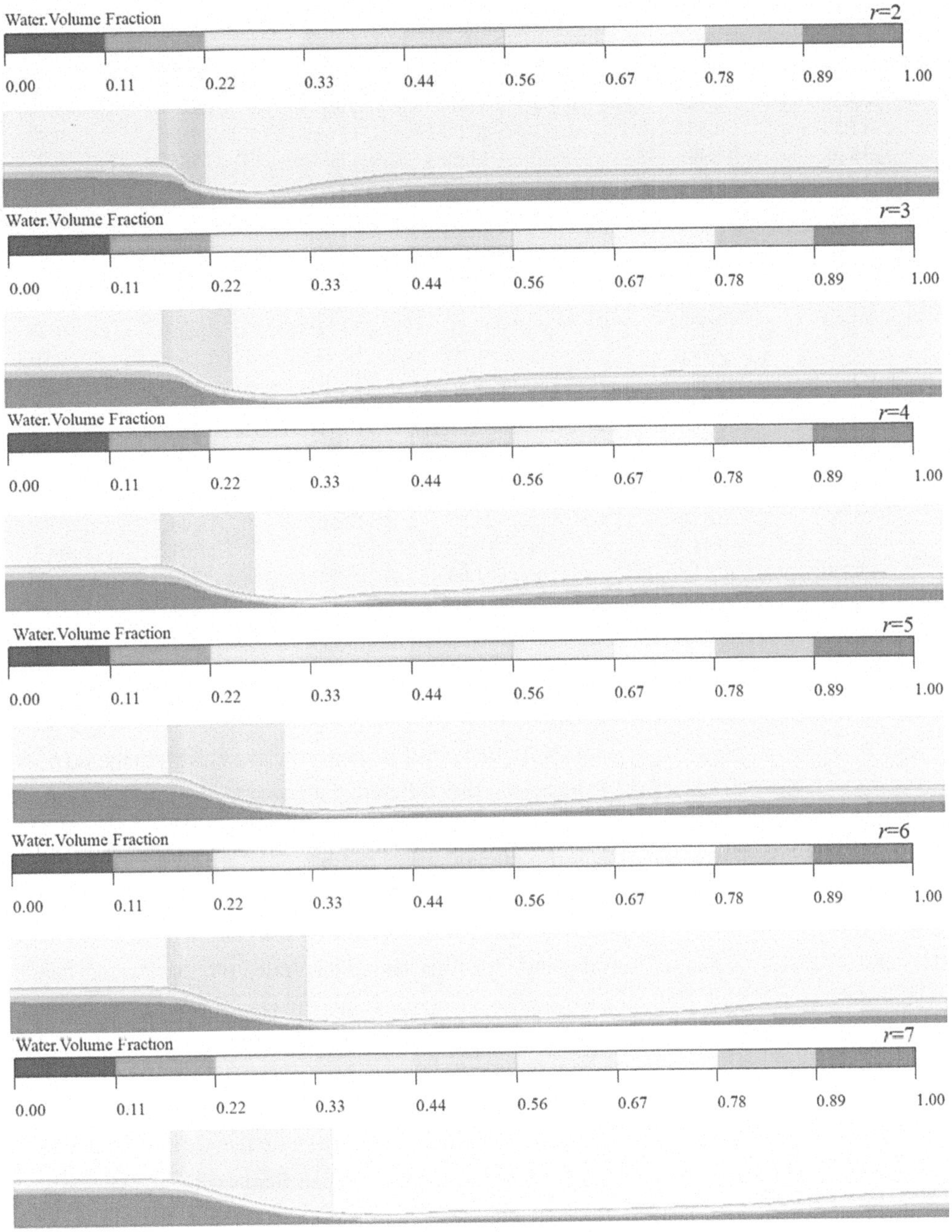

Figure 9 | Flow pattern of the longitudinal section of NACA airfoil pillar-shaped flumes when $Q = 15$ L/s.

the upstream section. It can also be seen that the flow velocity at the throat section increases when r increases, while the rate is decreasing. In addition, the flow velocity near the sidewall of the channel is also less than the flow velocity near the airfoil pillar. At the outlet section, it can be seen that, similar to the inlet section, the water flow is still divided into two streams and the water depth becomes smaller. The maximum velocity is located in the middle and upper parts of the flow. The water depth on both sides of the channel is obviously lower than the water depth in the center of the channel, and this phenomenon is more noticeable when r is small.

Table 4 | Comparison of measured and calculated discharge

r	h	Q_1 (L·s^{-1})	Q_2 (L·s^{-1})	|Error| %	r	h	Q_1 (L·s^{-1})	Q_2 (L·s^{-1})	|Error| %
2	8.54	6.98	6.93	0.72	5	8.96	6.86	6.60	3.79
	11.52	11.02	11.53	4.63		12.08	11.01	10.98	0.27
	13.79	15.01	15.67	4.40		14.60	15.03	15.17	0.93
	15.64	19.08	19.41	1.73		16.71	18.98	19.08	0.53
	17.18	22.98	22.78	0.87		18.63	23.02	22.96	0.26
	18.64	26.88	26.16	2.68		20.59	27.02	27.24	0.81
3	8.73	7.04	6.78	3.69	6	9.02	6.97	6.51	6.60
	11.76	11.03	11.28	2.27		12.17	11.05	10.84	1.90
	14.13	15.04	15.42	2.53		14.78	15.01	15.09	0.53
	16.10	19.02	19.26	1.26		16.95	19.03	19.06	0.16
	17.83	23.02	22.91	0.48		18.93	23.01	23.01	0.00
	19.53	26.96	26.75	0.78		20.87	27.04	27.17	0.48
4	8.83	7.02	6.65	5.27	7	9.07	7.04	6.43	8.66
	11.91	11.02	11.07	0.45		12.24	11.02	10.71	2.81
	14.34	15.02	15.18	1.07		14.92	15.04	15.01	0.20
	16.36	19.06	18.99	0.37		17.11	18.96	18.96	0.00
	18.23	22.94	22.84	0.44		19.18	23.02	23.01	0.04
	20.06	26.92	26.89	0.11		21.09	27.01	27.06	0.19

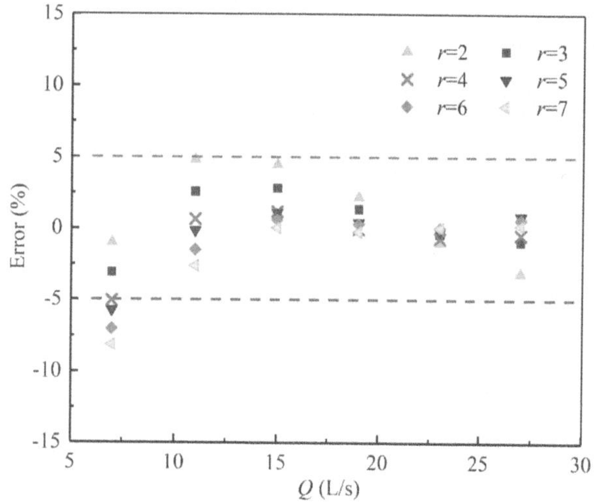

Figure 10 | The error between the calculated discharge and the measured discharge.

Backwater height

Note that, as a portable measuring flume, the cylindrical measuring flume is most suitable for small channels. In addition, due to the effect of side contraction or vertical contraction of the measuring flume, the upstream water level has to be larger than the normal depth of the original channel. Thus, the backwater phenomenon is likely to happen. Therefore, after installing the measuring flume, it is necessary to be sure if during measuring flow the backwater height of the channel may cause water to overflow the channel. Note that, in this work, the backwater height (H_b) is defined as the difference between the upstream water depth after installing the NACA airfoil pillar-shaped flume, and the original water depth.

It can be concluded from Figure 14 that, for the same r, the backwater height increases with an increase in the flow rate and, for the same flow rate, the backwater height increases as r increases. When $Q = 27$ L/s, the backwater heights of the NACA airfoil pillar-shaped flume with $r = 2$ and 7 are 5.27 and 7.72 cm, respectively. When $Q = 7$ L/s, the backwater heights with $r = 2$ and 7 are 1.97 and 2.50 cm, respectively. Therefore, it is recommended to use the NACA airfoil pillar-shaped flume with a small r value for small channel height, especially for large flow conditions.

In addition, it can be seen from Figure 14 that there is a good correlation between the backwater height and the flow rate. Note that the calculation formulas of the backwater height of the NACA airfoil pillar-shaped flume with different r values were obtained by fitting all data (Table 5). Through these formulas, the influence of the installation of the measuring flume on the channel depth can be predicted to facilitate the selection of a suitable NACA airfoil pillar-shaped flume.

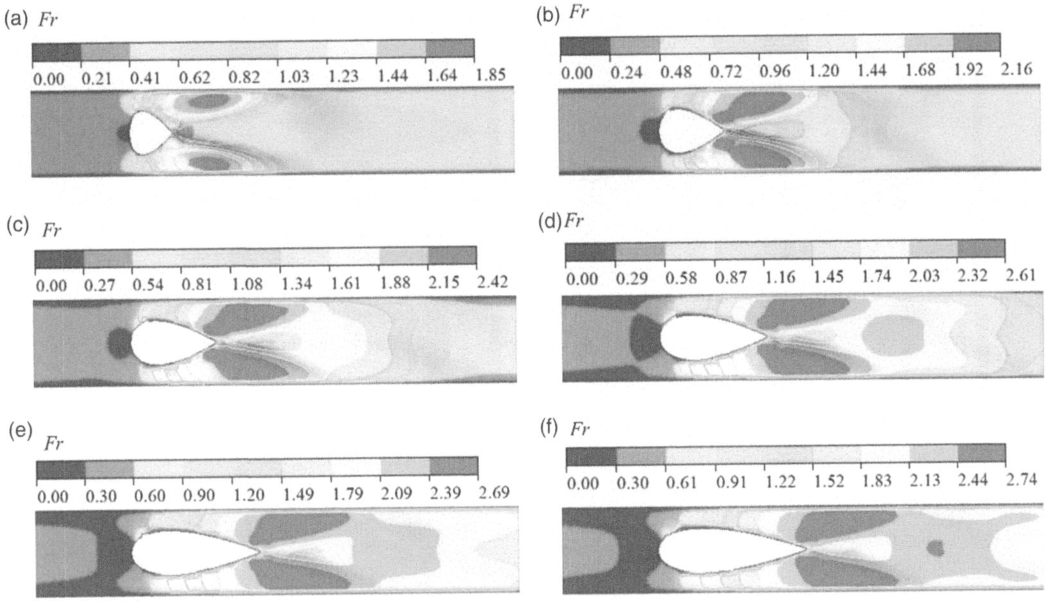

Figure 11 | Relationship between Froude number and discharge with different r: (a) $r=2$; (b) $r=3$; (c) $r=4$; (d) $r=5$; (e) $r=6$; (f) $r=7$.

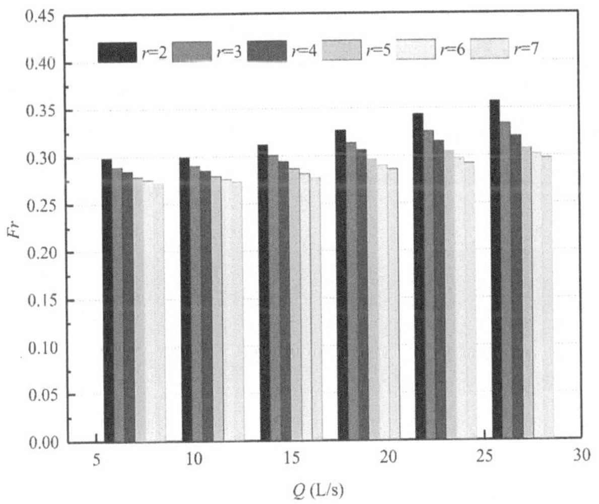

Figure 12 | Relationship between Froude number and discharge with different r.

Critical submergence

The critical submergence degree refers to the ratio of the downstream and upstream water depths. Note that for a certain contraction ratio, flow rate, and channel bottom slope, the critical submergence degree is used when the downstream water depth of the measuring trough just begins to affect the upstream water depth. This parameter reflects the range of water depths that can ensure the flume to be free flow in downstream. In this paper, the downstream water depth was adjusted by a gate at the end of the channel and the critical submergence degree of the NACA airfoil pillar-shaped flume was measured when the critical discharge condition was reached.

Table 6 shows the critical submergence degree of the NACA airfoil pillar-shaped flume under various working conditions. When $r=2$, the critical submergence range of the NACA airfoil pillar-shaped flume is from 0.76 to 0.84. When $r=7$, the critical submergence degree of the NACA airfoil pillar-shaped flume is in the range of 0.88–0.89. Variation range of the critical submergence degree of the NACA airfoil pillar-shaped flumes with respect to r is shown in Figure 15. This figure illustrates that the critical submergence degree is positively correlated with r so that it increases when r increases. Therefore, to provide larger ranges of free flow, the NACA airfoil pillar-shaped flume with larger r values should be used.

Energy loss

The viscosity of water is the root cause of the energy loss of water flow. Due to the existence of viscosity, each layer of the water flow will generate frictional resistance during the

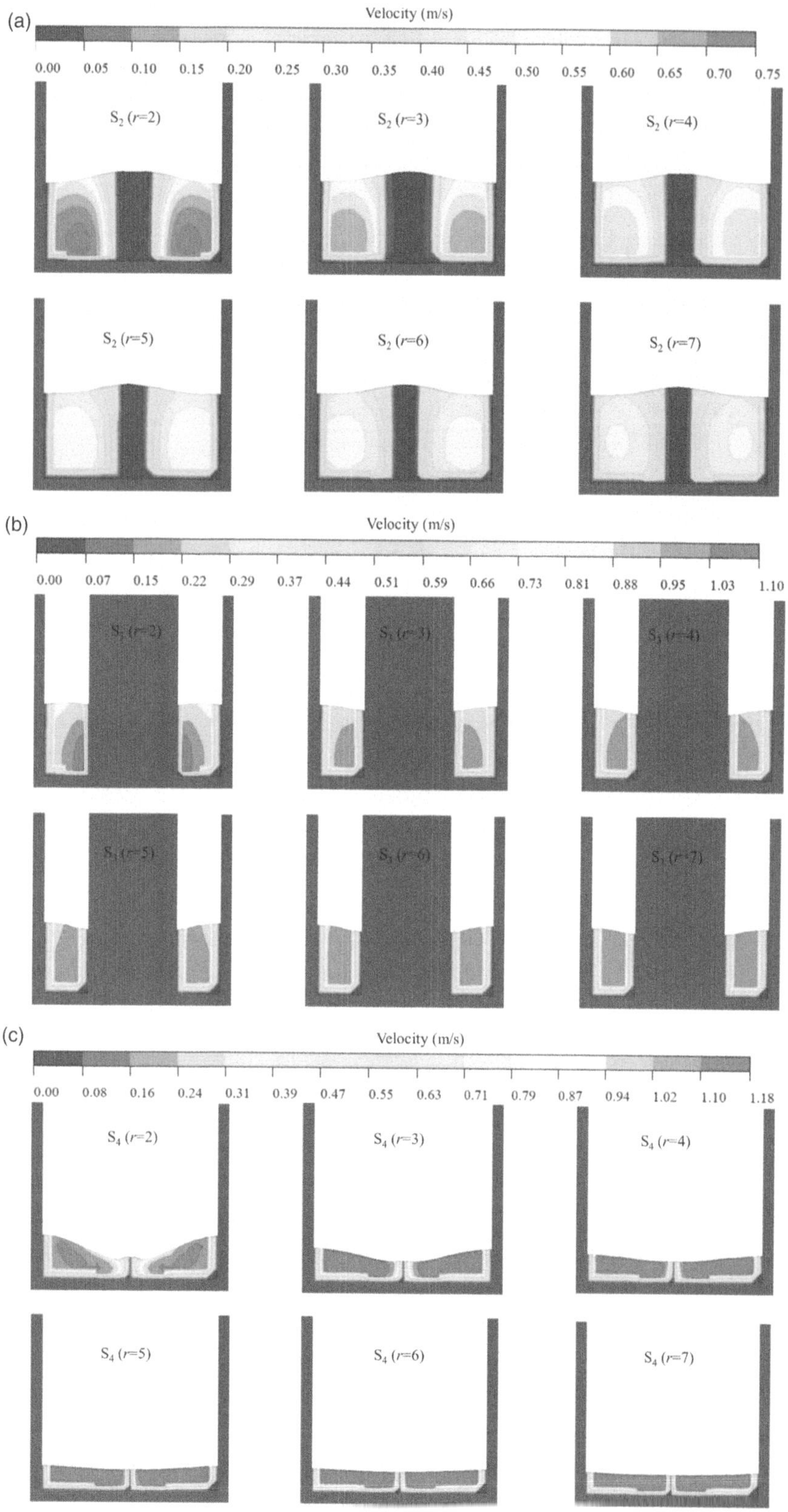

Figure 13 | Velocity distribution of the inlet section (a), throat section (b), and outlet section (c) of six NACA airfoil pillar-shaped flumes.

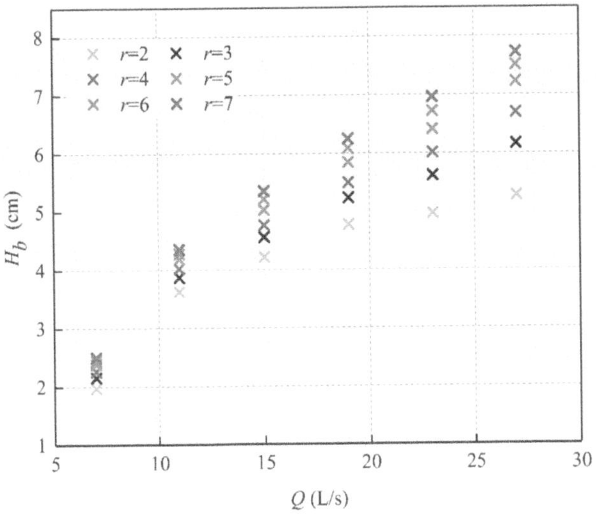

Figure 14 | The backwater height of NACA airfoil pillar-shaped flumes under all working conditions.

Table 5 | Calculation formulas for backwater height of NACA airfoil pillar-shaped flumes

r	Equation	R^2	r	Equation	R^2
2	$H_b = 0.0237\ln(Q) + 0.1401$	0.9636	5	$H_b = 0.0344\ln(Q) + 0.1951$	0.9948
3	$H_b = 0.0286\ln(Q) + 0.1401$	0.9890	6	$H_b = 0.0363\ln(Q) + 0.2052$	0.9971
4	$H_b = 0.0315\ln(Q) + 0.1800$	0.9934	7	$H_b = 0.0378\ln(Q) + 0.2127$	0.9981

Table 6 | Critical submergence of NACA airfoil pillar-shaped flume under all conditions

r	Q (L/s)	S	r	Q (L/s)	S	r	Q (L/s)	S
2	6.98	0.76	3	7.04	0.78	4	7.02	0.80
	11.02	0.83		11.03	0.83		11.02	0.82
	15.01	0.82		15.04	0.84		15.02	0.84
	19.08	0.84		19.02	0.83		19.06	0.83
	22.98	0.83		23.02	0.84		22.94	0.85
	26.88	0.84		26.96	0.85		26.92	0.85
5	6.86	0.86	6	6.97	0.87	7	7.04	0.88
	11.01	0.83		11.05	0.85		11.02	0.89
	15.03	0.83		15.01	0.85		15.04	0.88
	18.98	0.84		19.03	0.86		18.96	0.89
	23.02	0.85		23.01	0.87		23.02	0.89
	27.02	0.86		27.04	0.88		27.01	0.89

flow process, causing the loss of energy of the moving liquid. There are two kinds of local hydraulic phenomena, the water level drop and water level jump (Figure 16), which can cause large energy losses. In the process of measuring flow, the shape and the size of the solid boundary change when the water flows excessively in the flume and the streamline is bent. Note that these characteristics represent a non-uniform abrupt flow. In addition, the head loss and local head loss will occur along the course. However, in the place of local head loss, the main flow is separated from the boundary and a vortex forms in the separation area. In the vortex region, the turbulence intensifies and a constant energy exchange exists between the mainstream and the vortex region. Also, a lot of mechanical energy is consumed by the friction and violent collision between the particles. Therefore, the local head loss is much larger than the head loss along the same length of a flow segment. In this study, the length of the measuring flume is relatively short. Thus, the head loss is neglected along the flow segment.

The total head in a cross-section, H_{total} (m), is defined as:

$$H_{total} = \frac{\sum z_i a_i + \sum \frac{p_i}{\rho g} a_i + \sum \frac{\partial(u_i^2 + v_i^2 + w_i^2)}{2g} a_i}{\sum a_i} \tag{18}$$

where z_i is the elevation of a cell from the channel bottom (m), p_i is the pressure of a cell (Pa), and u_i, v_i, and w_i (m/s) are the velocities for a cell in the x, y, and z directions, respectively. a_i is the volume fraction of water in a cell, and α is the kinetic energy correction coefficient, which is set to $\alpha = 1$.

In this study, the head loss is defined as the difference between the total head of the upstream and downstream sections:

$$H_{\text{loss}} = H_u - H_d \tag{19}$$

where, H_{loss} is the head loss (m), H_{u} is total head of upstream sections (m), and H_{d} is total head of downstream sections (m).

Figure 17 illustrates the variation trend of the head loss of NACA airfoil pillar-shaped flumes at critical outflow under all working conditions. Figure 17 shows that when r is constant, the head loss increases with an increase in the flow rate. While, when the flow rate is constant, the head loss shows a decreasing trend as r increases. For plain irrigation areas with poor water head conditions, it is

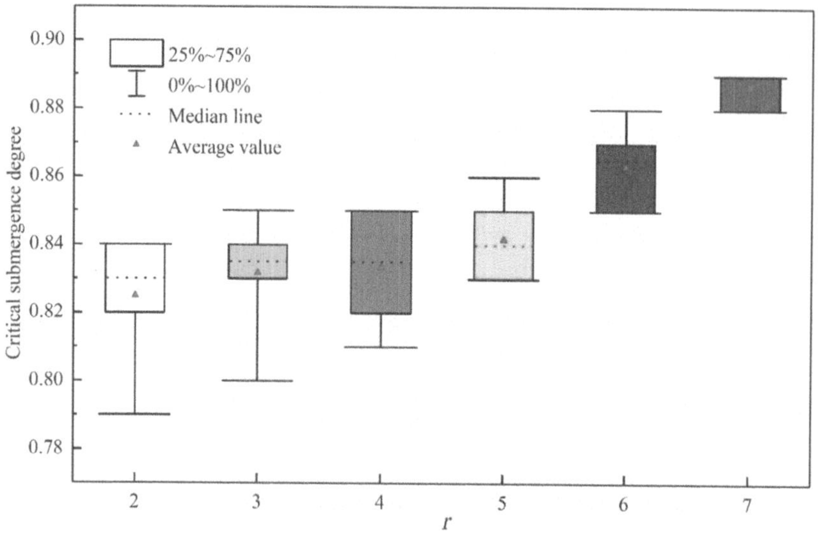

Figure 15 | Variation range of critical submergence degree of NACA airfoil pillar-shaped flumes.

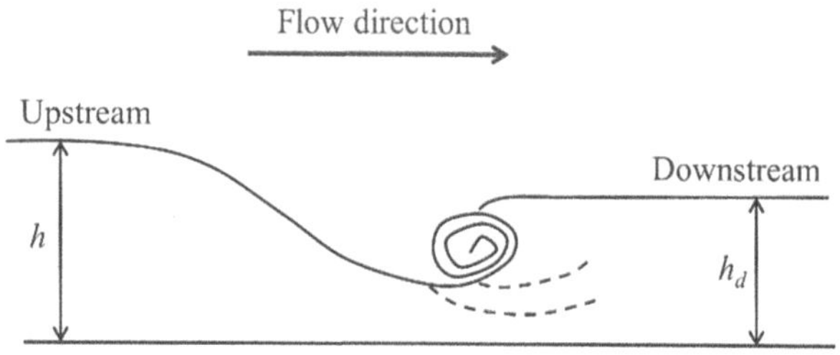

Figure 16 | Schematic diagram of water fall and water jump.

recommended to use NACA airfoil pillar-shaped flumes with r in the range of 5–7.

CONCLUSIONS

Open channel flow measurement is mostly used for measuring flow in different applications such as the optimal

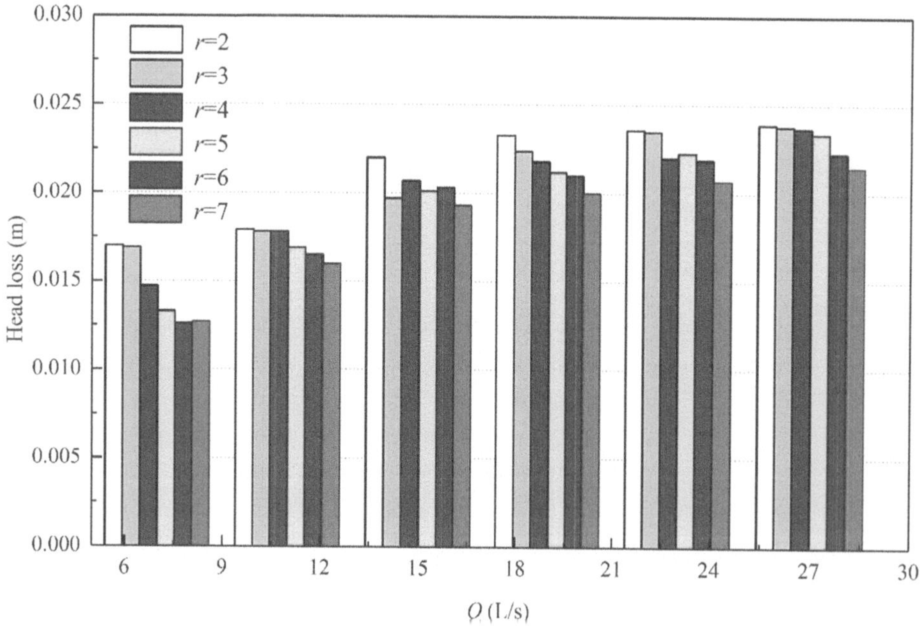

Figure 17 | Variation trend of head loss of NACA airfoil pillar-shaped flumes at critical outflow under all working conditions.

allocation of water resources, flood discharge forecasting, and water-saving irrigation. In this study, experimental and numerical simulations have been carried out on six NACA airfoil pillar-shaped flumes with different wing lengths to further discuss the abilities of these flumes in flow measurement. For the numerical calculations, the standard k-ε three-dimensional turbulence model and VOF method were used. In addition, the Buckingham Pi theorem along with the incomplete self-similarity theory were used for the theoretical flow calculation. Following conclusions can be drawn as a result of this study:

(1) The standard k-ε model and the VOF method can accurately simulate the flow in the flume. The average error of the experimental and numerical water surface line is 4.83%. It provides a theoretical basis for further research on the internal flow field distribution and the optimization and promotion of NACA airfoil pillar-shaped flumes.
(2) Dimensional analysis and self-similarity theory are used to derive the relationship between water level and flow and a formula suitable for different wing lengths is obtained, in which R^2 is 0.9981. The average error between the calculated and measured flows is 1.72%, which meets the requirements of open channel flow measurement.
(3) The flow pattern, Froude number, flow velocity distribution, backwater height, critical submergence degree, and energy loss of six NACA airfoil pillar-shaped flumes were compared. It was found that the backwater height is inversely proportional to *r*. However, the critical submergence and energy loss are directly proportional to *r*. In terms of portability and backwater height, NACA airfoil pillar-shaped flumes with smaller *r* values should be considered. For plain irrigation areas or channels with poor water head conditions, it is recommended to use NACA airfoil pillar-shaped flumes with *r* in the range of 5–7. This provides a theoretical reference for the selection of NACA airfoil pillar-shaped flumes in small rectangular channels.

ACKNOWLEDGEMENTS

The authors gratefully acknowledge the support provided by the National Key Research and Development Program of China (No. 2017YFC1501204), the Program for Science and Technology Innovation Talents in Universities of Henan Province (No. 19HASTIT043), the Outstanding Young Talent Research Fund of Zhengzhou University (1621323001), the Program for Guangdong Introducing Innovative and Enterpreneurial Teams (No. HG-GCKY-01-002), National Natural Science Foundation of China (No. 51909242, 52009125), Key scientific research projects of colleges and universities in Henan Province (No. 21A570007), China Postdoctoral Science Foundation funded project (No. 2020TQ0285).

DATA AVAILABILITY STATEMENT

All relevant data are included in the paper or its Supplementary Information.

REFERENCES

Aydin, M. C. 2012 CFD simulation of free-surface flow over triangular labyrinth side weir. *Advances in Engineering Software* **45**, 159–166.

Bridgman, P. W. 1922 *Dimensional Analysis*. Yale University Press, Oxford University Press, New Haven.

Carollo, F. G., Di Stefano, C., Ferro, V. & Pampalone, V. 2016 New stage-discharge equation for the SMBF flume. *Journal of Irrigation and Drainage Engineering* **142** (5).

Cox, A. L., Thornton, C. I. & Abt, S. R. 2013 Supercritical flow measurement using a large Parshall flume. *Journal of Irrigation and Drainage Engineering* **139**, 655–662.

Dabrowski, W. & Polak, U. 2012 Improvements in flow rate measurements by flumes. *Journal of Hydraulic Engineering-Asce* **138**, 757–763.

Das, R., Nayek, M., Das, S., Dutta, P. & Mazumdar, A. 2017 Design and analysis of 0.127 m (5 ") Cutthroat flume. *Ain Shams Engineering Journal* **8**, 295–303.

Ferro, V. 2002 Discussion of 'simple flume for flow measurement in open channel' by Zohrab Samani and Henry Magallanez. *Journal of Irrigation and Drainage Engineering* **128**, 129–131.

Ferro, V. 2018 Discussion of 'three simple flumes for flow measurement in open channels' by Zohrab Samani. *Journal of Irrigation and Drainage Engineering* **144**, 1323.

Hager, W. H. 1985 Modified venturi channel. *Journal of Irrigation and Drainage Engineering* **111**, 19–35.

Hirt, C. W. & Nichols, B. D. 1981 Volume of fluid (VOF) method for the dynamics of free boundaries. *Journal of Computational Physics* **39**, 201–225.

Hu, H., Huang, J., Qian, Z., Huai, W. & Yu, G. 2014 Hydraulic analysis of parabolic flume for flow measurement. *Flow Measurement and Instrumentation* **37**, 54–64.

Jesson, M., Sterling, M. & Baker, D. 2017 Application of ISO4359 for discharge calculation in a narrow flume. *Flow Measurement and Instrumentation* **54**, 283–287.

Karakouzian, M., Chavez, A., Hayes, D. & Nazari-Sharabian, M. 2019 Bulbous pier: an alternative to bridge pier extensions as a countermeasure against bridge deck splashing. *Fluids* **4** (3), 4030140.

Li, X., Zhang, T. & Li, G. 2004 The improved Parshall flume. *Chinese Journal of Scientific Instrument* **25**, 747–750, 69.

Li, X., Jin, L., Engel, B. A., Yang, Z., Wang, W., He, W. & Wang, Y. 2020 Influence of the structure of cylindrical mobile flumes on hydraulic performance characteristics in U-shaped channels. *Flow Measurement and Instrumentation* **72**.

Liu, H., Zhao, R., Huang, J. & Zhang, L. 2014 Experimental research on airfoil pillar-shaped measuring-flume. *China Rural Water and Hydropower* **06**, 170–172 (in Chinese).

Manekar, V. L., Porey, P. D. & Ingle, R. N. 2007 Discharge relation for cutthroat flume under free-flow condition. *Journal of Irrigation and Drainage Engineering* **133**, 495–499.

Nazari-Sharabian, M., Karakouzian, M. & Hayes, D. 2020a Flow topology in the confluence of an open channel with lateral drainage pipe. *Hydrology* **7** (3), 7030057.

Nazari-Sharabian, M., Nazari-Sharabian, A., Karakouzian, M. & Karami, M. 2020b Sacrificial piles as scour countermeasures in river bridges A numerical study using FLOW-3D. *Civil Engineering Journal-Tehran* **6** (6), 1091–1103.

Qi, Q., Huo, Q., Ye, Q. & Zhang, J. 2014 3-D Numerical simulation of the flow field of cutthroat flume. In: *Materials Science, Civil Engineering and Architecture Science, Mechanical Engineering and Manufacturing Technology, Pts 1 and 2* (H. W. Liu, ed.). Trans Tech Publications, Zurich, Switzerland.

Ramamurthy, A. S. & Tadayon, R. 2008 Numerical simulation of flows in cut-throat flumes. *Journal of Irrigation and Drainage Engineering* **134**, 857–860.

Saadatnejadgharahassanlou, H., Zeynali, R. I., Gharehbaghi, A., Mehdizadeh, S. & Vaheddoost, B. 2020 Three dimensional flow simulation over a sharp-crested V-Notch weir. *Flow Measurement and Instrumentation* **71**, 101684.

Samani, Z. 2017 Three simple flumes for flow measurement in open channels. *Journal of Irrigation and Drainage Engineering* **143**.

Thornton, C. I., Smith, B. A., Abt, S. R. & Robeson, M. D. 2009 Supercritical flow measurement using a small Parshall flume. *Journal of Irrigation and Drainage Engineering* **135**, 683–692.

Torres, A. F. & Merkley, G. P. 2008 Cutthroat measurement flume calibration for free and submerged flow using a single equation. *Journal of Irrigation and Drainage Engineering* **134**, 521–526.

Wang, S. & Wen, J. 1990 Iverstigation on the potential flow model of the hydraulic characteristics of the airfoil weir. *Journal of Hohai University* **04**, 93–97 (in Chinese).

Xiao, Y., Wang, W., Hu, X. & Zhou, Y. 2016 Experimental and numerical research on portable short-throat flume in the field'. *Flow Measurement and Instrumentation* **47**, 54–61.

Ziaei, A. N., Nikou, N. S. R., Beyhaghi, A., Attarzadeh, F. & Khodashenas, S. R. 2019 Flow simulation over a triangular labyrinth side weir in a rectangular channel. *Progress in Computational Fluid Dynamics* **19**, 22–34.

First received 25 January 2021; accepted in revised form 23 March 2021. Available online 1 April 2021

doi: 10.2166/ws.2021.155

Discharge estimation for trapezoidal open channels applying fuzzy transformation method to a flow equation

N. Samarinas* and C. Evangelides

Department of Hydraulics and Transportation Engineering, Faculty of Engineering, Aristotle University of Thessaloniki, 54124 Thessaloniki, Greece
*Corresponding author. E-mail: smnikiforos@topo.auth.gr

ABSTRACT

The aim of this paper is to implement fuzzy logic theory in order to estimate the discharge for open channels, which is a well-known physical problem affected by many factors. The problem can be solved by the Manning equation but the parameters present uncertainties as to their true-real values. Especially, the Manning *n* roughness coefficient, which is an empirically derived coefficient, presents quite a high variation for different substrates. With the help of fuzzy logic and utilizing a fuzzy transformation method, it is possible to include the uncertainties of the problem in the calculation process. In this case, it is feasible to estimate the discharge, putting more emphasis on different uncertainty rates of the Manning roughness coefficient, while the rest of the parameters remain with constant or zero uncertainty level. By taking different α-cut levels, it is shown that the methodology gives realistic and reliable results, presenting with great accuracy the variations of the water discharge for trapezoidal open channels. In this way, a possible underestimation or overestimation of the actual physical condition is avoided, by helping engineers and researchers to obtain a more comprehensive view of the real physical conditions, and thus make better management plans.

Key words: discharge, fuzzy logic, Manning equation, open channel, transformation method

HIGHLIGHTS

- Novel use of fuzzy logic in the Manning equation.
- A new approach to estimate the discharge in open natural channels.
- Includes Manning's roughness coefficient uncertainties in the calculation process.
- Comparison between the uncertainty parameters of the Manning equation.
- Provides accurate information for better management plans and decisions.

INTRODUCTION

It is widely known that ambiguities and uncertainties in the values of parameters appear in many problems of mechanical and hydraulic engineering which are introduced in the corresponding mathematical models. Normally, these uncertainties arise from incomplete or imprecise information or from identification problems. They mainly refer to uncertain model parameters, sometimes also uncertain system inputs or uncertain initial or boundary conditions. These uncertainties greatly influence the numerical results, and they should be taken seriously into consideration in calculations.

The discharge estimation of water into open channels is a well-known and complicated problem for artificial channels such as irrigation systems, water supply systems etc., but even more complicated for natural open channels such as rivers, streams etc., due to their high variability. As yet, the majority of researchers are still using Manning's roughness coefficient in terms of hydrodynamics, hydraulics, hydrology and water resource engineering (Choo *et al.* 2018).

The Manning n is a coefficient which represents the roughness or friction applied to the flow by the channel and is used in the Manning equation for estimating the discharge in open channel flow. Manning's equation has a big advantage because it requires only two mathematical parameters, the water surface slope S and the roughness n from the bed load to estimate the discharge. However, it presents a certain drawback and does not work well due to a high sensitivity of the coefficient n which is discharge-dependent and applies only to a specific flow condition (Geem & Kim 2018). Furthermore, the Manning coefficient is influenced by channel slope, flow velocity, Reynolds number, Froude number and sediment load. Zhang *et al.* (2010)

investigated the potential effects of sediment load on the Manning coefficient on a fixed flume bed under wide ranges of hydraulics and sediment loads. Their work showed that the Manning coefficient decreased with increases in Reynolds number and Froude number. Also, although a qualitative description of movable bed roughness was clearly defined a long time ago (Yen 2002), its precise value in an unsteady sediment transport process is still difficult to determine (Huang 2007). In general, the procedure for selecting n values is quite complex and usually requires advanced judgment and expertise (Bahramifar *et al.* 2013). Moreover, one of the best methods is to have field observations from an experienced hydraulic engineer (Choo *et al.* 2018), but such a method is both time-consuming and costly. In this regard, and because it is an empirically derived coefficient, which is dependent on and affected by many factors (Chow 1959), the scientific community has shown in the past great interest in various methods for selecting appropriate values of roughness coefficient for discharge estimation in an open channel (Cowan 1956; Limerinos 1970; Shiono *et al.* 1999; Mailapalli *et al.* 2008; Khatua *et al.* 2012; Dash & Khatua 2016; Pradhan & Khatua 2018). Additionally, and because it is difficult to obtain a general equation to provide accurate estimation of the Manning coefficient, due to a lack of knowledge of some physical processes associated with channel formation and maintenance, new procedures for the estimation of Manning's n coefficient have been developed using innovative methods such as gene expression programming (Azamathulla *et al.* 2013), adaptive neuro-fuzzy inference system (ANFIS) (Şamandar 2011; Bahramifar *et al.* 2013; Moharana & Khatua 2014; Al-Husseini 2015) and machine learning techniques (Mohanta *et al.* 2018).

More specifically, the fuzzy set theory proposed by Zadeh (1965) proved to be a powerful tool for modelling uncertainties and for processing subjective information in mathematical models, giving a big boost to simulation procedures dealing with fuzzy parameterized models. In recent years, fuzzy parameterized models have become quite popular in engineering sciences and help in analyzing systems in view of the uncertainty of the parameters. Discharge estimation in open channels is a hydraulic problem characterized by ambiguities in all its parameters, but in this work greater attention is given to the Manning roughness coefficient for different substrates, of which due to complexities of physical processes the exact value is often more sensitive and presents the greatest ambiguity of all factors. In this regard, it is easily understood that the ambiguity of the roughness coefficient greatly affects the discharge results.

Understanding the need for accurate prediction of the Manning roughness coefficient and including its uncertainties, in this article, the reduced transformation method proposed by Hanss (2002, 2005) is implemented in order to transform the Manning equation in a fuzzy environment giving the final discharge estimation in a fuzzy form. The method can generally be considered as an advanced and extended version of the so-called vertex method proposed by Dong & Shah (1987) but with certain advantages. In contrast to classic fuzzy arithmetic, fuzzy arithmetic based on the transformation method does not exhibit the effect of overestimation, which is a critical point when we deal with the Manning equation. The method uses fuzzy numbers to solve the model and is generally used to simulate and analyze systems with uncertain parameters. In this way, the Manning equation parameters can be treated as fuzzy numbers with the potential to give different uncertainty rates to parameters, thus including the uncertainties in the results. The fuzzy transformation method works well by using triangular fuzzy numbers which are simple to apply, with easy and intuitive interpretation, based on numerical results as well as graphic representation. The method is applied in four different open channel types of trapezoidal profile. Also, two different scenarios were considered. In the first scenario all the parameters of the problem present uncertainties and in the second the roughness coefficient is the only uncertain parameter, accepting that all other parameters are accurate as to their values with zero uncertainty rate.

METHODS

Preliminaries

When the surface of a flow is open to the atmosphere, in other words when there is only atmospheric pressure on the surface, the flow is named open channel flow. Water flows in rivers and streams are obvious examples of open channel flow in natural channels.

Classified according to its origin, a channel may be either natural or artificial. Geomorphology of natural channels concerns their shape and structure. Natural channel sections are in general very irregular, usually varying from approximately parabolic to approximately trapezoidal (Chow 1959).

On the other hand, artificial open channels are built for some specific purpose, such as irrigation, water-power development, etc., and are usually designed with regular geometric shapes such as rectangular or trapezoidal. Figure 1 presents the construction of a trapezoidal channel and a side view.

The Manning equation is the most-used equation to analyze open channel flows. It is a semi-empirical equation for simulating water flows in channels and culverts where the water is open to the atmosphere, i.e. not flowing under pressure. It was introduced by the Irish engineer Robert Manning in 1891 as an alternative to the Chezy equation and can be evaluating as follows (Manning 1891):

$$V = \frac{1}{n} R^{2/3} S_0^{1/2} \tag{1}$$

where V is the average velocity (ms^{-1}), R is the hydraulic radius (m), S_0 is the channel slope, and n is the Manning coefficient.

Also,

$$R = \frac{A}{P} \quad \text{with} \quad P = b + 2y\sqrt{1+z^2} \quad \text{and} \quad A = (b + zy)y \tag{2}$$

where P is the wetted perimeter (m), A is the channel cross-sectional area (m^2), b is the bottom width (m), y is the water width (m) and z is the side slope.

The final discharge Q (m^3s^{-1}) is calculated from the equation:

$$Q = V \cdot A \tag{3}$$

To estimate the discharge in a reliable way, first it is important to understand and analyze the problem in individual steps to accurately identify the gaps appearing in the problem that are creating uncertainties. Then, the fuzzy methodology seems to be the most appropriate method of approaching the problem by enabling the uncertainties to be included in the calculations in a simple way. Figure 2 presents the problem approach and its solution.

Fuzzy methodology

Definition 1. In fuzzy logic we consider a set X and a subset P of X which consists of a set of ordered pairs of the form:

$$P = \{[x, \mu_P(x)]; x \in [0, 1]\} \tag{4}$$

where $\mu_P(x)$ is the value of the membership function of x in P.

Definition 2. A domain of a fuzzy set P is the set of elements x whose membership function has positive values:

$$\text{supp}(P) = \{x; \mu_P(x) > 0\} \tag{5}$$

Definition 3. Consider a fuzzy set $\tilde{P} \in F(R)$. Then the α–cuts of $\tilde{P}$ are:

$$[\tilde{P}]^\alpha = [P_\alpha^-(x), \; P_\alpha^+(x)] \tag{6}$$

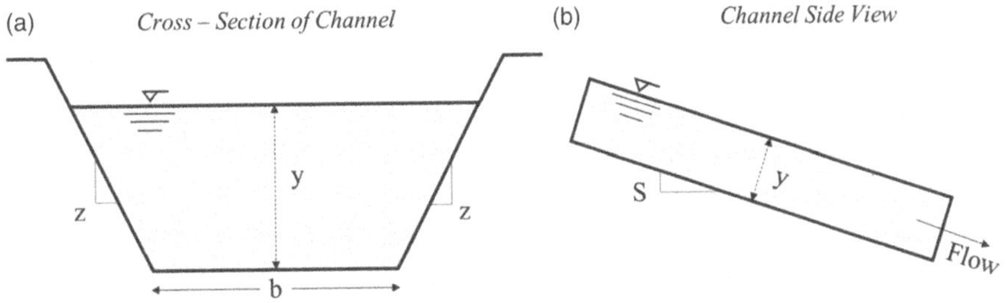

Figure 1 | Construction for (a) trapezoidal profile and (b) channel side view.

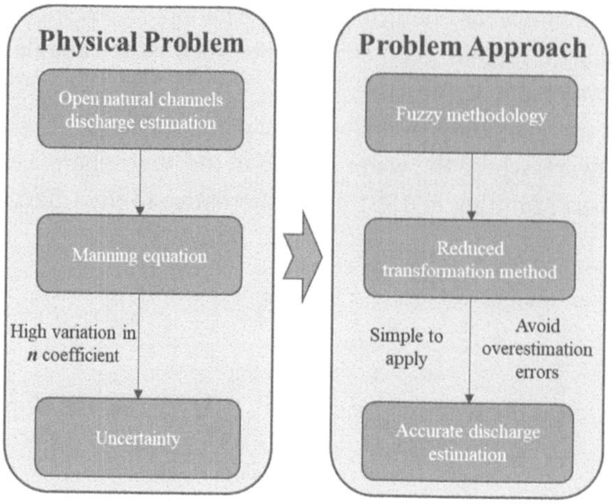

Figure 2 | Methodology for identifying the problem and its solution.

According to the theorem of Negoiţă & Ralescu (1975) and Goetschel & Voxman (1986) the membership function and the α-cut form of a fuzzy number $\tilde{P}$ are equivalent and more specifically the α-cuts (Equation (6)) represent $\tilde{P}$, provided that the two functions are monotonous ($P_\alpha^-(x)$ increasing, $P_\alpha^+(x)$ decreasing) and $P_1^- \leq P_1^+$, for $\alpha = 1$.

Definition 4. Among the large number of existing fuzzy numbers, some types are of particular importance and are used extensively (Zimmermann 1992). Triangular, trapezoidal, and Gaussian fuzzy numbers are in the foreground.

A triangular fuzzy number has the form (Hanss 2005):

$$\tilde{p} = \mathrm{tfn}(\bar{x}, a_\mathrm{l}, a_\mathrm{r}) \tag{7}$$

where the parameter $\bar{x}$, denotes the modal value of the fuzzy number, and a_l and a_r are the left-hand and right-hand worst-case deviations from the modal value.

In Figure 3(a) the straight section $(\alpha_\mathrm{l}(\alpha) - \alpha_\mathrm{r}(\alpha))$ is called the interval of confidence of the fuzzy number, while the ordinate α is called the level of presumption in accordance with Kaufmann & Gupta (1991). Due to its rather simple membership function of the linear type, the triangular fuzzy number is one of the most frequently used fuzzy numbers (Hanss 2005).

A trapezoidal fuzzy number (Figure 3(b)), which is a generalization of the triangular fuzzy number, has the form:

$$\tilde{p} = \mathrm{tpfn}(a_\mathrm{l}, b_\mathrm{l}, b_\mathrm{r}, a_\mathrm{r})$$

A Gaussian fuzzy number (Figure 3(c)) has the form:

$$\tilde{p} = \mathrm{gfn}(\bar{x}, \sigma_\mathrm{l}, \sigma_\mathrm{r})$$

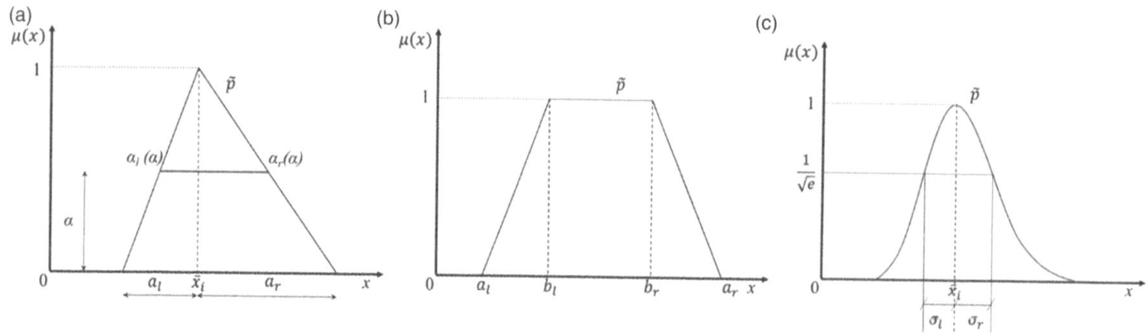

Figure 3 | Fuzzy numbers $\tilde{p}$: (a) triangular form, (b) trapezoidal form and (c) Gaussian form.

As in the triangular fuzzy number again here, the modal value is denoted by the parameter $\bar{x}$, and σ_l and σ_r denote the left-hand and right-hand spreads, corresponding to the standard deviations of the Gaussian distribution.

Reduced transformation method

According to Hanss (2002), given a problem with a number of n independent parameters which are assumed to be uncertain, the parameters can be represented by fuzzy numbers $\tilde{p}_i$, $i = 1, 2, \ldots, n$, each of them decomposed (Figure 4) into a set P_i of $m+1$ intervals $\hat{X}_i^{(j)}$, $i = 1, 2, \ldots, n$, of the form:

$$P_i = \{X_i^{(0)}, X_i^{(1)}, \ldots, X_i^{(m)}\}, \quad i = 1, 2, \ldots, n \tag{8}$$

with

$$X_i^{(j)} = [a_i^{(j)}, b_i^{(j)}],\ a_i^{(j)} \leq b_i^{(j)}, \quad i = 1, 2, \ldots, n, \quad j = 1, 2, \ldots, m \tag{9}$$

For the decomposition of a parameter into α-cuts, the μ-axis is subdivided into m segments, equally spaced by $\Delta\mu = 1/m$. The $m+1$ levels of membership μ_j are given by:

$$\mu_j = \frac{j}{m}, \quad j = 1, 2, \ldots, m \tag{10}$$

The intervals of each level of membership μ_j, $j = 0, 1, \ldots, m$, can now be transformed into arrays $\hat{X}_i^{(j)}$, of the following form:

$$\hat{X}_i^{(j)} = (\overbrace{(a_i^{(j)}, \beta_i^{(j)}), (a_i^{(j)}, \beta_i^{(j)}), \ldots, (a_i^{(j)}, \beta_i^{(j)})}^{2^{i-1}\text{pairs}}) \tag{11}$$

with

$$a_i^{(j)} = \underbrace{(a_i^{(j)}, \ldots, a_i^{(j)})}_{2^{n-1}\text{ elements}}, \qquad \beta_i^{(j)} = \underbrace{(\beta_i^{(j)}, \ldots, \beta_i^{(j)})}_{2^{n-1}\text{ elements}} \tag{12}$$

Note that $a_i^{(j)}$ and $\beta_i^{(j)}$ are the lower and upper bounds of the interval $\hat{X}_i^{(j)}$ at the membership level μ_j for the ith uncertain model parameter. Each of the transformed arrays $X_i^{(j)}$ has a total number of $(m+1-j)^n$ entries (Klimke 2003). Assuming the

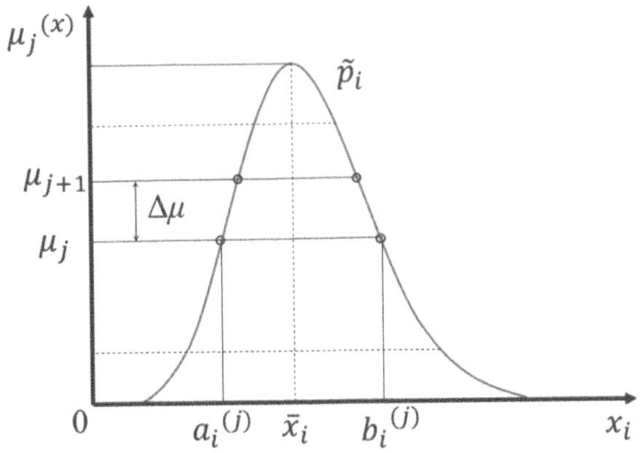

Figure 4 | Decomposition of the ith uncertain parameter into intervals.

system to be simulated is expressed by an arithmetical expression F of the form:

$$\tilde{q} = F(\tilde{p}_1,\ \tilde{p}_2,\ \ldots,\ \tilde{p}_n) \tag{13}$$

its evaluation is then carried out by evaluating the expression separately at the positions of the arrays using the conventional arithmetic for crisp numbers. Thus, if the output $\tilde{q}$ of the system can be expressed in its decomposed and transformed form by the arrays $Z^{(j)} = [a^{(j)}, b^{(j)}]$, $j = 1, 2, \ldots, m$, the kth element ${}^k\hat{z}^{(j)}$ of the array $Z^{(j)}$ is given by:

$${}^k\hat{z}^{(j)} = F({}^k\hat{x}_1^{(j)}, {}^k\hat{x}_2^{(j)}, \ldots, {}^k\hat{x}_n^{(j)}) \tag{14}$$

where ${}^k\hat{z}^{(j)}$ denotes the kth element of the array $\hat{X}_i^{(j)}$. Finally, the fuzzy-valued result $\tilde{q}$ of the problem can be obtained in its decomposed form:

$$Z^{(j)} = [a^{(j)}, b^{(j)}], \qquad j = 1, 2, \ldots, m \tag{15}$$

by re-transforming the arrays $Z^{(j)}$ according to the recursive formulae:

$$\alpha^{(j)} = \min_k (a^{(j+1)},\ {}^k\hat{z}^{(j)}) \ \text{ and } \ b^{(j)} = \max_k (b^{(j+1)},\ {}^k\hat{z}^{(j)}), \quad j = 1, 2, \ldots, m-1 \tag{16}$$

and

$$\alpha^{(m)} = \min_k ({}^k\hat{z}^{(j)}) = \max_k ({}^k\hat{z}^{(j)}) = b^{(m)} \tag{17}$$

The reduction of fuzzy arithmetic to multiple crisp-number operations means that the transformation method can be implemented quite easily in an appropriate software environment.

Fuzzy mathematical model

Moving from classical logic to fuzzy set theory, the five (four for rectangular profile, $z = 0$) above parameters lead to the following rational expressions:

$$\left.\begin{aligned} P(\tilde{b}, \tilde{y}, \tilde{z}) &= \tilde{b} + 2\tilde{y}\sqrt{1+\tilde{z}^2} \\ A(\tilde{b}, \tilde{y}, \tilde{z}) &= (\tilde{b} + \tilde{z}\tilde{y})\tilde{y} \\ R(\tilde{b}, \tilde{y}, \tilde{z}) &= \frac{(\tilde{b} + \tilde{z}\tilde{y})\tilde{y}}{\tilde{b} + 2\tilde{y}\sqrt{1+\tilde{z}^2}} \end{aligned}\right\} V(\tilde{b}, \tilde{y}, \tilde{z}, \tilde{S}_0, \tilde{n}) = \left[\frac{1}{\tilde{n}} \cdot \left(\frac{(\tilde{b} + \tilde{z}\tilde{y})\tilde{y}}{\tilde{b} + 2\tilde{y}\sqrt{1+\tilde{z}^2}}\right)^{2/3} \cdot \tilde{S}_0^{1/2}\right]$$

Therefore, the final expression of the discharge equation (Equation (3)) in analytical form will be:

$$Q(\tilde{b}, \tilde{y}, \tilde{z}, \tilde{S}_0, \tilde{n}) = \left[\frac{1}{\tilde{n}} \cdot \left(\frac{(\tilde{b} + \tilde{z}\tilde{y})\tilde{y}}{\tilde{b} + 2\tilde{y}\sqrt{1+\tilde{z}^2}}\right)^{2/3} \cdot \tilde{S}_0^{1/2}\right] \cdot (\tilde{b} + \tilde{z}\tilde{y})\tilde{y} \tag{18}$$

RESULTS AND DISCUSSION

Definition of problem scenarios

In order to implement the methodology we will use two scenarios which refer to different open channel types with different uncertainty rates in the parameters of the problem. A necessary step first is to define the open channel substrates as well as the values and corresponding uncertainty rates of the Manning equation parameters.

In this regard, recommended Manning's n values for four different types of natural channels are given in Table 1. Chow (1959) suggested the three values of minimum, normal and maximum of n for each kind of channel.

In Table 2 the cases taken into consideration and the uncertainty rates (Ur) that are accepted for the Manning roughness coefficient which were obtained according to Table 1 are listed.

As can be easily observed, all the parameters of the problem remain with constant values for all cases, except for the n coefficient. This way comparison of the different scenarios is possible, showing the importance of the uncertainty of the n factor in the calculations. As is easily understood, the computational process is quite complex and time-consuming. However, the transformation method can be implemented quite easily in an appropriate software environment. In this regard, computer algorithms were developed in the MATLAB environment and specifically in this work the algorithm as presented by Klimke (2003) for the reduced transformation method was used.

The two different scenarios taken into consideration are as follows:

- **Scenario 1:** This scenario applies the transformation method to estimate the discharge for the case 1a (as a computational example) of Table 2 and for α-cut level 0. In this scenario all parameters present uncertainties.
- **Scenario 2:** This scenario follows the same procedure as in scenario 1 but takes into consideration only the uncertainty rate of Manning's roughness coefficient. Therefore, the parameters b, y, z, S_0 have zero uncertainty rate (Ur 0%) and the parameter n has the uncertainty rate as shown in Table 2. Similarly in this case the transformation method was applied keeping the same values for all parameters.

Scenario implementation

The main purpose of the following calculations is the presentation of the methodology in the physical problem with real values in the parameters. The transformation method is applied in order to estimate the discharge for case 1a of Table 2, as an example, and for α-cut level 0.

Starting from the initial and simple calculations and according to Equations (1)–(3) the following is obtained:

$$A = 122.16\ \mathrm{m}^2,\ \ P = 104.32\ \mathrm{m},\ \ R = 1.17\ \mathrm{m},$$

$$V = 0.7\ \mathrm{ms}^{-1},\ \ Q = 86.70\ \mathrm{m}^3\mathrm{s}^{-1}$$

Table 1 | Recommended Manning's n values for natural channels

Type of channel and description		Min	Normal	Max
1. Main channels				
a.	Clean, straight, full stage, no riffles or deep pools with stones and weeds	0.03	0.035	0.04
b.	Sluggish reaches, weedy, deep pools	0.05	0.07	0.08
2. Mountain streams, no vegetation in channel, banks usually steep, trees and brush along banks submerged at high stages				
a.	bottom: gravels, cobbles, and few boulders	0.03	0.04	0.05
b.	bottom: cobbles with large boulders	0.04	0.05	0.07

Table 2 | Recommended uncertainty rates (Ur %) for all parameters

Trapezoidal profile											
Case	**b (m)**	**%**	**y (m)**	**%**	**z**	**%**	**n**	**%**	**S_0**	**%**	
1a	100	5	1.20	5	1.5	5	0.035	10	0.0005	4	
1b	100	5	1.20	5	1.5	5	0.07	30	0.0005	4	
2a	100	5	1.20	5	1.5	5	0.04	20	0.0005	4	
2b	100	5	1.20	5	1.5	5	0.05	30	0.0005	4	

The next step involves the transformation of the parameters of the problem into fuzzy triangular numbers. Thus, according to the uncertainty rates, presented in Table 2, of parameters and from Equation (4), the following triangular fuzzy numbers arise:

$$\tilde{b} = \text{tfn}(100,\ 5,\ 5),\ \tilde{y} = \text{tfn}(1.2,\ 1.14,\ 1.26),\ \tilde{n} = \text{tfn}(0.035,\ 0.0315,\ 0.0385),$$

$$\tilde{z} = \text{tfn}(1.5, 1.425, 0.0385, 1.575),\ \tilde{S}_0 = \text{tfn}(0.0005,\ 0.00048,\ 0.00052)$$

Having now transformed the parameters into triangular fuzzy numbers, it is possible to create the intervals of confidence for the fuzzy numbers $\tilde{b}, \tilde{y}, \tilde{z}, \tilde{n}$ and $\tilde{S}_0$ for the α-cut $= 0$ as follows:

$$X_1 = [95,\ 105],\ X_2 = [1.14,\ 1.26],\ \ X_3 = [1.425,\ 1.575],\ X_4 = [0.0315,\ 0.0385] \text{ and}$$
$$X_5 = [0.00048,\ 0.00052]$$

The transformation will convert those intervals into arrays $\hat{X}_i$ (with $i = 1, \ldots, n$) each of length $2^{n=5} = 32$ for α-cut $= 0$ using Equations (11) and (12).

The final results are given in Table 3, replacing the appropriate values in Equation (18). From these results and considering the minimum and maximum values of $\hat{Q}$, the final interval $Q = [67.32,\ 111.95]$ is obtained.

A similar approach is utilized for all α-cuts of case 1a and for all cases in general as well as for both scenarios. The results for all cases and scenarios are listed in Table 4.

Figure 5 shows the final discharge estimations for both scenarios and for all cases at all α-cut levels. It is very useful that the graphs are presented in this form as the variation of the final discharge estimation is more obvious for the same case (i.e. 1a, 1b, etc.) in the two different scenarios.

Observing the final intervals of Table 4 as well as their graphical representation in Figure 5, first it is important to highlight that applying scenario 1, where all the parameters have a degree of uncertainty, we would expect to see a big difference in the discharge results in relation to scenario 2, where only the n coefficient is uncertain. However, it was clear that although the variation of the discharge decreases with scenario 2, the overall problem uncertainty remains at high levels for all cases. According to the results, the differences in terms of the variation are approximately in the range of 5% to 15% for the

Table 3 | Final calculations of $\hat{Q}$ for case 1a

	$\hat{Q}(\hat{X}_1, \hat{X}_2, \hat{X}_3, \hat{X}_4, \hat{X}_5)$	Value		$\hat{Q}(\hat{X}_1, \hat{X}_2, \hat{X}_3, \hat{X}_4, \hat{X}_5)$	Value
1	Q(95, 1.14, 1.425, 0.0315, 0.00048)	82.28	17	Q(105, 1.14, 1.425, 0.0315, 0.00048)	85.64
2	Q(95, 1.14, 1.425, 0.0315, 0.00052)	90.93	18	Q(105, 1.14, 1.425, 0.0315, 0.00052)	94.64
3	Q(95, 1.14, 1.425, 0.0385, 0.00048)	97.23	19	Q(105, 1.14, 1.425, 0.0385, 0.00048)	101.20
4	Q(95, 1.14, 1.425, 0.0385, 0.00052)	107.45	20	Q(105, 1.14, 1.425, 0.0385, 0.00052)	111.83
5	Q(95, 1.14, 1.575, 0.0315, 0.00048)	82.36	21	Q(105, 1.14, 1.575, 0.0315, 0.00048)	85.73
6	Q(95, 1.14, 1.575, 0.0315, 0.00052)	91.01	22	Q(105, 1.14, 1.575, 0.0315 0.00052)	94.73
7	Q(95, 1.14, 1.575, 0.0385, 0.00048)	97.34	23	Q(105, 1.14, 1.575, 0.0385, 0.00048)	101.31
8	Q(95, 1.14, 1.575, 0.0385, 0.00052)	107.56	24	Q(105, 1.14, 1.575, 0.0385, 0.00052)	111.95
9	Q(95, 1.26, 1.425, 0.0315, 0.00048)	67.32	25	Q(105, 1.26, 1.425, 0.0315, 0.00048)	70.06
10	Q(95, 1.26, 1.425, 0.0315, 0.00052)	74.39	26	Q(105, 1.26, 1.425, 0.0315, 0.00052)	77.43
11	Q(95, 1.26, 1.425, 0.0385, 0.00048)	79.55	27	Q(105, 1.26, 1.425, 0.0385, 0.00048)	82.80
12	Q(95, 1.26, 1.425, 0.0385 0.00052)	87.91	28	Q(105, 1.26, 1.425, 0.0385, 0.00052)	91.50
13	Q(95, 1.26, 1.575, 0.0315, 0.00048)	67.39	29	Q(105, 1.26, 1.575, 0.0315, 0.00048)	70.14
14	Q(95, 1.26, 1.575, 0.0315 0.00052)	74.46	30	Q(105, 1.26, 1.575, 0.0315, 0.00052)	77.50
15	Q(95, 1.26, 1.575, 0.0385, 0.00048)	79.64	31	Q(105, 1.26, 1.575, 0.0385, 0.00048)	82.89
16	Q(95, 1.26, 1.575, 0.0385, 0.00052)	88.00	32	Q(105, 1.26, 1.575, 0.0385, 0.00052)	91.59

Table 4 | Final discharge Q ranges for all cases and for all α-cuts

		Case			
		Scenario 1			
α-cut		**1a**	**1b**	**2a**	**2b**
0	Q	[67.32, 111.95]	[28.48, 71.97]	[54.00, 110.20]	[39.87, 100.75]
0.25		[71.70, 104.98]	[31.46, 62.65]	[58.65, 99.96]	[44.05, 87.71]
0.5		[76.38, 98.47]	[34.84, 55.03]	[63.80, 90.95]	[48.82, 77.04]
0.75		[81.37, 92.39]	[38.79, 48.69]	[69.51, 82.97]	[54.31, 68.17]
1		[86.70, 86.70]	[43.35, 43.35]	[75.86, 75.86]	[60.69, 60.69]
		Scenario 2			
0	Q	[78.82, 96.33]	[33.34, 61.93]	[63.22, 94.83]	[46.68, 86.70]
0.25		[80.65, 93.73]	[35.38, 55.93]	[65.97, 89.25]	[49.54, 78.31]
0.5		[82.57, 91.26]	[37.69, 51.00]	[68.96, 84.29]	[52.77, 71.40]
0.75		[84.58, 88.92]	[40.32, 46.86]	[72.25, 79.85]	[56.45, 65.61]
1		[86.70, 86.70]	[43.35, 43.35]	[75.86, 75.86]	[60.69, 60.69]

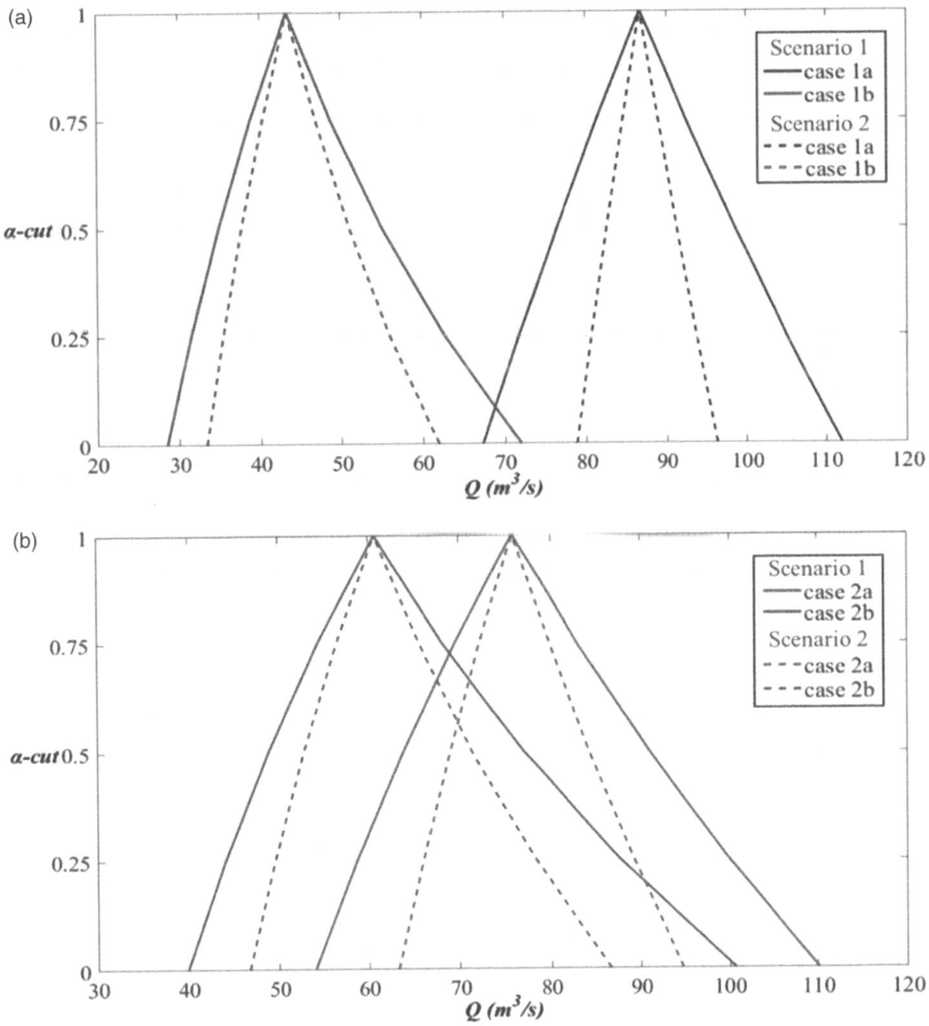

Figure 5 | Discharge estimations for both scenarios for cases (a) 1a and 1b, (b) 2a and 2b.

four different substrates. This fact shows the great influence that the n coefficient has on the Manning equation and consequently on the final estimation of the discharge. It is also noticeable that having all the parameters of the problem in uncertain mode, the problem becomes quite complex as the number of parameters for the case of a trapezoidal channel profile is large ($n = 5$) and the total degree of uncertainty of the problem reaches the point where the results are difficult to interpret. Even though the transformation method was applied to the problem, which gives the opportunity to avoid overestimation errors and gives satisfactory results, it turned out that a complex problem consisting of many uncertain parameters requires advanced judgment in terms of understanding the results and making final decisions.

In a more flexible context, and considering the results, it is suggested in the first stage of the study to take into consideration only the n coefficient as an uncertain parameter. However, if it is absolutely necessary to treat other parameters of the problem in a fuzzy way, then it is proposed that the uncertainty rate given to each parameter should be based on the engineer's or researcher's skills and expertise. The uncertainty rate should be wisely defined without extreme definition values.

Nonetheless, we must admit that the presented methodology was applied to an ideal problem of trapezoidal cross-section open channel flow and its theoretical approach in the investigated problem. This methodology has a limitation in its application to natural open channels (i.e. streams, rivers) in which the cross-section is irregular and variable. In these cases, field measurements and reliable cross-sectional data of slopes, depth, substrates etc. along the flow are considered necessary for the accurate application of the methodology and the extraction of correct conclusions.

CONCLUSIONS

The main goal of this work was to implement an already known fuzzy method, namely the reduced transformation method, to a practical problem of hydraulic engineering described by the Manning equation. The variation of Manning's roughness coefficient is quite high and accurate values cannot usually be obtained. It should be noted that the purpose of this work was not to find an optimal value of the Manning's roughness coefficient for each type of channel, as has been done so far in the literature, but to include mainly the uncertainties of the n coefficient in the calculations. Using the transformation method in its reduced form, the normal value of the coefficient for each case is introduced and variations are considered, thus achieving satisfactory accuracy in estimating the discharge. Finally, as the results show, the uncertainty of the roughness coefficient cannot be neglected since its influence on the results is quite high. Fuzzy methodology and more specifically the transformation method is suitable for engineering problems, described by uncertain parameters, because it is simple to apply and avoids overestimation errors. This way, better management plans and decisions can be carried out.

DATA AVAILABILITY STATEMENT

All relevant data are included in the paper or its Supplementary Information.

REFERENCES

Al-Husseini, T. R. 2015 Estimation of Manning's roughness coefficient for Al-Diwaniya River. *Eng. Sust. Dev.* **19** (4), 170–184.

Azamathulla, H. M., Ahmad, Z. & Ghani, A. Ab. 2013 An expert system for predicting Manning's roughness coefficient in open channels by using gene expression programming. *Neural Comput. Appl.* **23**, 1343–1349. doi:10.1007/s00521-012-1078-z.

Bahramifar, A., Shirkhani, R. & Mohammadi, M. 2013 An ANFIS-based approach for predicting the Manning roughness coefficient in alluvial channels at the bank-full stage. *Int. J. Eng. Trans. B Appl.* **26**, 177–186.

Choo, Y. M., Yun, G. S. & Choo, T. H. 2018 A research on the estimation of coefficient roughness in open channel applying entropy concept. *Environ. Earth Sci.* **77**, 624. doi:10.1007/s12665-018-7809-4.

Chow, V. T. 1959 *Open-Channel Hydraulics*. McGraw-Hill, Singapore.

Cowan, W. L. 1956 Estimating hydraulic roughness coefficients. *Agric. Eng.* **37**, 473–475.

Dash, S. S. & Khatua, K. K. 2016 Sinuosity dependency on stage discharge in meandering channels. *J. Irrig. Drain. Eng.* **142**, 04016030. doi:10.1061/(asce)ir.1943-4774.0001037.

Dong, W. & Shah, H. C. 1987 Vertex method for computing functions of fuzzy variables. *Fuzzy Sets Syst.* **24**, 65–78. doi:10.1016/0165-0114(87)90114-X.

Geem, Z. W. & Kim, J. H. 2018 Application of computational intelligence techniques to an environmental flow formula. *Int. J. Fuzzy Log. Intell. Syst.* **18**, 237–244. doi:10.5391/IJFIS.2018.18.4.237.

Goetschel, R. & Voxman, W. 1986 Elementary fuzzy calculus. *Fuzzy Sets Syst.* **18**, 31–43.

Hanss, M. 2002 The transformation method for the simulation and analysis of systems with uncertain parameters. *Fuzzy Sets Syst.* **130**, 277–289. doi:10.1016/S0165-0114(02)00045-3.

Hanss, M. 2005 *Applied Fuzzy Arithmetic: An Introduction with Engineering Applications*. Springer, Berlin, Germany.

Huang, S. L. 2007 Effects of using different sediment transport formulae and methods of computing Manning's roughness coefficient on numerical modeling of sediment transport. *J. Hyd. Res.* **45** (3), 347–356. doi:10.1080/00221686.2007.9521768.

Kaufmann, A. & Gupta, M. M. 1991 *Introduction to Fuzzy Arithmetic Theory and Applications*. Van Nostrand Reinhold, New York, USA.

Khatua, K. K., Patra, K. C. & Mohanty, P. K. 2012 Stage–discharge prediction for straight and smooth compound channels with wide floodplains. *J. Hydraul. Eng.* **138** (1), 93–99. doi:10.1061/(asce)hy.1943-7900.0000491.

Klimke, A. 2003 An efficient implementation of the transformation method of fuzzy arithmetic. In: *22nd International Conference of the North American Fuzzy Information Processing Society, NAFIPS 2003*. IEEE, Piscataway, NJ, USA, pp. 468–473. doi:10.1109/NAFIPS.2003.1226830.

Limerinos, J. T. 1970 *Determination of the Manning Coefficient from Measured Bed Roughness in Natural Channels*. US Department of the Interior, Geological Survey, Water Resources Division, Washington, DC, USA.

Mailapalli, D. R., Raghuwanshi, N. S., Singh, R., Schmitz, G. H. & Lennartz, F. 2008 Spatial and temporal variation of Manning's roughness coefficient in furrow irrigation. *J. Irrig. Drain. Eng.* **134**, 185–192. doi:10.1061/(asce)0733-9437(2008)134:2(185).

Manning, R. 1891 On the flow of water in open channels and pipes. *Trans. Inst. Civil Eng. Ireland* **20**, 161–207.

Mohanta, A., Patra, K. C. & Sahoo, B. B. 2018 Anticipate Manning's coefficient in meandering compound channels. *Hydrology* **5**, 47. doi:10.3390/hydrology5030047.

Moharana, S. & Khatua, K. K. 2014 Prediction of roughness coefficient of a meandering open channel flow using Neuro-Fuzzy Inference System. *Measurement* **51**, 112–123. doi:10.1016/j.measurement.2014.01.026.

Negoiță, C. V. & Ralescu, D. A. 1975 *Applications of Fuzzy Sets to Systems Analysis*. Birkhauser, Basel, Switzerland. doi:10.1007/978-3-0348-5921-9.

Pradhan, A. & Khatua, K. K. 2018 Assessment of roughness coefficient for meandering compound channels. *KSCE J. Civ. Eng.* **22**, 2010–2022. doi:10.1007/s12205-017-1818-9.

Şamandar, A. 2011 A model of adaptive neural-based fuzzy inference system (ANFIS) for prediction of friction coefficient in open channel flow. *Sci. Res. Essays.* **6**, 1020–1027.

Shiono, K., Al-Romaih, J. S. & Knight, D. W. 1999 Stage–discharge assessment in compound meandering channels. *J. Hydraul. Eng.* **125**, 66–77. doi:10.1061/(asce)0733-9429(1999)125:1(66).

Yen, B. C. 2002 Open channel flow resistance. *J. Hydraul. Eng.* **128**, 20–39. doi:10.1061/(asce)0733-9429(2002)128:1(20).

Zadeh, L. A. 1965 Fuzzy sets. *Inf. Control* **8**, 338–353. doi:10.1016/S0019-9958(65)90241-X.

Zhang, G. H., Luo, R. T., Cao, Y., Shen, R. C. & Zhang, X. C. 2010 Impacts of sediment load on Manning coefficient in supercritical shallow flow on steep slopes. *Hydrol. Proc.* **24**, 3909–3914. doi:10.1002/hyp.7892.

Zimmermann, H. J. 1992 *Fuzzy Set Theory and Its Applications*, 2nd edn. Kluwer Academic, Dordrecht, The Netherlands.

First received 31 January 2021; accepted in revised form 11 May 2021. Available online 24 May 2021

Study on optimal start-up time and water usage volume of blowdown residue discharge of mesh filter

Kai Shi, Zhenji Liu, Yan Xie and Man Li

ABSTRACT

The water volume used for blowdown of a mesh filter is one of the important indicators to evaluate the economic benefit of a mesh filter. The amount of water used for drainage of the mesh filter can be controlled by setting the start-up time of the blowdown residue discharge of the mesh filter. Through the indoor prototype test, the sediment of the Manas riverbed in Xinjiang, China, was used as the filter medium to test the mesh filter and analyze the water usage volume at each stage. The test focused on the analysis of the trend of the pressure difference between the inlet and outlet of the mesh filter over the filtration time. The results showed that as the filtration time was prolonged, the inlet and outlet pressure difference of the mesh filter showed an S-shaped curve with time. The pressure difference between the inlet and outlet can be divided into four stages, i.e., the start of filtering to the first inflection point of the abrupt change in the pressure difference, the first inflection point of the abrupt pressure difference change to the median point of filtering, the median point of filtering to the second inflection point of the abrupt change in the pressure difference, and the second inflection point of the abrupt pressure difference change to the end of filtering. At the same time, the water usage volume for blowdown was calculated under various working conditions. This paper analyzed the start-up time of blowdown at three different time points, i.e., after the completion of the entire filtration cycle of the traditional mesh filter, at the first inflection point of pressure difference, and at the second inflection point of pressure difference. The tests were performed under the condition that the filtration cycle of the mesh filter was ten days. The results have shown that compared with the traditional screen filter that starts the blowdown at the end of the entire filtration cycle, starting the blowdown at the first inflection point of pressure difference can save 200–300 m^3 of water resources for a single mesh filter, and starting the blowdown at the second inflection point of pressure difference can save 30–90 m^3 of water resources for a single mesh filter. Meanwhile, the working conditions in the analyses in this paper were consistent with the actual engineering operating conditions. It is recommended that the mesh filter in the micro-irrigation system use the first inflection point of pressure difference to control the start-up time of the blowdown residue discharge, which can not only save water resources, but also ensure that the mesh filter runs at the optimal filtering conditions.

Key words | inflection point of inlet and outlet pressure difference, mesh filter, S-shaped growth, start-up time of blowdown residue discharge, water usage volume of blowdown residue discharge

Kai Shi
Zhenji Liu (corresponding author)
Yan Xie
Man Li
College of Water Conservancy and Civil Engineering,
Shihezi University,
Shihezi 832000,
China
E-mail: *39080706@qq.com*

doi: 10.2166/ws.2020.262

HIGHLIGHTS

- The study analyzed the change rule of the inlet and outlet pressure difference over time during the entire filtration stage of the mesh filter.
- The research focused on analyzing the water usage volume when the blowdown was started at different filtration time points.

GRAPHICAL ABSTRACT

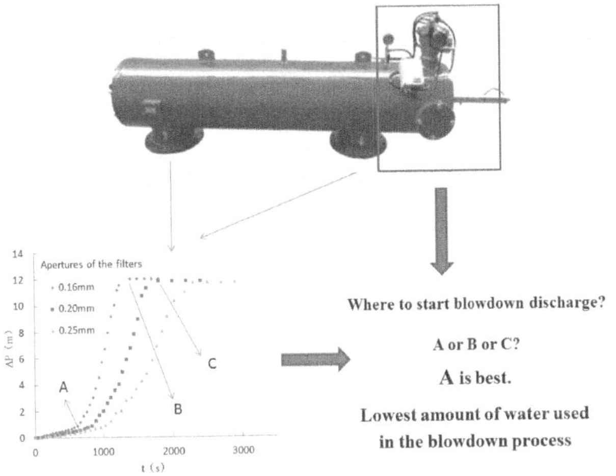

PREFACE

Water resources are an irreplaceable basic element in agricultural development, but the current shortage of water resources restricts the sustainable development of agricultural production (Hamududu & Ngoma 2020; Kulmatov *et al.* 2020; Li *et al.* 2020). Water conservation and efficient use of water resources have become the common goal of the world's agricultural development (de Vito *et al.* 2019). Micro-irrigation is an important technology for saving water resources for agricultural irrigation, and provides a strong technical prerequisite for the development of water-saving irrigation. As the core component of the micro-irrigation system, the filter mainly serves to remove mud particles and other suspended impurities from muddy water sources (Zong *et al.* 2015). Since the mesh filter has the advantages of good filtering effect, simple cleaning method and good cleaning effect, it has a wide range of applications. The studies on mesh filter have been focused on two key research directions, i.e., filtration process and cleaning process of mesh filters. Many scholars have studied the pressure difference of the filter during the filtration process, and the results have shown that the filter pressure difference is related to the inlet water flow, the sand content of the inlet water and the porosity of the filter mesh. In addition, the pressure difference between the inlet and outlet shows exponential change with time (Duran-Ros *et al.* 2010; Elbana *et al.* 2013; Bové *et al.* 2015; Bounoua *et al.* 2016). Secondly, models such as Gaussian process regression (GPR) have been applied to study the filters. The GPR model was used to establish a predictive model to predict the dissolved oxygen of the wastewater during the filtering process, analyze the turbidity of the effluent under different working conditions of the filter, and analyze the best parameters through regression (García-Nieto *et al.* 2020a, 2020b). Based on the artificial neural network (ANN), gene expression programming (GEP) and multivariable linear regression (MLR) models, Martí *et al.* used the test data of 769 filtration cycles to estimate dissolved oxygen content. The predicted results were consistent with the test results and the previously published results (Martí *et al.* 2013; García-Nieto *et al.* 2016). Mesquita *et al.* (2019) designed the diffuser and evaluated its hydraulic performance based on computational fluid dynamics (CFD) numerical simulation technology to maximize hydrodynamic efficiency (Mesquita *et al.* 2019). Meanwhile, research on backflush in the blowdown residue discharge stage is also a main research angle. The test method was used to measure the backflush time of the filter, and the backwashing process was studied by combining the backflush efficiency and the backflush pressure difference (Elbana *et al.* 2012; Zong *et al.* 2019). At the same time, some scholars proposed a hydraulic characterization method for the filter backflush process to evaluate the influence of sediment elements on filter design in micro-irrigation (de Deus *et al.* 2020). Many research results have been obtained on the mesh filter and have shown that the pressure difference between the inlet and outlet of the mesh filter has a consistent trend over time. However, there have been only few systematic theoretical studies on the trend of the pressure difference between the inlet and outlet of the mesh filter over time, the optimal start-up time of the filter's blowdown, and the required water volume for blowdown. Furthermore, the relationship between the pressure difference of the mesh filter and the amount of clogged impurities in the pores has rarely been studied. The clogging of impurities further determines the water usage volume of the mesh filter during blowdown, which in turn determines the economic benefit of the filter. Therefore, this paper took a commonly used micro-irrigation mesh filter as the research object, focused on the trend of the pressure difference between the inlet and outlet over time under different working conditions, and studied the inflection point of the pressure difference curves. Moreover, the required water volume for blowdown residue discharge was studied under various operation conditions. The required water volumes for blowdown were compared and analyzed when the blowdown was applied at different time points (including traditionally at the end of the entire filtration cycle and at the inflection points). Based on the results, the optimal blowdown start-up time of the mesh filter was determined. At this optimal start-up time, the mesh filter can operate under the best economic efficiency and the optimal working conditions.

FILTER STRUCTURE AND WORKING PRINCIPLE

The mesh filter is mainly composed of filter housing, a filtering device, a blowdown device and a blowdown-starting control device. The working principle diagram is shown in Figure 1 within which, Figure 1(a) is the working principle diagram of a vertical mesh filter, and Figure 1(b) is the working principle diagram of a horizontal mesh filter. The filter design parameters are shown in Table 1. The apertures of the filters used in this test were 0.25 mm, 0.20 mm, and 0.16 mm. The working principle of the mesh filter is as follows: during the filtering process, the sand-containing water for irrigation enters the water inlet of the filter and is filtered from the inside to the outside through the filter. Large particles of silt impurities and organic impurities are intercepted by the filter. Impurity particles with a size smaller than the pore of the filter screen and filtered clean water enter the designated irrigation system. When the pressure difference between the inside and outside of the filter

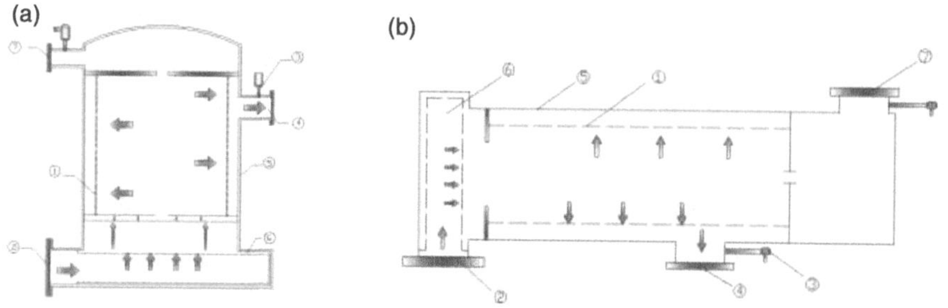

Figure 1 | Working principle diagram of (a) the vertical mesh filter and (b) the horizontal mesh filter: ① fine filter; ② water inlet; ③ precision electronic pressure gauge; ④ water outlet; ⑤ filter housing; ⑥ coarse filter; ⑦ sewage outlet.

Table 1 | Summary table of filter design parameters

Parameter category	Specification (mm)
Inlet diameter	150
Outlet diameter	150
Filter housing diameter	420
Filter mesh diameter	360
Import and export height difference	400

reaches the optimal start pressure difference for blowdown, the blowdown process is initiated to complete the filter cleaning work, and then the next filtration cycle is initiated. In the discharge process, the water carries the sediment impurities from the discharge pipe to the designated sedimentation system, and the mesh-filtered water continues to enter the irrigation water source for irrigation, avoiding the waste of water resources.

TEST DEVICE AND TEST METHOD

Test device

In this test, a universal micro-irrigation mesh filter produced by Shihezi Tianlu Water Saving Equipment Co., Ltd, Xinjiang, China, was used. The diagram of the test device is shown in Figure 2.

Test equipment

The test equipment is summarized in Table 2.

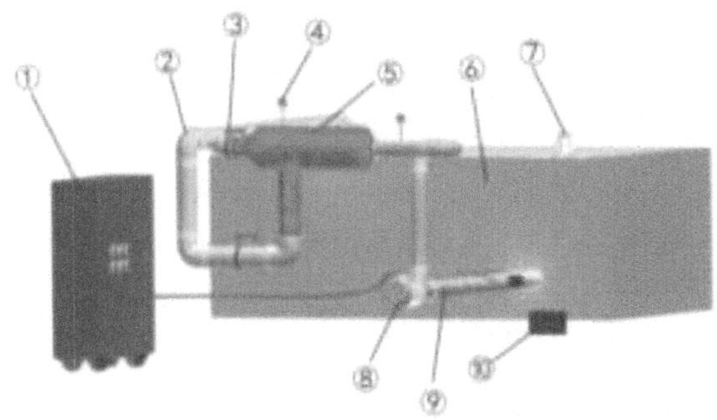

Figure 2 | Platform construction diagram of the test device: ① inverter cabinet; ② outlet pipe; ③ waste pipe; ④ pressure gauge; ⑤ filter; ⑥ reservoir; ⑦ mixing tank; ⑧ water pump; ⑨ inlet pipe; ⑩ flow meter.

Test methods and steps

The test device was installed and debugged according to Figure 2. This experiment used the orthogonal experiment method to arrange the experiment groups. Each experiment was conducted twice, and the average of the two tests was taken as the experiment result. The optimal flow rate of the commonly used mesh filter is in the range of 180–220 m^3/h. Thus, in this test, the inlet water flow rate was set to 180 m^3/h, 200 m^3/h, and 220 m^3/h, respectively. In order to allow enough time for the test to take water samples without prolonging the test time, the influent sand contents were set to four values, which were 0.045 g/L, 0.068 g/L, 0.082 g/L, and 0.125 g/L, respectively. During the filtering process, the artificially mixed riverbed sand was uniformly added to the sedimentation tank, and the water samples were taken at the inlet and outlet of the filtration system every five minutes. The purpose of this test is to determine the change of the pressure difference between the inlet and the outlet after reaching the peak value, so the filtration time should be appropriately extended after the pressure difference reaches the peak value. The blowdown process was initiated immediately after the filtration process was completed. After the filter screen was cleaned, the next filtration process was initiated. After the filtration and blowdown processes were completed, the sand content of the connected water sample was measured to complete all the tests. In this test, a reservoir was used to store and supply water for the filter. The water source of the reservoir was obtained by extracting groundwater. The water source did not contain organic impurities or suspended impurities, and its turbidity was less than 0.3NTU. The sieved sediment impurities were added uniformly at the junction of the reservoir and the water inlet of the filter. This test mainly studies the working conditions in which the inlet water only contains sediment impurities, because in the actual project, organic impurities and suspended impurities can be filtered by first-level system filtering and do not enter the filter.

Gradation of sand particles in the test

Studies have shown that the physical properties of the filter media such as bulk density, actual density, porosity and

Table 2 | Summary of test equipment

Equipment name	Model	Specification	Quantity	Remarks
Three-phase asynchronous motor	Y90L-04	Rated power 1.5 kW	1	Placed in at the mixing tank
Frequency control cabinet	DZB300B0015L4A		1	
Electronic pressure gauge	JDC600	Accuracy is 0.001 MPa	4	Located inside the water inlet, water outlet, sewage outlet and strainer
Pressure sensor	AS-131		4	Located inside the water inlet, water outlet, sewage outlet and strainer
Butterfly valve			3	Located at the water inlet, water outlet and sewage outlet respectively

sphericity have a certain impact on the performance of the filter. But the size and gradation of particles have more significant effect on the filter performance. Thus in this study, the riverbed sand of the Manas River Basin in Xinjiang Uygur Autonomous Region was used as the filter medium, and the particle gradation of the medium was analyzed. Through the particle size screening test, particles with a particle size smaller than 0.25 mm accounted for 30% of the total sediment, particles with a particle size smaller than 0.20 mm accounted for 25% of the total sediment, and particles with a particle size smaller than 0.16 mm accounted for 20% of the total sediment of the sand.

THEORETICAL ANALYSES OF PRESSURE DIFFERENCE AND WATER USAGE FOR BLOWDOWN

Theoretical analysis of pressure difference

The change of the pressure difference of the mesh filter with time was mainly affected by the filtration flow rate. The change of pressure difference was positively correlated with the quadratic square of the filtration flow rate. Filter clogging can be mainly divided into two stages, i.e., the clogging of the filter pores and the clogging of the inner surface of the filter. When the filter is completely clogged, the pressure difference between the inlet and outlet of the filter reaches the peak value. From analyzing the relevant results, the increase in pressure difference with filtration time during the filtration process is approximately exponentially increasing with time. When the pressure difference peak is reached, the filtration time is appropriately extended, and the change curve is approximately S-shaped. Origin was used to fit the pre-test multi-group test results, and the satisfied functional relationship is shown in Equation (1):

$$f(x) = \frac{A_1 - A_2}{1 + e^{\frac{x - x_0}{dx}}} + A_2 \tag{1}$$

where A_1 is the initial value, A_2 is the final value, x_0 is the center, and dx is a constant.

The mathematical expression of the filter pressure difference of the mesh filter with the filter time can be shown in Equation (2):

$$\Delta P = \frac{P_1 - P_2}{1 + e^{\alpha(t - t_0)}} + P_2 \tag{2}$$

where P_1 is the initial value of the pressure difference (m), P_2 is the final value of the pressure difference (m), t_0 is the center (s), α is a constant, ΔP is the pressure drop of the filter (m), and t is the filtration time (s).

Analysis of required water volume for blowdown

The change in the pressure difference between the inlet and outlet of the mesh filter is mainly caused by the accumulation of impurities in the mesh. The clogging of the mesh can be divided into two main stages, i.e., clogging of the pores and clogging of the inner surface of the mesh. When the clogging on the surface of the mesh reached a certain thickness, the pressure difference between the inlet and outlet of the filter reached the peak value. Therefore, in the first stage, the pores of the mesh filter were blocked, and only particles larger than the mesh were blocked inside the mesh surface. The interior of the mesh was calculated according to the principle of standard screening and mass conservation. The mathematical expression of the mass of the clogged impurities is shown in Equation (3):

$$M_1 = S_1 * Q_1 * t_1 * \gamma \tag{3}$$

where M_1 is the amount of blocked sand in the filter screen during the clogging stage (kg), S_1 is the influent sand content (g/L), Q_1 is the influent filtration flow rate (m^3/h), and t_1 is the time to reach an inflection point (s), and γ is the percentage of impurities greater than the pore size of the filter.

In the second stage, the clogging process was completed on the surface of the filter screen, that is, all particles were clogged inside the filter screen. The mathematical expression of the amount of clogged sand is shown in Equation (4):

$$M_2 = S_1 * Q_1 * (t_1 - t_2) \tag{4}$$

where M_2 is the amount of sand clogged inside the filter during the clogging stage (kg), and t_2 is the time to complete the filtration (s).

The principle of the blowdown of the mesh filter is mainly to use water flow to discharge sediment. The calculation expression of the single discharge time is shown in Equation (5):

$$t_2 = \frac{M}{S_* Q_2} \tag{5}$$

where t_2 is the single blowdown time (s), M is the total amount of clogged impurities (kg), S_* is the sediment carrying capability (g/L), and Q_2 is the blowdown flow rate (m^3/h).

According to calculation of the sediment carrying capacity of the water in the pressure irrigation pipe, the sediment carrying capacity of the pressure irrigation pipe flow can be calculated by Equation (6):

$$S_* = \frac{\lambda v^2}{0.000023\omega^{0.25} \cdot 8g} \tag{6}$$

where v is the average flow velocity of the pipe flow (m/s), g is the acceleration of gravity (m/s^2), and ω is the average sedimentation velocity of the sediment (mm/s).

The filter material is steel pipe or cast iron, and the drag coefficient along the pipeline can be calculated by Equation (7):

$$\lambda = \frac{0.02164}{d^{0.1}} \tag{7}$$

where d is the pipe diameter (m).

The consumed water volume by the mesh filter is shown in Equation (8):

$$V = Q_2 * t_2 * \beta \tag{8}$$

where β is the number of blowdowns in a filtration cycle.

TEST RESULTS AND ANALYSIS

Test results

Under various sandy soil conditions, the mesh filters with the apertures of 0.25 mm, 0.20 mm and 0.16 mm were tested. These filters are commonly used in field micro-irrigation. The trends of the pressure difference between the inlet and outlet (ΔP) of the mesh filter over the filtration time (t) of four different sand content conditions were plotted, as shown in Figures 3, 4, and 5.

Analysis of test results

Through orthogonal test analysis, when the sediment content in the inlet water was constant, the pressure difference between the inlet and outlet changed with time in the same pattern, showing an S-shaped growth trend. Each curve can be divided into four parts, i.e., the beginning of the test to the first inflection point of the sudden change in the pressure difference, the first inflection point of the sudden change in pressure difference to the median point of filtering, the median point of filtering to the second inflection point of the sudden change in pressure difference, and the second inflection point of the sudden change in pressure

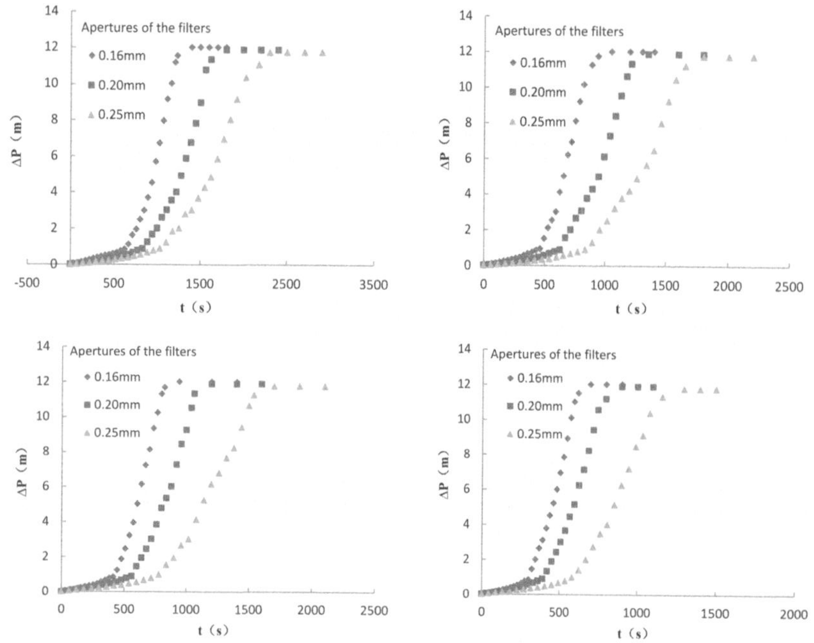

Figure 3 | Trend of pressure difference under the working condition of 180 m^3/h: (1) $S_1 = 0.045$ g/L, (2) $S_1 = 0.068$ g/L, (3) $S_1 = 0.082$ g/L, (4) $S_1 = 0.125$ g/L.

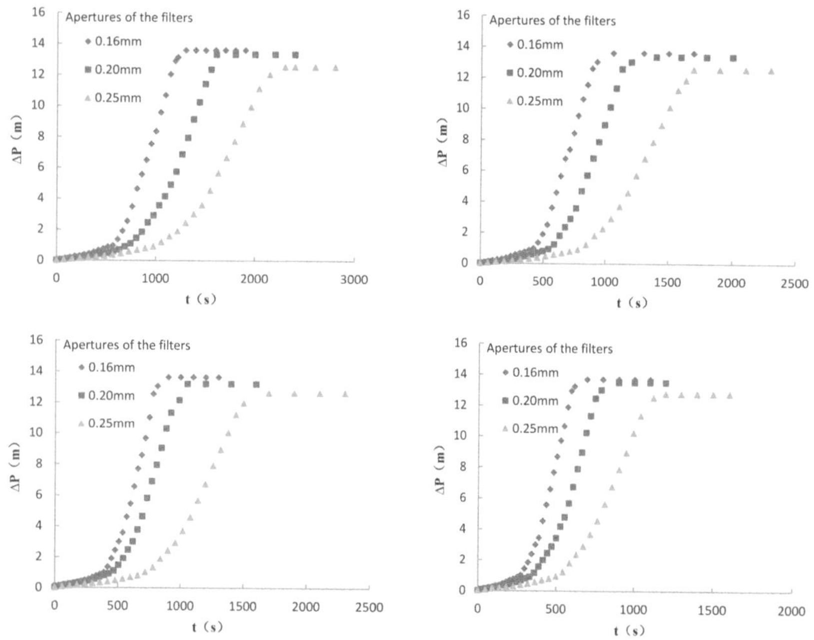

Figure 4 | Trend of pressure difference under the working condition of 200 m^3/h: (1) $S_1 = 0.045$ g/L, (2) $S_1 = 0.068$ g/L, (3) $S_1 = 0.082$ g/L, (4) $S_1 = 0.125$ g/L.

difference to the end of filtering. In the first stage, the pressure difference between the inlet and outlet of the mesh filter basically remained unchanged. In the second stage, the pressure difference between the inlet and outlet of the mesh filter changed at an accelerating rate with time, and the pressure difference between the inlet and outlet increased sharply. In the third stage, the changing rate of the pressure difference between the inlet and outlet of the mesh filter slowed down, and the pressure difference increased slowly. In the fourth stage, the pressure difference

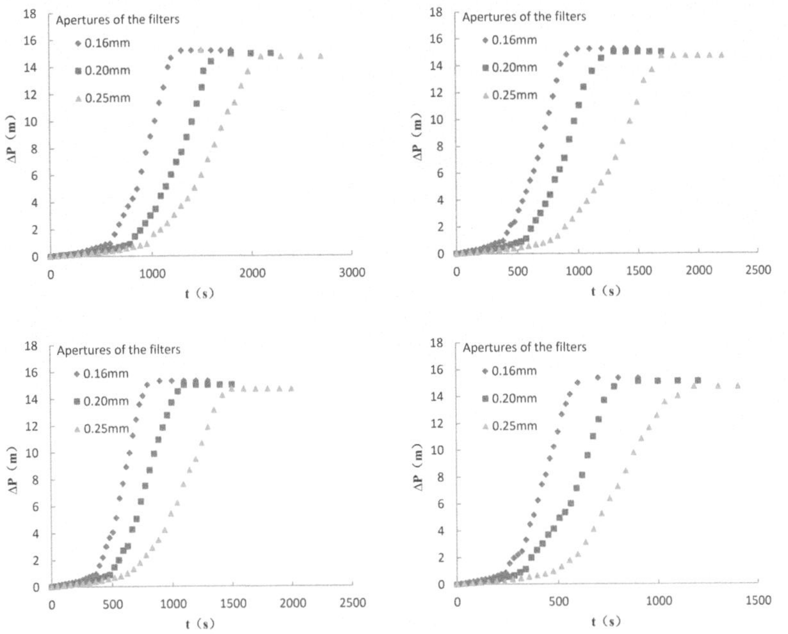

Figure 5 | Trend of pressure difference under the working condition of 220 m^3/h: (1) $S_1 = 0.045$ g/L, (2) $S_1 = 0.068$ g/L, (3) $S_1 = 0.082$ g/L, (4) $S_1 = 0.125$ g/L.

reached the peak value, and the pressure difference between the inlet and outlet of the mesh filter did not increase with time. Based on the analysis of the filtering principle of the mesh filter, the mesh filter clogging can be divided into two stages, i.e., clogging of the mesh filter pores and clogging of the mesh surface. Combining with the analysis of the curves in Figures 3–5, the clogging of the filter pores was almost completed at the first inflection point of the pressure difference, and it was difficult for the impurities of small particles to pass through the filter pores. Then the surface clogging of the filter started. When the second inflection point of the pressure difference was reached, the pores and the surface of the filter were completely clogged, and the contaminated water could not be filtered. At this time, the pressure difference of the filter screen reached the peak value. Origin was used to fit the raw data in each group using Equation (2), and the fitting correlation was higher than 99%. Therefore, it is considered that Equation (2) has a high correlation and can be used to calculate the pressure difference of similar types of network filters.

Water usage analysis

Under different working conditions, the water usage volumes by the mesh filter at three time points, i.e., the first inflection point of the pressure difference, the second inflection point of the pressure difference, and the start of blowdown residue discharge at the end of the filtration phase, were calculated based on the filtration test results. Taking a ten-day filtration cycle as the research object, the water usage in one filtration cycle was investigated at the flow rate of 180 m^3/h. The analysis focused on the time of single blowdown (t_2), the total number of blowdowns (β), the total time of the blowdowns (T), and the cumulative water volume for all the blowdowns (V) over the ten-day test cycle. The predicted results are shown in Tables 3–5.

Tables 3–5 show that when the screen clogging stage starts, the total mass of clogged sediment impurities rapidly increases. In addition, the required water volume for blowdown increases with the increase of the total mass of clogged sediment impurities inside the screen. As the mesh number of the filter increased, that is, the impurities smaller than the pore size of the filter accounted for a higher portion, the amount of the clogged sand in the filter increased, and the filtration time was shortened. As a result, more blowdowns were needed in the entire filtration cycle of the filter and the required water volume for blowdown was increased. When the inlet water flow rate was 180 m^3/h, the maximum water usage for blowdown was 1,476 m^3 at the

Table 3 | Analysis of water usage for blowdown at the filter aperture of 0.25 mm in ten days

		Blowdown at the first inflection point of the pressure difference				Blowdown at the second inflection point of the pressure difference				Blowdown at the end of the filtration cycle			
Q_1 m³/h	S_1 g/L	t_2 s	β	T s	V m³	t_2 s	β	T s	V m³	t_2 s	β	T s	V m³
180	0.045	11.15	835	9,305	465	30.61	376	11,498	575	39.84	298	11,869	593
180	0.068	14.65	960	14,060	703	35.57	480	17,073	854	44.87	393	17,621	881
180	0.082	15.31	1,108	16,955	848	41.10	508	20,888	1,044	52.31	411	21,523	1,076
180	0.125	17.50	1,477	25,846	1,292	48.06	665	31,938	1,597	56.60	576	32,603	1,630
200	0.045	11.68	885	10,338	517	34.31	376	12,889	644	42.86	309	13,225	661
200	0.068	15.41	1,014	15,623	781	37.31	508	18,962	948	52.81	376	19,838	992
200	0.082	16.66	1,131	18,839	942	45.81	508	23,284	1,164	64.50	376	24,231	1,212
200	0.125	16.62	1,728	28,718	1,436	49.86	720	35,897	1,795	68.85	540	37,179	1,859
220	0.045	12.49	910	11,372	569	22.85	576	13,163	658	34.13	411	14,044	702
220	0.068	15.32	1,122	17,185	859	41.74	508	21,214	1,061	55.95	393	21,972	1,099
220	0.082	15.40	1,346	20,723	1,036	44.80	576	25,803	1,290	61.93	432	26,753	1,338
220	0.125	15.90	1,986	31,590	1,579	54.82	732	40,137	2,007	66.31	617	40,922	2,046

Table 4 | Analysis of water usage for blowdown at the filter aperture of 0.20 mm in ten days

		Blowdown at the first inflection point of the pressure difference				Blowdown at the second inflection point of the pressure difference				Blowdown at the end of the filtration cycle			
Q_1 m³/h	S_1 g/L	t_2 s	β	T s	V m³	t_2 s	β	T s	V m³	t_2 s	β	T s	V m³
180	0.045	10.27	971	9,969	498	24.27	480	11,649	582	33.50	360	12,060	603
180	0.068	11.77	1,280	15,065	753	27.46	640	17,575	879	37.92	480	18,203	910
180	0.082	12.62	1,440	18,166	908	29.44	720	21,194	1,060	40.65	540	21,951	1,098
180	0.125	13.46	2,057	27,692	1,385	33.97	960	32,615	1,631	42.52	785	33,399	1,670
200	0.045	10.41	1,064	11,077	554	24.05	537	12,907	645	37.56	360	13,520	676
200	0.068	12.67	1,321	16,738	837	31.94	617	19,712	986	47.44	432	20,493	1,025
200	0.082	11.82	1,708	20,185	1,009	28.89	820	23,683	1,184	45.90	540	24,785	1,239
200	0.125	14.60	2,107	30,769	1,538	37.87	960	36,353	1,818	52.11	720	37,521	1,876
220	0.045	10.51	1,160	12,185	609	26.58	540	14,355	718	37.87	393	14,871	744
220	0.068	12.57	1,464	18,412	921	32.75	665	21,764	1,088	44.11	508	22,420	1,121
220	0.082	12.90	1,721	22,203	1,110	33.39	785	26,227	1,311	47.10	576	27,127	1,356
220	0.125	13.91	2,434	33,846	1,692	42.37	960	40,678	2,034	58.04	720	41,791	2,090

first inflection point, 1,656 m^3 at the second inflection point, and 1,698 m^3 at the end of the filtration cycle. When the inlet water flow rate was 200 m^3/h, the maximum water usage for blowdown was 1,641 m^3 at the first inflection point, 1,888 m^3 at the second inflection point, and 1,947 m^3 at the end of the filtration cycle. When the inlet water flow rate was 220 m^3/h, the maximum water usage for blowdown was 1,805 m^3 at the first inflection point, 2,092 m^3 at the second inflection point, and 2,129 m^3 at the end of the filtration cycle. Therefore, when blowdown

Table 5 | Analysis of water usage for blowdown at the filter aperture of 0.16 mm in ten days

		Blowdown at the first inflection point of the pressure difference				Blowdown at the second inflection point of the pressure difference				Blowdown at the end of the filtration cycle			
Q_1 m³/h	S_1 g/L	t_2 s	β	T s	V m³	t_2 s	β	T s	V m³	t_2 s	β	T s	V m³
180	0.045	10.26	1,036	10,634	532	18.97	617	11,709	585	25.13	480	12,061	603
180	0.068	8.59	1,870	16,069	803	25.75	720	18,540	927	30.40	617	18,760	938
180	0.082	9.28	2,087	19,377	969	24.03	919	22,088	1,104	36.93	617	22,789	1,139
180	0.125	12.31	2,400	29,538	1,476	26.84	1,234	33,125	1,656	35.38	960	33,969	1,698
200	0.045	9.41	1,256	11,815	591	19.87	665	13,206	660	30.13	455	13,700	685
200	0.068	9.40	1,899	17,854	893	25.03	815	20,402	1,020	41.56	508	21,123	1,056
200	0.082	10.14	2,123	21,530	1,077	25.50	960	24,479	1,224	37.96	665	25,228	1,261
200	0.125	10.52	3,119	32,821	1,641	30.61	1,234	37,779	1,889	49.60	785	38,959	1,948
220	0.045	9.36	1,389	12,997	650	22.11	665	14,692	735	31.51	480	15,123	756
220	0.068	9.32	2,107	19,640	982	26.08	864	22,537	1,127	40.29	576	23,208	1,160
220	0.082	9.26	2,556	23,683	1,184	28.52	960	27,381	1369	42.23	665	28,065	1,403
220	0.125	10.61	3,402	36,103	1,805	33.91	1,234	41,853	2,093	44.36	960	42,581	2,129

was started at the first inflection point of the pressure difference, although the number of blowdowns was increased, the overall water volume used for blowdown decreased by 200–300 m^3 in one filtration cycle. Considering the short duration of the filtration time, the total discharge of irrigation water is much larger than the total volume of backwashing water. Therefore, in the analysis of the water usage in the blowdown process, the relationship between the blowdown water usage and the total outflow of irrigation water was not considered and the relationship should not affect the economy of the blowdown process. By converting the ten-day filtration cycle accordingly, the analysis in this paper can be applied to actual projects with different filtration cycles.

DISCUSSION

The pressure difference between the inlet and outlet of the mesh filter is mainly caused by the clogging of the pores and the inner surface of the filter. As the clogging continues, the pressure difference between the inside and outside of the filter gradually increases. When the filter is completely clogged, the pressure difference between the inlet and outlet of the filter screen reaches the peak value, and normal filtration cannot be performed. In the actual projects and in many studies, the starting point of the blowdown of the filters was set at the second inflection point of the pressure difference. In this study, the trend of the pressure difference between the inlet and outlet during filtration was used to control the network. The water usage for blowdown was analyzed when the filter was activated at the first inflection point of pressure difference and the second inflection point of pressure difference. Before the first inflection point of pressure difference, the pores of the mesh filter screen were gradually clogged. The sediment impurities entering the filter inlet did not completely accumulate on the inner surface of the filter. Instead, only impurities larger than the pore size of the filter accumulated on the inner surface of the filter. If blowdown was initiated at this time, due to the large particle size of the impurity particles and the small cohesion force, the required sand-carrying force for blowdown was small. Thus the impurities can be transported to the outside of the filter screen to complete the blowdown within a short time. The inner surface of the filter screen was mainly blocked between the first inflection point of pressure change and the second inflection point of pressure difference. During this period, the sediment entering the filter inlet accumulated on the inner surface of the filter, and impurities with a particle size smaller than the pore size of the filter failed to pass through the filter. Due to the accumulation of smaller particles with higher viscosity on the inside of the filter screen, the adhesion of the impurities inside the filter screen was greatly increased. If blowdown was initiated at this time, the transport of impurities took a longer time. At the same time, the pressure difference and drainage characteristics of the filter in related studies are consistent with the conclusions of this article. Therefore, the water usage for the blowdown of the mesh filter can be calculated by Equation (8). Starting the blowdown of the mesh filter at the first inflection point of pressure difference can save water usage by the filter and ensure that the filter is running under the optimal filtering condition.

CONCLUSIONS

(1) The pressure difference between the inlet and outlet of the mesh filter showed a consistently S-shaped growth trend over time. Each curve can be divided into four stages, i.e., the beginning of the test to the first inflection point of the pressure difference, the first inflection point of pressure difference to the median of filtering, the median point of filtering to the second inflection point of pressure difference, and the second inflection point of pressure difference to the end of filtering. Equation (2) was used to fit and calculate the trend of the pressure difference between inlet and outlet over time with a fitting degree higher than 99%. Thus Equation (2) can be regarded as a highly accurate mathematical expression of the pressure difference of this mesh filter type.

(2) Through the analysis of the mass conservation principle and the variation law of the pressure difference between the inlet and the outlet, the mathematical expression of the water usage for the blowdown of the mesh filter was obtained. Equation (8) was used to calculate the number of blowdowns and the required water volume for blowdown under different conditions when the blowdown was started at three time points, i.e., the first inflection point of the pressure difference, the second inflection point of the pressure difference, and the completion of the entire filtration cycle.

(3) When the blowdown was started at the first inflection point of pressure difference, the number of blowdowns increased but the overall water usage for blowdown decreased. Compared with starting blowdown after the entire filtration process was completed, the water usage volume was smaller when the blowdown was started at the first inflection point of pressure difference. Starting the blowdown at the first inflection point, a single filter can save 200–300 m^3 of water on average in each filtration cycle of ten days. Therefore, starting the blowdown at the first inflection point can ensure that the mesh filter saves water resources and always runs under optimal conditions.

ACKNOWLEDGEMENTS

The authors would like to express their gratitude to the National Natural Science Foundation of China (Grant No.11662018). The authors would also like to thank the Shihezi 'Jintudi' Water Saving Equipment Co., Ltd of Xinjiang, China, for its help in developing this experiment at their installations.

FUNDING

This study was supported by the National Natural Science Foundation of China (Grant No. 11662018).

DATA AVAILABILITY STATEMENT

All relevant data are included in the paper or its Supplementary Information.

REFERENCES

Bounoua, S., Tomas, S., Labille, J., Molle, B., Granier, J., Haldenwang, P. & Izzati, S. N. 2016 Understanding physical clogging in drip irrigation: in situ, in-lab and numerical approaches. *Irrigation Science* **34** (4), 327–342. doi:10.1007/s00271-016-0506-8.

Bové, J., Arbat, G., Duran-Ros, M., Pujol, T., Velayos, J., de Cartagena, F. R. & Puig-Bargués, J. 2015 Pressure drop across sand and recycled glass media used in micro irrigation filters.

Biosystems Engineering **137**, 55–63. doi:10.1016/j.biosystemseng.2015.07.009.

de Deus, F. P., Mesquita, M., Testezlaf, R., de Almeida, R. C. & de Oliveira, H. F. E. 2020 Methodology for hydraulic characterisation of the sand filter backwashing processes used in micro irrigation. *MethodsX* **7**, 100962. doi:10.1016/j.mex.2020.100962.

de Vito, R., Pagano, A, Portoghese, I., Giordano, R., Vurro, M. & Fratino, U., 2019 Integrated approach for supporting sustainable water resources management of irrigation based on the WEFN framework. *Water Resources Management* **33** (4), 1281–1295. doi:10.1007/s11269-019-2196-5.

Duran-Ros, M., Arbat, G., Barragán, J., de Cartagena, F. R. & Puig-Bargués, J. 2010 Assessment of head loss equations developed with dimensional analysis for micro irrigation filters using effluents. *Biosystems Engineering* **106** (4), 521–526. doi:10.1016/j.biosystemseng.2010.06.001.

Elbana, M., de Cartagena, F. R. & Puig-Bargués, J. 2012 Effectiveness of sand media filters for removing turbidity and recovering dissolved oxygen from a reclaimed effluent used for micro-irrigation. *Agricultural Water Management* **111**, 27–33. doi:10.1016/j.agwat.2012.04.010.

Elbana, M., de Cartagena, F. R. & Puig-Bargués, J. 2013 New mathematical model for computing head loss across sand media filter for microirrigation systems. *Irrigation Science* **31** (3), 343–349. doi:10.1007/s00271-011-0310-4.

García-Nieto, P. J., García-Gonzalo, E., Arbat, G., Duran-Ros, M., de Cartagena, F. R. & Puig-Bargués, J. 2016 A new predictive model for the filtered volume and outlet parameters in micro-irrigation sand filters fed with effluents using the hybrid PSO–SVM-based approach. *Computers and Electronics in Agriculture* **125**, 74–80. doi:10.1016/j.compag.2016.04.031.

García-Nieto, P. J., García-Gonzalo, E., Puig-Bargués, J., Solé-Torres, C., Duran-Ros, M. & Arbat, G. 2020a A new predictive model for the outlet turbidity in micro-irrigation sand filters fed with effluents using Gaussian process regression. *Computers and Electronics in Agriculture* **170**, 105292. doi:10.1016/j.compag.2020.105292.

García-Nieto, P. J., García-Gonzalo, E., Puig-Bargués, J., Duran-Ros, M., de Cartagena, F. R. & Arbat, G. 2020b Prediction of outlet dissolved oxygen in micro-irrigation sand media filters using a Gaussian process regression. *Biosystems Engineering* **195**, 198–207. doi:10.1016/j.biosystemseng.2020.05.009.

Hamududu, B. H. & Ngoma, H. 2020 Impacts of climate change on water resources availability in Zambia: implications for irrigation development. *Environment Development and Sustainability* **22** (4), 2817–2838. doi:10.1007/s10668-019-00320-9.

Kulmatov, R., Mirzaev, J., Abuduwaili, J. & Karimov, B. 2020 Challenges for the sustainable use of water and land resources under a changing climate and increasing salinization in the Jizzakh irrigation zone of Uzbekistan. *Journal of Arid Land* **12** (1), 90–103. doi:10.1007/s40333-020-0092-8.

Li, J. S., Fei, L. J., Li, S., Shi, Z. X. & Liu, L. H. 2020 The influence of optimized allocation of agricultural water and soil resources on irrigation and drainage in the Jingdian Irrigation District, China. *Irrigation Science* **38** (1), 37–47. doi:10.1007/s00271-019-00649-z.

Martí, P., Shiri, J., Duran-Ros, M., Arbat, G., de Cartagena, F. R. & Puig-Bargués, J. 2013 Artificial neural networks vs. Gene Expression Programming for estimating outlet dissolved oxygen in micro-irrigation sand filters fed with effluents. *Computers and Electronics in Agriculture* **99**, 176–185. doi:10.1016/j.compag.2013.08.016.

Mesquita, M., de Deus, F. P., Testezlaf, R., da Rosa, L. M. & Diotto, A. V. 2019 Design and hydrodynamic performance testing of a new pressure sand filter diffuser plate using numerical simulation. *Biosystems Engineering* **183**, 58–69. doi:10.1016/j.biosystemseng.2019.04.015.

Zong, Q. L., Zheng, T. G., Liu, H. F. & Li, C. J. 2015 Development of head loss equations for self-cleaning screen filters in drip irrigation systems using dimensional analysis. *Biosystems Engineering* **133**, 116–127. doi:10.1016/j.biosystemseng.2015.03.001.

Zong, Q. L., Liu, Z. J., Liu, H. F. & Yang, H. F. 2019 Backwashing performance of self-cleaning screen filters in drip irrigation systems. *PloS ONE* **14** (12), e0226354. doi:10.1371/journal.pone.0226354.

First received 22 June 2020; accepted in revised form 30 September 2020. Available online 16 October 2020

Assessing desertification sensitivity map under climate change and agricultural practices scenarios: the island of Crete case study

G. Morianou, N. N. Kourgialas, V. Pisinaras, G. Psarras and G. Arambatzis

ABSTRACT

The aim of this study is the assessment of desertification risk for a typical Mediterranean island, in the frame of climate change and the application of good agricultural practices. Based on the MEDALUS Environmentally Sensitive Area Index (ESAI) approach, the sensitivity in desertification is estimated by employing 15 quantitative parameters divided in to four main quality indices: climate, vegetation, soil, and management quality. The methodology applied for a baseline scenario (current conditions), two future climate change scenarios (RCP 4.5 and RCP 8.5) and a soil quality improvement scenario. According to the results, more than 13% of the island's area is characterized as critically sensitive to desertification in the current conditions. This percentage will increase in the future under both the RCP 4.5 and the RCP 8.5 climate scenarios, where the critical areas will rise above to 15%. By applying the soil quality improvement scenario simultaneously with the climate change scenarios, a slight mitigation of desertification risk in the future could be achieved. The methodology developed in this study may be used to assess desertification process under various climate, soil, and land use management scenarios in regions of the Mediterranean Sea.

Key words | agricultural practices, climate change, Crete, desertification, ESAI, GIS

G. Morianou (corresponding author)
N. N. Kourgialas
G. Psarras
NAGREF – Hellenic Agricultural Organization (H.A.O.-DEMETER), Institute for Olive Tree Subtropical Crops and Viticulture,
Agrokipio, Chania,
Greece
E-mail: *morianou@elgo.iosv.gr*

V. Pisinaras
G. Arambatzis
NAGREF – Hellenic Agricultural Organization (H.A.O.-DEMETER), Soil and Water Resources Institute,
Thessaloniki,
Greece

HIGHLIGHTS

- This study investigates, using GIS, the effects of all the above approaches on the sensitivity of desertification risk for the Mediterranean island of Crete (Greece) for the period 2031–2060. This study also highlights the results of the comparison of the future desertification scenarios under the application of GAPs and the maintenance of the traditional-conventional agricultural practices.

doi: 10.2166/ws.2021.132

GRAPHICAL ABSTRACT

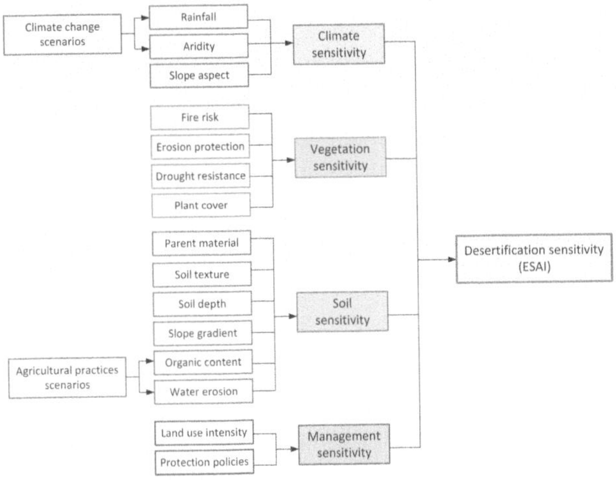

INTRODUCTION

Land degradation is the 'reduction or loss in the biological or economic productive capacity of the land caused by human activities and often magnified by the impacts of climate change' (UNCCD 1994). Desertification is a type of land degradation that occurs mainly in arid, semi-arid, and dry sub-humid environments, where water is the main limiting factor of land use performance in ecosystems. Environmental systems are generally in a state of dynamic equilibrium, thus a small change in climate or at the land use intensity can affect the desertification risk of an area (Kosmas *et al.* 1999; Baartman *et al.* 2007).

In the Mediterranean region, the desertification risk in combination with the expected climate changes pose a significant threat. Several research studies, support that large areas of the Mediterranean region have been recognized as vulnerable to desertification (Drake & Vafeidis 2004; Kepner *et al.* 2006; Sommer *et al.* 2011; Kourgialas *et al.* 2016a; Symeonakis *et al.* 2016; Morianou *et al.* 2018). In the Mediterranean region, desertification is mainly an outcome of the climatic conditions and the intensity of human activity on land (Conacher & Sala 1998; Geist & Lambin 2004; Reynolds *et al.* 2007). The Eastern Mediterranean area is particularly considered as one of the main 'hot spots for climate change' at a global scale, meaning that climatic factors in this region are especially susceptible to climate change, a fact that may increase the desertification risk (Türkeş 1999; Kourgialas 2021). In line with this, the establishment of simple modeling tools for assessing the spatiotemporal variability of such risks is very important (Eris *et al.* 2020). Overgrazing and deforestation are two of the main causes of land degradation. In arid regions, vegetation cover is particularly necessary to prevent soil erosion and nutrient loss. In addition, improper agricultural practices such as the long-term conservation tillage may also lead to land that is more vulnerable to desertification. Recent studies have suggested that global climate change will have a significant impact on both local and extended scale territories (Mickley *et al.* 2004; Weaver *et al.* 2009; Kourgialas *et al.* 2016b). The results of many climate model simulations, under various CO_2 emission scenarios, also showed that temperatures in the Mediterranean are projected to rise significantly, while precipitation is projected to decrease. These trends could significantly affect the risk of desertification at specific areas in the Mediterranean region (Giorgi & Bi 2005; Giorgi & Lionello 2008; Zanis *et al.* 2009). Furthermore, organic matter and erosion by water are two important factors in terms of soil quality and desertification risk. Organic matter affects both the chemical and physical properties of the soil (soil structure, moisture, and nutrient holding capacity and activity of soil organisms) (Montanaro *et al.* 2012). On the other hand, soil erosion, especially in arid and semi-arid lands, contributes to soil degradation, which in turn affects the sustainable agricultural land use and productivity (Lal 2001).

Over the past decades, various models and methodologies have been proposed to assess the desertification sensitivity of an area (Oldeman & GWJ 1997; Babaev & Kharin 1999; Kosmas *et al.* 1999; Brandt *et al.* 2003; DESERTLINKS 2004). The most commonly used methodology in the Mediterranean region is an indicator-based approach proposed by Kosmas *et al.* (1999), the Environmental Sensitivity Area Index (ESAI) methodology. The Environmental Sensitivity Area (ESA) is a simple and flexible model that uses a wide variety of indicators. In this system, four main factors of climate, soil, vegetation, and land management are produced and combined in a synthetic way to make the so-called ESAI (Kosmas *et al.* 1999; Basso *et al.* 2000). ESA methodology has been applied to quite a few case studies in Mediterranean region: Italy (Basso *et al.* 2000; SORRISO-VALVO 2005; Salvati & Zitti 2009; Ladisa *et al.* 2012; Salvati *et al.* 2015), Spain (Contador *et al.* 2009), Iran (Sepehr *et al.* 2007), Egypt (Bakr *et al.* 2012), and Algeria (Boudjemline & Semar 2018). In Greece, the methodology has been applied in Lesvos island by Kosmas *et al.* (1999) and Symeonakis *et al.* (2016) and in Crete by Morianou *et al.* (2018). The majority of the aforementioned studies assess the desertification risk in a one-time-step static system. Symeonakis *et al.* (2016) noted that the land degradation process is dynamic due to the time dimension and the indicators should, therefore, derive from a sequence of temporal steps. To fill in this gap, they developed a continuous monitoring system of environmental sensitivity to desertification of the island of Lesvos (Greece) and they examined two different time periods, one during 1990 and one during 2000. However, up to now, no studies have assessed the risk of desertification under future climate change scenarios.

This study tries to fill this gap by incorporating the desertification risk assessment methodology for the RCP 4.5 and RCP 8.5 climate change scenarios. More specifically, the main purpose of this work was to assess the environmentally sensitive areas to desertification of a typical Mediterranean island (Crete) under two future climate change scenarios and under the application of good agricultural practices (GAPs). Many studies have shown that soil quality improvement can be achieved by applying GAPs (Hernández *et al.* 2005; Moreno *et al.* 2009; Koubouris *et al.* 2017; Kavvadias *et al.* 2018; Michalopoulos *et al.* 2020). For instance, cover crops and the maintenance of weeds during winter contribute as good practices to protection of land from erosion and increase soil water infiltration and carbon storage (Xiloyannis *et al.* 2008). In addition, the application of organic matter from composted byproducts and tree pruning residues can increase organic carbon in the soil in the long-term (Montanaro *et al.* 2010, 2012). Thus, for the first time, this study investigates, using GIS, the effects of all the above approaches on the sensitivity of the desertification risk for the Mediterranean island of Crete (Greece) for the period 2031–2060. This study also compares future desertification scenarios under GAPs and the maintenance of the traditional-conventional agricultural practices.

MATERIALS AND METHODS

Case study

Crete, the study area, is an island in the Mediterranean Sea and is the largest island in Greece (35:20:27 N, 25:07:46 E). Crete is an island of the Mediterranean Sea with a total area of about 8,265 km^2. The western part of Crete includes the Chania and Rethymnon prefectures, while the eastern includes the prefectures of Heraklion and Lassithi. The elevations in Crete range from 0 m to 2,400 m mean sea level. In Crete, lowlands (<200 m) cover an area of 2,165 km^2, 26% of the total area, semi-mountainous areas (200–800 m) cover an area of 4,627 km^2 which is 56% of the total area, while the mountainous area (>800 m) is approximately equal to 1,473 km^2 which is 18% of the total area.

The climate of Crete is sub-humid Mediterranean with humid and relatively cold winters, and dry and warm summers. The annual rainfall ranges from 300 to 700 mm/year in the lowlands, from 700 to 1,000 mm/year in the semi-mountainous areas, while in the mountainous areas it reaches 2,000 mm/year. Significant rainfall differences are recorded between the western and eastern areas of the island during the wet seasons (Kourgialas *et al.* 2018).

According to Panagos *et al.* (2014), soils in Crete are generally characterized as poorly developed and shallow. The biggest part of the island consists of fine soils (clay, sandy-clay, clay-loam, silty-clay & silty-clay-loam) and medium soils (sandy-clay-loam, loam, silt-loam & silt). Coarse soils (sand, loamy-sand & sandy-loam) occur mainly in the western part of the island.

According to the CORINE Land Cover maps (2012), Crete is mostly covered by natural grasslands and pastures. Evergreen forests, coniferous forests, and Mediterranean maquis cover less than 5% of the island. The agricultural areas of the island cover about 3,205 km^2. These areas are mainly dominated by tree crops (olive, citrus, avocado), vineyards, and greenhouse vegetation cultivations. Tree crops cover about 72% of the total agricultural area in Crete (OPEKEPE 2017). Olive, avocado, and citrus trees thrive at the western part of the island due to the local climate conditions. On the other hand, the agricultural areas at the eastern part of the island are characterized by drier climate conditions and the dominant cultivations are olive trees and vineyards.

ESAI methodology

The areas environmentally sensitive to desertification can be assessed in relation to various parameters such as landscape, soil, geology, erosion, vegetation, climate, and human action (Kosmas *et al.* 1999). Each of these parameters is reclassified according to its behavior on desertification and weighting factors are assigned in each class.

In this study, the modified ESAI methodology proposed by Morianou *et al.* (2018) is used. This modification of the MEDALUS approach includes two additional parameters related to soil quality (water erosion and soil organic matter) compared to the original one. Fifteen layers belonging to four main sensitivity groups (climate, soil, vegetation management, and land management) were collected from various sources and their values were standardized between 1 (low sensitivity) and 2 (high sensitivity) (Tables 1–4). The four sensitivity groups were estimated using the following formula (Equation (1)):

$$\text{Sensitivity}_x = (\text{layer}_1 \times \text{layer}_2 \times \text{layer}_3 \times \ldots \text{layer}_n)^{1/n} \quad (1)$$

where: sensitivity$_x$, represents the computed value of each sensitivity and n represents the number of sub-indicators (layers) used to calculate each Sensitivity Index (SI).

Also, in this study, climate change scenarios (RCP 4.5 and RCP 8.5) as well as GAP scenarios were incorporated into climate and soil main sensitivity groups to investigate their effects on the final desertification sensitivity map. A flow chart of the proposed methodology is presented in Figure 1. In the following sections, analytical information for each involved sensitivity group index/parameter is given.

Climate sensitivity

In this study, the Climate Sensitivity Index was calculated from three climate indicators: rainfall, evapotranspiration, and aspect. Rainfall amount and its spatial distribution as well as the hydrological extreme events in the semi-arid and arid zones of the Mediterranean are the main climatic

Table 1 | Main indicators adopted classification scheme- scores and methods used in the GIS to estimate the Climate Sensitivity Index

	Parameter	Classes	Index	Data/source
Climate sensitivity	Rainfall (mm)	>650	1	Data from meteorological stations of the island
		280–650	1.5	
		<280	2	
	Aridity = Precipitation/PET	>0.65	1	PET: FAO-56 Penman–Monteith equation (Allen *et al.* (1998)), mean monthly min and max temperature from the meteorological stations of the island
		0.5–0.65	1.5	
		<0.5	2	
	Aspect	N, NE, NW	1	25 m – pixel DEM EU-DEM v1.1
		S, SE, SW	2	

Table 2 | Main indicators adopted classification scheme- scores and methods used in the GIS to estimate the Vegetation Sensitivity Index

	Parameter	Classes	Index	Data/source
Vegetation sensitivity	Drought resistance	Evergreen forests (except coniferous), mixed Mediterranean maquis-evergreen forests	1	Corine Land Cover (CLC) for the year (2012), source: http://www.data.gov.gr/dataset/xartes-kalypshs-ghs-corine-land-cover-gia-ta-eth-2006-and-2012
		Conifer forests, deciduous forests, and olives	1.2	
		Almonds, orchards, and vines	1.4	
		Perennial grasslands, pastures, and shrubland	1.7	
		Annual crops (annual grassland, cereals, maize, and sunflower), horticulture, and very low vegetated	2	
	Erosion protection	Evergreen forest (except conifers), mixed Mediterranean maquis-evergreen forest	1	
		Mediterranean maquis, conifer forests, perennial grasslands, pastures, olives, and shrubs	1.3	
		Deciduous forests (oak and mixed)	1.6	
		Almonds and orchards	1.8	
		Vines, horticulture, annual crops, very low vegetated, and bare soils	2	
	Fire risk	Almonds, orchards, vines, olives, irrigated annual crops, and horticulture	1	
		Perennial grasslands, pastures, cereals, annual grasslands, deciduous forests (oak and mixed), mixed Mediterranean maquis evergreen forests, very low vegetated, and shrublands	1.3	
		Mediterranean maquis	1.6	
		Pines and other conifer forests	2	
	Plant cover (%)	>40	1	
		10–40	1.8	
		<10	2	

attributes that contribute to the degradation of land (Kosmas *et al.* 1999). Annual precipitation was classified in three classes considering that annual precipitation of 280 mm is a critical threshold for plant growth (Sepehr *et al.* 2007). The aridity index was also used in this study. According to Allen *et al.* (2005), the aridity index is the ratio of the annual precipitation divided by the annual potential evapotranspiration (P/PET). Precipitation and evapotranspiration data were collected from 56 meteorological stations distributed all over the island providing sufficient coverage. Figure 2 depicts the spatial distribution of these stations. Monthly data has been collected during the last 40 years, provided by the Department of Water Resources Management of the Region of Crete, the National Observatory of Athens and the Hellenic Agricultural Organization – National Agricultural Research Foundation. Data gaps were filled using values from the nearby stations, while a data quality check was applied to ensure the homogeneity of the time-series (Alexandersson 1986; Vicente-Serrano *et al.* 2010). Aspect was calculated from a Digital Elevation Model (EU-DEM v1.1) at 25 m resolution scale. The input data that were used in the climate sensitivity group index are presented in Table 1.

Vegetation sensitivity

Vegetation cover is a very crucial factor affecting desertification and land degradation. The key indicators of desertification related to the existing natural or agricultural vegetation can be considered in relation to: (i) fire risk and ability to recover, (ii) erosion protection offered to the soil, (iii) drought resistance, and (iv) percentage plant cover. The vegetation sensitivity map was produced using the vegetation cover data according to the CORINE Land Cover 2012 (CLC 2012). The input data that were used in the vegetation sensitivity group index are presented in Table 2.

Table 3 | Main indicators adopted classification scheme – scores and methods used in the GIS to estimate the Soil Sensitivity Index

	Parameter	Classes	Index	Data/source
Soil sensitivity	Parent material	Shale, schist, basic, ultra-basic, conglomerates, unconsolidated, clays, and marl (with natural vegetation)	1	Geological maps were bought from the Institute of Geology & Mineral Exploration
		Limestone, marble, granite, rhyolite, ignibrite, gneiss, siltstone, sandstone, and Dolomite	1.7	
		Marl and pyroclastics	2	
	Texture	LSCL, SL, LS, and CL	1	The map produced by the Institute of Geology and Mineral Exploration was downloaded from the European Soil Data Centre (ESDAC) http://esdac.jrc.ec.europa.eu/content/soilmap-greece-edafologikos-xartis-ellados
		SC, SiL, and SiCL	1.2	
		Si, C, and SiC	1.6	
		S	2	
	Soil depth (cm)	>75	1	
		30–75	1.3	
		15–30	1.6	
		<15	2	
	Soil organic matter (%)	<0.5	2	The map downloaded from https://esdac.jrc.ec.europa.eu/content/octop-topsoil-organic-carbon-content-europe
		0.5–1	1.7	
		1–2	1.5	
		2–3	1.2	
		>3	1	
	Water erosion (tn/ha/yr)	<1	1	A map was produced by Panagos *et al.* 2012 https://esdac.jrc.ec.europa.eu/content/g2-soil-erosion-model-data-crete-greece-and-strymonas-greecebulgaria-ishmi-erzeni-albania
		1–5	1.3	
		5–20	1.7	
		>2	2	
	Slope gradient	<6	1	20X20 m – pixel DEM Hellenic Military Geographical Service
		6–18	1.2	
		18–35	1.5	
		>35	2	

Table 4 | Main indicators adopted classification scheme – scores and methods used in the GIS to estimate the Management Sensitivity Index

	Parameter	Classes	Index	Data/source
Management sensitivity	Intesity of land use	Woodlands, semi-natural areas, and natural grasslands	1	CORINE Land Cover (CLC) 2012
		Irrigated and non-irrigated vineyards, irrigated and non-irrigated fruit trees, irrigated and non-irrigated olive groves, annual crops associated with permanent crops, complex cultivation patterns, land principally occupied by agriculture with significant areas of natural vegetation, non-irrigated arable land	1.5	
		Open field herbaceous with spring–summer cycle, horticultures with spring–summer cycle, horticultures with spring–summer–autumn cycle, greenhouses	2	
	Protection policies	Reserves, parks, and archaeological areas	1	Sites of Community Importance (SCI) and Special Protected Areas (SPA) according to the EU Directive 92/43. Source: http://www.ypeka.gr/?tabid=504
		Woodlands, semi-natural, and coastal areas	1.5	
		Areas not subject to restrictions	2	

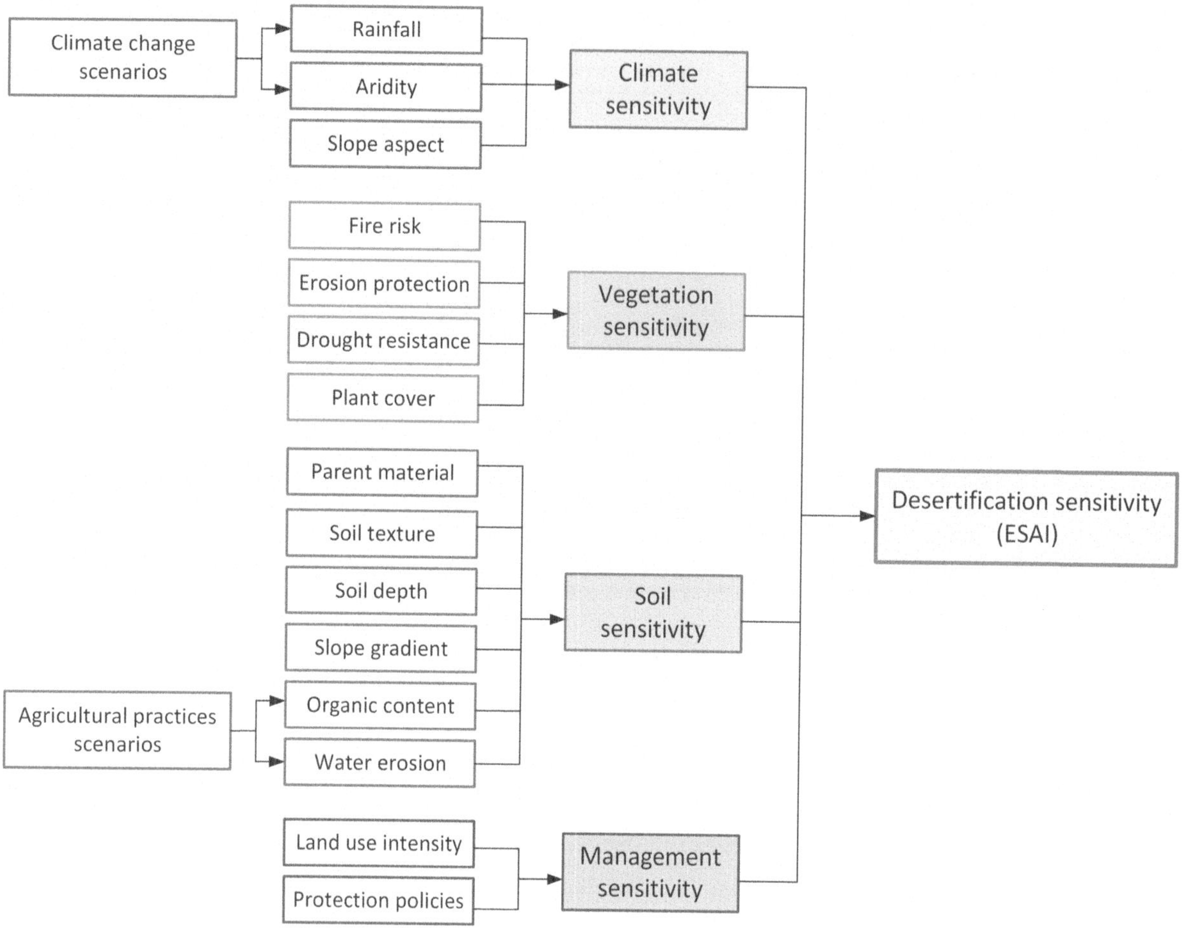

Figure 1 | Flowchart of the proposed methodological framework for the estimation of ESA to desertification.

Soil sensitivity

The soil sensitivity map was produced using the following soil data: texture, slope, parent material, soil depth, organic, and soil erosion caused by water. For the purpose of this study, soil data was derived from the European Soil Database (ESDAC). Soil erosion caused by water were derived from the study of seasonal and annual erosion assessments in Mediterranean agricultural areas using the G2 model (Panagos *et al.* 2014). The input data that were used in the soil sensitivity group index are presented in Table 3.

Management sensitivity

The definition of ESAs to desertification requires key indicators related to the physical environment and to the human-induced stress. The intensity of land use factor was produced according to the CORINE Land Cover 2012 (CLC 2012) and the protection policies factor that was produced using data from EU Directive 92/43 (NATURA 2000). The combination of these two parameters gives the final Management Sensitivity map. The input data that were used in the management sensitivity group index are presented in Table 4.

Climate change scenarios

Regarding the two studied climate change scenarios, RCP 8.5 scenario is characterized by increasing greenhouse gas emissions over time. The underlying scenario drivers and resulting development path are based on the A2 scenario detailed in Moss *et al.* (2008). This can be considered as a

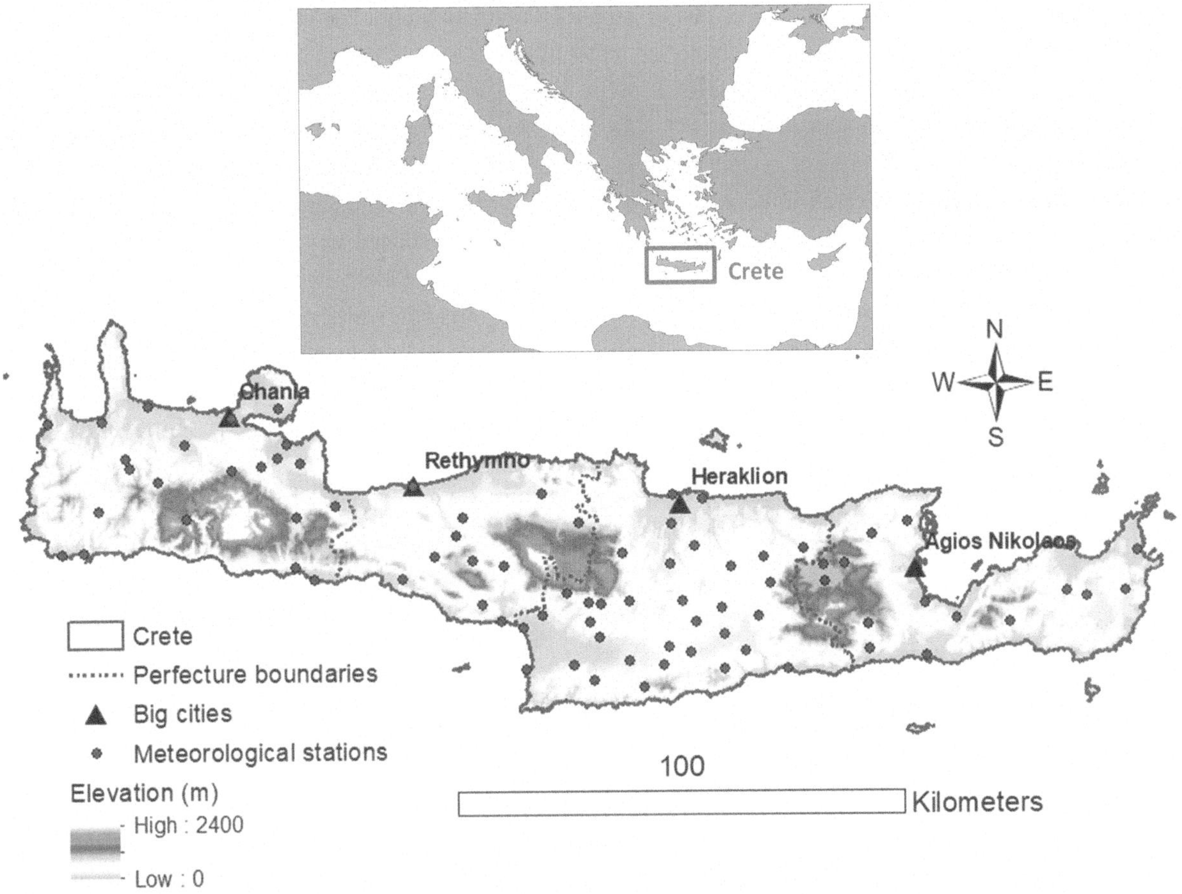

Figure 2 | The Island of Crete. Upper panel: location of Crete (red frame). Lower panel: the elevation map of Crete and the location of the meteorological stations.

non-mitigation business-as-usual scenario with high emissions. The RCP 8.5 (Moss *et al.* 2008) is developed by the MESSAGE modelling team and the Integrated Assessment Framework at the International Institute for Applied Systems Analysis, Austria. On the other hand, RCP 4.5 scenario is similar to the lowest-emission scenarios (B1) assessed in the Intergovernmental Panel on Climate Change AR4. It is a stabilization scenario where total radiative forcing is stabilized by around 2050 by using a range of technologies and strategies for reducing greenhouse gas emissions. This can be considered as a weak climate change mitigation scenario. The RCP 4.5 (Clarke *et al.* 2007; Moss *et al.* 2010) is developed by the MiniCAM modelling team at the Pacific Northwest National Laboratory's Joint Global Change Research Institute.

A lot of effort has been given by the scientific community in the last decades to produce high-resolution data from Regional Climate Models (RCMs) that are widely used to support climate change impact assessment studies. Nevertheless, and in the context of small-scale studies, these data are usually further processed with bias-correction methodologies to reduce the incorporated biases. This procedure requires adequate time-series of observed data. Due to data availability restrictions, RCM data that represents the climate conditions of the study had to be chosen. According to LIFE ADAPT2-CLIMA (2017), such data for Crete comes from the RCA4 RCM developed by the Swedish Meteorological and Hydrological Institute (Strandberg *et al.* 2015), driven by the Global Circulation Model named Hadley Centre Global Environmental Model, version 2 Earth System (Martin *et al.* 2010; Collins *et al.* 2013), which was developed by the Met Office Hadley Centre, hereafter referred as RCA4_MOHC.

Spatial downscaling can improve the resolution of climate data from RCMs and therefore produce a better representation of the climate conditions. RCA4_MOHC data (precipitation, and minimum and maximum air temperature) were downscaled from 0.11° to 100 m using a bilinear approach that was suggested and implemented in previous studies (Marke *et al.* 2011; Herrmann *et al.* 2016). Minimum and maximum air temperature data from RCA4_MOHC was used to calculate reference evapotranspiration (ETo) using the Hargreaves method (Hargreaves & Samani 1985). This method has been proposed by Allen *et al.* (1998) to be implemented as an alternative to Penman–Monteith when temperature data is the only input available. The period used in the context of the present study is 2031–2060, which is considered as representative of the near- to medium-distant future.

Agricultural practices scenarios

In terms of the desertification risk assessment, the factors that could be affected by the applied GAPs are related to soil sensitivity (soil organic matter and soil erosion). More specifically, a set of GAPs for improving soil conservation in tree crops areas includes: (i) application of organic matter by composted olive mill byproducts and olive tree pruning residue and (ii) reduction of water erosion using cover crops and/or introducing natural barriers in sloppy tree crops areas, also for this purpose, no tillage is strongly recommended. The above practices, as part of LIFE AgroCli-maWater project, were applied in two pilot river basins in Crete, one in the western side and one at the eastern side of the island during the period 2017–2019. It is important to mention that for each of the two studied river basins, 20 experimental sites/tree crop parcels were selected to apply the GAPs. These experimental sites were selected based on different criteria (geomorphology, soil characteristics, climate data, water availability, management regime of the parcels, as well as the already applied agricultural practices) to capture a wide range of alternative agricultural ecosystems on Crete, including climatic, soil, and cropping systems heterogeneity. The results of the above applied GAPs, as well as the results from other related studies and research projects in Crete (Kavvadias *et al.* 2018; Kavvadias & Koubouris 2019), indicate that organic matter in agricultural areas of the island can be increased up to 14% until 2060 by the application of organic matter by composted olive mill byproducts, tree pruning, and weed mowing residue application. In addition, the findings of the above studies/projects indicate that by establishing means of physical reduction of surface runoff along the contour lines in orchards located in slopy areas (slope >10%), the water erosion could be reduced up to 40%. For simulating the effects of the above GAPs, in the proposed methodology, the tree crops areas in Crete were isolated with a mask in GIS environment and the percentage of organic matter was increased by 14% (the baseline soil organic matter content was obtained from ESDAC database). In the same way, water erosion, in the agricultural areas with a slope of more than 10%, was reduced by 40%. The new map indicators of erosion and organic matter were then evaluated according to the ESA methodology and combined with the other soil factors to provide a new soil sensitivity map that incorporates GAPs implementation in tree crops areas of Crete.

Final desertification sensitivity map

After the combination of the environmental and land use parameters into the four sensitivity groups maps (climate, soil, vegetation management, and land management), the quantitative results of each factor are classified into three qualitative classes (low, moderate, and high quality) according to the scheme suggested by Sepehr *et al.* 2007 (Table 5).

Analytically, in the last ESAI stage, the final sensitivity to desertification of an area is evaluated from the

Table 5 | Sensitivity classes of the four sensitivity indices, according to the classification suggested by Sepehr *et al.* (2007)

Sensitivity index	Range	Sensitivity class
Climate Sensitivity Index	1	Low
	1.1–1.5	Moderate
	1.5–2	High
Vegetation Sensitivity Index	<1.13	Low
	1.13–1.38	Moderate
	>1.38	High
Soil Sensitivity Index	<1.13	Low
	1.13–1.45	Moderate
	>1.45	High
Management Sensitivity Index	1–1.3	Low
	1.3–1.5	Moderate
	>1.5	High

combination of the above mentioned four sensitivity maps in Equation (2) in a linear way:

$$\mathrm{ESAI} = (\mathrm{Sensitivity}_1 \times \mathrm{Sensitivity}_2 \times \mathrm{Sensitivity}_3 \times \ldots \mathrm{Sensitivity}_n)^{1/n} \tag{2}$$

The ESAI assigns equal weights to layer n, within each SI, as well as equal weights to each SI when computing the overall environmental sensitivity (ESAI). This way, the computations in Equation (2) are unaffected by the number of layers used in every SI, which means that a SI is not considered of less importance because less layers are used for its computations (Basso *et al.* 2000).

According to Kosmas *et al.* (1999), four general types of Environmentally Sensitive Areas (ESAs) to desertification can be distinguished based on the stage of land degradation.

Not Affected: Areas not affected by desertification are usually areas with deep to very deep, well-drained, coarse-textured or finer soils, under semi-arid or wetter climatic conditions, independent of vegetation.

Potential ESAs: Areas with nearly flat to gently sloping (slope <12%), moderately fine-textured, deep and mainly well-drained. Thc climatc is mainly characterized as dry sub-humid with rainfall greater than 650 mm. The dominant vegetation of these areas is usually olives and pines and, in some cases, evergreen oaks or annuals, with usually low fire risk (olives) or high fire risk (pines), high to moderate erosion protection, and mainly high resistance to drought. These areas are mainly under moderate land use intensity.

Fragile ESAs: Areas in which any change in the delicate balance of natural and human activity is likely to bring about desertification. For example, the impact of predicted climate change is likely to enhance reduction in the biological potential due to drought, causing areas to lose their vegetation cover and are subject to greater erosion.

Critical ESAs: Areas already highly degraded through past misuse, presenting a threat to the environment of the surrounding areas. For example, badly eroded areas subject to high runoff and sediment loss.

RESULTS AND DISCUSSION

The results/maps for the four sensitivity groups are shown in Figures 3–6. In terms of climate sensitivity (Figure 3), the majority of the island was mapped as having low to moderate climatic sensitivity, except for a part of the eastern island that was mapped as having high sensitivity under the current conditions (Figure 3(a)). This could be attributed to the fact that eastern areas (south Heraklion and Lasithi prefecture) receive less precipitation than the western part of the island (Koutroulis *et al.* 2011). Comparing the climate maps of the current situation (Figure 3(a)) with the two climate scenarios, RCP 4.5 (Figure 3(b)) and RCP 8.5 (Figure 3(c)), a significant part of the island seems to become more sensitive to desertification in terms of climate. It can also be noted that under the climate scenario RCP 4.5, there is higher sensitivity than under the scenario RCP 8.5. This is due to the fact that the climate model used gives less rain under the scenario RCP 4.5 compared to RCP 8.5 for the period 2031–2060. Since RCP 8.5 is a no-mitigation scenario, while RCP 4.5 is an average mitigation scenario, it is considered that, in general, the amount of rainfall for the Mediterranean region under the RCP 8.5 scenario is higher compared to the RCP 4.5. Nevertheless, this corresponds to an average trend which can differentiate depending on the RCM used and the specific area under study.

The amount of precipitation also affects the aridity index (precipitation/evapotranspiration), which gives lower values under climate scenario RCP 4.5. Specifically, at the eastern part of the island, in contrast to the general depiction presented by the climate maps, there seems to be an improvement in terms of climate sensitivity under both future climate scenarios. This is because, in recent years, the region of eastern Crete has long periods of drought, and in both climatic scenarios this region shows an increase in precipitation amounts.

Regarding the Vegetation Sensitivity Index, the areas with low drought resistance (e.g. agricultural areas), high fire risk (e.g. pine forests), and low percentage of vegetation cover are identified as highly sensitive. As Figure 4 indicates, the largest part of the island is characterized by moderate vegetation sensitivity except for some areas of Heraklion prefecture, which are characterized by high vegetation sensitivity. This is due to the fact that Heraklion prefecture includes large urbanization areas in the north and areas with intensive agricultural activity across the prefecture. Considering also that vegetation cover is a crucial factor for soil erosion control in sloppy areas, a considerable part

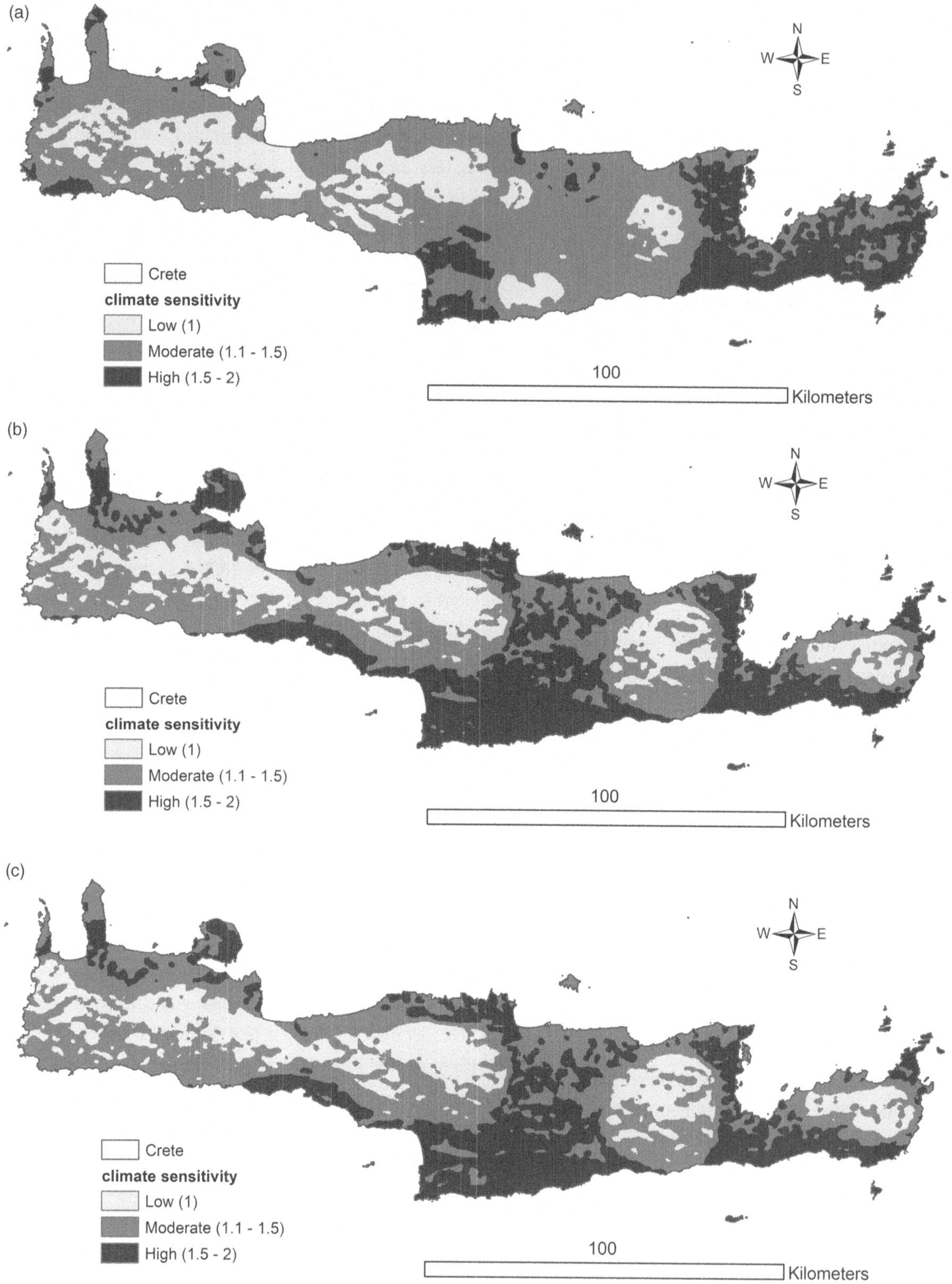

Figure 3 | (a) Climate sensitivity map of Crete (Greece) in the current conditions, (b) climate sensitivity map under the scenario RCP 4.5 and (c) climate sensitivity map under the scenario RCP 8.5.

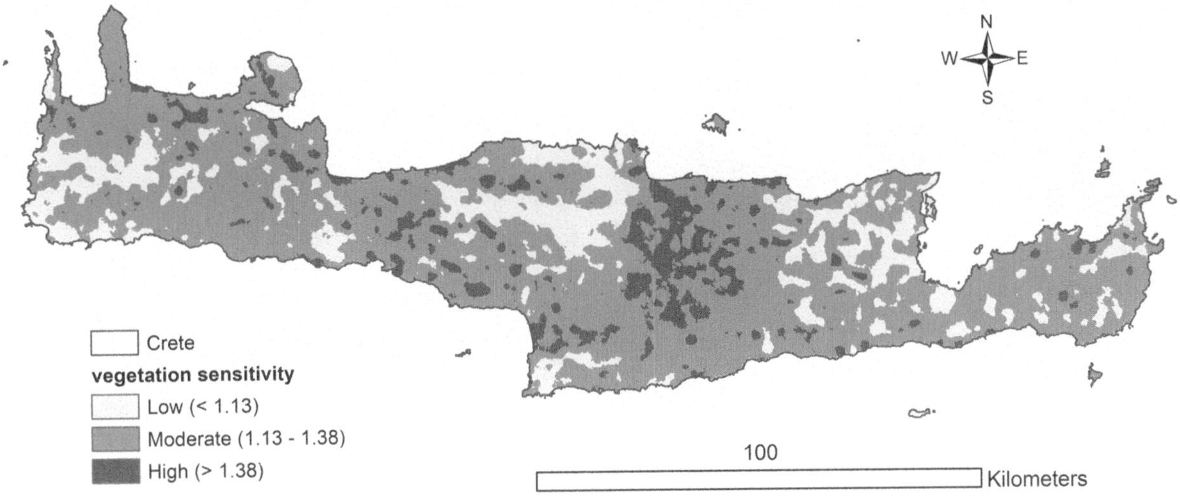

Figure 4 | Vegetation sensitivity map of Crete (Greece).

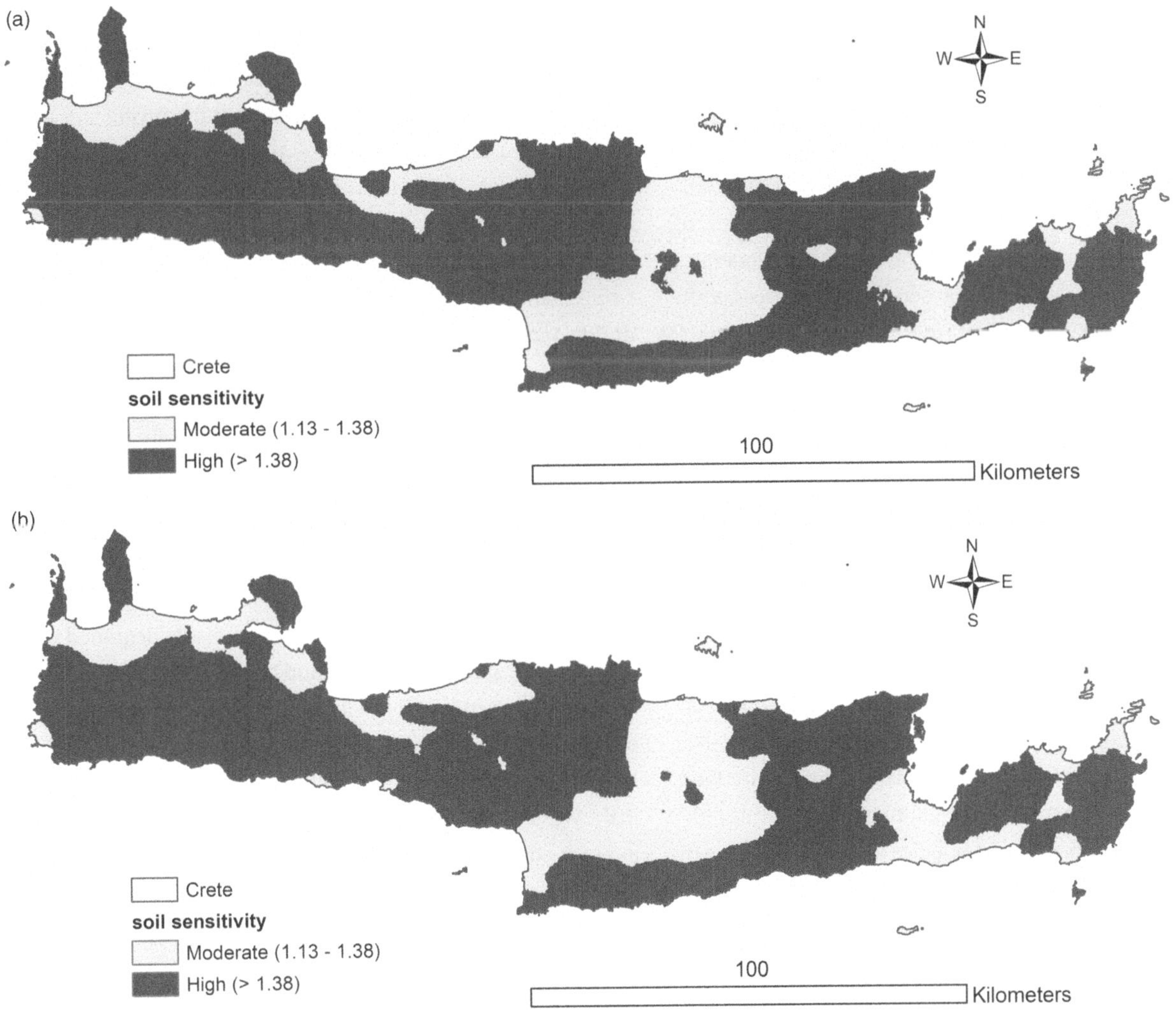

Figure 5 | (a) Soil sensitivity map of Crete (Greece) in the current conditions and (b) soil sensitivity map of Crete after the application of GAPs for the next 40 years.

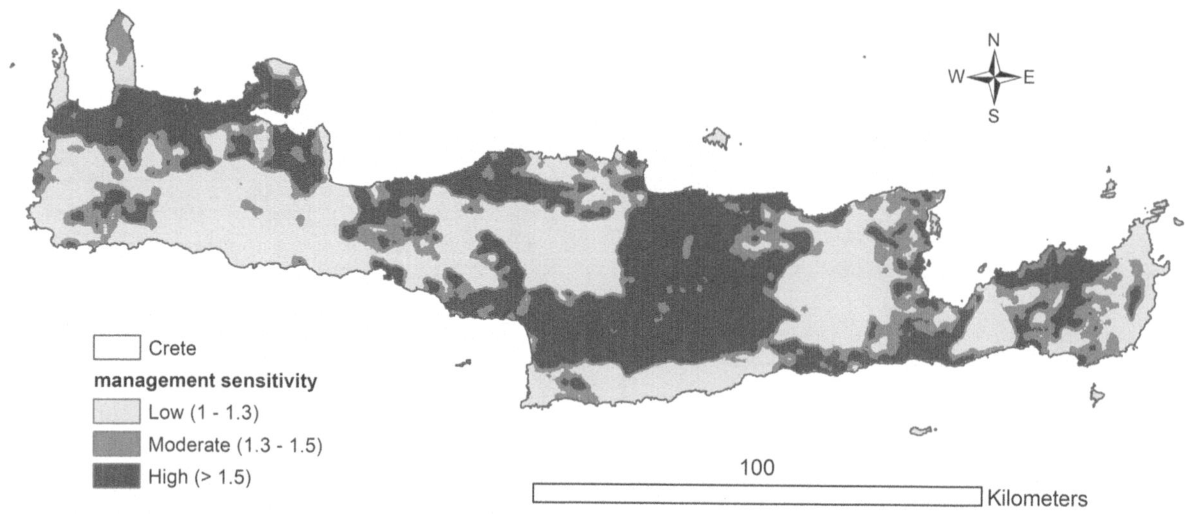

Figure 6 | Management sensitivity map of Crete (Greece).

of the island's natural lands are subject to very high erosion risk due to the animal breeding activities (overgrazing).

Figure 5(a) shows the resulting soil sensitivity map of Crete under current conditions. The areas with the most sensitive soils are found at the southwest part of the island, in the major part of Heraklion prefecture and in the eastern part of Lasithi prefecture. The rest of the island has moderate soil sensitivity. This is due to the presence of large areas with slopes greater than 18%, a high presence of soils with a depth of less than 30 cm, and an important presence of fine texture soils. The results of soil quality are also related to soil erosion from water and the poor soil organic content in agricultural areas.

Comparing the soil result maps, in the current condition (Figure 5(a)) and after the application of GAPs for 40 years (Figure 5(b)), a reduction of high sensitivity areas and an increase of the moderate sensitivity areas is observed. This is due to the increase of organic content in agricultural areas and the reduction of water erosion in areas with a slope of more than 10%. Therefore, it is observed that by the improvement of even two of the six soil sensitivity factors, a slight improvement in soil quality could be achieved.

In terms of the management policies dimension, Figure 6 shows that the north-western part of the island and the major part of Heraklion prefecture are considered very sensitive regarding land management. The two largest cities of Crete (Chania and Heraklion) are located here and therefore the human activity is more intense. The southern areas of Chania prefecture and the mountainous areas in the eastern part of the island are covered by pine or oak forests and are considered environmentally protected (low sensitivity). Large parts of these areas are also managed by environmental protection policies (e.g. NATURA 2000 areas and Important Bird Areas).

The environmental sensitivity to desertification of the Crete island according to the ESA approach in the current conditions (baseline scenario) is shown in Figure 7(a). According to these results and Table 6, the vast majority of the island is considered as fragile or critical for desertification. The most environmentally sensitive areas (critical areas) of Crete, in the current conditions, are found mainly in the southern and eastern parts of the island. More specifically, the results show that 29% of the island is fragile or critical to desertification (fragile: 16.24% and critical: 13.15%) (Figure 8). According to Morianou *et al.* (2018), these areas have badly degraded shallow soils, are poorly vegetated, and are very sensitive in terms of climate due to low rainfall and high air temperatures. The management policies are also quite poor in these areas. The Not Affected areas and the Potential ESAs are mainly located in the western and in the middle parts of the island (Figure 7(a)), in areas characterized by favorable climate (high annual rainfalls) and efficient land management.

By applying the ESAI methodology under two climate change scenarios, this study indicates that, in the future, a greater extent of the island will be characterized as critical to desertification rather than fragile and potential. This

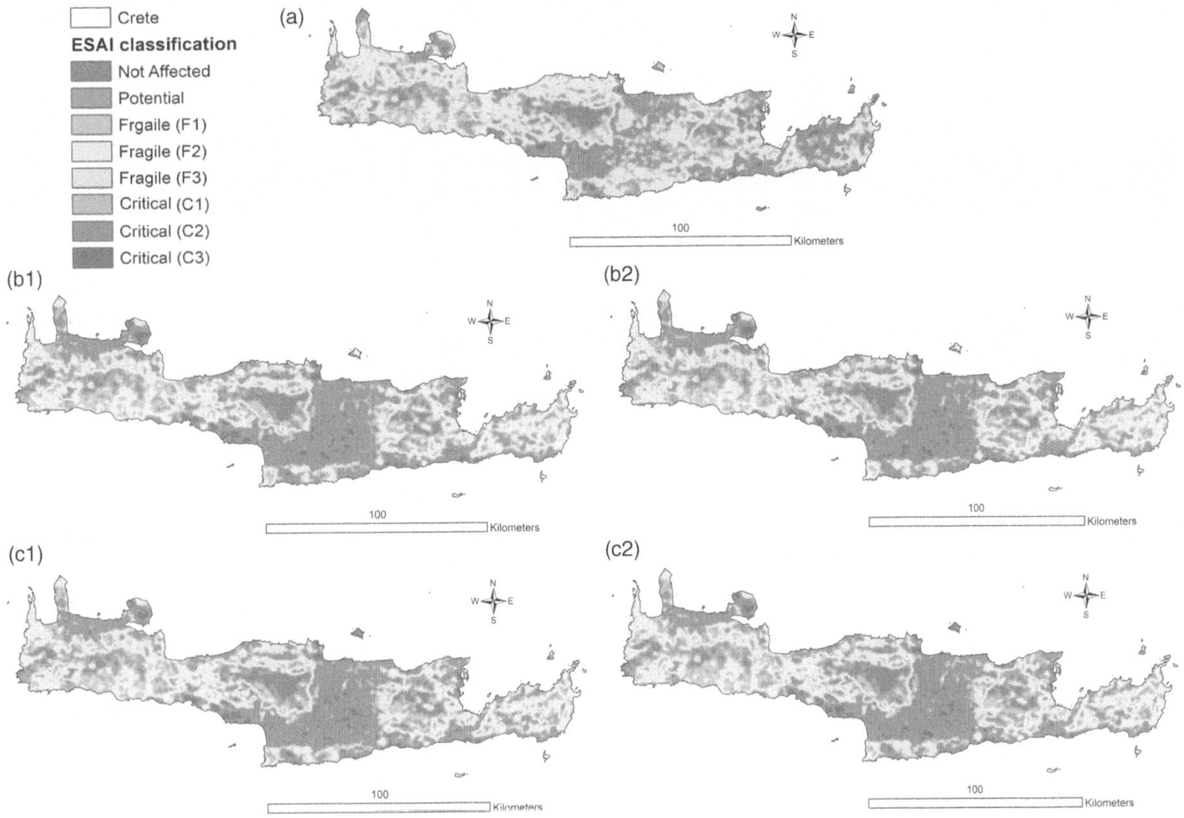

Figure 7 | (a) Map of ESA to desertification for Crete related to desertification risk in the current conditions (baseline scenario), (b1) under climate scenario RCP 4.5, (b2) under the climate scenario RCP 4.5 incorporating GAPs, (c1) under climate scenario RCP 8.5, (c2) under the climate scenario RCP 8.5 incorporating GAPs.

Table 6 | Sensitivity classes of the four sensitivity indices, according to the classification suggested by Sepehr *et al.* (2007)

ESA Index	ESAI classification
<1.17	Not Affected
1.17–1.22	Potential
1.23–1.26	Fragile (F1)
1.26–1.32	Fragile (F2)
1.33–1.37	Fragile (F3)
1.38–1.41	Critical (C1)
1.42–1.53	Critical (C2)
>1.53	Critical (C3)

seems likely to occur under both the average mitigation scenario (RCP 4.5) and the no-mitigation scenario (RCP 8.5), where the critical areas will rise to 15.76% and 15.09%, respectively, versus 13.15% for the baseline scenario (Figure 8). Unlike the general trend, some regions of eastern Crete show an improvement in terms of the desertification risk. This happens due to the potential increase in precipitation in the future and therefore the local increase of climate quality.

Comparing the effects on desertification risk between the two future climate change scenarios (RCP 4.5 and RCP 8.5), it is observed that under the climate scenario RCP 4.5 more areas are characterized as critical (15.76% VS 15.09%) and less as non-affected and potential to desertification than under the scenario RCP 8.5 (3.97% vs 4.28% and 8.60% vs 9.05%, respectively) (Figure 8). This is because the estimated amount of precipitation is lower in scenario RCP 4.5 compared to RCP 8.5.

Finally, the results obtained by the incorporation of the soil improvement scenario with the climate change scenarios showed that with the application of GAPs, the risk of desertification is slightly reduced. More specifically, the critical to desertification areas are reduced and, correspondingly, the percentage of areas not affected by desertification has increased with the application of GAPs. The fragile areas remain at the

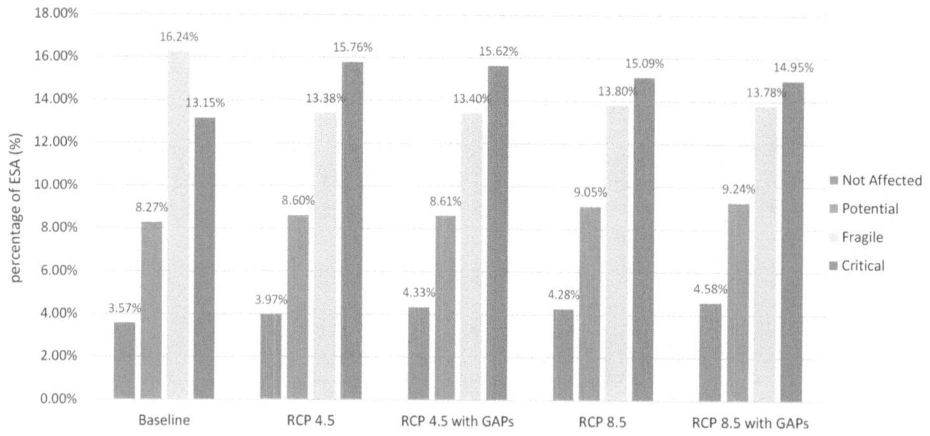

Figure 8 | Summarized histogram for the desertification risk assessment of Crete under the baseline scenario and the two climatic scenarios (with and without the application of GAPs).

same percentages with or without application of GAPs (Figures 7(b2), 7(c2) and 8).

The results of this study indicate that in the frame of climate change the desertification risk in the Mediterranean island of Crete could be higher. It also shows that by applying GAPs in cultivated areas a slight mitigation of desertification risk can be achieved. This is because GAPs affect only two of the 15 factors of ESAI methodology. These findings should alert local authorities considering measures to prevent or/and mitigate desertification in terms of land management and protection policies. In addition, the produced ESAI maps could become a valuable tool for the local authorities to inform farmers and convince them on the measures to be taken for the mitigation of desertification in their area.

This study, using the ESAI methodology, aims to assess the desertification risk on Crete and includes, as a new approach, the RCP 4.5 and RCP 8.5 climate change scenarios as well as the effects of the applied GAPs in tree crops, which cover about 72% of the total agricultural area on the island. Although ESAI methodology has been widely used and numerous scientific studies have been published based on it, the main limitations are the coarse assessments and the uncertainty in evaluating the final desertification map. In our case, apart from the model's inherent uncertainty, including the applied climate change scenarios, the uncertainty for some of the four main sensitivity groups/maps that are involved, such as soil and the scenario of how GAPs could affect it, can be significant. The uncertainty for the others (climate, vegetation management, and land management) is usually low, provided that the map information is accurate and up-to-date. Concerning the soil sensitivity group, there is uncertainty about soil organic matter stocks and trends because long-term soil monitoring networks, with a sufficient number of sampling sites across Europe, note contrasting soil organic matter trends and several assumptions on their relationship with soil type, geomorphology, land use, and climate (Panagos *et al.* 2013; Stolte *et al.* 2016). These sources of uncertainty could be reduced for small-scale field measurements, capturing as far as possible the heterogeneity of the studied system, as well as detailed climatological information. These approaches are included in the present study and contribute to overcoming the limitations of the proposed method.

CONCLUSIONS

The present study is an application of the MEDALUS – ESAI methodology in a typical Mediterranean island (Crete) for the assessment of the desertification risk under different climate and agricultural practices scenarios. As a step forward from the previous desertification study in Crete by Morianou *et al.* (2018), two future climate change scenarios were used to dynamically study the evolution of desertification sensitivity in the future. Going one step further, a scenario of soil quality improvement using GAPs is incorporated for the mitigation of the phenomenon.

In the current conditions (baseline scenario), a significant part of the island is characterized as fragile or critically sensitive to desertification. Human activities are the main cause – the non-proper changes of land use, overgrazing and deforestation are only some of these activities. The eastern part of Crete is more sensitive to desertification than the western part mainly due to the drier climate conditions.

By applying the ESAI methodology for two climate change scenarios, this study shows that the desertification risk of the island warrants attention for the future. The sensitivity of the island will increase greatly due to the increasing temperature and decreasing precipitation. Therefore, in this study a soil quality improvement scenario was also applied in combination with the two future scenarios. The scenario was based on the results of GAPs applied in the frame of LIFE + AgroClimaWater, to reduce the desertification sensitive areas in Crete. The results of this scenario show a slight mitigation of desertification risk in the future. However, it seems that good agricultural practices, individually, are not enough to mitigate the risk of desertification in the context of climate change. Therefore, the problem of land degradation must be addressed comprehensively by the authorities by applying policies for the prevention and mitigation of desertification in the Mediterranean region.

ACKNOWLEDGEMENTS

The authors acknowledge contribution of the LIFE + financial instrument of the European Union to project LIFE14 CCA/GR/000389 – AgroClimaWater.

DATA AVAILABILITY STATEMENT

All relevant data are included in the paper or its Supplementary Information.

REFERENCES

Alexandersson, H. 1986 A homogeneity test applied to precipitation data. *Journal of Climatology* **6** (6), 661–675.

Allen, R. G., Pereira, L. S., Raes, D. & Smith, M. 1998 Crop evapotranspiration – Guidelines for computing crop water requirements – FAO Irrigation and Drainage Paper 56, FAO, Rome, **300** (9), D05109.

Allen, R. G., Pereira, L. S., Smith, M., Raes, D. & Wright, J. L. 2005 FAO-56 dual crop coefficient method for estimating evaporation from soil and application extensions. *Journal of Irrigation and Drainage Engineering* **131** (1), 2–13.

Baartman, J. E., van Lynden, G. W., Reed, M., Ritsema, C. & Hessel, R. 2007 *Desertification and Land Degradation: Origins, Processes and Solutions. DESIRE Report Series: Scientific Report*, Wageningen: DESIRE.

Babaev, A. & Kharin, N. (eds) 1999 The monitoring and forecast of desertification processes. In: *Desert Problems and Desertification in Central Asia*. Springer, Berlin, pp. 59–75.

Bakr, N., Weindorf, D. C., Bahnassy, M. H. & El-Badawi, M. M. 2012 Multi-temporal assessment of land sensitivity to desertification in a fragile agro-ecosystem: environmental indicators. *Ecological Indicators* **15** (1), 271–280.

Basso, F., Bove, E., Dumontet, S., Ferrara, A., Pisante, M., Quaranta, G. & Taberner, M. 2000 Evaluating environmental sensitivity at the basin scale through the use of geographic information systems and remotely sensed data: an example covering the Agri basin (Southern Italy). *Catena* **40** (1), 19–35.

Boudjemline, F. & Semar, A. 2018 Assessment and mapping of desertification sensitivity with MEDALUS model and GIS–Case study: basin of Hodna, Algeria. *Journal of Water and Land Development* **36** (1), 17–26.

Brandt, J., Geeson, N. & Imeson, A. 2003 *A Desertification Indicator System for Mediterranean Europe. DESERTLINKS Project*. Available from: www.kcl.ac.uk/desertlinks.

Clarke, L., Lurz, J., Wise, M., Edmonds, J., Kim, S., Smith, S. & Pitcher, H. 2007 *Model Documentation for the Minicam Climate Change Science Program Stabilization Scenarios: Ccsp Product 2.1 A. Pacific Northwest National Laboratory, PNNL-16735*.

CLC 2012 *2012 CORINE Land Cover*. Available from: https://land.copernicus.eu/pan-european/corine-land-cover/clc-2012.

Collins, M., Knutti, R., Arblaster, J., Dufresne, J.-L., Fichefet, T., Friedlingstein, P., Gao, X., Gutowski, W. J., Johns, T. & Krinner, G. 2013 Long-term climate change: projections, commitments and irreversibility. In: *Climate Change 2013-The Physical Science Basis: Contribution of Working Group I to the Fifth Assessment Report of the Intergovernmental Panel on Climate Change*. Cambridge University Press, Cambridge, pp. 1029–1136.

Conacher, A. J. & Sala, M. 1998 *Land Degradation in Mediterranean Environments of the World: Nature and Extent, Causes and Solutions*. John Wiley and Sons Ltd, Chichester.

Contador, J. L., Schnabel, S., Gutierrez, A. G. & Fernandez, M. P. 2009 Mapping sensitivity to land degradation in Extremadura, SW Spain. *Land Degradation & Development* **20** (2), 129–144.

DESERTLINKS 2004 *Desertification Indicator System for Mediterranean Europe*. Available from: http://www.desire-his.eu/en/assessment-with-indicators/related-sitesthematicmenu-277/204-desertifi.

Drake, N. & Vafeidis, A. 2004 A review of European Union funded research into the monitoring and mapping of Mediterranean desertification. *Advances in Environmental Monitoring and Modelling* **1** (4), 1–51.

Eris, E., Cavus, Y., Aksoy, H., Burgan, H. I., Aksu, H. & Boyacioglu, H. 2020 Spatiotemporal analysis of meteorological drought over Kucuk Menderes River Basin in the Aegean Region of Turkey. *Theoretical and Applied Climatology* **142** (3), 1515–1530.

Geist, H. J. & Lambin, E. F. 2004 Dynamic causal patterns of desertification. *BioScience* **54** (9), 817–829.

Giorgi, F. & Bi, X. 2005 Regional changes in surface climate interannual variability for the 21st century from ensembles of global model simulations. *Geophysical Research Letters* **32** (13), L13701.

Giorgi, F. & Lionello, P. 2008 Climate change projections for the Mediterranean region. *Global and Planetary Change* **63** (2–3), 90–104.

Hargreaves, G. H. & Samani, Z. A. 1985 Reference crop evapotranspiration from temperature. *Applied Engineering in Agriculture* **1** (2), 96–99.

Hernández, A. J., Lacasta, C. & Pastor, J. 2005 Effects of different management practices on soil conservation and soil water in a rainfed olive orchard. *Agricultural Water Management* **77** (1–3), 232–248.

Herrmann, F., Kunkel, R., Ostermann, U., Vereecken, H. & Wendland, F. 2016 Projected impact of climate change on irrigation needs and groundwater resources in the metropolitan area of Hamburg (Germany). *Environmental Earth Sciences* **75** (14), 1104.

Kavvadias, V. & Koubouris, G. 2019 Sustainable soil management practices in olive groves. In: *Soil Fertility Management for Sustainable Development* (D. G. Panpatte & Y. K. Jhala, eds). Springer Singapore, Singapore, pp. 167–188.

Kavvadias, V., Papadopoulou, M., Vavoulidou, E., Theocharopoulos, S., Repas, S., Koubouris, G., Psarras, G. & Kokkinos, G. 2018 Effect of addition of organic materials and irrigation practices on soil quality in olive Groves. *Journal of Water and Climate Change* **9** (4), 775–785.

Kepner, W. G., Rubio, J. L., Mouat, D. A. & Pedrazzini, F. 2006 Desertification in the Mediterranean Region. In: *A Security Issue: Proceedings of the NATO Mediterranean Dialogue Workshop, Held in Valencia*, 2–5 December 2003, Spain. Springer Science & Business Media.

Kosmas, C., Ferrara, A., Briasouli, H. & Imeson, A. 1999 Methodology for mapping environmentally sensitive areas (ESAs) to desertification. In: *The Medalus Project Mediterranean Desertification and Land Use. Manual on Key Indicators of Desertification and Mapping Environmentally Sensitive Areas to Desertification* (K. Kosmas, M. Kirkby & N. Geeson, eds.), European Commission, pp. 31–47.

Koubouris, G., Kourgialas, N., Kavvadias, V., Digalaki, N. & Psarras, G. 2017 Sustainable agricultural practices for improving soil carbon and nitrogen content in relation to water availability–an adapted approach to Mediterranean olive groves. *Communications in Soil Science and Plant Analysis* **48** (22), 2687–2700.

Kourgialas, N. N. 2021 A critical review of water resources in Greece: the key role of agricultural adaptation to climate-water effects. *Science of The Total Environment* **775**, 145857.

Kourgialas, N. N., Dokou, Z., Karatzas, G. P., Panagopoulos, G., Soupios, P., Vafidis, A., Manoutsoglou, E. & Schafmeister, M. 2016a Saltwater intrusion in an irrigated agricultural area: combining density-dependent modeling and geophysical methods. *Environmental Earth Science* **17** (1), 1–13,15.

Kourgialas, N. N., Koubouris, G. C., Karatzas, G. P. & Metzidakis, I. 2016b Assessing water erosion in Mediterranean tree crops using geoinformatic techniques and field measurements; the effect of climate change. *Natural Hazards* **83**, 65–81. doi: 10.1007/s11069-016-2354-5.

Kourgialas, N. N., Anyfanti, I., Karatzas, G. P. & Dokou, Z. 2018 An integrated method for assessing drought prone areas – water efficiency practices for a climate resilient Mediterranean agriculture. *Science of the Total Environment* **625**, 1290–1300.

Koutroulis, A. G., Vrohidou, A.-E. K. & Tsanis, I. K. 2011 Spatiotemporal characteristics of meteorological drought for the island of Crete. *Journal of Hydrometeorology* **12** (2), 206–226.

Ladisa, G., Todorovic, M. & Liuzzi, G. T. 2012 A GIS-based approach for desertification risk assessment in Apulia region, SE Italy. *Physics and Chemistry of the Earth, Parts A/B/C* **49**, 103–113.

Lal, R. 2001 Soil degradation by erosion. *Land Degradation & Development* **12** (6), 519–539.

LIFE ADAPT2CLIMA 2017 *Deliverable C3: Future Projections on Climatic Indices with Particular Relevance to Agriculture for the Three Islands (Coarse Resolution) and for Each Agricultural Pilot Area (Fine Resolution) Project ADAPT2CLIMA LIFE14 CCA/GR/000928*. Available from: http://adapt2clima.eu/uploads/2017/ADAPT2CLIMA_DEL_C.3_Final_3.pdf (accessed 1 April 2020).

Marke, T., Mauser, W., Pfeiffer, A. & Zängl, G. 2011 A pragmatic approach for the downscaling and bias correction of regional climate simulations: evaluation in hydrological modeling. *Geoscientific Model Development* **4** (3), 759.

Martin, G., Milton, S., Senior, C., Brooks, M., Ineson, S., Reichler, T. & Kim, J. 2010 Analysis and reduction of systematic errors through a seamless approach to modeling weather and climate. *Journal of Climate* **23** (22), 5933–5957.

Michalopoulos, G., Kasapi, K., Koubouris, G., Psarras, G., Arampatzis, G., Hatzigiannakis, E., Kavvadias, V., Xiloyannis, C., Montanaro, G. & Malliaraki, S. 2020 Adaptation of Mediterranean olive groves to climate change through sustainable cultivation practices. *Climate* **8** (4), 54.

Mickley, L. J., Jacob, D. J., Field, B. & Rind, D. 2004 Effects of future climate change on regional air pollution episodes in the United States. *Geophysical Research Letters* **31** (24), L24103.

Montanaro, G., Celano, G., Dichio, B. & Xiloyannis, C. 2010 Effects of soil-protecting agricultural practices on soil organic carbon and productivity in fruit tree orchards. *Land Degradation & Development* **21** (2), 132–138.

Montanaro, G., Dichio, B., Bati, C. B. & Xiloyannis, C. 2012 Soil management affects carbon dynamics and yield in a Mediterranean peach orchard. *Agriculture, Ecosystems & Environment* **161**, 46–54.

Moreno, B., Garcia-Rodriguez, S., Cañizares, R., Castro, J. & Benítez, E. 2009 Rainfed olive farming in south-eastern Spain: long-term effect of soil management on biological indicators of soil quality. *Agriculture, Ecosystems & Environment* **131** (3–4), 333–339.

Morianou, G., Kourgialas, N., Psarras, G. & Koubouris, G. 2018 Mapping sensitivity to desertification in Crete (Greece), the risk for agricultural areas. *Journal of Water and Climate Change* **9** (4), 691–702.

Moss, R., Babiker, W., Brinkman, S., Calvo, E., Carter, T., Edmonds, J., Elgizouli, I., Emori, S., Erda, L. & Hibbard, K. 2008 *Towards New Scenarios for the Analysis of Emissions: Climate Change, Impacts and Response Strategies*. IPCC Expert Meeting Report, 19 September, Noordwijkerhout, The Netherlands.

Moss, R. H., Edmonds, J. A., Hibbard, K. A., Manning, M. R., Rose, S. K., Van Vuuren, D. P., Carter, T. R., Emori, S., Kainuma, M. & Kram, T. 2010 The next generation of scenarios for climate change research and assessment. *Nature* **463** (7282), 747–756.

NATURA 2000 *Natura 2000 Data and Maps*. Available from: http://ec.europa.eu/environment/nature/natura2000/data/index_en.htm.

Oldeman L. & GWJ v. L. 1997 Assessment of the status of human-induced soil degradation in China. *Proceedings Workshop Wageningen-China* (P. W. J. Uithol & J. J. R. Groot, eds.), Wageningen, December 1997, Report 84 (8), 101–111.

OPEKEPE 2017 *Greek Payment Authority of Common Agricultural Policy (C.A.P.) Aid Schemes*. Available from: http://www.opekepe.gr/

Panagos, P., Meusburger, K., Alewell, C. & Montanarella, L. 2012 Soil erodibility estimation using LUCAS point survey data of Europe. *Environmental Modelling & Software* **30** (2012), 143–145.

Panagos, P., Hiederer, R., Van Liedekerke, M. & Bampa, F. 2013 Estimating soil organic carbon in Europe based on data collected through a European network. *Ecological Indicators* **24**, 439–450.

Panagos, P., Christos, K., Cristiano, B. & Ioannis, G. 2014 Seasonal monitoring of soil erosion at regional scale: an application of the G2 model in Crete focusing on agricultural land uses. *International Journal of Applied Earth Observation and Geoinformation* **27**, 147–155.

Reynolds, J. F., Smith, D. M. S., Lambin, E. F., Turner, B., Mortimore, M., Batterbury, S. P., Downing, T. E., Dowlatabadi, H., Fernández, R. J. & Herrick, J. E. 2007 Global desertification: building a science for dryland development. *Science* **316** (5826), 847–851.

Salvati, L. & Zitti, M. 2009 Convergence or divergence in desertification risk? scale-based assessment and policy implications in a Mediterranean country. *Journal of Environmental Planning and Management* **52** (7), 957–971.

Salvati, L., Ferrara, C. & Corona, P. 2015 Indirect validation of the environmental sensitive area index using soil degradation indicators: a country-scale approach. *Ecological Indicators* **57**, 360–365.

Sepehr, A., Hassanli, A., Ekhtesasi, M. & Jamali, J. 2007 Quantitative assessment of desertification in south of Iran using MEDALUS method. *Environmental Monitoring and Assessment* **134** (1–3), 243.

Sommer, S., Zucca, C., Grainger, A., Cherlet, M., Zougmore, R., Sokona, Y., Hill, J., Della Peruta, R., Roehrig, J. & Wang, G. 2011 Application of indicator systems for monitoring and assessment of desertification from national to global scales. *Land Degradation & Development* **22** (2), 184–197.

Sorriso-Valvo, M. 2005 To desertification: an application to the Calabrian territory (Italy). *Geomorphological Processes and Human Impacts in River Basins* **23**, 299.

Stolte, J., Tesfai, M., Oygarden, L., Kvaerno, S., Keizer, J., Verheijen, F. & Hessel, R. 2016 *Soil Threats in Europe: Status, Methods, Drivers and Effects on Ecosystem Services: Deliverable 2.1 RECARE Project*. European Commission DG Joint Research Centre.

Strandberg, G., Bärring, L., Hansson, U., Jansson, C., Jones, C., Kjellström, E., Kupiainen, M., Nikulin, G., Samuelsson, P. & Ullerstig, A. 2015 *CORDEX Scenarios for Europe From the Rossby Centre Regional Climate Model RCA4*. 116 SMHI. SE-60176, Norrköping, Sweden.

Symeonakis, E., Karathanasis, N., Koukoulas, S. & Panagopoulos, G. 2016 Monitoring sensitivity to land degradation and desertification with the environmentally sensitive area index: the case of lesvos island. *Land Degradation & Development* **27** (6), 1562–1573.

Türkeş, M. 1999 Vulnerability of Turkey to desertification with respect to precipitation and aridity conditions. *Turkish Journal of Engineering and Environmental Sciences* **23** (5), 363–380.

UNCCD 1994 *United Nations Convention to Combat Desertification, Elaboration of an International Convention to Combat Desertification in Countries Experiencing Serious Drought and/or Desertification, Particularly in Africa (U.N. Doc. A/AC.241/27, 33 I.L.M. 1328*. United Nations, New York.

Vicente-Serrano, S. M., Beguería, S., López-Moreno, J. I., García-Vera, M. A. & Stepanek, P. 2010 A complete daily precipitation database for northeast Spain: reconstruction, quality control, and homogeneity. *International Journal of Climatology* **30** (8), 1146–1163.

Weaver, C., Liang, X.-Z., Zhu, J., Adams, P., Amar, P., Avise, J., Caughey, M., Chen, J., Cohen, R. & Cooter, E. 2009 A preliminary synthesis of modeled climate change impacts on US regional ozone concentrations. *Bulletin of the American Meteorological Society* **90** (12), 1843–1864.

Xiloyannis, C., Martinez Raya, A., Kosmas, C. & Favia, M. 2008 Semi-intensive olive orchards on sloping land: requiring good

land husbandry for future development. *Journal of Environmental Management* **89** (2), 110–119.
Zanis, P., Kapsomenakis, I., Philandras, C., Douvis, K., Nikolakis, D., Kanellopoulou, E., Zerefos, C. & Repapis, C. 2009 Analysis of an ensemble of present day and future regional climate simulations for Greece. *International Journal of Climatology: A Journal of the Royal Meteorological Society* **29** (11), 1614–1633.

First received 28 January 2021; accepted in revised form 23 April 2021. Available online 6 May 2021

Consumptive water footprints, water use efficiencies and productivities of rice under alternate wetting and drying for Kharagpur, West Bengal, India

A. Biswas, D. R. Mailapalli and N. S. Raghuwanshi

ABSTRACT

An experimental study was carried out with medium duration rice variety (IR 36) during kharif and rabi seasons of 2015/16 and 2016/17 to investigate the effect of alternate wetting and drying (AWD) practice on water use efficiency, productivity, and consumptive water footprints of rice. The performance of AWD practice was compared with the conventionally (CON) irrigated rice using non-weighing lysimeters. The study resulted that by managing the alternate wetting and drying up to 15 cm below the ground level, a significant reduction in water input (26–29% in kharif and 22–27% in rabi season) could be achieved under AWD. A reduction in evapotranspiration (about 6% in both kharif and rabi seasons) was also observed under AWD. Reduction in consumptive water footprint (about 2–3% in kharif and 2–5% in rabi) was obtained under AWD. Reductions in blue water footprints (7% in kharif and 4–5% in rabi) was also observed under AWD. On average, crop water use efficiency was significantly enhanced by 27–33% and 20–29% in the respective kharif and rabi seasons under AWD practice. Significant improvement in total water productivity by 29–37% and 23–35% in the respective two seasons exhibited the superiority of AWD over CON during the two years of field experiments.

Key words | alternate wetting and drying, consumptive water footprint, crop water use efficiency, lysimeter, water productivity

A. Biswas
D. R. Mailapalli (corresponding author)
N. S. Raghuwanshi
Agricultural & Food Engineering Department,
IIT Kharagpur,
Kharagpur, 721302, West Bengal,
India
E-mail: *mailapalli@agfe.iitkgp.ac.in*

HIGHLIGHTS

- Substantial reduction (22–29%) in water input was under AWD.
- Rice evapotranspiration decreased by 6% under AWD.
- AWD reduced rice consumptive water footprint by 2–5%.
- A substantial increase (20–33%) in crop water use efficiency was under AWD.
- AWD improved total water productivity considerably by 23–37%.

INTRODUCTION

In Asia, irrigated agriculture accounts for about 90% of total diverted freshwater, and more than 50% of this is used to irrigate rice (Hatiye *et al.* 2015). However, water is frequently being wasted as a result of inefficient management, resulting in low water productivity (Yao *et al.* 2012; Amarasingha *et al.* 2015). It is estimated that about 2 million ha of Asia's irrigated dry-season rice and 13 million ha of its irrigated wet-season rice will experience water scarcity by 2025 (Bouman & Tuong 2001; Amarasingha *et al.* 2015). Therefore, major studies (Yadav *et al.* 2011; Amarasingha *et al.* 2015; Sharda *et al.* 2017) suggested few efficient irrigation water

doi: 10.2166/ws.2021.068

management practices that could save a significant quantity of water during rice cultivation. One such water-saving irrigation technique is an alternate wetting-and-drying (AWD) method (Carrijo *et al.* 2017), in which water is applied to the field a number of days after the disappearance of ponded water (Amarasingha *et al.* 2015; Carrijo *et al.* 2017). This means that the rice fields are not kept continuously submerged (i.e. never letting the ponded water disappear) as practiced in conventional irrigation, but are allowed to dry intermittently during the rice-growing stage (Sriphirom *et al.* 2018). A 'Safe' AWD allows the water level in the rice field to drop down to 15 cm below the soil surface before the crop is irrigated again (Amarasingha *et al.* 2015; Carrijo *et al.* 2017).

Well-designed AWD water management conditions revealed higher rice water productivity compared to continuously flooded soil conditions in most of the previous instances (Gaydon *et al.* 2012; Ye *et al.* 2013; Amarasingha *et al.* 2015; Thakur *et al.* 2018). This AWD technique also proved its ability to save up to 30% irrigation water and improve irrigation water use efficiency by 5–35% depending upon the crop cultivar, soil characteristics, climate, and crop water (Cabangon *et al.* 2011; Liang *et al.* 2013; Djaman *et al.* 2018). However, research is still lacking on knowledge about water consumption per unit production of rice under this AWD technique. Recognizing the significance of the above facts discussed, quantification of consumptive water footprint (CWF) of rice under AWD is of utmost importance as it is the reliable way to evaluate real water consumption during rice growth periods.

The CWF of any crop denotes the ratio of the total evapotranspiration during that crop growth period to the quantity of that crop obtained (Cao *et al.* 2018). The CWF combines blue and green water footprints (Cao *et al.* 2018). Therefore, it reflects the amount of freshwater resources utilized per unit quantity of agricultural produces from a particular management system. Hence, the lower CWF of a crop under any water management will reveal its efficiency to produce more biological yield with less volume of freshwater (Lovarelli *et al.* 2016). Therefore, the research on the CWF of rice under AWD irrigation has turned out to be a hot research topic.

In view of the importance of CWF, the previous researches on CWF of crop production were carried out in several regional levels from an irrigation district (Cao *et al.* 2014a; Suttayakul *et al.* 2016; Shrestha *et al.* 2017), a river basin (Bocchiola *et al.* 2013; Denis *et al.* 2016; Roux *et al.* 2017), a city-level region (Marano & Filippi 2015; Lu *et al.* 2016; Chu *et al.* 2017), a country (Cao *et al.* 2014b; Wang *et al.* 2015; Zhuo *et al.* 2016) to the global standpoint (Mekonnen & Hoekstra 2011, 2014; Lovarelli *et al.* 2016). The previous studies on CWF were satisfactorily addressed in terms of research scales and crop types (Cao *et al.* 2018). Mekonnen & Hoekstra (2011) documented CWF values of 126 crops and derived crop produces for 10 years (1996–2005) at multiple regional levels. However, most of the previous studies used different types of models like the water balance model, hydrological model, and crop model to estimate CWF of crop production (Cao *et al.* 2018). Although the modeling assessment of CWF is proficient in economy and time to estimate costs for large scale, field evaluation is essential for validation of results obtained from modeling work (Johannessen *et al.* 2015). Meanwhile, CWF assessment for a specific farm area and real-time estimation of CWF-related parameters are more precise than the model simulated results and can provide a more truthful reference for farmers.

Very few researchers have started to investigate field measurement methods for CWF of rice and its viability in current years. Kar *et al.* (2015) studied for the first time, CWF of rice under intermittent irrigation management in the East coast region of India. Cao *et al.* (2018) estimated CWF of rice production under traditional irrigation practice in East China. The above relevant literature principally paid attention to the measurement of rice CWF at the farm level. However, more exploration on such farm level assessment of CWF of rice under different irrigation practices is of utmost importance to provide accurate information on real water consumption of rice under different irrigation management to the local farmers. Recognizing the importance of the above fact, the present study was conducted to investigate the impact of AWD irrigation practice on CWF of rice at farm level during monsoon (kharif) and post-monsoon season (rabi) of Eastern India. Along with CWF, the variations of water input, yield, water use efficiency, and water productivities under AWD were also explored and documented in the present study.

MATERIALS AND METHODS

Study area

The field experiments were carried out in the Research Farm of Agricultural and Food Engineering Department, Indian Institute of Technology (IIT), Kharagpur, India. The study site is located at 22.32°N latitude and 87.31°E longitude. The average annual rainfall in the study area is around 1,400 mm, 78% of which occurs during the monsoon period (June to September). Winter (December to mid-February) is brief but chilly (10° to 25 °C); summers (March to June) are hot (25° to 40 °C) and relative humidity ranges from 50 to 95%. The climate is mostly sub-humid and subtropical. The soil is of sandy loam type and grouped under alfisol based on the USDA Soil Taxonomy. Table 1 presents the physical and chemical properties of the soil at the experimental site.

Field experiments

Experimental plot design

The field experiments were carried out during the kharif (July-October) and rabi seasons (January-April) of 2015/16 and 2016/17 in 6 (3 m × 3 m) plots under conventional (CON) and alternate wetting and drying (AWD) practices. These two water treatments (CON and AWD) were replicated three times. The high yielding and popularly cultivated IR 36 rice cultivar was selected for this study. Rice seedlings of 25 days old were transplanted manually by planting 5 seedlings per hill and maintained a grid spacing of 20 cm × 20 cm for both CON and AWD practices. All plots of both treatments were mechanically weeded at 20 days after transplantation (DAT) and manually weeded at 30 and 45 DAT. The recommended dose of N: P: K (120:50:60 kg/ha) was considered (Adhikari *et al.* 2011) for the experiments under both water treatments. The N-fertilizer was applied in the form of urea (46% N) in four splits; basal (during transplanting), 20, 40, and 60 DAT whereas the full fertilizer doses of phosphorus in the form of Single Super Phosphate (18% P_2O_5) and potassium as Muriate of Potash (62.5% K_2O) were applied in one-split as basal dose (during transplanting).

Irrigation management

Groundwater was used to irrigate the experimental plots of both treatments. In CON practice, a maximum of 2–3 cm standing water was maintained during 0–20 DAT for the crop to establish, and 5 cm was maintained during 21–95 DAT. Similar to CON practice, the AWD plots were initially maintained at a ponding depth of 2–3 cm during 0–20 DAT. After that, the AWD plots were irrigated to a depth of 5 cm above the ground surface when the standing water level depleted to 15 cm below the ground surface (i.e. -ve water level). The ponding depth was measured with a measuring scale installed in each plot of both water treatments. In the case of AWD, 'Safe' AWD guidelines were followed (Amarasingha *et al.* 2015) in this study. To record the water level below the ground surface, a perforated PVC pipe of 30 cm long and 15 cm diameter were installed in the AWD plots. The diameter was made large enough for easy visibility, freehand movement inside the tube to clean during clogging of soil. All plots were accommodated with an outlet 5 cm above the soil surface after 20 DAT to facilitate 5 cm of ponding depth in rain events. The excess rainwater in all plots was channelled out and collected in a constructed farm pond, and utilized for irrigation purposes in both treatment plots. All plots were surrounded by 50 cm

Table 1 | Selected physical and chemical properties of soil at the experimental site

Depth (cm)	Sand (%)	Silt (%)	Clay (%)	BD (g/cm^3)	pH	EC (dS/m)	TN (%)	OC (%)
0–20	59.70 (3.91)	25.80 (2.65)	14.50 (3.22)	1.59 (0.07)	5.70	0.60	0.05	0.5
20–40	53.60 (2.43)	21.00 (2.86)	25.40 (3.81)	1.48 (0.05)	5.80	0.70	0.02	0.2
40–80	53.20 (3.31)	19.70 (3.07)	27.10 (3.29)	1.51 (0.04)	5.91	0.84	0.01	0.1

BD, Bulk density; EC, Electrical conductivity; TN, Total nitrogen; OC, Organic carbon.

wide bunds to prevent lateral water seepage and nutrient diffusion between the treatments, followed by 50 cm wide irrigation channels. The bunds of all plots were plugged based on the procedure reported by Patil *et al.* (2011) for controlling the lateral flow from plot-to-plot. Each plot was provided by the bund height of 80 cm to control the overland flow from the adjacent irrigation and drainage channels.

Data collection

The data were grouped into meteorological, water, and crop-related data. The meteorological data were collected from the meteorological station of IIT Kharagpur. Daily standing water depth, crop evapotranspiration (ET_c), evaporation (E), and deep percolation (DP) readings were estimated during the crop growth period of both seasons using the field lysimeters. Irrigation and rainfall were taken as inputs and DP and ET_C as outputs from the soil reservoir. The ET_C of rice under both water managements were measured by using the volume balance approach (Cao *et al.* 2018; Shekhar *et al.* 2020a). In this study, four bottom open and two bottom closed non-weighing lysimeters of $1.25\ m \times 1.25\ m \times 1\ m$ dimensions were installed in the experimental plots. The bottom of each lysimeter was kept at 800 mm below the ground surface and the water going beyond 800 mm was considered as DP loss. A combination of two bottom open and one bottom closed lysimeter was used to measure ET_c, E, and DP losses from the plots of each water treatment. Crop evapotranspiration together with deep percolation ($ET_c + DP$) was measured using a bottom open lysimeter in which paddy was grown (Figure 1(a)). Deep percolation together with evaporation ($E + DP$) was measured using another bottom open lysimeter in which there was no crop (Figure 1(b)), and E was measured using a bottom closed lysimeter without paddy (Figure 1(c)). The water level in the lysimeters was held at the same level as in the experimental plots and the rate of decrease of water level inside the lysimeters was recorded daily. Then, the values of daily E, ET_c, and DP at the study area were calculated during both kharif and rabi seasons (2015–17) as follows in Equations (1)–(3):

$$ET_C = \Delta H_{Ai} - (\Delta H_{Bi} - \Delta H_{Ci}) \tag{1}$$

$$DP = \Delta H_{Bi} - \Delta H_{Ci} \tag{2}$$

$$E = \Delta H_{Ci} \tag{3}$$

where, ΔH_A, ΔH_B, ΔH_C is the difference between the two consecutive ponding depth readings in the lysimeters A, B, and C, respectively, in a given time step (i). In this study, a daily time step was used.

The crop-related information was collected during the crop growing periods. The information related to tillering, panicle initiation, flowering, and harvesting were collected at different growth stages in both seasons. For obtaining

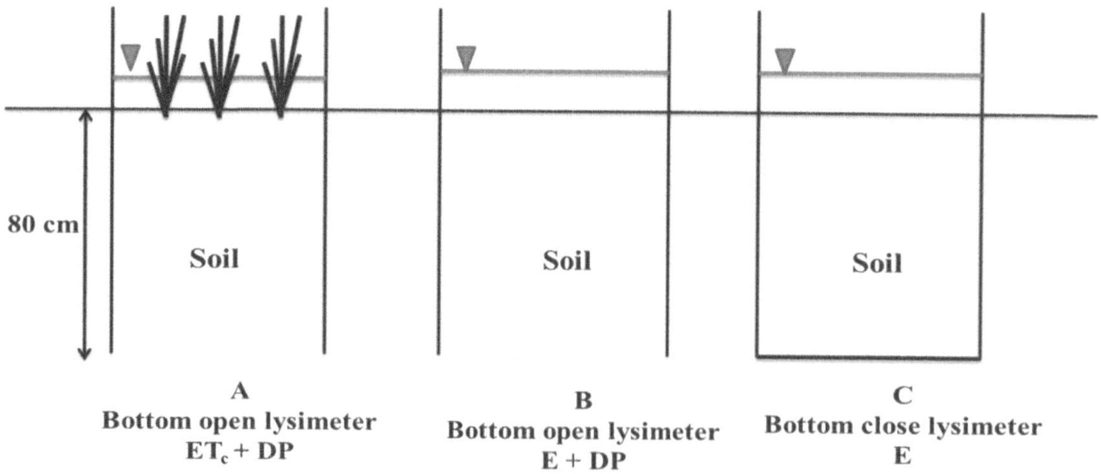

Figure 1 | Field measurement of ET_C, E, and DP in the paddy field during rice growing season.

grain yield, rice plants were first harvested from a 1 m × 1 m sampling area marked at three randomly selected spots in each plot. The grains were then separated from the straws using a threshing machine and the average grain yield (grain weight) was recorded for each plot.

Crop water use efficiency concept. Crop water use efficiency (CWUE) is defined as the ratio of the water that is effectively taken up by the plant to the total water supplied (TWS); that is, that was not lost to drainage, bare soil evaporation, or interception (Ragab 2015). Mathematically, it can be defined as Equation (4):

$$CWUE = \frac{ET_c\,(mm)}{TWS\,(mm)} \tag{4}$$

Crop water productivity concept. Crop water productivity (CWP) is a measure of water productivity expressed in terms of grain yield (Y, kg/ha) per unit of the amount of water consumed by the crop in the process of evapotranspiration (m^3/ha). Mathematically, it is denoted as Equation (5):

$$CWP = \frac{Y\,(kg/ha)}{ET_C\,(m^3/ha)} \tag{5}$$

Total water productivity concept. Total water productivity (TWP) is also a measure of water productivity stated in terms of grain yield (Y, kg/ha) per unit of total water supplied (TWS) including irrigation and rainfall (m^3/ha) (Bouman & Tuong 2001). Mathematically, it can be stated as Equation (6):

$$TWP = \frac{Y\,(kg/ha)}{TWS\,(m^3/ha)} \tag{6}$$

Consumptive water footprint (CWF) concept. Consumptive water footprint refers to the ET_C during the crop growth period (Cao *et al.* 2018). Initially, the water footprint was developed by Hoekstra & Hung (2002). These authors defined water footprint (WF) as 'the total annual volume of fresh water needed to produce the goods and services related to certain consumption pattern'. Generally, WF has three components, namely green, blue and grey. But, CWF of crop production consists of green and blue water components only. Mathematically, it (m^3/t) is expressed as Equation (7):

$$CWF = Green\,WF + Blue\,WF \tag{7}$$

Green WF (GWF) is expressed as the volume of rainwater consumed (W_g, m^3/ha) to crop yield (Y, t/ha) and Blue WF (BWF) is defined as the volume of irrigation water consumed (W_b, m^3/ha) to crop production (Y, t/ha) as presented in Equations (8) and (9).

$$GWF = \frac{W_g\,(m^3/ha)}{Y\,(t/ha)} \tag{8}$$

$$BWF = \frac{W_b\,(m^3/ha)}{Y\,(t/ha)} \tag{9}$$

Specifically, green water (GW) refers to the minimum of effective rainfall (the difference between total rainfall and actual ET_C) and crop ET_C. If ET_C is higher than the effective rainfall (P_{eff}) during crop growth, GW is equal to the effective rainfall (Equation (10a)); on the contrary, if ET_C is less than P_{eff}, GW will be equal to ET_C (Equation (10b)) (Bocchiola *et al.* 2013).

$$\text{If } ET_C \geq P_{eff},\ GW = P_{eff} \tag{10a}$$

$$\text{If } ET_C < P_{eff},\ GW = ET_C \tag{10b}$$

Blue water (BW) refers to the ground or surface water volume that is supplied to the field through irrigation. From the agricultural point of view, BW is dependent on GW. If GW is insufficient, BW is used. Particularly, if ET_C is higher than P_{eff}, BW is potentially equal to the difference between ET_C and P_{eff} (Bocchiola *et al.* 2013) (Equation (11a)). Otherwise, BW is zero (Equation (11b)).

$$\text{If } ET_C \geq P_{eff},\ BW = ET_C - P_{eff} \tag{11a}$$

$$\text{If } ET_C < P_{eff},\ BW = 0 \tag{11b}$$

Data analysis

The data (yield, CWUE, CWP, TWP, GWF, BWF, and CWF) were statistically analyzed by the analysis of variance (ANOVA). Duncan's multiple range test (DMRT) was employed to assess the differences between the treatment means at the 1, 5, and 10% significance levels. All statistical analyses were performed using SPSS 16.0 software.

RESULTS

Water balance components and CWUE

Table 2 presents the values of the water balance components of CON and AWD practices during both seasons. The DP loss under AWD was reduced by 31 and 27% in kharif seasons of two respective years; whereas, in the rabi season, it was reduced by 27 and 26% for the respective years. The ET_C under AWD was reduced by 6% during both kharif and rabi seasons of two respective years (Figure 2). The total water input (irrigation + rainfall) for AWD practice was significantly ($p < 0.01$) reduced by 26 and 29% in kharif seasons and by 22 and 27% in rabi seasons of the two respective years (Figure 2).

Table 2 reveals the variation of the CWUE under AWD and CON during both kharif and rabi seasons. The CWUE in AWD was found to be increased by 27 and 33%, respectively, during the kharif seasons, whereas it was 20 and 29% higher for rabi seasons of the respective two years (Figure 2). The improvement of CWUE in AWD was found to be significant at a 95% level of confidence ($p < 0.05$) during both kharif and rabi seasons (Table 2).

Water productivity

Table 3 presents the variation of the grain yield, CWP, and TWP under AWD and CON during kharif and rabi seasons of 2015/16 and 2016/17 periods. The grain yield was found to be lower in AWD practice and the reduction was observed as about 4 and 3% in respective kharif seasons; whereas in rabi season, it was about 4 and 2% for the respective two years (Figure 3). However, this yield reduction in AWD during both kharif and rabi was not statistically ($p > 0.05$) significant (Table 3). The CWP improved by 2 and 3%, respectively in AWD during respective kharif seasons, whereas in rabi season, the percentage increase in CWP under AWD was 2 and 5% for the two respective years (Figure 3). But, this enhancement of CWP under AWD was found to be statistically insignificant ($p > 0.05$) during both seasons (Table 3). Similarly, TWP increased by 29 and 37%, respectively in AWD during the respective two years' kharif season, whereas in rabi season, the percen-

Table 2 | Water balance components and CWUE under AWD and CON during kharif and rabi seasons of 2015/16 and 2016/17

		2015/16		2016/17	
Parameter	**Treatment**	**Kharif**	**Rabi**	**Kharif**	**Rabi**
Irrigation (mm)	CON	1,070 (±16.35)	1,453 (±17.10)	510 (±7.21)	1,430 (±23.11)
	AWD	710 (±5.69)	1,100 (±6.00)	50 (±0.00)	1,010 (±13.45)
ANOVA		***	***	***	***
Rainfall (mm)	CON	313	173	1,060	121
	AWD	313	173	1,060	121
ET_c (mm)	CON	447	485	433	497
	AWD	420	456	408	466
DP (mm)	CON	785	1,087	858	1,006
	AWD	540	796	629	745
CWUE (%)	CON	32 (±1.49)	30 (±1.11)	28 (±0.64)	32 (±1.47)
	AWD	41 (±1.63)	36 (±1.41)	37 (±0.00)	41 (±2.23)
ANOVA		**	**	**	**

***indicates a strong significant difference at $p < 0.01$ level (high level of significance) in AWD as compared to CON.
**indicates a moderate significant difference at $p < 0.05$ level (medium level of significance) in AWD as compared to CON; value in parenthesis indicates standard deviation.

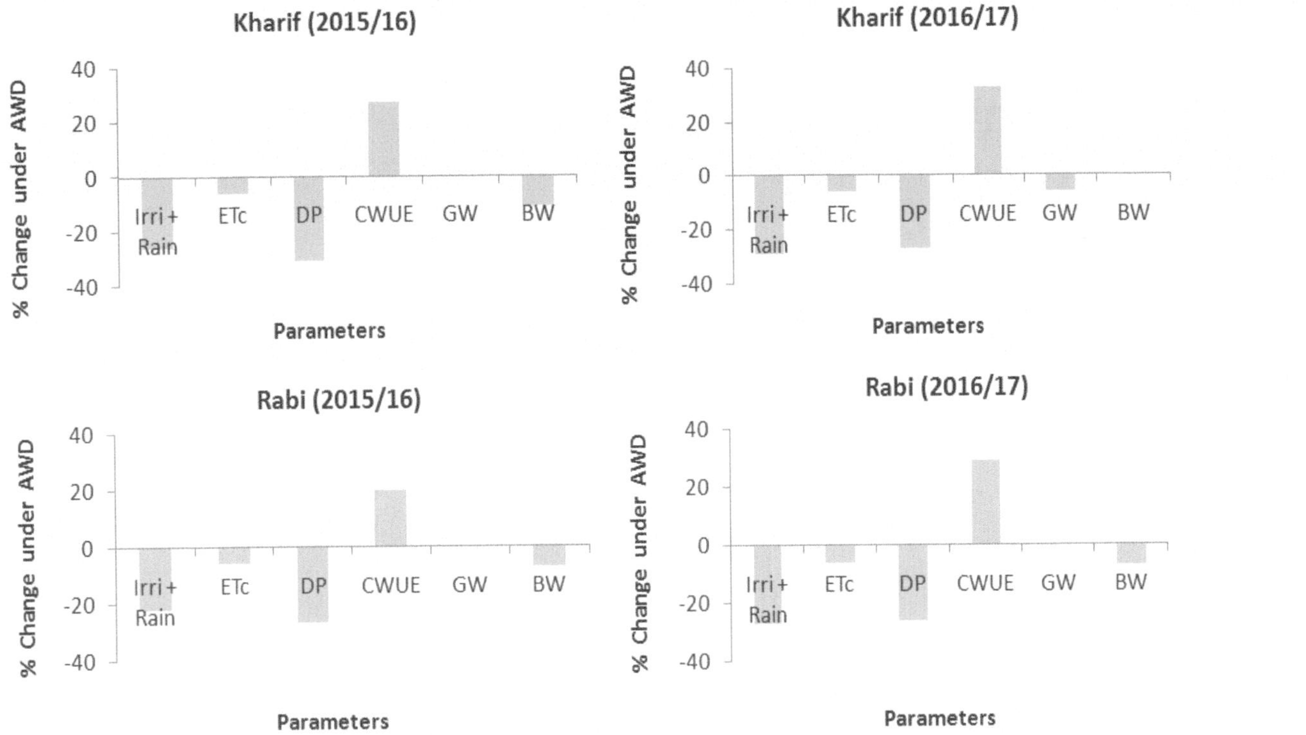

Figure 2 | Variation (%) of different water balance parameters under AWD along with changes (%) of CWUE, GW, and BW during both kharif and rabi seasons of two experimental years.

tage improvement in TWP was found as 23 and 35 for two respective years (Figure 3). This improvement under AWD was found to be statistically significant at a 95% level of confidence ($p < 0.05$) during both seasons of each year (Table 3).

Water footprints

Table 3 presents the observed data of green water (GW) and blue water (BW) for both irrigation practices. The use of GW under both irrigation managements was found as same during kharif of 2015/16 and rabi of both experimental years. However, the scenario of kharif 2016/17 showed about a 6% reduction in GW use under AWD. The BW usage under AWD was declined by about 11% in kharif of 2015/16; whereas in 2016/17 (kharif), ET_C of both practices was fulfilled by GW only. In the case of the rabi season, about a 7% decrease in BW utilization under AWD for both years (Figure 2).

The GWF during kharif was increased by 6% in 2015/16 and reduced by 3% under AWD in 2016/17; whereas during rabi season, the percentage increase was 4 and 2% for 2015/16 and 2016/17 respectively (Figure 3). But, no substantial change ($p > 0.05$) of GWF was found between CON and AWD during both kharif and rabi seasons (Table 3). In the case of BWF, the reduction under AWD was estimated at about 6% in 2015/16 kharif and 3 and 5% in the respective two rabi seasons (Figure 3). However, there was no change of BWF between AWD and CON practices in the kharif season of 2016/17. Such reductions of BWF under AWD were also statistically insignificant ($p > 0.05$) in this study (Table 3). Similarly, CWF under AWD declined by 1 and 3% in the respective two kharif seasons; while in rabi season, the reduction percentage was about 2 and 5% for the respective two years (Figure 3). However, this reduction of CWF under AWD was not statistically significant ($p > 0.05$) (Table 3).

DISCUSSION

Instead of following the CON method, the adoption of the AWD method showed a beneficial effect in terms of water-saving in this study. However, following AWD practice is tricky during the monsoon (kharif) season, and

Table 3 | Green water, blue water, yield, water productivities, and water footprints under AWD and CON for both kharif and rabi seasons

		2015/16		2016/17	
Parameter	**Treatment**	**Kharif**	**Rabi**	**Kharif**	**Rabi**
GW (mm)	CON	200	91	433	42
	AWD	200	91	408	42
BW (mm)	CON	247	394	0	455
	AWD	220	365	0	424
Yield (t ha^{-1})	CON	3.65 (±0.58)	4.07 (±0.55)	3.51 (±0.64)	4.15 (±0.48)
	AWD	3.50 (±0.72)	3.92 (±0.64)	3.39 (±0.60)	4.08 (±0.47)
ANOVA		ns	ns	ns	ns
CWP (kg m^{-3})	CON	0.82 (±0.13)	0.84 (±0.11)	0.81 (±0.15)	0.84 (±0.10)
	AWD	0.83 (±0.17)	0.86 (±0.14)	0.83 (±0.15)	0.88 (±0.10)
ANOVA		ns	ns	ns	ns
TWP (kg m^{-3})	CON	0.26 (±0.04)	0.25 (±0.04)	0.22 (±0.04)	0.27 (±0.02)
	AWD	0.34 (±0.07)	0.31 (±0.06)	0.31 (±0.05)	0.36 (±0.02)
ANOVA		**	**	**	**
GWF (m^3 kg^{-1})	CON	0.559 (±0.09)	0.227 (±0.03)	1.261 (±0.23)	0.103 (±0.01)
	AWD	0.590 (±0.13)	0.237 (±0.04)	1.228 (±0.22)	0.105 (±0.01)
ANOVA		ns	ns	ns	ns
BWF (m^3 kg^{-1})	CON	0.687 (±0.11)	0.980 (±0.13)	0.000 (±0.00)	1.105 (±0.12)
	AWD	0.647 (±0.14)	0.948 (±0.16)	0.000 (±0.00)	1.046 (±0.12)
ANOVA		ns	ns	ns	ns
CWF (m^3 kg^{-1})	CON	1.246 (±0.20)	1.206 (±0.16)	1.261 (±0.23)	1.208 (±0.14)
	AWD	1.237 (±0.27)	1.184 (±0.20)	1.228 (±0.22)	1.151 (±0.13)
ANOVA		ns	ns	ns	ns

(ns) indicates no significant difference.
**indicates a moderate significant difference at $p < 0.05$ level (medium level of significance) in AWD as compared to CON; values in parenthesis indicate standard deviation.

maintaining depletion of water may be challenging. However, the study area experienced some dry periods during monsoon and supplementary irrigation was provided when the water level in AWD pipe reached the threshold (i.e. 15 cm below ground surface). Table 2 shows a significant ($p < 0.01$) reduction in water use (irrigation + rainfall) under AWD for kharif and rabi seasons of both years. This reduction can be attributed to the decline in the number of irrigations under AWD. Similar types of results were also registered by other researchers (Yao *et al.* 2012; Tan *et al.* 2013; Lampayan et al. 2015a, 2015b; Pan *et al.* 2017; Pascual & Wang 2017; Djaman *et al.* 2018). Zhang *et al.* (2012) obtained a substantial reduction of 25% in irrigation water application under AWD. Lampayan *et al.* (2015a, 2015b) reviewed that a 38% reduction in water input under AWD management would be possible if AWD was implemented properly. Umesh *et al.* (2017) documented that properly implemented AWD management in the rice field could save water up to 30%. Djaman *et al.* (2018) registered a considerable quantity of water saving by over 27% with this AWD. Similarly, Sriphirom *et al.* (2018) showed the possibility of 19% water saving using the AWD technique during the tillering stage of rice growth. This reduction in water input under AWD had an impact on the different outflow parameters like ET_C and DP in the field during rice growth. Table 2 and Figure 2 represented the reductions in both ET_C and DP under AWD in this study. These reductions can be attributed to the significant decline in irrigation water under AWD. Alberto *et al.* (2011) documented the possibility of ET_C reduction by about 11% in aerobic rice fields. Tan *et al.* (2013) also registered similar types of findings in their study where intermittent irrigation was found superior over traditional methods in terms of percolation reductions by over 15%. Recently, Shekhar

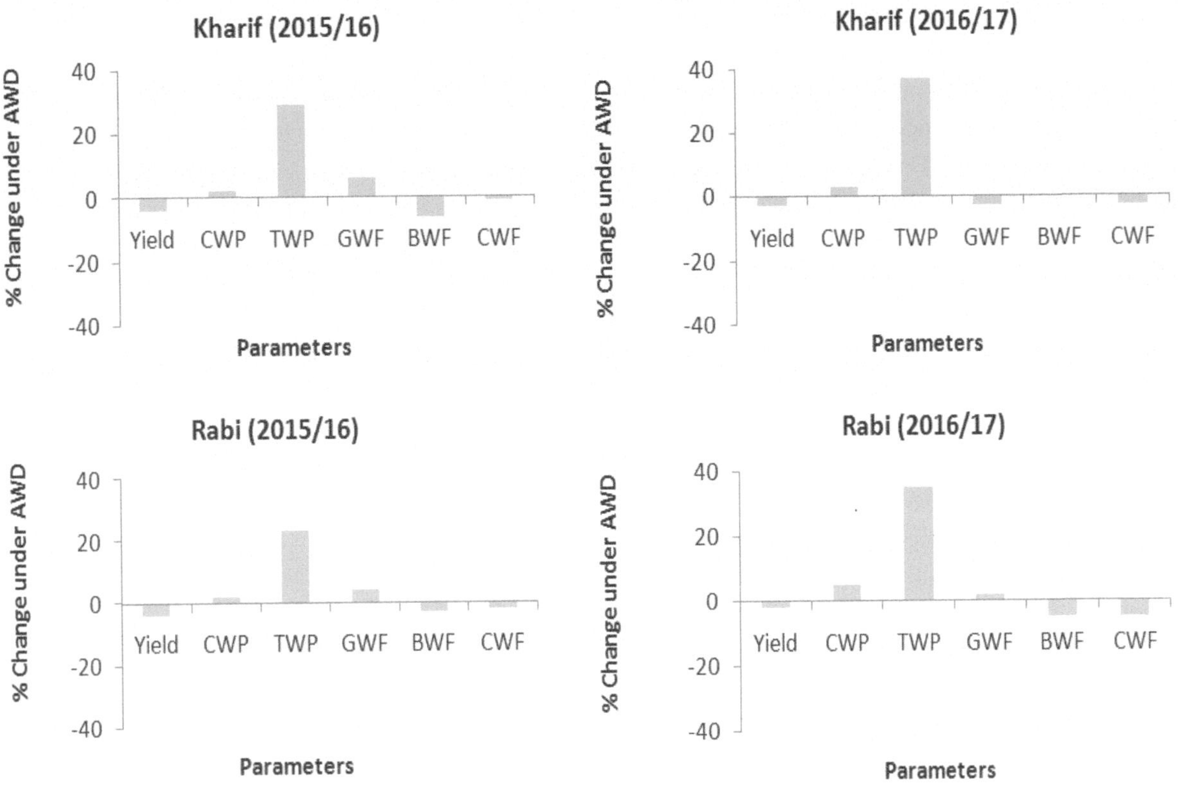

Figure 3 | Change (%) of yield, water productivities and footprints under AWD for both kharif and rabi seasons of two experimental years.

et al. (2020b) reported a 13% reduction in ET_C under AWD practice as compared to CON in their study. Therefore, the AWD management had a positive impact on CWUE in rice cultivation. In the present study, CWUE was significantly ($p < 0.05$) enhanced in AWD during all experimental seasons (Table 2; Figure 2). This improvement could be attributed to a reduction in ET_c and a substantial ($p < 0.01$) decrease in water input under AWD. Wang *et al.* (2016) observed a similar kind of improvement of WUE under a mild AWD approach and documented that the reduced unproductive tillers, increased root growth, and enhanced harvest index attributed to the higher WUE under moderate AWD than CON irrigation. Yang *et al.* (2017) also registered 21–30% enhanced WUE under moderate AWD in their study. Zhou *et al.* (2017) also obtained significantly higher WUE under AWD management compared to traditional irrigation practices in their study.

The current research work demonstrated that the AWD with significantly ($p < 0.01$) reduced water input than CON and had a positive impact on rice yield and productivities (Table 3; Figure 3). The adoption of AWD did not reduce yield significantly in the present study (Table 3). Hence, statistically similar grain yield using significantly ($p < 0.01$) reduced water application and decreased ET_C resulted in enhanced TWP and CWP respectively under AWD in the current study. Other researchers (Yadav *et al.* 2011; Yao *et al.* 2012; Lampayan *et al.* 2015a, 2015b; Pan *et al.* 2017; Djaman *et al.* 2018) also documented similar results in their studies. Thakur *et al.* (2014) registered about 3% higher yield under mild AWD compared to the conventional method. In the study of Lampayan et al. (2015a, 2015b), total water productivity was found to be higher under AWD compared to CON. Wang *et al.* (2016) also obtained better yield and productivity in moderate AWD than traditional practice in their study and reported that reduced unproductive tillers, increased root growth, and enhanced harvest index attributed to the higher yield under moderate AWD than CON irrigation. Pan *et al.* (2017) reported that there was no significant reduction of yield under the 'safe' AWD method (explain) compared to the farmer's water management practice. The study also documented that the 'safe' AWD was advantageous over the farmer's water management practice to enhance water productivity significantly with a reduction in the number of irrigations. According to Zhou *et al.* (2017), rice yield also improved under moderate AWD practice. Such kind of positive reports of AWD on rice yield and productivity also strengthened the outcomes of the present study.

Since CWF of any crop refers to the real water consumption by crop, only ET_C was considered for farm-level water footprint calculation in the current research work. Figure 2 and Table 3 clearly show the variations of BW usage between the AWD and CON method. The change of BW and ET_C coupled with little reduction of grain yield in AWD made the little decline in BWF and CWF under AWD in this study. But, Kar *et al.* (2015) obtained a significant ($p < 0.05$) decrease in CWF under intermittent irrigation method in their study. However, the present study illustrated the statistical similarity of different water footprints between AWD and CON. According to Kar *et al.* (2015), about 5 and 6% reduction in CWF with 3 and 7% decrease in ET_C was possible under moderate and severe AWD respectively. The current farm level investigation also reported similar kinds of reduction percentages of ET_C and CWF under AWD (Figure 3). Hence, it was inferred that the CWF of rice to a small extent could be influenced by AWD management practice.

CONCLUSIONS

The conventional method of rice production has very low water use efficiency, resulting in low productivity, high water footprints, and huge losses of water from the field. Thus, in order to increase the production and water use efficiency and to reduce the water footprints of rice, we need to adopt advanced water-saving irrigation techniques. In this study, the effect of AWD practice on yield, CWUE, water productivity, and water footprints was assessed in comparison to conventional practice. The AWD practice was able to provide significant enhancement of CWUE (27–33% in kharif and 20–29% in rabi) with a considerable reduction in water input (26–29% in kharif and 22–27% in rabi) and deep percolation (27–31% in kharif and 26–27% in rabi). Total water productivity of AWD was found to be significantly improved (29–37% in kharif and 23–35% in rabi) in association with a reduction in ET_C (6% in both kharif and rabi), BWF (6% in kharif and 3–5% in rabi) and CWF (1–3% in kharif and 2–5% in rabi). Therefore, positive responses were achieved under AWD over two years' field experiments in the present study. Since water management in AWD involves wetting and drying phases, water application needs to be adjusted to the local demand as well as local environmental features to obtain more productivity with less water footprint. To fulfil the challenging amount of rice productivity under water-scarce situation in future, water footprints of rice under such kinds of non-conventional water environment should be investigated thoroughly in different environments.

DATA AVAILABILITY STATEMENT

All relevant data are included in the paper or its Supplementary Information.

REFERENCES

Adhikari, B., Bag, M. K., Bhowmick, M. K. & Kundu, C. 2011 *Status Paper on Rice in West Bengal.* Rice Knowledge Management Portal (RKMP), Directorate of Rice Research, Rajendranagar, Hyderabad. Available from: http://www.rkmp.co.in/.

Alberto, M. C. R., Wassmann, R., Hirano, T., Miyata, A., Hatano, R., Kumar, A., Padre, A. & Amante, M. 2011 Comparisons of energy balance and evapotranspiration between flooded and aerobic rice fields in the Philippines. *Agricultural Water Management* **98**, 1417–1430. https://doi/10.1016/j.agwat.2011.04.011.

Amarasingha, R. P. R. K., Suriyagoda, L. D. B., Marambe, B., Gaydon, D. S., Galagedara, L. W., Punyawardena, R. & Howden, M. 2015 Simulation of crop and water productivity for rice (*Oryza sativa* L.) using APSIM under diverse agro-climatic conditions and water management techniques in Sri Lanka. *Agricultural Water Management* **160**, 132–143. https://doi.org/10.1016/j.agwat.2015.07.001.

Bocchiola, D., Nana, E. & Soncini, A. 2013 Impact of climate change scenarios on crop yield and water footprint of maize in the Po valley of Italy. *Agricultural Water Management* **116**, 50–61. https://doi.org/10.1016/j.agwat.2012.10.009.

Bouman, B. A. M. & Tuong, T. P. 2001 Field water management to save water and increase its productivity in irrigated lowland rice. *Agricultural Water Management* **49**, 11–30. https://doi.org/10.1016/S0378-3774(00)00128-1.

Cabangon, R. J., Castillo, E. G. & Tuong, T. P. 2011 Chlorophyll meter-based nitrogen management of rice grown under alternate wetting and drying irrigation. *Field Crops Research* **121**, 136–146. https://doi.org/10.1016/j.fcr.2010.12.002.

Cao, X., Wu, P., Wang, Y. & Zhao, X. 2014a Assessing blue and green water utilisation in wheat production of China from the perspectives of water footprint and total water use. *Hydrology and Earth System Sciences* **18**, 3165–3178. http://dx.doi.org/10.5194/hess-18-3165- 2014.

Cao, X., Wu, P., Wang, Y. & Zhao, X. 2014b Water footprint of grain product in irrigated farmland of China. *Water Resources Management* **28** (8), 2213–2227. http://dx.doi.org/10.1007/s11269-014-0607-1.

Cao, X., Wu, M., Shu, R., Zhuo, L., Chen, D., Shao, G., Guo, X., Wang, W. & Tang, S. 2018 Water footprint assessment for crop production based on field measurements: a case study of irrigated paddy rice in East China. *Science of the Total Environment* **610–611**, 84–93. https://doi.org/10.1016/j.scitotenv.2017.08.011.

Carrijo, D. R., Lundy, M. E. & Linquist, B. A. 2017 Rice yields and water use under alternate wetting and drying irrigation: a meta-analysis. *Field Crops Research* **203**, 173–180. https://doi.org/10.1016/j.fcr.2016.12.002.

Chu, Y., Shen, Y. & Yuan, Z. 2017 Water footprint of crop production for different crop structures in the Hebei southern plain, North China. *Hydrology and Earth System Sciences* **21**, 3061–3069. http://dx.doi.org/10.5194/hess-21-3061-2017.

Denis, D., Kumar, M., Srivastava, S., Suryavanshi, S., Denis, A., Singh, R., Yadav, A. & Mishra, H. 2016 A high resolution assessment of water footprint of wheat to understand yield and water use heterogeneity. *Water Resources Management* **30** (8), 2641–2649. http://dx.doi.org/10.1007/s11269-016-1305-y.12-14.

Djaman, K., Mel, V. C., Diop, L., Sow, A., El-Namaky, R., Manneh, B., Saito, K., Futakuchi, K. & Irmak, S. 2018 Effects of alternate wetting and drying irrigation regime and nitrogen fertilizer on yield and nitrogen use efficiency of irrigated rice in the Sahel. *Water* **10**, 711. https://doi.org/10.3390/w10060711.

Gaydon, D. S., Probert, M. E., Buresh, R. J., Meinke, H., Suriadi, A., Dobermann, A., Bouman, B. & Timsina, J. 2012 Rice in cropping systems-modelling transitions between flooded and non-flooded soil environments. *European Journal of Agronomy* **39**, 9–24. https://doi.org/10.1016/j.eja.2012.01.003.

Hatiye, S. D., Hari Prasad, K. S., Ojha, C. S. P., Kaushika, G. S. & Adeloye, A. J. 2015 Evaluating irrigation scheduling efficiency of paddy rice and berseem fodder crops in sandy loam soil. *Irrigation and Drainage Systems Engineering* **4**, 147. https://doi.org/10.4172/2168-9768.1000147.

Hoekstra, A. Y. & Hung, P. Q. 2002 Virtual water trade. A quantification of virtual water flows between nations in relation to international crop trade. *Value of Water Research Report Series* **11**, 166.

Johannessen, S. C., Macdonald, R. W., Burd, B., van Roodselaar, A. & Bertold, S. 2015 Local environmental conditions determine the footprint of municipal effluent in coastal waters: a case study in the Strait of Georgia, British Columbia. *Science of the Total Environment* **508**, 228–239. https://doi.org/10.1016/j.scitotenv.2014.11.096.

Kar, G., Kumar, A., Singh, R., Sikka, A. K., Panigrahi, S., Sahoo, N., Swain, S. & Sahoo, H. N. 2015 Farm level water footprints, water productivity and nitrogen use efficiency in irrigated rice under different water and nitrogen management. *Journal of the Indian Society of Soil Science* **63** (3), 266–275. https://doi.org/10.5958/0974-0228.2015.00036.5.

Lampayan, R. M., Rejesus, R. M., Singleton, G. R. & Bouman, B. A. 2015a Adoption and economics of alternate wetting and drying water management for irrigated lowland rice. *Field Crops Research* **170**, 95–108. http://dx.doi.org/10.1016/j.fcr.2014.10.013.

Lampayan, R. M., Samoy-Pascual, K. C., Sibayan, E. B., Ella, V. B., Jayag, O. P., Cabangaon, R. J. & Bouman, B. A. M. 2015b Effects of alternate wetting and drying (AWD) threshold level and plant seedling age on crop performance, water input, and water productivity of transplanted rice in Central Luzon, Philippines. *Paddy and Water Environment* **13**, 215–227. https://doi.org/10.1007/s10333-014-0423-5.

Liang, X. Q., Chen, Y. X., Nie, Z. Y., Ye, Y. S., Liu, J., Tian, G. M., Wang, G. H. & Tuong, T. P. 2013 Mitigation of nutrient losses via surface runoff from rice cropping systems with alternate wetting and drying irrigation and site-specific nutrient management practices. *Environmental Science and Pollution Research* **20**, 6980–6991. https://doi.org/10.1007/s11356-012-1391-1.

Lovarelli, D., Bacenetti, J. & Fiala, M. 2016 Water footprint of crop productions: a review. *Science of the Total Environment* **548–549**, 236–251. https://doi.org/10.1016/j.scitotenv.2016.01.022.

Lu, Y., Zhang, X., Chen, S., Shao, L. & Sun, H. 2016 Changes in water use efficiency and water footprint in grain production over the past 35 years: a case study in the North China Plain. *Journal of Cleaner Production* **116**, 71–79. http://dx.doi.org/10.1016/j.jclepro.2016.01.008.

Marano, R. P. & Filippi, R. A. 2015 Water footprint in paddy rice systems. Its determination in the provinces of Santa Fe and Entre Ríos, Argentina. *Ecological Indicators* **56**, 229–236. http://dx.doi.org/10.1016/j.ecolind.2015.03.027.

Mekonnen, M. M. & Hoekstra, A. Y. 2011 The green, blue and grey water footprint of crops and derived crop products. *Hydrology and Earth System Sciences* **15** (5), 1577–1600. https://doi.org/10.5194/hess-15-1577-2011.

Mekonnen, M. M. & Hoekstra, A. Y. 2014 Water footprint benchmarks for crop production: a first global assessment. *Ecological Indicators* **46**, 214–223. http://dx.doi.org/10.1016/j.ecolind. 2014.06.013.

Pan, J., Liu, Y., Zhong, X., Lampayan, R. M., Singleton, G. R., Huang, N., Liang, K., Peng, B. & Tian, K. 2017 Grain yield, water productivity and nitrogen use efficiency of rice under different water management and fertilizer-N inputs in South

China. *Agricultural Water Management* **184**, 191–200. http://dx.doi.org/10.1016/j.agwat.2017.01.013.

Pascual, V. J. & Wang, Y. M. 2017 Utilizing rainfall and alternate wetting and drying irrigation for high water productivity in irrigated lowland paddy rice in southern Taiwan. *Plant Production Science* **20** (1), 24–35. http://dx.doi.org/10.1080/1343943X.2016.1242373.

Patil, M. D., Das, B. S. & Bhadoria, P. B. 2011 A simple bund plugging technique for improving water productivity in wetland rice. *Soil and Tillage Research* **112** (1), 66–75. https://doi.org/10.1016/j.still.2010.11.010.

Ragab, R. 2015 *A Note on Water use Efficiency and Water Productivity*. Available from: http://www.water4crops.org/wp.../08/RR_Water-use-efficiency-and-water-productivity.

Roux, B., Laan, M., Vahrmeijer, T., Bristow, K. & Annandale, J. 2017 Establishing and testing a catchment water footprint framework to inform sustainable irrigation water use for an aquifer under stress. *Science of the Total Environment* **599**, 1119–1129. http://dx.doi.org/10.1016/j.scitotenv.2017.04.170.

Sharda, R., Mahajan, G., Siag, M., Singh, A. & Chauhan, B. S. 2017 Performance of drip-irrigated dry-seeded rice (*Oryza sativa* L.) in South Asia. *Paddy and Water Environment* **15**, 93–100. https://doi.org/10.1007/s10333-016-0531-5.

Shekhar, S., Mailapalli, D. R., Raghuwanshi, N. S. & Das, B. S. 2020a Hydrus-1D model for simulating water flow through paddy soils under alternate wetting and drying irrigation practice. *Paddy and Water Environment* **18** (1), 73–85.

Shekhar, S., Tamilarasan, R., Mailapalli, D. R. & Raghuwanshi, N. S. 2020b Estimation of evapotranspiration for paddy under alternate wetting and drying irrigation practice. *Irrigation and Drainage*. https://doi.org/10.1002/ird.2536.

Shrestha, S., Chapagain, R. & Babel, M. S. 2017 Quantifying the impact of climate change on crop yield and water footprint of rice in the Nam Oon Irrigation Project, Thailand. *Science of the Total Environment* **599–600**, 689–699. https://doi.org/10.1016/j.scitotenv.2017.05.028.

Sriphirom, P., Chidthaisong, A. & Towprayoon, S. 2018 Rice cultivation to cope with drought situation by alternate wet and dry (AWD) water management system: case study of Ratchaburi province, Thailand. *Chemical Engineering Transactions* **63**, 139–144. https://doi.org/10.3303/CET1863024.

Suttayakul, P., Aran, H., Suksaroj, C., Mungkalasiri, J., Wisansuwannakorn, R. & Musikavong, C. 2016 Water footprints of products of oil palm plantations and palm oil mills in Thailand. *Science of the Total Environment* **542**, 521–529. https://doi.org/10.1016/j.scitotenv.2015.10.060.

Tan, X., Shao, D., Liu, H., Yang, F., Xiao, C. & Yang, H. 2013 Effects of alternate wetting and drying irrigation on percolation and nitrogen leaching in paddy fields. *Paddy and Water Environment* **11** (1–4), 381–395. https://doi.org/10.1007/s10333-012-0328-0.

Thakur, A. K., Mohanty, R. K., Patil, D. U. & Kumar, A. 2014 Impact of water management on yield and water productivity with system of rice intensification (SRI) and conventional transplanting system in rice. *Paddy and Water Environment* **12** (4), 413–424. https://doi.org/10.1007/s10333-013-0397-8.

Thakur, A. K., Mandal, K. G., Mohanty, R. K. & Ambast, S. K. 2018 Rice root growth, photosynthesis, yield and water productivity improvements through modifying cultivation practices and water management. *Agricultural Water Management* **206**, 67–77. https://doi.org/10.1016/j.agwat.2018.04.027.

Umesh, M. R., Chittapur, B. M. & Angadi, S. 2017 Alternate wetting and drying (AWD) irrigation for rice to enhance water productivity and sustainable production: a review. *Journal of Farm Sciences* **30** (4), 441–449.

Wang, Y., Wu, P., Engel, B. & Sun, S. 2015 Comparison of volumetric and stress-weighted water footprint of grain products in China. *Ecological Indicators* **48**, 324–333. https://doi/10.1016/j.ecolind.2014.08.014.

Wang, Z., Zhang, W., Beebout, S. S., Zhang, H., Liu, L., Yang, J. & Zhang, J. 2016 Grain yield, water and nitrogen use efficiencies of rice as influenced by irrigation regimes and their interaction with nitrogen rates. *Field Crops Research* **193**, 54–69. http://dx.doi.org/10.1016/j.fcr.2016.03.006.

Yadav, S., Gill, G., Humphreys, E., Kukal, S. S. & Walia, U. S. 2011 Effect of water management on dry seeded and puddled transplanted rice. Part 1. Crop performance. *Field Crops Research* **120**, 112–122. https://doi.org/10.1016/j.fcr.2010.09.002.

Yang, J., Zhou, Q. & Zhang, J. 2017 Moderate wetting and drying increases rice yield and reduces water use, grain arsenic level, and methane emission. *The Crop Journal* **5** (2), 151–158. https://doi.org/10.1016/j.cj.2016.06.002.

Yao, F., Huang, J., Cui, K., Nie, L., Xiang, J., Liu, X., Wu, W., Chen, M. & Peng, S. 2012 Agronomic performance of high-yielding rice variety grown under alternate wetting and drying irrigation. *Field Crops Research* **126**, 16–22. https://doi.org/10.1016/j.fcr.2011.09.018.

Ye, Y., Liang, X., Chen, Y., Liu, J., Gu, J., Guo, R. & Li, L. 2013 Alternate wetting and drying irrigation and controlled-release nitrogen fertilizer in late-season rice. Effects on dry matter accumulation, yield, water and nitrogen use. *Field Crops Research* **144**, 212–224. https://doi.org/10.1016/j.fcr.2012.12.003.

Zhang, Y., Tang, Q., Peng, S., Xing, D., Qin, J., Laza, R. C. & Punzalan, B. R. 2012 Water use efficiency and physiological response of rice cultivars under alternate wetting and drying conditions. *Scientific World Journal* **2012**, 10. http://dx.doi.org/10.1100/2012/287907.

Zhou, Q., Ju, C. X., Wang, Z. Q., Zhang, H., Liu, L. J., Yang, J. C. & Zhang, J. H. 2017 Grain yield and water use efficiency of super rice under soil water deficit and alternate wetting and drying irrigation. *Journal of Integrative Agriculture* **16** (5), 1028–1043. https://doi.org/10.1016/S2095-3119(16)61506-X.

Zhuo, L., Mekonnen, M. M., Hoekstra, A. Y. & Wada, Y. 2016 Inter- and intra-annual variation of water footprint of crops and blue water scarcity in the Yellow River Basin (1961–2009). *Advances in Water Resources* **87**, 21–41. http://dx.doi.org/10.1016/j.advwatres.2015.11.002.

First received 3 January 2021; accepted in revised form 24 February 2021. Available online 11 March 2021

Mitigation of the effects of climate change in the agricultural sector of Cyprus, through optimization of benefit

Y. Economidou, M. K. Doula and A. A. Zorpas

ABSTRACT

Water scarcity has a significant impact on agricultural productivity, and it usually constitutes the main criterion for the continuation of agricultural activity. The Cyprus agri-sector faces increasing challenges since water resources are very limited and particularly susceptible to climate change. As total profit highly depends on cultivation patterns, the present study aims to define the optimal pattern of agricultural crops in Cyprus. A linear programming model is set up to maximize the net profit of annual and permanent crops. As for the constraints, these include limitations on land and water availability under four different scenarios, provided that production secures current consumption of (1) domestic products, (2) domestic and imported products, (3) domestic products but under a 40% decrease in water availability and (4) domestic products but limited to only traditional, tropical and subtropical crops, under a 40% decrease in water availability. The results indicate that when using an optimal cultivation pattern, economic benefit increases by 120%, 62%, 20%, and 48%, based on Scenarios 1, 2, 3 and 4, respectively. Furthermore, the results are considered useful and directly applicable for policy makers to redesign their strategy in light of water scarcity.

Key words | agriculture, climate change, Cyprus, irrigation, linear programming

Y. Economidou (corresponding author)
M. K. Doula
A. A. Zorpas
Open University of Cyprus, Faculty of Pure and Applied Science, Environmental Conservation and Management,
Laboratory of Chemical Engineering and Engineering Sustainability,
Giannou Kranidioti, 33, P.O. Box 12794, Latsia, 2252, Nicosia,
Cyprus
E-mail: *yianna.economidou@gmail.com*

HIGHLIGHTS

- Water scarcity in Cyprus impacts the agri-sector, leading to high economic losses.
- Linear programming can provide optimal cultivation patterns to maximize economic benefit.
- If Cypriot agriculture shifts to less water-intensive/higher-added-value crops, an increase in economic benefit is achieved, even with significant water reduction.
- Linear Programming results provide a valuable tool for policy makers, in light of climate change.

INTRODUCTION

The agricultural sector is considered one of the most dynamic and important sectors for economic development on the entire planet, playing a vital role in food security. A dedicated Sustainable Development Goal (SDG2), based on specific key performance indicators to be reached by 2030, provides a comprehensive approach to tackling food insecurity while promoting sustainable agriculture. In most countries, the agri-sector represents a huge fraction of total manufacturing benefit and is significantly important for gross domestic product (GDP). At the same time, however,

doi: 10.2166/ws.2021.118

the agri-sector is responsible for major environmental problems, such as greenhouse gas (GHG) emissions and excessive water demands. In contrast, in various eastern Mediterranean countries and in Cyprus, Greece (Crete region), Sicily, Spain, Morocco, and Tunisia, water scarcity is considered to be a critical problem, with adverse effects on the agri-sector (Bonneau *et al.* 2017; Luque *et al.* 2017; Tsangas *et al.* 2020).

In general, the agri-sector is the one sector of the economy that will likely be hit the hardest by climate change since it directly depends on environmental and climatic factors, such as precipitation, temperature, sunlight, water consumption and deforestation (Doula *et al.* 2019; Cradock-Henry *et al.* 2020; Ioannou *et al.* 2020; Tsangas *et al.* 2020). The magnitude of these effects is strongly related to the existing agricultural approaches and practices applied (Navarro-Pedreño *et al.* 2018; Recanati *et al.* 2018; Soudejani *et al.* 2019).

According to the European Environment Agency (EEA 2008), climate change affects crop yields, the growing season and the timing of the crop life cycle. While climate change is projected to have some positive effects through improved crop-growing conditions in northern Europe (Peltonen-Sainio *et al.* 2009), southern Europe is expected to experience adverse effects, including declining crop yields, decreasing soil (Almendro-Candel *et al.* 2018), organic matter (Escolano *et al.* 2018), water resources, as well as desertification (Behrens *et al.* 2010). Especially, the Mediterranean region is expected to face the most adverse effects on natural conditions for crop cultivation, leading to higher economic losses (Jones *et al.* 2003; Porter & Semenov 2005; Norrant & Douguédroit 2006; EEA 2008; Iglesias *et al.* 2010). Furthermore, taking into account the report released by the Intergovernmental Panel on Climate Change (IPCC 2007), the entire Mediterranean ecosystem is susceptible to climate variation since the availability of water is limited, as well as due to the increase of desertification risk (Meléndez-Pastor *et al.* 2017). The quantity and rate of water, oxygen, and nutrient absorption by plants depend on the ability of the roots to absorb the soil solution and the ability of the soil to supply it to the roots. Low hydraulic conductivity, as one of the main soil properties affected by climate conditions, may reduce the availability of free water supply and oxygen to the roots, which has a direct negative affect on agricultural yield (Escolano *et al.* 2018).

In particular, the agricultural sector of Cyprus, which has always been an economic sector of great importance for the island, faces increasing challenges due to climate change (Marcou *et al.* 2011). The island has a semi-arid climate and limited water resources that depend mainly on rainfall. However, rainfall is unevenly distributed, with temporal and geographical variations. Forecasts from regional climate model simulations for the twenty-first century highlight the vulnerability of Cyprus to climate change by projecting an increase in temperature and a decrease in rainfall by the end of the century (Zachariadis 2016). According to universal projections, these conditions affect the physical properties of soil, which are vital for the agri-sector (Escolano *et al.* 2018).

Water is a fundamental resource for agricultural productivity with water footprint being a very crucial indicator for the agri-sector, in general (Goudouva & Zorpas 2017). The water scarcity problem in Cyprus has already had a negative impact on the island's agri-sector and in many cases, the degree of irrigation is the main condition for the continuation of an agricultural activity. The aim of the agricultural sector under the pressure of climate change should be water saving, as well as the allocation of this limited and valuable resource to crops, which could potentially be in a position to bring the maximum economic benefit (Gruda *et al.* 2019). The allocation of crops is also linked to land use planning and management, which in turn has an impact on many socio-economic sectors (Escolano *et al.* 2018). Unless suitable land management procedures are implemented, more frequent and more severe droughts will cause soil water retention mechanisms to collapse, resulting in the onset of erosion and desertification (JRC 2012).

Within this context, the purpose of the present paper is to define the optimum cultivation pattern of the main agricultural crops in Cyprus, for maximizing the net economic profit, under limited water use constraints.

MATERIALS AND METHODS

A literature review shows that optimization methods are very effective tools for water resources management, and in particular for finding optimum agricultural cultivation patterns (Sarker & Ray 2009; Mansourifar *et al.* 2013). Numerous other empirical studies have revealed that linear programming (LP) is one of the best approaches for optimization, because of its simplicity and applicability (Singh *et al.* 2001; Zare & Koch 2014).

In the present paper, a linear programming model was formulated with the objective of finding an optimal cropping pattern that results in a maximum net profit, consisting of the following components: the decision variables, which quantify the decisions that need to be taken, the objective function, which quantifies the measure of performance to be maximized or minimized, and a set of constraints, which reflect the physical limitations of some or all decision variables. The LP model was formulated and solved using the MS Excel Solver add-in (Simplex method).

Decision variables

In accordance with the Statistical Service of Cyprus, the decision variables were defined as the cultivated areas of the main crops produced on the island (Statistical Service of Cyprus 2019). The crops were divided into two categories, which do not compete with each other, the annual (i) and the permanent (j) crops. An additional scenario was also formulated considering only the traditional, tropical, and subtropical crops (k) that grow on the island.

The annual crops (i) include eight different types of crops: Wheat, Barley, Tomatoes, Cucumbers, Watermelons, Sweet Melons, Carrots and Potatoes, with a total existing cultivated area of 24,900 hectares. The permanent crops (j) include 18 types: Olives, Carobs, Almonds, Apples, Pears, Cherries, Peaches, Avocado, Oranges, Grapefruit, Lemons, Bananas, Kiwi, Plums, Figs, Artichokes, Wine Grapes and Table Grapes, with a total existing cultivated area of 23,685 hectares. The traditional, tropical and subtropical crops (k) include nine types: Olives, Carobs, Wheat, Avocado, Bananas, Kiwi, Plums, Figs and Artichokes, with a total existing cultivated area of 21,419 hectares.

Objective function

The objective function for each crop category aims to maximize the total annual net profit, denoted by Z, for the selected crops, according to the mathematical equations presented below.

For the annual crops, i (1–8):

$$\text{Max}\, Z_i = \sum_{i=1}^{8} A_i(R_i - E_i)$$

For the permanent crops, j (1–18):

$$\text{Max}\, Z_j = \sum_{j=1}^{18} A_j(R_j - E_j)$$

For the traditional, tropical and subtropical crops, k (1–9):

$$\text{Max}\, Z_k = \sum_{k=1}^{9} A_k(R_k - E_k)$$

The term R (R_i or R_j or R_k) denotes the total Revenues for each type of crop per hectare, based on the expected sale price. The term E (E_i or E_j or E_k) denotes the total Expenses for each type of crop and includes all variable costs per hectare, such as fertilizers, irrigation, mechanical work, electricity, external labor and more. The term A (A_i or A_j or A_k) denotes the optimal cultivated area by each type of crop that results in a net profit maximization. These are the decision variables, which will constitute our proposals.

Constraints

In the present study, land and water constraints have been used. For the land constraints, both maximum and minimum limits have been set. The maximum limit was used based on the assumption that the total area of the existing cultivated land will not change. The minimum limit on the cultivated land has been introduced to ensure that domestic production covers the current consumption needs of the products.

Regarding irrigation water, the basic assumption of the model is that total irrigation water consumed should not exceed the total quantity currently consumed by the existing crops. Furthermore, it was assumed that the efficiency of the irrigation networks in Cyprus is 85%.

Scenarios

Four scenarios were applied to the model, to determine various proposals for the agricultural sector of Cyprus. In the first scenario, it was assumed that (a) the total area of cultivated land per crop category (annual and permanent) does not exceed the total cultivated land of the existing crops, per category, (b) the minimum areas of crops should ensure that domestic production covers the current consumption needs in domestic products, with the remaining needs of the population to be covered by imports, and (c) total irrigation water does not exceed current consumption in irrigation water.

The constraints in the second scenario are the same as the ones in the first scenario in terms of maximum land areas and maximum water consumption. What differs is the assumption that the minimum areas of crops should ensure the total consumption of all products. More specifically, all products (both domestic and imported) that are currently consumed should be produced in Cyprus.

The constraints in the third scenario are the same as the ones in the first scenario in terms of maximum and minimum areas. The difference is in the assumption that irrigation water is reduced by 40%.

The main assumption of the fourth scenario is that under the pressure of climate change, only traditional, tropical and subtropical crops can grow and thrive in Cyprus. Also, similarly to the first and third scenarios, the minimum areas of crops should ensure that domestic production covers the current consumption needs in domestic products, with the remaining needs of the population to be covered by imports. Furthermore, water availability is reduced by 40%, due to climate change. It is noted that under this scenario, the proposed crops could potentially cover the total existing area cultivated by all other current crops. However, for direct comparison, the maximum area used is the one taken up by the traditional, tropical and subtropical crops that are currently cultivated.

RESULTS AND DISCUSSION

Scenario 1 – annual crops

The results of the LP model for the annual crops under Scenario 1, for which (a) the minimum cultivated areas should cover the current consumption needs in domestic products, supplementing the remaining needs with imports, and (b) the total irrigation water does not exceed current consumption in irrigation water, are presented in Table 1.

Table 1 | Scenario 1 – results of the linear programming model for the annual crops

		Irrigation needs per hectare	Revenue – Expenses per hectare	Cultivated Area		Net Profit		
				A_i		$Z = A_i(R_i - E_i)$		
i	Crop	IRR_i (m^3/ha)	$R_i - E_i$ (€/ha)	Existing Area (ha)	Proposed Area (ha)	With Existing Area (€ mil.)	With Proposed Area (€ mil.)	Incr./(Decr.) (%)
1	Wheat	–	430	8,678	**21,766**	3.7	**9.4**	*151%*
2	Barley	–	287	10,953	–	3.1	–	*(100%)*
3	Tomatoes	6,394	45,832	261	**2,192**	12.0	**100.5**	*740%*
4	Cucumbers	4,761	4,211	190	**188**	0.8	**0.8**	*(1%)*
5	Watermelons	4,988	16,406	436	**435**	7.2	**7.1**	*(0.1%)*
6	Sweet Melons	5,192	17,564	146	**146**	2.6	**2.6**	*(0.1%)*
7	Carrots	4,244	12,506	57	**54**	0.7	**0.7**	*(6%)*
8	Potatoes	3,035	8,261	4,179	**119**	34.5	**1.0**	*(97%)*
Total				**24,900**	**24,900**	**65**	**122**	***89%***

Note: values in bold type specify (1) all Total values and (2) Proposed Cultivated Areas and Net Profit that corresponds to Proposed Cultivated Areas.

According to the model, the total net profit of annual crops may increase by 89% if the cultivations of tomatoes and wheat are increased by 740% and 150%, respectively. On the other hand, barley should be entirely omitted from the cultivation pattern, and the rest of the crops should be produced in the minimum possible quantity to cover domestic consumption. These results were expected since the net profit for the tomato crops, which is €45,832/ha, is multiple times higher than the net profit of the rest.

In the case of wheat, the low net profit of the crop is offset by the fact that it requires no irrigation water, however in the case of barley, even though it also has no irrigation requirements, its low net profit makes it unprofitable as a crop. It is worth noting that for the cultivation of potato, which today is the main exported agricultural product for Cyprus, even though its irrigation water requirements are not prohibitively high (3,035 m^3/ha), its low net profit (€8,261/ha) compared with the rest of the irrigated crops results in a reduction of the crop.

Scenario 1 – permanent crops

The proposed quantities of the permanent crops under Scenario 1 are shown in Table 2.

The crops that should be increased are figs (by 4,500%) and to a lesser extent wine grapes (by 83%). The rest of the crops should be produced in the minimum possible quantity just to cover domestic consumption, whereas the cultivation of carob trees is omitted completely.

The explanation of the results under this scenario is similar to the explanation provided in the case of the annual crops. Figs have the highest net profit (€46,000/ha) and relatively low water needs; as a result, they are ranked as the top crop with the highest productivity. In the case

Table 2 | Scenario 1 – results of the linear programming model for the permanent crops

		Irrigation needs per hectare	Revenue – Expenses per hectare	Cultivated Area		Net Profit		
				A_j		$Z = A_j(R_j - E_j)$		
j	Crop	IRR_j (m^3/ha)	$R_j - E_j$ (€/ha)	Existing Area (ha)	Proposed Area (ha)	With Existing Area (€ mil.)	With Proposed Area (€ mil.)	Incr./(Decr.) (%)
1	Olives	4,307	5,285	10,830	**1,678**	57.2	**8.9**	(85%)
2	Carobs	–	927	982	–	0.9	–	(100%)
3	Almonds	3,540	3,874	2,195	**1,978**	8.5	**7.7**	(10%)
4	Apples	6,821	19,886	372	**371**	7.4	**7.4**	(0%)
5	Pears	6,821	25,414	66	**66**	1.7	**1.7**	(0%)
6	Cherries	6,821	27,390	227	**225**	6.2	**6.2**	(1%)
7	Peaches	6,821	14,146	297	**297**	4.2	**4.2**	(0%)
8	Avocado	8,468	15,363	77	**77**	1.2	**1.2**	0%
9	Oranges	8,468	2,239	1,090	**603**	2.4	**1.3**	(45%)
10	Grapefruit	8,468	474	413	**113**	0.2	**0.1**	(73%)
11	Lemons	8,468	3,023	363	**153**	1.1	**0.5**	(58%)
12	Bananas	12,559	7,047	211	**207**	1.5	**1.5**	(2%)
13	Kiwi	6,821	1,593	2	**2**	0.0	**0.0**	0%
14	Plums	6,821	9,233	383	**383**	3.5	**3.5**	(0%)
15	Figs	6,821	28,061	163	**7,471**	4.6	**209.6**	4,483%
16	Artichokes	4,405	14,190	93	**93**	1.3	**1.3**	(0%)
17	Wine Grapes	–	2,302	5,313	**9,732**	12.2	**22.4**	83%
18	Table Grapes	3,082	6,149	608	**237**	3.7	**1.5**	(61%)
Total				**23,685**	**23,685**	**118**	**279**	**136%**

Note: values in bold type specify (1) all Total values and (2) Proposed Cultivated Areas and Net Profit that corresponds to Proposed Cultivated Areas.

of grapes, the model has picked wine grapes, when compared with table grapes, even if they have a lower net profit, due to the fact that they do not require any water, which makes them more productive relative to water.

It is interesting to note the results of olive and carob trees, which are considered traditional crops and part of the Cypriot rural environment. In the case of carob, which is a non-irrigated crop, the very low net profit for the raw product (€927/ha) is the reason for its removal from the proposed cultivation pattern. In the case of olive, even though its net profit is relatively high (€5,285/ha) and its water requirements are relatively low (4,307 m^3/ha), it is not ranked by the model as the crop with the highest economic productivity. Under Scenario 1, citrus, another main exported product, is also negatively affected. Its high water requirements, combined with its low net profit, make it a non-profitable crop.

Scenario 2 – annual crops

Under Scenario 2, for which (a) the minimum areas of crops should ensure the total consumption of all products, including imported, and (b) the total irrigation water should not exceed current consumption, the total net profit of annual crops increases by 87% (see Table 3). Also, in this case, the tomato and wheat crops have the greatest increase, 722% and 150%, respectively. Furthermore, the potato crops are reduced under this scenario, for the reasons mentioned earlier.

Scenario 2 – permanent crops

The results of the permanent crops under Scenario 2 are indicated in Table 4.

In this case, there is also an increase in the total net profit (48%), but to a lesser extent, compared with Scenario 1. Again, figs and wine grapes are increased by 1,090% and 21%, respectively, whereas the rest of the crops are increased in the minimum possible quantity to cover consumption needs. In Scenario 2, carobs are also omitted completely and citrus trees are reduced to the minimum necessary quantities.

Scenario 3 – annual crops

The results of the LP model for the annual crops under Scenario 3, for which (a) the minimum areas of crops should cover current consumption needs in domestic products, and (b) total irrigation water is reduced by 40%, are indicated in Table 5.

The results show that even if the available quantities of water are decreased by 40%, which is a possible future

Table 3 | Scenario 2 – results of the linear programming model for the annual crops

		Irrigation needs per hectare	Revenue – Expenses per hectare	Cultivated Area		Net Profit		
				A_i		$Z = A_i(R_i - E_i)$		
i	Crop	IRR_i (m^3/ha)	$R_i - E_i$ (€/ha)	Existing Area (ha)	Proposed Area (ha)	With Existing Area (€ mil.)	With Proposed Area (€ mil.)	Incr./(Decr.) (%)
1	Wheat	–	430	8,678	**21,732**	3.7	**9.3**	*150%*
2	Barley	–	287	10,953	–	3.1	–	*(100%)*
3	Tomatoes	6,394	45,832	261	**2,145**	12.0	**98.3**	*722%*
4	Cucumbers	4,761	4,211	190	**191**	0.8	**0.8**	*0%*
5	Watermelons	4,988	16,406	436	**442**	7.2	**7.2**	*1%*
6	Sweet Melons	5,192	17,564	146	**148**	2.6	**2.6**	*1%*
7	Carrots	4,244	12,506	57	**83**	0.7	**1.0**	*45%*
8	Potatoes	3,035	8,261	4,179	**160**	34.5	**1.3**	*(96%)*
Total				**24,900**	**24,900**	**65**	**121**	***87%***

Note: values in bold type specify (1) all Total values and (2) Proposed Cultivated Areas and Net Profit that corresponds to Proposed Cultivated Areas.

Table 4 | Scenario 2 – results of the linear programming model for the permanent crops

		Irrigation needs per hectare	Revenue – Expenses per hectare	Cultivated Area		Net Profit		
				A_j		$Z = A_j(R_j - E_j)$		
j	Crop	IRR_j (m³/ha)	$R_j - E_j$ (€/ha)	Existing Area (ha)	Proposed Area (ha)	With Existing Area (€ mil.)	With Proposed Area (€ mil.)	Incr./(Decr.) (%)
1	Olives	4,307	5,285	10,830	**1,680**	57.2	**8.9**	*(84%)*
2	Carobs	–	927	982	–	0.9	–	*(100%)*
3	Almonds	3,540	3,874	2,195	**8,920**	8.5	**34.6**	*306%*
4	Apples	6,821	19,886	372	**944**	7.4	**18.8**	*154%*
5	Pears	6,821	25,414	66	**317**	1.7	**8.1**	*381%*
6	Cherries	6,821	27,390	227	**546**	6.2	**14.9**	*140%*
7	Peaches	6,821	14,146	297	**318**	4.2	**4.5**	*7%*
8	Avocado	8,468	15,363	77	**131**	1.2	**2.0**	*70%*
9	Oranges	8,468	2,239	1,090	**620**	2.4	**1.4**	*(43%)*
10	Grapefruit	8,468	474	413	**115**	0.2	**0.1**	*(72%)*
11	Lemons	8,468	3,023	363	**228**	1.1	**0.7**	*(37%)*
12	Bananas	12,559	7,047	211	**396**	1.5	**2.8**	*88%*
13	Kiwi	6,821	1,593	2	**16**	0.0	**0.0**	*712%*
14	Plums	6,821	9,233	383	**483**	3.5	**4.5**	*26%*
15	Figs	6,821	28,061	163	**1,939**	4.6	**54.4**	*1,090%*
16	Artichokes	4,405	14,190	93	**94**	1.3	**1.3**	*1%*
17	Wine Grapes	–	2,302	5,313	**6,446**	12.2	**14.8**	*21%*
18	Table Grapes	3,082	6,149	608	**492**	3.7	**3.0**	*(19%)*
Total				**23,685**	**23,685**	**118**	**175**	***48%***

Note: values in bold type specify (1) all Total values and (2) Proposed Cultivated Areas and Net Profit that corresponds to Proposed Cultivated Areas.

Table 5 | Scenario 3 – results of the linear programming model for the annual crops

		Irrigation needs per hectare	Revenue – Expenses per hectare	Cultivated Area		Net Profit		
				A_i		$Z = A_i(R_i - E_i)$		
i	Crop	IRR_i (m³/ha)	$R_i - E_i$ (€/ha)	Existing Area (ha)	Proposed Area (ha)	With Existing Area (€ mil.)	With Proposed Area (€ mil.)	Incr./(Decr.) (%)
1	Wheat	–	430	8,678	**22,919**	3.7	**9.9**	*164%*
2	Barley	–	287	10,953	–	3.1	–	*(100%)*
3	Tomatoes	6,394	45,832	261	**1,039**	12.0	**47.6**	*298%*
4	Cucumbers	4,761	4,211	190	**188**	0.8	**0.8**	*(1%)*
5	Watermelons	4,988	16,406	436	**435**	7.2	**7.1**	*(0.1%)*
6	Sweet Melons	5,192	17,564	146	**146**	2.6	**2.6**	*(0.1%)*
7	Carrots	4,244	12,506	57	**54**	0.7	**0.7**	*(6%)*
8	Potatoes	3,035	8,261	4,179	**119**	34.5	**1.0**	*(97%)*
Total				**24,900**	**24,900**	**65**	**70**	***8%***

Note: values in bold type specify (1) all Total values and (2) Proposed Cultivated Areas and Net Profit that corresponds to Proposed Cultivated Areas.

scenario under the pressure of climate change, the total net profit could potentially be increased by 8%. The pattern includes the same prevailing crops as in the previous scenarios, tomatoes and wheat, which are increased by 298% and 164%, respectively.

Scenario 3 – permanent crops

Similarly, for the permanent crops under Scenario 3 (see Table 6), there is an increase in the total net profit by 26%, and figs and wine grapes also prevail.

Scenario 4 – traditional, tropical and subtropical crops

The results of the LP model for the annual crops under Scenario 4, which is an extreme case of extended water scarcity and droughts, under which only traditional, tropical and subtropical crops can grow when water availability is reduced by 40%, are presented in Table 7. In addition, the minimum cultivated areas should ensure that domestic production covers current consumption of existing domestic production, with the remaining needs of the population to be covered by imports.

The total net profit under Scenario 4 is increased by 48%. In this case, there is also an increase in figs by 1,604% and for the first time, carob trees appear in the cultivation pattern. The rest of the crops are limited to the minimum possible areas, just to cover the consumption of current production.

The overall results of the linear programming model are indicated in Table 8. In summary, there is an increase in the total net profit for all categories of crops, under all four

Table 6 | Scenario 3 – results of the linear programming model for the permanent crops

		Irrigation needs per hectare	Revenue – Expenses per hectare	Cultivated Area A_j		Net Profit $Z = A_j(R_j - E_j)$		
j	Crop	IRR_j (m^3/ha)	$R_j - E_j$ (€/ha)	Existing Area (ha)	Proposed Area (ha)	With Existing Area (€ mil.)	With Proposed Area (€ mil.)	Incr./(Decr.) (%)
1	Olives	4,307	5,285	10,830	**1,678**	57.2	**8.9**	*(85%)*
2	Carobs	–	927	982	–	0.9	–	*(100%)*
3	Almonds	3,540	3,874	2,195	**1,978**	8.5	**7.7**	*(10%)*
4	Apples	6,821	19,886	372	**371**	7.4	**7.4**	*(0%)*
5	Pears	6,821	25,414	66	**66**	1.7	**1.7**	*(0%)*
6	Cherries	6,821	27,390	227	**225**	6.2	**6.2**	*(1%)*
7	Peaches	6,821	14,146	297	**297**	4.2	**4.2**	*(0%)*
8	Avocado	8,468	15,363	77	**77**	1.2	**1.2**	*0%*
9	Oranges	8,468	2,239	1,090	**603**	2.4	**1.3**	*(45%)*
10	Grapefruit	8,468	474	413	**113**	0.2	**0.1**	*(73%)*
11	Lemons	8,468	3,023	363	**153**	1.1	**0.5**	*(58%)*
12	Bananas	12,559	7,047	211	**207**	1.5	**1.5**	*(2%)*
13	Kiwi	6,821	1,593	2	**2**	0.0	**0.0**	*0%*
14	Plums	6,821	9,233	383	**383**	3.5	**3.5**	*(0%)*
15	Figs	6,821	28,061	163	**2,422**	4.6	**68.0**	*1,386%*
16	Artichokes	4,405	14,190	93	**93**	1.3	**1.3**	*(0%)*
17	Wine Grapes	–	2,302	5,313	**14,781**	12.2	**34.0**	*178%*
18	Table Grapes	3,082	6,149	608	**237**	3.7	**1.5**	*(61%)*
Total				**23,685**	**23,685**	**118**	**149**	**26%**

Note: values in bold type specify (1) all Total values and (2) Proposed Cultivated Areas and Net Profit that corresponds to Proposed Cultivated Areas.

Table 7 | Scenario 4 – results of the linear programming model for the traditional, tropical and subtropical crops

		Irrigation needs per hectare	Revenue – Expenses per hectare	Cultivated Area		Net Profit		
				A_k		$Z = A_k(R_k - E_k)$		
k	**Crop**	IRR_k **(m³/ha)**	$R_k - E_k$ **(€/ha)**	**Existing Area (ha)**	**Proposed Area (ha)**	**With Existing Area (€ mil.)**	**With Proposed Area (€ mil.)**	**Incr./(Decr.) (%)**
1	Olives	4,307	5,285	10,830	**1,678**	57.2	**8.9**	*(85%)*
2	Carobs	–	927	982	**16,142**	0.9	**15.0**	*1,544%*
3	Wheat	–	430	8,678	**60**	3.7	**0.0**	*(99%)*
4	Avocado	8,468	15,363	77	**77**	1.2	**1.2**	*0%*
5	Bananas	12,559	7,047	211	**207**	1.5	**1.5**	*(2%)*
6	Kiwi	6,821	1,593	2	**2**	0.003	**0.003**	*0%*
7	Plums	6,821	9,233	383	**383**	3.5	**3.5**	*(0%)*
8	Figs	6,821	28,061	163	**2,777**	4.6	**77.9**	*1,604%*
9	Artichokes	4,405	14,190	93	**93**	1.3	**1.3**	*(0.4%)*
Total				**21,419**	**21,419**	**74**	**109**	***48%***

Note: values in bold type specify (1) all Total values and (2) Proposed Cultivated Areas and Net Profit that corresponds to Proposed Cultivated Areas.

Table 8 | Summary of results of the linear programming model for all scenarios

	Annual Crops			Permanent Crops			All Crops		
Scenario	**With Exist. Area (€ mil.)**	**With Prop. Area (€ mil.)**	***Incr./ (Decr.) (%)***	**With Exist. Area (€ mil.)**	**With Prop. Area (€ mil.)**	***Incr./ (Decr.) (%)***	**With Exist. Area (€ mil.)**	**With Prop. Area (€ mil.)**	***Incr./ (Decr.) (%)***
1	65	122	*89%*	118	279	*136%*	183	401	*120%*
2	65	121	*87%*	118	175	*48%*	183	295	*62%*
3	65	70	*8%*	118	149	*26%*	183	218	*20%*
4							74	109	*48%*

scenarios. The most considerable increase is observed in Scenario 1, a smaller increase in Scenario 2, and the smallest increase in Scenario 3, as was expected due to reduced water availability. Under the fourth scenario, the combination of traditional, tropical and subtropical crops results in an increase in the total net profit by 48%, even with water availability reduced by 40%.

Limitations

Linear programming is a simple and effective tool that can easily determine the optimal solution to a problem under specific conditions and constraints. However, the reliability and accuracy of the model is as good as its assumptions.

For the simulation of the problem, not all categories of crops produced in Cyprus have been selected, due to data unavailability. As a result, crops with an even higher profitability may not have been included. In addition, in the formulation of the model constraints, no upper limits have been set for the cultivated area, which exist due to variations in the soil morphology and microclimatic conditions – not all crops can grow everywhere. Furthermore, the model does not take into account any variations in the product sale prices, which may arise if the existing market equilibrium is disturbed. If there is a change in prices, for example, if the surplus of products does not follow the export route, there is a possibility of revenue loss from lower market prices due to oversupply. Finally, it should be noted that over the years, a specific cultivation structure has been formed, based on the knowledge, skills and dedication of the rural population, something that is difficult to change.

Despite the model's limitations, it can be assumed that the relatively high increase in the net profit from the theoretical application of the various scenarios provides some cushion, in case the adopted cultivation pattern deviates to some degree from the theoretical one.

CONCLUSIONS

The results of the LP model show that the use of an optimal cultivation pattern maximizes the economic benefit, compared with the current cultivation pattern. The cultivations of tomato, wheat, figs and wine grapes prevail in the proposed cultivation patterns, as they all present high net profit compared with their irrigation water requirements, displacing other cultivations with lower profitability.

Exports should be limited to products with significantly higher profitability, based on the local conditions of Cyprus, such as cultivations of tomatoes, figs and wine grapes. The value of olives and carobs does not seem to be recognized by the model, since the revenue figures are based on the selling prices of the raw products, without taking into account the potential of the final processed products. For example, the carob tree is considered as an indigenous species of the island, fully adapted to the local conditions, and according to the new consumer trends towards superfoods, the prospect of the processed product is promising.

The prospect of the carob tree is recognized in the scenario of extended water scarcity, high temperatures and low crop productivity (Scenario 4), which is possible, according to the literature review, in a time span of ten to 30 years from now. Under these circumstances, most of the crops that are cultivated currently, such as citrus, vegetables, even grapes, may not be sustained under future climatic conditions and as a consequence, Cyprus may not be in a position to be self-sustained in food production. However, if a shift is made towards more traditional, tropical and subtropical crops, an increase in economic benefit could be achieved, even with a significant reduction in water availability. It should be noted though that this scenario does not take into account the fact that the cost of imports may increase, since all products consumed today may be imported rather than produced domestically.

Summing up, to tackle the challenges of climate change and protect agricultural production and income, Cypriot agriculture should, in the medium term, make a shift towards less water-intensive and higher added-value crops, such as figs, wine grapes, tomatoes and wheat. In the long term, the agricultural sector should make a gradual shift to species which are better suited for dry conditions and allow for profit maximization, without additional burden on water consumption. In order to achieve these goals, appropriate incentives should be provided to the farmers, for example through funding mechanisms, such as the Rural Development Program, for the cultivation and processing of such varieties, especially in the superfood category, that have a promising prospect in the future.

DATA AVAILABILITY STATEMENT

All relevant data are included in the paper or its Supplementary Information.

REFERENCES

Almendro-Candel, M. B., Lucas, I. G., Navarro-Pedreño, J. & Zorpas, A. A. 2018 Physical properties of soils affected by the use of agricultural waste. In: *Agricultural Waste and Residues* (A. Aladjadjiyan, ed.), IntechOpen, London, UK, pp. 9–28. doi: 10.5772/intechopen.77993.

Behrens, A., Georgiev, A. & Carraro, M. 2010 *Future Impacts of Climate Change across Europe*, CEPS Working Document No. 324. Centre for European Policy Studies, Brussels, Belgium.

Bonneau, V., Copigneaux, B., Probst, L. & Pedersen, B. 2017 *Industry 4.0 in Agriculture: Focus on IoT Aspects*. European Commission, Brussels, Belgium.

Cradock-Henry, N. A., Blackett, P., Hall, M., Johnstone, P., Teixeira, E. & Wreford, A. 2020 Climate adaptation pathways for agriculture: insights from a participatory process. *Environmental Science and Policy* **107**, 66–79. doi: 10.1016/j.envsci.2020.02.020.

Doula, M. K., Zorpas, A. A., Inglezakis, V. J., Navarro, J. P. & Bilalis, D. J. 2019 Optimization of heavy polluted soil from olive mill waste through the implementation of zeolites. *Environmental Engineering and Management Journal* **18** (6),

1297–1309. Available from: http://eemj.eu/index.php/EEMJ/article/view/3890 (accessed: 22 January 2021).

EEA 2008 *Impacts of Europe's Changing Climate – 2008 Indicator-Based Assessment*, EEA Report No.4/2008. EEA, Copenhagen, Denmark.

Escolano, J. J., Pedreño, J. N.,, Lucas, I. G., Candel, M. B. A. & Zorpas, A. A. 2018 Decreased organic carbon associated with land management in Mediterranean environments. In: *Soil Management and Climate Change: Effects on Organic Carbon, Nitrogen Dynamics, and Greenhouse Gas Emissions* (M. A. Muñoz & R. Zornoza, eds), Elsevier, London, UK, pp. 1–13. doi: 10.1016/B978-0-12-812128-3.00001-X.

Goudouva, G. T. & Zorpas, A. A. 2017 Water footprint determination by quarry operation in island regions. *Desalination and Water Treatment* **86**, 271–276. doi: 10.5004/dwt.2017.20814.

Gruda, N., Bisbis, M. & Tanny, J. 2019 Influence of climate change on protected cultivation: impacts and sustainable adaptation strategies – a review. *Journal of Cleaner Production* **225**, 481–495. doi: 10.1016/j.jclepro.2019.03.210.

Iglesias, A., Quiroga, S. & Schlickenrieder, J. 2010 Climate change and agricultural adaptation: assessing management uncertainty for four crop types in Spain. *Climate Research* **44** (1), 83–94. doi: 10.3354/cr00921.

Ioannou, A., Gohari, G., Papaphilippou, P., Panahirad, S., Akbari, A., Dadpour, M. R., Krasia-Christoforou, T. & Fotopoulos, V. 2020 Advanced nanomaterials in agriculture under a changing climate: the way to the future? *Environmental and Experimental Botany* **176**, 104048. doi: 10.1016/j.envexpbot.2020.104048.

IPCC 2007 *Climate Change 2007 – The Physical Science Basis: Working Group I Contribution to the Fourth Assessment Report of the IPCC*. Cambridge University Press, Cambridge, UK. Available from: https://www.cambridge.org/cy/academic/subjects/earth-and-environmental-science/climatology-and-climate-change/climate-change-2007-physical-science-basis-working-group-i-contribution-fourth-assessment-report-ipcc?format=PB&isbn=9780521705967 (accessed: 22 January 2021).

Jones, P. D., Lister, D. H., Jaggard, K. W. & Pidgeon, J. D. 2003 Future climate impact on the productivity of sugar beet (*Beta vulgaris* L.) in Europe. *Climatic Change* **58**, 93–108. doi: 10.1023/A:1023420102432.

JRC 2012 *The State of Soil in Europe: A Contribution of the JRC to the European Environment Agency's Environment State and Outlook Report – SOER 2010*. JRC Publications Repository, Publications Office of the European Union, Luxembourg, Publications Office of the European Union, Luxembourg. Available from: https://publications.jrc.ec.europa.eu/repository/handle/JRC68418 (accessed: 22 January 2021).

Luque, A., Estela Peralta, M., de las Heras, A. & Córdoba, A. 2017 State of the Industry 4.0 in the Andalusian food sector. *Procedia Manufacturing* **13**, 1199–1205. doi: 10.1016/j.promfg.2017.09.195.

Mansourifar, M., Almassi, M., Borghaee, A.-M. & Moghadassi, R. 2013 Optimization crops pattern in variable field ownership. *World Applied Sciences Journal* 21 (4), 492–497.

Marcou, M., Stylianou, A., Bruggeman, A., Zoumides, C., Pashiardis, S., Hadjinicolaou, P., Lange, M.A., Zachariadis, T. & Michaelides, A. 2011 *Economic Impact of Climate Change on the Cypriot Agricultural Sector*, Working Paper. MANRE, Nicosia, Cyprus.

Meléndez-Pastor, I., Pedreño, J. N., Lucas, I. G. & Zorpas, A. A. 2017 A model for evaluating soil vulnerability to erosion using remote sensing data and a fuzzy logic system. In: *Modern Fuzzy Control Systems and Its Applications* (S. Ramakrishnan, ed.), InTech, Rijeka, Croatia, pp. 437–453. doi: 10.5772/67989.

Navarro-Pedreño, J., Almendro-Candel, M. B., Gómez Lucas, I., Jordán Vidal, M. M., Bech Borras, J. & Zorpas, A. A. 2018 Trace metal content and availability of essential metals in agricultural soils of Alicante (Spain). *Sustainability* **10** (12), 4534. doi: 10.3390/su10124534.

Norrant, C. & Douguédroit, A. 2006 Monthly and daily precipitation trends in the Mediterranean (1950–2000). *Theoretical and Applied Climatology* **83** (1–4), 89–106. doi: 10.1007/s00704-005-0163-y.

Peltonen-Sainio, P., Jauhiainen, L. & Laurila, I. P. 2009 Cereal yield trends in northern European conditions: changes in yield potential and its realisation. *Field Crops Research* **110** (1), 85–90. doi: 10.1016/j.fcr.2008.07.007.

Porter, J. R. & Semenov, M. A. 2005 Crop responses to climatic variation. *Philosophical Transactions of the Royal Society B: Biological Sciences* **360** (1463), 2021–2035. doi: 10.1098/rstb.2005.1752.

Recanati, F., Arrigoni, A., Scaccabarozzi, G., Marveggio, D., Meliá, P. & Dotelli, G. 2018 LCA towards sustainable agriculture: the case study of cupuaçu jam from agroforestry. *Procedia CIRP* **69**, 557–561. doi: 10.1016/j.procir.2017.11.003.

Sarker, R. & Ray, T. 2009 An improved evolutionary algorithm for solving multi-objective crop planning models. *Computers and Electronics in Agriculture* **68** (2), 191–199. doi: 10.1016/j.compag.2009.06.002.

Singh, D. K., Jaiswal, C. S., Reddy, K. S., Singh, R. M. & Bhandarkar, D. M. 2001 Optimal cropping pattern in a canal command area. *Agricultural Water Management* **50** (1), 1–8. doi: 10.1016/S0378-3774(01)00104-4.

Soudejani, H. T., Kazemian, H., Inglezakis, V. J. & Zorpas, A. A. 2019 Application of zeolites in organic waste composting: a review. *Biocatalysis and Agricultural Biotechnology* **22**, 101396. doi: 10.1016/j.bcab.2019.101396.

Statistical Service of Cyprus 2019 *Areas, Production and Yield of Main Agricultural Products 2009–2017*. Available from: https://www.mof.gov.cy/mof/cystat/statistics.nsf/agriculture_51main_gr/agriculture_51main_gr?OpenForm&sub=1&sel=2 (accessed: 16 April 2020).

Tsangas, M., Gavriel, I., Doula, M., Xeni, F. & Zorpas, A. A. 2020 Life cycle analysis in the framework of agricultural strategic development planning in the Balkan region. *Sustainability* **12** (5), 1813. doi: 10.3390/su12051813.

Zachariadis, T. 2016 Climate change impacts. In: *Climate Change in Cyprus: Review of the Impacts and Outline of an*

Adaptation Strategy, Springer, Cham, Switzerland, pp. 25–49. doi: 10.1007/978-3-319-29688-3_3.

Zare, M. & Koch, M. 2014 Optimization of cultivation pattern for maximizing farmers' profits under land- and water constraints by means of linear-programming: an iranian case study. In: *ICHE 2014, Proceedings of the 11th International Conference on Hydroscience & Engineering* (R. Lehfeldt & R. Kopmann, eds), Bundesanstalt für Wasserbau, Karlsruhe, Germany, pp. 141–150. Available from: https://hdl.handle.net/20.500.11970/99425 (accessed: 19 March 2020).

First received 29 January 2021; accepted in revised form 9 April 2021. Available online 20 April 2021

Examining the potential of an irrigation work to improve sustainability in a rural area

Athanasios Ragkos and Vasileios Ambas

ABSTRACT

Irrigation works aim to increase the efficiency of water use and economic benefits for farmers. This study adopts a broader view and investigates their potential to contribute to the achievement of other sustainability objectives. In particular, the paper employs a multi-objective programming (MOP) model, which examines the possibilities of simultaneously achieving four conflicting objectives with the upgrade of an irrigation network in a rural area in Greece. The four objectives are maximization of economic result (economic sustainability) and of employment (social sustainability) as well as the minimization of agrochemical use and irrigation water consumption (environmental sustainability). The compromise is sought through different cropping patterns either by restructuring existing crops (Scenario 1) or by also introducing new crops (Scenario 2). The results show that solutions in Scenario 2 perform much better in all dimensions of sustainability; however, large increases in economic performance and employment come with lower environmental gains. A cost-benefit analysis shows that very few solutions yield positive Net Present Value and the investment could be halted if benefits relating to social and environmental sustainability are disregarded. Results are discussed in conjunction with the proposal of a new governance scheme, which could assume broader roles in supporting sustainable development.

Key words | cost-benefit analysis, crop restructuring, governance, multi-objective programming, public-private partnerships, technical and economic data

Athanasios Ragkos (corresponding author)
Hellenic Agricultural Organization – ELGO-DIMITRA,
Agricultural Economics Research Institute,
Terma Alkmanos,
115 28 Athens,
Greece
E-mail: *ragkos@agreri.gr*

Vasileios Ambas
Department of Agriculture,
University of Western Macedonia,
Terma Kontopoulou,
531 00 Florina,
Greece

HIGHLIGHTS

- Achievement of sustainable development through irrigation works.
- Quantification of social and environmental benefits from an irrigation work.
- Examination of simultaneous achievement of conflicting objectives.
- New insights into governance of irrigation networks for sustainable development through public-private partnerships.

doi: 10.2166/ws.2021.106

INTRODUCTION

According to the Sustainability Assessment of Food and Agriculture Systems of the Food and Agriculture Organization (FAO 2014), there are four dimensions in sustainable development. *Economic resilience* deals with issues relating to local economic development and operational profits of businesses. Complementary to this, *social wellbeing* includes equity, labor rights and quality in order to satisfy human needs and aspirations for a better life. *Environmental integrity* examines the management of water resources, biodiversity and other environmental goods and services, within an ecosystem approach across the whole food and agriculture sector. Apart from these three dimensions, good *governance* has also been noted as an integral element of sustainability – although not always examined as a separate dimension – and highlights the need of coordinating efforts and setting effective procedures towards sustainability. These dimensions were reflected in the 17 sustainable development goals proposed by the United Nations in 2015 and adopted by all United Nations Member States (United Nations 2015).

Sustainable development of rural areas constitutes one of the key policy goals in the European Union (EU). The new European Union (EU) Common Agricultural Policy (CAP) objectives for 2021–2027 focus on broader issues of interest to rural areas including fair income, vibrant rural areas and environmental care (https://ec.europa.eu/; European Commission 2020) and seek to implement measures that will – *inter alia* – encourage innovation in rural areas and improve farm infrastructure and amenities for rural dwellers. In this context, public infrastructure – including irrigation works – can serve as a tool for sustainable development (Jägermeyr *et al.* 2017). Irrigation works predominantly aim to increase of water use efficiency, which will also permit the design of more efficient irrigation plans, combined with deeper knowledge of crop needs (Kourgialas *et al.* 2019). Irrigated agriculture is linked to increased farm incomes but irrigation works can also contribute to sustainability through promoting environmental balance, the provision of ecosystem services, employment and social equity, especially in areas where the small percentage of irrigated land causes social conflicts for access to it.

Since irrigation works can serve objectives related to sustainable development, effective governance of irrigation water use and allocation is required to valorize the expected results of an irrigation investment. In Greece, the competent bodies for the management of local irrigation networks are the Local Land Reclamation Organizations (TOEB in Greek). TOEB are independent collective non-profit organizations governed by public law, under the supervision of Regional Authorities (see Dercas 2020) for a detailed presentation of the evolution of the framework). Their specific responsibilities are described in their Statutes and typically include the smooth operation and management of irrigation and land reclamation works, compilation of irrigation programmes, setting up and collection of irrigation fees as well as drafting studies for necessary reparation, improvement and/or expansion works. Despite being collective bodies, TOEB have not generally undertaken broader roles in Greek rural areas, such as proposing strategies for agricultural development, orientation towards innovation and improvement of human resources involved in the farming sector.

The construction of an irrigation work is accompanied by its financial appraisal, which demonstrates its feasibility and the conditions under which it is beneficial and profitable. Research regarding water management and irrigation project appraisal is rich (for instance see Latinopoulos 2005; Tillery & Jones 2007; Panagopoulos *et al.* 2014) and is mainly based on the Cost-Benefit Analysis (CBA), which is the established methodology (Hoehn & Randall 1987; Hanley & Spash 1993). As Tsakiris *et al.* (2017) pointed out, however, CBA cannot always capture indirect socio-economic benefits of irrigation works and dams. Indeed, environmental benefits are sometimes neglected due to market failures, while the potential to achieve broader sustainable development objectives is not reflected in economic performance indicators. In addition, the costs of irrigation works can be exaggerated or not estimated inaccurately (Psychoudakis *et al.* 1995). Therefore, appraisal with purely economic criteria could lead to underestimation of their real contribution in quality of life, environment and job creation.

In order to overcome this shortcoming, the financial appraisal is often supplemented with a socioeconomic and environmental analysis, which incorporates social and environmental issues. Yet, a considerable body of literature focuses on the environmental effects of irrigation water management in terms of water consumption and does not consider impact on the development process in rural areas. Latinopoulos (2008) examined the environmental, social and economic effects of irrigation water pricing on rural areas with multi-objective techniques and found that higher water prices could compromise economic and social sustainability objectives. Another significant part of relevant research analyzes the relationship between irrigation water consumption and economic performance (Latinopoulos & Mylopoulos 2005; Georgiou & Papamichail 2008). In the same context, Michailidis *et al.* (2009) described the process of economic analysis of an irrigation system by incorporating risk analysis and allowing for variability in cost elements using the real options method. Berbel *et al.* (2011) concluded that decision making about water management needs to take into account complex interactions among various factors, which cannot be tackled by an analysis of cost effectiveness alone. In this context, Niu *et al.* (2016) – with a two-stage fuzzy stochastic programming method – and Xu *et al.* (2019) – with an equilibrium model – analysed the importance of equity in irrigation water distribution to achieve sustainability. Capitanio *et al.* (2015) emphasized the importance of irrigation for the sustainable development of Italian agriculture under the sustainability objectives of CAP 2014–2020. General and partial equilibrium models have also been used to analyze the economics of water allocation (Dudu & Chumi 2008).

The purpose of this study is to showcase how an irrigation work can serve to achieve broader objectives related to the sustainability of rural areas. In particular, the paper describes the development of a multi-objective programming (MOP) model that seeks to compromise conflicting objectives regarding the maximization of farm incomes and employment (economic and social sustainability) and the minimization of agrochemical use and water consumption (environmental sustainability). The method has been used before in agricultural economic analysis. Mosleh *et al.* (2017) used MOP and suitability analysis to propose the optimal cropping pattern for a rural area in Iran. Groot *et al.* (2018) combined MOP with other methods to explore multifunctional landscapes is many parts of the world. The study of Xevi & Khan (2005) focused on water use in agriculture in contrast to the achievement of environmental objectives. Galan-Martin *et al.* (2017) applied MOP along with life cycle assessment and water footprint concepts to examine the optimal allocation of irrigated and rainfed wheat production in Spain. Li *et al.* (2018) applied interval MOP in a Chinese irrigation district to examine how water savings can be achieved simultaneously with higher yields and economic performance.

In this paper, the analysis focuses on a Greek community for which technical and economic indicators are derived for main crops and for new crops that could be introduced in the cropping pattern. This way, the paper approaches irrigation works as opportunities for development. In order to emphasize this aspect of irrigation works, the paper also discusses the role of irrigation governance in the coordination of efforts to more towards higher levels of overall sustainability in rural areas.

MATERIALS AND METHODS

Study area

Characteristics

The study area is the Community of Limnochori in the Regional Unit of Florina, Region of Western Macedonia (RWM), Greece (Figure 1) (590–600 m a.s.l.). The population of the Community and of the whole Municipal Unit was 257 and 2.952 persons in 2011, with significant decreasing trends compared to previous decades (ELSTAT 2011). Unemployment constitutes one of the main problems of the whole Region, rating to 22.3% and harshly affecting especially younger people (ELSTAT 2011). RWM was ranked 11th among the 13 Greek Regions in terms of GDP (2,1% of national GDP) but 5th in terms of GDP per capita (14.800 €) (59% of EU average) in 2018 (Eurostat 2020), mainly due to the operation of coal-based electricity plants. Agriculture is the main source of employment and income in the study area, followed by employment in energy production in the electric power plants. However, RWM is actually in the course of a serious transition, as the coal-based energy production will be significantly

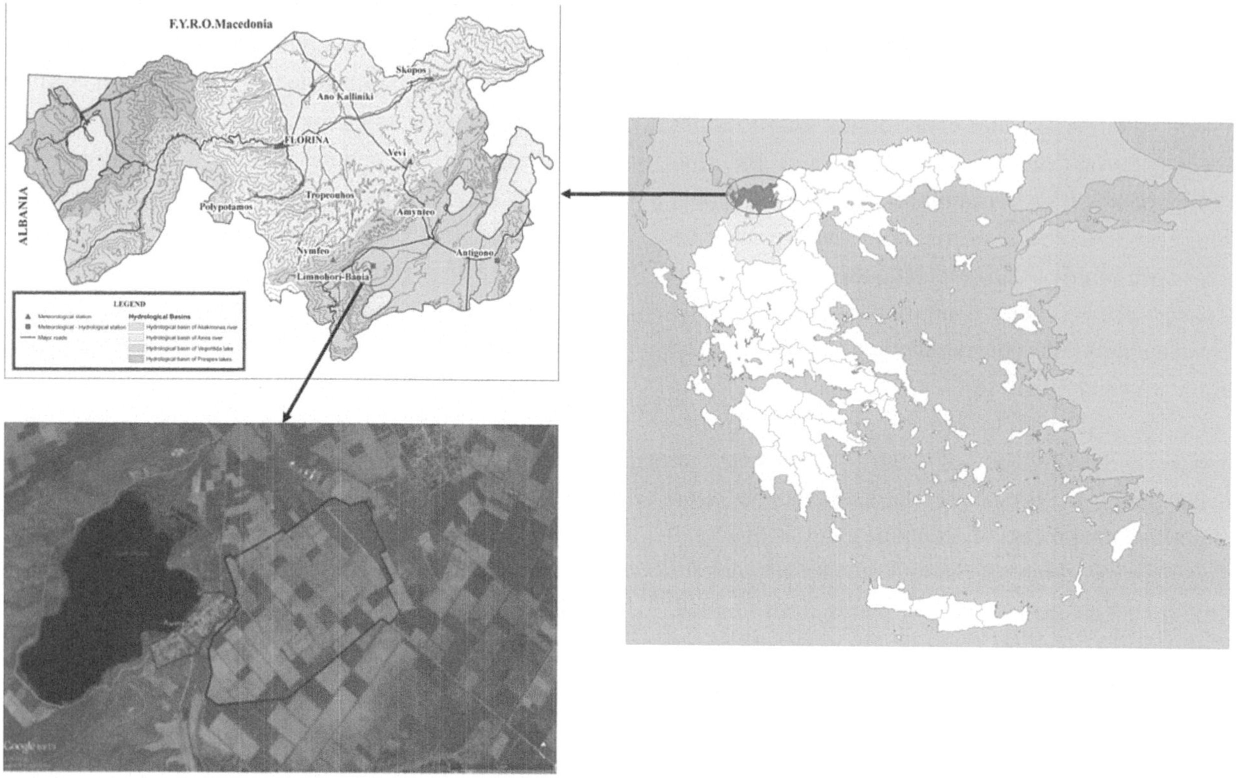

Figure 1 | Map of the study area (Limnochrori, Regional Unit of Florina, Western Macedonia, Greece). *Note*: The blue and pink polygons denote the irrigated area in the present situation and the additional irrigated area in the future situation respectively. The full color version of this figure is available in the online version of this paper, at http://dx.doi.org/10.2166/ws.2021.106.

reduced (Karlopoulos & Sidiropoulos 2020). Therefore, interventions to induce income generation and employment in other sectors of the economy is crucial.

Agricultural production in the study area is characterized by the predominance of few arable crops (maize, winter cereal, lucerne, potatoes, peas for livestock) which account for more than 90% of the total cultivated land, while sugar beet production has been significantly reduced in the past few years. The cropping pattern is the result of a long process of adjustment under the influence of climate and soil conditions but also of the Common Agricultural Policy, in the form of income support or of direct subsidies on the volume or acreage of production in previous periods. Through this process, necessary infrastructure has been formed and specific farming practices have prevailed, formulating a fairly intensive production pattern with high requirements in irrigation water, pesticides, fertilizers and other inputs. One of the problems in the farming sector of the study area is disguised unemployment, as the existing cropping pattern has low human labor requirements, with significant seasonal variation and peaks.

Irrigation network in the study area

The existing irrigation network covers 213,5 ha. It uses open canals and is supplied with water from Lake Zazari nearby. It was constructed in 1985 with an annual recoverable water quantity from the lake of 1,190,000 m^3 for a four-month irrigation period (Evangelidis 2017a). Since then, however, its capacity has been reduced, due to evaporation from the open canals and the use of old irrigation methods (mainly sprinklers) resulting in significant water losses. Also, about 30 ha are actually not irrigated due to network damages. Due to this situation, an upgrade and repair of the irrigation network has been planned and designed. Optimization of water use will be achieved through a standard closed pressure fully automated network, with innovative technologies that will ensure proper operation as well as gradual replacement of sprinklers by drip irrigation. This will lead to less water losses and higher water availability for farmers, which will allow expansion of the total irrigated area by 64 ha to a total of 276,7 ha, without increasing the amount of water recovered from the lake annually. Specific works that will be undertaken are described in the relevant technical and economic studies (Evangelidis 2017a, 2017b) and include:

- Replacement/reconstruction of the existing hydrant inside lake Zazari with new equipment, in the same spot and with the same capacity of 1,190,000 m^3 per year
- A 78 m water pipeline from the hydrant to the pumping station
- Construction of a pumping station for network supply, in a municipal area 13 m from the shore of the lake
- A 360 m water pipeline from the pumping station to the limit of the irrigated area
- Replacement of existing canals by a pipeline network (approximately 11 km)
- A 3.183 m water pipeline from the pumping station to the limit of the dilatation of the irrigated area
- A pipeline network for the irrigation of additional 64 ha without increasing water recovery from the lake. The total irrigated area after construction will thus be 276,7 ha

Methodological approach

Model development

The analysis in this paper is based on an application of multi-objective programming (MOP), an optimization method that examines the possibilities of simultaneously achieving conflicting objectives. The model in this study is specified to include more objectives reflecting all dimensions of sustainability. The multiobjective problem for p objectives ($Z_1, Z_z \ldots Z_p$) is formulated as follows (Cohon 1978; Romero & Rehman 1989; Zeleny 2012)

$$\text{maximize } Z(x_1, x_2 \ldots x_j) = [Z_1(x_1, x_2 \ldots x_j), Z_2(x_1, x_2 \ldots x_j), \ldots, Z_p(x_1, x_2 \ldots x_j)]$$

$$\text{subject to } (x_1, x_2 \ldots x_j) \in F_d \tag{1}$$

where $x_1 \ldots x_j$ are the different activities through which maximization can be achieved (variables). Among different methods to solve the multi-objective problem, this study applies the ϵ-constraint method (Romero & Rehman 1989), which is expressed as follows

$$\text{maximize } Z_h(x_1, x_2 \ldots x_j) \tag{2}$$

$$\text{subject to } (x_1, x_2 \ldots x_j) \in F_d$$

$$Z_\kappa(x_1, x_2 \ldots x_j) \leq L_k$$

where $k = 1, 2, \ldots, h-1, h+1, \ldots, p$

In this model, Z_h is one of the p objectives which is set as the objective function for optimization (minimization or maximization). The remaining p-1 objectives (Z_k) are included as constraints in the model specification. This specification is a linear programming (LP) problem – which is specified in Equation (3)

$$\max(\min) \sum_{j=1}^{M} c_j x_j = Z$$

$$\text{subject to } \sum_{j=1}^{M} a_{i_j} x_j \leq A_i$$

$$\sum_{j=1}^{M} \beta_{ij} x_j \leq B_\iota$$

$$x_j \geq 0 \tag{3}$$

In the above LP problem, Z stands for the Z_h objective, while c_j is the vector of coefficients of M variables (activities), whose vector is denoted by x_j and stands for the contribution of each activity x_j to the objective function. α_{ij} are the requirements per unit of x_j for constraint A_i, while the set of the Z_k alternatives is expressed as a set of constraints, where β_{ij} are the coefficients of variables x_j and B_i is the available quantity of the specific constraint. This LP problem is specified similarly and solved for each one of the *p* conflicting objectives.

The solution of each one of the LP problems is included in a *pxp* table, the pay-off matrix, where columns represent the Z_p objectives and rows LP solutions. Therefore, the elements of the pay-off matrix are the values of each optimized objective Z_h and the value of Z_k objectives included as constraints in the corresponding model specification. The elements of the diagonal of the pay-off matrix (M_p) indicate an optimal solution to the MOP problem; that is, the infeasible solution where all conflicting objectives are optimized, while the difference between the minimum (m_p) and the maximum (M_p) of each Z_p objective define the range within each objective can range.

Based on the pay-off matrix, the MOP model is transformed to a right-hand-side parametric model (RHS-PP). The mathematical expression of the RHS-PP model is the same as the LP model (Equation (3)); however, the right-hand-side value of a constraint (A_i or B_i) is allowed to vary within an acceptable range, which yields multiple optimal solutions according to the availability of the constraint. In MOP, *p-1* RHS-PP models are specified, where one Z_h objective is the objective function in all specifications, while the remaining Z_k objectives are set as constraints and in each specification one of them is the parametrized (varying) constraint. The available quantity Bi of each Z_k constraint is allowed to vary between m_p and M_p

$$m_p \leq Z_k \leq M_p \tag{4}$$

This guarantees feasibility and non-inferiority for each solution (Ragkos & Psychoudakis 2009). For each specification, this procedure yields a set of alternative optimal solutions according to the availability of the varying constraints, all of which constitute an approximation of the non-inferior set of solutions (Cohon 1978). The final set of solutions reflects different options, each one of which compromises the conflicting objectives at different levels.

In the MOP application in this paper, the possibilities of achieving four conflicting objectives (Z_p) are examined.

- *Z1. Maximization of economic result* (Gross margin = Gross revenue − Variable expenses). This objective reflects higher economic sustainability, linked to higher economic performance of farms. For the calculation of the gross margin, hired labor expenses are not included in variable expenses, as the study does not discern between sources of farm labor (family or hired).
- *Z2. Maximization of employment* (hours of farm labor). This relates to the social sustainability of the sector, as more jobs in the farming sector would render it a more alluring alternative. Considering that young people seem to consider farm employment as an option to overcome high unemployment rates, this objective also reflects opportunities to keep population in the area or even to attract young people to return. Since the increase of employment is a general objective, the model does not discern between hired and family labor.
- *Z3. Minimization of pesticide and fertilizer use (agrochemicals)* (value of inputs per ha). The reduction in the use of agrochemical inputs relates to environmental sustainability. Residues of pesticides pollute water reserves and soil, while the excessive use of nitrogen fertilizers has led to degradation of surface and ground water (Pretty *et al.* 2000). In this study we include the value of pesticides and not physical units (e.g. kg of agrochemicals and fertilizers), because their expression in € provides an option for their simultaneous consideration – as it is difficult to add their quantities. Since farming practices and input prices are stable in the model, changes in the value of pesticides reflect changes in their total quantities used.
- *Z4. Minimization of irrigation water use.* Irrigation water saving constitutes one of the main objectives relating to the environmental sustainability of the production system. As described above, the new irrigation network will permit the irrigation of more hectares by reducing water losses and increasing irrigation efficiency.

It should be noted that the MOP problem presented here assumes that there are no changes in the production practices before and after the construction of the irrigation work. The compromise of the conflicting objectives is, therefore, sought only through changes in the cropping pattern, with crop restructuring which leads to optimization of each objective or to different levels of simultaneous achievement of all objectives.

Using the results of the ϵ-constraint method, the trade-offs between objectives can be calculated (Cohon 1978). Trade-offs represent the shadow price of a parametrized constrained input i.e. the additional amount of the objective function that can be derived by increasing the availability of the constrained input by one unit.

Common constraints in all model specifications include:

- *Land constraints.* The irrigated land supported by the irrigation network in the existing situation was 212.7 ha. The new network will allow the irrigation of 64 additional hectares, which are actually cultivated with winter cereal (wheat and barley, arid crops). The model includes constraints regarding the maximum acreage of each crop, taking into account agronomic and market factors
- *Labor constraints,* which simulate the requirements for human labor of all crops
- *Capital constraints,* which refer to the availability of variable capital for purchased inputs

The paper examines two scenarios in order to demonstrate the potential effects of the irrigation work on sustainability.

- *Scenario 1 – Optimization without introducing new crops.* Scenario 1 involves the restructure of the existing crops (maize/silage, lucerne, potatoes, livestock peas) and the expansion of their cultivation in the additional 64 irrigated hectares. The optimal solutions in this scenario show if the irrigation work could have positive effects on the development of the area without significant changes in the cropping pattern.
- *Scenario 2 – Crop restructuring (optimization) with the introduction of two new crops with more dynamic potential.* These two crops are

  - Chamomile, which is a non-irrigated crop with high labor requirements and relatively low investment costs. Cultivation is quite extensive, as only small quantities of nitrogen fertilizers are applied (usually no more than 100 kg/ha) and pesticides are not necessary. The acreage of aromatic plants in the broader region had expanded during the last few years, with similar development trends for other activities across the value chain (standardization and processing units, retail sales).
  - Fresh peas, an irrigated crop that requires intensive use of agrochemicals (fertilizers, herbicides and insecticides) but also human and machine labor. One of the main opportunities for this crop is contractual production, already occurring in the area.

The new cropping pattern in this case would require a mid-term development strategy for the area in order to facilitate the transition.

Cost-benefit analysis

Based on the results of MOP, a CBA of the construction of the irrigation work was conducted. For each solution, the Net Present Value (NPV) and the Internal Rate of Return (IRR) were calculated, which are two of the most commonly used indicators for the appraisal of construction works. In this paper, net cash flows were discounted with a 3% rate – which is considered satisfactory for a public investment – and the NPV was calculated over a period of 50 years. The IRR stands for the discount rate, which returns zero NPV and expresses the marginal return on invested capital, which can be compared to bank and loan interest rates.

The annual economic benefits from the investment (irrigation work) were calculated as the difference between the economic output (gross margin) in the optimal existing solution and the future situation. This approach was used because this difference would reflect the difference between the existing and future situation attributable to irrigation only. Indeed, a comparison between future optimal and existing sub-optimal organization would include the difference between actual and optimal organization in the existing situation, which is not related to irrigation but to other factors. Therefore, the optimization of the existing situation provides an alternative to the irrigation work, as is common for CBA assessments. The optimal existing solution was derived by means of an LP model (Formula 3) where land constraints were set close to the existing acreage of each crop and showed how the existing cropping pattern could be rearranged to achieve optimal economic performance. The future benefits (gross margin, Z1) were calculated from the selected optimal solutions of the MOP model.

Data

Data for the LP and MOP models were derived combining surveys and secondary (published) sources. Technical and economic indicators for production inputs (value of fertilizers and pesticides), labor requirements, product yields and market prices were obtained from the Region of Western Macedonia (2009) and were then updated with interviews with farmers and verified by two local experts. Irrigation water requirements were calculated based on Ambas (2010), using a method that considered the needs of all crops, annual rainfall – based on meteorological data from a local station – and evapotranspiration. In more detail, Ambas (2010) calculated evapotranspiration with five different methods and estimated parameters of each method, yielding a total of 26 values of evapotranspiration. The main evaporation method considered was the FAO56 Penman-Monteith method with clear-sky radiation estimated using spectral analysis and solar radiation from raw data. Effective rainfall was calculated by means of the Soil Conservation Service method. The indicators used in the MOP model (c_{ij} and β_{ij}) for the four objectives are presented in Table 1.

Construction costs of the irrigation work were calculated to 3,427,419.35 €, including all surcharges foreseen by Greek law except VAT (technical works were budgeted to 2,441,475.24 €, 439,465.54 € (18%) were general expenses and contractor profit, 432,141.12 € (15) were unforeseen expenses and 114,337.46 € were budget revision). Annual costs raised to a total of 41,7887.24 €/year including maintenance costs (0.25% of construction costs per year), electricity costs and management/operation costs (wages for two persons employed full-time 18,564 €/year) (Evangelidis 2017b).

Table 2 | Optimal cropping pattern under the existing situation

Crops	**Acreage (ha)**
Maize	77.5
Silage	19.4
Potatoes	13.8
Lucerne	69.2
Peas	0
Cereal	96.8
TOTAL	**276.7**
Objectives	
Gross margin (X1000€)	439.5
Agrochemicals (X1000€)	59.5
Labor (Labor Units[a])	12.5
Water (X1000 m^3)	844.3[b]

[a]1 Labor Unit = 1,720 h.

[b]Water consumption here represents crop requirements and not the total consumption.

RESULTS

Table 2 presents the basic LP solution, where the existing situation (before the construction of the irrigation work) was optimized. Non-irrigated cereals cover 96.8 ha, therefore they are cultivated even in areas that can be irrigated. This is due to the fact that the available water is used for crops with higher gross margin per ha, which are more demanding in irrigation. In addition, the reduced capacity of the existing network leads to losses and inefficient use of the annual

Table 1 | Variable coefficients (c_{ij} and β_{ij}) of the four conflicting objectives considered in the MOP model

	Z1 (max) Gross margin (€/ha/year)	**Z2 (max) Labor (h/ha/year)**	**Z3 (min) Agrochemicals (€/ha/year)**	**Z4 (min) Irrigation water (m^3/ha/year)**
Existing crops				
Maize	1,536	100	280	4,150
Silage	1,664	80	325	4,150
Potatoes	3,572	160	640	5,834
Lucerne	2,822	120	125	5,230
Dry peas (livestock)	344	120	175	3,033
Cereal	450	20	140	0
New crops				
Chamomile	4,800	1,300	50	0
Fresh peas	2,482	450	175	3,576

recoverable water quantity. In irrigated land, maize, silage and lucerne account for more than 90%. Note that water consumption in Table 2 expresses the total requirements of crops and not the actual consumption, which should be considerably higher due to water losses from evaporation and reduced irrigation efficiency in the existing situation.

Table 3 presents cropping patterns and the pay-off matrices for both scenarios. Differences between the highest and the lowest values of the four objectives indicate the possibilities of increasing or decreasing each one of them. Gross margin could be increased from 471.4 thousand € to 572.1 thousand € (21.3%) and employment from 15.5 to 17.7 LU (2.2 more jobs or an increase of 14.2%). Furthermore, a decrease from 79.6 to 63.8 thousand € for the value of agrochemicals (19.8%) and considerable irrigation water savings (from 1,184.4 to 1,067.5 thousand m^3 or by 9.9%) could occur. The acreage of irrigated crops was higher for solutions maximizing gross profit and employment, mainly because potatoes were included with an acreage of 41.5 ha in both optimal cropping patterns. Livestock peas (8%) and winter cereal (13%) were included in the solutions relating to environmental sustainability (minimizing agrochemical and irrigation water use), along with maize (and silage), which occupied half of the total available land. The acreage of lucerne was relatively stable across all four cropping patterns (21–25%).

In Scenario 2, which considers the inclusion of new crops (chamomile and fresh peas) (Table 3), the ranges of the values of the four objectives were not significantly different, as increases of 21.4% and 13.7% can be achieved for gross margin (482.7–639.7 thousand €) and employment (25.6–28.9 LU) respectively and savings of 9.9% and 19.8% could occur for irrigation water (975.6–1,184.4 thousand m^3) and agrochemicals (55.6–78.9 thousand €). Cropping patterns varied across the four solutions, as only the two new crops (camomile and fresh peas) were included in all solutions. Gross margin was maximized mainly with the inclusion of maize, lucerne and potatoes (76.3%) and when dry peas for livestock feeds were excluded. For the maximization of labor, the acreage of maize was increased

Table 3 | Pay-off matrices in the two scenarios – results of multi-objective programming

	Scenario 1. Existing crops				Scenario 2. Existing and new crops			
	Z1 (max) Gross margin	Z2 (max) Labor	Z3 (min) Agrochemicals	Z4 (min) Irrigation water	Z1 (max) Gross margin	Z2 (max) Labor	Z3 (min) Agrochemicals	Z4 (min) Irrigation water
Non-irrigated crops (ha)	**27.7**	**36.0**	**17.1**	**36.0**	**23.9**	**47.1**	**11.9**	**47.1**
Winter cereal	27.7	36.0	17.1	36.0	12.8	36.0	0.8	36.0
Chamomile	–	–		–	11.1	11.1	11.1	11.1
Irrigated crops (ha)	**249.1**	**240.8**	**259.6**	**240.8**	**252.9**	**229.6**	**264.8**	**229.7**
Lucerne	69.2	69.2	57.6	69.2	69.2	69.2	46.2	52.6
Maize	110.7	138.4	138.4	110.7	100.5	121.7	138.4	110.7
Dry peas (livestock)	0.0	22.1	22.1	22.1	0.0	22.1	22.1	22.1
Potatoes	41.5	11.1	41.5	11.1	41.5	0.0	41.5	0.0
Silage	27.7	0.0	0.0	27.7	25.1	0.0	0.0	27.7
Peas (fresh)	–	–	–	–	16.6	16.6	16.6	16.6
Pay-off matrices								
Gross margin (X1000€)	**572.1**	*471.4*	538.9	474.7	**639.7**	500.6	593.8	*482.7*
Agrochemicals (X1000€)	*79.6*	**63.8**	79.2	65.2	77.2	**55.6**	*78.9*	59.4
Labor (LU)	16.7	15.9	**17.7**	*15.5*	28.1	26.1	**28.9**	*25.6*
Water (X1000m^3)	1,177.9	1,067.5	*1,184.4*	**1,067.5**	*1,184.4*	993.5	*1,184.4*	**975.6**

even more (138.4 ha – 50% of total land) along with lucerne and potatoes (31.7%). With regards to environmental sustainability, the use of agrochemicals was minimized by increasing the acreage of cereal (57.0%), lucerne (25.0%) and livestock peas (8.0%), while irrigation water consumption was minimized by increasing silage (10.0%) and winter cereal (13.0%) apart from maize (40.0%).

The comparison of the two pay-off matrices illustrates how the inclusion of new dynamic crops in the cropping pattern could have significant impact on all dimensions of sustainability. Indeed, maximum gross margin and employment were higher by 11.8% and 63.6% respectively in Scenario 2, while in the same scenario additional reduction of 12.9% and 8.6% could be achieved in the value of agrochemicals and water consumption.

The results of multi-objective programming yielded 21 optimal solutions in total for both scenarios (10 for S1 and 11 for S2). Table 4 presents ten selected plans among the 21 optimal solutions yielded in both scenarios. Each one of these plans compromises the four objectives and reflects varying levels of economic, social and environmental sustainability. Solutions are ranked according to the level of irrigation water use and each successive solution in each scenario (S1.1-S1.9 and S2.1-S2.11) represents higher gross margin. A key finding is that increases in economic and social sustainability were achieved mainly with the inclusion of potatoes and high acreage of maize and a reduction in winter cereals. On the contrary, potatoes were excluded for livestock peas, maize and/or silage when environmental objectives were achieved to higher degrees. Cropping patterns differed across solutions. As irrigation water use increased, the acreage of winter cereal was reduced, dry livestock peas were excluded, and the cultivation of lucerne was maximized.

Based on the results of the MOP model, for both scenarios, Table 5 presents the trade-offs between objectives Z2–Z4 and gross margin (objective Z1). As expected, trade-offs increase for lower levels of input (agrochemicals, labor, water) use. It is interesting to notice that the shadow price (trade-off) of labor in Scenario 1 is considerably higher compared to Scenario 2, as in Scenario 1 labor use is generally much lower. When it comes to water, the use of less water increases its shadow price significantly, which reflects the importance of this input.

Figure 2 illustrates the various contributions of the irrigation work that can be considered in its appraisal. The

Table 4 | Optimal cropping pattern with the new irrigation network for the developed scenarios

	S2.1	S2.3	S2.5	S1.1	S1.5	S2.8	S1.9	S1.6	S2.9	S2.11
Non-irrigated crops (ha)	**47.1**	**47.1**	**47.1**	**36**	**36**	**47.1**	**36**	**27.7**	**20.2**	**23.9**
Winter cereal	36	36	36	36	36	36	36	27.7	9.1	12.8
Chamomile	11.1	11.1	11.1	–	–	11.1	–	–	11.1	11.1
Irrigated crops (ha)	**229.7**	**229.7**	**229.7**	**240.8**	**240.7**	**229.7**	**240.8**	**249.1**	**256.6**	**252.9**
Lucerne	52.6	11.1	69.2	69.2	38.7	69.2	69.2	69.2	69.2	69.2
Maize	110.7	110.7	138.4	138.4	110.7	81.9	110.7	110.7	138.4	100.5
Dry peas (livestock)	22.1	22.1	0	22.1	22.1	0	0	0	0	0
Potatoes	0	41.5	5.5	11.1	41.5	41.5	33.2	41.5	32.4	41.5
Silage	27.7	27.7	0	0	27.7	20.5	27.7	27.7	0	25.1
Peas (fresh)	16.6	16.6	16.6	–	–	16.6	–	–	16.6	16.6
OBJECTIVES										
Gross margin (X1000€)	483	514	538	471	497	614	546	572	622	640
Agrochemicals (X1000€)	59	81	58	64	81	74	75	80	72	77
Labor (LU)	26	27	26	16	16	27	16	16	29	29
Water (X1000m^3)	976	1,001	1,028	1,068	1,086	1,088	1,129	1,178	1,184	1,184

Note: S1 and S2 stand for Scenario 1 and Scenario 2 respectively and numbers after the dots the sequence number of each solution in each scenario. In this table, solutions are presented according to the volume of water consumption.

Table 5 | Trade-offs between gross margin and use of agrochemicals, labor and irrigation water

Agrochemicals		Labor		Irrigation water	
Value of agrochemicals (thousand €)	Trade-offs (€/€)	Hours of labor (Thousand hours)	Trade-offs (€/hour)	Use of irrigation water (thousand m³)	Trade-offs (€/m³)
Scenario 1					
78.2	2.40	28.8	14.66	1,168	0.27
74.0	6.24	28.1	31.34	1,111	1.16
63.8	7.02	26.7	80.75		
Scenario 2					
77.2	2.40	49.1	2.94	1,184	0.27
73.2	6.24	29.6	4.39	1,165	1.09
57.2	9.01	28.8	5.54	1,063	1.25
55.6	12.00	26.7	17.76		

NPV was marginally positive in only two solutions (S2.9 and S2.11), while the IRR was less than 4% – or even negative – in all solutions. This shows that the expected economic performance of the investment was low – especially when new crops were not considered – despite the fact that gross margin was higher than the optimal existing situation in all solutions. This is presented in Figure 3, where also other contributions were taken into account. Indeed, all scenarios entailed higher levels of employment in the area, which was more than double for Scenario 2. Furthermore, in all solutions there was more irrigation water use compared to the existing situation, without, however, increasing the quantity recovered from the lake and also ensuring the irrigation of additional areas. Nevertheless, in only two out of ten solutions (S2.1 and S2.5) was the use of agrochemicals lower than the optimal existing situation and these solutions performed badly in terms of NPV and IRR.

DISCUSSION

The MOP model yielded a set of non-inferior solutions that reflect different levels of achievement of conflicting objectives. The sustainable development pattern based on these objectives is in line with the CAP Specific Objectives for 2021–2027 (https://ec.europa.eu/). Depending on the choice of the specific optimal solution, objectives regarding environmental care, fair income for farmers and competitiveness are met. However, the choice of the best solution is subject to internal and external factors. Internal factors are related to strengths and weaknesses of the farming sector in

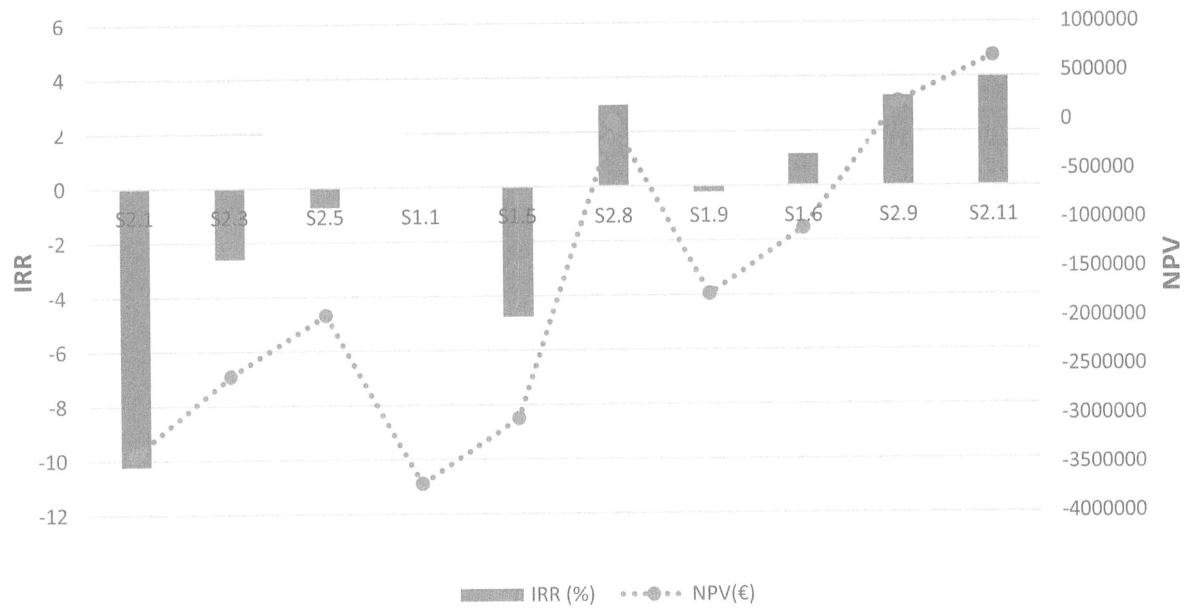

Figure 2 | Cost-benefit analysis indicators in selected MOP solutions. *Note*: S1 and S2 stand for Scenario 1 and Scenario 2 respectively and numbers after the dots the sequence number of each solution in each scenario. In this figure, solutions are presented according to the volume of water consumption.

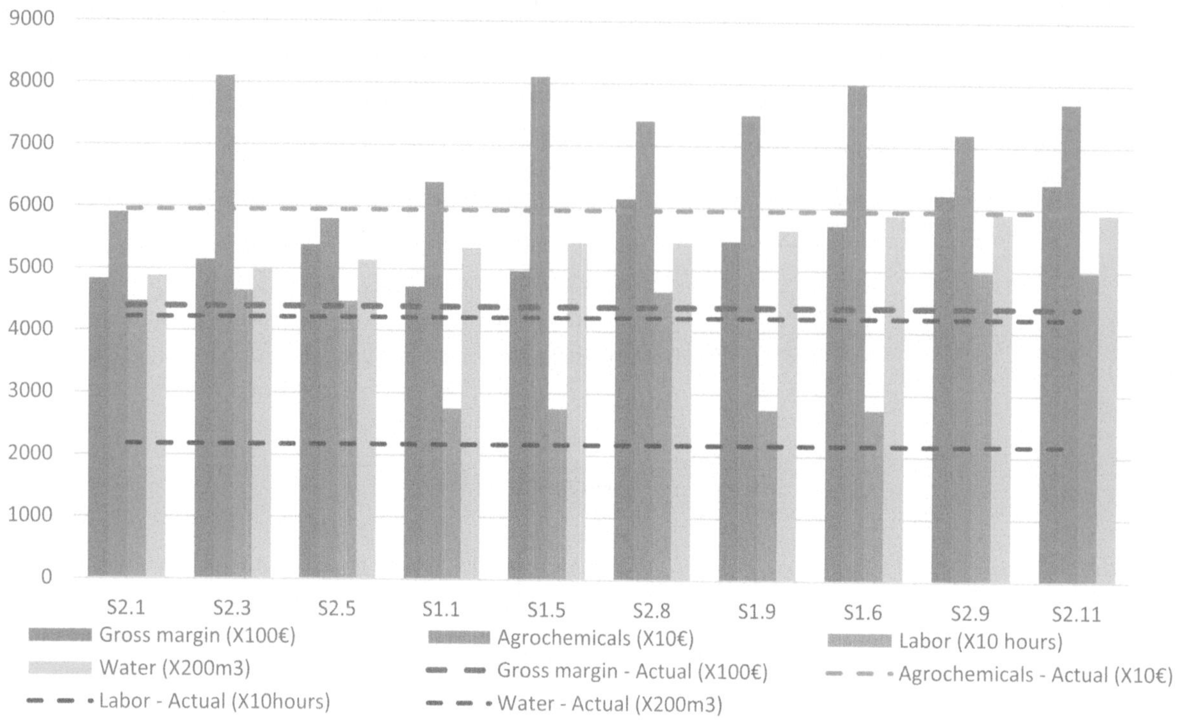

Figure 3 | Achievement of social and environmental sustainability objectives in selected MOP solutions. *Note:* S1 and S2 stand for Scenario 1 and Scenario 2 respectively and numbers after the dots the sequence number of each solution in each scenario. In this figure, solutions are presented according to the volume of water consumption.

the area and primarily include locals' preferences and aspirations – not only of farmers. For example, reactions could be expected if property issues arize, especially for newly irrigated land. In addition, the sustainable development process should ensure that conditions are met, including access to capital, training and advice. In any case, the views of the affected population should be explored, in order to valorize the potential of the irrigation project to induce a new development process in the area. On the other hand, external factors represent the influence of general policy objectives, the evolution of farm policies and income support schemes for specific crops, the economic environment, other employment opportunities, environmental policies in force and other socio-economic challenges. Therefore, this choice is also contingent upon broader policy objectives.

The increase of the irrigated areas will only have marginal effects on the development process, if it is not accompanied by strategic planning and effective governance. There is a need for a new improved scheme not only to manage water but also to provide guidance to farmers and assist the transition to crop restructuring towards higher levels of sustainability. In this context, this study considers the perspective of redefining the role of TOEB by introducing alternative and more flexible – and potentially more efficient – forms of administration and management of irrigation projects such as public-private partnerships (PPPs) (Warner *et al.* 2008). These are government initiatives for cooperation in which the role of the state entity in the operation and maintenance of irrigation projects is transferred to another body through a management agreement. PPPs could assume the form of financially autonomous government agencies and involve a professional Water Users Association or even private companies (Kleemeier & Lockwood 2012; Lamaddalena & Khadra 2012; Aarnoudse *et al.* 2018). This increases the participation of farmers, who become co-responsible for the management and operation of the project.

PPPs become increasingly popular as they ensure important benefits, mainly because they counterbalance bureaucratic issues and public funding requirements for operation. In particular, PPPs can be cost efficient in determining irrigation fees – as a result of consensus of involved stakeholders – based on objective criteria and acceptable collection mechanisms. In addition, they can act as intermediate actors between government and farmers with increased flexibility and can be more responsive to local needs, as they motivate the active participation of local actors. This could be expected to reduce social conflicts due to competition for access to irrigation water, especially when the participation of farmers is active, substantially influencing the decision-making process. A PPP could also be effective in undertaking advisory and supporting roles for farmers in the process of transition towards more sustainable development, thus supporting crop restructuring to achieve benefits as reported in this paper. This role corresponds to the fourth dimension of sustainability, as PPP could also undertake broader roles in local societies, relating to agricultural development as well as training, advice and communication. Despite these benefits, PPPs in irrigation water management could be proven inefficient under the influence of numerous factors, including cases where farmers' involvement is not active (Gastineau 2006; Zhanq & Tariq 2020).

An important aspect that affects the successful operation of irrigation works is the calculation of irrigation fees. This study shows that the imposition of lower or higher irrigation fees could affect the gross margin of farms and – to some extent – economic sustainability of farm production in the area, which would also affect other dimensions of sustainable development. Therefore, one of the main duties of the collective governance scheme in charge of irrigation management would not only be the effective calculation of irrigation fees but also the option to readjust them dynamically. In Europe, there is accumulated practical and academic experience from a variety of methods and schemes (Berbel *et al.* 2007). In Greece, the efficient calculation of irrigation water fees can play an important role in its efficient use and can also be affected by policies in force (Manos *et al.* 2006).

CONCLUSIONS

This study described the development of a MOP model to examine the extent to which an irrigation work could contribute to sustainable development and induce economic, social and environmental improvements. The results showed that there is significant potential for increases in all dimensions of sustainability. The CBA confirmed that the consideration of only the economic performance in the appraisal could neglect other important benefits that are not directly reflected in financial indicators. The advantage of the MOP approach is that the solutions it provides allow consolidation of many of these benefits without excessive data requirements or analysis.

In order to achieve sustainable results, the study pinpoints the importance of effective governance, which should be based on inclusive participatory approaches of farmers and other affected stakeholders. Thus, the human factor is of paramount importance for the success of such a project, which will ultimately determine the exact conditions for the realization of the project and the exploitation of the advantages that this entails. However, the proposed PPP scheme constitutes an alternative to the existing TOEB schemes and should aim to support their operation and to redefine their objectives, by adding broader roles in their portfolios. PPP could therefore provide them with flexibility and better managerial capacities. A particular function of the PPP governance scheme could also include ongoing monitoring and ex-post evaluation of sustainable development in the area. This would also show the degree of adaptation of farmers to the optimal cropping pattern that would be chosen.

The methodology presented in this study can be easily applied in other settings and even for larger irrigation works and areas, with larger affected populations. However, this approach is subject to two limitations. First, farming practices in the model represent the observed practices in the area, but with this specification, other practices (such as organic production or integrated pest management) are not considered. This option should be taken into account in future research. Second, the expected environmental gains from more efficient water use and less agrochemicals have been found to have non-market values (Ragkos & Theodoridis 2016), which can be included in CBA frameworks through the application of non-market valuation techniques.

ACKNOWLEDGEMENT

This research was partially funded by the Region of Western Macedonia through the project 'Appraisal of irrigation infrastructure works and new methods for irrigation water pricing'.

DATA AVAILABILITY STATEMENT

Data cannot be made publicly available; readers should contact the corresponding author for details.

REFERENCES

Aarnoudse, E., Closas, A. & Lefore, N. 2018 Water user associations: a review of approaches and alternative management options for Sub-Saharan Africa. In: *Working Paper 180, International Water Management Institute*, Colombo, Sri Lanka. doi:10.5337/2018.210.

Ambas, V. 2010 *Research and Estimation of Meteorological Parameters with Direct Impact on Agriculture*. Ph.D. Thesis, Aristotle University of Thessaloniki, Thessaloniki, Greece.

Berbel, J., Calatrava, J. & Garrido, A. 2007 Water pricing and irrigation: a review of the European experience. *Irrigation Water Pricing Policy: The Gap Between Theory and Practice*. CAB International, Oxon, pp. 295–327.

Berbel, J., Martin-Ortega, J. & Mesa, P. 2011 A cost-effectiveness analysis of water-saving measures for the water framework directive: the case of the Guadalquivir River Basin in Southern Spain. *Water Resources Management* **25** (2), 623–640.

Capitanio, F., Di Falco, S., Zucaro, R. & Zilberman, D. 2015 Italian agriculture in the context of climate change: the role of irrigation for sustainable development of rural areas. *Rivista di Studi Sulla Sostenibilita* **2**, 131–152.

Cohon, J. L. 1978 *Multiobjective Programming and Planning*. New York Academic Press, New York, NY.

Dercas, N. 2020 Agricultural Water Management in Greece. In: *Water Resources Management in Balkan Countries* (A. M. Negm, G. Romanescu & M. Zelenakova, eds). Springer, Cham, Switzerland, pp. 421–453.

Dudu, H. & Chumi, S. 2008 Economics of irrigation water management: a literature survey with focus on partial and general equilibrium models. *World Bank Policy Research Working Paper* (4556).

ELSTAT (Hellenic Statistical Authority) 2011 *Results of the Population Census 2011*. Retrieved on 15/1/2021 from: www.statistics.gr.

Eurostat 2020 GDP per capita in EU regions. News release. Retrieved on 24/1/2021 from https://ec.europa.eu/eurostat/documents/2995521/10474907/1-05032020-AP-EN.pdf/81807e19-e4c8-2e53-c98a-933f5bf30f58.

European Commission 2020 https://ec.europa.eu/info/food-farming-fisheries/key-policies/common-agricultural-policy/future-cap_en.

Evangelidis, C. 2017a *Investigation of the Replacement of Surface Irrigation Network in Limnochori: Agrotechnical Report (Phase 2)*. Thessaloniki, Greece (in Greek).

Evangelidis, C. 2017b *Investigation of the Replacement of Surface Irrigation Network in Limnochori: Cost-Benefit Analysis (Phase 4)*. Thessaloniki, Greece (in Greek).

Food and Agriculture Organization (FAO) 2014 *SAFA Sustainability Assessment of of Food And and Agriculture Systems Guidelines*. FAO, Rome, Italy. ISBN 978-92-5-108485-4.

Galan-Martin, A., Vaskan, P., Anton, A., Esteller, L. J. & Guillen-Gosalbez, G. 2017 Multi-objective optimization of rainfed and irrigated agricultural areas considering production and environmental criteria: a case study of wheat production in Spain. *Journal of Cleaner Production* **140**, 816–830.

Gastineau, P. 2006 A water users association in Madagascar: Why does it fail. In: *Building the European Commons: From Open Fields to Open Source-European Regional Meeting of the International Association for the Study of Common Property*, Brescia, Italy.

Georgiou, P. E. & Papamichail, D. M. 2008 Optimization model of an irrigation reservoir for water allocation and crop planning under various weather conditions. *Irrigation Science* **26** (6), 487–504.

Groot, J. C., Yalew, S. G. & Rossing, W. A. 2018 Exploring ecosystem services trade-offs in agricultural landscapes with a multi-objective programming approach. *Landscape and Urban Planning* **172**, 29–36.

Hanley, N. & Spash, C. L. 1993 *Cost – Benefit Analysis and the Environment*. Edward Elgar Publishing, Cheltenham, Gloucestershire, UK.

Hoehn, J. P. & Randall, A. 1987 A satisfactory benefit cost indicator from contingent valuation. *Journal of Environmental Economics and Management* **14** (3), 226–247.

Jägermeyr, J., Pastor, A., Biemans, H. & Gerten, D. 2017 Reconciling irrigated food production with environmental flows for Sustainable Development Goals implementation. *Nature Communications* **8** (1), 1–9.

Karlopoulos, E. & Sidiropoulos, A. 2020 *Regional Strategy for the Transition of Western Macedonia to Zero Lignite Dependence*. A Practical Guide to the smooth transition of regions with high dependence on solid fossil fuels to a new production model – The case of Western Macedonia. Retrieved on 24/1/2021 from: https://ekpaa.ypeka.gr/wp-content/uploads/2020/07/D4.1_%CF%83%CF%84%CF%81%CE%B1%CF%84%CE%B7%CE%B3%CE%B9%CE%BA%CE%AE_%CE%BC%CE%B5%CF%84%CE%B1%CE%B2%CE%B1%CF%83%CE%B7%CF%82.pdf (in Greek).

Kleemeier, E. & Lockwood, H. 2012 *Public-Private Partnerships for Rural Water Services*. IRC, The Hague, The Netherlands.

Kourgialas, N. N., Koubouris, G. C. & Dokou, Z. 2019 Optimal irrigation planning for addressing current or future water scarcity in Mediterranean tree crops. *Science of the Total Environment* **654**, 616–632.

Lamaddalena, N. & Khadra, R. 2012 *Regional Assessment – Water Users' Associations in the SWIM-SM Project Countries.* Sustainable Water Integrated Management (SWIM) - Support Mechanism, WUAs Expert Regional Workshop (23–24 April, 2012, Athens, Greece). Retrieved online from http://www.swim-sm.eu/files/ASSESSMENT_WUAs_FINAL.pdf.

Latinopoulos, D. 2005 Derivation of irrigation water demand functions through linear and non-linear optimisation models: application to an intensively irrigated area in northern Greece. *Water Science and Technology: Water Supply* **5** (6), 75–84.

Latinopoulos, D. 2008 Estimating the potential impacts of irrigation water pricing using multicriteria decision making modelling. An application to Northern Greece. *Water Resources Management* **22** (12), 1761–1782.

Latinopoulos, D. & Mylopoulos, Y. 2005 Optimal allocation of land and water resources in irrigated agriculture by means of goal programming: application in Loudias river basin. *Global Nest. The International Journal* **7** (3), 264–273.

Li, M., Fu, Q., Singh, V. P. & Liu, D. 2018 An interval multi-objective programming model for irrigation water allocation under uncertainty. *Agricultural Water Management* **196**, 24–36.

Manos, B., Bournaris, T., Kamruzzaman, M., Begum, M., Anjuman, A. & Papathanasiou, J. 2006 Regional impact of irrigation water pricing in Greece under alternative scenarios of European policy: a multicriteria analysis. *Regional Studies* **40** (9), 1055–1068.

Michailidis, A., Mattas, K., Tzouramani, I. & Karamouzis, D. 2009 A socioeconomic valuation of an irrigation system project based on real option analysis approach. *Water Resources Management* **23** (10), 1989–2001.

Mosleh, Z., Salehi, M. H., Fasakhodi, A. A., Jafari, A., Mehnatkesh, A. & Borujeni, I. E. 2017 Sustainable allocation of agricultural lands and water resources using suitability analysis and mathematical multi-objective programming. *Geoderma* **303**, 52–59.

Niu, G., Li, Y. P., Huang, G. H., Liu, J. & Fan, Y. R. 2016 Crop planning and water resource allocation for sustainable development of an irrigation region in China under multiple uncertainties. *Agricultural Water Management* **166**, 53–69.

Panagopoulos, Y., Makropoulos, C., Gkiokas, A., Kossida, M., Evangelou, L., Lourmas, G., Michas, S., Tsadilas, C., Papageorgiou, S., Perleros, V. & Drakopoulou, S. 2014 Assessing the cost-effectiveness of irrigation water management practices in water stressed agricultural catchments: the case of Pinios. *Agricultural Water Management* **139**, 31–42.

Pretty, J. N., Brett, C., Gee, D., Hine, R. E., Mason, C. F., Morison, J. I. L., Raven, H., Rayment, M. D. & van der Bijl, G. 2000 An assessment of the total external costs of UK agriculture. *Agricultural Systems* **65** (2), 113–136.

Psychoudakis, A., Papoutsi-Psychoudaki, S. & McFarquhar, A. M. M. 1995 An economic assessment of an irrigation project affecting a Greek wetland. *Wetlands Ecology and Management* **3** (4), 225–232.

Ragkos, A. & Psychoudakis, A. 2009 Minimizing adverse environmental effects of agriculture: a multi-objective programming approach. *Operational Research* **9** (3), 267–280.

Ragkos, A. & Theodoridis, A. 2016 Valuation of environmental and social functions of the multifunctional Cypriot agriculture. *Land Use Policy* **54**, 593–601.

Region of Western Macedonia 2009 *Rural Development Indicators*. Region of Western Macedonia, Kozani, Greece.

Romero, C. & Rehman, T. 1989 *Multiple Criteria Analysis for Agricultural Decisions*. Elsevier, Amsterdam, The Netherlands.

Tillery, J. A. & Jones, V. P. E. 2007 Modeling and cost-benefit analysis of restoration alternatives for the lakefield watershed. In: *Proceedings of the 2007 Georgia Water Resources Conference*, March 27–29, 2007. University of Georgia, Athens, GA.

Tsakiris, G., Safiolea, E. & Tsakiris, V. 2017 On the economic appraisal of water utility projects in the framework of cohesion policy 2014–2020. *Water Utility Journal* **15**, 81–90.

United Nations General Assembly (UNGA) 2015 *Transforming Our World: The 2030 Agenda for Sustainable Development*. UNGA, New York, NY, USA.

Warner, M., Kahan, D. & Lehel, S. 2008 *Market-Oriented Agricultural Infrastructure: Appraisal of Public-Private Partnerships. Agricultural Management, Marketing and Finance*. Occasional paper 23. Food and Agriculture Organization, Rome.

Xevi, E. & Khan, S. 2005 A multi-objective optimisation approach to water management. *Journal of Environmental Management* **77** (4), 269–277.

Xu, Z., Yao, L., Zhou, X., Moudi, M. & Zhang, L. 2019 Optimal irrigation for sustainable development considering water rights transaction: a Stackelberg-Nash-Cournot equilibrium model. *Journal of Hydrology* **575**, 628–637.

Zeleny, M. 2012 *Linear Multiobjective Programming (Vol. 95)*. Springer Science & Business Media, New York, NY.

Zhang, X. & Tariq, S. 2020 Failure mechanisms in international water PPP projects: a public sector perspective. *Journal of Construction Engineering and Management* **146** (6), 04020055.

First received 25 January 2021; accepted in revised form 30 March 2021. Available online 8 April 2021

Enhancing urban and sub-urban riparian areas through ecosystem services and ecotourism activities

G. Gkiatas, I. Kasapidis, P. Koutalakis, V. Iakovoglou, A. Savvopoulou, I. Germantzidis and G. N. Zaimes

ABSTRACT

Riparian ecosystems in urban environments are usually degraded. Their sustainable management can reduce water pollution, mitigate climate change while providing recreation opportunities. This study assessed the current condition of the streams/torrents and their riparian areas in the city of Drama in Greece. In addition, practices to improve their functionality and enhance community awareness were recommended. A spatial database of the wider urban setting including suburbs and rural areas (weather, land cover, soils, roads, protected areas, riparian areas, etc.) was developed within GIS. The datasets were then inputted in a hydrologic model to simulate the water balance and stream discharge in the main urban streams and torrents. Field measurements of stream water discharge, and vegetation and stream channel conditions were undertaken using two visual protocols. The results of the protocols indicate that most streams/torrents and their riparian areas are of low ecological quality. Nature-based solutions were recommended to improve their ecological quality. In addition, the implementation of Eco-Routes was recommended to enhance the awareness of their importance and to promote their sustainable management. Overall, the urban streams/torrents and riparian ecosystems of Drama are degraded and innovative management practices should be implemented to conserve them and feature their ecosystem services.

Key words | Eco-Routes, GIS, hydrologic models, riparian, SWAT, visual protocols

G. Gkiatas (corresponding author)
I. Kasapidis
P. Koutalakis
V. Iakovoglou
A. Savvopoulou
I. Germantzidis
G. N. Zaimes
UNESCO Chair Con-E-Ect, Department of Forest and Natural Environment Sciences,
International Hellenic University,
1st km Drama-Mikrohoriou, 66100 Drama, Greece
E-mail: *george.giatas@hotmail.com*

HIGHLIGHTS

- The urban streams/torrents and riparian areas of Drama City are degraded.
- Spatial databases should be utilized by the authorities for urban planning.
- Hydrologic models are essential for the sustainable management of urban water resources.
- Nature-based solutions can restore the streams/torrent and their riparian areas.
- Eco-Routes can enhance the awareness on the importance of riparian areas.

doi: 10.2166/ws.2021.114

GRAPHICAL ABSTRACT

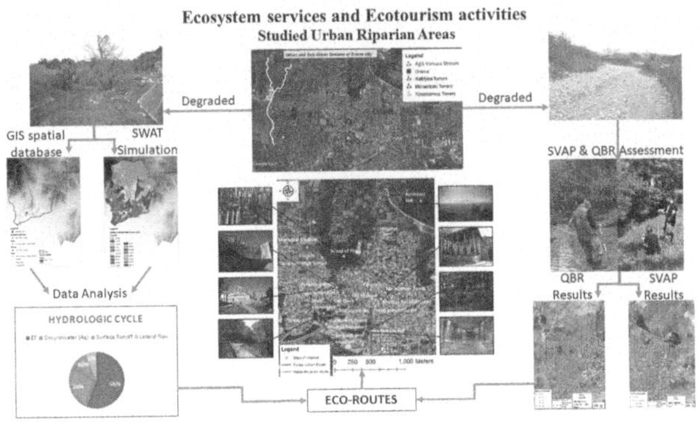

INTRODUCTION

The conservation and protection of riparian areas is a main priority for most regions of the world because of the many ecosystem services they offer (Zaimes & Iakovoglou 2021). In addition, riparian areas have been heavily utilized and are considered as some of the most degraded ecosystems. Climate change is further compounding the pressures on riparian areas (Zaimes 2020). This is particularly true for the Mediterranean region whose inhabitants have been dependent on and have utilized riparian areas for thousands of years. The anthropogenic activities that negatively impact the riparian areas include river transportation (e.g. boats), energy generation, urbanization and agriculture that alter the channel's morphology and hydrologic regime and/or eliminate, shrink, degrade or reduce of the riparian vegetation diversity (Balzan *et al.* 2020).

Urban sprawl is expected to continue since it has been projected that by 2050, close to 70% of the world's population will be living in urban environments (Monk *et al.* 2019). Many urban areas are along streams, rivers and deltas highlighting that the anthropogenic pressure of urbanization on the riparian areas will further increase. Urbanization can have the most aggressive anthropogenic alterations on the structure and functionality of riparian areas (Iakovoglou *et al.* 2013). The hydrologic regimes are significantly altered in urban watersheds. The pathways that rainwater follows (e.g., evapotranspiration, surface runoff, infiltration, lateral flow) are altered due to the extensive impervious surfaces, elimination of natural vegetation and the urban water infrastructures (Paul & Meyer 2001). Overall, rainwater moves substantially quicker out of the urban watershed because urban drainage systems convey rainwater efficiently and rapidly which can create flash flood events and afterwards 'hydrologic drought' of the riparian areas by lowering the water table. The alteration of the hydrologic regime impacts the soil, vegetation, and their ability to remove pollutants. Riparian areas are important corridors and the degradation of their urban segments can lead to the fragmentation of their continuum from the headwaters to the deltas and the disconnection of the stream/rivers to their floodplains (Hudson 2002).

In contrast the sustainable management of urban riparian areas can help revitalize ecologically and socioeconomically urban ecosystems (Wu *et al.* 2017). This has been recognized worldwide and is why most urban strategies include the maintenance or restoration of riparian areas. The ecosystem services that they can offer in urban settings include: improved stream water quality, protection from flood events, enhanced aesthetics, expanded recreational opportunities, and climate change mitigation. To provide these services, riparian restoration efforts typically include fluvial and geomorphic restoration, active introduction of native plants and re-establishment of the natural floodplain (González *et al.* 2015).

The reasons behind the many ecosystem services that riparian areas can offer are because they are ecological hotspots that support high biodiversity, disproportionately larger than their extent in the watershed (Monk *et al.* 2019). They are highly productive and biodiverse habitats because they have characteristics that are a mixture of both terrestrial and aquatic ecosystems. This is why riparian areas are called semi-aquatic, transition zones between the aquatic and terrestrial ecosystems. The greater presence of groundwater and the frequent flooding, leads to unique soils and geomorphologic formations that affect the vegetation that is primarily hydrophilic and differentiates them from the adjacent ecosystems (Zaimes & Iakovoglou 2021). As transition zones they have biophysical, ecological and biota gradients as you move from the aquatic to the terrestrial ecosystems leading to unique environmental conditions.

In the Mediterranean region, climate change is leading to increasing temperatures and evapotranspiration, while rainfalls are more intense but at the same time less frequent, especially in the summer (Cramer *et al.* 2018). Consequently, drought frequency and strength are expected to increase, while the hydrological cycle will intensify (more flash floods are expected). Overall, these alterations will reduce water availability in the region (Fader *et al.* 2020). Streams and rivers in the Mediterranean might shift from perennial to intermittent and from intermittent to ephemeral which could also alter vegetation communities. Overall, riparian areas of the region appear to be very sensitive to climate change impacts (Zaimes 2020), and can be indicators for natural and human systems vulnerability to climate change (Capon *et al.* 2013).

To sustain the healthy function and structure of riparian ecosystems in relation to their services, is an even greater challenge today due to climate change. Safeguarding these ecosystems through proper management (e.g., ecotourism) becomes a major priority (Iakovoglou *et al.* 2013). 'Ecotourism' is a fast-growing trend in the tourism industry that can provide increased economic growth while also sustaining ecosystems (Ghermandi *et al.* 2020). The increased diversity and services of the riparian ecosystems in and adjacent to residential areas provide unique ecotourism opportunities that can potentially increase the welfare of the urban area. The sustainable implementation of ecotourism in urban areas can help to educate citizens on the importance of riparian areas and their benefits. This should promote their enhanced protection and conservation, improve the quality of life of the urban dwellers and help the local economy through an increase in the number of visitors.

Urban riparian areas in Greece have not been studied extensively, especially in regard to maintaining and enhancing sustainable cities and helping reduce climate change impacts (Iakovoglou *et al.* 2013). The aim of this study was to assess the characteristics of the urban streams and riparian areas and based on these, recommend management practices to improve their utility. Firstly, the stream flows for reaches that run through the city of Drama were measured and afterwards utilized in a hydrologic model. The urban riparian areas were assessed via GIS mapping and the implementation of two visual protocols. These data led to suggestions on how to enhance the functionality and ecosystem services of the urban riparian areas. Finally, ecotourism paths were suggested in order to enhance the city's ecotourism opportunities and improve the quality of life of its residents.

METHODS

Study area

The city of Drama, which is the capital of Regional Unit of Drama, is located in the region of Eastern Macedonia and Thrace, in Northern Greece (see Figure 1(a)). It was built on the foothills of the Falakro Mountain in a verdant area with abundant water sources. The city was originally named Ydrama or Dyrama by the ancient Greeks because of the abundant water sources. The city of Drama has an area of 59.37 km^2, located at an altitude of 118 m above sea level and with a population of approximately 44,800 people (Population Census of 2011). The climate in Drama is warm and temperate. The Köppen-Geiger climate classification is Csa (typical Mediterranean climate). The average temperature in Drama is 14.2 °C and the yearly precipitation 520 mm. As Drama is an inland city, this has led to the promotion of alternative tourism and not the usual Greek tourism activities of coastal areas. This is the reason why during the Christmas period, the city hosts the Dream Town event (that is a Christmas market and a fun park) that attracts many tourists. Moreover, every year there are additional events such as Drama International Short Film Festival, Wine Festival 'Draminognosia', Dramaica Youth Festival. The main sources of revenue are agriculture

(a)

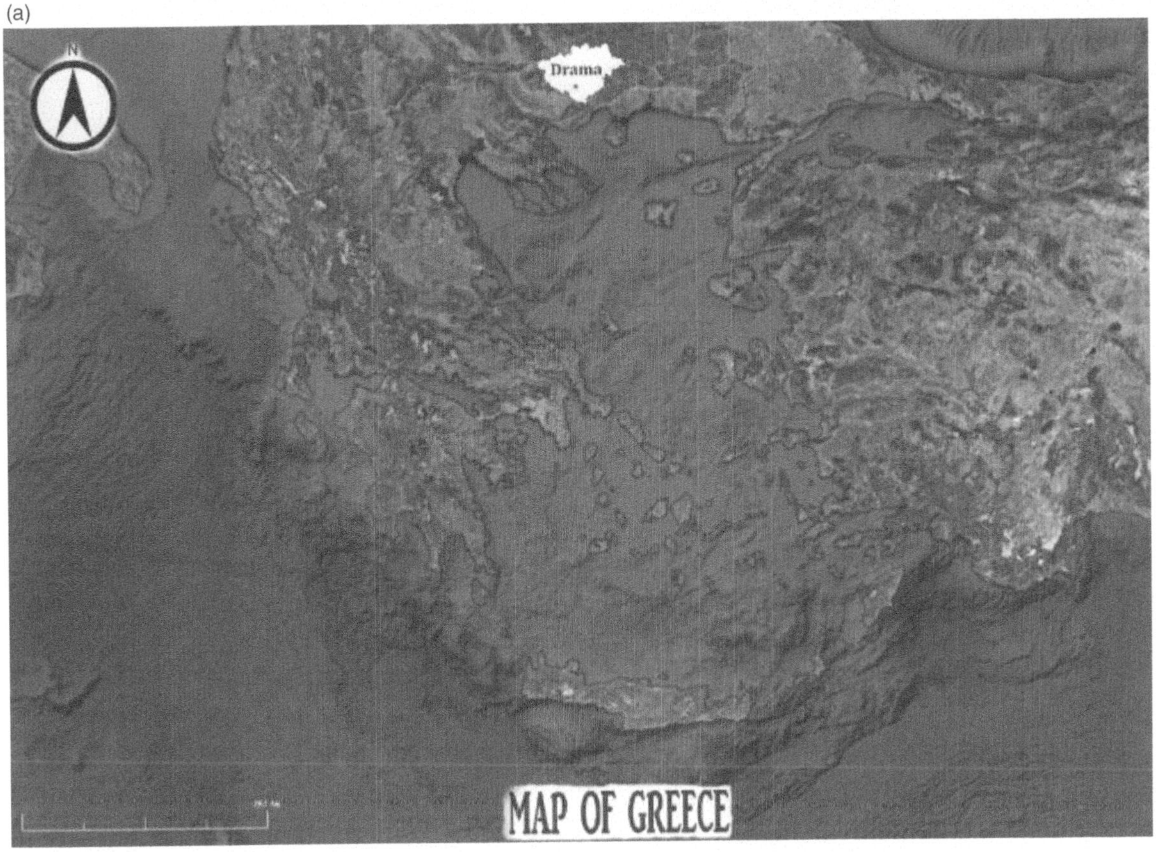

(b)

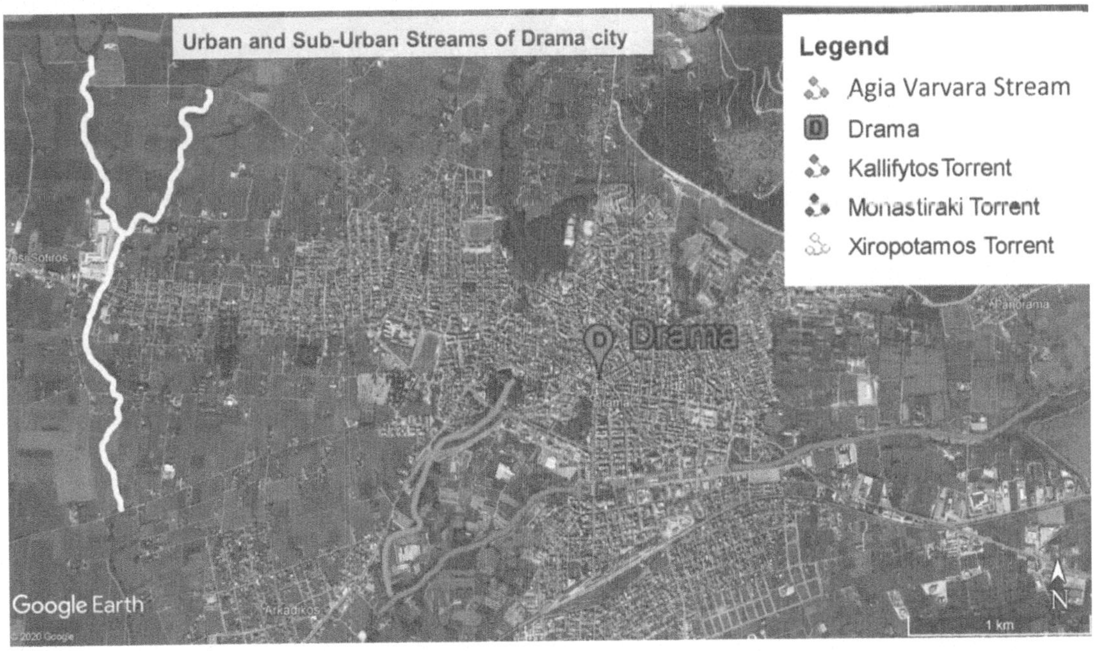

Figure 1 | (a) The location of Regional Unit of Drama (in white) in Greece. (b) The four studied urban stream and torrent reaches of Agia Varvara (in green), Monastiraki (in blue), Kallifytos (in red) and Xiropotamos (in yellow). Please refer to the online version of this paper to see this figure in colour: http://dx.doi.org/10.2166/ws.2021.114.

(crops and livestock), marble mining and winemaking. Recently there have been efforts to exploit the unique natural ecosystems that surround the city for ecotourism opportunities. Such an example is the environmentally friendly development of the recreational area of Agia Varvara Springs. In this area the natural riparian vegetation has been maintained in many reaches while the city has constructed eco-friendly paths that make this a 'hot spot' for leisure and recreational activities, concerts and festivals especially during spring and summer. In addition, restaurants, coffee shops and hotels have also been developed that attract many local residents and visitors. In order to further promote tourism and also enhance the quality of life that could lead to an increase in the number of residents, the city plans to further utilize other urban riparian areas with the sustainable exploitation of their ecosystem services. In this study four different (ephemeral to perennial) reaches, typical of the Mediterranean, were studied: (a) Agia Varvara Springs, (b) the Monastiraki torrent, (c) Xiropotamos torrent and (d) Kallifytos torrent, which run through the city (see Figure 1(b)).

Visual protocols

The two visual protocols were utilized: (a) the Ecological Quality of Riparian Habitat (QBR) protocol (Munné *et al.* 2003) that focus on the riparian vegetation and (b) the Stream Visual Assessment Protocol (SVAP) (Bjorkland *et al.* 2001) that focus on the stream/torrent channel condition.

The QBR protocol was developed in Spain for Mediterranean riparian areas. Its assessment consists of four different sections: (a) total vegetation cover, (b) vegetation cover structure, (c) cover quality and (d) river channel alterations (Munné *et al.* 2003). Each section can have a score from 0 to 25 points. The sum of these four different scores provides the final score that can range from 0 to 100. The categories of the protocol based on the final score are: (a) riparian habitat in natural condition, greater or equal than 95, (b) low disturbance, good quality from 75 to 90, (c) disturbance important, fair quality from 55 to 70, (d) strong alteration, poor quality from 30 to 50, and (e) extreme degradation, bad quality less or equal than 25.

The SVAP protocol was developed in the United States and was modified for the Greek and other similar Euro-Mediterranean riparian areas (Savopoulou *et al.* 2017). Fourteen characteristics are used in order to evaluate the field plots within a score that ranges from 0 to 10. The characteristics evaluated include: (a) channel condition, (b) hydrologic alteration, (c) riparian zone condition, (d) bank stability, (e) water existence, (f) water appearance, (g) presence of livestock shed, (h) instream fish cover, (i) pools, (j) insect/invertebrate habitat, (k) canopy cover, (l) presence of manure, (m) presence of biological wastewater treatment, and (n) presence of garbage. The final score of a plot is calculated as the mean of the above 14 characteristics. Based on the SVAP score there are the following categories: (a) poor condition with mean values less than 6, (b) moderate condition with mean values ranging from 6.1 to 7.4, (c) good condition with mean values ranging from 7.5 to 8.9 and (d) excellent condition with mean values greater than 9 (Koutalakis *et al.* 2020).

Field measurements were carried out in the reaches of all four studied stream/torrents. In each of them, the total length was measured and further sub-divided into intervals of approximately 150 m that were the locations of the plots. This is the reason why the total number of the plots was different in each reach. The total number of measured plots was 113 and distributed to 50 plots in the Kallifytos torrent, 9 plots in the Monastiraki torrent, 17 plots in the Agia Varvara stream and 37 in the Xiropotamos torrent.

The SWAT model

The Soil and Water Assessment Tool (SWAT) is a continuous-time, semi-distributed, process-based river basin model (Krysanova & Srinivasan 2015). It was developed by the United States Department of Agriculture and specifically the Agricultural Research Service to evaluate the impacts of alternative management decisions on water resources and nonpoint-source pollution in watersheds (Arnold *et al.* 2012). The SWAT model has been utilized in many different environments, while there is an extensive documentation and a variety of assisted pre- and post-processing software (Gassman *et al.* 2014). The SWAT uses a large number of input parameters but the necessary input data are: (a) the digital elevation map (DEM), (b) the land use map, (c) the soil type map, (d) the slope map and (e) weather data (Geza & McCray 2008). Furthermore, the model divides the watershed into hydrologic response units (HRUs) which are unique combinations of land use, slope and soil type, thus, each sub-watershed may contain multiple HRUs (Reddy & Reddy 2015). The ArcGIS interface (known as ArcSWAT) was used in the current study (Olivera *et al.* 2006). The first step was the watershed delineation by using the DEM focused on Drama City and its surroundings, an area of approximately 24,243 ha. This step created the reaches, the longest flow paths, the sub-watersheds, and topographic statistics such as the minimum and maximum elevation which were 52 m and 2,174 m, respectively (see Figure 2). The next step developed the slope map based on three categories (0–20%, 20–40% and >40%) that were derived from the topographic statistics (see Figure 2). The soil map of the area was based on the data acquired from the European Union World Soil Harmonised Database (see Figure 3). The dominant soil types of the area are: (a) Dystric Cambisol, (b) Calcaro-chromic Vertisol, (c) Calcaric Fluvisol, (d) Calcaric Leptosol and (e) Chromic Luvisol. The land use map (see Figure 3) was derived from the European Union CORINE land cover types. The land uses of the area included: (a) Urban Residential-High Density (URHD), (b) Urban Residential-Low Density (URLD), (c) Urban Residential-Medium Density (URMD), (d) Urban Industrial (UIDU), (e) Urban Commercial (UCOM), (f) Agricultural Land-Generic (AGRL), (g) Agricultural Land-Row Crops (AGRR), (h) Pasture (PAST), (i) Cropland/Grassland MOSAIC (CRGR), (j) Agricultural Land-Close-grown (AGRC), (k) Forest-Mixed (FRST), (l) Forest-Deciduous (FRSD), (m) Range-Brush (RNGB), (n) Forest-Evergreen (FRSE), (o) Grassland (GRAS), (p) Shrubland (SHRB) and (q) Barren (BARR) (r) Barren or sparsely vegetation (BSVG). The combination of the above input data created 259 HRUs for the six sub-watersheds and the report of the HRU coverage in each sub-watershed.

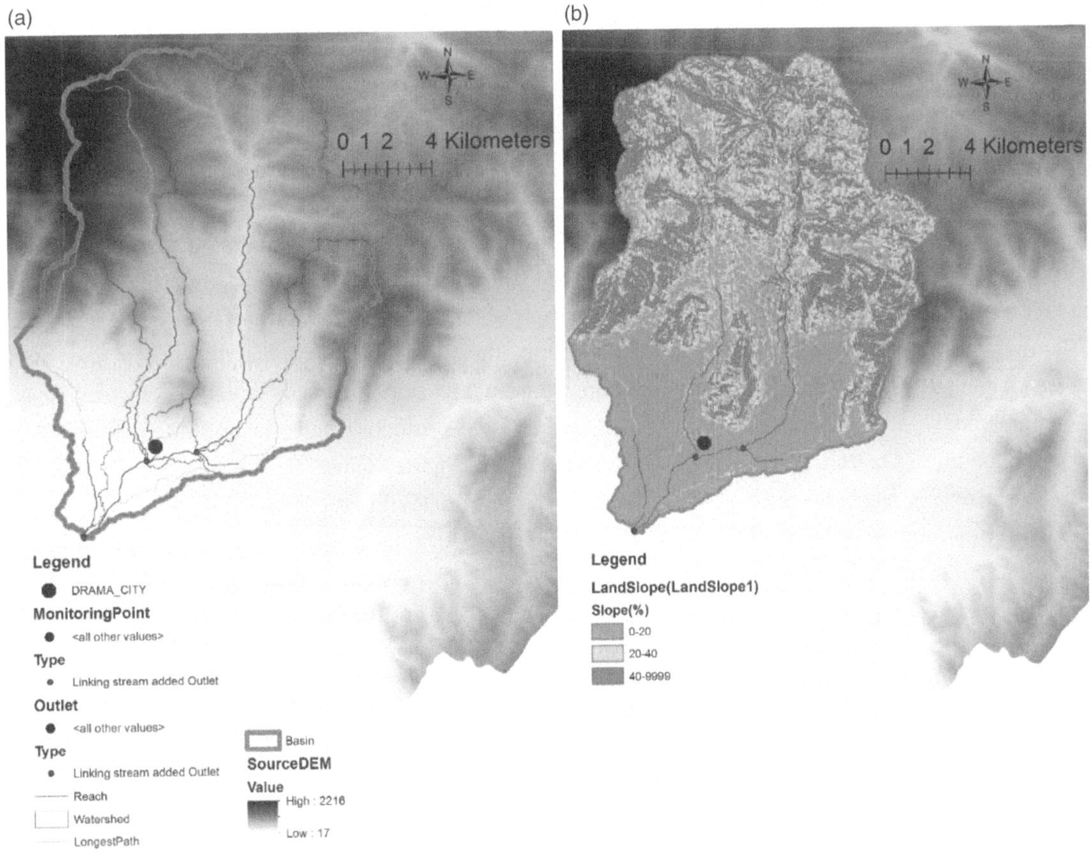

Figure 2 | (a) The hydrographic network (reaches) along with the longest path, the basin, the sub-watershed and the DEM based on the SWAT model. (b) The slope categories of the study area based on the SWAT model.

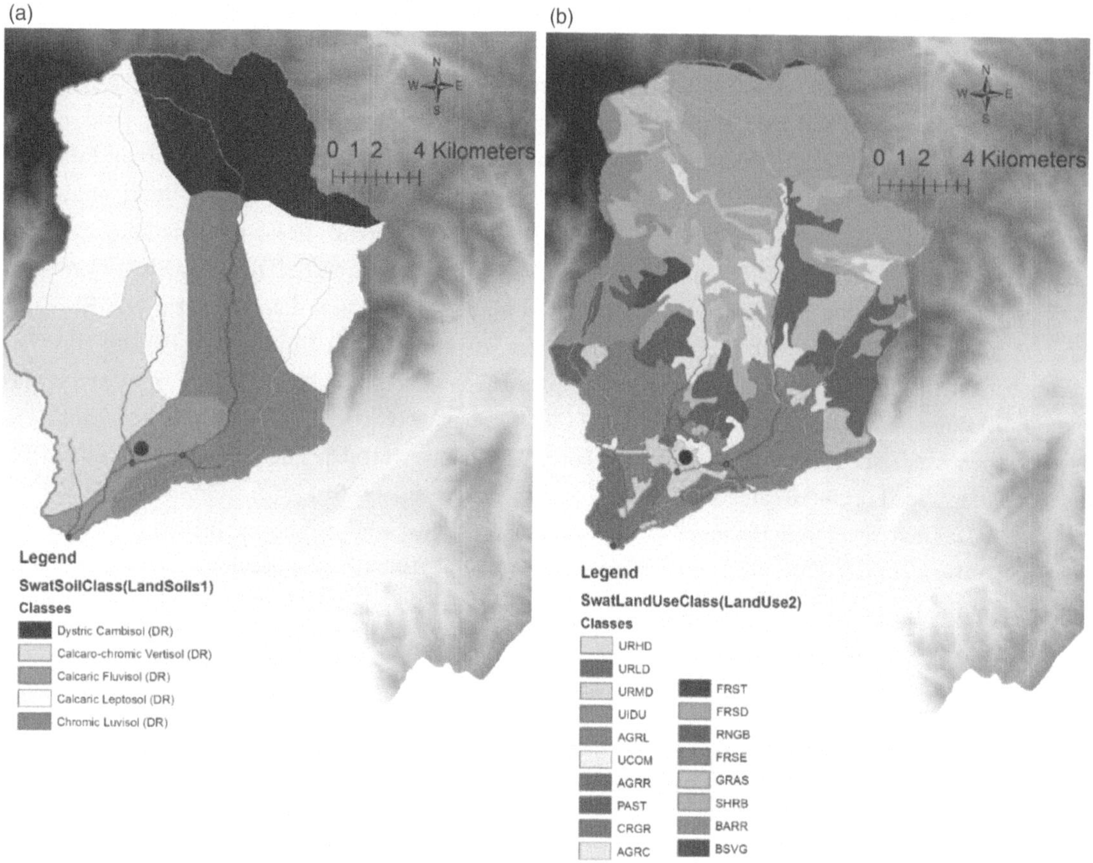

Figure 3 | The soil types (a) and the land uses (b) of the study area based on the SWAT model.

Finally, the daily weather data (precipitation, low and high temperature, relative humidity, solar radiation and wind velocity) utilized in the model were obtained from the weather station located in the Campus of the International Hellenic University in Drama. The simulated results were produced by using default initial parameters with no calibration data because of the short period of time. On the other hand, field measurements via a current meter allowed us to measure the water quantity in order to determine the hydrologic conditions of the studied streams/torrents and to compare them to the results generated by the SWAT model.

GIS data

The project included the creation of a spatial database focused on the urban area of Drama based on the free software QGIS; a powerful tool for spatial analysis (Pujari & Bhosale 2017). The GIS data included the digitized maps of the hydrographic network, the catchment areas, the land cover, the soil map, the road network, the protected areas, the riparian areas, the weather stations, etc. (see Figure 4). In addition, the digitization included the urban and suburban riparian areas in the study area (see Figure 5) as well as the visual protocol results for both QBR and SVAP assessments (see Figures 6 and 7). The base maps were: (a) the OpenStreetMap (Wiki Free Word Map) and (b) the KTIMATOLOGIO SA map (Hellenic Mapping and Cadastral Organization). The spatial database was crucial in order to develop the possible routes for ecotourism activities and to prepare the best management plans as future activities.

RESULTS AND DISCUSSION

The QBR protocol showed that the environmental conditions of the riparian vegetation adjacent to the torrent

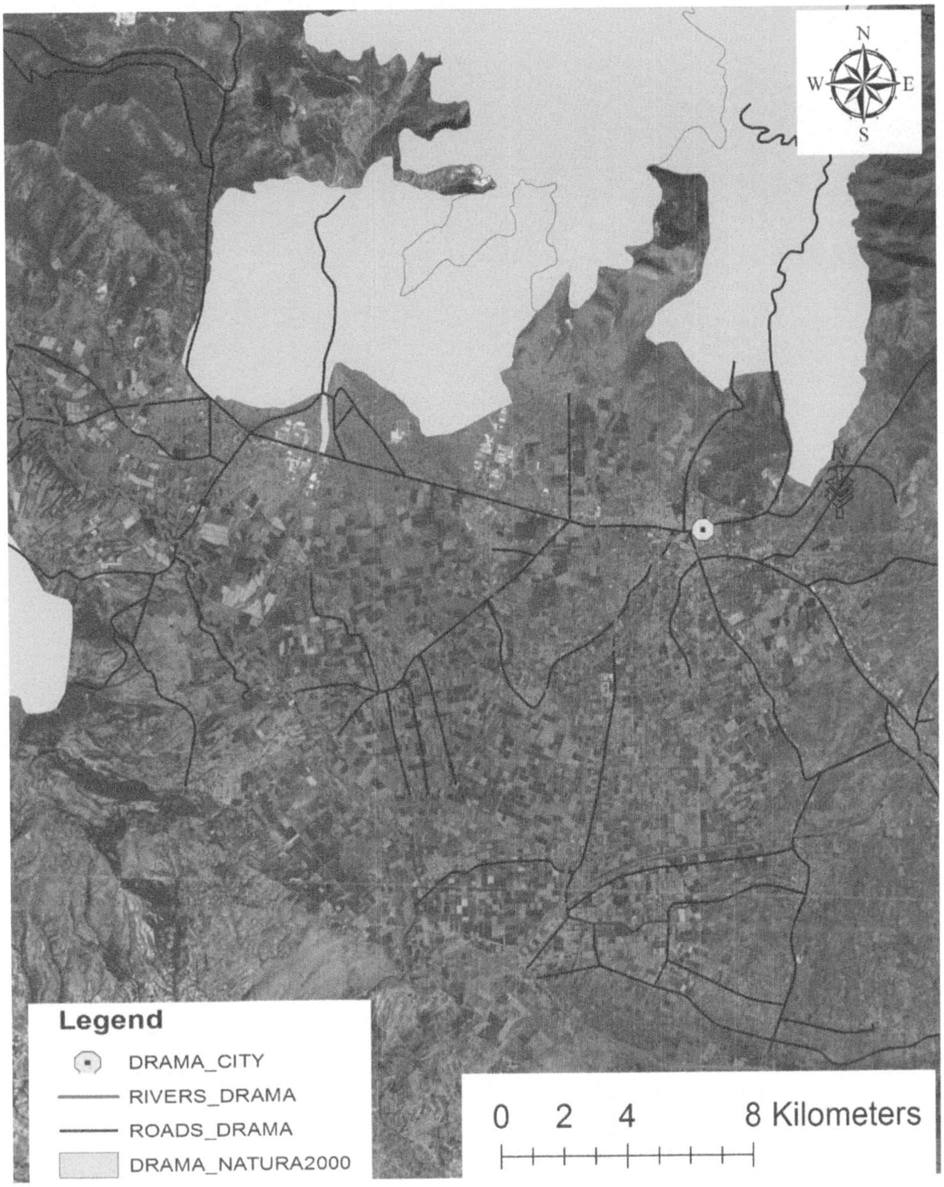

Figure 4 | An example of the digitized spatial database for the area surrounding Drama City.

and stream urban and sub-urban reaches are degraded (see Figure 6). Specifically, no plots of natural conditions were recorded along the Kallifytos torrent and only 4 plots were of good quality. The vegetation was of fair quality in 16 of the plots while 18 plots were of poor quality and 12 plots of bad quality. In the Monastiraki torrent, 4 plots were characterized as natural, 3 plots were of good quality, 1 of fair and 1 of poor quality. In the Agia Varvara stream 7 plots were of good quality, 3 plots were fair, 6 plots were poor and only 1 was of bad quality. Finally, along the Xiropotamos torrent, 3 plots were in natural conditions, 18 were plots of good quality, 11 plots were of fair quality, while 5 plots were of poor quality.

The SVAP protocol assessment of all reaches resulted in no plots in excellent condition (see Figure 7). The Kallifytos torrent had 10 plots in poor conditions, while 30 of the plots were moderate and 10 were good. Along the Monastiraki torrent, 2 plots were poor, 1 was moderate and 6 (the

Figure 5 | The digitized boundaries of the riparian areas in the urban and sub-urban areas in Drama City.

majority) were in good condition. For the Agia Varvara stream only 1 was in good condition, 6 plots were poor and 10 plots were in moderate conditions. Finally, the Xiropotamos torrent had the worst results based on the SVAP, with 29 plots in poor conditions and only 8 plots in moderate condition. These QBR and SVAP results show that the ecological conditions of the urban and sub-urban reaches of the torrents/streams in the area of Drama are highly degraded.

The hydrologic model SWAT was utilized to complement the results of the ecological quality assessment based on the two visual protocols. The model was developed for the greater area of Drama that included both the urban and sub-urban streams and torrents around the city. The simulated annual average hydrologic budget based on SWAT model (see Figure 8) estimated that the 418 mm of yearly precipitation was distributed as following: 10% surface runoff, 6% lateral flow, 28% groundwater that correspond to the total water yield while the remaining 56% was returned to the atmosphere as evapotranspiration (ET).

The simulated results of discharge from the entire watershed of the study area, for the period 2011–2019, based on the SWAT model are depicted in Figure 9. Furthermore, Figure 10 shows the field measurements of Agia Varvara stream in the city of Drama. Field measurements were done in two different cross-sections. The endpoint of the watershed for the SWAT model was located at the village of Kalos Agros (see Figure 11). The simulated discharge resulted in an average flow of 1.06 m^3/s. As this point is a confluence of many smaller streams and torrents, this value was expected to be higher than the field measurements taken in Agia Varvara which had average values of 0.55 m^3/s and 0.49 m^3/s at the two measured cross-sections. The simulated results also highlighted great peaks of runoff after heavy rainfall in the surrounding mountains. These peaks were recorded as high-water events in ephemeral torrents, and should be managed to mitigate soil erosion, tree debris, floods and even failures in infrastructures (e.g., bridges, basements/cellars etc.) that have been recorded in the past. Overall, in these torrents and streams discharge

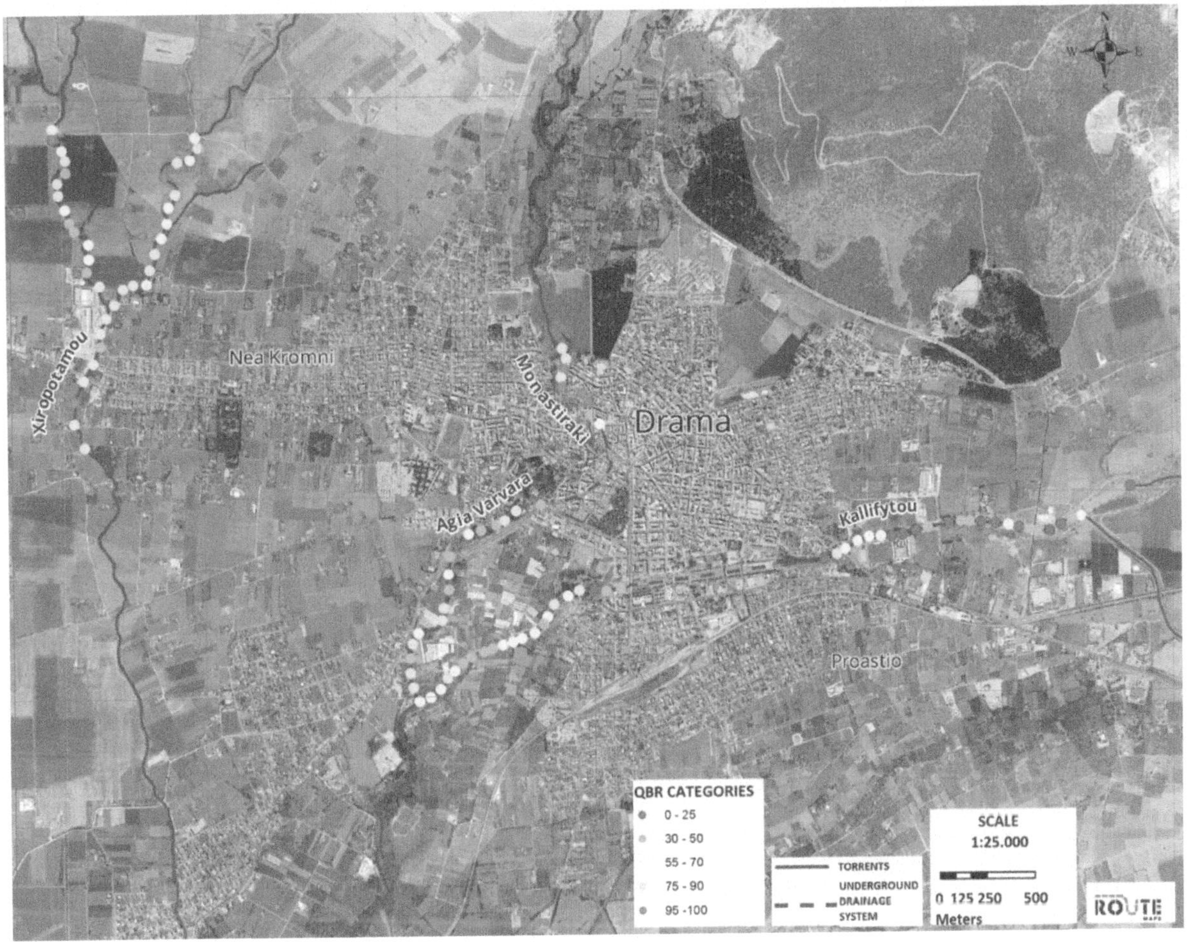

Figure 6 | The characterization of each plot based on the five QBR categories (represented with different colours) for the four reaches of the study torrents/streams in Drama City.

appears typically as flash floods and can cause serious damage to infrastructure and also the loss of human lives. Flash floods are the result of land-use changes in the watershed that enhance surface runoff and decrease infiltration such as the increase of impervious surfaces in urban setting. Ecosystem-based approaches such as maintaining and re-establishing healthy riparian areas in the urban areas and also at the watershed scale can help mitigate flash floods. One of the main services of riparian areas is acting as sinks to store rain and flood waters. Measures to improve the riparian areas in the urban and sub-urban areas are imperative based on the results of the two protocols QBR and SVAP. The nature-based solutions can include planting of native riparian species and reshaping of the stream banks. Of course, additional measures might be required such as detention and/or retention ponds, installation of an early warning systems and signs that inform the public areas that they should not cross streams and torrents during storms to avoid human injuries or even loss of life.

Another important measure that can improve the sustainable management of the urban and sub-urban riparian areas is awareness. These awareness activities should focus on presenting to the public the many ecosystems services that riparian areas can offer to humans. This was done by developing ecotourism paths along the riparian and forested areas in the urban area to be constructed in the future. Specifically, two different routes were suggested the: (a) Water-Riparian and (b) Forest-Urban (see Figure 12). The two visual protocols, the SWAT model, the GIS data and potential sites of interest were utilized in order to determine the exact routes. The protocol results identified riparian

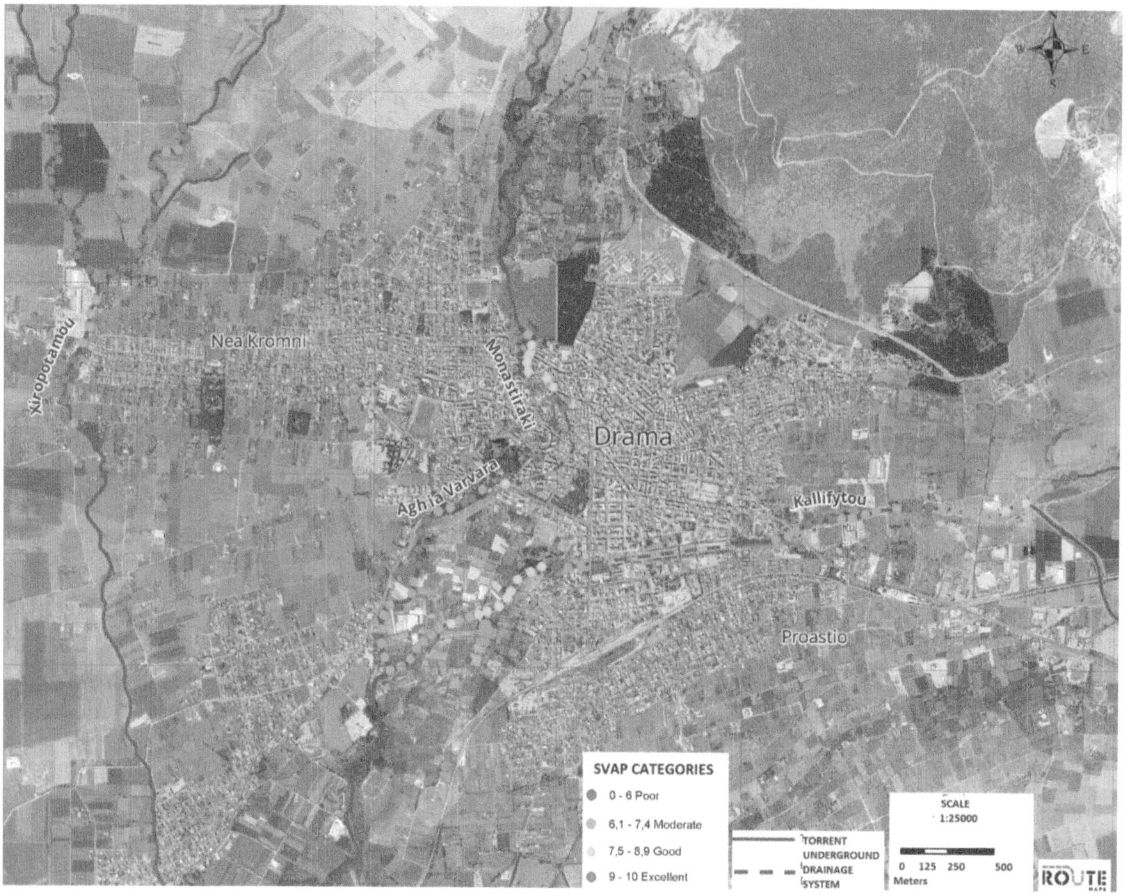

Figure 7 | The categorization of each plot based on the four SVAP categories (represented with different colours) for the four reaches of the study torrents/streams in Drama City.

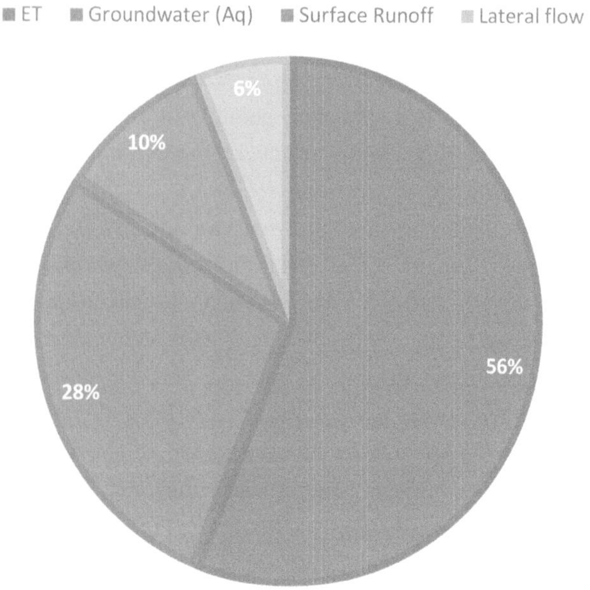

Figure 8 | The simulated hydrologic budget of the study area in Drama City based on SWAT model.

areas that in good condition. This would showcase healthy riparian areas and what the potential services could be if all the riparian areas throughout the city were in this condition. The SWAT data provided information on when and where flash floods could happen in the city. This allowed ecotourism routes (Eco-Routes) that are safe from such events to be selected and/or to showcase areas that frequently get flooded. To further enhance awareness five signs and a kiosk will be placed along the routes that are nature-based and environmentally friendly. These signs and the kiosk will provide information on the importance of the ecosystem services that riparian areas offer with real life examples. These signs will also highlight how the microclimatic condition due to the existence of riparian vegetation can alleviate some of the extreme temperatures in cities due to climate change. Finally, in addition to being informative and educational these two routes will

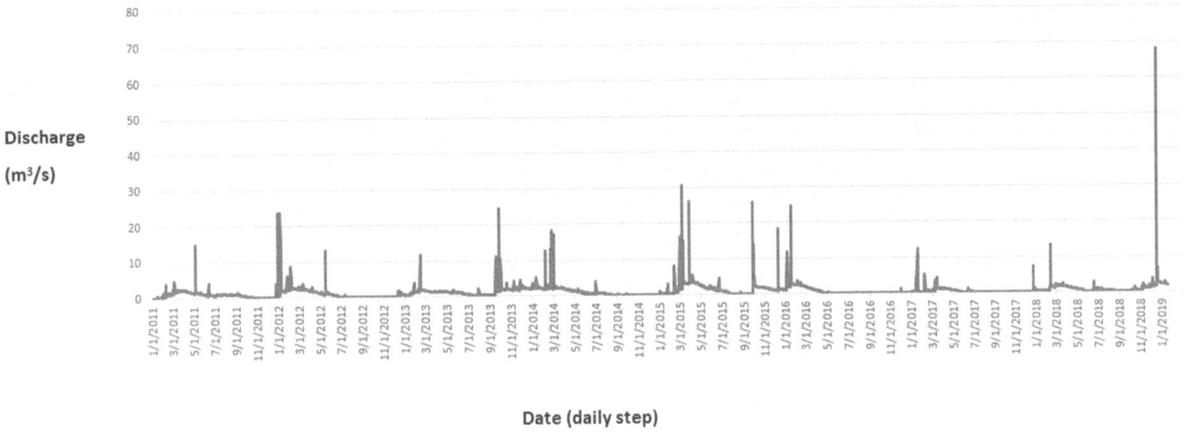

Figure 9 | The simulated discharge of all urban torrents/streams of the study area in Drama City based on SWAT.

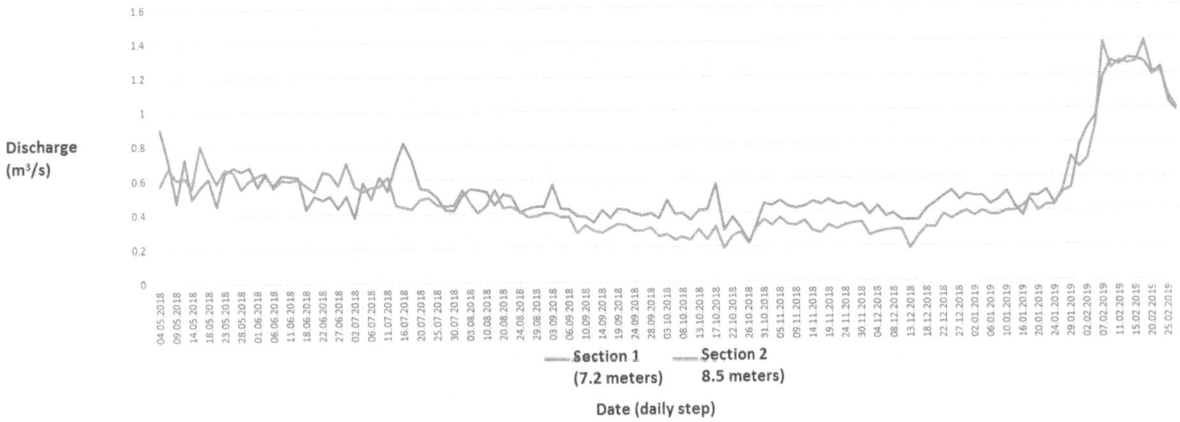

Figure 10 | The field measurements of the water discharge for the two measured cross-sections in the Agia Varvara stream in Drama City.

provide the citizens of the town and also visitors leisure and relaxation opportunities.

CONCLUSIONS

Riparian areas are unique ecosystems that if properly managed can be great assets for urban areas. In most cases riparian areas in urban settings are degraded. Visual protocols, stream discharge measurements, the development of a geodatabase and the use of SWAT can provide an assessment of the waterways and adjacent riparian areas ecological and hydrologic conditions in an urban setting. Most of the riparian areas along the urban and sub-urban reaches in Drama City based on the results of the assessment using the two visual protocols indicated degraded environmental conditions. Ecosystem-based approaches and nature-based solutions are recommended to be adopted to improve the riparian areas, which would greatly benefit the city of Drama. This could be based on the practices implemented in a small reach of Agia Varvara stream. Such activities could include replanting native riparian vegetation and alleviating negative anthropogenic activities. Measures for flood mitigation should be implemented, based on the SWAT results, the discharge measurements and past flash flood experiences in the city. Riparian areas if managed properly, along with other engineering works (e.g. blue-grey-green infrastructure) could help mitigate flash floods. Finally, understanding the important services riparian areas can offer in urban ecosystems is a key element of their sustainable management. This is why the establishment of Eco-Routes, which have been suggested to the Municipality of Drama,

Figure 11 | The results of the SWAT model. The stream network, the sub-watersheds and the entire watershed draining through the city of Drama.

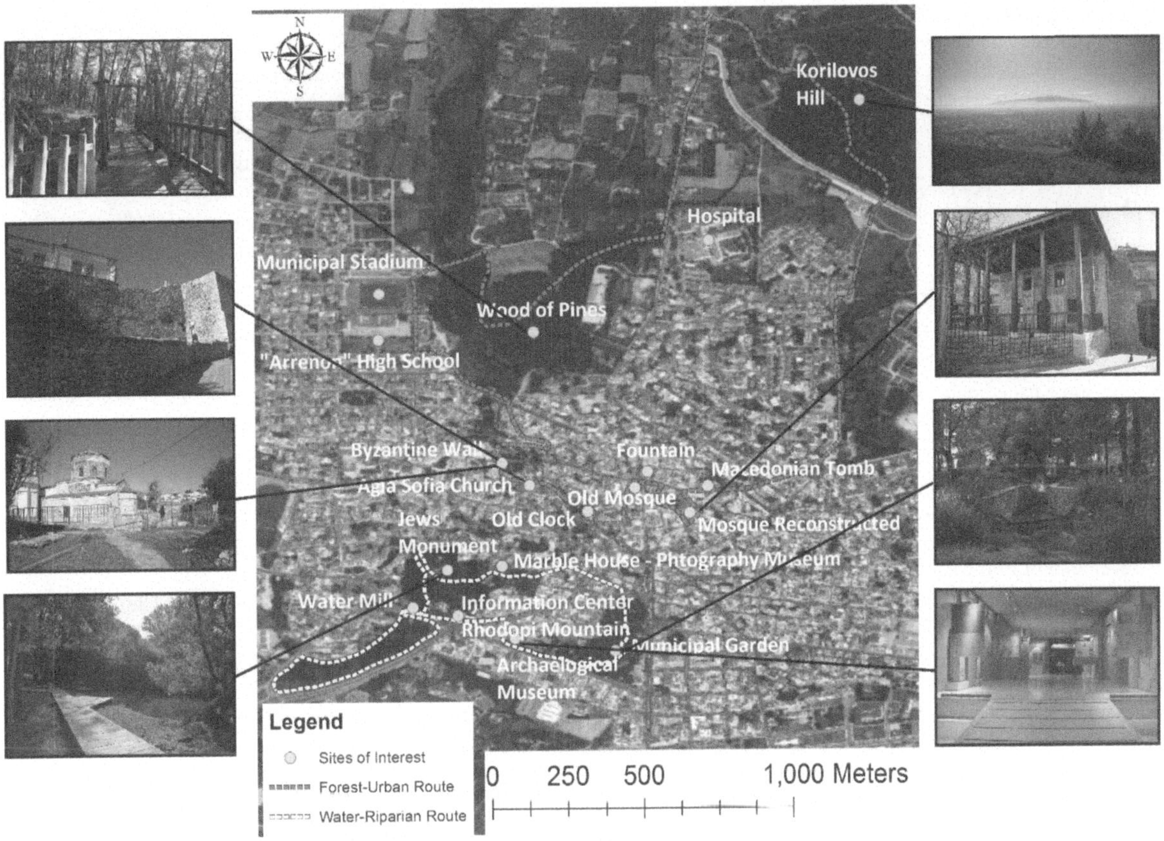

Figure 12 | The suggested two routes (Forest-Urban and Water-Riparian) for ecotourism activities.

are recommended that will be educational and informative. Visiting these routes will highlight the environmental benefits of restoring of riparian areas along with the cultural heritage of the city of Drama. Riparian areas are important for rejuvenating urban ecosystems and with their proper assessment that will lead to the implementation of effective practices (e.g., nature-based solutions, Eco-Routes) their sustainable management can be achieved in urban riparian areas of Greece and the Mediterranean.

ACKNOWLEDGEMENTS

This project was funded by the European Regional Development Fund in the framework of the implementation of the 'Sustainable Urban Development Strategy of the Municipality of Drama' of the R.O.P. A.M.T. 2014–2020. This publication reflects the views only of the authors, and the Commission cannot be held responsible for any use which may be made of the information contained therein.

DATA AVAILABILITY STATEMENT

Data cannot be made publicly available; readers should contact the corresponding author for details.

REFERENCES

Arnold, J. G., Moriasi, D. N., Gassman, P. W., Abbaspour, K. C., White, M. J., Srinivasan, R., Harmel, S. D., van Griensven, A., Van Liew, M. W., Kannan, N. & Jha, M. K. 2012 SWAT: Model use, calibration, and validation. *Transactions of the ASABE* **55** (4), 1491–1508.

Balzan, M. V., Hassoun, A. E. R., Aroua, N., Baldy, V., Bou Dagher, M., Branquinho, C., Dutay, J. C., El Bour, M., Médail, F., Mojtahid, M., Morán-Ordóñez, A., Roggero, P. P., Rossi Heras, S., Schatz, B., Vogiatzakis, I. N., Zaimes, G. N.

& Ziveri, P. 2020 Ecosystems. In: *Climate and Environmental Change in the Mediterranean Basin – Current Situation and Risks for the Future* (W. Cramer, J. Guiot & K. Marini, eds). First Mediterranean Assessment Report Union for the Mediterranean, Plan Bleu, UNEP/MAP, Marseille, France, p. 151, in press.

Bjorkland, R., Pringle, C. M. & Newton, B. 2001 A stream visual assessment protocol (SVAP) for riparian landowners. *Environmental Monitoring and Assessment* **68** (2), 99–125.

Capon, S. J., Chambers, L. E., Mac Nally, R., Naiman, R. J., Davies, P., Marshall, N., Pittock, J., Reid, M., Capon, T., Douglas, M., Catford, J., Baldwin, D. S., Stewardson, M., Roberts, J., Parsons, M. & Williams, S. E. 2013 Riparian ecosystems in the 21st century: hotspots for climate change adaptation? *Ecosystems* **16** (3), 359–381.

Cramer, W., Guiot, J., Fader, M., Garrabou, J., Gattuso, J. P., Iglesias, A., Lange, M. A., Lionello, P., Llasat, M. C., Paz, S., Peñuelas, J., Snoussi, M., Toreti, A., Tsimplis, M. N. & Xoplaki, E. 2018 Climate change and interconnected risks to sustainable development in the Mediterranean. *Nature Climate Change* **8** (11), 972–980.

Fader, M., Giupponi, C., Burak, S., Dakhlaoui, H., Koutroulis, A., Lange, M. A., Llasat, M. C., Pulido-Velazquez, D. & Sanz-Cobeña, A. 2020 Water. In: *Climate and Environmental Change in the Mediterranean Basin – Current Situation and Risks for the Future* (W. Cramer, J. Guiot & K. Marini, eds). First Mediterranean Assessment Report Union for the Mediterranean, Plan Bleu, UNEP/MAP, Marseille, France, p. 57, in press.

Gassman, P. W., Sadeghi, A. M. & Srinivasan, R. 2014 Applications of the SWAT model special section: overview and insights. *Journal of Environmental Quality* **43** (1), 1–8.

Geza, M. & McCray, J. E. 2008 Effects of soil data resolution on SWAT model stream flow and water quality predictions. *Journal of Environmental Management* **88** (3), 393–406.

Ghermandi, A., Camacho-Valdez, V. & Trejo-Espinosa, H. 2020 Identifying cultural services of natural and heritage sites for locals, domestic and international visitors to a coastal region in Mexico through social media analysis. *Tourism Management* **77**, 104002.

González, E., Sher, A. A., Tabacchi, E., Masip, A. & Poulin, M. 2015 Restoration of riparian vegetation: a global review of implementation and evaluation approaches in the international, peer-reviewed literature. *Journal of Environmental Management* **158** (C), 85–94.

Hudson, H. R. 2002 Linking the physical form and processes of rivers with ecological response. In: *Proc. of the International Symposium 'The Structure, Function and Management Implications of Fluvial Sedimentary Systems'*, 2–6 September 2002, Alice Springs, Australia.

Iakovoglou, V., Zaimes, G. N. & Gounaridis, D. 2013 Riparian areas in urban settings: two case studies from Greece. *International Journal of Innovation and Sustainable Development* **7**, 271–288.

Koutalakis, P., Tzoraki, O., Gkiatas, G. & Zaimes, G. N. 2020 Using UAV to capture and record torrent bed and banks, flood debris, and riparian areas. *Drones* **4**, 77.

Krysanova, V. & Srinivasan, R. 2015 Assessment of climate and land use change impacts with SWAT. *Regional Environmental Change* **15**, 431–434.

Monk, W. A., Compson, Z. G., Choung, C. C., Korbel, K. L., Rideout, N. K. & Baird, D. J. 2019 Urbanisation of floodplain ecosystems: weight-of-evidence and network meta-analysis elucidate multiple stressor pathways. *Science of the Total Environment* **684**, 741–752.

Munné, A., Prat, N., Solà, C., Bonada, N. & Rieradevall, M. J. A. C. M. 2003 A simple field method for assessing the ecological quality of riparian habitat in rivers and streams: QBR index. *Aquatic Conservation: Marine and Freshwater Ecosystem* **13**, 147–163.

Olivera, F., Valenzuela, M., Srinivasn, R., Choi, J., Cho, H., Koka, S. & Agrawal, A. 2006 ArcGIS-SWAT: A geodata model and GIS interface for SWAT. *Journal of the American Water Resources Association* **42** (2), 295–309.

Paul, M. J. & Meyer, J. L. 2001 Streams in the urban landscape. *Annual Review of Ecology and Systematics* **32** (1), 333–365.

Pujari, M. S. S. & Bhosale, M. S. 2017 Watershed protection and management using QGIS: study area of Borgoan Village, Sangli, Maharashtra. A review. *International Journal for Research in Applied Science and Engineering Technology* **5** (1), 219–221.

Reddy, A. S. & Reddy, M. J. 2015 Evaluating the influence of spatial resolutions of DEM on watershed runoff and sediment yield using SWAT. *Journal of Earth System Science* **124** (7), 1517–1529.

Savopoulou, A., Giatas, G., Pagonis, G., Iakovoglou, V. & Zaimes, G. N. 2017 Visual protocols and GIS as preliminary investigative tools to locate potential ecoengineering in streams and riparian areas. *Procedia Environmental Science, Engineering and Management* **4**, 227–234.

Wu, Y., Liu, J., Shen, R. & Fu, B. 2017 Mitigation of nonpoint source pollution in rural areas: from control to synergies of multi ecosystem services. *Science of the Total Environment* **607–608**, 1376–1380.

Zaimes, G. N. 2020 Mediterranean riparian areas – climate change implications and recommendations. *Journal of Environmental Biology* **41**, 957–965.

Zaimes, G. N. & Iakovoglou, V. 2021 Assessing riparian areas of Greece – an overview. *Sustainability* **13** (1), 309.

First received 31 January 2021; accepted in revised form 6 April 2021. Available online 16 April 2021

IWA Publishing's authorised EU representative for General Product Safety Regulations is Diane D'Arras, 15 rue Duret, 75116 Paris, France, e-mail: safety@iwap.co.uk.

Printed and bound by CPI Group (UK) Ltd, Croydon, CR0 4YY
05/06/2026
02128743-0002